# Topics in Biodiversity and Conservation

**Series Editor**

David L. Hawksworth, Ashtead, Surrey, UK

Springer's book series, Topics in Biodiversity and Conservation, brings together some of the most exciting and topical papers in biodiversity and conservation research. The result is a series of useful themed collections covering issues such as the diversity and conservation of specific habitats or groups of organisms, and the key dilemma of resource use versus conservation.
Based on Springer's popular journal, Biodiversity and Conservation, the series provides access to selected peer-reviewed papers which represent the cutting edge of current research to provide a valuable overview of progress in each topic addressed. With their diversity of case studies and depth of investigation, these collections will be of particular interest for courses including biodiversity and/or conservation issues, and to advanced students and researchers working in related fields.

More information about this series at http://www.springer.com/series/7488

Anurag Chaurasia • David L. Hawksworth
Manoela Pessoa de Miranda
Editors

# GMOs

## Implications for Biodiversity Conservation and Ecological Processes

Springer

*Editors*
Anurag Chaurasia
Indian Institute of Vegetable Research
Indian Council of Agricultural Research
Varanasi, Uttar Pradesh, India

David L. Hawksworth
Comparative Plant and Fungal Biology
Royal Botanical Gardens Kew
Kew, Richmond, UK

Manoela Pessoa de Miranda
Biosafety and Biosecurity
United Nations Environment Programme
Montreal, QC, Canada

ISSN 1875-1288 ISSN 1875-1296 (electronic)
Topics in Biodiversity and Conservation
ISBN 978-3-030-53182-9 ISBN 978-3-030-53183-6 (eBook)
https://doi.org/10.1007/978-3-030-53183-6



This Springer imprint is published by the registered company Springer Nature Switzerland AG.
The registered company address is: Gewerbestrasse 11, 6330 Cham, Switzerland

# Foreword

The very mention of GMOs has come to elicit fear in many people, the issues of risks as opposed to benefits are too rarely weighed, and generalizations may be too broad and not soundly based. The modification of crops and livestock is a process that has been going on since the origins of subsistence agriculture 6,000 or more years ago. In order to produce the amount of food to satisfy people's needs, there has been a drive to improve yield, quality, and resistance to pests and diseases. These are issues of more importance now than ever in world history.

Traditionally, improvements were achieved through selective breeding of plants and animals, and in the case of plants this has also included hybridization between different species. As our understanding of genetics increased, the achievements of plant breeding accelerated dramatically, especially through the latter half of the last century. This resulted in what became known as the Green Revolution, with unprecedented rises in crop production and pest and disease resistance. It was fuelled by series of crop-based internationally supported Crop Genetic. Research Institutes. This was an exciting period which I was privileged to be a part of as Director of the International Rice Research Institute in Manila. Local rice breeding stations were also established to breed and select races suited to their particular areas. The improved yields of rice made a major contribution to the alleviation of hunger in the Indian subcontinent and other parts of South-East Asia in particular.

Our capacity to transfer genes has progressed enormously since those of plant breeders using genetic variants of the same species to develop improved races. It is now possible to not only transfer genes which impart particular benefits from one species to another, but even to edit existing genomes using CRISPR technology. In both cases, the inserted or edited genes become an integral part of the genome and can be passed to future generations. It is therefore necessary to proceed cautiously and for funding agencies to support basic research to elucidate the various processes occurring in genetically engineered cells, and for regulatory authorities to put in place appropriate safeguards to protect not only human health, but also other organisms and so the ecosystems on which we ultimately depend.

Along with other scientists in India, I also wish to see benefits emerging from these new technologies accruing to and not harming resource-poor small farmers. We should learn from issues that have already arisen from the use of some Genetically Modified Organisms (GMOs) and adopt the Precautionary Principle on a case-by-case basis.

I am very pleased to see that the editors have now put together an authoritative volume which explains what is involved in genetic modification today, and the steps being taken to address concerns. Most importantly, these include contributions reviewing the various legal and regulatory systems now or being put in place to address concerns and minimize the risk of adverse effects.

This is a major contribution to a rational approach to an emotive issue, which serves to inform the current debates. Further, as in so many areas of science, it also shows the variety of situations for which genetically modified organisms have been developed. It emphasizes the need for appropriate regulation which will help to minimize risks and maximize benefits from new technologies.

Founder Chairman, M. S. Swaminathan Research Foundation, Chennai, Tamil Nadu, India
Ex-Member of Parliament
(Rajya Sabha) New Delhi, India

M. S. Swaminathan

# Preface

Whether Genetically Modified Organisms (GMOs) should be allowed to enter our supply chains, and if so under what circumstances, is a topic that has been hotly debated since the early 1990s. Whether or not they are safe, or pose a risk to human health or the environment, not only remains a topical issue today but is rapidly changing and increasingly challenging due to unprecedented advances in gene technology now proceeding apace. These complex issues are under discussion at fora ranging from intergovernmental conventions to national governments and regulatory authorities to consumer groups and producers. Our aim with this book is to showcase both the benefits and risks in diverse applications, and to explore the current situation regarding regulation and risk assessment both internationally and in a selection of individual countries.

A genetically modified organism (GMO), or a living modified organism (LMO), is any organism having a novel combination of genetic material created through genetic engineering or manipulating an organism's genetic makeup by introducing or eliminating specific genes using rapidly advancing biotechnological tools and techniques – now embracing CRISPR (clustered regularly interspaced short palindromic repeats) enabling sections of DNA to be edited, the so-called synthetic biology with the potential to design what could be regarded as new organisms, and engineered gene-drives aiming to change the genetic make-up in existing wild populations.

The potential benefits of biotechnology and GMOs to address global challenges are boundless. At a time when the world battles the devastating and multi-faceted effects of the Covid-19 pandemic, turning to innovative technologies in search of opportunities to improve food security, address environmental issues such as climate change and strengthen healthcare systems becomes more appealing and even necessary. As with any technological development, the risk and benefits of GMOs must be carefully weighed in an open dialogue that transcends the science–policy interface to also take into account cultural and ethical values.

Similar to classical animal and plant breeding, which have done so much to enable the Earth to support today's world population, to date, the new and emerging biotechnologies still depend primarily on naturally occurring genes as their key raw materials. However, most agricultural and other genetic biodiversity is located in the tropics, while the tools of modern biotechnology are generally held under strict patents and licensing agreements by private sector companies headquartered in temperate zones. This inevitably raises ethical questions as to how GMOs could better contribute to food security, especially the food deficit of many less developed countries, to really fulfil the United Nations' Sustainable Development Goal 2 on 'Zero Hunger'. Another question is that of unintended, and indeed possibly unforeseen, consequences of releasing GMOs into ecosystems. This has led to unprecedented radical and emotionally charged global debates embracing almost every section of society: scientists, governments, policy makers, producers, consumers and the public. There is also the issue that it may require many years, even decades, of critical observations to become aware of the effects of any release or escape into our ecosystems.

While the first genetically engineered crop was tobacco in 1983, widespread introductions of GM crops and products date from 1996. Twenty-four years on, some 190 million hectares of land are devoted to GM crops across 26 countries and involving almost 17 million farmers. There is consequently a huge ongoing experiment of possible GMO introductions into ecosystem subject to ongoing research observation. Field environmental data from released GMOs other than crops, such as ones of bacteria, fish and insects, however, are still in their infancy. Impacts of diverse GMOs on the biodiversity and ecological processes in soil are also still particularly inadequately known, as are those of gene drive organisms that might alter wild populations in unforeseen ways affecting the ecosystems in which they occur.

These new technologies have the potential for impacts that might parallel or exceed those of the Green Revolution that took off in the 1960s, in which selective breeding programmes produced high-yielding cultivars of major staples, most spectacularly in rice and wheat. We are fortunate in having one of the key players in that Revolution, M S Swaminathan FRS, contribute a Foreword to this volume, in which he recognizes the benefits that can emerge from the new GMO revolutions but cautions that these should not be to the detriment of resource-poor small farmers. He also draws attention to the importance of adopting a Precautionary Principle approach, and it is pleasing to see that this is being emphasized in many countries as a part of their risk assessment regulations. We are conscious, however, that even the most robust risk assessments may not uncover all potential adverse effects. Long-term careful monitoring following any release is therefore a key need so that any unforeseen ecological consequences can be recognized as quickly as possible while they can still be contained.

Researchers dealing with GMOs and other products of the new technologies, and in particular countries, are not always aware of the issues being confronted by those working on diverse different organisms and actions being taken to minimize risks through regulation. We have therefore brought together a series of contributions from researchers and regulators around the world presenting contrasting views on a range of topics: GM crops and their impacts, GM insects, GM vertebrates, ecological risk assessments, gene drive approaches, and governance and regulation. Governance and regulations are critical to how decisions under international agreements, in particular the Cartagena Protocol on Biosafety to the Convention of Biological Diversity, are implemented. We have therefore also provided national perspectives of the current situation from 22 countries across Africa, Asia, Australasia, Europe and the Americas reflecting their legal perspectives which reveal a range of approaches and rates of progress.

We trust that this volume will inform and broaden the perspectives of researchers and those involved in risk assessments and the development of national regulatory frameworks. Through the approach taken in this volume, we aim to support progress towards scientifically sound decision-making on this very contentious topic and further contribute to the framing of internationally acceptable principles focused on conserving and sustainably utilizing biodiversity – across national borders to the benefit of all of us and the global environment.

Varanasi, Uttar Pradesh, India — Anurag Chaurasia
Kew, Richmond, United Kingdom — David L. Hawksworth
Montreal, QC, Canada — Manoela Pessoa de Miranda
6 May 2020

# Contents

# Part I
# Introduction

# GMOs, Biodiversity and Ecosystem Processes

**Muhammad Amjad Nawaz, Kirill S. Golokhvast, Aristides M. Tsatsakis, Hon-Ming Lam, and Gyuhwa Chung**

**Abstract** Potential impacts of genetically modified (GM) crops on biodiversity is a controversial topic of public interest that comes under scrutiny in the Convention on Biological Diversity. The commercialization of GM crops has delivered global agronomic, economic and social benefits, but it has also raised concerns on the risks to human and environmental health. The current state of knowledge reveals that farmland biodiversity is jeopardized by intensive agricultural practices. The monoculture practice used in the cultivation of GM crops has increased the risk of the emergence of herbicide tolerance and insecticide resistance between weed and insect pest species. This, in turn, may interrupt the food web at different trophic levels. To avoid an overreliance on GM crops, alternative weed control and insect pest management strategies should be considered.

**Keywords** Biodiversity · Bioethics and biodiversity · Ecosystem functioning · Food web · GMO and public perception · Genetically modified plants · GMO regulation · Protecting biodiversity

---

M. A. Nawaz
Laboratory of Bio-Economics and Biotechnology, Department of Bio-Economics and Food Safety, School of Economics and Management, Far Eastern Federal University, Vladivostok, Russian Federation

K. S. Golokhvast (✉)
Pacific Geographical Institute FEB RAS, Vladivostok, Russian Federation

N.I. Vavilov All-Russian Institute of Plant Genetic Resources, Saint Petersburg, Russia

A. M. Tsatsakis
Department of Toxicology and Forensic Sciences, School of Medicine, University of Crete, Heraklion, Crete, Greece

H.-M. Lam
School of Life Sciences and Centre for Soybean Research of the State Key Laboratory of Agrobiotechnology, The Chinese University of Hong Kong, Hong Kong, SAR, China

G. Chung (✉)
Department of Biotechnology, Chonnam National University, Yeosu, Republic of Korea
e-mail: chung@chonnam.ac.kr

 
A. Chaurasia et al. (eds.), *GMOs*, Topics in Biodiversity and Conservation 19,
https://doi.org/10.1007/978-3-030-53183-6_1

# Introduction

## *Genetically Modified Organisms*

Genetically modified organisms (GMO) are any organisms with their genetic material altered in an unnatural, intended and targeted way via recombinant DNA technology to enable them to perform specific functions such as enhanced productivity and resistance to diseases and pests. Recombinant DNA constructs that often consist of the gene(s) of interest, promoter/enhancer, terminator and marker genes (Mertens 2008) are introduced into plants mainly by two well-developed techniques: biolistic transformation and *Agrobacterium tumefaciens*-mediated transformation. Although there are some other transformation technologies, such as protoplast transformation via electroporation or polyethylene glycol, they are of limited success (Dandekar and Fisk 2005).

With a mission to feed an increasing world population, modern agriculture is constantly developing innovative farming practices to increase the efficient use of arable land, energy and water resources to meet the global needs for food, animal feed and fibre. However, increasing evidence has shown that neither conventional farming nor GM technology alone can completely solve the problem (FAO 2017).

## *The History of GM Food and Feed*

The first commercial plantation of GM crops started in 1994, and within 2 years the area under GM crop cultivation reached 1.66 million hectares (Brookes and Barfoot 2018). In 2017, 2.15 billion hectares were planted with four GM crops: soybean (48.37%), maize (29.76%), cotton (15.81) and canola (6.04%). GM crops have delivered agronomic, economic and social benefits to farmers, and a growing trend in GM crop cultivation in developing Asian countries can be anticipated. The global area under GM crops has already reached a record high of 5.8 billion acres; involving 24 countries in which 19 are developing countries (ISAAA 2017). At the same time, the production of GM crops is a major industry in many developed countries, especially in the USA, which has an articulated biotechnology-based economic strategy. Within the giant biotechnology industry, GM crops alone (Carlson 2016) generated ~40% of US biotechnology revenue (128 billion of 324 billion USD).

In 2014, global farm income amounted to 17.74 billion USD, and the cumulated global farm income since the first commercialization of GM crops has reached 150.3 billion USD (Brookes and Barfoot 2018). The global acceptance of GM food and feed is largely contingent upon regulations, approval and public trust in governing authorities. Until 2014, 27 GM crops and 357 GM events have been approved across 65 nations, suggesting that the global biotechnology industry is expanding and not limited to the four major GM crops discussed above (Babar et al. 2020; Lucht 2015). Approximately 70–90% of GM crops are used as feeds (Van Eenennaam 2013).

## *Concerns over GMOs and the Public's Perception*

Unlike most widely used modern technologies, such as electronics, communication and medicine, there is a serious distrust among consumers regarding GM foods. Before the products of GM crops reach consumers, farmers who will determine whether GM crops will be grown will make the first and most important decision. The selection criteria of the farmers are mainly centred on farming income, the availability of a local or global market where the farmers' produce could be sold, and government regulatory policies. In some occasions, governments could impose strong interventions to overrule the decision of farmers, as in the case of soybean farming in Romania (Otiman et al. 2008).

So far, the public has not built up overall confidence in GM foods and some have demanded mandatory labelling. However, this varies greatly by countries and even within a country over labelling requirements; results have been mixed relative to the acceptance of food containing GM ingredients. GM food and feed remains a contentious issue involving governments, policy makers, farmers, biotechnology companies (mainly those that produce seeds of GM crops) and consumers. The consumer attitude on both sides of the Atlantic, that is, North America and Europe, has been strongly diverged since the arrival of the United States (US) GM soybean in Europe (EU) in 1996 (Lucht 2015). This seems to be directly correlated with the more stringent regulations and approval processes adopted in the EU compared with the much-relaxed regulations in the USA (Fig. 1). Relative to the USA, the EU has approved fewer GM foods or feeds to be cultivated or sold in the EU. The EU has approved GM cotton, maize, oilseed rape, soybean, sugar beet and swede-rapa, and there is an EU register of authorized GMOs (https://webgate.ec.europa.eu/dyna/gm_register/index_en.cfm). In addition to some government policies, some confusing reports released by the media and some Non-Governmental Organizations (NGOs) have given rise to negative public sentiments on GM crops (Frewer et al. 2002). During 2011 and 2013, groups of anti-GMO protestors destroyed GM wheats in Australia and the Golden Rice in the Philippines, respectively (Zhang et al. 2016). On the other hand, in 2016, an organization supporting "precision agriculture" stepped in and published a letter signed by more than one hundred Nobel laureates in support of GM crops (http://supportprecisionagriculture.org/nobel-laureate-gmo-letter_rjr.html).

**Fig. 1** Differences in GMO regulatory concepts between the USA and Europe

Considering this complex debate, certain factors are responsible for developing consumers' attitudes towards the acceptance or rejection of GM food, notably: (1) consumer perception of risks versus benefits, (2) understanding of the basic principle of GM technology, (3) source of information about GM foods and feed, (4) uncertainty/distrust about scientific soundness of the technology, (5) personal attitude and food preferences, (6) regulations and policies in the home country of the consumer, and (7) the role of NGOs and media. A comprehensive overview on the public acceptance of GM crops is provided by Lucht (2015). Consumers in different societies have different levels of knowledge about GM food and feed. However, there are several common questions being asked. For example, is GM food safe? Is it true that GM plant production coupled with intensive use of herbicides harm wildlife and non-target species? Will gene flow be a risk to the environment? Could the GMOs outcross to produce super-weeds? Are the effects of GMOs on the environment acceptable or unacceptable? (Tsatsakis et al. 2017a). Consumers have also expressed concerns related to the expression of GM food-driven DNA and RNA in the human body. However, this is not the focus of this chapter and is discussed in detail in a recent review by Nawaz et al. (2019).

## Biodiversity and GMOs

### *Defining Biodiversity – The Diversity of Diversities*

The potential impact of GMOs on biodiversity is a complex topic of great interest, particularly in relation to biological conservation and ecosystem resilience. While there is now an increasing amount of literature on the possible environmental implications of GM plants on biodiversity at different trophic levels, it is essential to first define biodiversity. Diversity is a multifaceted concept, spanning from molecular (genetic) diversity at the intracellular level to the organismal and supra-organismal levels, encompassing the variety of biological life within species, populations (both intra- and inter-), communities, ecosystems and biomes (Magurran 2003). The Convention on Biological Diversity (1992) defined the term "biological diversity" as "the variability among living organisms from all sources, including inter alia, terrestrial, marine and other aquatic ecosystems and the ecological complexes of which they are part; this includes diversity within species, between species and of ecosystems" (Lovei et al. 2010; www.cbd.int).

The long struggle of defining biodiversity has not reached a perfect definition of diversity, nor the scale, index or formula for quantifying it. Harper and Hawksworth (1994) discuss the history of the use of the term, but in practice, there exists a "diversity of diversities" (Juhasz-Nagy 1993). This expands the description of diversity to all the scales described above. Since we are discussing the impacts of GMOs on biodiversity, it is important to consider that genetic diversity is essential for individual organisms and populations to adapt to new environments and to develop favourable traits to withstand the changing climatic conditions. A single change at the DNA level could bring changes to the structure of genetic diversity. In relation to population genetics, the term "genetic diversity" refers to the total number of genetic characters in the genetic makeup of the species and could be defined as the presence of genetic differences between and within species. In order to preserve biodiversity, it is important to create, maintain and develop genetic diversity; that is what nature does: by competition, predation and evolution (Bøhn and Amundsen 2004).

## *Impacts of GMOs on Biodiversity*

With the rise in GM crop production and area under GM crop cultivation in the last 24 years, the debate on the risks associated with GM crop cultivation has also intensified, with more and more studies being published both in support of and against GM crops (Prakash et al. 2011). GM plants increase the crop's genetic diversity when considered in the contexts of improving underutilized crops by increasing their yield and/or nutritional values, and making them suitable for large-scale cultivation. However, the development of GM crop plants often uses a limited set of high-performing breeding lines, which results in a reduction in the diversity of cultivars being planted on farmland (Sneller 2003).

The structure of genetic diversity is a consequence of gene flow and is characterized by demographic factors and the life history of commercially cultivated crops (Lu and Yang 2009). The genes introduced by genetic engineering are mostly targeted to insect pest resistance and herbicide tolerance (Verma 2013) while product quality enhancement and agronomic traits are rapidly being added to the list of approved and commercially cultivated crops (USDA-APHIS-BRS database; www.aphis.usda.gov). There are possibilities of gene flow to the wild and related species (Andersson and de Vicente 2010). In the worst-case scenario, this could lead to the complete genetic extinction of wild populations. However, GM crop-specific examples in this regard are scarce while such a phenomenon has been observed in conventional crops: for example, sorghum/shattercane/Johnson grass; sugarbeet/sea beet and other wild *Beta* spp. in Europe; sunflower/wild annual sunflower in the USA (Papa and Gepts 2004). The structure of genetic diversity could be influenced by many factors, for example, migration, mutation, genetic drift and selection, all of which contribute towards the modification of gene frequencies. Tsatsakis et al. (2017b) discussed the potential disturbance in the structure of genetic diversity due to GMOs. The GM plants can affect the biodiversity at the crop, farm and landscape scales (Carpenter 2011) (see below). In summary, the main risks to biodiversity associated with GM plants are: (1) changes in the structure of genetic diversity, (2) intensification of farming, (3) increased pressure on biodiversity, (4) increase in herbicide use, (5) reduction in the diversity of non-farm biota, (6) development of herbicide resistance in weed species, (7) changes in arable weed populations, and (8) spatial and temporal spread of GM traits to non-target species. Some of these risks are common to intensive agriculture where non-GM crops are cultivated.

## *Risks to Plant Diversity at the Crop and Farm Levels*

Intensive agriculture is among the main drivers behind the loss of biodiversity at the crop, farm and landscape levels. Half of the global food demands are met with only four major crops: wheat, maize, rice and potato, implying that 50% of the biota occurring on farmlands (i.e. plants, insects, pests, farm animals, birds, fungi and microbes of farmland and water) is directly related to the diversity of these four crop species. The genetic and varietal diversity of the dominant plant species can limit, maintain or increase the biodiversity in the local landscape. However, this phenomenon is related to crop plant (and species) because in some countries the approved GM varieties are then bred with high yielding cultivars/lines tolerant to other stresses. With the increasing spread of GM crop cultivation across the globe, it is speculated that the genetic diversity of each of these main crops will be decreased as each breeding programme is targeted at improving one particular trait at a time and hence it uses a limited set of breeding material. Many large-scale whole-genome sequencing studies have revealed that domestication and targeted breeding have reduced/altered the levels of genetic diversity in the improved cultivars and landraces of cultivated species compared to their wild relatives, for example, rice, soybean, wheat, maize and potato (Xu et al. 2012; Lam et al. 2010; Fu and Somers 2009; Diez et al. 2013; Hardigan et al. 2017). The renewed interests in the utilization of the wild relatives of crops in breeding programs have resulted in the identification and subsequent breeding of adaptive ancestral traits back into the cultivated species, thus confirming that interspecific gene flow is possible (Nawaz et al. 2018). It is important to understand that many of the cultivated crop species originated from "centres of diversity" in different regions of the biosphere. Often these centres of diversity contain many wild relatives having interspecific gene transfer abilities via different routes, such as pollination. This poses the risk of transferring the transgenes into local flora. If GM crop cultivation spreads to more countries, which is evident based on recent reports (Mathur et al. 2017), it will eventually increase the chances of transgene transfers to closely related species. This will soon become a worldwide phenomenon (Mertens 2008).

At the level of individual crop species, the transfer of transgenes is possible through pollination, seed-mediated and vegetative propagule-mediated gene transfers (Dick et al. 2008). The gene transfer itself can be either crop-to-crop, crop-to-wild or crop-to-microorganisms (Papa and Gepts 2004). Each type of gene flow is accompanied by potential impacts on biodiversity. The gene transfer can be a rare hybridization event between a GM crop and its wild relative and can generate hybrids with increased fitness (Darmency 2000). The resulting changes in somaclonal variations and pleiotropic effects of the transgene could lead to changes in DNA methylation patterns, thus altering gene expressions. An example is a transgenic potato with altered carbohydrate metabolism (Becker et al. 1998). The persistence of transgenes in wild hybrids is influenced by a strong selection for the acquired trait and hybrid vigour. Many such cases of unintended effects have been observed, such as larger flower sizes with better pollen donation abilities of non-GM plants and the altered bolting pattern of GM sugar beet X Swiss chard hybrids (Ellstrand 2003; Bergelson et al. 1998; Mertens 2008). Therefore hybrid vigour, selection pressure, fitness cost and heterosis are some of the factors which determine the success and persistence of transgene introgression into wild relatives, as well as pleiotropy and the insertion site of the transgene in the recipient species' genome (Chèvre et al. 2000; Campbell et al. 2016). It is essential to understand the ecological significance of the transgenic plants as a single inserted trait may result in the production of a crop plant, which could be novel to the existing ecological network. An unexpected consequence of the spread of transgenes into wild species could also result in more invasive weeds (Vrbnicanin et al. 2017).

At the level of the farmland, the local biotic diversity is severely suppressed due to the use of selective herbicides (e.g. glyphosate) when herbicide-resistant crops are planted. GM plants with herbicide resistance and insect tolerance may interact negatively with organisms present on the farm, leading to the loss of native species (Schutte et al. 2017). During the initial stage of the introduction of herbicide-resistant crops, the density of weeds will decrease (intended goal), while at the same time there will also be an increased pressure on surrounding biota ranging from non-target weeds to non-target insects, farm animals and even amphibians and aquatic animals in the local water bodies (unintended result) (Marshall 2001; Tsatsakis et al. 2017a). Weed communities in the farmland area are directly affected by intensive herbicide sprays where less tolerant weeds will completely disappear (due to the depletion of weed seed bank) while moderately tolerant species will survive, evolve more resistance, grow and spread.

More than 240 weed species have been reported as herbicide-resistant (Heap 2014). The question to be addressed then is: what are the risks to farmland biodiversity because of the development of herbicide resistance in weeds and the complete removal of weeds? The first and foremost risk is to insects living and/or foraging on these weeds. With a change in weed density, the foraging behaviour of insects will be changed (Capinera 2005). For examples, wolf spiders feeding on crickets will experience a reduced food supply, the populations of monarch butterflies will be reduced, and there will be less frequent visits by insect pollinators (Wrinn et al. 2012; Schutte et al. 2017; Boyle et al. 2019). The changed insect behaviour may also be an indication of the development of secondary pests. The reduction in pollinator populations can significantly affect the non-GM crops growing near GM crops, the yields of which largely rely on pollination by insects (Nicholls and Altieri 2012). Secondarily, this can result in reduced bird populations owing to the reduction in insect populations on the farms. Furthermore, aquatic life residing within the farmland water bodies is exposed to increasing concentrations of residual herbicides, and the soil fungal and bacterial communities will also be at risk from herbicide exposure, although is a poorly researched area. A shift in the activities and the composition of soil microbiota can also be the after-effect of modified nutrient uptake by herbicide-resistant GM plants (Tappeser et al. 2014). The other main group of GM crops, the Bt (*Bacillus thuringiensis*) crops, negatively affect the insect populations which directly feed on them and the microbial communities residing in the root zone of Bt crops (Clark et al. 2005; see also chapters "Impacts of Genetically Engineered Crops on the Soil Microbiome, Biological Processes and Ecosystem Services" and "Environmental Analytical and Ecotoxicological Aspects of *Bt* Maize in the Pannonian Biogeographical Region of the European Union").

A report funded by the European Commission to review the results of GMO-related experiments was published by the Research Directorate-General of the European Commission in 2001 (Kessler and Economidis 2001). This report summarized the impacts of 81 projects they had funded before year 2000. This review found that genetic sequences of plant origin were not transferred to bacteria tested (project BIOT-CT91-0282) and the fitness of GM plants was not significantly different from non-GM plants but the possibility for gene transfer to related species was dependent on the crop species (BIOT-CT91-0298) (Kessler and Economidis 2001). Under experimental conditions, dispersal of potato transgenes was not possible, while gene transfer between alfalfa and non-cultivated relatives could occur (BAP-0371/0384/0408/0423). Incorporation of insecticidal proteins in the diet of beneficial insects (at concentrations higher than that expressed in GM plants) showed significant effects on insect behaviour and physiology, but the report stated that these results needed validation under field conditions (BIO4-CT96-0365).

Other experimental projects reviewed in this report were either not completed or did not include a statement which could be categorized as either pro- or anti-GMO. Subsequently, another report from the Directorate-General for Research and Innovation Biotechnologies, Agriculture and Food was published in 2010 reviewing the next decade of EU-funded GMO research (2001–10). Compared to the previous report, this one was more comprehensive and covered 50 projects and experiments involving 400 scientific research groups working on GMOs (European Commission 2010). It contained a chapter that focused on the environmental impacts of GMOs and outlined the conclusions from all related projects. Overall, this second report concluded that no significant effects of Bt cotton were found in non-target species tested. However, the cotton project showed concerns over the development of Bt resistance in bollworms. On the other hand, the studies on rice and potato did not report any risk for comparative assessment. Furthermore, no direct significant effect of Bt maize was reported on lacewings (Rodrigo-Simon et al. 2006). However, the study did not address any indirect risks. Most of these projects reported no effects on tri-trophic levels but indicated that pollinator behaviour might have been affected. Overall, no significant negative impacts were reported but it is also worth noting that these authors reported a few negative effects on arthropod biodiversity near Bt crops (European Commission 2010).

Similarly, a report by the Division on Earth and Life Studies, Board on Agriculture and Natural Resources of The National Academies of Sciences, Engineering and Medicine (NAS 2016) presented a balanced opinion regarding the benefits and risks associated with GM plants and did not rule out the risks associated with GM plants to biodiversity (NAS 2016). Collectively, risks associated with the cultivation of GM crops provide relevant indicators for monitoring changes in biodiversity at the crop and farmland levels.

### *Protecting Biodiversity*

Overreliance on GM crops in general, and herbicide-resistant GM crops in particular, discourages the application of alternative weed control and insect pest management strategies. Since the very beginning of agriculture, all the developments have aimed at achieving higher yields and increasing farm income, with less focus placed on integrating the externalities of agriculture such as its negative impacts on climate change and biodiversity. For example, the use of herbicide-resistant GM crops has become a problem for biodiversity conservation and works against sustainable agriculture (Lovei et al. 2010). Therefore, to counteract the negative effects of GM crops, certain measures, such as the implementation of long-term, ecosystem-based weed management strategies instead of relying on a single herbicide, should be adopted to reduce the selection pressure on weed species to develop herbicide resistance. However, these alternative strategies are not easy to implement because current herbicide-resistant GM crops are developed to work with modern mechanized farming with large-scale monoculture. It is efficient and straightforward for farmers to grow, and resulted in high yields of uniform-quality crops. The implementation of integrated pest management strategies may entail the use of guaranteed weed-free seeds, crop rotation and intercropping, which will require a major change in the current mainstream farming practices.

## GMOs and Ecosystem Processes

A community of diverse species living and interacting in conjunction with non-living components in their environment forms an ecosystem (Schulze et al. 2002). In the context of GMOs, it is important to consider the farmland itself as an ecosystem that provides us with the necessities for life: that is, food, animal feed and forage, pharmaceuticals, bioenergy and shelter. Furthermore, these ecosystems also regulate soil and water quality, help in the cycling and regulation of the flow of natural substances such as water and carbon, biodiversity maintenance, regulation of microclimate, detoxification of noxious chemicals as well as cultural services; these are ecosystem processes (sometimes referred to as “functional biodiversity”; see Harper and Hawksworth 1994). Farmland ecosystems contain and rely mostly on natural ecosystem services but are also at the same time, heavily dependent on human activities, that is, the types of agricultural practices. Ecosystem maintenance is the sum total of all processes involved in energy and matter transfer within and between ecosystems.

The introduction of GM plants into the farmland ecosystem often could not sufficiently anticipate the long-term risks associated with their release, mostly because of the complexity of ecosystem processes. Major concerns are disturbances in species diversity mainly via hybridization and gene flow (Tsatsakis et al. 2017a, b). The development of insecticide resistance among insect pests, the emergence of new viral pathogens, the rise of super-weeds and altered agronomic practices have received much attention (Mertens 2008). To maintain the productivity, stability and nutrient recycling ability of an ecosystem, the diversity of diversities is critical. Intensive agricultural practices, particularly increased applications of broad-spectrum herbicides and insecticides associated with GM crops, in conjunction with interspecific complementarity, decreased herbivore populations, changed foraging behaviours and greater use of evermore limiting natural resources, are pushing ecosystems in an undesirable direction. This trend enables invasion by an alien species that happens to have certain traits such as the ability to withstand a broad range of environmental conditions, high dispersal ability, aided by the reduced herbivory risks due to biocide use (Mertens 2008). Apart from such short-term risks, long-term risks such as species displacement and extinction as a result of GM crops are also important considerations. Once a species is displaced from an ecosystem, the multitude of species involved directly and indirectly with it through the food web will experience changes in their respective richness and abundance as a consequence. However, such an observation is yet to be made on a large scale (Szenasi et al. 2014).

Farmland ecosystems, just like any other ecosystems, consist of a complex web of organisms interacting with one another under various combinations of biotic and abiotic conditions. An imbalance in this complex food web can result in the upset of many symbiotic associations and tri-trophic interactions (Lovei et al. 2010). Tri-trophic interaction disturbance has been speculated by different reports but the large-scale scientific investigations in this regard are scarce. Changes in community structure, species interaction and biodiversity have a pronounced impact on agricultural intensification whether it is GMO- or non-GMO-related (Lohaus et al. 2013). However, some studies concluded that no adverse effect was seen on non-target arthropods in Bt Corn (*Cry1Ac*) (Guo et al. 2014). Additionally, a recent dataset from different genetically modified maize events (coleopteran resistant, coleopteran and lepidopteran resistant, lepidopteran+herbicide tolerant and coleopteran resistant and herbicide tolerant) and associated controls comprising 363,555 arthropod individuals has been made available with a conclusion that GM plants cause no short-term direct harm to arthropods (ground-dwelling and plant canopy dwelling) (Szenasi et al. 2014; Palinkas et al. 2018). A study based on this dataset reported that the structure of the food webs remained stable at a larger scale. However, the detailed results found differences in average trophic links as well as in characteristic path lengths of GM and non-GM food webs (Palinkas et al. 2017). These examples suggest that impacts of agriculture on food webs is evident and there are certain risks associated with the cultivation of GM crops. However, long-term studies may better explain the reported differences in trophic links/trophic groups because the continued use of a GMO is likely to change the ecological balance and genetic diversity with the passage of time.

With increased pressure from selective herbicide applications, a single weed species could be removed from the farmland food web and hence impart risks to tri-trophic levels. Similarly, exposure to Bt toxins released into farmland soil can affect the balance in the predator and parasitoid populations, depending on the levels of Bt toxins released, the target insect species, their prey, and the local root zone and soil microbiota (Hilbeck and Otto 2015 and references therein). As Bt toxins may also be expressed in different plant tissues other than the target tissue to be protected against insect pests, the risks to non-target organisms cannot be avoided as "the introduction of a natural Cry protein into a plant can significantly enhance its toxicity towards both target and non-target species" (Latham et al. 2017). The Bt toxins ingested by those non-target insects or herbivores will be passed onto the next trophic levels in additive, synergistic and/or antagonistic ways (Hilbeck et al. 1998).

## The Links Between Bioethics, Biodiversity and Ecosystems

Understanding the risks of GMOs on biodiversity and ecosystems is important because anthropogenic activities related to intensive agriculture and industrialization have already impacted ecosystem substantially. Nearly half of the land occupied by plants across the globe now belongs to farmland ecosystems, which provide us with benefits such as food and animal feed, but they also have large ecological footprints (Kanianska 2016). It is our moral obligation to protect biodiversity, without which ecosystems may not be maintained. Thus, the Convention on Biological Diversity has made this one of their specific objectives. The unintended effects of the release of transgenes into the ecosystem must be dealt with on a case-by-case basis and contextualized according to the environment into which they are released. Misinformation and doubts created by popular media and inadequate scientific experiments on both sides of the debate, by giant biotech industries and environmental groups alike, have raised more hype and fear both for and against the consumption of GM food and feed (Arntzen et al. 2003). In this uncertain situation, a sound policy developed by a competent authority on the conservation of biodiversity is needed to avoid conflicting interpretations and to make policy decisions based on the soundest scientific evidence available. In this regard, only balanced and proactive risk-benefit analyses as well as strategies to mitigate risks could lead towards a more balanced and constructive decision-making process. In this era where novel gene manipulation techniques such as CRISPR/Cas 9 are being introduced and established, new guiding principles are needed in addition to "comparative assessment" and "substantial equivalence" (Agapito-Tenfen et al. 2018).

## Conclusions

The rise in cultivated areas for GM crops worldwide indicates that this technology is now largely accepted by the global farming community, mainly due to increased farm economic output. GM crops have been contributing towards the fight against hunger and feeding the ever-increasing world population. The impact of GM crops on biodiversity and ecosystem functioning have been studied for the last two decades and the findings have been incontrovertible. The cultivations of herbicide-resistant and insecticide-tolerant crops have raised concerns on the reduction in the diversity in the local biota. However, these effects are not only due to GM plants but are the collective results of intensive agricultural practices. Nonetheless, the most prominent changes in biodiversity are linked to changes in herbicide application regimes associated with herbicide-resistant crops. Overall, intensive agriculture has converted natural habitats into farmlands and disturbed the local biodiversity. Within farmland ecosystems, food webs have been affected at different trophic levels. There is therefore a dire need to address the resulting destruction of biodiversity.

**Acknowledgements** This work was supported by Hong Kong Research Grants Council Area of Excellence Scheme (AoE/M-403/16). JY Chu copy-edited the text.

## References

Agapito-Tenfen SZ, Okoli AS, Bernstein MJ, Wikmark O, Myhr AI (2018) Revisiting risk governance of GM plants: the need to consider new and emerging gene-editing techniques. Front Plant Sci 9:1874. https://doi.org/10.3389/fpls.2018.01874

Andersson MS, de Vicente MC (2010) Gene flow between crops and their wild relatives. Evol Appl 3(4):402–403. https://doi.org/10.1111/j.1752-4571.2010.00138.x

Arntzen CJ, Coghlan A, Johnson B, Peacock J, Rodemeyer M (2003) GM crops: science, politics and communication. Nat Rev Genet 4:839–843

Babar U, Nawaz MA, Arshad U, Azhar MT, Atif RM, Golokhvast KS, Tsatsakis AM, Shcerbakova K, Chung G, Rana IA (2020) Transgenic crops for the agricultural improvement in Pakistan: a perspective of environmental stresses and the current status of genetically modified crops. GM Crops Food 11(1):1–29. https://doi.org/10.1080/21645698.2019.1680078

Becker R, Marty B, Ulrich A (1998) ExperimentalleVerifizierung von Veränderungenrisikorelevanterökologischer Parameter beitransgenen Kartoffelnmit Veränderungenim Phosphat- und Kohlenhydrat metabolismus. Landesumweltamt Brandenburg, Müncheberg

Bergelson J, Purrington CB, Wichmann G (1998) Promiscuity in transgenic plants. Nature 395:25

Bøhn T, Amundsen PA (2004) Ecological interactions and evolution: forgotten parts of biodiversity? Bioscience 54:804–805

Boyle JH, Dalgleish HJ, Puzey JR (2019) Monarch butterfly and milkweed declines substantially predate the use of genetically modified crops. PNAS 116(8):3006–3011. https://doi.org/10.1073/pnas.1811437116

Brookes G, Barfoot P (2018) Farm income and production impacts of using GM crop technology 1996–2015. GM Crop Food 8:156–193

Campbell LG, Lee D, Shukla K, Waite TA, Bartsch D (2016) An ecological approach to measuring the evolutionary consequences of gene flow from crops to wild or weedy relatives. Appl Plants Sci 4(3):apps.1500114. https://doi.org/10.3732/apps.1500114

Capinera JL (2005) Relationships between insect pests and weeds: an evolutionary perspective. Weed Sci 53(6):892–901

Carlson R (2016) Estimating the biotech sector's contribution to the US economy. Nat Biotechnol 34:247e255

Carpenter JE (2011) Impact of GM crops on biodiversity. GM Crops 2:7–23. https://doi.org/10.4161/gmcr.2.1.15086

Chèvre AM, Eber F, Darmency H, Fleury A, Picault H, Letanneur JC, Renard M (2000) Assessment of interspecific hybridization between transgenic oilseed rape and wild radish under normal agronomic conditions. Theor Appl Genet 100:1233–1239
Clark BW, Phillips TA, Coats JR (2005) Environmental fate and effects of *Bacillus thuringiensis* (Bt) proteins from transgenic crops: a review. J Agric Food Chem 53:4643–4653
Dandekar AM, Fisk HJ (2005) Plant transformation: agrobacterium-mediated gene transfer. Methods Mol Biol 286:35–46
Darmency H (2000) Unpredictability of transgene flow between oilseed rape and wild relatives. Xieme Colloque international sur la biologie des mauvaises herbes, Dijon, pp 597–603
Dick CW, Hardy OJ, Jones FA, Petit RJ (2008) Spatial scales of pollen and seed mediated gene flow in tropical rain forest trees. Trop Plant Biol 1(1):20–33
Diez CM, Gaut BS, Meca E, Scheinvar E, Montes-Hernandez S, Equiarte LE, Tenaillon MI (2013) Genome size variation in wild and cultivated maize along altitudinal gradients. New Phytol 199:264–276. https://doi.org/10.1111/nph.12247
Ellstrand NC (2003) Current knowledge of gene flow in plants: implications for transgene flow. Philos Trans R Soc B Biol Sci 358:1163–1170
European Commission (2010) A decade of EU-funded GMO research. Directorate-General for Research and Innovation Biotechnologies, Agriculture, Food. European Union, Luxembourg, Belgium. ISBN:978-92-79-16344-9. https://doi.org/10.2777/97784
FAO (2017) The future of food and agriculture – trends and challenges. FAO, Rome. ISBN:978-92-5-109551-5
Frewer LJ, Miles S, Marsh R (2002) The media and genetically modified foods: evidence in support of social amplification of risk. Risk Anal 22:701–711. https://doi.org/10.1111/0272-4332.00062
Fu YB, Somers DJ (2009) Genome-wide reduction of genetic diversity in wheat breeding. Crop Sci 49:161–168. https://doi.org/10.2135/cropsci2008.03.0125
Guo Y, Feng Y, Ge Y, Tetreau G, Chen X, Dong X, Shi W (2014) The cultivation of Bt corn producing Cry1Ac toxins does not adversely affect non-target arthropods. PLoS One 9(12):e114228. https://doi.org/10.1371/journal.pone.0114228
Hardigan MA, Laimbeer PE, Newton L et al (2017) Genome diversity of tuber-bearing Solanum uncovers complex evolutionary history and targets of domestication in the cultivated potato. PNAS 114:E9999–E10008. https://doi.org/10.1073/pnas.1714380114
Harper JL, Hawksworth DL (1994) Biodiversity: measurement and estimation. Phil Trans R Soc Lond B 345:5–12
Heap I (2014) Global perspective of herbicide-resistance weeds. Pest Manag Sci 70:1306–1315. https://doi.org/10.1002/ps.3696
Hilbeck A, Otto M (2015) Specificity and combinatorial effects of *Bacillus thuringiensis* cry toxins in the context of GMO environmental risk assessment. Front Environ Sci 3:71. https://doi.org/10.3389/fenvs.2015.00071
Hilbeck A, Baumgartner M, Fried PM, Bigler F (1998) Effects of transgenic *Bacillus thuringiensis*-cornfed prey on mortality and development time of immature *Chrysoperlacarnea* (Neuroptera: Chrysopidae). Environ Entomol 27:480–487
ISAAA (2017) Global status of commercialized biotech/GM crops in 2017: biotech crop adoption surges as economic benefits accumulate in 22 years. ISAAA brief no. 53. ISAAA, Ithaca
Juhasz-Nagy P (1993) Notes on compositional diversity. Hydrobiologia 249:173–182
Kanianska R (2016) Agriculture and its impact on land-use, environment, and ecosystem services. In: Almusaed A (ed) The influence of land use and anthropogenic impacts of landscape creation. IntechOpen. https://doi.org/10.5772/63719
Kessler C, Economidis I (2001) A review of results. EC-sponsored research on safety of genetically modified organisms. European Commission. European Communities. ISBN:92-894-1527-4
Lam HM et al (2010) Resequencing of 31 wild and cultivated soybean genomes identifies patterns of genetic diversity and selection. Nat Genet 42:1053–1059. https://doi.org/10.1038/ng.715

Latham JR, Love M, Hilbeck A (2017) The distinct properties of natural and GM cry insecticidal proteins. Biotech Genet Eng Rev 33(1):62–96. https://doi.org/10.1080/02648725.2017.135729 5
Lohaus K, Vidal S, Thies C (2013) Farming practices change food web structures in cereal aphid-parasitoid-hyperparasitoid communities. Oecologia 171(1):249–259
Lovei GL, Bøhn T, Hilbeck A (2010) Ecosystem services and genetically modified organisms. Third World Network, Penang. ISBN:978-967-5412-13-4
Lu BR, Yang C (2009) Gene flow from genetically modified rice to its wild relatives: assessing potential ecological consequences. Biotechnol Adv 27:1083–1091
Lucht JM (2015) Public acceptance of plant biotechnology and GM crops. Viruses 7:4254–4281. https://doi.org/10.3390/v7082819
Magurran AE (2003) Measuring biological diversity. Blackwell Publishing, Oxford, 256p
Marshall EJP (2001) Biodiversity, herbicides and non-target plants. In: BCPC conference weeds. BCPC, Farnham, pp 855–862
Mathur V, Javid L, Kushrestha S, Mandal A, Reddy AA (2017) World cultivation of genetically modified crops: opportunities and risks. In: Lichtfouse E (ed) Sustainable agriculture reviews, vol 25. Springer, Cham
Mertens M (2008) Assessment of environmental impacts of genetically modified plants BfN – Skripten 217. New York, Federal Agency for Nature Conservation
National Academy of Sciences, Engineering, and Medicine (2016) Genetically engineered crops: experiences and prospects. The National Academies Press, Washington, DC. https://doi.org/10.17226/23395
Nawaz MA, Yang SH, Chung G (2018) Wild soybeans: an opportunistic resource for soybean improvement. In: Grillo O (ed) Rediscovery of landraces as a resource for the future. IntechOpen. https://doi.org/10.5772/intechopen.74973
Nawaz MA, Mesnage R, Tsatsakis AM, Golokhvast KS, Yang SH, Antoniou MN, Chung G (2019) Addressing concerns over the fate of DNA derived from genetically modified food in the human body: a review. Food Chem Toxicol 124:423–430. https://doi.org/10.1016/j.fct.2018.12.030
Nicholls CI, Altieri MA (2012) Plant biodiversity enhances bees and other insect pollinators in agroecosystems. A review. Agron Sustain Dev 33:257–274
Otiman IP, Badea EM, Buzdugan L (2008) Roundup ready soybean, a Romanian story. Bull UnivAgricSci Vet Med Cluj-Napoca Anim Sci Biotechnol 65:352–357
Palinkas Z, Kiss J, Zalai M, Szenasi A, Dorner Z, North S, Woodward G, Balog A (2017) Effects of genetically modified maize events expressing Cry34Ab1, Cry35Ab1, Cry1F and CP4 EPSPS proteins on arthropod complex food webs. Ecol Evol 7(7):2286–2293
Palinkas Z, Zalai M, Szensai A, Dorner Z, Kiss J, North S, Woodward G, Balog A (2018) Arthropods dataset from different genetically modified maize events and associated controls. Sci Data 5(180019). https://doi.org/10.1038/sdata.2018.19
Papa R, Gepts P (2004) Gene flow between crops and their wild progenitors. Encycl Plant Crop Sci:488–491. https://doi.org/10.1081/E-EPCS120017095
Prakash D, Verma S, Bhatia R, Tiwary BN (2011) Risks and precautions of genetically modified organisms. ISRN Ecol 369573:13. https://doi.org/10.5402/2011/369573
Rodrigo-Simon A, de Maagd RA, Avilla C, Bakker PL, Molthoff JW, Gonzalez-Zamora J, Ferre J (2006) Lack of detrimental effects of Bacillus thuringiensis Cry toxins on the insect predator green lacewing: a toxicological, histopathological, and biochemical approach. Appl Environ Microbiol 72:1595–1603
Schulze ED, Beck E, Müller-Hohenstein K (2002) Pflanzenökologie. Springer, Heidelberg
Schutte G, Eckerstorfer M, Rastelli V et al (2017) Herbicide resistance and biodiversity: agronomic and environmental aspects of genetically modified herbicide-resistant plants. Environ Sci Eur 29:5. https://doi.org/10.1186/s12302-016-0100-y
Sneller CH (2003) Impact of transgenic genotypes and subdivisions on diversity within elite North American soybean germplasm. Crop Sci 43:409–414. https://doi.org/10.2135/cropsci2003.4090

Szenasi A, Palinkas Z, Zalai M, Schmitz OJ, Balog A (2014) Short-term effects of different genetically modified maize varieties on arthropod food web properties: an experimental field assessment. Sci Rep 4:5315. https://doi.org/10.1038/srep05315

Tappeser B, Reichenbecher W, Teichmann H (2014) Agronomic and environmental aspects of the cultivation of genetically modified herbicide-resistant plants. A BfNFOEN-EAA-Joint paper, BfN-Skripten. http://www.bfn.de/fileadmin/MDB/documents/service/skript362.pdf

Tsatsakis AM, Nawaz MA, Tutelyan VA, Golokhvast KS, Kalantzi O-I, Chung DH et al (2017a) Impact on environment, ecosystem, diversity and health from culturing and using GMOs as feed and food. Food Chem Toxicol 107:108–121

Tsatsakis AM, Nawaz MA, Kouretas D, Balias G, Savolainen K, Tutelyan VA, Chung G (2017b) Environmental impacts of genetically modified plants: a review. Environ Res 156:818–833

Van Eenennaam AL (2013) GMOs in animal agriculture: time to consider both costs and benefits in regulatory evaluations. J Anim Sci Biotechnol 4:37. https://doi.org/10.1186/2049-1891-4-37

Verma SR (2013) Genetically modified plants: public and scientific perceptions. ISRN Biotechnol. https://doi.org/10.5402/2013/820671

Vrbnicanin S, Bozic D, Pavlovic D (2017) Gene flow from herbicide-resistant crops to wild relatives. In: Pacanoski Z (ed) Herbicide resistance in weeds and crops. IntechOpen. https://doi.org/10.5772/67645

Wrinn KM, Evans SC, Rypstra AL (2012) Predator cues and an herbicide affect activity and emigration in an agrobiont wolf spider. Chemosphere 87(4):390–396. https://doi.org/10.1016/j.chemosphere.2011.12.030

Xu X, Liu X, Ge S et al (2012) Resequencing 50 accessions of cultivated and wild rice yields markers for identifying ergonomically important genes. Nat Biotech 30:105–111. https://doi.org/10.1038/nbt.2050

Zhang C, Wohlhueter R, Zhang H (2016) Genetically modified foods: a critical review of their promise and problems. Food Sci Human Wellness 5:116–123. https://doi.org/10.1016/j.fshw.2016.04.002

# Part II
# GM Crops and Their Impacts

# Impact of GM Crops on Farmland Biodiversity

**Tlou S. Masehela, Bruce Maseko, and Eugenia Barros**

**Abstract** The potential impact of genetically modified (GM) crops on farmland biodiversity has been a topic of interest since the adoption of the Cartagena Protocol on Biodiversity and the first commercial release of GM crops in 1996. The adoption and use of genetically modified (GM) crops have not only revolutionised the farming landscapes but also continue to fuel debates on the benefits versus risks to both the environment and human health. Farmlands provide diverse habitats for numerous fauna and flora. At the same time, these lands are also used for various agricultural practices such as the rearing of livestock and crop production. The impacts of agriculture on biodiversity and the environment have been detailed in many studies, and recommendations for management towards their sustainable use point to an intricate balance between the needs for biodiversity conservation while ensuring agricultural productivity – for sustainable food production. This chapter details the different pressures and impacts (positive and negative) experienced by farmland biodiversity resulting from the cultivation of GM crops. Here, we present some examples – on a case-by-case basis – to illustrate not only the complexities of some of the issues (i.e. benefits, risks and uncertainties) but highlight the need to understand and account for various dynamics that results in trade-offs at a farm management level, socio-economic implications as well as conservation priorities.

**Keywords** Genetically modified crops · Farmland · Biodiversity · Agricultural production · Biotech crops · Modern agriculture

T. S. Masehela (✉)
South African National Biodiversity Institute, Kirstenbosch Research Centre, Cape Town, South Africa
e-mail: T.Masehela@sanbi.org.za

B. Maseko
AfricaBio, The Innovation Hub, Pretoria, South Africa

E. Barros
Eugenia Barros Biosciences and Consulting, Sandton, Gauteng, South Africa

 
A. Chaurasia et al. (eds.), *GMOs*, Topics in Biodiversity and Conservation 19,
https://doi.org/10.1007/978-3-030-53183-6_2

# Introduction

The use of genetically modified (GM) seeds continues to have a profound impact on agricultural productivity since their first commercial release in 1996 (Brookes and Barfoot 2017).

The global area planted of biotech crops has been on a steady increase over the past 23 years with some countries reporting increases or decreases in a hectare of biotech crops during the period (ISAAA 2017). Drought, pest and disease pressures, invasive species, land-use changes are some of the multiple drivers that contribute to the reported fluctuations in hectare (Tesfahun 2018; Cogato et al. 2019). Developing countries such as Brazil, India and China grow more biotech crops than some of the industrialised countries (ISAAA 2017). Four main biotech crops include canola, maize, cotton and soybean with improved insect and herbicide protection traits.

Despite enhanced and efficiency gains at farm level attributed to the planting of GM crops, their adoption has been slow in Africa and Europe, mainly due to regulatory hurdles and perceived risks of GM crops on the environment. There has been several biosafety concerns and potential risk of GM crop especially on the environment. This chapter focuses on the potential impact of GM crops on farmland biodiversity. It explores some of the environmental concerns associated with direct and indirect impact on GM crops on farmland biodiversity.

## *Farmland Biodiversity*

Farmlands also referred to as agricultural lands, are typically devoted to various agricultural practices such as the rearing of livestock and production of crops – mainly to produce food for humans. At the same time, farmland also provides habitat for numerous species. Sustainable use of farmland requires an intricate balance between the needs of biodiversity conservation and increased agricultural productivity (Erisman et al. 2016). For example, the rearing of livestock and production of crops – mainly to produce food for humans. At the same time, farmland also provides habitat for numerous species. In farmlands, biodiversity includes all animals and plants that exist and interact at different ecosystem levels to perform different functions. In essence, all species at different levels can be linked to essential ecosystem services such as pollination of crops (Kremen et al. 2007), the breaking down of organic matter in the soil and nutrient cycling (Dominati et al. 2010), and seed dispersal in some instances (Sekercioglu 2006). All farmlands' biodiversity is critical to maintaining a healthy and well-functioning ecosystem. Furthermore, greater biodiversity in farmlands can contribute to their resilience (Oliver et al. 2015).

Over the years, maintaining the balance for competing demands on land uses has become a global concern (Lambin and Meyfroidt 2011). Across all continents, there is increasing competition between ecological goals for biodiversity management and conservation (McShane et al. 2011), ever-increasing production needs to meet the human food demand (Rosegrant et al. 2001), and in some instances land needed for housing infrastructure and economic development opportunities (Lambin and Meyfroidt 2011). The scale of these needs as well as their trends and activities are bound to differ from region to region, as well as the rate at which they take place. However, it is widely known that when it comes to agricultural practices that are highly productive, a large ecological footprint is almost-always experienced (Laurance et al. 2014).

## *Context of Farmland Biodiversity*

The Convention on Biological Diversity (CBD) describes biological diversity (biodiversity) as the variability among living organisms from all sources including, among other things, terrestrial, marine and other aquatic ecosystems and the ecological complexes of which they are part of: this includes diversity within species, between species and of the ecosystems. Biodiversity is essential for ecosystem productivity and function, whereby the ecosystem's ability to adapt to change over time or even maintain resilience to environmental change is of great importance (CBD 2000). Therefore, a balance must be found on farmlands which allow biodiversity to thrive to support agricultural production without impacting on ecosystem functions and the services it provides. The Millennium Ecosystem Assessment (MEA) outlines various categories of services associated with biodiversity and ecosystems and also described benefits (goods and services) people obtain from the ecosystem (MEA 2005). All four categories, (1) provisioning services; (2) regulating services; (3) cultural services and (4) supporting services, can be linked to farmland biodiversity. These services are necessary for the functioning and productivity of all ecosystems (Costanza et al. 1997).

Agricultural activities associated with farmlands often threaten the very same biodiversity responsible for supporting various ecosystems and their associated services (Díaz et al. 2006). The intensification and specialisation of farming have led to a simplification of agricultural landscapes and a loss of both natural and semi-natural habitats (Erisman et al. 2016). This is further supported by work from Sub-Saharan Africa which indicates that the massive growth of agriculture is directly associated with increasing threats to biodiversity at all time scales (FAO 2017). The Rural Investment Support for Europe (RISE Foundation) highlights that farmland biodiversity in Europe has declined drastically over the last decades. Their data shows that 76% of species and 70% of habitats related to agriculture currently have unfavourable conservation status. In other regions, marginal farmland is being neglected to allow for the natural process of succession for recovery. Unfortunately, this has resulted in both the loss of farmland habitats and associated species (Plieninger et al. 2014).

The interactions between modern agriculture and biodiversity in most parts of the world have also become challenging to research, understand and manage.

Interestingly, these ecological problems seem to be associated with agriculture in general (Raven 2010) and not just the cultivation of crops derived from a particular technology. This is because crops do not damage the environment simply because they are derived from a particular technology (e.g. Genetic Modification), as some farming practices, such as the incorrect and at times overuse of chemicals for weed control can impact negatively on the environment. Such problems are similar for conventional and GM crops.

## *Agricultural Practices on Farmlands and Surrounding Landscapes*

Extensive literature globally has recognised the rapid global human population growth and also attributed the challenges associated with this trend (Cohen 2003; Lutz et al. 2004; Godfray et al. 2010). Top of the agenda is the land that will be essential to agricultural production to meet a likely rise in demand for food (Gibbs et al. 2010; Tilman et al. 2011). In light of the above, there is no doubt that the pressure is bound to increase the conversion of land in particular, for those areas that have not been previously used for agriculture. At the same time, previous and current agricultural land might intensify due to increased activities for mainly food production. In any of the scenarios, farmlands, as well as agroecosystem would have to somehow adapt to rapid changes in associated activities while some losses or reductions in their biodiversity should be expected.

As early as the late 1990s, there were indications that the expansion of land in agricultural use takes place all the time (FAO 2003). This increase was firmly attributed to countries that combine growing needs for food and employment with limited access to technology packages that could increase intensification of cultivation on land already in agricultural use, for example, sub-Saharan Africa, South America and East Asia (FAO 2003). Does this mean that the use of technology is the answer to slowing down or curbing the expansion of land for agriculture, for the benefit of conservation and biodiversity management at different landscapes?

The International Service for the Acquisition of Agri-biotech Applications (ISAAA) already indicated GM crops to have helped lower agriculture's ecological footprint, reduced the quantity and range of agrochemical applications, and reduced pressure on the natural environment and indirectly contributing to the conservation of biodiversity. Could this be one of the approaches that are indicative of the sentiments that modern agriculture and biodiversity protection are compatible? Alternatively, an indication that modern farming practices are and will continue to evolve to facilitate the preservation of farmland biodiversity?

It has been more than 20 years since the introduction of genetically modified (GM) crops, also called biotech crops, and their commercialisation has no doubt revolutionised global agriculture (ISAAA 2017). The development and use of GM crops have made significant, positive contributions to sustainability in agriculture around the world. Yet, uncertainties surrounding their safety for the environment and impact on biodiversity continue to increase (Singh et al. 2006; Bawa and Anilakumar 2013). Although most research has found GM crops to increase crop productivity, and at times on smaller margins of land – which in turn conserves natural habitats/vegetation (Brookes and Barfoot 2017), there is still healthy scepticism which points to their risks and potential harm to the environment and ecosystems (Paull 2017). For example, the environmental harm associated with chemical use in GM crops, increased cultivation area (leading to habitat loss), genetic flow (subsequently contamination) to wild relatives, harm to non-target organisms and many others (Andow and Zwahlen 2006).

## *Pressures Associated with Agriculture in Farmlands*

Climate change is a risk factor that has a profound impact on agricultural production. It influences crop production, water supply, input supplies and several components of farming systems (Adams et al. 1998). Climate change is the critical driver of food insecurity in the developing countries since it affects agricultural productivity and several other components of the food value system, including storage, access and use (Campbell et al. 2016). Climate change-induced risks on agrarian production are directly linked to a food security risk for the smallholder farmers who directly depend on agriculture for their livelihood (FAO 2016).

Smallholder farmers in rain-fed areas with inadequate farming systems are most vulnerable because of their dependence on climate and natural resources (Pereira 2017). Viable crops are not guaranteed due to limited financial, inadequate infrastructure and access to improved seed varieties (Ehui and Pender 2005). The impact of climate change at the farm level includes drought that leads to crop failure and loss of arable land (Morton 2007). As a result, this increases uncertainty to the availability of water in many regions, especially drought-prone countries (Liu et al. 2018). Excessive heat increases the threat of new pests, diseases and weeds, which leads to crop damage and yield loss (Oerke 2006), and compromises the farmer's ability to produce quality crops sustainably. The El Niño-induced drought severely affected agricultural productivity in several regions across the globe (Niang et al. 2014). Shortly, after the drought, the fall armyworm (*Spodoptera frugiperda*) invaded sub-Saharan Africa leading to heavy maize yield losses (Day et al. 2017).

The potential use of GM crops to mitigate against drought and potential damage of fall armyworm in Africa has been proposed (Abrahams et al. 2017; Prasanna et al. 2018). However, the potential impact of GM crops on biodiversity has been the subject of scientific and societal debates globally. After more than two decades of commercial cultivation of GM crops, there is a substantial body of literature reviews addressing the potential beneficial impacts of GM crops on the environment, especially in the context of farmland diversity (Icoz and Stotzky 2008; Carpenter 2011). In general, the decline of farmland biodiversity is a challenge globally and has an impact on many economic sectors, especially the agricultural industry. Farmland biodiversity is critical to maintaining healthy ecosystems that enable farmers to produce food sustainably while presenting natural resources.

GM crops have been reported to contribute to climate change mitigation and adaptation (Ortiz-Bobea and Tack 2018). Drought-tolerant maize varieties have been reported to perform better across several countries in eastern and southern Africa (Setimela et al. 2018). A stacked gene maize product MON89034 which expresses two Cry proteins have been reported to offer better protection against the control of lepidopteran pests in South Africa (Van den Berg et al. 2013). Improved GM maize varieties have the potential to be used as part of the integrated pest management to control fall armyworm. GM crops can help farmers adapt and become more resilient, mitigate and adapt better to climate change. Recent breakthroughs in new breeding innovations continue to introduce improved crop varieties with improved ability to withstand insect pest, disease and the effects of climate change. This chapter will review the knowledge on the impact of GM crops on farmland biodiversity.

## Case Studies (Evidence-Based) on GM Crops Impact on Farmlands and Surrounding Landscapes

The adoption of first-generation GM crops worldwide, where crops were genetically modified for resistance to insect pests and herbicides, saw an increase in agricultural land use, raising concerns on the potential risk on soil biodiversity and other environmental effects by the conversion of extensive forested lands into farmland. Similarly, scientific and public debates over the environmental impact of GM crops became complex and fiercely intense. In other publishing and media platforms, the debates often became extremely emotional. Behind all this, was one question: are GM crops safe for the environment? Lately, scientists are also beginning to ask how GM crops can help protect our habitats and ecosystems.

Noting these trends and interactions in different platforms, it is once more an indication that assessing the impacts of GM crops is a complex exercise. Multiple factors need to be considered focusing on both the potential risks (negative impacts) and benefits (positive impacts). Following some of the definitions by Carpenter (2011), our use and subsequent application of the term "impact" in this section encompasses any impacts at the farmland scale level, inclusive of all organisms. However, examples presented in Table 1 demonstrate and support the different impacts and are limited to the availability of scientific evidence and literature.

Studies by Toft (2004), Beckie et al. (2011), Carpenter (2011), Brookes and Barfoot (2017) and Romeis et al. (2019) demonstrate that the adoption of transgenic crops that confer tolerance to a herbicide has resulted in less herbicide being applied in the field to control weeds which in turn has less impact on non-target sites and soil biodiversity. Similarly, insect-resistant GM crops expressing the proteins from the soil bacterium *Bacillus thuringiensis*, Bt proteins, are beneficial to the

**Table 1** Summary of studies that document the impacts of GM crops on the environment and farmlands

| Crop type | Trait/event | Countries covered | Year(s) of study | Documented area of impact | Documented impact (+/–) | Level of impact (primary/secondary) | Reference |
|---|---|---|---|---|---|---|---|
| Bt crops in general | Insect resistance | Multiple | Not specified | Reduced insecticide use in Bt crops can enhance the conservation of natural enemies | Positive | Primary | Romeis et al. (2019) |
| Canola | Herbicide-resistant | Canada | Not specified | Reduced environmental impacts of herbicides | Positive | Primary | Beckie et al. (2011) |
| Maize | Insect resistance | South Africa | 3 years | Plant and arthropod diversity | Positive | Primary | Botha et al. (2015) |
| Cotton | Insect resistance | USA and Australia | Not specified | Diversity of beneficial insects | Positive | Primary | Carpenter et al. (2002) |
| Cotton | Insect resistance | Philippines | Not specified | Insect abundance and diversity | Positive | Primary | Yorobe and Quicoy (2006) |
| GM crops in general | Not specific | Multiple | 16 years | Land under cultivation | Positive | Secondary | Brookes and Barfoot (2013) |
| GM crops in general | Not specific | Multiple | Not specified | Multiple | Positive | Primary and secondary | Carpenter (2011) |
| Sugar beet | Insect resistance | Denmark | Not specified | Increase in populations of insects, spiders and other arthropods contributed to an increase in bird life | Positive | Primary and secondary | Toft (2004) |
| GM crops in general | Not specific | Multiple | 20 years | Significant reduction in the environmental impact associated with insecticide and herbicide | Positive | Primary | Brookes and Barfoot (2017) |
| Maize | Herbicide | South Africa | 3 years | Minimum tillage in a communal area of Kwazulu-Natal | Positive | Secondary | Gouse et al. (2006) |
| Maize | Insect resistance | South Africa | 2 years | Arthropod diversity | Neutral | Primary | Truter et al. (2014) |

environment because they require less insecticide (if any at all) for pest control. Bt toxins are confined within the plant and are very specific to the pests that attack them and have been shown to have no direct effect on natural insect predators and parasites they have been tested against. With a decrease in the use of insecticides, beneficial insects and microscopic organisms of all kinds can grow among the Bt crops creating a natural environment where they can help control the growth of secondary pests.

A large-scale practice in agriculture mainly increasing cultivation area has led to a postulated decrease in biodiversity worldwide. In most instances, this has also led to the fragmentation and degradation of natural habitat bordering field margins. Often enough, the idea of high abundance and richness diversity in arthropods associated with GM crops may seem unrealistic. However, a few studies have shown this to be possible (Table 1). In particular, patterns of arthropod species in and around different GM crops showed that these areas are not species-poor ecosystems. In most instances, richness in arthropod diversity was evident (Carpenter et al. 2002; Yorobe and Quicoy 2006; Carpenter 2011; Botha et al. 2015).

Another requirement of traditional agriculture is the ploughing and turning over of the soil through some tillage as a way to control weeds. Herbicide-resistant crops have contributed significantly to the adoption of no-tillage or limited tillage with great benefit to the environment. As the soil is not agitated, soil erosion is reduced because the root systems can hold the soil together, also contributing to a decrease in contamination of waterways with agrochemicals from runoff soil of cultivated fields. The minimum disturbance to the soil has seen the restoration of natural populations of organisms, like soil bacteria and especially mycorrhizal fungi, and earthworms (Gouse et al. 2006; Buiatti et al. 2013; Alori and Babalola 2018). Besides, other fauna such as ants and birds benefit from soil-dwelling organisms.

## Positive Attributes of GM Crops in Farmlands

### *Restoring Biodiversity of Farmlands*

Herbicide and insect-resistant GM crops reduce the need for herbicide applications and eliminate insecticide applications. This contributes to a decrease in $CO_2$ emissions, conservation of soil, water retention, a saving of fossil fuels, lessens erosion through the use of no-till, restores agricultural ecosystems and prevents contamination of water sources. All these factors contribute to a reduction in the agricultural environmental footprint.

## *Coexistence: Farmland Biodiversity and GM Crops*

The potential adverse effects associated with GM crops and biodiversity (the environment) remain a pertinent issue even though some positive impacts have been demonstrated through different studies. In some instances, the release of any GMO into the environment possesses the general difficulty in predicting the occurrence and extent of long-term environmental effects for various reasons. For example, there might not be enough resources or time to carry out the necessary research and monitoring required for the intended study design or research. In other cases, environmental protection goals and assessment endpoints are not set or defined, therefore defining and contextualising harm and addressing various pathways for potential hard become difficult to impossible.

From the current literature, the presence of GM crops and their cultivation has shown that both herbicide and insect-resistant GM crops can contribute positively to the reduction in the agricultural environment footprint and lead to the positive management of farmland biodiversity.

# GM Crops Impacts Arising from Farm and Landscape Management Practices

## *Development of Resistance*

Development of resistance to insecticides and herbicides by some target species is a regular occurrence with conventional chemical insecticides and herbicides. Indeed, there can be an "arms race" to discover novel pesticides while pest organisms are being inadvertently selected for resistance to those already being applied. This is also a possible occurrence with Bt-expressing crops and transgenic herbicide-resistant crops resulting in both instances in the additional application of the particular insecticide or herbicide. However, this is a natural phenomenon as pests and diseases are likely to develop resistance irrespective of whether the crop was developed using traditional breeding or GM technology. This is more a farm management issue than a GM crop issue, and there are some strategies to prevent or at least delay the development of resistance.

The planting of 'refuge areas' in which the GM crop is grown near a non-GM crop provides a source of non-resistant target species and thus dilutes the pressure for resistant genes to be selected (IUCN 2007). Unfortunately, the adoption of the refuge strategy is not always followed as it was the case of Bt maize (MON810) in some areas of South Africa. Non-compliance to the refuge strategy contributed to the accelerated development of resistance in those areas, with another contributing factor being a variation in the insecticidal Bt protein content in GM plants due to variations in local environmental conditions (Kunert 2011). Large amounts of Bt protein need to be produced continuously in a Bt plant to limit the growth of the insect larvae and the potential build-up of resistance (Chilcutt and Tabashnik 2004). New strategies are now available that should prevent the development of resistance, which includes more efficient Bt proteins as well as combining two or more types of Bt proteins. Another strategy to restrict the development of resistance is the rotation of crops.

## *Gene Flow*

GM crops, like conventional crops, have the potential to crossbreed with non-GM crops or related wild species growing nearby. However, both GM and non-GM crop plants cannot survive without human intervention, and the GM traits only increase plant fitness in their natural habitats and do not survive well in the wild. Traditional domesticated crops are more susceptible to insect pests and disease than native species that naturally produce toxins to keep pests and pathogens under control. If some weeds acquire resistance to the introduced toxins, the resistance can be expected to soon be diluted from the weeds because it offers them little or no increased fitness. Gene flow is species-specific, and to date, there are no reports of a trait being successfully established in a wild plant. As a result, the best way to evaluate transgenic crops is on a case-by-case basis. Should the trait offer no advantage to the wild populations, it is unlikely to survive, and so no long-term effects or damage are impacted on the biodiversity of wild species (Buiatti et al. 2013).

A case of hybridisation between a GM grass and a wild species was reported for creeping bentgrass, *Agrostis stolonifera*, commonly used in golf courses. It was genetically modified to have a herbicide-resistant gene that enabled spraying the golf course with a common herbicide to kill the weeds but leaving the grass unharmed. This grass is a perennial and wind-pollinated, and the herbicide-resistant gene was found to have been transferred through hybridisation to a closely related wild grass, *A. gigantea*, some nine miles from its origin (Landry 2015). As the GM grass is unlikely to encounter the herbicide in the wild, it seems improbable that it would have an advantage over wild grasses. However, this highlights the importance of understanding the reproductive behaviour of the GM plant and the function of the genetic modification before releasing it into the wild. Most cultivated GM crops are not perennial and do not have wild relatives nearby (Hopkin 2006).

In most countries where GM crops have been introduced, they have no wild relatives, and thus studies that explore GM, non-GM crops and farmland biodiversity co-existence are very limited. Most only explore GM versus non-GM maize on a case-by-case scenario (Friesen et al. 2003; Devos et al. 2005; Groenewald and Groenewald 2009).

## Concluding Remarks

Interactions among GM crops and farmland biodiversity likely differ among regions and landscapes. This means that any assessment seeking to make observations and document these aspects should be specific to the crop, region and landscape. Although this then implies the need for sound knowledge, skills and resources, the assessments would be more thorough, and the scope can be focused on key issues through prioritisation as the need arises. It is to be anticipated that farmlands' biodiversity conditions can have trade-offs with socio-economic as well as conservation priorities. However, these cannot be established unless gaps in baseline data are addressed, sufficient data is accumulated and thoroughly interrogated, and concrete conclusions reached on the respective aspects. Furthermore, any assessment on GM interaction with or on farmland biodiversity needs to account for both negative and positive impacts.

## References

Abrahams P, Bateman M, Beale T et al (2017) Fall armyworm: impacts and implications for poverty in sub-Saharan Africa: pathways out of the spiral. Agric Econ 32:225–242

Adams RM, Hurd BH, Lenhart S, Leary N (1998) Effects of global climate change on agriculture: an interpretative review. Clim Res 11(1):19–30

Alori ET, Babalola OO (2018) Microbial inoculants for improving crop quality and human health in Africa. Front Microbiol 9:2213. https://doi.org/10.3389/fmicb.2018.02213

Andow DA, Zwahlen C (2006) Assessing environmental risks of transgenic plants. Ecol Lett 9(2):196–214

Bawa AS, Anilakumar KR (2013) Genetically modified foods: safety, risks and public concerns – a review. J Food Sci Technol 50(6):1035–1046

Beckie HJ, Harker KN, Legere A, Morrison MJ, Seguin-Swartz G, Falk KC (2011) GM canola: the Canadian experience. Farm Policy J 8(1):43–49

Botha M, Siebert SJ, Van den Berg J, Maliba BG, Ellis SM (2015) Plant and arthropod diversity patterns of maize agro-ecosystems in two grassy biomes of South Africa. Biodivers Conserv 24(7):1797–1824

Brookes G, Barfoot P (2013) Key environmental impacts of global genetically modified (GM) crop use 1996–2011. GM Crop Food 4(2):109–119

Brookes G, Barfoot P (2017) Environmental impacts of genetically modified (GM) crop use 1996–2015: impacts on pesticide use and carbon emissions. GM Crop Food 8(2):117–147. https://doi.org/10.1080/21645698.2017.1309490

Buiatti M, Christou P, Pastore G (2013) The application of GMOs in agriculture and in food production for a better nutrition: two different scientific points. Gene Nutr 8:255–270

Campbell BM, Vermeulen SJ, Aggarwal PK, Corner-Dolloff C, Girvetz E, Loboguerrero AM, Ramirez-Villegas J, Rosenstock T, Sebastian L, Thornton PK (2016) Reducing risks to food security from climate change. Glob Food Sec 11:34–43

Carpenter JE (2011) Impact of GM crops on biodiversity. GM Crops 2(1):7–23

Carpenter J, Felsot A, Goode T, Hammig M, Onstad D, Sankula S (2002) Comparative environmental impacts of biotechnology-derived and traditional soybean, corn, and cotton crops. Council for Agricultural Science and Technology, Ames. www.cast-science.org

CBD (2000) Cartagena protocol on biosafety to the convention on biological diversity. Montreal, Canada
Chilcutt CF, Tabashnik BE (2004) Contamination of refuges by *Bacillus thuringiensis* toxin genes from transgenic maize. Proc Natl Acad Sci U S A 101:7526–7529
Cogato A, Meggio F, Migliorati M, Marinello F (2019) Extreme weather events in agriculture: a systematic review. Sustainability 11(9):2547
Cohen JE (2003) Human population: the next half century. Science 302(5648):1172–1175
Costanza R, d'Arge R, De Groot R, Farber S, Grasso M, Hannon B, Limburg K, Naeem S, O'neill RV, Paruelo J, Raskin RG (1997) The value of the world's ecosystem services and natural capital. Nature 387(6630):253
Day R, Abrahams P, Bateman M, Beale T, Clottey V, Cock M, Colmenarez Y, Corniani N, Early R, Godwin J (2017) Fall armyworm: impacts and implications for Africa. Outlook Pest Manag 28:196–201
Devos Y, Reheul D, De Schrijver A (2005) The co-existence between transgenic and non-transgenic maize in the European Union: a focus on pollen flow and cross-fertilization. Environ Biosaf Res 4(2):71–87
Díaz S, Fargione J, Chapin FS III, Tilman D (2006) Biodiversity loss threatens human well-being. PLoS Biol 4(8):e277
Dominati E, Patterson M, Mackay A (2010) A framework for classifying and quantifying the natural capital and ecosystem services of soils. Ecol Econ 69(9):1858–1868
Ehui S, Pender J (2005) Resource degradation, low agricultural productivity, and poverty in sub-Saharan Africa: pathways out of the spiral. Agric Econ 32:225–242
Erisman JW, van Eekeren N, de Wit J, Koopmans C, Cuijpers W, Oerlemans N, Koks BJ (2016) Agriculture and biodiversity: a better balance benefits both. AIMS Agric Food 1(2):157–174
FAO (2003) World agriculture: towards 2015/2030 an FAO perspective. FAO, Rome
FAO (2016) Climate change and food security: risks and responses. FAO, Rome
FAO (2017) The future of food and agriculture trends and challenges. FAO, Rome
Friesen LF, Nelson AG, Van Acker RC (2003) Evidence of contamination of pedigreed canola (*Brassica napus*) seedlots in western Canada with genetically engineered herbicide resistance traits. Agron J 95(5):1342–1347
Gibbs HK, Ruesch AS, Achard F, Clayton MK, Holmgren P, Ramankutty N, Foley JA (2010) Tropical forests were the primary sources of new agricultural land in the 1980s and 1990s. Proc Natl Acad Sci 107(38):16732–16737
Godfray HCJ, Beddington JR, Crute IR, Haddad L, Lawrence D, Muir JF, Pretty J, Robinson S, Thomas SM, Toulmin C (2010) Food security: the challenge of feeding 9 billion people. Science 327(5967):812–818
Gouse M, Piesse J, Thirlle C (2006) Output and labour effects of GM maize and minimum tillage in a communal area of Kwazulu-Natal. J Dev Perspect 2(2):192–207
Groenewald S, Groenewald J-H (2009) Current practice and potential impact of in-field separation strategies for GM and non-GM maize. BSA DSD 09-01:1–22
Hopkin M (2006) Escaped GM grass could spread bad news. Nature News http://www.nature.com/news/2006/060807/full/060807-17.html
Icoz I, Stotzky G (2008) Review: fate and effects of insect-resistant Bt crops in soil ecosystems. Soil Biol Biochem 40:559–586
ISAAA (2017) Global status of commercialized biotech/GM crops in 2017: biotech crop adoption surges as economic benefits accumulate in 22 years, ISAAA brief no. 53. ISAAA, Ithaca
IUCN (2007) The World Conservation Union, Current knowledge of the impacts of genetically modified organisms on biodiversity and human health – information paper
Kremen C, Williams NM, Aizen MA, Gemmill-Herren B, LeBuhn G, Minckley R, Packer L, Potts SG, Roulston TA, Steffan-Dewenter I, Vázquez DP (2007) Pollination and other ecosystem services produced by mobile organisms: a conceptual framework for the effects of land-use change. Ecol Lett 10(4):299–314
Kunert KJ (2011) How effective and safe is Bt-maize in South Africa? S Afr J Sci 107:9–10

Lambin EF, Meyfroidt P (2011) Global land use change, economic globalization, and the looming land scarcity. Proc Natl Acad Sci 108(9):3465–3472

Landry H (2015) Challenging evolution: how GMOs can influence genetic diversity. Special Edition: genetically modified organisms and our food. http://sitn.hms.harvard.edu/flash/2015/challenging-evolution-how-gmos-can-influence-genetic-diversity/

Laurance WF, Sayer J, Cassman KG (2014) Agricultural expansion and its impacts on tropical nature. Trends Ecol Evol 29(2):107–116

Liu WB, Sun FB, Lim WH, Zhang J, Wang H, Shiogama H, Zhang YQ (2018) Global drought and severe drought-affected population in 1.5 and 2°C warmer worlds. Earth Syst Dynam 9:267–283

Lutz W, Sanderson WC, Scherbov S (2004) The end of world population growth in the 21st century: new challenges for human capital formation and sustainable development. Earthscan

McShane TO, Hirsch PD, Trung TC, Songorwa AN, Kinzig A, Monteferri B, Mutekanga D, Van Thang H, Dammert JL, Pulgar-Vidal M, Welch-Devine M (2011) Hard choices: making trade-offs between biodiversity conservation and human well-being. Biol Conserv 144(3):966–972

MEA (2005) Ecosystems and human well-being: current state and trends. Island Press, Washington, DC

Morton JF (2007) The impact of climate change on smallholder and subsistence agriculture. Proc Natl Acad Sci U S A 104:19680–19685

Niang I, Ruppel OC, Abdrabo MA, Essel A, Lennard C, Padgham J, Urquhart P (2014) Africa. In: Barros VR, Field CB, Dokken DJ, Mastrandrea MD, Mach KJ, Bilir TE, Chatterjee M, Ebi KL, Estrada YO, Genova RC, Girma B, Kissel ES, Levy AN, MacCracken S, Mastrandrea PR, White LL (eds) Climate change 2014: impacts, adaptation, and vulnerability. Part B: Regional aspects. Contribution of Working Group II to the Fifth Assessment Report of the Intergovernmental Panel on Climate Change. Cambridge University Press, Cambridge, pp 1199–1265

Oerke EC (2006) Crop losses to pests. J Agric Sci 144(1):31–43

Oliver TH, Heard MS, Isaac NJ, Roy DB, Procter D, Eigenbrod F, Freckleton R, Hector A, Orme CDL, Petchey OL, Proença V (2015) Biodiversity and resilience of ecosystem functions. Trends Ecol Evol 30(11):673–684

Ortiz-Bobea A, Tack J (2018) Is another genetic revolution needed to offset climate change impacts for US maize yields? Environ Res Lett 13:124009

Paull J (2017) Genetically modified organisms (GMOs) as invasive species. J Environ Prot Sustain Dev 4:31–37

Pereira L (2017) Climate change impacts on agriculture across Africa. In: Oxford research encyclopedia of environmental science. Oxford University Press, New York

Plieninger T, Hui C, Gaertner M, Huntsinger L (2014) The impact of land abandonment on species richness and abundance in the Mediterranean Basin: a meta-analysis. PLoS One 9(5):e98355

Prasanna BM, Huesing JE, Virginia RE, Peschke M (eds) (2018) Fall armyworm in Africa: a guide for integrated pest management, 1st edn. CIMMYT, Mexico, CDMX

Raven PH (2010) Does the use of transgenic plants diminish or promote biodiversity? New Biotechnol 27(5):528–533

Romeis J, Naranjo SE, Meissle M, Shelton AM (2019) Genetically engineered crops help support conservation biological contral. Biol Control 130:136–154

Rosegrant MW, Paisner MS, Meijer S, Witcover J (2001) 2020 global food outlook: trends, alternatives, and choices, vol 11. International Food Policy Research Institute, Washington, DC

Sekercioglu CH (2006) Increasing awareness of avian ecological function. Trends Ecol Evol 21(8):464–471

Setimela PS, Zaman-Allah M, Ndoro OF (2018) Performance of elite maize varieties tested on-farm trials in eastern and southern Africa. CIMMYT

Singh OV, Ghai S, Paul D, Jain RK (2006) Genetically modified crops: success, safety assessment, and public concern. Appl Microbiol Biotechnol 71(5):598–607

Tesfahun W (2018) Climate change mitigation and adaptation through biotechnology approaches – a review. Cogent Food Agric 4:1. https://doi.org/10.1080/23311932.2018.1512837

Tilman D, Balzer C, Hill J, Befort BL (2011) Global food demand and the sustainable intensification of agriculture. Proc Natl Acad Sci 108(50):20260–20264
Toft J (2004) Denmark national report: co-existence bypassing issues. Precautionary expertise for GM crops. http://technology.open.ac.uk/cts/national/denmark%20national%report%2OPEG.pdf
Truter J, Van Hamburg H, Van Den Berg J (2014) Comparative diversity of arthropods on Bt maize and non-Bt maize in two different cropping systems in South Africa. Environ Entomol 43(1):197–208
Van den Berg J, Hilbeck A, Bøhn T (2013) Pest resistance to Cry1Ab Bt maize: field resistance, contributing factors and lessons from South Africa. Crop Prot 54:154–160
Yorobe JM, Quicoy CB (2006) Economic impact of Bt corn in the Philippines. Philipp Agric Sci 89(3):258

# GM Crops: Resistance Development and Impact on Biodiversity

**Lombardo Luca, Trenti Massimiliano, and Zelasco Samanta**

**Abstract** Despite a consolidated increase in their employment, herbicide-tolerant (HT) and insect-resistant (IR) GM varieties have commonly been suspected to represent a threat to biodiversity. In this chapter, we analyze the major concerns related to the potential persistence and invasiveness of GM plants, selection of superweeds and resistant insects, effects on non-target organisms including vertical and horizontal gene flow, with the acquisition of antibiotic resistance and biological contamination. Mitigation measures to reduce the ecological impact on biodiversity are briefly considered.

**Keywords** Herbicide-tolerant GM crops · Insect-resistant GM crops · Biodiversity · Environment · Herbicide-resistant weeds · Non-target organisms

## Introduction

Twenty-four years after the first commercial cultivation of genetically modified (GM) crops in 1996, there are about 18 million biotech plant farmers in 30 countries, for a total of 189.8 million hectares worldwide. However, 99% of GM crops are represented by only four species: soybean (50%), maize (31%), cotton (13%), and canola (5%) (ISAAA 2017), and 70–80% of harvested genetically engineered (GE) biomass is destined for feeding food-producing animals (Flachowsky et al. 2012). Nearly 47% of cultivated GM crops (88.7 Mha) have been engineered for

L. Luca (✉)
Center Agriculture Food Environment (C3A), University of Trento, Trento, Italy

Research and Innovation Centre, Fondazione Edmund Mach, San Michele all'Adige, Italy
e-mail: luca.lombardo@unitn.it

T. Massimiliano
Department of Biotechnology, University of Verona, Verona, Italy

Z. Samanta
CREA Research Centre for Olive, Citrus and Tree Fruit, Rende, Italy



A. Chaurasia et al. (eds.), *GMOs*, Topics in Biodiversity and Conservation 19,
https://doi.org/10.1007/978-3-030-53183-6_3

herbicide tolerance, ~12% (23.3 Mha) for insect resistance, and ~41% for expressing both (stacked) traits. Less than 1% is represented by nutritionally improved and virus-resistant varieties (ISAAA 2017; Lombardo and Grando 2020).

## *Herbicide-Tolerant GM plants*

As the vast majority of herbicide-tolerant (HT) crops involve glyphosate tolerance, glyphosate, *N-(phosphonomethyl)-glycine*, is the most widely used herbicide worldwide. Glyphosate inhibits the chloroplast enzyme 5-enolpyruvylshikimate-3-phosphate synthase (EPSPS), arresting the biosynthesis of folates and aromatic amino acids (phenylalanine, tyrosine, and tryptophan) in the shikimate pathway. Since this enzyme is present only in plants and microorganisms, it has been considered safe for humans and animals; nevertheless, the International Agency for Research on Cancer (IARC) classified it as probably carcinogenic to humans (Group 2A) in March 2015.

The approaches to develop glyphosate tolerance in plants via genetic engineering are based on: (1) the introduction of genes encoding enzymes (e.g., glyphosate *N*-acetyltransferase -gat- and glyphosate oxidase -gox-) degrading the herbicide into nontoxic compounds (Castle et al. 2004; Pedotti et al. 2009); (2) the insertion of a gene coding for a glyphosate-insensitive EPSPS, derived from *Agrobacterium sp.* (*CP4-EPSPS*; Meilan et al. 2002), *Pseudomonas fluorescens* (*G2-EPSPS;* Liu et al. 2015), or *Ochrobactrum anthropic* (a mutant EPSP; Tian et al. 2011); or (3) the overproduction of the unmodified target protein permitting normal metabolism to occur (Rao 2015). Coexpression of *gat* and *G2-EPSPS* genes has also been exploited (Dun et al. 2014; Guo et al. 2015). The CP4-EPSPS alone makes soybean approximately 50-fold less sensitive to glyphosate (Nandula et al. 2007). Further, HT GE crops have been obtained by introducing the *Streptomyces-derived pat* or *bar* gene into plants. Both genes are quite similar and encode the enzyme phosphinothricin acetyl transferase (PAT), which inactivates the l-form of phosphinothricin (l-PPT), thus conferring tolerance to glufosinate, a broad-spectrum synthetic herbicide composed of a mixture of the d- (with no biological activity) and l- isomer (the active ingredient). l-phosphinothricin targets glutamine synthetase (GS), leading to the accumulation of lethal levels of ammonia in plants (Dayan et al. 2015).

Several commercially available crops have been engineered to express the *dmo* (dicamba mono-oxygenase) gene from *Stenotrophomonas maltophilia* (syn. *Pseudomonas maltophilia*), that confers tolerance to the herbicide dicamba (2-methoxy-3,6-dichlorobenzoic acid) by converting it to 3,6-dichlorosalicylic acid (3,6-DCSA), a compound with no herbicidal activity. Dicamba functions by increasing plant growth to an unsustainable level.

Another GM HT trait on the market at present is sulfonylurea resistance, obtained through the insertion of the *gm-hra* gene coding for a modified acetolactate synthase (ALS) enzyme conferring resistance to ALS inhibitors.

Cotton transformed to express the aryloxyalkanoate di-oxygenase 12 (AAD-12) protein is resistant to the 2,4-D herbicide, as this enzyme catalyzes the side chain degradation of 2,4-dichlorophenoxyacetic acid (Tan et al. 2005). A second gene of the *aad* family, *aad-1*, encodes an enzyme that detoxifies aryloxyphenoxypropionate (FOPs) herbicides—a group of auxins and acetyl CoA carboxylase (ACCase) inhibitors—, via an α-ketoglutarate-dependent dioxygenase reaction (Wright et al. 2010).

Bromoxynil-resistant cotton was one of the first transgenic HT crops (Stalker et al. 1996). Bromoxynil inhibits photosynthesis by binding the D1 protein of photosystem II. An oilseed rape, Westar Oxy-235, was created for commercial use and is tolerant to bromoxynil (3,5-dibromo- 4-hydroxybenzonitrile) thanks to the *bxn* gene from *Klebsiella ozaenae*, which encodes a nitrilase enzyme that detoxifies the herbicide. However, it was withdrawn from the market in 2002, and thus, no bromoxynil-tolerant varieties are currently available.

The single-gene transgenic events currently approved for commercial release are listed in Table 1.

Because intense selective pressure created by the prolonged exposure to a single herbicide generally leads to the development of more and more resistant weeds (Evans et al. 2015), the new tendency is toward the stacking of multiple genes into a single genotype-gene pyramiding (Table 2). For example, in 2016, Monsanto developed a variety of seeds called Roundup 2 Xtend which was designed to withstand both Roundup (the commercial name of glyphosate) and dicamba.

A different approach toward HR is the employment of oligonucleotide-directed mutagenesis (ODM), a highly site-specific targeted gene correction, gene knockout, allelic replacement, and/or genetic modification of plant genomes (Beetham et al. 1999; Lombardo and Zelasco 2016). This technology works through the introduction of complementary chimeric DNA/RNA oligonucleotide (CO) from 20 to 80 nucleotides in size, differing for at least one nucleotide from the endogenous gene sequence. The resulting helical distortion induced by the mismatch is subsequently recognized by the cell's DNA repair machinery and the base pair corrected using the DNA sequence of the chimera as a template. ODM has been exploited to develop the marketed sulfonylurea-tolerant GM canola, branded SU Canola™, to survive the application of a specific imidazolinone herbicide (Beyond™) inhibiting the ALS enzyme (also called acetohydroxyacid synthase AHAS), through an insensitive mutated form of the acetolactate synthase (ALS) enzyme.

## *Insect-Resistant GM plants*

The currently available biotech insect-resistant (IR) crops differ in their end uses (food, feed, or cultivation) and include cotton (*Gossypium hirsutum*), eggplant (*Solanum melongena*), maize (*Zea mays*), tomato (*Lycopersicon esculentum*), poplar (*Populus* spp.), potato (*Solanum tuberosum*), rice (*Oryza sativa*), soybean (*Glycine max*), and sugarcane (*Saccharum* spp.) with 47, 1, 200, 2,1, 30, 3, 6, and 1

**Table 1** List of the commercially approved transgenic crops transformed with a single HT gene, including the mechanism of action

| Gene | Gene source | Product | Function | Crop |
|---|---|---|---|---|
| *cp4 epsps (aroA:CP4)* | *Agrobacterium tumefaciens* strain CP4 | Herbicide-tolerant form of 5-enolpyruvulshikimate-3-phosphate synthase (EPSPS) enzyme | Decreases binding affinity for glyphosate, thereby conferring increased tolerance to glyphosate herbicide | Alfalfa – *Medicago sativa* |
| | | | | Argentine canola – *Brassica napus* |
| | | | | Cotton – *Gossypium hirsutum* |
| | | | | Creeping bentgrass – *Agrostis stolonifera* |
| | | | | Maize – *Zea mays* |
| | | | | Potato – *Solanum tuberosum* |
| | | | | Soybean – *Glycine max* |
| | | | | Sugar beet – *Beta vulgaris* |
| | | | | Wheat – *Triticum aestivum* |
| *gat4621/gat 4601* | *Bacillus licheniformis* | Glyphosate N-acetyltransferase enzyme | Catalyzes the inactivation of glyphosate, conferring tolerance to glyphosate herbicides | Argentine canola – *Brassica napus* |

| Gene | Gene source | Product | Function | Crop |
|---|---|---|---|---|
| *bar* | *Streptomyces hygroscopicus* | Phosphinothricin N-acetyltransferase (PAT) enzyme | Eliminates herbicidal activity of glufosinate (phosphinothricin) herbicides by acetylation | Argentine canola – *Brassica napus* |
| | | | | Chicory – *Cichorium intybus* |
| | | | | Cotton – *Gossypium hirsutum* |
| | | | | Maize – *Zea mays* |
| | | | | Rice – *Oryza sativa* |
| | | | | Soybean – *Glycine max* |
| *als* | *Arabidopsis thaliana* | Herbicide-tolerant enzyme acetolactate synthase (als) | Allows the synthesis of essential amino acids in the presence of sulfonylurea herbicides | Flax – *Linum usitatissimum* |
| *pat* | *Streptomyces viridochromogenes* | Phosphinothricin N-acetyltransferase (PAT) enzyme | Eliminates herbicidal activity of glufosinate (phosphinothricin) herbicides by acetylation | Maize – *Zea mays* |
| | | | | Polish canola – *Brassica rapa* |
| | | | | Soybean – *Glycine max* |
| | | | | Sugar beet – *Beta vulgaris* |
| *pat (syn)* | Synthetic form of *pat* gene derived from *Streptomyces viridochromogenes* strain Tu 494 | Phosphinothricin N-acetyltransferase (PAT) enzyme | Eliminates herbicidal activity of glufosinate (phosphinothricin) herbicides by acetylation | Argentine canola – *Brassica napus* |
| | | | | Maize – *Zea mays* |

(continued)

**Table 1** (continued)

| Gene | Gene source | Product | Function | Crop |
|---|---|---|---|---|
| *bxn* | *Klebsiella pneumoniae* subsp. *Ozaenae* | Nitrilase enzyme | Eliminates herbicidal activity of oxynil herbicides (e.g., bromoxynil) | Argentine canola – *Brassica napus* |
| | | | | Cotton – *Gossypium hirsutum* |
| | | | | Tobacco – *Nicotiana tabacum* |
| *surB* | *Nicotiana tabacum* | Herbicide-tolerant acetolactate synthase (ALS) enzyme | Confers tolerance to sulfonylurea herbicides and other acetolactate synthase (ALS)-inhibiting herbicides | Carnation – *Dianthus caryophyllus* |
| *S4-HrA* | *Nicotiana tabacum* cv. Xanthi | Herbicide-tolerant acetolactate synthase (ALS) enzyme | Allows the plant to synthesize essential amino acids in the presence of sulfonylurea herbicides | Cotton – *Gossypium hirsutum* |
| *mepsps* | *Zea mays* | Modified 5-enolpyruvylshikimate-3-phosphate synthase (EPSPS) enzyme | Confers tolerance to glyphosate herbicides | Maize – *Zea mays* |
| *2mepsps* | *Zea mays* | 5-Enolpyruvyl shikimate-3-phosphate synthase enzyme (double mutant version) | Decreases binding affinity for glyphosate, thereby increasing tolerance to glyphosate herbicide | Cotton – *Gossypium hirsutum* |
| | | | | Maize – *Zea mays* |
| *aad-1* | Synthetic form of the aad-1 gene from *Sphingobium herbicidovorans* | Aryloxyalkanoate dioxygenase 1 (AAD-1) protein | Detoxifies 2,4-D herbicide by side-chain degradation and degrades the R-enantiomers of aryloxyphenoxypropionate | Maize – *Zea mays* |
| *epsps grg23ace5* | Synthetic gene; similar to epsps grg23 gene from soil bacterium *Arthrobacter globiformis* | Modified 5-enolpyruvylshikimate-3-phosphate synthase (EPSPS) protein or EPSPS ACE5 protein | Confers tolerance to glyphosate herbicides | Maize – *Zea mays* |

| Gene | Gene source | Product | Function | Crop |
|---|---|---|---|---|
| *csr1-2* | *Arabidopsis thaliana* | Modified acetohydroxyacid synthase large subunit (AtAHASL) | Confers tolerance to imidazolinone herbicides | Soybean – *Glycine max* |
| *dmo* | *Stenotrophomonas maltophilia* strain DI-6 | Dicamba mono-oxygenase enzyme | Confers tolerance to the herbicide dicamba (2-methoxy-3,6-dichlorobenzoic acid) by using dicamba as a substrate in an enzymatic reaction | Soybean – *Glycine max* |
| *aad-12* | *Delftia acidovorans* | Aryloxyalkanoate di-oxygenase 12 (AAD-12) protein | Catalyzes the side chain degradation of 2,4-D herbicide | Soybean – *Glycine max* |
| | | | | Maize – *Zea mays* |
| *avhppd-03* | Oats (*Avena sativa*) | p-Hydroxyphenylpyruvate dioxygenase | Tolerance to mesotrione herbicide | Soybean – *Glycine max* |
| *hppdPF W336* | *Pseudomonas fluorescens* strain A32 | Modified p-hydroxyphenylpyruvate dioxygenase (hppd) enzyme | Confers tolerance to HPPD-inhibiting herbicides (such as isoxaflutole) by reducing the specificity for the herbicide's bioactive constituent | Soybean – *Glycine max* |
| | | | | Cotton – *Gossypium hirsutum* |
| *goxv247* | *Ochrobactrum anthropi* strain LBAA | Glyphosate oxidase | Confers tolerance to glyphosate herbicides by degrading glyphosate into aminomethylphosphonic acid (AMPA) and glyoxylate | Argentine canola – *Brassica napus* |
| | | | | Maize – *Zea mays* |
| | | | | Polish canola – *Brassica rapa* |
| | | | | Sugar beet – *Beta vulgaris* |

Source: ISAAA GM Approval Database, 2019

**Table 2** Commercially approved transgenic crops transformed with multiple HT genes

| Gene | Crop |
|---|---|
| *gat4601 + bar* | Argentine canola – *Brassica napus* |
| *cp4 epsps (aroA:CP4)+ goxv247* | Argentine canola – *Brassica napus* |
| | Maize – *Zea mays* |
| | Polish canola – *Brassica rapa* |
| | Sugar beet – *Beta vulgaris* |
| *pat (syn) + cp4 epsps (aroA:CP4)* | Argentine canola – *Brassica napus* |
| | Maize – *Zea mays* |
| *bar + cp4 epsps (aroA:CP4)* | Argentine canola – *Brassica napus* |
| | Cotton – *Gossypium hirsutum* |
| *bar + cp4 epsps (aroA:CP4)+ goxv247* | Argentine canola – *Brassica napus* |
| *pat (syn) +aad-12 +pat* | Cotton – *Gossypium hirsutum* |
| *cp4 epsps (aroA:CP4) + pat* | Cotton – *Gossypium hirsutum* |
| | Maize – *Zea mays* |
| *cp4 epsps (aroA:CP4)+pat+aad-12* | Cotton – *Gossypium hirsutum* |
| | Soybean – *Glycine max* |
| *aad-12 + pat* | Cotton – *Gossypium hirsutum* |
| | Soybean – *Glycine max* |
| *dmo + bar+ cp4 epsps (aroA:CP4)* | Cotton – *Gossypium hirsutum* |
| *2mepsps +bar* | Cotton – *Gossypium hirsutum* |
| *dmo + bar* | Cotton – *Gossypium hirsutum* |
| *pat + meps* | Maize – *Zea mays* |
| *pat + aad-1* | Maize – *Zea mays* |
| *pat +cp4 epsps (aroA:CP4) + aad-1* | Maize – *Zea mays* |
| *zm-hra +gat4621* | Maize – *Zea mays* |
| *zm-hra +gat4621 + pat* | Maize – *Zea mays* |
| *aad-1 + cp4 epsps (aroA:CP4)* | Maize – *Zea mays* |
| *pat (syn) +mepsps* | Maize – *Zea mays* |
| *dmo + pat* | Maize – *Zea mays* |
| *2mepsps + pat* | Maize – *Zea mays* |
| *cp4 epsps (aroA:CP4) + pat + goxv247* | Maize – *Zea mays* |
| *aad-12 + 2mepsp + pat* | Soybean – *Glycine max* |
| *gm-hra + cp4 epsps (aroA:CP4)* | Soybean – *Glycine max* |
| *dmo +cp4 epsps (aroA:CP4)* | Soybean – *Glycine max* |
| *gm-hra + gat4601* | Soybean – *Glycine max* |
| *2mepsps+ hppdPF W336* | Soybean – *Glycine max* |
| *2mepsps + hppdPF W336 + pat* | Soybean – *Glycine max* |
| *pat + dmo + cp4 epsps (aroA:CP4)* | Soybean – *Glycine max* |
| *pat + avhppd-03* | Soybean – *Glycine max* |

Source: ISAAA GM Approval Database 2019

approved events, respectively. The main strategy to develop insect-resistant plants via genetic engineering has been based on the use of the Bt Cry genes from the soil bacterium *Bacillus thuringiensis*. Cry proteins are solubilized in the insect midgut where intestinal proteases process the formed protoxins and cleave the C or N terminus. This activates the toxins which recognize binding sites on the midgut brush border membrane surface and form ion channels or pores in the epithelial membrane, leading to cell lysis and eventually death (Lombardo et al. 2016). Different Bt toxin genes have been stacked using conventional breeding by cross-hybridization and selection involving transgenic donor(s), giving rise to the main commercial biotech lines WideStrike™, Bollgard® II, Bollgard® III, and Twinlink™ for cotton; Agrisure®, Duracade™, Viptera™, Herculex™ RW, YieldGard™ VT Triple SmartStax™ Pro x Enlist™, and Genuity® VT Triple Pro™ for maize. A variant is represented by the Agrisure®, Duracade™, and Viptera™ maize lines by Syngenta with stacked genes coding for the Bt Cry1Ab delta-endotoxin and the vegetative insecticidal protein Vip3Aa20.

Alternative strategies use protease inhibitors (PIs), proteins which are abundantly present in plant reproductive, storage, and vegetative tissues. They are constitutively expressed or wound-induced, making them an important strategy of natural plant defense against phytophagous insects (Zhu-Salzman and Zeng 2015). Protease inhibitors bind tightly to the enzyme's active site of digestive protease, lowering the hydrolysis rate of proteases on the protein substrate. They act in a substrate-like manner resulting in a stable complex, which is more stable than the enzyme-substrate and enzyme-product complexes, leading to hypersecretion of the digestive enzymes to compensate the inhibition. This over-induction of protein synthesis causes an inability for the insect to acquire essential amino acids, with severe delayed development, maturation, and procreation (Smigocki et al. 2013). Following catalysis mechanisms, five classes of proteases (serine, threonine, cysteine, aspartic, and metallo catalytic types) have been described (Zhu-Salzman and Zeng 2015). As the two major classes in the digestive systems of phytophagous insects are the serine (predominant in Lepidoptera and Diptera) and cysteine proteases (Homoptera and Coleoptera orders class) (Smigocki et al. 2013), transgenic plants expressing inhibitors of these two proteinase types have been developed to combat insect pests (Hilder et al. 1987; Srinivasan et al. 2006) but have even been found to act against nematodes (Vishnudasan et al. 2005; Chan et al. 2010), bacteria (Khadeeva et al. 2009), and viruses (Gutierrez-Campos et al. 1999). This strategy has been used for the development of the transgenic cotton SGK 321, containing genes encoding a Bt toxin and a seed-expressed Bowman-Birk type trypsin proteinase inhibitor from cowpea (CpTI), commercially available in China since 1999. Other plant protein antimetabolites including lectins (Vandenborre et al. 2011), alpha-amylase inhibitors (Campbell et al. 2011), chitinases (Wang et al. 2005), and biotin-binding proteins (Cooper et al. 2009) altering the digestive system of phytophagous pests are still at the experimental stage in GM plants.

Table 3 describes insect resistance-encoding genes usually inserted in commercially approved GM crops.

**Table 3** List of genes inserted in commercially available IR crops, with gene source, product, and function

| Gene | Gene source | Product | Function |
|---|---|---|---|
| Trait: *Coleopteran insect resistance* | | | |
| *cry34Ab1/cry35Ab1* | *Bacillus thuringiensis* strain PS149B1 | Cry34Ab1 delta-endotoxin | Confers resistance to coleopteran insects particularly corn rootworm by selectively damaging their midgut lining |
| *cry3A* | *Bacillus thuringiensis* subsp. tenebrionis | cry3A delta-endotoxin | Confers resistance to coleopteran insects by selectively damaging their midgut lining |
| *cry3Bb1* | *Bacillus thuringiensis* subsp. *kumamotoensis* | Cry3Bb1 delta-endotoxin | Confers resistance to coleopteran insects particularly corn rootworm by selectively damaging their midgut lining |
| *dvsnf7* | Western corn rootworm (*Diabrotica virgifera virgifera*) | Double-stranded RNA transcript containing a 240 bp fragment of the WCR Snf7 gene | RNAi interference resulting to down-regulation of the function of the targeted Snf7 gene leading to Western Corn Rootworm mortality |
| *mcry3A* | Synthetic form of cry3A gene from *Bacillus thuringiensis* subsp. *tenebrionis* | Modified Cry3A delta-endotoxin | Confers resistance to coleopteran insects particularly corn rootworm pests by selectively damaging their midgut lining |
| Trait: *Hemipteran insect resistance* | | | |
| *mCry51Aa2* | *Bacillus thuringiensis* | Modified Bt Cry51Aa2 protein | Confers resistance to hemipteran insects *Lygus hesperus* and *L. lineolaris* by selectively damaging their midgut lining |
| Trait: *Lepidopteran insect resistance* | | | |
| *cry1A* | *Bacillus thuringiensis* | delta-endotoxin of the Cry1A group | Confers resistance to lepidopteran insects by selectively damaging their midgut lining |
| *cry1A.105* | *Bacillus thuringiensis* subsp. *kumamotoensis* | Cry1A.105 protein which comprises the Cry1Ab, Cry1F and Cry1Ac proteins | Confers resistance to lepidopteran insects by selectively damaging their midgut lining |
| *cry1Ab* | *Bacillus thuringiensis* subsp. *kurstaki* | Cry1Ab delta-endotoxin | Confers resistance to lepidopteran insects by selectively damaging their midgut lining |

(continued)

**Table 3** (continued)

| Gene | Gene source | Product | Function |
|---|---|---|---|
| *cry1Ab* (truncated) | Synthetic form of Cry1Ab from *Bacillus thuringiensis* subsp. *kumamotoensis* | Cry1Ab delta-endotoxin | Confers resistance to lepidopteran insects by selectively damaging their midgut lining |
| *cry1Ab-Ac* | Synthetic fusion gene derived from *Bacillus thuringiensis* | Cry1Ab-Ac delta-endotoxin (fusion protein) | Confers resistance to lepidopteran insects by selectively damaging their midgut lining |
| *cry1Ac* | *Bacillus thuringiensis* subsp. *kurstaki* strain HD73 | Cry1Ac delta-endotoxin | Confers resistance to lepidopteran insects by selectively damaging their midgut lining |
| *cry1C* | Synthetic gene derived from *Bacillus thuringiensis* | Cry1C delta-endotoxin | Confers resistance to lepidopteran insects, specifically Spodoptera |
| *cry1F* | *Bacillus thuringiensis* subsp. aizawai | Cry1F delta-endotoxin | Confers resistance to lepidopteran insects by selectively damaging their midgut lining |
| *cry1Fa2* | Synthetic form of cry1F gene derived from *Bacillus thuringiensis* subsp. aizawai | Modified Cry1F protein | Confers resistance to lepidopteran insects by selectively damaging their midgut lining |
| *cry2Ab2* | *Bacillus thuringiensis* subsp. kumamotoensis | Cry2Ab delta-endotoxin | Confers resistance to lepidopteran insects by selectively damaging their midgut lining |
| *cry2Ae* | *Bacillus thuringiensis* subsp. dakota | Cry2Ae delta-endotoxin | Confers resistance to lepidopteran insects by selectively damaging their midgut lining |
| *cry9C* | *Bacillus thuringiensis* subsp. *tolworthi* strain BTS02618A | Cry9C delta-endotoxin | Confers resistance to lepidopteran insects by selectively damaging their midgut lining |
| *mocry1F* | Synthetic form of cry1F gene from *Bacillus thuringiensis* subsp. aizawai | Modified Cry1F protein | Confers resistance to lepidopteran insects by selectively damaging their midgut lining |
| *pinII* | *Solanum tuberosum* | Protease inhibitor protein | Enhances defense against insect predators by reducing the digestibility and nutritional quality of the leaves |

(continued)

**Table 3** (continued)

| Gene | Gene source | Product | Function |
|---|---|---|---|
| *vip3A(a)* | *Bacillus thuringiensis* strain AB88 | VIP3A vegetative insecticidal protein | Confers resistance to feeding damage caused by lepidopteran insects by selectively damaging their midgut lining |
| *vip3Aa20* | *Bacillus thuringiensis* strain AB88 | Vegetative insecticidal protein (vip3Aa variant) | Confers resistance to feeding damage caused by lepidopteran insects by selectively damaging their midgut |
| Trait: *Multiple insect resistance* | | | |
| API | *Sagittaria sagittifolia* (arrowhead) | Arrowhead protease inhibitor protein A or B | Confers resistance to a wide range of insect pests |
| CpTI | *Vigna unguiculata* | Trypsin inhibitor | Confers resistance to a wide range of insect pests |
| *ecry3.1Ab* | Synthetic form of Cry3A gene and Cry1Ab gene from *Bacillus thuringiensis* | Chimeric (Cry3A-Cry1Ab) delta-endotoxin protein | Confers resistance to coleopteran and lepidopteran insects by selectively damaging their midgut lining |

Source: ISAAA GM Approval Database 2019

## *Virus-Resistant GM Plants*

Genetically modified plants that resist viruses exploit a pathogen-derived resistance (PDR) plant system. PDR-based strategies in transgenic plants provide the insertion of genes that encode particular viral proteins (coat and movement proteins, replicases) or the insertion of sequences coding for nucleic acids that prevent the expression of viral proteins through base-pairing interactions commonly referred to as post-transcriptional gene silencing (PTGS). The molecular mechanisms behind viral protein-mediated resistance are not fully understood (due to protein accumulation and/or small RNA accumulation), and are different for different viruses (Hull 2014). It is known, however, that PTGS is triggered by double-stranded RNAs which are cleaved into RNA duplexes of approximately 21 to 28 nucleotides by the Dicer enzyme (a ribonuclease), a member of the RNase III family. These short interfering siRNAs (including guide RNAs) are incorporated into a nuclease complex, called RNA-induced silencing complex (RISC), which degrades any mRNA sharing sequence homology (Ahlquist 2002). The target mRNA is not silenced until the virus vector infects the plant. The first commercial sale of transgenic virus-resistant crops obtained by employing the PDR concept dates back to 1995, in the USA, with virus-resistant yellow crookneck squash line designated ZW-20 (Federal Register, pp. 64187-64189) by Asgrow Co. (Kalamazoo, MI, USA). This squash was modified to contain the coat protein genes of watermelon mosaic virus 2 (WMV2), zucchini yellow mosaic virus (ZYMV), and cucumber mosaic virus (CMV). Additional GE varieties which exhibit coat protein-mediated resistance, that are approved for food and commercial planting in the USA, are the papaya ringspot virus (PRSV)-resistant "Rainbow" and "SunUp" papaya varieties (Fuchs and Gonsalves 2007); plum pox virus (PPV)-resistant "HoneySweet" plum released by the USDA; and Monsanto's "NewLeaf™ Plus" and NewLeaf™ Y potatoes with resistance to potato leafroll virus (PLV) and potato virus Y (PVY).

## Impact of GM Crops on Biodiversity

The actual impact of GM crops on biodiversity strongly depends on the technologies and the management systems that they replace. It is possible that switching to biotech crops may lead to the degradation of natural resources and the displacement of local varieties and, as a consequence, to an increase of crop vulnerability due to the higher uniformity of GM varieties (Steinbrecher 1996; Wolfenbarger and Phifer 2000; Kolady and Lesser 2012). However, this may happen even by switching to non-GE higher income crops. Moreover, some authors suggest that the progressive development of more and more GE varieties, possibly with GE traits incorporated into a large number of landraces, could avert these issues (Bowman et al. 2003; Sneller 2003; Qaim et al. 2005).

On the other hand, GM crops reduce the need for insecticide application and mechanical tillage for weed eradication, with positive effects for aerial and telluric species (Carpenter 2011; Fernandez-Cornejo et al. 2012). In the same regard, the development of glyphosate- and glufosinate-resistant varieties favored the decrease rate of application of more toxic herbicides for both animals and man (Giesy et al. 2000; Williams et al. 2000). Several aspects must be taken into account when analyzing the impact of GM crops on biodiversity; several different possible consequences of GM crop use are reported in the following sections.

### *Evolution of Herbicide-Resistant Weeds*

After a promising reduction in each of the first 6 years (1996–2001) following the introduction of HT GE crops, the use of agrochemicals (specifically glyphosate) in HT GE cropland compared to non-GE counterparts began to rise in the seventh year as a consequence of the onset of tougher-to-control herbicide-resistant "superweeds", and because of a concomitant decline in the volume of agrochemicals applied to conventional crops (Brookes and Barfoot 2017). The herbicidal activity of glyphosate was discovered in 1970 (Baird et al. 1971) and in 1974 it was marketed by Monsanto as Roundup. Since the first mass cultivation of genetically engineered glyphosate-tolerant crops in 1996, glyphosate use has risen almost 15-fold (from 51 to about 750 million kg), with HT GM crops accounting for about 56% of global glyphosate use (Benbrook 2016). This vast adoption has led to evolved glyphosate resistance in several important weeds. Naturally evolved glyphosate resistance was first discovered in annual ryegrass (*Lolium rigidum*) in an apple orchard in Australia in 1996 (Powles et al. 1998), but the first case of a glyphosate-resistant (GR) *Conyza canadensis* weed was found in a continuous GR soybean in 2000 (Van Gessel 2001). From eight reported resistant species in 2004, 43 weed species are now glyphosate-resistant (Heap 2019). According to the 2016 University of Illinois Plant Clinic Herbicide Resistance Report, the presence of glyphosate-resistant weeds occurs in 76.8% of the 593 crop fields from ten Midwestern states. The HT types include PPO inhibitors resistance in 62.5% and "stacked" resistance in 49% of the cases.

Mechanisms of resistance to glyphosate are better understood than for any other herbicide. These mechanisms include adaptive gene amplification, amino acid substitutions, vacuole sequestration, reduced translocation (the most common mechanism; Duke and Cedeira 2010), and rapid necrosis response (Sammons and Gaines 2014). Other potential mechanisms comprise the enhanced conversion of glyphosate to its relatively innocuous metabolite aminomethylphosphonic acid (AMPA; Reddy et al. 2008) and endophyte-mediated evolution of herbicide resistance (Handayani et al. 2017). In particular, *epsps* gene duplication and amino acid substitutions in EPSPS synthase represent two mechanisms of rapid genetic evolution under high selection pressure. In *epsps* gene duplication, increased transcript levels of *epsps* correspond to a marked resistance to glyphosate. This mechanism has been reported in 8 weed species (Gaines et al. 2010; Salas et al. 2012; Jugulam et al. 2014; Lorentz et al. 2014; Nandula et al. 2014; Chen et al. 2015; Wiersma et al. 2015; Malone et al. 2016; Molin et al. 2017; Dillon et al. 2017; Ngo et al. 2018), for a total of 17 amino acid variants (Gaines and Heap 2019) thus representing one of the major mechanisms enabling resistance evolution. Interestingly, Koo et al. (2018) described the occurrence of an extra ring chromosome harboring *epsps* in glyphosate-resistant *Amaranthus tuberculatus* as an evolutionary driver of herbicide resistance. It should however be stressed that weed resistance to herbicides is an extremely common natural phenomenon regarding 255 species and 163 different herbicides (Heap 2019) and may be considered more a consequence of the overuse of herbicides than the use of HT GM crops.

A collateral effect of the adoption of HT crops and the appearance of HT weeds is the reduced weed diversity because, whenever the herbicide is sprayed, the HT weed is more likely to result more persistent and/or invasive than the non-HT one.

## *Evolution of Bt-Resistant Insects*

The evolution of Bt resistance in herbivore insects has been first documented for diamondback moths (*Plutella xylostella*) and cabbage loopers (*Trichoplusia ni*) consistently exposed to Bt-containing pesticides (Chapman and Burke 2006). Interestingly, these pests showed reduced fitness in the absence of the toxin and resistance in wild populations rapidly declined under such conditions. Since 2002, field-evolved resistance to Bt crops has been reported for five major insect pests across different countries (Dively et al. 2016). Resistance to Bt corn-derived Cry1F and Cry1Ab was described in the fall armyworm *Spodoptera frugiperda* in the USA and Brazil (Matten et al. 2008; Dively et al. 2016). In 2007, the resistance evolved against Cry1F in fall armyworm pushed Dow AgroSciences and Pioneer Hi-Bred International to withdraw TC1507 maize from local commercial use in Puerto Rico (Storer et al. 2012). Among the other lepidopterans, the stem borer *Busseola fusca* evolved resistance to Bt corn producing Cry1Ab in two different areas in South Africa (Van Rensburg 2007; Kruger et al. 2009), *Helicoverpa zea* and *Helicoverpa armigera* evolved resistance to Bt cotton producing Cry1Ac in 8 years in the south-eastern USA and in 8 years in China (Luttrel and Ali 2007; Liu et al. 2010). In November 2009, the Monsanto Company declared that *Pectinophora gossypiella* could develop resistance to Bt cotton producing Cry1Ac in four districts in India (Abbas 2018). Growing resistance to Bt toxins (Cry3Bb and mCry3A) expressed in maize was also observed for western corn rootworm, *Diabrotica* var. *virgifera* (LeConte), in the USA (Shrestha et al. 2018). Different studies have indicated that a cross-resistance phenomenon can also occur among closely related Cry toxins, particularly those selected in single-gene Bt crops (Dively et al. 2016). *H. zea* showed cross-resistance to Cry1A proteins (1Ab, 1Ac, Cry1A.105) (Anilkumar et al. 2008; Welch et al. 2015), and *S. frugiperda* and *Ostrinia nubilalis* to Cry1A.105, Cry1Ab, and Cry1Ac proteins (Hernandez-Rodriguez et al. 2013). Santos-Amaya et al. (2015) found that Cry1F-resistant *S. frugiperda* selected in a Bt corn was also highly resistant to Cry1Ac and Cry1f toxins from Bt cotton, while a laboratory-selected strain of *S. frugiperda* resistant to Cry1A.105 and Cry2Ab2 showed cross-resistance to Cry1F (Yang et al. 2014).

Similarly, the onset of insects resistant to PIs has also been reported (Reeck et al. 1997; Girard et al. 1998). At first, plant PIs were considered good candidates and a number of transgenic plants expressing PIs-genes could be found in the literature (Duan et al. 1996; Gatehouse et al. 1997; Delledonne et al. 2001; Ribeiro et al. 2006; Zavala et al. 2008), but currently very few are commercially available. Insofar as PIs are naturally occurring proteins released in response to a physical injury or biological attack, a co-evolutionary process has been occurring between phytophagous insects and their host plants leading to sophisticated and flexible physiological responses to dietary PIs. Multiple adaptive strategies are adopted by insect herbivores such as overproduction of existing digestive proteases, increased expression of inhibitor-insensitive protease isoforms, and activation of proteases that hydrolyze and thus detoxify plant inhibitors Many insects are able to combine multiple strategies to circumvent the effects of PIs in very short order (Zhu-Salzman and Zeng 2015), making them less economically viable as a resistance strategy.

Evolved resistance to Bt crops involves the increased use of traditional insecticides with a higher mortality rate for the broader range of non-target species and higher toxicity for animals and humans. As an opposite effect, Bt resistance would allow pests to colonize other areas in which Bt-based insecticides are applied, such as organic farms and orchards, where their eradication would be hampered by the prohibition of the use of synthetic pesticides, implying higher costs for pest control and possibly modification of the trophic chain. Once again these risks are tightly coupled with the selective pressure generated by the cultivation of GM plants, whereas the transgene products of most currently marketed GM plants are constitutively expressed throughout the plant organs and the entire growing season, resulting in a longer persistence when comparing with "classical" pesticide-based pest-control strategies (Lövei et al. 2007). To date, there is no scientific evidence supporting differential selective pressure between the GM vs. non-GM strategies, and in fact resistance to around 300 insecticide compounds, including Bt pesticides, has been observed in at least 590 insect species, up to 2014 (Sparks and Nauen 2015).

## *Invasiveness of GM Plants*

To date first-generation GM crops resulted non-invasive in non-agricultural areas. Crawley et al. (1993) firstly evaluated the invasiveness of GM glufosinate-resistant canola in four habitat types in the UK, finding no increase in invasive ability, as substantiated by successive trials (Hall et al. 2005; Simard et al. 2005). A long-term study on 4 different IR and HT GM crops (a transgenic potato expressing either the insecticidal Bt or pea lectin, glufosinate-tolerant maize and oilseed-rape, and glyphosate-tolerant sugar beet), monitoring 12 different natural habitats over a period of 10 years was conducted in the United Kingdom. None of the crops, GM or conventional, increased in abundance at any of the sites and no significant differences in average recruitment between conventional and GM crops were found. Population size declined after the first year as a result of increased competition from native perennial plants. The few cases of increased GM plant survival that were significant in the short-term did not translate into long-term differences in invasiveness (Crawley et al. 2001). Further studies have confirmed these results (Beckie et al. 2006; Beckie and Owen 2007).

## *Vertical Gene Flow*

Vertical gene flow refers to the transfer of genes among different populations of the same or closely related species, including wild or weedy relatives, through reproductive processes. Such transfer can occur via pollen, seeds, or vegetative propagules, the relative importance of which varies according to plant species. Reproductive biology factors such as species fertility (i.e., male fertile or -sterile receptor plants), sexual compatibility, allogamous or autogamous pollination, wind and/or insect vectors, fertility pollen viability and longevity, synchrony of flowering or pollen production and environmental factors such as wind speed and direction, temperature, humidity, topography, presence of vegetation, and relative density of donor and receptor plant populations affect the likelihood of pollen-mediated gene flow in a specific region or environment (Warwick et al. 2009; Sanchez et al. 2016).

Hybridization with wild and weedy relatives has been described in 44 cultivated crops, including 12 of the 13 most widely cultivated crops (Ellstrand et al. 1999). In a study of transgene flow via pollen-mediated crop-to-crop and crop-to-wild relatives gene flow from six major crops discriminated low risk (potato, wheat, and barley), medium to high risk (sugar beet and maize) and high risk (oilseed rape) crops (Eastham and Sweet 2002). However, despite the fact that gene flow between GM cultivars with different IR and HT traits has led to the occurrence of doubly resistant volunteers, no altered weedy or invasiveness potential has been to date observed (Simard et al. 2005).

A particular circumstance could be represented by the "de-domestication" of the GM crop itself, namely the establishment of feral populations from cultivated varieties, as suggested by Lu and Snow (2005).

Since herbicide tolerance (Gealy et al. 2003) and insect resistance (Sachs et al. 1998; Zhang et al. 2000) are mostly inherited as dominant Mendelian traits, they can easily spread by cross-pollination between GM and non-GM plants that can occur at distances of several kilometers (Rieger et al. 2002; Watrud et al. 2004; Busi et al. 2008). Furthermore, although the stable or inheritable integration of a transgene into a related genome (introgression) beyond the first generation might be prevented or slowed by hybrid sterility, the presence of heterotic and abundant F1 weed-crop hybrids might enhance the probability of introgression of transgenic alleles. Finally, the degree of gene introgression from an allopolyploid crop (e.g., *Brassica napus*) to one of its relatives may depend on the genome on which the gene is located (Tomiuk et al. 2000). Nevertheless, the flow of HT transgenes to weedy relatives has not been described in the principal GM crops (soybean, cotton, and maize), as a likely absence of sexually compatible weedy species in their growing environments (Duke and Cedeira 2010).

The potential hazards related to gene flow are in particular related to the possible change in weediness of these non-target organisms, where either an increase or a reduction of the fitness through the acquisition of transgenic traits via hybridization is possible. Decreased weediness might produce a major reduction in weed biomass, that in agro-ecosystems may even play an important role in maintaining ecosystem services (Gaba et al. 2016; Bretagnolle and Gaba 2015) with possible depressing effects on species that feed on it. However, in the case of HT GM crops the hazards might appear only in the agro-ecosystems where the herbicide is applied. The ecological risk may be further amplified by seed-mediated gene flow allowing the long-distance movement of transgenes and ensuring a high survival rate. Also in this case the survival rate depends on some specific characteristics of the species such as seed vitality, germination ability and natural and induced dormancy mechanisms as well as agronomic (tillage, herbicide distribution) and environmental conditions.

Vertical transgene flow has been described from IR and HT *Brassica napus* to its wild relative *B. rapa* (Linder and Schmitt 1995; Halfhill et al. 2004; Warwick et al. 2008) and less frequently to wild radish (*Raphanus raphanistrum*; Gueritaine et al.

2002; Warwick et al. 2003), and to *S. arvensis* and *E. gallicum* (Warwick et al. 2003). Other examples involve the escape of transgenes to weedy species from glyphosate-resistant *Agrostis stolonifera* (Reichman et al. 2006), from glyphosate- and gluphosinate-resistant *Beta vulgaris* (Alibert et al. 2005; Darmency et al. 2007), from HT rice (Chen et al. 2004) and from IR *Cucurbita pepo* (Spencer and Snow 2001). Snow et al. (2003) demonstrated first that a transgene derived from a crop has the potential to increase the fitness of wild plants, and thus increase in frequency in wild populations, while Halfhill et al. (2004) observed that Bt *B. napus x B. rapa* transgenic hybrids have a high potential to produce transgenic seeds in backcrosses.

## *Horizontal Gene Flow*

Horizontal gene flow (HGF) or lateral gene flow is the movement of genetic information between sexually unrelated and incompatible organisms and is known to have occurred across the boundaries of species, genera, and even kingdoms. Despite being an extremely rare phenomenon, horizontal gene flow might theoretically apply to GM crops through the transfer of pesticide, herbicide, or antibiotic resistance to unrelated organisms or through the escape of agrobacterial vectors. Horizontal gene transfer from transgenic plants to bacteria has been proven in laboratory-scale trials (de Vries et al. 2004; Simpson et al. 2007; Rizzi et al. 2008; Pontiroli et al. 2009). However, entire transgenes with the regulatory portions of the DNA have never been found to be horizontally transferred (Rao 2015). Gene flow conferring pesticide and herbicide resistance would pose the same risks described for vertical gene flow, namely possible increased (or decreased) weed or pest burden and adverse effects on species communities, eventually resulting in possible ecological imbalances. One of the steps essential for a successful gene transfer is the stabilization of the foreign DNA into the recipient organism. This could occur through different recombination mechanisms. In line with current scientific knowledge on frequency rates, double homologous recombination scenarios are considered the most relevant. The efficiency of double homologous recombination in bacteria depends on nucleotide sequence identity and mostly on the length of the non-homologous DNA. Homologous recombination becomes increasingly inefficient with decreasing length of sequences with high identity and reaches under *in vitro* conditions a plateau phase with about 1 kb of sequences with high identity at both stretches of DNA. Recombination efficiencies also decrease with increasing size (from 1 to 2 kb the non-homologous insert in a double HR event; Gennaro et al. 2017).

Antibiotic resistance from GM crops may relate to the fact that most of the transgene constructs for developing GM plants, even only for research purpose, include antibiotic resistance genes like *nptII* (resistance to kanamycin), *hpt* (resistance to hygromycin B), *aad* (resistance to streptomycin and spectinomycin), and *bla* (resistance to ampicillin). Given the evidence that transgene DNA from GM plants can persist days to months (Widmer et al. 1997; Paget et al. 1998; Hay et al. 2002; Zhu et al. 2010) and even years (Gebhard and Smalla 1999) in the soil, thus becoming accessible to bacteria cells, the HGF-mediated spread of diseases in plants (if the recipient organism is a pathogenic bacteria infecting plants) as well as in plant-eating animals and men, cannot be *a priori* excluded (Pontiroli et al. 2007; Keese 2008). Notwithstanding, HGF of recombinant genes from GM plants to bacteria has never been shown under field conditions with GM plants used in agriculture (Badosa et al. 2004; Ma et al. 2011), but in a 2012 study, genetically engineered plasmids containing a synthetic version of the *bla* gene were found in six Chinese rivers (Chen et al. 2012), likely deriving from genetic engineering applications as the source of the ampicillin resistance. This issue can however be easily bypassed through the employment of several alternatives to antibiotic resistance marker genes now available (Breyer et al. 2014).

Eventually, as genetic transformation is usually mediated via agrobacterial vectors, residual engineered agrobacteria with all the diagnostic genes (*BtCry1Ac*, *API-A* – arrowhead double-headed proteinase inhibitors A, *NptII*, *Vir-G*-regulatory protein VirG- and ChvE-glutamine amidotransferase-like protein chyE) were detected in the soil after 24 months of GM plant cultivation in a greenhouse trial (Yang et al. 2006). As HGF by specific mechanisms such as conjugation, transduction, and transformation is a quite common phenomenon in bacteria (Dröge et al. 1998), further studies are necessary to better understand any potential hazard.

## *Impact on Non-target Organisms*

Some transgenic traits such as the pesticidal toxins expressed by Bt genes may affect non-target species as well as the crop pests. However, the experimental results are not always univocal. The paradigmatic case in the literature is represented by the monarch butterfly (*Danaus plexippus*). An increased mortality was shown by larvae fed on milkweed (*Asclepias curassavica*) leaves dusted with pollen from Bt corn expressing Cry1A protein. Toxicological studies to assess exposure and population-level effects determined that larval exposure to pollen on a population-wide basis was low, given the proportion of larvae in maize fields during pollen shed, the proportion of fields planted in Bt maize, and the levels of pollen within and around maize fields that exceeded the toxicity threshold (Losey et al. 1999). A 2-year study demonstrated that the risk to monarch butterfly populations is 0.6% of the total of monarch butterflies breeding in the North American Corn Belt (Prakash et al. 2011). Six other laboratory field studies showed that the density of Bt toxin in Bt corn pollen is not enough to cause any harm to the insect larvae (Sears et al. 2001). Moreover, the effects of standard pesticide applications on monarch butterfly populations may be more detrimental than the endogenous production of Bt toxins in a GM crop (Chapman and Burke 2006). Another related example involves the increased mortality and delayed development of lacewings (*Chrysoperla carnea*) when reared on Bt corn-fed insects. Also in this case the amount of Bt toxin fed to the insects was greater by over 30 times that expected to be encountered in the field (Chapman and

Burke 2006). Therefore, as stated by Pimentel and Raven (2000) the effect on the survival of butterfly populations of Bt corn pollen dusting their larval food plants appears to be relatively insignificant compared with other factors.

Numerous other studies showed that non-target organisms are not affected by exposition to Bt toxins in similar or higher amounts than those produced by the Bt crops (Mendelsohn et al. 2003; Zwahlen et al. 2003; Ferry et al. 2005; Lu et al. 2010; Schuler et al. 2013). Naranjo et al. (2005) collected the results of 11 field studies conducted in the United States and Australia that focused on the longer-term assessment of potential non-target effects of transgenic Bt cotton and hybrid corn plants, producing five insecticidal proteins active against lepidopteran and coleopteran pests. These studies considered a wide taxonomic breadth of non-target arthropods. In all cases, the experimental trials that included a comparison between conventional agronomic management and Bt crops, adopting either selective and/or wide range insecticides and/or in their absence, confirmed the highly selective activity of the most widely used δ-endotoxins from *B. thuringiensis*. The extant cultivars of transgenic Bt cotton and corn pose a relatively low risk to non-target arthropods (Naranjo et al. 2005).

Conversely, in Mexico, after 20 years of using Bt cotton, drastic changes in the composition of insect pest species were observed. The Lepidoptera complex represented only up to 5% of the reported pests while the sucking insect pests comprised around 73%. Very low population levels in the different cotton regions were observed for *P. gossypiella*, *H. virescens*, and *Bucculatrix thurberiella* while *H. zea* and *S. exigua* are currently considered pests of secondary importance in all cotton areas (Rocha-Munive et al. 2018). Other authors observed an increase of non-target secondary pests following the reduction of the target pests in Bt crops, suggesting that secondary pests can occupy the resources previously used by target insects (Tian et al. 2015; Rocha-Munive et al. 2018). However, fluctuations in non-target populations in Bt cotton fields also occur because the pest density may have consequences on the abundance of predators and parasitoids (Romeis et al. 2006). Recently, Pellegrino et al. (2018) performed a meta-analysis of the peer-reviewed literature from 1996 to 2016 regarding the environmental safety of GM maize cultivation including hazards for human health. The analysis, considering the abundance of non-target organisms reported in studies on plants expressing resistance to Coleoptera (35%) and Lepidoptera (65%), concluded that Anthocoridae, Aphididae, Araneae, Carabidae, Chrysopidae (adults and larvae), Coccinellidae (adults and larvae), Nabidae, Nitidulidae, and Staphylinidae were not affected by the cultivation of Bt crops. Only the Braconidae taxa was significantly decreased by 31.5%. By contrast, the increase reported in the abundance of the Cicadellidae taxa was not supported by sensitivity analysis. Braconidae taxa was mostly represented by *M. cingulum* (98% of observations), belonging to the functional guild of parasitoids. This phenomenon can be interpreted as an indirect effect of the decrease of its target insect (*O. nubilalis*) caused by the Bt maize. Similarly, no effect of Bt maize on 26 non-target arthropods has been highlighted in a meta-analysis based on 13 field trials in Spain (Comas et al. 2014), while Marvier et al. (2007), analyzing 42 field experiments, indicated that non-target invertebrates are generally more abundant in Bt cotton and maize fields, except for a decrease in Hymenoptera number.

Additionally, as engineered toxins remain in the soil with GM plant debris and are actively exuded from the roots of some, but not all, Bt crops, effects may be noted (Saxena et al. 2004). The effects on soil decomposers and other telluric communities have been evaluated, but contrasting results have been reported. Decomposer communities and degradation speed did not differ between three Bt corn varieties (2x Cry1Ab, 1x Cry3Bb1), their untransformed corresponding near-isolines, and three conventional varieties in a 9-month litter bag field study (Hönemann et al. 2008). Field comparisons of GM and non-GM maize revealed sporadic decreases in the biomass of amoebae, earthworms, flagellates and ciliates, as well as of nematodes with no difference or small difference in nematode community composition (Pellegrino et al. 2018). In a 2013 study, Souza et al. (2013) reported no relevant impact of HR soybean crop on soil microbial communities. In accordance, several studies describe Bt plants responsible for minor or no effects on soil micro-organisms (Mocali, 2010; Zhou et al. 2016), while arbuscular mycorrhizal fungal (AMF) community, spore abundance, and root colonization did not change in Bt versus non-Bt maize, in soil under field conditions (Cheeke et al. 2013). On the other hand, transgenic plants have been found to significantly affect non-target bacterial and fungal populations and the structure of microbial communities (Turrini et al. 2015; Lu et al. 2018). However, the huge number of variables affecting soil microbial communities made most of the studies not comparable.

Watkinson et al. (2000) modeled the effects of the introduction of a GM HT sugar beet on the population dynamics of the annual weed *Chenopodium album*, whose seeds are an important source of food for farmland birds, thus extending their research on the consequent impact on the seed-eating bird skylark (*Alauda arvensis*). They predicted that weed populations might be reduced to low levels or practically eradicated at field scale level, with possible severe effects on the local land use by birds. Similarly, Hawes et al. (2003) observed that the reduction in weed biomass under GMHT management in beet and spring oilseed rape, compared with conventional treatments, had a detrimental effect on herbivores, pollinators, and natural enemies abundance while detritivores increased in number. Contrasting results have been obtained by Haughton and collaborators (2003) on carabids that feed on weed seeds, whose number was lower in GM HT beet and spring oilseed rape orchards, but higher in GM HT maize fields when compared to the non-GM counterparts, while Bohan et al. (2007) found that invertebrate dynamics under GMHT management are similar to conventional management in spring- and winter-sown oilseed rape.

## *Virus-Resistant GM Crops*

Potential effects on biodiversity caused by virus-resistant transgenic plants relate to the possible interactions of the inserted viral sequences with other viruses via heteroencapsidation, recombination, and synergism (Fuchs and Gonsalves 2007). Heteroencapsidation is the encapsidation of the genome of one virus by the coat protein (CP) of another virus that may occur naturally or as a result of the expression of the CP subunits in transgenic plants. Consequently, a virus could infect an otherwise non-host plant or could be transformed from vector-non-transmissible to vector-transmissible. Recombination refers to the exchange of genetic material between transcripts of a viral transgene and the genome of a challenge virus during replication in a transgenic plant cell resulting in the formation of a chimeric virus with theoretical new abilities to break existing biological barriers for infection. Synergism refers to the interaction of a viral protein product with another challenge virus that can result in an aggravation of host symptom severity and an increase in virus titer that neither virus can cause independently. It has been demonstrated that extending the transgene construct with additional viral sequences extends the breadth of the resistance (Bucher et al. 2006). The spread of more pathogenic viruses with a broader spectrum of hosts could pose a threat to ecological balance, but to date this phenomenon has not been observed in GM plants.

## Mitigation Strategies

The refuge strategy has been the primary approach used worldwide to delay pest resistance to Bt crops. The deployment of "refuges" consists of plots with non-Bt plants near GM crops. For the refuge strategy to be most effective, insect resistance should be recessive (Carrière et al. 2010), and in this way heterozygous offspring from homozygous susceptible insects from the refuge and resistant insects from the Bt crop field will be sensitive to a Cry toxin present in Bt crops. This approach is sometimes called the "high dose refuge strategy" because it works best if the toxin dose for Bt plant-eating insects is high enough to kill them all. The US Environmental Protection Agency guidelines specify that high-dose Bt plants should kill at least 99.99% of susceptible insects in the field (Tabashnik et al. 2013a). Nevertheless, if two heterozygous insects mate, ¼ of their progeny will be resistant. For this reason, in addition to the refuge strategy, the stacking of two or more *cry* genes that have different modes of action has been widely used to delay the evolution of resistance to Bt crops, greatly reducing the probability of having a pest with double mutation (Rocha-Munive et al. 2018). Therefore, in theory, three key factors favor the success of the refuge strategy: first, recessive inheritance of resistance; second, low resistance allele frequency; and third, abundant refuges of non-Bt host plants near Bt crops. This strategy will still work for dominant inheritance of resistance in case of bigger refuges. Two additional factors predicted to delay resistance are fitness costs and incomplete resistance. The first one occurs when the fitness of resistant insects is lower than those susceptible on non-Bt host plants (Tabashnik et al. 2013a). The incomplete resistance occurs when resistant insects developing on non-Bt crops are at an advantage compared with resistant insects on Bt plants, in the development complementation (Tabashnik et al. 2013a). Field outcomes support theoretical predictions that factors delaying resistance include recessive inheritance of resistance, low initial frequency of resistance alleles, abundant refuges of non-Bt host plants.

A recent biotechnological strategy based on modification and/or truncation or combination of the three domains of the largest group of Cry toxins has been suggested as a valid alternative for enhancing Bt toxin activity. The creation of modified toxins (Cry1AbMod and Cry1AcMod) led to an improved biological activity against six Cry-resistant Lepidoptera species (Tabashnik et al. 2011, 2013b). The V171C mutant of Cry1Ab toxin exhibits a 25-fold increase in toxicity against gypsy moth *Lymantria dispar* (Alzate et al. 2010). Modified Cry3A toxin showed higher activity against the western corn rootworm *Diabrotica virgifera*, and mulberry longhorn beetle *Apriona germari* (Walters et al. 2008; Guo et al. 2012). The fusion of the binding domains of Bt toxins to each other or with domains of other anti-nutritional factors such as the ricin protein, represents another approach that increase the potential number of interactions at the molecular level in target insects, resulted in increased toxicity to a range of insect pests (Mehlo et al. 2005; Walters et al. 2010). Given that insects can probably evolve resistance to modified Bt toxins, pyramiding strategies, combining natural and engineered Bt toxins, represents a promising approach, broadening the options for pest control (Tabashnik et al. 2013b). Nevertheless, the broader spectrum of toxicity of truncated Bt toxins could more greatly affect non-target organisms, with potential impacts on biodiversity (Lombardo et al. 2016).

Mitigation measures have been proposed to contain the environmental risks of GM crops related to their impact on biodiversity. A specific strategy relies on transgene containment. There are several proposed methods for transgene containment in plants, such as physical containment (in greenhouses, growth rooms, and bioreactors), partial genome incompatibility, harvesting before flowering, parthenocarpy, stenospermocarpy, reduced shattering, inhibition of seed dormancy, apomixis, plastid transformation (transplastomic approach), cleistogamy, induced triploidy, conditional lethality, male sterility, inducible promoters, complete sterility by nonflowering, transgene excision, transgene mitigation (TM), inteins, and auxotrophy (Kausch et al. 2010; Liu et al. 2013). However, none of the strategies currently available blocks all avenues for transgene spread (de Maagd and Boutilier 2009). Eventually, genetic use restriction technologies (GURTs) could represent an effective method for the environmental containment of transgenic seeds (variety-specific V-GURTs) or genes (trait-specific T-GURTs). GURTs regroup a series of experimental methods aimed at restricting the unauthorized use of genetic material by controlling gene expression through the insertion in plants of a "genetic switch" activated (or inactivated) by an external –chemical or physica l– inducer to prevent germination in V-GURTs, or to turn on/off a value-added trait such as tolerance to herbicides or biotic and abiotic stresses and pest resistance in the case of T-GURTs.

Strong protests all over the world raised against these technologies that would have forced farmers to purchase new seeds every year, preventing the practice of seed saving estimated to account for between 15% and 20% of the world's food supply involving 1.4 billion people, so that the Convention on Biological Diversity (CBD) Decision V/5 section III of the Fifth Conference of the Parties (COP5) held in Nairobi in June 2000 imposed a *de facto* global moratorium on GURTs (Lombardo 2014).

## Conclusions

Twenty-four years after their first commercial cultivation, GM crops cover around 190 million hectares worldwide, for the most part in the form of insect-resistant and herbicide-tolerant plants.

The scientific evidence reported here, in addition to the general concerns regarding GM crop's impact on biodiversity, entails some degree of uncertainty and ambiguity. Several potential risks have been highlighted, but only few cases represent actual hazards, especially if compared to the conventional management systems aimed at increasing farmers' productions by encouraging the expansion of intensive plantation, providing frequent tillage and heavy water and pesticide use, and increased the marginalization of traditional low-input systems with generally negative consequences for the environment like erosion, decreased biodiversity (considerable reduction in both the diversity and total number of flora and fauna), significant loss of wildlife habitat and water bodies pollution. There is more and more evidence that first-generation GM crops are non-invasive in non-agricultural areas. More controversial seems to be the issue about vertical gene flow because many factors affect pollen and seed migration. Nevertheless, the relatively low number of cases of hybridization among GM plants and wild and/or weedy relatives reported in the literature do not seem to represent a particular threat for biodiversity. In the same regard, despite the fact that horizontal gene flow of recombinant genes from GM plants to bacteria has been experimentally observed in the laboratory, it has never been shown under field conditions with GM plants used in agriculture, and entire transgenes with the regulatory portions of the DNA have never been found to be horizontally transferred.

No or minor effects have been reported on non-target Coleoptera and Lepidoptera, while discording results have been obtained for soil micro-organisms. The selection of HR weeds and toxin-tolerant insects due to the selection pressure, to date, has not had any relevant impact on biodiversity, considering however that this is a common phenomenon in non-GM agroenvironments where conventional pesticides and herbicides are used.

Finally, GM crops have demonstrated to represent a resource in the agricultural field, contributing in combination with appropriate agronomic management practices, to a more sustainable use of soil in the long term and to reduce the use of agrochemicals that are among the leading causes of cancer onset in the world.

Moreover, the development of increasingly targeted new approaches represents a promising alternative to limit unintended effects on non-target organisms and increase crop yield, that in the near future could contribute, at least in part, to meet the increase in production demand.

## References

Abbas MST (2018) Genetically engineered (modified) crops (*Bacillus thuringiensis* crops) and the world controversy on their safety. Egypt J of Biol Pest Control 28:52. https://doi.org/10.1186/s41938-018-0051-2

Ahlquist P (2002) RNA-dependent RNA polymerases, viruses, and RNA silencing. Science 296(5571):1270–1273

Alibert B, Sellier H, Souvré A (2005) A combined method to study gene flow from cultivated sugar beet to ruderal beets in the glasshouse and open field. Eur J Agron 23:195–208

Alzate O, Osorio C, Florez AM, Dean DH (2010) Participation of valine 171 in alpha-Helix 5 of *Bacillus thuringiensis* Cry1Ab delta-endotoxin in translocation of toxin into *Lymantria dispar* midgut membranes. Appl Environ Microbiol 76:7878–7880

Anilkumar KJ, Rodrigo-Simón A, Ferré J, Pusztai-Carey M, Sivasupramaniam S, Moar WJ (2008) Production and characterization of *Bacillus thuringiensis* Cry1Ac-resistant cotton bollworm *Helicoverpa zea* (Boddie). Appl Environ Microbiol 74:462–469

Badosa E, Moreno C, Montesinos E (2004) Lack of detection of ampicillin resistance gene transfer from Bt176 transgenic corn to culturable bacteria under field conditions. FEMS Microbiol Ecol 48(2):169–178. https://doi.org/10.1016/j.femsec.2004.01.005

Baird DD, Upchurch RP, Homesley WB, Franz JE (1971) Introduction of a new broad spectrum post emergence herbicide class with utility for herbaceous perennial weed control. Proc North Central Weed Control Conf 26:64–68

Beckie HJ, Owen MDK (2007) Herbicide-resistant crops as weeds in North America. CAB Rev Persp Agric Veter Sci Nutr Natl Resour 2(44):1–22

Beckie HJ, Harker KN, Hall LM, Warwick SI, Légère A, Sikkema PH, Clayton GW, Thomas AG, Leeson JY, Séguin-Swartz G, Simard MJ (2006) A decade of herbicide-resistant crops. Can J Plant Sci 86:1243–1264

Beetham PR, Kipp PB, Sawycky XL, Arntzen CJ, May GD (1999) A tool for functional plant genomics: chimeric RNA/DNA oligonucleotides cause in vivo gene-specific mutations. Proc Natl Acad Sci U S A 96(15):8774–8778. https://doi.org/10.1073/pnas.96.15.8774

Benbrook M (2016) Trends in glyphosate herbicide use in the United States and globally. Environ Sci Eur 28(1):3. https://doi.org/10.1186/s12302-016-0070-0

Bohan DA, Hawes C, Haughton AJ, Denholm I, Champion GT, Perry JN, Clark SJ (2007) Statistical models to evaluate invertebrate-plant trophic interactions in arable systems. Bull Entomol Res 97(3):265–280. https://doi.org/10.1017/S0007485307004890

Bowman D, May O, Creech J (2003) Genetic uniformity of the US upland cotton crop since the introduction of transgenic cottons. Crop Sci 43:515–518

Bretagnolle V, Gaba S (2015) Weeds for bees? Agron Sustain Dev 35(3):891–909

Breyer D, Kopertekh L, Reheul D (2014) Alternatives to antibiotic resistance marker genes for in vitro selection of genetically modified plants – scientific developments, current use, operational access and biosafety considerations. Crit Rev Plant Sci 33(4):286–330. https://doi.org/10.1080/07352689.2013.870422

Brookes G, Barfoot P (2017) GM crops: global socio-economic and environmental impacts 1996–2015. PG Economics Ltd, Dorchester, UK, 201 pp

Bucher E, Lohuis D, van Poppel PMJA, Geerts-Dimitriadou C, Goldbach R, Prins M (2006) Multiple virus resistance at a high frequency using a single transgene construct. J Gen Virol 87:3697–3701
Busi R, Yu Q, Barrett-Lennard R, Powles S (2008) Long distance pollen-mediated flow of herbicide resistance genes in *Lolium rigidum*. Theor Appl Genet 117:1281–1290
Campbell PM, Reiner D, Moore AE, Lee RY, Epstein MM, Higgins TJ (2011) Comparison of the α-amylase inhibitor-1 from common bean (*Phaseolus vulgaris*) varieties and transgenic expression in other legumes. Post-translational modifications and immunogenicity. J Agric Food Chem 59(11):6047–6054
Carpenter JE (2011) Impacts of GM crops on biodiversity. GM Crops 2(1):1–17
Carrière Y, Crowder DW, Tabashnik BE (2010) Evolutionary ecology of insect adaptation to Bt crops. Evol Appl 3:561–573
Castle LA, Daniel LS, Gorton R, Patten PA, Chen YH, Bertain S, Cho HJ, Duck N, Wong J, Liu D, Lassner MW (2004) Discovery and directed evolution of a glyphosate tolerance gene. Science 304:1151–1154. https://doi.org/10.1126/science.1096770
Chan YL, Yang AH, Chen JT, Yeh KW, Chan MT (2010) Heterologous expression of taro cystatin protects transgenic tomato against *Meloidogyne incognita* infection by means of interfering sex determination and suppressing gall formation. Plant Cell Rep 29:231–238
Chapman MA, Burke JM (2006) Letting the gene out of the bottle: the population genetics of genetically modified crops. New Phytol 170:429–443
Cheeke TE, Cruzan MB, Rosenstiel TN (2013) Field evaluation of arbuscular mycorrhizal fungal colonization in *Bacillus thuringiensis* toxin-expressing (Bt) and non-Bt maize. Appl Environ Microb 79:4078–4086
Chen LJ, Lee DS, Song ZP, Suh HS, Lu BR (2004) Gene flow from cultivated rice (*Oryza sativa*) to its weedy and wild relatives. Ann Bot 93:67–73
Chen J, Jin M, Ziu ZG, Guo C, Chen Z-L, Shen ZQ, Wang XW, Li JW (2012) A survey of drug resistance *blá* genes originating from synthetic plasmid vectors in six Chinese rivers. Environ Sci Technol 45:3448–3454
Chen J, Huang H, Zhang C, Wei S, Huang Z, Chen J, Wang X (2015) Mutations and amplification of *EPSPS* gene confer resistance to glyphosate in goosegrass (*Eleusine indica*). Planta 242:859–868
Comas C, Lumbierres B, Pons X, Albajes R (2014) No effects of *Bacillus thuringiensis* maize on non-target organisms in the field in southern Europe: a meta-analysis of 26 arthropod taxa. Transg Res 23:135–143. https://doi.org/10.1007/s11248-013-9737-0
Cooper SG, Douches DS, Grafius EJ (2009) Combining engineered resistance, avidin, and natural resistance derived from *Solanum chacoense* bitter to control Colorado potato beetle (*Coleoptera: Chrysomelidae*). J Econ Entomol 102:1270–1280
Crawley MJ, Hails RS, Rees M, Kohn D, Buxton J (1993) Ecology of transgenic oilseed rape in natural habitats. Nature 363:620–623
Crawley MJ, Brown SL, Hails RS, Kohn DD, Rees M (2001) Transgenic crops in natural habitats. Nature 409:682–683. https://doi.org/10.1038/35055621
Darmency H, Vigouroux Y, De Garambe T, Gestat R-MM, Muchembled C (2007) Transgene escape in sugar beet production fields: data from six years farm scale monitoring. Environ Biosaf Res 6:197–206
Dayan FE, Owens DK, Corniani N, Silva FML, Watson SB, Howell JL, Shaner DL (2015) Biochemical markers and enzyme assays for herbicide mode of action and resistance studies. Weed Sci 63:23–63. https://doi.org/10.1614/WS-D-13-00063.1
de Maagd RA, Boutilier K (2009) Efficacy of strategies for biological containment of transgenic crops. Plant Research International BV, Wageningen, p 58
de Vries J, Herzfeld T, Wackernagel W (2004) Transfer of plastid DNA from tobacco to the soil bacterium *Acinetobacter* sp by natural transformation. Mol Microbiol 53:323–334
Delledonne M, Allegro G, Belenghi B, Balestrazzi A, Picco F, Levine A, Zelasco S, Calligari P, Confalonieri M (2001) Transformation of white poplar (*Populus alba* L) with a novel

*Arabidopsis thaliana* cysteine proteinase inhibitor and analysis of insect pest resistance. Mol Breed 7:35–42
Dillon A, Varanasi VK, Danilova TV, Koo D-H, Nakka S, Peterson DE, Tranel PJ, Friebe B, Gill BS, Jugulam M (2017) Physical mapping of amplified 5-enolpyruvylshikimate-3-phosphate synthase gene copies in glyphosate-resistant *Amaranthus tuberculatus*. Plant Physiol 173:1226–1234
Dively GP, Venugopal PD, Finkenbinder C (2016) Field-evolved resistance in corn earworm to Cry proteins expressed by transgenic sweet corn. PLoS One 11:e0169115
Dröge M, Pühler A, Selbitschka W (1998) Horizontal gene transfer as a biosafety issue: a natural phenomenon of public concern. J Biotechnol 64(1):75–90
Duan X, Li X, Xue Q (1996) Transgenic rice plants harboring an introduced potato proteinase inhibitor II gene are insect resistant. Nat Biotechnol 14:494–498
Duke SO, Cedeira AL (2010) Transgenic crops for herbicide resistance. In: Kole C, Michler C, Albert Gand Hall TC (eds) Transgenic crop plants. Springer, Berlin/Heidelberg, pp 133–166
Dun B, Wang X, Lu W, Chen M, Wei W, Shuzhen P, Wang Z, Zhang B, Lin M (2014) Development of highly glyphosate-tolerant tobacco by coexpression of glyphosate acetyltransferase *gat* and *EPSPS G2-aroA* genes. Crop J 2(2-3):164–169. https://doi.org/10.1016/j.cj.2014.03.003
Eastham K, Sweet J (2002) Genetically modified organisms: the significance of gene flow through pollen transfer. Environ Issues Rep 28:1–75
Ellstrand NC, Prentice HC, Hancock JF (1999) Gene flow and introgression from domesticated plants into their wild relatives. Ann Rev Ecol Syst 30:539–563
Evans JA, Tranel PJ, Hager AG, Schutte B, Wu C, Chatham LA, Davis AS (2015) Managing the evolution of herbicide resistance. Pest Manag Sci 72(1):74–80. https://doi.org/10.1002/ps.4009
Fernandez-Cornejo J, Hallahan C, Nehring R, Wechsler S, Grube A (2012) Conservation tillage, herbicide use, and genetically engineered crops in the United States: the case of soybeans. AgBioForum 15(3):231–241
Ferry N, Mulligan EA, Majerus ME, Gatehouse AM (2005) Bitrophic and tritrophic effects of Bt Cry3A transgenic potato on beneficial, non-target, beetles. Transg Res 16:795–812
Flachowsky G, Schafft H, Meyer U (2012) Animal feeding studies for nutritional and safety assessments of feeds from genetically modified plants: a review. J Verbraucherschutz Lebensmittelsicherh (J Consum Prot Food Saf) 7:179–194
Fuchs MF, Gonsalves D (2007) Safety of virus-resistant transgenic plants two decades after their introduction: lessons from realistic field risk assessment studies. Annu Rev Phytopathol 45:173–202
Gaba S, Gabriel E, Chadoeuf J, Bonneu F, Bretagnolle V (2016) Herbicides do not ensure for higher wheat yield, but eliminate rare plant species. Sci Rep 6:30112
Gaines TA, Heap (2019) IMMutations in herbicide-resistant weeds to EPSP synthase inhibitors. Online. http://www.weedscience.com. Accessed Jan 2019
Gaines TA, Zhang W, Wang D, Bukun B, Chisholm ST, Shaner DL, Nissen SJ, Patzoldt WL, Tranel PJ, Culpepper AS, Grey TL, Webster TM, Vencill WK, Sammons RD, Jiang J, Preston C, Leach JE, Westra P (2010) Gene amplification confers glyphosate resistance in *Amaranthus palmeri*. Proc Natl Acad Sci U S A 107(3):1029–1034
Gatehouse AMR, Davison GM, Newell CA, Merryweather A, Hamilton WDO, Burgess EPJ, Gilbert RJC, Gatehouse JA (1997) Transgenic potato plants with enhanced resistance to the tomato moth, *Lacanobia oleracea*: growth room trials. Mol Breed 3:49–63
Gealy DR, Mitten DH, Rutger JN (2003) Gene flow between red rice (*Oryza sativa*) and herbicide-resistant rice (*O. sativa*): implications for weed management. Weed Technol 17:627–645
Gebhard F, Smalla K (1999) Monitoring field releases of genetically modified sugarbeets for persistence of transgenic plant DNA and horizontal gene transfer. FEMS Microbiol Ecol 28:261–272
Gennaro A, Gomes A, Herman L, Nogue F, Papadopoulou N, Tebbe C (2017) Technical report on the explanatory note on DNA sequence similarity searches in the context of the assessment of horizontal gene transfer from plants to microorganisms. EFSA (European Food Safety Authority) supporting publication 14: 7, 11 pp. https://doi.org/10.2903/sp.efsa.2017.EN-1273

Giesy JP, Dobson S, Solomon KR (2000) Ecotoxicological risk assessment for Roundup herbicide. Rev Environ Contam Toxicol 167:35–120

Girard C, Le Métayer M, Bonadé-Bottino M, Pham-Delègue MH, Jouanin L (1998) High level of resistance to proteinase inhibitors may be conferred by proteolytic cleavage in beetle larvae. Insect Biochem Mol Biol 28:229–237

Gueritaine G, Sester M, Eber F, Chevre AM, Darmency H (2002) Fitness components of progeny of hybrids between transgenic oilseed rape (*Brassica napus*) and wild radish (*Raphanus raphanistrum*). Mol Ecol 11:1419–1426

Guo CH, Zhao ST, Ma Y, Hu JJ, Han XJ, Chen J, Lu MZ (2012) *Bacillus thuringiensis* Cry3Aa fused to a cellulase-binding peptide shows increased toxicity against the longhorned beetle. Appl Microbiol Biotechnol 93:1249–1256

Guo B, Guo Y, Hong H, Jin L, Zhang L, Chang RZ, Lu W, Lin M, Qiu LJ (2015) Co-expression of G2-EPSPS and glyphosate acetyltransferase GAT genes conferring high tolerance to glyphosate in soybean. Front Plant Sci 6(847):1–11

Gutierrez-Campos R, Torres-Acosta JA, Saucedo-Arias LJ, Gomez-Lim MA (1999) The use of cysteine proteinase inhibitors to engineer resistance against potyviruses in transgenic tobacco plants. Nat Biotechnol 17:1223–1226

Halfhill MD, Zhu B, Warwick SI, Raymer PL, Millwood RJ, Weissinger AK, Stewart CN Jr (2004) Hybridization and backcrossing between transgenic oilseed rape and two related weed species under field conditions. Environ Biosaf Res 3(2):73–81

Hall LM, Rahman MH, Gulden RH, Thomas AG (2005) Volunteer oilseed rape: will herbicide-resistance traits assist ferality. In: Gressel J (ed.) Crop Ferality and Volunteerism. Taylor and Francis pp 59-80

Handayani VDS, Tanno Y, Yamashita M, Tobina H, Ichihara M, Ishida Y, Sawada H (2017) Influence of weed management measures on glyphosate resistance and endophyte infection in naturalized Italian ryegrass (*Lolium multiflorum*): glyphosate resistance and endophytes. Weed Biol Man 17(2):84–90. https://doi.org/10.1111/wbm.12122

Haughton AJ, Champion GT, Hawes C, Heard MS, Brooks DR, Bohan DA, Clark SJ, Dewar AM, Firbank LG, Osborne JL, Perry JN, Rothery P, Roy DB, Scott RJ, Woiwod IP, Birchall C, Skellern MP, Walker JH, Baker P, Browne EL, Dewar AJ, Garner BH, Haylock LA, Horne SL, Mason NS, Sands RJ, Walker MJ (2003) Invertebrate responses to the management of genetically modified herbicide-tolerant and conventional spring cropsII within-field epigeal and aerial arthropods. Philos Trans R Soc Lond B Biol Sci 358(1439):1863–1877

Hawes C, Haughton AJ, Osborne JL, Roy DB, Clark SJ, Perry JN, Rothery P, Bohan DA, Brooks DR, Champion GT, Dewar AM, Heard MS, Woiwod IP, Daniels RE, Young MW, Parish AM, Scott RJ, Firbank LG, Squire GR (2003) Responses of plants and invertebrate trophic groups to contrasting herbicide regimes in the Farm Scale Evaluations of genetically modified herbicide-tolerant crops. Philos Trans R Soc Lond B Biol Sci 358(1439):1899–1913

Hay I, Morency MJ, Séguin A (2002) Assessing the persistence of DNA in decomposing leaves of genetically modified poplar trees. Can J For Res 32:977–982

Heap I (2019) The international survey of herbicide resistant weeds. Available: www.weedscience.org. Accessed Jan 2019

Hernandez-Rodriguez CS, Hernandez-Martinez P, Van Rie J, Escriche B, Ferré J (2013) Shared midgut binding sites for Cry1A105, Cry1Aa, Cry1Ab, Cry1Ac and Cry1Fa proteins from *Bacillus thuringiensis* in two important corn pests, *Ostrinia nubilalis* and *Spodoptera frugiperda*. PLoS One 8(7):e68164. https://doi.org/10.1371/journal.pone.0068164

Hilder VA, Gatehouse AMR, Sheerman SE, Barker RF, Boulter D (1987) A novel mechanism of insect resistance engineered into tobacco. Nature 330:169–163

Hönemann L, Zurbrügg C, Nentwig W (2008) Effects of Bt-corn decomposition on the composition of the soil meso- and macrofauna. Appl Soil Ecol 40(2):203–209. https://doi.org/10.1016/j.apsoil.2008.04.006

Hull R (2014) Plant virology, 5th edn. Academic, 1120 pp

ISAAA (2017) Global status of commercialized biotech/gm crops in 2017: biotech crop adoption surges as economic benefits accumulate in 22 years, ISAAA brief no 53. ISAAA, Ithaca
ISAAA GM Approval Database. http://www.isaaa.org/gmapprovaldatabase/default.asp. Accessed Jan 2019
Jugulam M, Niehues K, Godar AS, Koo DH, Danilova T, Friebe B, Sehgal S, Varanasi VK, Wiersma A, Westra P, Stahlman PW, Gill BS (2014) Tandem amplification of a chromosomal segment harboring 5-enolpyruvylshikimate-3-phosphate synthase locus confers glyphosate resistance in *Kochia scoparia*. Plant Physiol 166:1200–1207
Kausch AP, Hague J, Oliver M, Li Y, Daniell H, Mascia P, Stewart CN (2010) Genetic modification in dedicated bioenergy crops and strategies for gene confinement. In: Mascia PN, Scheffran CN Jr, Widholm JM (eds) Plant biotechnology for sustainable production of energy and co-products 66. Springer, Berlin/Heidelberg, pp 299–315
Keese P (2008) Risks from GMOs due to horizontal gene transfer. Environ Biosaf Res 7:123–149. https://doi.org/10.1051/ebr:2008014
Khadeeva NV, Kochieva EZ, Tcherednitchenko MY, Yakovleva EY, Sydoruk KV, Bogush VG, Dunaevsky YE, Belozersky MA (2009) Use of buckwheat seed protease inhibitor gene for improvement of tobacco and potato plant resistance to biotic stress. Biochemistry (Moscow) 74(3):260–267
Kolady DE, Lesser W (2012) Genetically-engineered crops and their effects on varietal diversity: a case of Bt eggplant in India. Agric Hum Value 29:3–15
Koo DH, Jugulam M, Putta K, Cuvaca IB, Peterson DE, Currie RS, Friebe B, Gill BS (2018) Gene duplication and aneuploidy trigger rapid evolution of herbicide resistance in common waterhemp. Plant Physiol 176(3):1932–1938. https://doi.org/10.1104/pp.17.01668
Kruger M, Van Renzburg JJ, Van den Berg J (2009) Prospective on the development of the stem borer resistance to Bt maize and refuge compliance at the Vaalharts irrigation scheme in South Africa. Crop Port 28:684–689
Linder CR, Schmitt J (1995) Potential persistence of escaped transgenes: performance of transgenic oil-modified Brassica seeds and seedlings. Ecol Appl 5:1056–1068
Liu F, Xu Z, Zhu YC, Huang F, Wang Y, Li H, Li H, Gao C, Zhou W, Shen J (2010) Evidence of field evolved resistance to *Cry1Ac* expressing Bt cotton in *Helicoverpa armigera* in northern China. Pest Manag Sci 66:155–161
Liu W, Yuan JS, Stewart CN Jr (2013) Advanced genetic tools for plant biotechnology. Nat Rev Genet 14:781–793
Liu YY, Zhang Y, Liu Y, Lu W, Wang G (2015) Metabolic effects of glyphosate on transgenic maize expressing a G2-EPSPS gene from *Pseudomonas fluorescens*. J Plant Biochem Biotech 24(2):233–241
Lombardo L (2014) Genetic use restriction technologies: a review. Plant Biotech J 12:995–1005
Lombardo L, Grando MS (2020) Genetically modified plants for nutritionally improved food: a promise kept? Food Rev Int 36(1):58–76. https://doi.org/10.1080/87559129.2019.1613664
Lombardo L, Zelasco S (2016) Biotech approaches to overcome the limitations of using transgenic plants in organic farming. Sustainability 8(5):497
Lombardo L, Coppola G, Zelasco S (2016) New technologies for insect-resistant and herbicide-tolerant plants. Trend in Biotech 34:149–157
Lorentz L, Gaines TA, Nissen SJ, Westra P, Strek H, Dehne HW, Ruiz-Santaella JP, Beffa R (2014) Characterization of glyphosate resistance in *Amaranthus tuberculatus* populations. J Agric Food Chem 62:8134–8142
Losey JE, Rayor LS, Carter ME (1999) Transgenic pollen harms monarch larvae. Nature 399(6733):214. https://doi.org/10.1038/20338
Lövei GL, Bøhn T, Hilbeck A (2007) Biodiversity, ecosystem services and genetically modified organisms. In: Traavik T, Lim LC (eds) Biosafety first: holistic approaches to risk and uncertainty in genetic engineering and genetically modified organisms. Tapir Academic Press, Trondheim, pp 161–180

Lu BR, Snow AA (2005) Gene flow from genetically modified rice and its environmental consequences. BioSci 55(8):669–678

Lu Y, Wu K, Jiang Y, Xia B, Li P, Feng H, Wyckhuys KA, Guo Y (2010) Mirid bug outbreaks in multiple crops correlated with wide-scale adoption of Bt cotton in China. Science 328:1151–1154

Lu GH, Hua XM, Liang L, Wen ZL, Du MH, Meng FF, Pang YJ, Qi JL, Tang CY, Yang YH (2018) Identification of major rhizobacterial taxa affected by a glyphosate-tolerant soybean line via shotgun metagenomic approach. Genes 9(4):E214. https://doi.org/10.3390/genes9040214

Luttrel RG, Ali MI (2007) Exploring selection for Bt resistance in heliothines: results of laboratory and field studies. In: Boyd S, Huffman M, Richter D, Robertson B (eds) Proceedings of the 2007 Beltwide cotton conferences, New Orleans, LA, January 9–12. National Cotton Council of America, Memphis, pp 1073–1086

Ma BL, Blackshaw RE, Roy J, He T (2011) Investigation on gene transfer from genetically modified corn (*Zea mays* L) plants to soil bacteria. J Environ Sci Health B46(7):590–599. https://doi.org/10.1080/03601234.2011.586598

Malone JM, Morran S, Shirley N, Boutsalis P, Preston C (2016) *EPSPS* gene amplification in glyphosate-resistant *Bromus diandrus*. Pest Manag Sci 72:81–88

Marvier M, McCreedy C, Regetz J, Kareiva P (2007) A meta-analysis of effects of Bt cotton and maize on nontarget invertebrates. Science 316:1475–1477

Matten SR, Head GP, Quemada HD (2008) How governmental regulation can help the integration of Bt crops within IPM programs. In: Romois J, Shelton AM, Kennedy GG (eds) Integration of insect resistant genetically modified crops within IPM programs. Springer, New York, pp 27–39

Mehlo L, Gahakwa D, Nghia PT, Loc NT, Capell T, Gatehouse JA, Gatehouse AM, Christou P (2005) An alternative strategy for sustainable pest resistance in genetically enhanced crops. Proc Natl Acad Sci U S A 5:7812–7816

Meilan R, Han KH, Ma C, Di Fazio SP, Eaton J, Hoien EA, Stanton BJ, Crockett RP, Taylor ML, James RR, Skinner JS, Jouanin L, Pilate G, Strauss SH (2002) The CP4 transgene provides high levels of tolerance to Roundup® herbicide in field-grown hybrid poplars. Can J Res 32:967–976. https://doi.org/10.1139/x02-015

Mendelsohn M, Kough J, Vaituzis Z, Matthews K (2003) Are Bt crops safe? Nat Biotechnol 21:1003–1009

Mocali S (2010) Bt plants and effects on soil micro-organisms. Agric Vet Sci Nutr Nat Resour 5(036):1–19. https://doi.org/10.1079/PAVSNNR20105036

Molin WT, Wright AA, Lawton-Rauh A, Saski CA (2017) The unique genomic landscape surrounding the *EPSPS* gene in glyphosate resistant *Amaranthus palmeri*: a repetitive path to resistance. BMC Genomics 18:91. https://doi.org/10.1186/s12864-016-3336-4

Nandula VK, Reddy KN, Rimando AM, Duke SO, Poston DH (2007) Glyphosate-resistant and-susceptible soybean (*Glycine max*) and canola (*Brassica napus*) dose response and metabolism relationships with glyphosate. J Agric Food Chem 55:3540–3545

Nandula VK, Wright AA, Bond JA, Ray JD, Eubank TW, Molin WT (2014) EPSPS amplification in glyphosate-resistant spiny amaranth (*Amaranthus spinosus*): a case of gene transfer via interspecific hybridization from glyphosate-resistant Palmer amaranth (*Amaranthus palmeri*). Pest Manag Sci 70:1902–1909

Naranjo SE, Head G, Gbalen P (2005) Field studies assessing arthropod non target effects in Bt transgenic crops. Introduc Environ Ent 34(5):1178–1180

Ngo TD, Malone JM, Boutsalis P, Gill G, Preston C (2018) EPSPS gene amplification conferring resistance to glyphosate in windmill grass (*Chloris truncata*) in Australia. Pest Manag Sci 74(5):1101–1108. https://doi.org/10.1002/ps.4573

Paget E, Lebrun M, Freyssinet G, Simonet P (1998) The fate of recombinant plant DNA in soil. Eur J Soil Biol 34:81–88

Pedotti M, Rosini E, Molla G, Moschetti T, Savino C, Vallone B, Pollegioni L (2009) Glyphosate resistance by engineering the flavoenzyme glycine oxidase. J Biol Chem 284:36415–36423. https://doi.org/10.1074/jbc.M109.051631

Pellegrino E, Bedini S, Nuti M, Ercoli L (2018) Impact of genetically engineered maize on agronomic, environmental and toxicological traits: a meta-analysis of 21 years of field data. Sci Rep 8:3113. https://doi.org/10.1038/s41598-018-21284

Pimentel DS, Raven PH (2000) Bt corn pollen impacts on nontarget Lepidoptera: Assessment of effects in nature. Proc Natl Acad Sci 97 (15):8198–8199

Pontiroli A, Simonet P, Frostegard A, Vogel TM, Monier JM (2007) Fate of transgenic plant DNA in the environment. Environ Biosaf Res 6:15–35

Pontiroli A, Rizzi A, Simonet P, Daffonchio D, Vogel TM, Monier JM (2009) Visual evidence of horizontal gene transfer between plants and bacteria in the phytosphere of transplastomic tobacco. Appl Environ Microbiol 75:3314–3322

Powles SB, Lorraine-Colwill DF, Dellow JJ, Preston C (1998) Evolved resistance to glyphosate in rigid ryegrass (*Lolium rigidum*) in Australia. Weed Sci 46:604–607

Prakash D, Verma S, Bhatia R, Tiwary BN (2011) Risks and precautions of genetically modified organisms. ISRN Ecol:1–13. https://doi.org/10.5402/2011/369573

Qaim M, Yarkin C, Zilberman D (2005) Impact of biotechnology on crop genetic diversity. In: Cooper J, Lipper LM, Zilberman D (eds) Agricultural biodiversity and biotechnology in economic development, Natural resource management and policy 27. Springer, Boston, pp 283–307

Rao VS (2015) Transgenic herbicide resistance in plants. CRC Press, Boca Raton, 458 pp

Reddy KN, Rimando AM, Duke SO, Nandula VK (2008) Aminomethylphosphonic acid accumulation in plant species treated with glyphosate. J Agric Food Chem 56:2125–2130

Reeck GR, Kramer KJ, Baker JE, Kanost MR, Fabrick JA, Behnke CA (1997) Proteinase inhibitors and resistance of transgenic plants to insects. In: Carozzi N, Koziel M (eds) Advances in insect control: the role of transgenic plants to insects. Taylor & Francis, London, pp 157–183

Reichman J, Watrud L, Lee EH, Burdick CA, Bollman MA, Storm MJ, King GA, Mallory-Smith C (2006) Establishment of transgenic herbicide-resistant creeping bentgrass (*Agrostis stolonifera* L) in nonagronomic habitats. Mol Ecol 15:4243–4255

Ribeiro APO, Pereira EJG, Galvan TL, Picanco MC, Picoli EAT, da Silva DJH, Fári MG, Otoni WC (2006) Effect of eggplant transformed with oryzacystatin gene on *Myzus persicae* and *Macrosiphum euphorbiae*. J Appl Entomol 130(841):84–90

Rieger MA, Lamond M, Preston C, Powles SB, Roush RT (2002) Pollen-mediated movement of herbicide resistance between commercial canola fields. Science 296:2386–2388

Rizzi A, Pontiroli A, Brusetti L, Borin S, Sorlini C, Abruzzese A, Sacchi GA, Vogel TM, Simonet P, Bazzicalupo M, Nielsen KM, Monier JM, Daffonchio D (2008) Strategy for *in situ* detection of natural transformation-based horizontal gene transfer events. Appl Environ Microbiol 74:1250–1254

Rocha-Munive MG, Soberón M, Castañeda S, Niaves E, Scheinvar E, Eguiarte LE, Mota-Sánchez D, Rosales-Robles E, Nava-Camberos U, Martínez-Carrillo JL, Blanco CA, Bravo A, Souza V (2018) Evaluation of the impact of genetically modified cotton after 20 years of cultivation in Mexico. Front Bioeng Biotechnol 6:82. https://doi.org/10.3389/fbioe.2018.00082

Romeis J, Meissle M, Bigler F (2006) Transgenic crops expressing *Bacillus thuringiensis* toxin and biological control. Nat Biotech 24:63–71

Sachs ES, Benedict JH, Stelly DM, Taylor JF, Altman DW, Berberich SA, Davis SK (1998) Expression and segregation of genes encoding CryIA insecticidal proteins in cotton. Crop Sci 38:1–11

Salas RA, Dayan FE, Pan Z, Watson SB, Dickson JW, Scott RC, Burgos NR (2012) *EPSPS* gene amplification in glyphosate-resistant Italian ryegrass (*Lolium perenne* ssp*multiflorum*) from Arkansas. Pest Manag Sci 68:1223–1230

Sammons RD, Gaines TA (2014) Glyphosate resistance: state of knowledge. Pest Manag Sci 70(9):1367–1377. https://doi.org/10.1002/ps.3743

Sanchez MA, Cid P, Navarrete H, Aguirre C, Chacón G, Salazar E, Prieto H (2016) Outcrossing potential between 11 important genetically modified crops and the Chilean vascular flora. P Biotech J 14:625–637
Santos-Amaya OF, Rodrigues JVC, Souza TC, Tavares CS, Campos SO, Guedes RN, Pereira EJ (2015) Resistance to dual-gene Bt maize in *Spodoptera frugiperda*: selection, inheritance, and cross-resistance to other transgenic events. Sci Rep 5(18243). https://doi.org/10.1038/srep18243
Saxena D, Stewart CN, Altosaar I, Shu Q, Stotzky G (2004) Larvicidal Cry proteins from *Bacillus thuringiensis* are released in root exudates of transgenic *B thuringiensis* corn, potato, and rice but not of *B thuringiensis* canola, cotton, and tobacco. Plant Phys Biochem 42:383–387
Schuler TH, Potting RPJ, Denholm I, Clark SJ, Clark AJ, Stewart CN, Poppy GM (2013) Tritrophic choice experiments with Bt plants, the diamondback moth (*Plutella xylostella*) and the parasitoid *Cotesia plutellae*. Trans Res 12:351–361
Sears MK, Hillmich RL, Stanley-Horn DE, Oberhauser KS, Pleasants JM, Mattila HR, Siegfried BD, Dively GP (2001) Impact of Bt corn pollen on monarch butterfly populations: a risk assessment. Proc Natl Acad Sci U S A 98:11937–11943
Shrestha RM, Dunbar MW, French BW, Gassmann AJ (2018) Effects of field history on resistance to Bt maize by western corn rootworm, *Diabrotica virgifera virgifera* LeConte (Coleoptera: Chrysomelidae). PLoS One 13(7):e0200156
Simard MJ, Légère A, Séguin-Swartz G, Nair H, Warwick S (2005) Fitness of double vs. single herbicide-resistant canola. Weed Sci 53:489–498
Simpson DJ, Fry JC, Rogers HJ, Day MJ (2007) Transformation of *Acinetobacter baylyi* in non-sterile soil using recombinant plant nuclear DNA. Environ Biosaf Res 6:101–112
Smigocki AC, Ivic-Haymes S, Li H, Savic J (2013) Pest protection conferred by a beta vulgaris serine proteinase inhibitor gene. PLoS One 8(2):e57303
Sneller CH (2003) Impact of transgenic genotypes and subdivision on diversity within elite North American soybean germplasm. Crop Sci 43:409–414
Snow AA, Pilson D, Rieseberg LH, Paulsen MJ, Pleskac N, Reagon MR, Wolf DE, Selbo SM (2003) A Bt transgene reduces herbivory and enhances fecundity in wild sunflowers. Ecol Appl 13(2):279–286
Souza RA, Babujia LC, Silva AP, Guimaraes MF, Arias CA, Hungria M (2013) Impact of the *ahas* transgene and of herbicides associated with the soybean crop on soil microbial communities. Transg Res 22:877–892
Sparks TC, Nauen R (2015) IRAC: mode of action classification and insecticide resistance management. Pest Biochem Phys 121:122–128
Spencer LJ, Snow AA (2001) Fecundity of transgenic wild-crop hybrids of Curcubita pepo (Cucurbitaceae): implications for crop-to-wild gene flow. Heredity 86(6):694–702
Srinivasan A, Giri AP, Gupta VS (2006) Structural and functional diversities in Lepidopteran serine proteases. Cell Mol Biol Lett 11:132–154
Stalker DM, Kiser JA, Baldwin G, Coulombe B, Houck CM (1996) Cotton weed control using the BXN system. In: Duke SO (ed) Herbicide resistant crops: agricultural, environmental, economic, regulator, and technical aspects. CRC Press, Boca Raton, pp 93–105
Steinbrecher R (1996) From green revolution to gene revolution: the environmental risks of genetically engineered crops. Ecologist 26:240–273
Storer NP, Kubiszak ME, King JE, Thompson GD, Santos AC (2012) Status of resistance to Bt maize in *Spodoptera frugiperda*: lessons from Puerto Rico. J Invertebr Pathol 110:294–300
Tabashnik BE, Huang F, Ghimire MN (2011) Efficacy of genetically modified Bt toxins against insects with different genetic mechanisms of resistance. Nat Biotechnol 29:1128–1131
Tabashnik BE, BreÂvault T, Carrière Y (2013a) Insect resistance to Bt crops: lessons from the first billion acres. Nat Biotech 31:510–521
Tabashnik BE, Fabrick JA, Unnithan GC, Yelich AJ, Masson L, Zhang J, Bravo A, Soberón M (2013b) Efficacy of genetically modified Bt toxins alone and in combinations against pink bollworm resistant to Cry1Ac and Cry2Ab. PLoS One 8:e80496

Tan S, Evans RR, Dahmer ML, Singh BK, Shaner DL (2005) Imidazolinone tolerant crops: history, current status and future. Pest Manag Sci 61:246–257
Tian YS, Xu J, Xiong AS, Zhao W, Fu XY, Peng RH, Yao QH (2011) Improvement of glyphosate resistance through concurrent mutations in three amino acids of the Ochrobactrum 5-enopyruvylshikimate-3-phosphate synthase. Appl Environ Microbiol 77:8409–8414. https://doi.org/10.1128/AEM.05271-11
Tian JC, Yao J, Long LP, Romeis J, Shelton AM (2015) Bt crops benefit natural enemies to control non-target pests. Sci Rep 5:16636. https://doi.org/10.1038/srep16636
Tomiuk J, Hauser TP, Bagger-Jorgensen R (2000) A- or C-chromosomes, does it matter for the transfer of transgenes from *Brassica napus*? Theor Appl Genet 100:750–754
Turrini A, Sbrana C, Giovannetti M (2015) Belowground environmental effects of transgenic crops: a soil microbial perspective. Res Microbiol 166:121–131
University of Illinois Plant Clinic Herbicide Resistance Report (2016). Available at: http://bulletin.ipm.illinois.edu/?p=3821
Van Gessel MJ (2001) Glyphosate–resistant horseweed from Delaware. Weed Sci 49:703–705
Van Rensburg JJ (2007) First report of field resistance by stem borer, *Busseolafusca* (Fuller) to Bt transgenic maize. S Afr J Plant Soil 24:147–151
Vandenborre G, Smagghe G, Van Damme EJ (2011) Plant lectins as defense proteins against phytophagous insects. Phytochemistry 72:1538–1550. https://doi.org/10.1016/j.phytochem.2011.02.024
Vishnudasan D, Tripathi MN, Rao U, Khurana P (2005) Assessment of nematode resistance in wheat transgenic plants expressing potato proteinase inhibitor (PIN2) gene. Transg Res 14:665–675
Walters FS, Stacy CM, Lee MK, Palekar N, Chen JS (2008) An engineered chymotrypsin/cathepsin G site in domain I renders *Bacillus thuringiensis* Cry3A active against Western corn rootworm larvae. Appl Environ Microbiol 74:367–374
Walters FS, deFontes CM, Hart H, Warren GW, Chen JS (2010) Lepidopteran-active variable-region sequence imparts coleopteran activity in eCry31ab, an engineered *Bacillus thuringiensis* hybrid insecticidal protein. Appl Environ Microbiol 76:3082–3088
Wang J, Chen Z, Du J, Sun Y, Liang A (2005) Novel insect resistance in *Brassica napus* developed by transformation of chitinase and scorpion toxin genes. Plant Cell Rep 24:549–555
Warwick SI, Simard MJ, Légère A, Beckie HJ, Braun L, Zhu B, Stewart CN Jr (2003) Hybridization between transgenic *Brassica napus* Land its wild relatives: *Brassica rapa* L, *Raphanus raphanistrum* L, *Sinapis arvensis* L, and *Erucastrum gallicum* (Willd) OE Schulz. Theor Appl Genet 107(3):528–539. https://doi.org/10.1007/s00122-003-1278-0
Warwick SI, Légère A, Simard MJ, James T (2008) Do escaped transgenes persist in nature? The case of an herbicide resistance transgene in a weedy *Brassica rapa* population. Mol Ecol 17(5):1387–1395. https://doi.org/10.1111/j.1365-294X.2007.03567.x
Warwick SI, Beckie HJ, Hall LM (2009) Gene flow, invasiveness, and ecological impact of genetically modified crops. Ann N Y Acad Sci 1168:72–99
Watkinson AR, Freckleton RP, Robinson RA, Sutherland WJ (2000) Predictions of biodiversity response to genetically modified herbicide-tolerant crops. Science 289(5484):1554–1557
Watrud LS, Lee EH, Fairbrother A, Burdick C, Reichman JR, Bollman M, Storm M, King G, Van de Water PK (2004) Evidence for landscape-level, pollen-mediated gene flow from genetically modified creeping bentgrass with CP4 EPSPS as a marker. Proc Natl Acad Sci U S A 101:14533–14538
Welch KL, Unnithan GC, Degain BA, Wei J, Zhang J, Li X, Tabashnik BE, Carrière Y (2015) Cross-resistance to toxins used in pyramided Bt crops and resistance to Bt sprays in *Helicoverpa zea*. J Invertebr Pathol 132:149–156
Widmer F, Seidler RJ, Donegan KK, Reed GL (1997) Quantification of transgenic plant marker gene persistence in the field. Mol Ecol 6:1–7

Wiersma AT, Gaines TA, Preston C, Hamilton JP, Giacomini D, Buell CR, Leach JE, Westra P (2015) Gene amplification of 5-enol-pyruvylshikimate-3-phosphate synthase in glyphosate-resistant *Kochia scoparia*. Planta 241:463–474

Williams GM, Kroes R, Munro IC (2000) Safety evaluation and risk assessment of the herbicide roundup and its active ingredient, glyphosate, for humans. Regul Toxicol Pharmacol 31:117–165

Wolfenbarger LL, Phifer PR (2000) The ecological risks and benefits of genetically engineered plants. Science 290:2088–2093

Wright TR, Shan G, Walsh TA, Lira JM, Cui C, Song P, Zhuang M, Arnold NL, Lin G, Yau K, Russell SM, Cicchillo RM, Peterson MA, Simpson DM, Zhou N, Ponsamuel J, Zhang Z (2010) Robust crop resistance to broadleaf and grass herbicides provided by aryloxyalkanoate dioxygenase transgenes. Proc Natl Acad Sci U S A 107:20240–20245. https://doi.org/10.1073/pnas.1013154107

Yang MS, Mi D, Ewald D, Wang Y, Liang HY, Zhen ZX (2006) The survival and escape of Agrobacterium tumefaciens in triploid hybrid lines of Chinese white poplar transformed with two insect-resistant genes. Acta Ecol Sin 26:3555–3561. https://doi.org/10.1016/s1872-2032(06)60055-3

Yang F, Kerns DL, Head GP, Leonard BR, Levy R, Niu Y, Huang F (2014) A challenge for the seed mixture refuge strategy in Bt maize: impact of cross-pollination on an ear-feeding pest, corn earworm. PLoS One 9(11):e112962. https://doi.org/10.1371/journal.pone.0112962

Zavala JA, Giri AP, Jongsma MA, Baldwin IT (2008) Digestive duet: midgut digestive proteinases of *Manduca sexta* ingesting *Nicotiana attenuata* with manipulated trypsin proteinase inhibitor expression. PLoS One 3(4):e2008

Zhang BH, Guo TL, Wang QL (2000) Inheritance and segregation of exogenous genes in transgenic cotton. J Genet 79:71–75

Zhou D, Xu L, Gao S, Guo J, Luo J, You Q, Que Y (2016) Cry1Ac transgenic sugarcane does not affect the diversity of microbial communities and has no significant effect on enzyme activities in rhizosphere soil within one crop season. Front Plant Sci 8(7):265. https://doi.org/10.3389/fpls.2016.00265

Zhu B, Ma BL, Blackshaw RE (2010) Development of real time PCR assays for detection and quantification of transgene DNA of a *Bacillus thuringiensis* (Bt) corn hybrid in soil samples. Transg Res 19:765. https://doi.org/10.1007/s11248-009-9353-1

Zhu-Salzman K, Zeng R (2015) Insect response to plant defensive protease inhibitors. Annu Rev Entomol 60:233–252. https://doi.org/10.1146/annurev-ento-010814-020816

Zwahlen C, Nentwig W, Bigler F, Hilbeck A (2003) Tritrophic interactions of transgenic *Bacillus thuringiensis* corn, *Anaphothripsobscurus* (Thysanoptera: Thripidae), and the predator *Orius majusculus* (Heteroptera: Anthocoridae). Environ Entomol 29:846–850

# Impact of Genetically Modified Crops on the Biodiversity of Arbuscular Mycorrhizal Fungi

**Roxana Paula Colombo, Laura Fernandez Bidondo, Vanesa Analía Silvani, and Alicia Margarita Godeas**

**Abstract** Plants, including genetically modified (GM) ones, are able to shape the structure of soil microbiomes by modifying the soil properties and the rhizosphere chemistry by the exudation of different compounds, thus influencing the soil microbial dynamics. Among the rhizosphere microorganisms beneficial to the plant community, the arbuscular mycorrhizal (AM) fungi are considered as indicators of soil health, playing important roles in plant growth by improving mineral nutrition, stabilizing soil aggregates, and protecting their host plants against pathogens. In the context of a sustainable use of soils and conservation of the biodiversity of these beneficial fungi, a better understanding of the effects of GM plants on the mycorrhizal symbiosis is thus relevant. This chapter seeks to understand the effects of GM crops on AM fungi by reviewing the existing bibliography and to lay the foundation for further research works that enrich the current knowledge on this topic. Although to date most scientific investigations have found no significant effect of GM plants on AM fungal colonization, some found a negative effect of GM plants on the initial stages of the establishment of the mycorrhizal symbiosis. Thus, it is necessary to carry out further investigations on a long temporal scale and under different agro-ecological conditions, focusing on AM fungal biodiversity and on the impact of each GM crop on these beneficial microorganisms before authorizing release.

**Keywords** Arbuscular mycorrhizal fungi · Non-target microorganism · Soil health indicators · Rhizosphere effect · Intraradical effect · Long-term testing · Sustainable agriculture · Lack of monitoring · Impact on biodiversity

R. P. Colombo · V. A. Silvani · A. M. Godeas (✉)
Instituto de Biodiversidad y Biología Experimental y Aplicada, Universidad de Buenos Aires-CONICET, Buenos Aires, Argentina
e-mail: godeas@bg.fcen.uba.ar

L. Fernandez Bidondo
Laboratorio de Micología Molecular, Departamento de Ciencia y Tecnología, Universidad Nacional de Quilmes, Quilmes, Buenos Aires, Argentina
e-mail: laura.bidondo@unq.edu.ar



A. Chaurasia et al. (eds.), *GMOs*, Topics in Biodiversity and Conservation 19,
https://doi.org/10.1007/978-3-030-53183-6_4

## Introduction

Since 1996, the global area cultivated with genetically modified (GM) crops has increased, reaching 13% of the world's arable surface (189.8 million hectares) in 2017. The crops most commonly subjected to genetic modifications are soybean, cotton, maize, and canola, occupying 82%, 68%, 30%, and 25%, respectively, of the planted agricultural surface (ISAAA 2017). At present, approximately 78% of the global area under GM crop production is in the USA, Brazil, and Argentina, with the greatest area occupied by cultivars bearing herbicide resistance traits, followed by Bt crops and by a combination of both herbicide-resistant and Bt-tolerant crops. The most common trait used in GM crops is insecticidal toxin production (ISAAA 2017).

The rhizosphere is inhabited by certain chemotactic microorganisms that are influenced by the chemical compounds exuded from plant roots. These compounds include metabolites, such as sugars, amino acids, and carboxylic acids, as well as diverse secondary metabolites, whose production is under plant genetic control. The rhizosphere effect could be considered as a plant's strategy to select and recruit specific microorganisms that could help directly or indirectly in the nutrient mining process. The rhizosphere is principally colonized by fungi and bacteria, which in turn have their own nutrient strategies and biology to be symbionts, pathogens, or saprotrophs. The main role of soil fungi is to participate in organic matter decomposition, thus contributing to plant nutrition (Bridge and Spooner 2001). The soil fungal community includes arbuscular mycorrhizal (AM) fungi, which are specialized root endo-symbionts that are intimately associated with a great diversity of plant species (Smith and Read 2008). As other soil microorganisms, AM fungi may be exposed to the new traits expressed by GM plants even when they are not the target of genetic modification (Castaldini et al. 2005; Giovannetti et al. 2005). In addition, GM plants have the potential to influence microbial dynamics by modifying the rhizosphere's chemistry through the exudation of new compounds, which could also either inhibit or promote AM fungal symbiosis (Bruinsma et al. 2003).

This chapter seeks to provide information to better understand the effects of GM crops on the different development stages of AM fungi and their biodiversity. To this end, we summarized the results globally reported in scientific publications since 1993 to the present, aiming to detect the points that have not yet been thoroughly investigated and thus propose further studies.

## AM Fungi

AM fungi (phylum *Glomeromycota*) form mutualistic associations with more than 80% of terrestrial plant species from different taxonomic groups and varied habitats (Smith and Read 2008). Although these fungi exhibit little host specificity and a single plant can be simultaneously colonized by more than one species of AM fungi,

Öpik et al. (2010) found certain degree of AM fungal specificity when grouping plant species by their taxonomic families.

The origin of these fungi dates back over 400 million years (Humphreys et al. 2010). Ordovician and Devonian fossil records reveal the presence of AM symbioses, supporting the hypothesis that AM fungi played a crucial role in the colonization of land by plants (Redecker 2000). Because of their obligate biotrophy, AM fungi need carbohydrates from the host plant to complete their life cycle. In exchange, they uptake and transfer soil nutrients and water to plants, thus improving plant fitness, protecting them against biotic and abiotic stresses, and enhancing the quality of soils (Smith and Read 2008).

AM fungal life cycle can be divided into different phases (Bonfante and Genre 2010), presymbiotic phase, root colonization, active symbiotic phase, and extraradical phase (Fig. 1):

(a) Presymbiotic phase: after the germination of AM fungal propagules (spores, extraradical mycelium, mycorrhized roots), the hyphal apex grows and branches as a response elicited by plant-derived strigolactones, among other volatile and diffusible compounds of the rhizosphere that act as intermediaries of the molecular dialog between plants and AM fungi. During hyphal growth toward the root, hyphae also release signal molecules that trigger a host symbiotic program.

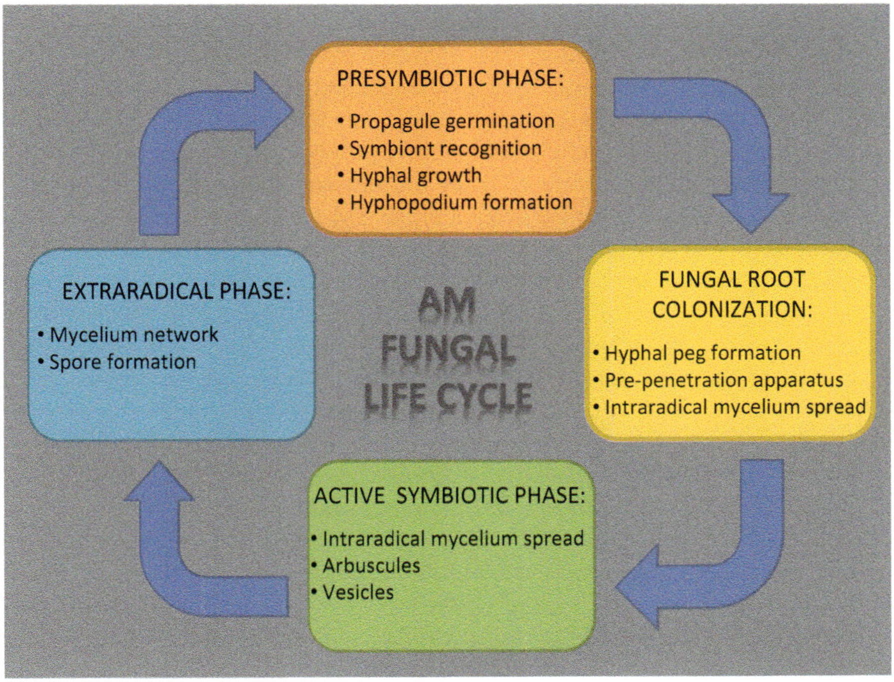


**Fig. 1** Typical arbuscular mycorrhizal fungal life cycle

(b) Fungal root colonization: when the AM fungal hyphae contact the root surface, a hyphopodium is differentiated and a penetration peg is formed to facilitate entry into the root. Simultaneously, root cells in the vicinity of the penetration peg develop a subcellular structure that resembles a tube called prepenetration apparatus (PPA). Intracellular fungal colonization moves behind the PPA from the epidermis to the inner cortex, to finally spread to the apoplast along the longitudinal axis of the root.
(c) Active symbiotic phase: eventually, fungal hyphae penetrate the host cells and form highly branched arbuscules or coils (another typical AM fungal structure). Each hyphal branch is surrounded by a plant-derived periarbuscular membrane, which physically separates the fungus from the cell cytoplasm. The apoplastic interface between the fungal and plant membrane plays a key role in the bidirectional exchange of nutrients between both symbionts. Next, certain AM fungal species develop storage structures within the roots, called vesicles.
(d) Extraradical phase: the establishment of the symbiosis allows the completion of the AM fungal life cycle. This involves the formation of a network of extraradical mycelium (ERM) that colonizes the surrounding soil and is responsible for the uptake of mineral nutrients (mainly phosphorus (P), nitrogen, and other micronutrients) and translocation to the host plant (Jansa et al. 2003). Finally, new spores develop at the apex of hyphae in the ERM and sometimes in the apoplast of roots, depending on the AM fungal species. AM fungi then produce abundant asexual multinucleate spores with different polymorphic deoxyribonucleic acid (DNA)-sequence variants and different numbers of nuclei among sister spores as a result of nuclear migration and mitosis (Marleau et al. 2011).

The biodiversity of a particular AM fungal community can be studied by different techniques. Some traditional techniques include the analysis of AM root colonization or the isolation and identification of spores from a soil of interest. Molecular techniques involve the isolation of DNA and the use of diverse molecular techniques for the molecular characterization of the AM fungal community. Regarding the identification of AM fungi, morphological characterization is based on observations of the ultrastructure and ontogeny of spores, whereas molecular characterization frequently considers different regions of DNA ribosomal genes. Based on these characterizations, approximately 244 AM fungal taxa and 348 to over 1600 virtual taxa have so far been described, if environmental sequences are contemplated (Ohsowski et al. 2014). Considering morphological and molecular phylogenetic analyses, until now *Glomeromycota* are divided in four orders and 11 families: *Glomerales* (*Glomeraceae* and *Claroideoglomeraceae*), *Diversisporales* (*Diversisporaceae*, *Acaulosporaceae*, *Entrophosporaceae*, *Pacisporaceae*, and *Gigasporaceae*), *Paraglomerales* (*Paraglomeraceae*), and *Archaeosporales* (*Archaeosporaceae*, *Ambisporaceae*, and *Geosiphonaceae*) (Krüger et al. 2012). However, since the classification of *Glomeromycota* is in continuous change and regrouping and numerous AM fungal species are still undescribed, AM fungal biodiversity in many ecosystems around the world has not been completely characterized yet.

## AM Fungal Biodiversity in Natural Environments

Arbuscular mycorrhizal fungi are ubiquitous and commonly found along different geographic locations and biomes, from tropical to temperate forests, grasslands, and deserts and even extreme ecosystems (Brundrett 1991). Thus, they are considered a cosmopolitan group of microorganisms. However, some evidence suggests that certain families and species of AM fungi occupy only certain continents and climatic zones (Öpik et al. 2010). For instance, members of *Acaulosporaceae* and *Gigasporaceae* are mainly reported from tropical areas of Africa and South America, whereas members of *Glomeraceae* dominate the temperate and boreal/austral habitats of Europe, Asia, and Oceania (Chaudhary et al. 2018; Stürmer et al. 2018). Members of *Archaeosporaceae* and *Paraglomeraceae* are found in all continents except the Oceanian and Madagascan realms and Antarctica, whereas *Glomeraceae* and *Claroideoglomeraceae* are found mainly in extreme natural environments (e.g., geothermal sites, hypersaline soils, or high-altitude habitats) (Appoloni et al. 2008; Silvani et al. 2016). In soils of Antarctica, only members of *Glomeraceae* and *Acaulosporaceae* have been reported.

It is widely accepted that the distribution of AM fungal species is influenced by latitudinal gradients and that species richness decreases with latitude at the global scale from tropical to boreal/austral zones (Kivlin et al. 2011; Davison et al. 2015). The ecological factors that affect the composition of the AM fungal community are very complex. Among abiotic factors, soil properties, habitat fragmentation, and seasonality play a key role in influencing the structure of the AM fungal community (Hausmann and Hawkes 2009). Given that AM fungi establish symbiotic associations with more than 80% of plant species, ecological surveys aimed to study the global AM fungal diversity primarily concern plant hosts and plant-defined biomes. These studies have noticed that plant community assemblages affect directly the AM fungal communities from a location (Garcia de Leon et al. 2016) and that although AM fungi are thought to be predominantly generalists, certain AM fungal taxa are more host specific than others (Smith and Read 2008). Other factors such as dispersal agents, environmental filtering, interactions within the soil biota, and intrinsic AM fungal traits may also contribute to their diversity patterns (Lekberg et al. 2007; Öpik and Davison 2016). Some phenotypic and functional traits of AM fungi, such as dispersal capacity or P translocation efficiency, may significantly impact plant host fitness and soil colonization (Hart and Reader 2005) and subsequently the occurrence of AM fungal species in an area.

Natural ecosystems possess a higher plant and microbial diversity than agricultural fields. The low diversity of plants in agroecosystems, especially in monocultures, promotes the homogenization of lands, induces modifications of the microbial communities, and diminishes their level of biodiversity (Andreote and Pereira e Silva 2017). A considerable number of studies have demonstrated a decline in AM fungal diversity or a replacement of AM species in agroecosystems when compared with natural ecosystems (Oehl et al. 2004; Bedini et al. 2007). In contrast to human-impacted habitats, which have a great proportion of cultured and described AM fungal taxa (Ohsowski et al. 2014), natural habitats and wild plants harbor AM fungal communities composed mainly of uncultured and undescribed taxa (or AM fungal species known only through DNA sequences, without morphological description). Therefore, information on the diversity of AM fungi and their functionality in natural environments could definitely contribute to a better understanding and prediction of AM fungal responses to anthropogenic environmental disturbance in a changing ecosystem. This information should include both marginal soils used for agriculture and land systems in which GM crops are newly incorporated into the cropping system.

As mentioned above, a large number of AM fungal species are yet to be discovered in natural habitats around the world. Therefore, the conservation of these natural environments is primary for the identification of new species of *Glomeromycota* and to maintain the reservoirs of their germplasms for future biotechnological interest under a sustainable program.

## Impact of the Agricultural Management of Soils on AM Fungal Diversity

Biodiversity surveys of AM fungal communities in agroecosystems have to consider the effects of different factors such as soil compaction by tillage equipment, chemical fertilization, pesticide application, crop variety, type of soil, and climate. The possible consequences of the management practices on the AM fungal community structure and mycorrhizal development have been widely investigated (Oehl et al. 2004). Many studies using traditional methodologies have demonstrated a reduction of AM fungal inoculum potential in agricultural soils, as well as of AM fungal abundance and species richness (Schalamuk and Cabello 2010). Recently, studies based on pyrosequencing technologies have provided new approaches to study the ecology of microbial communities in a wide range of environments (Unterseher et al. 2011; Colombo et al. 2014). In concordance with that found by traditional methods, these studies have found that the anthropogenic disturbance of soils has reduced the AM fungal species richness of arable, plantation, and pasture systems (Lumini et al. 2009; Moora et al. 2014; Xiang et al. 2014). Verbruggen et al. (2010), for example, showed that *Gigasporaceae* and *Acaulosporaceae* are vulnerable to threats from agricultural practices, whereas Hijri et al. (2006) observed the absence of *Acaulospora* and *Scutellospora* sequences in maize roots from an intensive conventional monoculture. In contrast, they also reported a high diversity of AM fungi under low application of fertilizers and pesticides. Many researchers agree that the most important factors that drive AM fungal community diversity in agricultural soils are intensive high-input farming crop and tillage, low plant diversity, excessive use of pesticides (particularly fungicides), and fertilization (Jansa et al. 2006). In contrast, some authors have suggested that organic farming is able to sustain greater AM fungal diversity (Manoharan et al. 2017).

Some GM plants pose inherent agricultural management, such as herbicide application, which directly or indirectly affects the function and diversity of soil and rhizosphere microbial communities, including AM fungi (Turrini et al. 2015). For instance, Druille et al. (2013) observed that glyphosate reduces the spore viability and root colonization of AM fungi.

In addition to agricultural management, some soil properties like pH, texture, organic matter, macronutrient content, and humidity are main abiotic factors that seem to determine the biodiversity of AM fungal communities (Hannula et al. 2010). It is expected that GM crop cultivation could modify these edaphic properties due to changes in root exudates, their physiology, and radical architecture or indirectly through effects on the soil biota (Liu 2010; Chen et al. 2017). The rhizosphere microbiome can also be affected by other important factors such as temperature and water availability in soils. However, Icoz et al. (2008) and Seres et al. (2014) recorded no significant differences in these soil parameters between Bt maize and isogenic maize plots. Regarding the effects of GM crops on soil chemistry, Liu (2010) found minor changes in soil N and P contents when growing Bt and glyphosate herbicide-tolerant GM corn crops instead of conventional crops, whereas Liang et al. (2015) observed no differences in N and C concentrations in the rhizosphere soil of transgenic soybean varieties. Regarding soil pH values, Cruz-Gutiérrez et al. (2015) and Liang et al. (2015) found that these values remained unaffected by some GM plants (soybean varieties and transgenic Mexican lime), whereas Donegan et al. (1999) recorded an increase in soil pH with genetically engineered alfalfa inoculated with recombinant *Sinorhizobium meliloti*. Thus, since results on the potential alterations in edaphic specificities due to GM plants that might influence the biology of AM fungi are still controversial, more works should address this issue under different field trials. A better understanding of the effects of agricultural practices with new GM crops on the mycorrhizal symbiosis is relevant to improve agricultural production and the sustainable use of soils.

## GM Crops and AM Fungi

Since the end of the twentieth century, with the start of the manipulation of organisms by genetic modification and interspecific gene transfer, biotechnology has had a great impact on the spread of GM plants. These GM crops have been commercialized for more than 20 years, and in the last years the offer has significantly increased not only because of the improvements in genetic engineering but also because of the need to scale up agricultural production (Qaim 2009). In addition, the land area destined to agricultural crops increased from 1.7 million hectares in 1996 to 189.8 million hectares in 2017. This increase has been more significant in developing countries (ISAAA 2017; Pellegrino et al. 2018). The list of GM crops approved for commercial release does not exceed 30 species in total, and genetic modifications almost exclusively point to making crops resistant to herbicides or pests (insects, viral and fungal pathogens), delaying the ripening and browning of fruits, modifying the colors of flowers, or enhancing nutritional values (ISAAA 2017).

Although GM crops provide numerous agricultural and economic benefits, during 2017, some countries totally stopped planting GM crops due to the high costs of cultivation and the preference of manufacturers for non-GM processed food and feed (ISAAA 2017). The cultivation of GM crops is far from being openly accepted since many studies have warned about their potential impact on the environment. These studies have focused mainly on (1) gene transfer from GM to wild plants, (2) the transference of antibiotic resistance to natural microbial populations, and (3) the impact of GM proteins on nontarget organisms (Giovannetti 2003; Turrini et al. 2015). These nontarget organisms include AM fungi, which are among the key functional groups of soil microbial communities. Despite this and the importance of AM fungi, few studies have assessed the effects of GM plants on this group of obligate biotrophic fungi (Liu 2010; Hannula et al. 2014).

Miller (1993) defined two genetic factors of host plants that could influence the establishment of the mycorrhizal symbiosis. The first factor is that the colonization susceptibility of AM fungi would be directly related to the resistance of host plants to nonspecific pathogens due to the production of pathogenesis-related proteins and/or signaling and recognition molecules, whereas the second factor would be the changes induced by unmet physiological needs, specifically at the nutrient and water levels.

Before their release for large-scale cultivation, all GM crops should be monitored for their detrimental effects on nontarget soil microorganisms. In this context, AM fungi are good candidates for this type of analysis as they are indicators of a normal rhizosphere structure and functionality and because the interaction between GM plants and the AM fungal community is critical to the ecology of agricultural soils (Liu 2010; Liu and Du 2008). However, the number of scientific studies on the interaction between GM plants and AM fungi, even taking into account experimental researches at field or greenhouse scale and reviews, barely reaches 50 (Table 1). Besides, these investigations have been carried out in a few countries, focusing on very few crops (Fig. 2a, d). When comparing the results of these studies with those published by the International Service for the Acquisition of Agri-biotech Applications (ISAAA) in 2017, it can be seen that the crops studied, the nature of the genetic modification, and the institutions where the research works were carried out do not match with the countries where the greatest amount of GM crops are cultivated (in terms of the number of released species and cultivated area) or with the most widely distributed genetic modifications. In addition, the overwhelming majority of these scientific trials have studied the effect of only two genetic traits: the expression of Bt proteins and chitinases in few host plants of agronomic or scientific interest, corn and tobacco, respectively (Fig. 2). This bias has been indicated in the reviews by Bruinsma (2003), Liu (2010), Hannula et al. (2014), and Mohandass and Muthukumar (2017).

Based on the scientific works analyzed in this chapter, the cultivation of GM plants seems to have no direct or significant effect on the establishment of AM symbiosis. However, some studies (about 19%) report negative effects on the

**Table 1** Studies examining the effect of different species of genetically modified plants on arbuscular mycorrhizal fungi and the countries where they were carried out

| Plant | Country | Genetic improvement | Effect on AM fungi | References |
|---|---|---|---|---|
| Corn | USA | Resistance to insects | No effect | Blackwood and Buyer (2004) |
| Corn | Italy | Resistance to insects | Negative | Castaldini et al. (2005) |
| Corn | USA | Resistance to insects | Negative | Cheeke et al. (2010) |
| Corn | USA | Resistance to insects | Negative | Cheeke et al. (2012) |
| Soybean | USA | Resistance to insects | No effect | Cheeke et al. (2013) |
| Corn | China | Resistance to insects | Negative | Chen (2017) |
| Corn | Argentina | Enhance drought tolerance | No effect | Colombo et al. (2017) |
| Mexican lime | Mexico | Unspecified | Positive | Cruz Gutierrez (2015) |
| Rice | Denmark | Resistance to insects | No effect | de Vaufleury (2007) |
| Corn | Switzerland | Resistance to insects | No effect | Fließbach et al. (2012) |
| Pea | Canada | Resistance to fungi | No effect | Gill Kahlon et al. (2017) |
| Wheat | Italy | Resistance to fungi | No effect | Girlanda et al. (2008) |
| Soybean | Germany | High amylopectin levels | No effect | Gschwendtner et al. (2010) |
| Potato | Netherlands | High starch content | No effect | Hannula (2010) |
| Potato | Netherlands | High starch content | No effect | Hannula (2012) |
| Aspen | Germany | Modification of phytohormone balance | No effect | Kaldorf et al. (2002) |
| Tomato | Australia | Resistance to glyphosate | No effect | Knox et al. (2008) |
| Cotton | Australia | Resistance to insects | No effect | Knox et al. (2008) |
| Corn | Australia | Resistance to insects and glyphosate | No effect | Knox et al. (2008) |
| Corn | Netherlands | Resistance to insects | No effect | Kuramae et al. (2013) |
| Corn | China | Enhance in methionine content | No effect | Liang et al. (2015) |
| Cotton | Spain | Resistance to pathogens | No effect | Medina et al. (2003) |
| Flax | Spain | Resistance to pathogens | No effect | Medina et al. (2003) |
| Rice | Switzerland | Resistance to fungi | No effect | Meyer (2013) |
| Potato | USA | Resistance to pathogens | No effect | Newhouse (2007) |
| Cotton | USA | Resistance to herbicide | No effect | Powell (2007) |
| Corn | China | Resistance to insects | No effect | Ren (2006) |
| Corn | Hungary | Resistance to insects | Negative | Seres et al. (2014) |
| Potato | Argentina | Resistance to fungi | No effect | Stephan et al. (2019 ) |
| Woodland tobacco | United Kingdom | Resistance to fungi | Positive | Tahiri-Alaoui et al. (1994) |
| Corn | China | Resistance to insects | No effect | Tan (2011) |
| Aubergine | Italy | Resistance to fungi | Negative | Turrini et al. (2004a) |
| Corn | Italy | Resistance to insects | No effect | Turrini et al. (2004a) |

(continued)

**Table 1** (continued)

| Plant | Country | Genetic improvement | Effect on AM fungi | References |
|---|---|---|---|---|
| Aubergine | Italy | Resistance to fungi | No effect | Turrini et al. (2004b) |
| Corn | Italy | Resistance to insects | Negative | Turrini et al. (2008) |
| Corn | Netherlands | Resistance to insects | No effect | Verbruggen (2012) |
| Tobacco | Switzerland | Resistance to fungi | No effect | Vierheilig et al. (1993) |
| American elm | Switzerland | Resistance to pathogens | No effect | Vierheilig et al. (1995) |
| Tobacco | Poland | Resistance to pathogens | No effect | Wróbel-Kwiatkowska et al. (2012) |
| Aubergine | China | Resistance to fungi | Negative | Yang et al. (2002) |
| Corn | China | Resistance to insects | No effect | Zeng et al. (2014) |
| Corn | China | Resistance to insects | No effect | Zeng et al. (2015) |

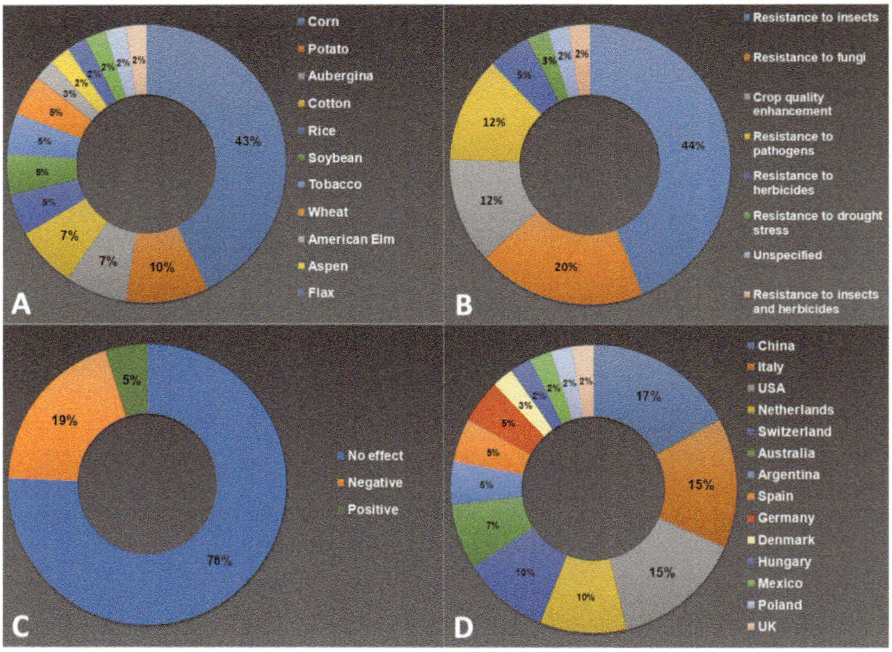


**Fig. 2** (**a**) Percentual representation of genetically modified plants species; (**b**) type of genetic improvements, tested for their effect on arbuscular mycorrhizal fungi (**c**); and (**d**) the countries where they were carried out. Data obtained from a survey of different scientific works (see Table 1)

establishment of the symbiosis, especially in the presymbiotic phase (lower hyphal growth), a delay at the initiation of fungal root colonization, and a decrease in the proportion of arbuscules, and only two studies (5% of the scientific reports) have described the positive effects of GM crops on AM fungal colonization (Tahiri-Alaoui et al. 1994) and biodiversity (Cruz Gutierrez 2015).

## *GM Crops Resistant to Abiotic Stressful Conditions*

Plants are continuously exposed to adverse environmental changes, such as decreases in water availability, increases in soil salinity, variation in season, soil compaction, and soil pollution. Under these conditions, the production of reactive oxygen species (ROS) increases, causing oxidative damage and cellular injury. To respond to such stressful conditions, plants have developed different mechanisms. To mitigate drought stresses, for example, they tend to increase the extension of their roots and/or to establish symbiotic association with AM fungi (Bompadre et al. 2015). Also under stress conditions, mycorrhizal plants display enhanced fitness due to a higher activity of antioxidant enzymes and an improved capacity to detoxify ROS. In addition, a higher photosynthetic efficiency and an increased production of endogenous cytokines in AM plants are responsible for the short-term relief during leaf senescence (Bompadre et al. 2015). Thus, the establishment of AM fungal symbiosis could also enhance the resistance of host plants to several extreme abiotic conditions, protecting them from oxidative stress.

The expression of certain genes also confers resilience to stressed plants, enhancing their capacity to adapt or tolerate environmental pressures. For example, the overexpression of the transcription factor Hahb-4 (HD-Zip family) from *Helianthus annuus*, regulated by water availability, abscisic acid, and soil salinity, inhibits the perception of ethylene or jasmonic acid, delaying plant senescence under drought conditions. This has led to a conserved heterologous response in transgenic *Arabidopsis thaliana* and *Zea mays* plants, resulting in tolerance to drought and salinity (Manavella et al. 2008). However, only one study by Colombo et al. (2016) examined the interaction between drought-tolerant GM corn and AM fungi. These authors detected no effect of the transgenic corn lines tested on AM fungal colonization, and the results were related to the fact that the overexpression of the Hahb-4 transcription factor does not produce or release metabolites into the rhizosphere. However, as mycorrhizal colonization entails an energy cost to the host plant, it is favored by nutrient- and water-stressful conditions. It is expected that when that cost is offset, AM fungal colonization levels will diminish.

In the cases of GM plants resistant to herbicides, Knox et al. (2008) and Powell (2007) found no effects on the community of AM fungi. It should be noted that most studies conducted to elucidate the effect of GM crops expressing abiotic stress resistance genes on AM fungi drew their conclusions only by studying root colonization and that none of them studied their effect on the diversity of these beneficial microorganisms.

## *GM Crops Resistant to Negative Biotic Interactions*

Crop protection and disease control are supported mainly by agrochemical pesticides, which present residual toxicity and induce pesticide-resistant pathogens. In that context, GM cultivars could aid to reduce losses caused by pests and diminish the input of chemical supplies in agroecosystems. The commercial use of GM crops resistant to the attack of predators and pathogens is increasingly widespread (ISAAA 2017). Regarding the soil microbial ecology, studies on the potential risks of these GM plants focus mainly on the impact of transgenic proteins, expressed by roots or incorporated through decaying plant biomass, accumulated in the rhizosphere, and potentially affecting nontarget microorganisms (Turrini et al. 2015). Phenotypic changes in GM plants (pleiotropic effects) and in their root exudate profiles could modify rhizosphere microbial communities.

Different studies that have evaluated the effects of GM plants on AM fungal infectivity by assessing root colonization have found both neutral and negative interactions. Results vary regarding the genes expressed, their expression levels, and the plant species involved (Castaldini et al. 2005; Liu 2010; Newhouse et al. 2007; Turrini et al. 2004b, 2015; Vierheilig et al. 1995).

The impact of these GM plants on AM fungi has also been estimated by spore counts and molecular methods. However, a few studies have explored the effect of GM crops on the biodiversity of AM fungi. Generally, the results of these studies have shown negative or null effects of GM plants on AM fungi (Fig. 3), either on the diversity of their communities or on their ability to effectively develop symbiosis (Bruinsma et al. 2003; Liu 2010; Hannula et al. 2014; Mohandass and Muthukumar 2017).

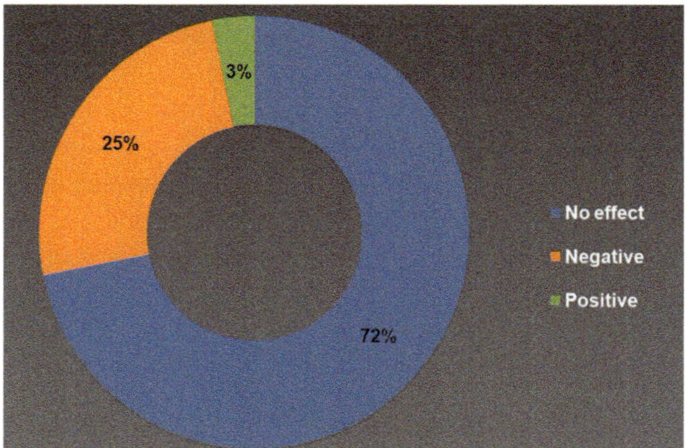


**Fig. 3** Percentual representation of the effect of genetically modified crops resistant to negative biotic interactions on arbuscular mycorrhizal fungi. Data obtained from a survey of different scientific works

## Conclusions and Future Prospects

Although many works have found no significant effect of GM cultivation on AM fungal colonization, it is necessary and urgent to carry out more investigations on a long temporal scale and different agroecological conditions. Possibly, researchers have not observed adverse effects in an immediate term due to the enormous plasticity of AM fungi, which allows them to adapt to any environment and host condition. Although intraradical colonization levels may not be affected in the short term, very few studies have evaluated which AM fungal species are actually colonizing GM crops, and thus the possible long-term effect on the biodiversity of these beneficial organisms still needs to be established. Particularly, in countries where the cultivation of GM plants has been widely adopted, there are very few reports or even no reports at all on their impact on AM fungi.

On the other hand, exhaustive genetic studies should be done to detect possible variation and selection of AM genes leading to undesirable effects on their fitness and, consequently, on their fungal functional traits. A drastic change at the genetic level could lead to an improper selection of less efficient AM fungal genotypes during the symbiosis. These variations in AM fungal life history traits could influence the nutritional status and growth of host cultivars and, consequently, their productivity.

Investigations based on putative modifications of the microbiome associated with AM fungi should equally be considered as it is well known that some rhizosphere microorganisms, especially some groups of bacteria, are intimately related to AM fungal structures, whereby a change in these microorganisms could have a direct effect on the survival of AM fungi and the effectiveness of the AM symbiosis.

These neglected issues could nevertheless be largely overcome with better sustainable agricultural management practices, such as crop rotation or intercropping with other AM plant species, to improve soil quality and to maintain microbial soil diversity. In addition, further research should include environmental impact studies of the food crops of interest in controlled conditions.

In summary, in order to prevent any possible adverse influence of GM plants on AM fungal communities, it should be mandatory to carry out studies of the environmental impact on the crop of interest under controlled conditions before authorizing its release. When GM plants are established as environmentally safe to AM fungi and other nontarget organisms, in the context of a particular soil, they can be considered as a sustainable agricultural practice. Depending on the nature of the genetic modification, they would allow a reduction in the input of agrochemicals and the cultivation of crops in agriculturally marginal land and result in higher crop yields and crops with improved nutritional values.

## References

Andreote FD, Pereira e Silva MC (2017) Microbial communities associated with plants: learning from nature to apply it in agriculture. Curr Opin Microbiol 37:29–34

Appoloni S, Lekberg Y, Tercek MT et al (2008) Molecular community analysis of arbuscular mycorrhizal fungi in roots of geothermal soils in Yellowstone National Park (USA). Microb Ecol 56(4):649–659

Bedini S, Avio L, Argese E et al (2007) Effects of long term land use on arbuscular mycorrhizal fungi and glomalin related soil protein. Agric Ecosyst Environ 120:463–466

Blackwood CB, Buyer JS (2004) Soil microbial communities associated with Bt and non-Bt corn in three soils. J Environ Qual 33:832–836

Bompadre MJ, Fernandez Bidondo L, Silvani VA et al (2015) Combined effects of arbuscular mycorrhizal fungi and exogenous cytokinins on pomegranate (*Punica granatum*) under two contrasting water availability conditions. Symbiosis 65(2):55–63

Bonfante P, Genre A (2010) Mechanisms underlying beneficial plant-fungus interactions in mycorrhizal symbiosis. Nat Commun 1:48

Bridge P, Spooner BM (2001) Soil fungi: diversity and detection. Plant Soil 232:147–154

Bruinsma M, Kowalchuk GA, van Veen JA (2003) Effects of genetically modified plants on microbial communities and processes in soil. Biol Fertil Soils 37:329–337

Brundrett MC (1991) Mycorrhizas in natural ecosystems. Adv Ecol Res 21:171–213

Castaldini M, Turrini A, Sbrana C et al (2005) Impact of Bt corn on rhizospheric and soil eubacterial communities and on beneficial mycorrhizal symbiosis in experimental microcosms. Appl Environ Microbiol 71:6719–6729

Chaudhary VB, Cuenca G, Johnson NC (2018) Tropical-temperate comparison of landscape-scale arbuscular mycorrhizal fungal species distributions. Divers Distrib 24:116–128

Cheeke TE, Pace BA, Rosenstiel TN et al (2010) The influence of fertilizer level and spore density on arbuscular mycorrhizal colonization of transgenic Bt 11 maize (*Zea mays*) in experimental microcosms. FEMS Microbiol 75:304–312

Cheeke TE, Rosenstiel TN, Cruzan MB (2012) Evidence of reduced arbuscular mycorrhizal fungal colonization in multiple lines of Bt maize. Am J Bot 99:700–707

Cheeke TE, Cruzan MB, Rosenstiel TN (2013) Field evaluation of arbuscular mycorrhizal fungal colonization in *Bacillus thuringiensis* toxin-expressing (Bt) and non-Bt maize. Appl Environ Microbiol 79:4078–4086

Chen X, Zhang R, Wang F (2017) Transgenic Bt cotton inhibited arbuscular mycorrhizal fungus differentiation and colonization. Plant Soil Environ 63:62–69

Colombo RP, Fernandez Bidondo L, Silvani VA et al (2014) Diversity of arbuscular mycorrhizal fungi in soil from the Pampa Ondulada, Argentina, assessed by pyrosequencing and morphological techniques. Can J Microbiol 60:1–9

Colombo RP, Ibarra G, Fernandez Bidondo L et al (2017) Arbuscular mycorrhizal fungal association in genetically modified drought-tolerant corn. J Environ Qual 46:227–231

Cruz-Gutiérrez EJ, Gutiérrez-Espinosa MA, González-Chávez MC et al (2015) Identification of arbuscular fungi in transgenic Mexican lime (*Citrus aurantifolia*). AP Agro Productividad 8:38–43

Davison J, Moora M, Öpik M et al (2015) Global assessment of arbuscular mycorrhizal fungus diversity reveals very low endemism. Science 349:970–973

de Vaufleury A, Kramarz PE, Binet P et al (2007) Exposure and effects assessments of Bt-maize on non-target organisms (gastropods, microarthropods, mycorrhizal fungi) in microcosms. Pedobiologia 51:185–194

Donegan KK, Seidler RJ, Doyle JD, Porteous LA, Digiovanni G, Widmer F, Watrud LS (1999) A field study with genetically engineered alfalfa inoculated with recombinant *Sinorhizobium meliloti*: effects on the soil ecosystem. J Appl Ecol 36:920–936

Druille M, Cabello MN, Omacini M et al (2013) Glyphosate reduces spore viability and root colonization of arbuscular mycorrhizal fungi. Appl Soil Ecol 64:99–103

Fließbach A, Messmer M, Nietlispach B et al (2012) Effects of conventionally bred and *Bacillus thuringiensis* (Bt) maize varieties on soil microbial biomass and activity. Biol Fertil Soils 48:315–324

Garcia de Leon D, Moora M, Öpik M et al (2016) Symbiont dynamics during ecosystem succession: co-occurring plant and arbuscular mycorrhizal fungal communities. FEMS Microbiol Ecol 92(7):fiw097. https://doi.org/10.1093/femsec/fiw097. (Last entrance: 2019/04/04)

Gill Kahlon J, Jacobsen HJ, Cahill JF et al (2017) Antifungal genes expressed in transgenic pea (*Pisum sativum* L.) do not affect root colonization of arbuscular mycorrhizae fungi. Mycorrhiza 27:683–694

Giovannetti M (2003) The ecological risk of transgenic plants. Riv Biol Biol Forum 96:207–224

Giovannetti M, Sbrana C, Turrini A (2005) The impact of genetically modified crops on soil microbial communities. Riv Biol Biol Forum 98:393–417

Girlanda M, Bianciotto V, Cappellazzo GA et al (2008) Interactions between engineered tomato plants expressing antifungal enzymes and nontarget fungi in the rhizosphere and phyllosphere. FEMS Microbiol Lett 288:9–18

Gschwendtner S, Esperschütz J, Buegger F et al (2010) Effects of genetically modified starch metabolism in potato plants on photosynthate fluxes into the rhizosphere and on microbial degraders of root exudates. FEMS Microbiol Ecol 76:564–575

Hannula SE, de Boer W, van Veen JA (2010) In situ dynamics of soil fungal communities under different genotypes of potato, including a genetically modified cultivar. Soil Biol Biochem 42:2211–2223

Hannula SE, de Boer W, van Veen JA (2012) A 3-year study reveals that plant growth stage, season and field site affect soil fungal communities while cultivar and GM-trait have minor effects. PLoS One 7:e33819

Hannula SE, de Boer W, van Veen JA (2014) Do genetic modifications in crops affect soil fungi? A review. Biol Fertil Soils 50:433–446

Hart M, Reader RJ (2005) The role of the external mycelium in early colonization for three arbuscular mycorrhizal fungal species with different colonization strategies. Pedobiologia 49:269–279

Hausmann NT, Hawkes CV (2009) Plant neighborhood control of arbuscular mycorrhizal community composition. New Phytol 183:1188–1200

Hijri I, Sykorova Z, Oehl F et al (2006) Communities of arbuscular mycorrhizal fungi in arable soils are not necessarily low in diversity. Mol Ecol 15:2277–2289

Humphreys CP, Franks PJ, Rees M et al (2010) Mutualistic mycorrhiza-like symbiosis in the most ancient group of land plants. Nat Commun 1:103

Icoz I, Saxena D, Andow D, Zwahlen C, Stotzky C (2008) Microbial populations and enzyme activities in soil in situ under transgenic corn expressing Cry proteins from *Bacillus thuringiensis*. J Environ Qual 37(2):647–662

ISAAA (2017) Global status of commercialized biotech/GM crops in 2017: biotech crop adoption surges as economic benefits accumulate in 22 years, ISAAA brief no. 53. ISAAA, Ithaca. http://www.isaaa.org/default.asp. (Last entrance: 2018/11/02)

Jansa J, Mozafar A, Frossard E (2003) Long-distance transport of P and Zn through the hyphae of an arbuscular mycorrhizal fungus in symbiosis with maize. Agronomie 23:481–488

Jansa J, Wiwmken A, Frossard E (2006) The effects of agricultural practices on arbuscular mycorrhizal fungi. In: Frossard E, Blum WEH, Warkentin BP (eds) Function of soils for human societies and the environment, Geological society, vol 266. Special Publications, London, pp 89–115

Kaldorf M, Fladung M, Muhs HJ et al (2002) Mycorrhizal colonization of transgenic aspen in a field trial. Planta 214:653–660

Kivlin SN, Hawkes CV, Treseder KK (2011) Global diversity and distribution of arbuscular mycorrhizal fungi. Soil Biol Biochem 43:2294–2303

Knox OGG, Nehl DB, Mor T et al (2008) Genetically modified cotton has no effect on arbuscular mycorrhizal colonisation of roots. Field Crop Res 109:57–60

Krüger M, Krüger C, Walker C et al (2012) Phylogenetic reference data for systematics and phylotaxonomy of arbuscular mycorrhizal fungi from phylum to species level. New Phytol 193:970–984
Kuramae EE, Verbruggen E, Hillekens R et al (2013) Tracking fungal community responses to maize plants by DNA- and RNA-based pyrosequencing. PLoS One 8:e69973
Lekberg Y, Koide RT, Rohr JR et al (2007) Role of niche restrictions and dispersal in the composition of arbuscular mycorrhizal fungal communities. J Ecol 95:95–105
Liang J, Meng F, Sun S et al (2015) Community structure of arbuscular mycorrhizal fungi in rhizospheric soil of a transgenic high-methionine soybean and a near isogenic variety. PLoS One 10(12):e0145001
Liu W (2010) Do genetically modified plants impact arbuscular mycorrhizal fungi? Ecotoxicology 19:229–238
Liu WK, Du LF (2008) Interactions between Bt transgenic crops and arbuscular mycorrhizal fungi: a new urgent soil ecology issue in agro-ecosystems. Acta Agric Scand Sect B Soil Plant Sci 58:187–192
Lumini E, Orgiazzi A, Borriello R et al (2009) Disclosing arbuscular mycorrhizal fungal biodiversity in soil through a land use gradient using a pyrosequencing approach. Environ Microbiol 12:2165–2179
Manavella PA, Dezar CA, Bonaventure G et al (2008) HAHB4, a sunflower HD-Zip protein, integrates signals from the jasmonic acid and ethylene pathways during wounding and biotic stress responses. Plant J 56:376–388
Manoharan L, Rosenstock NO, Williams A et al (2017) Agricultural management practices influence AMF diversity and community composition with cascading effects on plant productivity. Appl Soil Ecol 115:53–59
Marleau J, Dalpe Y, St-Arnaud M et al (2011) Spore development and nuclear inheritance in arbuscular mycorrhizal fungi. BMC Evol Biol 11:51
Medina MJH, Gagnon H, Piche Y et al (2003) Root colonization by arbuscular mycorrhizal fungi is affected by the salicylic acid content of the plant. Plant Sci 164:993–998
Meyer JB, Song-Wilson Y, Foetzki A et al (2013) Does wheat genetically modified for disease resistance affect root-colonizing pseudomonads and arbuscular mycorrhizal fungi? PLoS One 8:e53825
Miller RM (1993) Nontarget and ecological effects of transgenically altered disease resistance in crops – possible effects on the mycorrhizal symbiosis. Mol Ecol 2:327–335
Mohandass D, Muthukumar T (2017) An insight into genetically modified crop-mycorrhizal symbiosis. In: Singh DP et al (eds) Plant-microbe interactions in agro-ecological perspectives. https://doi.org/10.1007/978-981-10-5813-4_20. (Last entrance: 2019/04/04)
Moora M, Davison J, Öpik M et al (2014) Anthropogenic land use shapes the composition and phylogenetic structure of soil arbuscular mycorrhizal fungal communities. FEMS Microbiol Ecol 90:609–621
Newhouse AE, Schrodt F, Liang HY et al (2007) Transgenic American elm shows reduced Dutch elm disease symptoms and normal mycorrhizal colonization. Plant Cell Rep 26:977–987
Oehl F, Sieverding E, Mäder P et al (2004) Impact of long-term conventional and organic farming on the diversity of arbuscular mycorrhizal fungi. Oecologia 138:574–583
Ohsowski BM, Zaitsoff DP, Öpik M et al (2014) Where the wild things are: looking for uncultured Glomeromycota. New Phytol 204:171–179
Öpik M, Davison J (2016) Uniting species- and community-oriented approaches to understand arbuscular mycorrhizal fungal diversity. Fungal Ecol 24:106–113
Öpik M, Vanatoa A, Vanatoa E et al (2010) The online database MaarjAM reveals global and ecosystemic distribution patterns in arbuscular mycorrhizal fungi (Glomeromycota). New Phytol 188:223–241
Pellegrino E, Bedini S, Nuti M et al (2018) Impact of genetically engineered maize on. Sci Rep 8:3113. https://doi.org/10.1038/s41598-018-21284-2. (Last entrance: 2019/04/04)

Powell JR, Gulden RH, Hart MM et al (2007) Mycorrhizal and rhizobial colonization of genetically modified and conventional soybeans. Appl Environ Microbiol 73(13):4365–4367

Qaim M (2009) The economics of genetically modified crops. Annu Rev Resour Econ 1:665–694

Redecker D (2000) Specific PCR primers to identify arbuscular mycorrhizal fungi within colonized roots. Mycorrhiza 10:73–80

Ren X (2006) Effect of Bt transgenic rice (KMD) on soil bacterial community and rhizosphere AM fungi. Dissertation, Zhejiang University, Hang Zhou, China

Schalamuk S, Cabello M (2010) AMF propagules from tillage and no-tillage systems: possible impacts on Glomeromycota diversity. Mycologia 102:261–268

Seres A, Nagy KP, Saly P et al (2014) Arbuscular mycorrhizal fungi colonisation of Cry3 toxin-producing Bt maize and near isogenic maize. Plant Soil Environ 60:569–573

Silvani VA, Colombo RP, Scorza MV et al (2016) Arbuscular mycorrhizal fungal diversity in high-altitude hypersaline Andean wetlands studied by 454-sequencing and morphological approaches. Symbiosis 72:143–152

Smith SE, Read DJ (2008) Mycorrhizal symbiosis. Academic, London

Stephan BI, Colombo RP, Silvani VA et al (2019) Short-term effect of genetically modified potato on arbuscular mycorrhizal fungal communities. J Soil Sci Plant Nutr. https://doi.org/10.1007/s42729-019-00035-w. (Last entrance: 2019/04/30)

Stürmer S, Bever JD, Morton J (2018) Biogeography of arbuscular mycorrhizal fungi (Glomeromycota): a phylogenetic perspective on species distribution patterns. Mycorrhiza 28:587–603

Tahiri-Alaoui A, Grison R, Gianinazzi-Pearson VT et al (1994) The impact of the constitutive expression of chitinases in roots of transgenic tobacco on arbuscular mycorrhizal fungi. In: Abstract 406 of the 7th international symposium on molecular plant–microbe interactions, Edinburgh, 26 June–1 July, 1994

Tan F, Wang J, Chen Z et al (2011) Assessment of the arbuscular mycorrhizal fungal community in roots and rhizosphere soils of Bt corn and their non-Bt isolines. Soil Biol Biochem 43:2473–2479

Turrini A, Sbrana C, Nuti MP (2004a) Development of a model system to assess the impact of genetically modified corn and aubergine plants on arbuscular mycorrhizal fungi. Plant Soil 266:69–67

Turrini A, Sbrana C, Pitto L et al (2004b) The antifungal Dm-AMP1 protein from *Dahlia merckii* expressed in *Solanum melongena* is released in root exudates and differentially affects pathogenic fungi and mycorrhizal symbiosis. New Phytol 163:393–403

Turrini A, Sbrana C, Giovannetti M (2008) Experimental systems to monitor the impact of transgenic corn on keystone soil microorganisms. 16th IFOAM Organic World Congress, Modena, Italy

Turrini A, Sbrana C, Giovanetti M (2015) Belowground environmental effects of transgenic crops: a soil microbial perspective. Res Microbiol 166:121–131

Unterseher M, Jumpponen A, Opik M et al (2011) Species abundance distributions and richness estimations in fungal metagenomic—lesson learned from community ecology. Mol Ecol 20:275–285

Verbruggen E, Röling WFM, Gamper HA et al (2010) Positive effects of organic farming on below-ground mutualists: large-scale comparison of mycorrhizal fungal communities in agricultural soils. New Phytol 186:968–979

Verbruggen E, Kuramae EE, Hillekens R et al (2012) Testing potential effects of maize expressing the *Bacillus thuringiensis* Cry1Ab endotoxin (Bt maize) on mycorrhizal fungal communities via DNA and RNA-based pyrosequencing and molecular fingerprinting. Appl Environ Microbiol 78:7384–7392

Vierheilig H, Alt M, Neuhaus JM et al (1993) Colonization of transgenic *Nicotiana sylvestris* plants, expressing different forms of *Nicotiana tabacum* chitinase, by the root pathogen Rhizoctonia solani and by the mycorrhizal symbiont *Glomus mosseae*. Mol Plant-Microbe Interact 6:261–264

Vierheilig H, Alt M, Lange J et al (1995) Colonization of transgenic tobacco constitutively expressing pathogenesis related proteins by vesicular–arbuscular mycorrhizal fungus *Glomus mosseae*. Appl Environ Microbiol 61:3031–3034

Wróbel-Kwiatkowska M, Turnau K, Góralska K et al (2012) Effects of genetic modifications to flax (*Linum usitatissimum*) on arbuscular mycorrhiza and plant performance. Mycorrhiza 22:493–499

Xiang D, Verbruggen E, Hu Y (2014) Land use influences arbuscular mycorrhizal fungal communities in the farming–pastoral ecotone of northern China. New Phytol 204:968–978

Yang YF, Yuan HX, Liu YL et al (2002) Research on root microorganism community of "RCH" transgenic rice. Chin J Agric Econ China 10:29–31

Zeng H, Tan F, Zhang Y et al (2014) Effects of cultivation and return of *Bacillus thuringiensis* (Bt) maize on the diversity of the arbuscular mycorrhizal community in soils and roots of subsequently cultivated conventional maize. Soil Biol Biochem 75:254–263

Zeng H, Tan F, Shu Y et al (2015) The Cry1Ab protein has minor effects on the arbuscular mycorrhizal fungal communities after five seasons of continuous Bt maize cultivation. PLoS One 10:e014604

# GMOs – Impact on Non-target Arthropods

**Patricia Valle Pinheiro and Josias Correa de Faria**

**Abstract** Genetically modified (GM) plants have been adopted at unprecedented levels since they were first commercialized in 1996. At that time, two main objectives were set by the industry: tolerance to herbicides and resistance to insect pests. While there was no promise of productivity increase, GM crops certainly reduce crop losses by reducing weed and pest populations that otherwise compete with nutrients and the overall crop ability to yield at its genetic potential. At which cost for the environment? This review focuses on non-target herbivores, pollinators, natural enemies, and detritivores. A huge body of literature reporting laboratory and field studies have been published. Although the majority of these reports show neutral or "negligible" effects of GM crops or insecticidal proteins on non-target arthropods, some reported negative effects, while a few others reported positive effects. The massive adoption of insect-resistant GM crops is accepted in cases where no other pest control method is available or else the GM crop impact on the environment will be lower than that of pesticides. However, most of the studies published to date did not compare pesticides to GM crops, in terms of potential negative impacts. There are still many unanswered questions on how GM crops might interfere with multi-trophic interactions. Even so, it is certain that we currently know better the potential effects of GM crops on non-target arthropods than those of any other crop technology developed to date.

**Keywords** Genetically modified crops · Environmental costs · Pollinators · Natural enemies · Predators · Parasitoids · Detritivores · Non-target herbivores · Undesirable effects

P. V. Pinheiro · J. C. de Faria (✉)
EMBRAPA Rice and Beans, Santo Antônio de Goiás, GO, Brazil
e-mail: josias.faria@embrapa.br



A. Chaurasia et al. (eds.), *GMOs*, Topics in Biodiversity and Conservation 19,
https://doi.org/10.1007/978-3-030-53183-6_5

## Introduction

Pest control is one of the most impacting agricultural practices for crop production increase. There is a strong demand to increase crop production by 60–110% in the year 2050 due to population growth, meat, dairy, and biofuel consumption, besides food security (Ray et al. 2013). Most gains in production came from the so-called green revolution – increased use of higher yielding races of crops associated with increased use of pesticides, fertilizers, irrigation, and mechanization (Ray et al. 2013). Losses owing to pest, including arthropods, pathogens, and weeds, range from 20% to 40% annually (Oerke 2006; FAO 2015) and may reach 100% in some areas, depending on agricultural practices and climate conditions. As the world climate warms up, crop losses due to pests are expected to increase considerably (Deutsch et al. 2018). Among the current options for pest control, the intensive use of agrochemicals has taken the lead, followed by the adoption of resistant plant genotypes, obtained either by conventional plant breeding or by genetic engineering, whereas biological control has still taken a marginal position.

It is possible and desirable to combine these various tools into Integrated Pest Management (IPM) practices (Anderson et al. 2019). IPM aims to prevent, monitor, and control pest populations, once they reach economic damage levels. The biggest challenge is to reduce pest damage to crop yield and quality, and at the same time, causing minimum or zero negative impact to human health and the environment, including non-target organisms.

Although chemical control has made a significant contribution to food production, despite its well-known threats to the environment and health, an increasing number of commercially available chemical molecules are no longer efficient, due to the selection of resistant pest populations. Around 5 billion pounds of synthetic agrochemicals are applied to agriculture production yearly, costing more than 56 billion dollars in 2012 (US EPA 2015). Moreover, the intensive use of synthetic agrochemical molecules over the years has been related to the decline of pollinator populations, especially honeybees (Desneux et al. 2007; Potts et al. 2010; Goulson et al. 2015). At a lower level, biological control agents, including *Bacillus thuringiensis*, also cause negative impacts on non-target organisms (Flexner et al. 1986; Brimner and Boland 2003; Cardinale et al. 2003; Biondi et al. 2012). Thus, considering the options, when the selection of pest-resistant genotypes within the same species or genus is possible, this strategy is still the most environmentally sustainable method to reduce losses in crop yield or quality due to pest damage. However, in some cases no natural resistance to the target pest has been found in the germoplasm of the host plant. In such cases, an alternative strategy might be introducing exogenous genes into the plant genotype, obtaining transgenic, genetically modified (GM) or biotech crops.

With the promise to be target-specific, with reduced adoption costs and easy application, GM crops have become the fastest adopted crop technology in recent times (ISAAA 2017). Since the first commercial release in 1996, the cultivation of GM crops has increased from 1.7 to 185.1 million hectares, in 2017 (ISAAA 2017).

Herbicide tolerance (HT) has been the most adopted trait of GM crops, followed by insect resistance (IR). The major crops in which those traits have been introduced are soybean, maize, canola, and cotton, but there are other species of plants with GM cultivars approved, such as sugar beet and oilseed rape, as well as some minor crops, such as common bean, in Brazil, and eggplant in Philippines and Pakistan, for example. More recently, HT/IR traits have been stacked into soybean, maize and cotton and these GM crops have presented the highest increase in cultivated area, representing 41% of the total area cultivated with GM crops in the world, compared to 21% for IR GM crops. Among the top five GM crop adoption countries, Argentina is the first country to reach 100% of the soybean cultivated area with GM cultivars, followed by the USA and Brazil (about 94% adoption of GM cultivars), Canada, and India. The USA and Brazil are the countries with the largest area cultivated with GM crops (ISAAA 2017).

IR GM crops have incorporated the pesticides into plant genotypes, which makes it easier to apply the insect management strategy. However, merging plant genetics and pesticide technologies may have a negative impact on the biocontrol adoption, in cases where GM crops affect non-target organisms, such as predators and parasitoids, which are important biocontrol players. The risk of toxic effects to pollinators is a serious concern, because pollination is an essential ecosystem service. In the last two decades, a large body of literature has been produced on the impacts of GM crops on non-target organisms. Neutral to negative effects of GM crops on non-target organisms have been reported, as well as a few positive effects. Some authors have suggested "negligible" effects, but the meaning of this term is not well described. Several meta-analyses of data from the literature so far available have been published, using different methods for summarizing the literature and analyze the data. Most of them agree that cry toxins and protease inhibitors usually have negative effects on non-target arthropods, even if only indirectly (Romeis et al. 2006; Duan et al. 2008; Wolfenbarger et al. 2008; Lövei et al. 2009; Peterson et al. 2011). In this chapter, we review the extensive literature of risk assessment of GM crops to arthropods and microarthropods. Where authors described negative effects, we will take a closer look at the reports and their ecological significance, in light of the contrasting reports of similar studies.

## Insect-Resistant GM Crops

The first insect-resistant (IR) transgenic crops, released for commercialization in 1996, were maize, cotton, and potato plants genetically modified with genes from the gram-positive bacterium *Bacillus thuringiensis* (Bt), which encode crystalline inclusion proteins, the cry proteins. Although the toxicity of these proteins is highly specific to certain insect orders (Lepidoptera, Coleoptera, and Diptera), GM crops producing transgenic products might present different levels of toxicity to other arthropods, nematodes, and even to human cancer cells (Palma et al. 2014). Over 50 years before the development of Bt crops, *B. thuringiensis* formulations were being used for the biological control of insect pests (Koch et al. 2015), and proving an important tool for organic production. Once ingested by insects, cry proteins are solubilized and cleaved by midgut proteases. A potentially toxic part of the protein is then released. However, the toxin will only be active if the solubilized protein fragment is recognized by a particular protein receptor, to which it binds, leading to cell disruption and insect death (Palma et al. 2014). Other Bt proteins, also targeting the insect midgut, have been introduced into plants by genetic engineering, aiming to increase insect resistance. Among them are vegetative insecticidal proteins (vip) and cytolytic proteins (cyt) (Harrison and Bonning 2010), which are also toxic to lepidopteran insects, but with different modes of action, that, when combined, might delay the selection of resistant pest populations. While the cry proteins are mainly produced in the sporulation phase of the bacterium, vip and cyt proteins are produced during the vegetative and stationary phases, providing different active molecules for insect control. The majority of the IR transgenic crops currently approved express one or more Bt proteins, as described by a report of the International Service for the Acquisition of Agri-Biotech Application (ISAAA) (ISAAA 2017) (Table 1). There are also approved crop varieties expressing the proteins vip3A(a) and vip3Aa20 (Table 1), some of them with commercially available products.

A new generation of transgenic crops containing stacked genes has been developed, combining several different genes for insect resistance and herbicide tolerance in one crop variety. Concern has been raised on whether the different insecticidal proteins might interact, affecting non-target arthropods (Svobodová et al. 2017). For example, the co-expression of the Bt proteins Cyt2Aa3 and Vip3Aa29 inhibited the development of the lepidopterans *Chilo suppressalis*, and *Spodoptera exigua*, while the Cyt2Aa3 protein alone did not present the same effect (Yu et al. 2012). This result suggests that the synergism between Bt toxins or other insecticidal proteins might unexpectedly cause adverse effects on non-target arthropods. Therefore, for GM stacked crops, the potential synergistic effects of the stacked proteins should be evaluated.

Beyond Bt proteins, other insecticidal molecules, including endogenous plant defense proteins, from different origins, have been explored to develop insect-resistant GM crops, such as plant protease inhibitors (PIs). PIs are natural defense molecules accumulated constitutively in seeds and other storage tissues by plants to avoid insect herbivory. These proteins act in the insect metabolism by reducing protein digestion and amino acids availability, which are essential for insect growth and development. Plants produce a large variety of protease inhibitors (at least 13 gene families) that target the main protease families of herbivores and plant pathogens, such as viruses, bacteria, fungi, and nematodes (Rawlings et al. 2004). Therefore, PIs present a broad protective effect on plants. Among them are cysteine protease inhibitors (i.e., cathepsins B and L, cystatins, etc.) and serine protease inhibitors (i.e., trypsin inhibitor). Protease inhibitors are also produced in leaves as an induced defense, upon mechanical and insect damage (Green and Ryan 1972). Currently, there is no evidence of toxic effects of PIs on mammals. Three crop varieties expressing protease inhibitors for insect resistance, targeting Lepidoptera or a wide range of insect species, are currently approved for commercial cultivation. The

**Table 1** Genes introduced into genetically modified crops for insect resistance that have been approved for commercial release, in February 2019

| Gene | Gene Source | Product | Insect type controlled | Crop approved |
|---|---|---|---|---|
| cry1A | *Bacillus thuringiensis* | Delta-endotoxin of the cry1A group | Lepidoptera | Cotton and maize |
| cry1A.105 | *Bacillus thuringiensis* subsp. *kumamotoensis* | cry1A.105 protein which comprises the cry1Ab, cry1F and cry1Ac proteins | Lepidoptera | Maize and soybean |
| cry1Ab | *Bacillus thuringiensis* subsp. *kurstaki* | cry1Ab delta-endotoxin | Lepidoptera | Cotton, maize, rice, and sugarcane |
| cry1Ab (truncated) | Synthetic form of Cry1Ab from *Bacillus thuringiensis* subsp. *kumamotoensis* | cry1Ab delta-endotoxin | Lepidoptera | Maize and rice |
| cry1Ab-Ac | Synthetic fusion gene derived from *Bacillus thuringiensis* | cry1Ab-Ac delta-endotoxin (fusion protein) | Lepidoptera | Cotton |
| cry1Ac | *Bacillus thuringiensis* subsp. *kurstaki* strain HD73 | cry1Ac delta-endotoxin | Lepidoptera | Cotton, eggplant, maize, poplar, rice, soybean, tomato and sugarcane |
| cry1C | Synthetic gene derived from *Bacillus thuringiensis* | cry1C delta-endotoxin | Lepidoptera | Cotton |
| cry1F | *Bacillus thuringiensis* var. aizawai | cry1F delta-endotoxin | Lepidoptera | Cotton, maize and soybean |
| cry1Fa2 | Synthetic form of cry1F gene derived from *Bacillus thuringiensis* var. aizawai | Modified cry1F protein | Lepidoptera | Maize |
| cry2Ab2 | *Bacillus thuringiensis* subsp. *kumamotoensis* | cry2Ab delta-endotoxin | Lepidoptera | Cotton, maize and soybean |
| cry2Ae | *Bacillus thuringiensis* subsp. *dakota* | cry2Ae delta-endotoxin | Lepidoptera | Cotton and maize |
| cry9C | *Bacillus thuringiensis* subsp. *tolworthi* strain BTS02618A | cry9C delta-endotoxin | Lepidoptera | Maize |
| mocry1F | Synthetic form of cry1F gene from *Bacillus thuringiensis* var. aizawai | Modified cry1F protein | Lepidoptera | Maize |
| pinII | *Solanum tuberosum* | Serine protease inhibitor protein | Lepidoptera | Maize |

(continued)

**Table 1** (continued)

| Gene | Gene Source | Product | Insect type controlled | Crop approved |
|---|---|---|---|---|
| vip3A(a) | *Bacillus thuringiensis* strain AB88 | vip3A vegetative insecticidal protein | Lepidoptera | Cotton and maize |
| vip3Aa20 | *Bacillus thuringiensis* strain AB88 | Vegetative insecticidal protein (vip3Aa variant) | Lepidoptera | Maize |
| cry34Ab1 | *Bacillus thuringiensis* strain PS149B1 | cry34Ab1 delta-endotoxin | Coleoptera | Maize |
| cry35Ab1 | *Bacillus thuringiensis* strain PS149B1 | cry35Ab1 delta-endotoxin | Coleoptera | Maize |
| cry3A | *Bacillus thuringiensis* subsp. *tenebrionis* | cry3A delta-endotoxin | Coleoptera | Potato |
| cry3Bb1 | *Bacillus thuringiensis* subsp. *kumamotoensis* | cry3Bb1 delta-endotoxin | Coleoptera | Maize |
| dvsnf7 | Western corn rootworm (*Diabrotica virgifera virgifera*) | Double-stranded RNA transcript containing a 240 bp fragment of the WCR Snf7 gene | Coleoptera | Maize |
| mcry3A | Synthetic form of cry3A gene from *Bacillus thuringiensis* subsp. *tenebrionis* | Modified cry3A delta-endotoxin | Coleoptera | Maize |
| mCry51Aa2 | *Bacillus thuringiensis* | Modified Bt cry51Aa2 protein | Hemiptera | Cotton |
| API | *Sagittaria sagittifolia* (arrowhead) | Arrowhead protease inhibitor protein A or B | Multiple insect resistance | Poplar |
| CpTI | *Vigna unguiculata* | Trypsin inhibitor | Multiple insect resistance | Cotton |
| ecry3.1Ab | Synthetic form of Cry3A gene and Cry1Ab gene from *Bacillus thuringiensis* | Chimeric (cry3A-cry1Ab) delta-endotoxin protein | Coleopteran and lepidopteran | Maize |

Data available at http://www.isaaa.org/gmapprovaldatabase/

approved varieties express the following PIs: the serine protease inhibitor pinII from potato, the arrowhead protease inhibitor proteins A or B (API), and the cowpea trypsin inhibitor (CpTI) (Table 1).

The snowdrop (*Galanthus nivalis*) agglutinin, GNA, has been studied as a potential source of resistance to sap-sucking insects from the order Hemiptera. Insects from this order are less susceptible to Bt proteins than insects from the Lepidoptera, Coleoptera, and Diptera. Therefore, using GNA-transgenic crops might represent a promising strategy to control Hemipteran insects. Although there are several studies in the literature showing the potential effects of GNA on non-target organisms, currently no crop variety expressing this protein has been approved for commercial release. On the other hand, GM crops expressing modified Bt proteins have been developed for resistance to hemipterans (Liu et al. 2018). A cotton variety expressing the Bt cry51Aa2 protein, which confers resistance to the true bug *Lygus* spp., (Gowda et al. 2016) has been approved for commercial release (Table 1).

More recently, a new generation of IR plants has been engineered using the RNA interference (RNAi) mechanism to silence genes in the target insect pest that are essential for insect survival. This technique can be highly specific, if the target gene region is unique or not highly conserved across other non-target taxa. Using this approach, GM plants expressing double-stranded RNA (dsRNA) homologs that target transcripts of essential genes in the insect have been obtained. Insects ingest the dsRNA or their processed versions, small, single-stranded RNA, referred to as small interfering RNAs (siRNAs), when feeding on the plant, and have the expression of the target gene suppressed, leading to insect death (Baum et al. 2007). Currently, there is only one crop expressing dsRNA for insect resistance approved, the maize variety expressing the gene dvsnf7, which targets the gene Snf7 from the Western corn rootworm *Diabrotica virgifera* (Table 1). However, the RNAi mechanism is a promising tool for the development of insect-resistant crop varieties in the future.

The effect of insecticidal proteins expressed by GM crops has been evaluated both *in vivo* and *in vitro.* Although the vast majority of the risk assessment of GM crops to non-target organisms are conducted with insect-resistant (IR) GM crops, some studies also investigated the effect of herbicide-tolerant (HT) and virus-resistant (VR) GM crops on non-target organisms and a few others looked at potential negative effects of HT/IR stacked GM crops (Romeis et al. 2006; Duan et al. 2008; Lövei et al. 2009; Pinheiro et al. 2014; Seide et al. 2018).

## What Do We Know?

Since the first commercial release in 1996, the cultivation of GM crops has raised several questions on the implications of introducing organisms produced by genetic engineering (i.e., not naturally occurring) into agricultural ecosystems. Concerns about ecosystem balance and impact of GM crops on non-target organisms have been debated since then. Over 22 years of GM crops cultivation, the area cultivated and the adoption of GM crops have dramatically increased (ISAAA 2017). On the other hand, the extensive research and debate did not provide scientific evidence of ecological imbalance or negative effect on the arthropod community at the population level. Most of the results reported in the literature show neutral effects of GM crops on non-target arthropods and a few others show a negative impact (Table 2, Fig. 1). So, what do we know about these effects until now? We will discuss the main findings about the effects of GM crops on non-target herbivores, pollinators, non-target lepidopterans, natural enemies, such as predators and parasitoids and detritivore arthropods. This chapter is primarily focused on the molecules expressed by the currently approved transgenic crops, which express Bt cry and vip toxins, PIs and dsRNA.

**Table 2** Number of studies reported in the literature showing negative, neutral, or positive effects of genetically modified (GM) crops on non-target arthropods

| Arthropod guild | Negative | Neutral | Positive | Total |
|---|---|---|---|---|
| Arthropod community | 2 | 7 | 2 | 11 |
| Detritivore arthropods | 10 | 14 | 0 | 24 |
| Non-target herbivores | 5 | 22 | 6 | 33 |
| Non-target lepidopterans | 4 | 6 | 1 | 11 |
| Parasitoids | 13 | 24 | 2 | 39 |
| Pollinators | 8 | 23 | 1 | 32 |
| Predators | 8 | 23 | 1 | 32 |
| Total | 50 | 119 | 13 | 182 |
| % | 27.5 | 65.4 | 7.1 | |

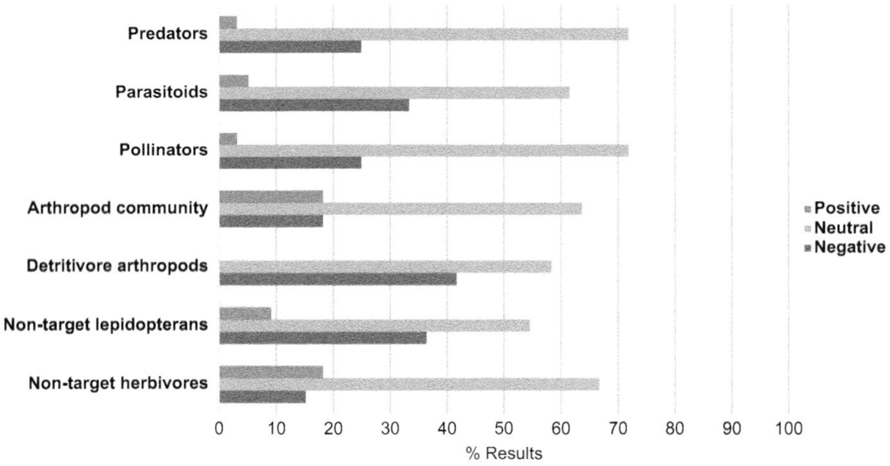


**Fig. 1** Percentage of negative, neutral, or positive effects of genetically modified (GM) crops or insecticidal toxins on non-target arthropods, as reported in the literature

## *Non-target Herbivores*

Genetically modified plants expressing insecticidal toxins for the control of specific insect pests might affect other species of arthropods, including other herbivores, which are non-target of the technology. Many different effects are possible when non-target herbivores are exposed to toxins produced in transgenic plants, from mortality or reduced/slowed development, to the uptake/accumulation of toxins, which can be transferred through the food chain to their natural enemies. Non-target herbivores can be affected by habitat loss, when weed density is altered in GM crop fields, usually by weed management using specific herbicides, affecting herbivore-weed connections on complex food webs (McPherson et al. 2003; Pálinkás et al. 2017). The population of non-target herbivores has been reported to fluctuate from one year to another in GM crop field, which is not necessarily related to a detrimental effect of the transgenic plant (McPherson et al. 2003; Chen et al. 2006; Pinheiro et al. 2014). A few reports claimed that crop resistance to one or more insect pest species will not necessarily promote outbreaks of non-target herbivores (Chen et al. 2006, 2012; Li et al. 2010). However, opposed to those potential detrimental effects of GM crops, in most cases, non-target herbivore populations might be benefitted by the reduction in pesticide applications, as well as by the decrease in intraspecific competition for resources (Bourguet et al. 2002; Wu and Guo 2003; Men et al. 2004; Lu et al. 2008; Bergé and Ricroch 2010). The biggest risk assessment question is how much safer GM plants are for the environment as a tool for insect pest management, in comparison with pesticides. The answer to that is mostly related to how specific the effect of transgenic plants is to the target species.

In 2017, the global area cultivated with insect-resistant GM crops reached 23.3 million hectares (ISAAA 2017), most of it represented by Bt crops, for the control of Lepidopteran pests (Hagenbucher et al. 2014a). The abundance of non-target herbivores that are not suppressed by the insecticidal trait has increased on Bt crops, most probably due to the reduction in insecticide applications and in resource competition, as indicated by studies around the world (Wu et al. 2002; Lu et al. 2010; Naranjo 2011). Specific insect resistance might lead to unexpected consequences in the complex multi-trophic interactions involving plants, herbivores, and natural enemies (Groot and Dicke 2002; Pálinkás et al. 2017). For example, cotton plants naturally activate induced defenses against chewing herbivores, such as caterpillars, producing insecticidal terpenoids (Bezemer et al. 2004). These compounds are toxic to insects with different feeding habits, such as sap-sucking insects. Thus, the plant defenses induced in cotton by caterpillars also protect the plant against aphids and other hemipteran pests. However, Bt cotton plants are less damaged by lepidopteran pests and therefore these plants produce lower amounts of insecticidal terpenoids. As a result, insects that do not induce terpenoid defenses in plants and are not affected by Bt cry toxins, such as aphids, are benefited by this system (Hagenbucher et al. 2013, 2014a, b).

Because aphids feed on the plant phloem sap, it was thought that they would not be able to ingest Bt toxins (Head et al. 2001; Raps et al. 2001). However, trace amounts of Bt cry toxins have been detected in aphids, indicating that they do ingest the toxins (Zhang et al. 2006; Burgio et al. 2007; Svobodová et al. 2017), with variable uptake abilities depending on the aphid species (Paula and Andow 2016). It is not yet clear whether aphids ingest very small amounts of Bt toxins or if the toxin is degraded in the aphid. Those small amounts of toxins are unlikely to harm natural enemies that consume aphids containing Bt toxins (Romeis and Meissle 2011). Additionally, there is no evidence that aphid survival is affected by Bt toxins expressed in transgenic crops (Lawo et al. 2009; Sujii et al. 2013; Zhao et al. 2016), except for reduced growth reported for *Myzus persicae* (Paula and Andow 2016).

Similar to that, transgenic potato plants expressing cysteine protease inhibitors (cystatins) for nematode resistance did not affect the survival and development of nymphs of the aphid *M. persicae* (Cowgill and Atkinson 2003). On the contrary, aphid populations have increased in Bt crops (Lumbierres et al. 2004; Liu et al. 2005; Pons et al. 2005; Fernandes et al. 2012; Hagenbucher et al. 2013), which might lead to serious crop losses, because aphids directly damage the plants by feeding and indirectly by transmitting plant viruses (Gray et al. 2014). A recent study showed that chemical inhibition of the cysteine protease cathepsin B in *M. persicae* restored the aphid capacity of transmitting a circulative virus, which had been impaired due to a host switch effect (Pinheiro et al. 2017). As we have begun to see, these interactions are more complex than thought not long ago and therefore, it is important to improve our understanding of the whole interaction.

Genetically modified plants expressing the Bt cry3B protein (and other proteins from this family) are resistant to chrysomelid herbivores. The main target species is the Western corn rootworm *D. virgifera*, a generalist herbivore and a key pest in maize and other crops. A comprehensive review by Devos et al. (2012) of the potential effects of a Bt cry3Bb1-maize line on non-target chrysomelids concluded that this Bt maize is not a threat to non-target herbivores from the same family as the target species, because they differ in feeding habits. For example, the exposure of non-target chrysomelidae larvae to Bt maize pollen deposited in host plants is minimal and the toxin presents low activity on adults. It remains to be investigated whether the resistance to *D. virgifera* might impact the arthropod community or/and non-target herbivores from other insect orders, as it has been shown for lepidopteran-resistant Bt crops (Pálinkás et al. 2017).

The small aquatic crustacean *Daphnia magna* is a model organism for toxicology evaluations. Although this arthropod is a primary consumer, not likely to be exposed to GM crops directly, it can be used as an indicator species to measure the toxic level of insecticidal proteins in surface water. In a laboratory study with 12 animal species including insect predators, detritivores, and pollinators, as well as birds and mammals, *D. magna* was the only species negatively affected by the Bt vip3Aa20 protein, even at very conservative concentrations (Raybould and Vlachos 2011). Although survival and reproduction were not affected, *D. magna* exposed to the protein *in vitro* grew more slowly than the controls. Later on, the authors compared the effect of the Bt vip3Aa20 protein to high concentrations of a protein test substance, the bovine serum albumin (BSA). They found out the same sporadic effects on growth and reproduction, indicating that the negative effect was not specific to Bt protein, but a result from non-toxic effects of high concentrations of protein consumption (Raybould et al. 2014).

Bt cry toxins have been detected in other non-target herbivores. For example, the rice leaf bug *Trigonotylus caelestialium*, a mirid bug, accumulates the Bt cry3Bb toxin when exposed to transgenic maize in field experiments (Rauschen et al. 2009). However, no difference on the density of this herbivore population could be related to the Bt maize variety. Bt toxin from GM oilseed rape plants was detected in larvae and feces of the turnip sawfly *Athalia rosae*, a tenthredinidae, but not in other developmental stages (Howald et al. 2003). The Tenthredinidae is a family of herbivorous hymenopterans, which are regular pests of brassicacean in Europe and Asia. Although the feeding behavior and susceptibility to Bt cry1Ac of tenthredinid larvae are similar to lepidopteran larvae, no detrimental effect was reported for *A. rosae* feeding on Bt oilseed rape plants. However, the detection of Bt toxin in the sawfly suggests that other trophic levels could be exposed to the toxin, when consuming the herbivore. Another study showed that Cry proteins are transferred to the next generation of the non-target lepidopteran *Chlosyne lacinia*, and to eggs and neonate larvae of the predator *Harmonia axyridis,* indicating another cascade effect of Bt toxins (Paula et al. 2014, 2015). The Bt cry1Ab protein has been detected in another non-target herbivores, the aphid *Rhopalosiphum maidis* and the moth *Ostrinia nubilalis* when the insects were fed *in vitro* with the purified protein, but not when the insects feed on Bt cry1Ab maize, suggesting that the non-target herbivores are less exposed to the toxins in plants (Head et al. 2001). Although these studies show toxin uptake by non-target herbivores, they have not investigated the effect on predators and parasitoids that prey on these herbivores, which is the major concern, regarding non-target arthropods.

GM plants expressing double-stranded RNA (dsRNA) homologs have been developed for insect pest management. Insects from the order Coleoptera present the highest sensitivity to ingested dsRNA (Swevers et al. 2013; Singh et al. 2017; Cooper et al. 2019). The efficiency of gene silencing by RNAi in target arthropods depends on several factors, including length of dsRNA fragments, dsRNA concentration and cellular uptake, timing and duration of exposure and life stage of the target organism (reviewed in Swevers et al. 2013; Scott et al. 2013; Roberts et al. 2015; Singh et al. 2017; Cooper et al. 2019). Beyond these factors, the potential of RNAi GM plants to harm non-target arthropods additionally depends on the exposure and susceptibility of non-target arthropods to environmental dsRNA, instability of dsRNA in the gut of the non-target arthropod and silencing of a transcript that will impact the organism survival (Roberts et al. 2015). There is little information on dsRNA persistence in the environment and whether dsRNA can be transferred through the food chain to different trophic levels. Until now, evidence has shown that dsRNA is quickly degraded in the soil and that uptake from the soil did not elicit an RNAi response (Dubelman et al. 2014). However, only a few studies have investigated the impact of RNAi GM crops on non-target arthropods to date (Nunes et al. 2013; Bachman et al. 2016). The only crop expressing dsRNA for insect resistance commercially approved to date is the maize variety expressing the gene dvsnf7 (Table 1). Our literature search recovered only one study on the potential effects of this dsRNA on non-target arthropods, but it did not include non-target herbivores (Bachman et al. 2016). There are several studies in the literature on the impact of other dsRNA on non-target arthropods, or on the development of protocols for risk assessment studies, but most of them have been performed *in vitro*, and the GM plants with these dsRNA have not yet been approved (Vélez et al. 2016; Haller et al. 2019).

## *Pollinators*

Insect pollinators provide essential ecosystem service and without this service, civilization as we know it might disappear. The majority of the pollinators are social and solitary bees, butterflies, beetles, wasps, and flies, but other insects also play a minor role in pollination, such as thrips and ants (Vanbergen and Initiative 2013; Eliyahu et al. 2015). Food crop production depends largely on pollination services, including fruits, vegetables, oil, seeds, and nut crops. Pollination is also critical for the reproduction of wild plants, directly related to biodiversity and ecosystem balance (Vanbergen and Initiative 2013). Over the last 20 years, pollinator populations have declined, due to a variety of factors, including the intensive use of pesticides, land-use intensification, climate change, and the spread of insect diseases (Vanbergen and Initiative 2013). The development and commercial release of GM plants has raised concerns about their impact on pollinators. Most of the risk assessment studies of GM crops on non-target organisms have been conducted with the honey bee *Apis mellifera*, considered the most important crop pollinator worldwide. Although most of the plant species carrying GM traits are considered as self-pollinated or depending very little on insect pollination, recent literature has shown that pollination highly contributes to an increase in crop productivity in soybean and cotton, among others (Milfont et al. 2013; Pires et al. 2014). In the environment, pollinators can be exposed to Bt proteins expressed in pollen or other plant parts from transgenic crops. The amount of insecticidal proteins expressed in the pollen of GM plants will depend on the type of promoter used in the transformation vector. The 35S, a constitutive gene expression promoter, is the most widely promoter used in plant transformation, especially for dicotyledonous plants. Plants transformed with the 35S present smaller amounts of the target protein in pollen, in comparison with plants transformed with pollen-specific promoters that have been used in the past. More recently, new alternatives have been studied, aiming to reduce transgene expression in non-photosynthetic tissues, such as roots, tissues, and flowering parts (Wang et al. 2016). The expression of Bt proteins has not been reported, and in some cases, not even studied, in plant secretions such as nectar and resins, also used by pollinators as a food source (Malone and Pham-Delègue 2001; Groot and Dicke 2002; Andow et al. 2008).

Extensive research has been conducted to evaluate the levels of exposure of insect pollinators to GM crops or their insecticidal proteins, risk of poisoning as well as the shortening in biodiversity and consequently, habitat loss (Potts et al. 2010). Most of the risk assessment studies with insect pollinators tested the effect of purified transgenic products mixed to artificial diets and offered to insects reared in the laboratory. A few of these *in vitro* studies showed negative effects on bees. The lectin GNA has been reported to cause significant larval mortality of honey bees (Hendriksma et al. 2012) and delayed development of solitary bees (*Osmia bicornis*) (Konrad et al. 2008). In comparison these insects were not as affected by Bt proteins (Hendriksma et al. 2012) or oryzacystatin-1 protein (Konrad et al. 2008). Serine protease inhibitor protein was also reported to cause significant larval mortality, plus delayed larval development and decreased adult body mass in honey bees (Brodsgaard et al. 2003). Another study showed that Bt cry1Ab did not affect survival, food consumption or learning capacities of honey bees, but reduced their foraging activity (Ramirez-Romero et al. 2005). Studies with two species of stingless bees showed contrasting results. In one of them, Bt cry1Ac protein did not affect survival and development of *Trigona sninipes* (Lima et al. 2013). In the other study, Bt cry1F and cry2Ab proteins delayed larval development of *Melipona quadrifasciata*, although larvae treated with cry2Ab presented higher survival rate, compared to the negative control. The major negative result found in this study was the lethal effect of the herbicide glyphosate on larvae and adults of *M. quadrifasciata*, after a few days of exposure. Surprisingly, glyphosate was more toxic to the stingless bees than the insecticide imidacloprid, used as a positive control (Seide et al. 2018). Glyphosate binds to and blocks the activity of the enzyme enolpyruvylshikimate-3-phosphate synthase (EPSPS), expressed only in plants and microorganisms and therefore, this herbicide was considered safe to animals. However, more recently, several studies have reported negative effects of glyphosate to honey bees (Gregorc and Ellis 2011; Herbert et al. 2014; Balbuena et al. 2015; Seide et al. 2018). That is especially important for the risk assessment of GM crops to pollinators due to the development of IR/HT stacked GM crops. Interestingly, the effect of pesticides (either herbicide or insecticide) on non-target arthropods is usually more negative than the insecticidal proteins expressed by GM crops (Oberhauser et al. 2001; Rose et al. 2007; Lima et al. 2016; Seide et al. 2018).

The majority of the *in vitro* studies reported in the literature showed that the effect of Bt cry and vip proteins (including a stacked Bt maize expressing three Bt toxins), protease inhibitor proteins, and *in vitro*-synthesized dsRNA on adults and larvae of bees is negligible. The parameters evaluated in these studies were survival of larvae and adults, food consumption, pupal weight, hypopharyngeal gland development, detoxification enzyme activity, midgut enzyme activity, midgut bacterial community diversity, hemolymph protein concentration, total hemocyte count, and learning capacity or flight ability (Arpaia and Metapontum 1996; Malone et al. 1999, 2001, 2004; Hanley et al. 2003; Babendreier et al. 2005; Ramirez-Romero et al. 2005; Han et al. 2010; Hendriksma et al. 2011, 2012; Raybould and Vlachos 2011; Niu et al. 2013; Bachman et al. 2016; Jia et al. 2017; Yi et al. 2018). Additionally, a meta-analysis of 25 independent laboratory studies showed that Bt proteins did not cause a negative impact on the survival of adults or larvae of bees (Duan et al. 2002).

Although *in vitro* studies are important as an indicator of risk, the level of exposure of non-target insects in these studies is usually higher than in the environment, either because of the high concentrations used in lab assays (Cowgill et al. 2002; Álvarez-Alfageme et al. 2007), or because other food sources are available in the environment, thus reducing the impact of the GM crops. Another caveat of *in vitro* studies is that, for social insects, evaluating the effect of potential threats on individuals is not a good measure of the impact on the colony. Using a robotic platform for continuous, multicolony monitoring of uniquely identified workers, researchers have shown that neonicotinoid exposure can dramatically affect bee's social behavior, by, for example, reducing the time bees spent with nursing and impairing their ability to warm the nest or to build insulating wax canopy (Crall et al. 2018). Based on this result, one might consider that evaluating the effect of GM crops on social insects individually, in laboratory assays, might misestimate the impact on the colony. However, field studies aimed to reduce this caveat, by evaluating the effect of exposing bee colonies to Bt pollen in the field, concluded that Bt plants pose no risk to survival, development and organ formation, colony performance, foraging behavior, or olfactory learning of honey bees (Arpaia and Metapontum 1996; Malone et al. 2001; Dai et al. 2012). Likewise, honey bee colonies placed in herbicide-tolerant GM canola fields were not affected in terms of survival, adult recovery, and pupal weight, compared to the colonies placed in non-transgenic canola fields (Huang et al. 2004).

The GFP (green fluorescent protein) gene was first isolated from the jellyfish, but it is commonly present in the genome of many other marine species. Because it is not present in the genome of arthropods or plants, this gene is frequently used as a reporter of expression. In RNAi studies, it has been used as an exogenous control for many arthropod species, including the honey bee. Arthropods are treated with *in-vitro*-synthesized dsRNA-GFP as a negative control in RNAi bioassays aiming to silencing target transcripts from the arthropod genome. As an exogenous gene, dsRNA-GFP is not expected to trigger RNAi response in arthropods. However, undesirable effects on gene expression, pigmentation, and developmental time have been reported for honey bees treated with dsRNA-GFP (Nunes et al. 2013). The authors reported the differential expression of about 1,400 genes in the honey bees treated with dsRNA-GFP, in comparison with the non-treated controls, because of either direct off-target effects or indirect downstream secondary effects. Off-target effects were reported for four non-target genes in the honey bee, leading to unspecific downregulation depending on the target dsRNA used and on the insect tissue (Jarosch and Moritz 2012). These results suggest that off-target effects of RNAi GM plants on non-target organisms might occur even in the absence of the target gene, and so, advanced studies should be performed to investigate the impact on the insect biology and behavior. Additionally, based on the specificity of the RNAi efficiency requirements, it is expected that the response of non-target organisms to RNAi GM crops will vary greatly. There are still several unanswered questions about the mechanisms of siRNA persistence and processing that might impact our understanding on how this mechanism can affect both target and non-target organisms.

## *Non-target Lepidopterans*

Non-target lepidopterans are those species which usually do not feed on the protected crop. These insects have another range of preferred hosts, mostly weed species. Non-target lepidopterans are also relevant insects for pollination services and are commonly studied as indicators of biodiversity. Therefore, they are another insect group commonly assessed to evaluate the risk of GM crops to pollinators.

They are, however, a special group of non-target insects because most of the insect-resistant transgenic plants are Bt crops specifically resistant to lepidopterans. A potential risk to non-target lepidoptera larvae is the ingestion of harmful amounts of Bt pollen (mostly maize) deposited on the leaves of their host plants in agricultural landscapes. A study of protected lepidopteran species in Hungary showed that the host plants of 30 out of 187 species grow in the borders of maize fields (Darvas et al. 2004). Larvae of these 30 species would likely be exposed to Bt pollen deposited on the leaves of their host plants. Because two of these 30 species feed on host plants that might be exposed to significant deposition of maize pollen, the Hungarian government proposed a ban on the cultivation of Bt maize in 2005 (Darvas et al. 2004; Perry et al. 2012).

The classic case study with non-target Lepidoptera is the monarch caterpillar, *Danaus plexippus*, which feeds exclusively on milkweed plants. In 1999, researchers at Cornell University found that monarch larvae exposed to milkweed leaves dusted with transgenic Bt corn pollen presented higher mortality and delayed development than those fed leaves dusted with untransformed corn pollen or leaves without pollen (Losey et al. 1999). These results raised a long discussion in the academic community, which produced a large body of work showing that Bt corn pollen did not pose any significant risk to monarch butterflies (Hellmich et al. 2001; Oberhauser et al. 2001; Pleasants et al. 2001; Sears et al. 2001; Stanley-Horn et al. 2001; Zangerl et al. 2001). Evidence showed that the amounts of Bt corn pollen used in the laboratory assays were much higher than the real exposure level the insects encounter in the environment. Although maize pollen can, under special weather conditions, be dispersed over long distances by wind, the amount of pollen deposited on milkweed leaves is not enough to cause significant larvae mortality in the environment on the population level. Moreover, under natural conditions, insects are able to avoid leaves covered with corn pollen, which they could not do in the laboratory assay. These discrepancies between results obtained at controlled vs. field conditions indicate the need to check the impact of GM crops on non-target arthropods at field conditions, or, in cases where adverse effects are observed in controlled experiments with high concentrations of the insecticidal protein, on a case-by-case analysis.

The effect of Bt crops on other non-target lepidopterans has been reported. No mortality was observed for larvae of the lepidopteran *Euchatias egle* after 48 h exposure to pollen from a Bt GM maize deposited on leaves of their host plant, milkweed (Jesse and Obrycki 2002). GM poplar expressing the Bt cry3A protein, which confers resistance to the crysomelidae *Plagiodera versicolora*, did not affect survival, exuviations index, pupation rate, or eclosion rate of the non-target Lepidoptera *Clostera anachoreta* (Zhang et al. 2011). In contrast, laboratory studies exposing first instar larvae of the Peacock butterfly *Inachis io* to several concentrations of Bt cry1ab maize pollen caused significant dose-dependent weight reduction and mortality of insects (Felke et al. 2010).

Based on mathematical models, the European Food Safety Authority (EFSA) panel on genetically modified organisms published a document recommending risk mitigation measures to reduce estimated mortality of non-target Lepidoptera of conservation concern associated with the ingestion of Bt maize pollen deposited on their host plants (European Food Safety Authority 2015). An isolation distance of 20 m around protected habitats from areas cultivated with Bt maize was recommended to reduce the risk to non-target Lepidoptera to a negligible level. For highly sensitive species, such as *Plutella xylostella*, the recommended isolation distance was of 30 m (European Food Safety Authority 2015).

## *Entomophagous Insects*

An important component of ecosystem balance is the natural occurrence of insect predation and parasitism. Herbivore insect populations are maintained at balanced levels by insect predators and parasitoids in the environment. In agricultural landscapes, predators and parasitoids have been frequently used as a tool for the Integrated Pest Management (IPM), either by preservation or by artificial introduction of these organisms in crop fields. Predators and parasitoids are critical in situations where the chemical control is no longer efficient for the management of insect pests. This is the case where insect populations are selected for resistance to transgenic crops, which has often been reported (Jurat-Fuentes et al. 2003; Sisterson et al. 2005; Tabashnik 2008, 2009, 2013; Tabashnik and Carrière 2017). It is then desirable that the pest management strategies, either by chemical control or by the use of insect-resistant GM plants, do not affect population and diversity of predators and parasitoids. For example, in a field study, Bt sweet corn and the foliar insecticide spinosad, which specifically target the pest species, were less toxic to predators than the broad spectrum pyrethroid lambda cyhalothrin (Musser and Shelton 2003). The intensive use of insecticides in non-Bt cotton fields to control *Helicoverpa armigera* in China has led to the reduction in predator populations, and therefore, increased aphid populations (Wu and Guo 2003).

The development and use of GM crops for insect resistance aims to provide an alternative for the intensive use of pesticides, which are harmful to the environment and to human health. They also aim to provide an alternative in cases where chemical control has become inefficient due to the selection of resistant insect populations. It is therefore reasonable that the potential detrimental effects of GM crops to non-target arthropods should be compared to their counterpart in chemical control. Some studies have done this comparison and the majority of the results indicate that insect-resistant GM crops have significantly lower negative impact on non-target organisms than pesticides (Naranjo 2005, 2011; Romeis et al. 2006). Some studies go beyond and estimate that using Bt crops benefits the biological control, providing suitable conditions for naturally occurring or introduced natural enemies (Ferry et al. 2006; Lu et al. 2012; Tian et al. 2015; Romeis et al. 2019).

There is an extensive literature showing laboratory and field studies in which the conclusions range from neutral to negative effects of GM crops on natural enemies, with a few studies showing positive results. However, a meta-analysis of data in the literature about the effects of GM crops on natural enemies showed that the effect of cry toxins and protease inhibitors on natural enemies is often non-neutral, and that parasitoids are usually more sensitive to transgenic toxins than predators (Lövei et al. 2009). Similar to pesticides, insecticidal proteins expressed in GM plants might affect natural enemies either directly or indirectly. Some predators and parasitoids feed alternatively on plant parts, such as pollen and nectar, or on weeds, when prey is scarce, or else as a supplemental food source at some of their life stages. Direct negative effects might occur when susceptible natural enemies ingest insecticidal proteins from transgenic plants. Indirect effects occur when susceptible predators and parasitoids feed on preys/hosts that were previously fed with transgenic plants and ingested insecticidal proteins, but usually only when the prey/host is susceptible to the insecticidal protein (Romeis et al. 2006). Another indirect effect on natural enemy populations is the decrease in weed abundance, due to weed management using HT GM crops. In a cascade, reducing the populations of natural enemies might lead to increased numbers of herbivores (McPherson et al. 2003). The review by Lövei et al. (2009) points to a bias toward certain species of predators and parasitoids and a limited number of insecticidal molecules. More recently, concern has been raised about the GM crops with stacked insecticidal genes, which expose non-target arthropods to multiple insecticidal proteins, most of them not yet tested for their potential synergism (Svobodová et al. 2017).

## *Predators*

Insecticidal molecules expressed in GM plants can be transferred via the food web from transgenic plants to herbivores and then to their natural enemies and accumulate in organisms to different degrees (Chen et al. 2005; Zhang et al. 2006; Pérez-Hedo et al. 2012; García et al. 2012; Zhou et al. 2014; Paula et al. 2015). The uptake/accumulation of insecticidal molecules varies with a number of factors, such as the type of molecule and the species, life stage, and the susceptibility level of the natural enemies (Zhou et al. 2014). For example, the concentration of cry3Bb1 toxin decreased through the trophic chain from maize to mites and then to adults of the predator beetle *Atheta coriaria* (Staphylinidae), but not from mites to *A. coriaria* larvae (García et al. 2012). That result was similar to the decline in Bt toxin concentration from Bt maize leaves to *S. littoralis* larvae and then to the larvae of the ground-dwelling predator *Poecilus cupreus* reported in another study (Alvarez-Alfageme et al. 2009). Bt cry1Ab toxins were also reported to accumulate in spiders that preyed *Nilaparvata lugens* reared on transgenic rice (Chen et al. 2005; Tian et al. 2013). However, the uptake of insecticidal proteins in herbivores does not always result in toxicity to predators. For example, although the Bt cry1Ac protein was detected in *N. lugens* reared on transgenic rice, no negative effect on the development or morphology of the predator *Propylea japonica* was observed after preying on Bt-reared *N. lugens* (Bai et al. 2006). The ladybug predator *Stethorus punctillum* also feeds specifically on spider mites that ingest high amounts of Bt toxins from Bt maize. However, the survival and development of this predator were not affected by preying spider mites reared on Bt maize, compared to non-Bt (Li and Romeis 2010). Several other reports showed that the concentration of Bt cry toxins was three to six folder lower in spiders than in their Bt-fed prey, demonstrating that the toxin did not accumulate in these predators (Tian et al. 2010, 2012; Meissle and Romeis 2012; Han et al. 2014). Even spiders consuming preys fed on stacked Bt crops, expressing six different Bt proteins, presented lower Bt concentrations than their food (Svobodová et al. 2017). In both scenarios, with or without accumulation of Bt toxins in the predator species through the trophic chain, no detrimental effect on the natural enemy was reported in these cases. Although the Bt cry1Ab protein was detected in the lepidopteran target species *Cnaphalocrocis medinalis* and on its predator, the spider *Pirata subpiraticus* (Lycosidae), the protein concentration did not increase over feeding time (Chen et al. 2009). Moreover, the authors showed that the binding receptor to the protein is missing in the predator, but not in the herbivore. Even so, feeding on Bt-reared prey increased the developmental time of *P. subpiraticus*, but it did not affect its survival and reproduction.

A debated case of detrimental effects of GM plants to natural enemies was presented for the green lacewing *Chrysoperla carnea*, a common predator in agricultural areas worldwide, which larvae feed on small soft-bodied insects, such as aphids. Adults of *C. carnea* feed on pollen, nectar, and honeydew. In 1998, studies from the Swiss Federal Research Station for Agroecology and Agriculture with Bt-expressing maize and purified Bt cry proteins showed that Bt maize harmed *C. carnae* larvae (Hilbeck et al. 1998a, b). These results raised for the first time the discussion of GM crops potentially harming natural enemies. Since then, independent groups have aimed to repeat the experiments with the green lacewing, but the results obtained in the initial reports have not been confirmed in follow-up studies. Instead, what the researchers found was that feeding *C. carnea* adults and larvae on pollen of Bt cry-toxin expressing crops mixed to artificial diets or on vip3A test substances did not negatively affect survival, development, and reproduction rates, when compared to non-Bt pollen (Li et al. 2008; Raybould and Vlachos 2011). In the follow-up studies by independent research groups, no negative effect on *C. carnea* was observed even at concentrations of the Bt toxin 10,000 higher than the observed in the body of susceptible Lepidoptera larvae (Romeis et al. 2004). Comprehensive reviews were published later by Dutton and colleagues (Dutton et al. 2003) and by Romeis and colleagues (Romeis et al. 2014) indicating that *C. carnea* larvae and adult will likely not be harmed by Bt maize and its insecticidal proteins based on the biology and ecology of this predator. Among their arguments is the fact that the preferred preys of *C. carnea* are aphids and other homopterans, which feed on the plant phloem-sap. As the amount of Bt proteins in the sap of transgenic maize is not relevant and aphids are not affected by feeding on Bt maize, it is not likely that *C. carnea* will encounter significant concentrations of Bt proteins when feeding on aphids. In fact, most of the studies showing indirect negative effects of Bt plants on *C. carnea* were conducted with susceptible Lepidoptera larvae as preys (Dutton et al. 2002; Lawo et al. 2010). Feeding on Bt-resistant Lepidoptera larvae did not affect the predator (Lawo et al. 2010). This suggests that the detrimental effects observed on *C. carnea* survival, longevity, and development are more related to the low quality of preys than to the toxic effect of the transgenic proteins (Bell et al. 2003; Lawo et al. 2010; Romeis et al. 2014).

Similar results have been reported for several other species of predators, such as the ladybugs *Coleomegilla maculata, Adalia bipunctata, Coccinella septempunctata,* and *Cicloneda sanguinea* (Duan et al. 2002; Lundgren and Wiedenmann 2002; Ahmad et al. 2006; Alvarez-Alfageme et al. 2011; Raybould and Vlachos 2011; Nakasu et al. 2013), the predatory mite *Euseius concordis* (de Castro et al. 2013), the carabid species, *Harpalus caliginosus* and *H. pensylvanicus* (Ahmad et al. 2006), the generalist predatory bugs *Orius insidiosus* and *Cyrtorhinus lividipennis* (Al-Deeb et al. 2001; Raybould and Vlachos 2011; Chen et al. 2015) showing none or negligible effects of directly exposing the predators to a variety of Bt cry and vip toxins. However, although *A. bipunctata* survival was not significantly affected by Bt cry3Bb protein, Bt cry1Ab cause significant larval mortality (Schmidt et al. 2009). Similarly, direct exposure of the ladybug *Cheilomenes sexmaculatus* to Bt cry1Ab and cry1Ac reduced larval survival and adult emergence compared to non-Bt controls, although preying Bt-fed aphids did not affect the development of this predator (Dhillon and Sharma 2009). Moreover, an increased concentration of purified cry toxins negatively affected the reproduction of *E. concordis* (de Castro et al. 2013). Those high concentrations, however, do not represent the amount of toxin to which the predator would be exposed in the field. A commonly used protocol for risk assessment of GM crops to non-target organisms evaluates the susceptibility of non-target arthropods by directly exposing them to high concentrations of the toxin, as a first trial, in laboratory tests. Only if toxicity is observed, lower concentrations, similar to the amount expressed in the GM plant, are evaluated, and a next step would be to verify the real threat to non-target organisms in field assays. An example is the study with five species of arthropod predators, to which an *in vitro*-synthesized DvSnf7-derived dsRNA was offered at a concentration >tenfold the maximum expected environmental concentration. DvSnf7-dsRNA targets the gene Snf7 from the Coleopteran *Diabrotica virgifera.* No adverse effects were observed for life table parameters, such as survival, growth, and development of any species tested (Bachman et al. 2016). In cases like that, field assays with the GM plant are not necessary.

Studies with other species of predators feeding on Bt-resistant preys also resulted in none or negligible effects on life table parameters, as exemplified by the reports that follow. Some lepidopteran species, such as *P. xylostella* and *Trichoplusia ni*, present populations that differ in their susceptibility to Bt toxins. When the predator beetles *Pterostichus madidus* were fed on larvae of Bt-resistant and Bt-susceptible *P. xylostella* reared on Bt or non-Bt canola, no difference was observed for survival, weight gain, and reproduction of the predator (Ferry et al. 2006). Interestingly, in choice tests, the predator avoided Bt-fed susceptible prey, but not Bt-fed resistant prey, suggesting that the most relevant impact of Bt toxins on predators is the decrease in prey quality. Similarly, survival, development time, adult weight, and fecundity of the predator *Coleomegilla maculata* did not differ between beetles fed on Bt-resistant *T. ni* larvae reared on either Bt Cry2Ab- and Cry1Ac-expressing cotton or non-Bt cotton (Li et al. 2011). Likewise, feeding on non-target herbivores usually cause no detrimental effect on predators (Zhang et al. 2008). For example, no effect on predatory capacity was observed for the predatory mite *Neoseiulus californicus* (Phytoseiidae), neither for *A. coriaria*, feeding on phytophagous mites *Tetranychus urticae* that were reared on lepidopteran-resistant Bt cotton (de Castro et al. 2013) and coleopteran-resistant Bt maize, respectively (García et al. 2012). Similar to that, predation and attack behavior of the water bug *Microvelia horvathi* were not affected when feeding on *Entomobrya griseoolivata* reared on Bt rice (Bai et al. 2005). The survival and development of *C. lividipennis* were not affected by feeding on *N. lugens* reared on Bt rice (Bernal et al. 2002a). Contrary to these results showing no impact on predatory capacity, a choice test conducted with another phytoseiid predatory mite, *Phytoseiulus persimilis*, showed that predators preferred non-Bt fed preys, when offered *T. urticae* fed on Coleoptera-resistant (cry3Bb) Bt eggplants versus its non-Bt isoline (Rovenská et al. 2005).

GNA protein, which has been reported as toxic to pollinators (Konrad et al. 2008; Hendriksma et al. 2012), has also been reported as detrimental to oviposition period, fecundity and dry weight of *C. carnea* adults, when mixed to an artificial diet (Li et al. 2008). However, the development and survival of the ladybug *Adalia bipunctata* was not affected by preying on the green peach aphid *M. persicae* reared on GNA-expressing transgenic plants, which are toxic to aphids (Down et al. 2003).

The ultimate goal of a pest management strategy is to reduce pest populations to non-damaging levels. It is then, natural, that the intentional decrease in herbivore (target species) populations by GM crops will reduce the offer of preys/hosts to natural enemies, similarly to the effect of pesticides. In the case of GM crops, this might be a reasonable concern for specialist predators and parasitoids, but usually not for generalists, which can consume alternative preys/hosts. Several field studies investigated the impact of GM crops on the abundance and diversity of arthropods. A few of them specifically looked at communities of entomophagous species with different feeding behaviors (Riddick et al. 1998). Specialist entomophagous insects are obviously less abundant in fields cultivated with GM crops resistant to their prey/host species, such as the case of the carabid beetle *Lebia grandis*, which prey specifically the Colorado potato beetle, *Leptinotarsa decemlineata* (Riddick et al. 1998). On the other hand, the populations of generalist predators are usually not affected by the GM crop and the potential decrease in abundance of their preferred preys/hosts (Riddick et al. 1998; Al-Deeb et al. 2001; Al-Deeb and Wilde 2003; Pilcher et al. 2005; Árpás et al. 2005; Ahmad et al. 2006; Tian et al. 2010; Bai et al. 2012; Guo et al. 2014; Zhao et al. 2016). For example, two multi-season field evaluations of arthropod communities showed that the GM maize expressing the Bt cry3Bb1 toxin (resistant to *D. virgifera*) and the lepidopteran-resistant Bt cry1Ab GM maize cause negligible effects on the community of staphylinid beetles, which consume a wide range of herbivore species (Balog et al. 2010; Svobodová et al. 2016). Similarly, a 3-year field evaluation of population density and dynamic of the generalist predator spider *Hylyphantes graminicola* showed no difference in Lepidoptera-resistant Bt rice versus its non-Bt isoline plots (Han et al. 2014). Also in a 3-year field study, no impact of Bt rice lines on population dynamic of the mirid bug *C. lividipennis* was reported (Chen et al. 2007). Another long-term field studies showed a fluctuation in the population of some non-target arthropods, including generalist predator species, in the GM crop field, with the decline of some species in a year and the rise of other species over time (McPherson et al. 2003; Naranjo 2005; Balog et al. 2010; von Burg et al. 2011; Pinheiro et al. 2014). This variation suggests that the decline in the population of preferred preys/hosts displaced the most abundant predators, favoring other species (Naranjo 2005). Two independent studies showed that the population of the generalist predatory bug *Nabis* spp. decreased in the Bt crop field, in comparison with the non-transgenic field (Daly and Buntin 2005; Whitehouse et al. 2005). A possible explanation for these latter results, given by the authors, is that the number of nabids was lower in Bt maize due to reduced availability of their lepidopteran prey. In another field study, significantly less green lacewing adults and rove beetles were found in HT/GM soybean and Bt maize plots, respectively, compared to their non-transgenic isoline (Jasinski et al. 2003).

Although most of the field studies indicate none or negligible effects of GM crops on the population of natural enemies, recent transcriptomics studies point to negative effects to spiders at biochemical and physiological levels. In the spiders *Ummeliata insecticeps* and *Pardosa pseudoannulata*, preying upon fruit flies reared on cry1Ab-containing medium resulted in altered activity of three key metabolic enzymes, in comparison with spiders preying on cry1Ab-free fruit flies (Zhou et al. 2014). In another study, the differential expression of genes related to chitin formation in spiderlings of *P. pseudoannulata* preying upon *N. lugens* fed on Bt versus non-Bt rice indicates that the Bt toxin may impair the formation of new cuticles during molting, contributing to the delayed development of spiderlings (Wang et al. 2017). These results need to be interpreted carefully, performing the adequate comparisons with other GM and conventional lines as a baseline for comparison. But the results certainly raise questions on how the non-target population is affected over time and whether the ecological functions of these predators/parasitoids are impaired.

## *Parasitoids*

The success of parasitism depends on a combination of factors such as host finding and acceptance. Parasitoids find their hosts mainly by chemical signs emitted by herbivore-injured plants, which parasitoids recognize and are attracted to. Plants specifically produce and emit organic volatiles in response to different stresses, such as mechanical injury, pathogen replication, and senescence. Parasitoids can be sensitive to changes in pollen, nectar and honeydew composition, as well as to changes in volatile emissions and to host quality (Price et al. 1980; Dicke and Sabelis 1987; Turlings et al. 1991; Romeis et al. 2003; Beale et al. 2006; Tompkins et al. 2010). Genetic transformation can lead to inadvertent effects in the plant, such as pleiotropic effects and other biochemical changes (Gutiérrez-Campos et al. 2001; Lorenc-Kukuła et al. 2005; Arpaia et al. 2017). It is, therefore, reasonable to consider that genetic transformation aiming to introduce and express foreign genes might interfere with the metabolism of plants and, therefore, with target and non-target herbivores, which in turn might affect parasitoid performance. Similar to what we have seen in the previous section for predators, parasitoids might perceive host herbivores fed on GM plants as poor quality hosts, when herbivores are sub-lethally harmed by the effect of insecticidal proteins (Bernal et al. 2002b). In these cases, reports showed that parasitoids prefer hosts fed on non-transgenic plants, and this behavior is more related to host quality than to potential toxic effects caused by the GM plant (Tomov et al. 2003). Moreover, it is suggested that when host susceptibility to transgenic plants reduces their suitability to parasitism, damage to crops and reproduction success of pests are also reduced (Tomov et al. 2003). Again, such as for predators, population density of generalist parasitoids is likely to be maintained, even in the lack of their preferred host. In contrast, for specialists, reducing their host population will reduce the parasitoid population, in most cases (Hellmich et al. 2008; Pálinkás et al. 2017).

For example, *Cotesia marginiventris*, a microhymenoptera from the family Braconidae, is a generalist solitary endoparasitoid that develops into lepidopteran larvae from the family Noctuidae, the main target of Bt technology. The attractiveness of *C. marginiventris* did not differ among conventional and Bt maize injured by its host *S. frugiperda* or artificially injured (Desneux et al. 2010), as expected for a generalist. However, a similarly weak attractiveness of the parasitoid was registered toward frass derived from hosts fed on Bt maize tissue, versus hosts fed with non-Bt tissue treated with antibiotics, to eliminate host gut bacteria. In comparison, *C. marginiventris* was strongly attracted to host frass derived from non-Bt, non-treated maize. Altogether, the results indicate that Bt toxins might affect the bacterial community present in the host's gut and frass, altering the composition of volatiles that attract the parasitoid, and so reducing the parasitoid response toward host frass derived from Bt maize (Desneux et al. 2010). Other studies have reported negative effects of Bt crop-fed hosts on the development, reproduction, and longevity of *C. marginiventris* and *C. floridanum* (Baur and Boethel 2003), usually when the host is susceptible to Bt.

On the other hand, none or negligible negative effects have been reported on parasitoids developing on Bt-resistant/tolerant hosts, or on hosts that are non-target of Bt crops, in agreement with the effects reported for predators. An example is the Lepidoptera *S. eridania*, which is tolerant to the Bt cry1Ac protein. No adverse effect was reported for the egg parasitoid *Telenomus remus* (Platygastridae) developing on eggs of *S. eridania* reared on a Bt cry1Ac-soybean line, in comparison with its non-transgenic isoline (Bortolotto et al. 2014). Similarly, some populations of the diamondback moth *P. xylostella*, which is usually highly susceptible to Bt toxins, have been selected for resistance to these toxins. *P. xylostella*-resistant larvae damage the brassica plants more than susceptible ones and that leads to stronger attraction of the parasitoid *Cotesia plutellae* (Schuler et al. 2003). Interestingly, the parasitoids *C. plutellae* and *Diadegma insulare* are able to develop in Bt-resistant *P. xylostella*, better than on Bt-susceptible hosts, and did not distinguish between Bt and non-Bt plants or the volatile compounds produced by them, as long as plants are damaged at similar levels by their host caterpillars (Schuler et al. 2004; Liu et al. 2011). In another example, transgenic Bt rice did not affect host preference of *Anisopteromalus calandrae* (Pteromalidae), a generalist parasitoid of stored-product coleopteran (Sun et al. 2015). In the same direction, transgenic Bt cotton did not affect the quality of aphids as hosts of the braconid *Lysiphlebus testaceipes* (Hagenbucher et al. 2014a). In fact, aphid populations are benefitted on Bt cotton, due to the suppression of lepidopteran pests, which reduces herbivore competition and also the amount of terpenoids produced in cotton as an induced defense against tissue-chewing insects. Honeydew produced by aphids reared on Bt cotton plants and offered to parasitoids as a food source did not affect the survival of *L. testaceipes* and *Eretmocerus eremicus* (Hagenbucher et al. 2014a, b). It is likely that insecticidal toxins that do not affect the herbivore host will not affect their parasitoids either.

Parasitoid size and fitness usually correlate to host suitability and food quality. For example, adult size of *C. marginiventris* and brood size of another braconid, *Parallorhogas pyralophagus,* were reduced when they developed on their lepidopteran hosts *S. frugiperda* and *Eoreuma loftini*, respectively, fed on Bt maize (Bernal et al. 2002b; Ramirez-Romero et al. 2007). Mean size and longevity of female parasitoids *Eulophus pennicornis* were significantly reduced when they developed on their lepidopteran host, *Lacanobia oleracea,* fed on GNA-containing diet, although no significant alteration in fecundity was observed for these smaller females (Bell et al. 2001). A recent study reported negative effects on the size of antennae, forewings, tibia, as well as morphological body parts asymmetry of the braconids *C. sesamia* and *C. flavipes* developing in two lepidopteran hosts, the native *Sesamia calamistis* and a recent East Africa invader, *Chilo partellus*, fed on Bt maize. The effects were more severe on *C. flavipes*, the interaction of which with its host *C. partellus* is evolutionarily more recent (Ndolo et al. 2018). The size of the aphelinid parasitoid *Aphelinus abdominalis* developing on *Macrosiphum euphorbiae* aphids reared on GNA-expressing potato was also reduced, although western blot analysis suggested that parasitoids excreted most of the GNA ingested (Couty et al. 2001). This size effect reported on parasitoids is probably due to detrimental effects caused by the insecticidal proteins on the susceptible hosts.

Purified GNA protein mixed to artificial diets negatively affected pollinators and predators, as we saw in the previous sections. Most of the studies with parasitoids looked at indirect effects of GNA protein, by exposing parasitoids to hosts fed on GNA-expressing transgenic plants or plant parts. No detrimental effects of GNA-expressing transgenic maize, tomato, and potato plants were reported on parasitism rate, development, longevity, and fecundity for parasitoid-host interactions such as *C. flavipes* developing on its lepidopteran host *Diatraea saccharalis* as well as for two other parasitoids, the solitary endoparasitoid *Meteorus gyrator* and the gregarious ectoparasitoid *E. pennicornis,* developing on their lepidopteran host *L. oleracea* (Bell et al. 1999, 2001; Couty and Poppy 2001; Sétamou et al. 2002; Wakefield et al. 2006). In fact, significantly more adults of *E. pennicornis* developed on GNA-fed hosts, compared with the controls, when female parasitoids were exposed to third instar hosts (Bell et al. 1999). GNA-binding glycoproteins were detected in the gut and other tissues of larval *M. gyrator*, but not in adults (Wakefield et al. 2006). Bt cry toxins have also been detected in newly emerged parasitoids, although no significant effect on parasitoid fecundity was observed (Gao et al. 2010; Liu et al. 2011).

Some field studies evaluated the effect of insect-resistant GM crops on the abundance and diversity of parasitoids. In one of them, significant effects on aphid-parasitoid food webs were found in a 2-year field study comparing transgenic disease-resistant wheat lines and non-transgenic controls, but the effects were inconsistent and then attributed to differences between years and to genetic variation among wheat varieties (von Burg et al. 2011). In another study, a higher number of *O. nubilalis* larvae parasitized by the tachinids *Lydella thompsoni* and *Pseudoperichaeta nigrolineata* was reported in Bt cry1Ab maize, as compared to its non-transgenic isoline (Bourguet et al. 2002), which could be considered as a positive result. In contrast, another study showed lower parasitism rate of aphids by Aphidiine parasitoids in transgenic cotton expressing Bt cry1Ac and CpTI (Cowpea trypsin inhibitor), compared to the non-transgenic cotton plots (Yao et al. 2016). The reduced parasitism rate in Bt cotton did not affect the management of aphid populations, probably because the abundance of other natural enemies, such as ladybugs, lacewings, and spiders, did not differ between treatments. It is likely that the observed fluctuation in the population of parasitoids was compensated by the rise in the population of other species of natural enemies. Parasitism rate can be affected by sex ratio, if the reduced number of females is generated in the population. Although significantly fewer females emerged from *Sitophilus zeamais* reared on Bt maize, no difference was reported for sex ratio between parasitoids developing in host reared on Bt vs. non-Bt maize (Hansen et al. 2012). Bt crops did not alter the sex ratio of other parasitoid species, such as *P. pyralophagus* and *Anagrus nilaparvatae* (Bernal et al. 2002b; Gao et al. 2010).

Feeding the parasitic wasp *Pediobius foveolatus* with very high concentrations of an *in vitro*-synthesized DvSnf7-derived dsRNA did not affect its survival, growth, and development (Bachman et al. 2016). Although the effects on the food chain were not evaluated in this case, it is likely that the dsRNA generated in plants will be highly specific to the target species and probably will not be available to entomophagous arthropods, due to the highly degradable nature of RNA.

## Detritivores

Detritivore arthropods are an important group of soil-dwelling organisms that contribute to important ecological services, by carrying out processes in the soil, such as nutrient cycling and decomposition of organic matter. The abundance and diversity of detritivores is usually taken as an indicative of disturbance in agricultural systems. Arthropods presenting this feeding habit might be exposed to insecticidal proteins produced in transgenic plants by feeding on plant residues containing the toxin. Some of these insecticidal proteins can be exudated through the roots into the rizhosphere and so soil-dwelling arthropods would be exposed to the toxin in the soil (Saxena et al. 2002, 2004). Bt toxins released by root exudates or plant biomass persist in the soil for at least 200 days, depending on several factors, such as temperature (Saxena and Stotzky 2002; Zwahlen et al. 2007). Bt toxins are protected from biodegradation by binding to soil surface-active particles (Saxena and Stotzky 2002). In a field study conducted after a long-term (9 years) cultivation of Bt maize, Bt toxins were detected in the soil only in one out of four field sites (Gruber et al. 2012). Bt toxins are highly labile and might decompose rapidly in the soil, depending on the amount of toxin introduced and the degradation rate (Hopkins and Gregorich 2003; Ahmad et al. 2005; Gruber et al. 2012). Evidence showed that these toxins were not taken up by non-Bt crops from soil or water in hydroponic culture (Saxena and Stotzky 2002). Similarly, protease inhibitor proteins from transgenic plants also remain active in the soil for a short period (57 days) and affected the carbon content and soil-dwelling organisms that are directly exposed to the toxin, such as Collembola and nematodes (Donegan et al. 1997).

*Folsomia candida*, a microarthropod from the Class Collembola, is a commonly used sentinel species for risk assessment studies, because of its sensitivity to pesticides and other environmental pollutants (Fountain and Hopkin 2005). The majority of the risk assessment studies showed that Bt crops, as well as an *in vitro*-synthesized dsRNA, caused none or negligible effects on the survival, reproduction, growth, abundance, and quality as preys of detritivores such as *F. candida* and others (Yu et al. 1997; Al-Deeb et al. 2001; Bourguet et al. 2002; Candolfi et al. 2004; Bai et al. 2005, 2011; Griffiths et al. 2006; Bakonyi et al. 2006; Priestley and Brownbridge 2009; Chang et al. 2011; Yuan et al. 2013; Bachman et al. 2016). In contrast, a synthetic insecticide and a Bt insecticide caused a significant negative effect on the community of plant and soil-dwelling arthropods, compared to the Bt cry1Ab maize (Candolfi et al. 2004). Similarly, the survival of the bulb mite *Rhizoglypus robini* was not affected by contact and ingestion of the Bt cry3Aa protein, but it was significantly affected by the insecticide fipronil, which is another example of pesticides being more toxic to non-target organisms than GM crops.

A laboratory study showed that *F. candida* had their reproduction and catalase activity reduced when feeding on some, but not all, Bt rice varieties tested, which were offered to the arthropods in diets mixed with yeast (Yuan et al. 2011). In another study, collembolan detritivore species were either not negatively affected or were more abundant in GM herbicide-tolerant crop fields, compared to conventional crop herbicide treatment (Bitzer et al. 2002; Brooks et al. 2003). These studies indicate that some direct and indirect effects of insecticidal proteins produced by transgenic crops on detritivore arthropods might be species-specific (Bakonyi et al. 2006; Chang et al. 2011).

## Concluding Remarks (And What We Don't Know)

Genetically modified crops are a tool for Integrated Pest Management with promising outcomes for reducing the use of synthetic pesticides, and so, reducing the negative impact of these molecules on the environment and benefitting the biological control, either natural or introduced. Over the last 23 years, studies on risk assessment of GM crops for non-target organisms have been predominantly conducted with GM crops for insect resistance (IR). With the increasing adoption of HT/IR stacked traits GM crops, it becomes important to look at whether these traits can have an interactive effect on non-target organisms. The benefits of reducing pesticide use in GM crops are clear. However, both IR and HT GM crops will certainly have an impact on non-target populations, as it would be expected from any other kind of human interference in the environment. It is not reasonable to expect that a technology will arise that will not cause any impact on the ecosystem balance. Importantly, all technologies should be applied with care and responsibility, taking the risk mitigation measures and following the recommendations for the safe application.

No other technology has been so extensively studied such as GM crops, in terms of environmental impact. Negative effects have been reported mostly for insecticidal toxins not introduced in approved GM crops, or in *in vitro* assays, with purified toxins. The majority of the studies show neutral effects on arthropod species or communities. However, most of these studies looked at direct toxic effects and only a few of them investigated the indirect impact on ecological functions. As we have seen, there are still many gaps in our understanding of how non-target arthropods are affected by GM crops. In general, there is no difference in the diversity and abundance of arthropod communities in Bt versus non-Bt fields. But do we know whether the ecological functions are maintained? Recent reports show that transgenic crops impact spiders at the physiological and biochemical levels, as well as the social behavior of honey bees. Could the ecological functions of non-target arthropods be altered as well by a change in biochemical processes? How natural selection pressures will affect community structure at a long-term? Although extensive research has been conducted, to our knowledge, there is no long-term experiment reported in the literature.

We know that some non-target herbivores are able to uptake and/or accumulate toxins expressed in GM crops. However, we know very little of how these insecticidal toxins are processed, excreted, or passed on by non-target arthropods. What are the mechanisms involved in the accumulation of toxins in non-target arthropods? What mechanisms confer resistance to some non-target species? Are bacterial endosymbionts of insects affected by insecticidal proteins? Do these proteins affect complex multi-trophic interactions, such as pathogen transmission by arthropods?

Also, there are few studies comparing the impacts of insecticides versus the use of GM insect-resistant crops. However, based on the extensive literature published in the short period of 23 years, we can say that we probably know much better the effect of GM crops on non-target organisms than that of insecticides, for example. Moreover, it seems that the strict requirements for GM crop approval are reasonable, as negative effects might occur and should continue to be evaluated on a case-by-case basis.

## References

Ahmad A, Wilde GE, Zhu KY (2005) Detectability of coleopteran-specific Cry3Bb1 protein in soil and its effect on nontarget surface and below-ground arthropods. Environ Entomol 34:385–395. https://doi.org/10.1603/0046-225X-34.2.385

Ahmad A, Wilde GE, Whitworth RJ, Zolnerowich G (2006) Effect of corn hybrids expressing the coleopteran-specific cry3Bb1 protein for corn rootworm control on aboveground insect predators. J Econ Entomol 99:1085–1095

Al-Deeb MA, Wilde GE (2003) Effect of Bt corn expressing the Cry3Bb1 toxin for corn rootworm control on aboveground nontarget arthropods. Environ Entomol 32:1164–1170. https://doi.org/10.1603/0046-225X-32.5.1164

Al-Deeb MA, Wilde GE, Higgins RA (2001) No effect of Bacillus thuringiensis corn and Bacillus thuringiensis on the predator Orius insidiosus (Hemiptera: Anthocoridae). Environ Entomol 30:625–629. https://doi.org/10.1603/0046-225X-30.3.625

Álvarez-Alfageme F, Martínez M, Pascual-Ruiz S, Castañera P, Diaz I, Ortego F (2007) Effects of potato plants expressing a barley cystatin on the predatory bug Podisus maculiventris via herbivorous prey feeding on the plant. Transgenic Res 16:1–13. https://doi.org/10.1007/s11248-006-9022-6

Alvarez-Alfageme F, Ortego F, Castañera P (2009) Bt maize fed-prey mediated effect on fitness and digestive physiology of the ground predator Poecilus cupreus L. (Coleoptera: Carabidae). J Insect Physiol 55:143–149. https://doi.org/10.1016/j.jinsphys.2008.10.014

Alvarez-Alfageme F, Bigler F, Romeis J (2011) Laboratory toxicity studies demonstrate no adverse effects of Cry1Ab and Cry3Bb1 to larvae of Adalia bipunctata (Coleoptera: Coccinellidae): the importance of study design. Transgenic Res 20:467–479. https://doi.org/10.1007/s11248-010-9430-5

Anderson JA, Ellsworth PC, Faria JC, Head GP, Owen MDK, Pilcher CD, Shelton AM, Meissle M (2019) Genetically engineered crops: importance of diversified integrated pest management for agricultural sustainability. Front Bioeng Biotechnol 7. https://doi.org/10.3389/fbioe.2019.00024

Andow DA, Hilbeck A, Tuat NV (2008) Environmental Risk Assessment of Genetically Modified Organisms: challenges and opportunities with Bt cotton in Vietnam. CABI

Arpaia S, Metapontum A (1996) Ecological impact of Bt-transgenic plants [Bacillus thuringiensis]. 1: assessing possible effects of CryIIIB toxin on honey bee (Apis mellifera L.) colonies. J Genet Breed Italy

Arpaia S, Birch ANE, Kiss J, van Loon JJA, Messéan A, Nuti M, Perry JN, Sweet JB, Tebbe CC (2017) Assessing environmental impacts of genetically modified plants on non-target organisms: the relevance of in planta studies. Sci Total Environ 583:123–132. https://doi.org/10.1016/j.scitotenv.2017.01.039

Árpás K, Tóth F, Kiss J (2005) Foliage-dwelling arthropods in Bt-transgenic and isogenic maize: a comparison through spider web analysis. Acta Phytopathol Entomol Hung 40:347–353. https://doi.org/10.1556/APhyt.40.2005.3-4.16

Babendreier D, Kalberer NM, Romeis J, Fluri P, Mulligan E, Bigler F (2005) Influence of Bt-transgenic pollen, Bt-toxin and protease inhibitor (SBTI) ingestion on development of the hypopharyngeal glands in honeybees. Apidologie 36:585–594. https://doi.org/10.1051/apido:2005049

Bachman PM, Huizinga KM, Jensen PD, Mueller G, Tan J, Uffman JP, Levine SL (2016) Ecological risk assessment for DvSnf7 RNA: a plant-incorporated protectant with targeted activity against western corn rootworm. Regul Toxicol Pharmacol 81:77–88. https://doi.org/10.1016/j.yrtph.2016.08.001

Bai Y, Jiang M, Cheng J (2005) Impacts of transgenic cry1Ab rice on two collembolan species and predation of Microvelia horvathi (Hemiptera: Veliidae). Kun Chong Xue Bao Acta Entomol Sin 48:42–47

Bai YY, Jiang MX, Cheng JA, Wang D (2006) Effects of Cry1Ab Toxin on Propylea japonica (Thunberg) (Coleoptera: Coccinellidae) Through Its Prey, Nilaparvata lugens Stål (Homoptera: Delphacidae), Feeding on Transgenic Bt Rice. Environ Entomol 35:1130–1137. https://doi.org/10.1603/0046-225X-35.4.1130

Bai Y, Yan R, Ke X, Ye G, Huang F, Luo Y, Cheng J (2011) Effects of transgenic Bt rice on growth, reproduction, and superoxide dismutase activity of Folsomia candida (Collembola: Isotomidae) in laboratory studies. J Econ Entomol 104:1892–1899

Bai Y-Y, Yan R-H, Ye G-Y, Huang F, Wangila DS, Wang J-J, Cheng J-A (2012) Field response of aboveground non-target arthropod community to transgenic Bt-Cry1Ab rice plant residues in postharvest seasons. Transgenic Res 21:1023–1032. https://doi.org/10.1007/s11248-012-9590-6

Bakonyi G, Szira F, Kiss I, Villányi I, Seres A, Székacs A (2006) Preference tests with collembolas on isogenic and Bt-maize. Eur J Soil Biol 42:S132–S135. https://doi.org/10.1016/j.ejsobi.2006.06.005

Balbuena MS, Tison L, Hahn M-L, Greggers U, Menzel R, Farina WM (2015) Effects of sublethal doses of glyphosate on honeybee navigation. J Exp Biol 218:2799–2805. https://doi.org/10.1242/jeb.117291

Balog A, Kiss J, Szekeres D, Szénási Á, Markó V (2010) Rove beetle (Coleoptera: Staphylinidae) communities in transgenic Bt (MON810) and near isogenic maize. Crop Prot 29:567–571. https://doi.org/10.1016/j.cropro.2009.12.020

Baum JA, Bogaert T, Clinton W, Heck GR, Feldmann P, Ilagan O, Johnson S, Plaetinck G, Munyikwa T, Pleau M, Vaughn T, Roberts J (2007) Control of coleopteran insect pests through RNA interference. Nat Biotechnol 25:1322–1326. https://doi.org/10.1038/nbt1359

Baur ME, Boethel DJ (2003) Effect of Bt-cotton expressing Cry1A(c) on the survival and fecundity of two hymenopteran parasitoids (Braconidae, Encyrtidae) in the laboratory. Biol Control 26:325–332. https://doi.org/10.1016/S1049-9644(02)00160-3

Beale MH, Birkett MA, Bruce TJA, Chamberlain K, Field LM, Huttly AK, Martin JL, Parker R, Phillips AL, Pickett JA, Prosser IM, Shewry PR, Smart LE, Wadhams LJ, Woodcock CM, Zhang Y (2006) Aphid alarm pheromone produced by transgenic plants affects aphid and parasitoid behavior. Proc Natl Acad Sci U S A 103:10509–10513. https://doi.org/10.1073/pnas.0603998103

Bell HA, Fitches EC, Down RE, Marris GC, Edwards JP, Gatehouse JA, Gatehouse AMR (1999) The effect of snowdrop lectin (GNA) delivered via artificial diet and transgenic plants on Eulophus pennicornis (Hymenoptera: Eulophidae), a parasitoid of the tomato moth Lacanobia oleracea (Lepidoptera: Noctuidae). J Insect Physiol 45:983–991. https://doi.org/10.1016/S0022-1910(99)00077-3

Bell HA, Fitches EC, Marris GC, Bell J, Edwards JP, Gatehouse JA, Gatehouse AM (2001) Transgenic GNA expressing potato plants augment the beneficial biocontrol of Lacanobia oleracea (Lepidoptera; Noctuidae) by the parasitoid Eulophus pennicornis (Hymenoptera; Eulophidae). Transgenic Res 10:35–42

Bell HA, Down RE, Fitches EC, Edwards JP, Gatehouse AMR (2003) Impact of genetically modified potato expressing plant-derived insect resistance genes on the predatory bug Podisus maculiventris (Heteroptera: Pentatomidae). Biocontrol Sci Technol 13:729–741. https://doi.org/10.1080/09583150310001606543

Bergé JB, Ricroch AE (2010) Emergence of minor pests becoming major pests in GE cotton in China: what are the reasons? What are the alternatives practices to this change of status? GM Crops 1:214–219. https://doi.org/10.4161/gmcr.1.4.13421

Bernal CC, Aguda RM, Cohen MB (2002a) Effect of rice lines transformed with Bacillus thuringiensis toxin genes on the brown planthopper and its predator Cyrtorhinus lividipennis. Entomol Exp Appl 102:21–28. https://doi.org/10.1046/j.1570-7458.2002.00921.x

Bernal J, Griset J, Gillogly (2002b) Impacts of developing on Bt maize-intoxicated hosts on fitness parameters of a stem borer parasitoid. J Entomol Sci 37:27–40

Bezemer TM, Wagenaar R, van Dam NM, van Der Putten WH, Wäckers FL (2004) Above- and below-ground terpenoid aldehyde induction in cotton, Gossypium herbaceum, following root and leaf injury. J Chem Ecol 30:53–67. https://doi.org/10.1023/B:JOEC.0000013182.50662.2a

Biondi A, Mommaerts V, Smagghe G, Viñuela E, Zappalà L, Desneux N (2012) The non-target impact of spinosyns on beneficial arthropods. Pest Manag Sci 68:1523–1536. https://doi.org/10.1002/ps.3396

Bitzer RJ, Buckelew LD, Pedigo LP (2002) Effects of transgenic herbicide-resistant soybean varieties and systems on surface-active springtails (Entognatha: Collembola). Environ Entomol 31:449–461. https://doi.org/10.1603/0046-225X-31.3.449

Bortolotto OC, Silva GV, de Freitas BA, Pomari AF, Martinelli S, Head GP, Carvalho RA, Barbosa GC (2014) Development and reproduction of Spodoptera eridania (Lepidoptera: Noctuidae) and its egg parasitoid Telenomus remus (Hymenoptera: Platygastridae) on the genetically modified soybean (Bt) MON 87701×MON 89788. Bull Entomol Res 104:724–730. https://doi.org/10.1017/S0007485314000546

Bourguet D, Chaufaux J, Micoud A, Delos M, Naibo B, Bombarde F, Marque G, Eychenne N, Pagliari C (2002) Ostrinia nubilalis parasitism and the field abundance of non-target insects in transgenic Bacillus thuringiensis corn (Zea mays). Environ Biosafety Res 1:49–60

Brimner TA, Boland GJ (2003) A review of the non-target effects of fungi used to biologically control plant diseases. Agric Ecosyst Environ 100:3–16. https://doi.org/10.1016/S0167-8809(03)00200-7

Brodsgaard HF, Brodsgaard CJ, Hansen H, Lövei GL (2003) Environmental risk assessment of transgene products using honey bee (Apis mellifera) larvae. Apidologie 34:139–145. https://doi.org/10.1051/apido:2003003

Brooks DR, Bohan DA, Champion GT, Haughton AJ, Hawes C, Heard MS, Clark SJ, Dewar AM, Firbank LG, Perry JN, Rothery P, Scott RJ, Woiwod IP, Birchall C, Skellern MP, Walker JH, Baker P, Bell D, Browne EL, Dewar AJG, Fairfax CM, Garner BH, Haylock LA, Horne SL, Hulmes SE, Mason NS, Norton LR, Nuttall P, Randle Z, Rossall MJ, Sands RJN, Singer EJ, Walker MJ (2003) Invertebrate responses to the management of genetically modified herbicide–tolerant and conventional spring crops. I. Soil-surface-active invertebrates. Philos Trans R Soc B Biol Sci 358:1847–1862. https://doi.org/10.1098/rstb.2003.1407

Burgio G, Lanzoni A, Accinelli G, Dinelli G, Bonetti A, Marotti I, Ramilli F (2007) Evaluation of Bt-toxin uptake by the non-target herbivore, Myzus persicae (Hemiptera: Aphididae), feeding on transgenic oilseed rape. Bull Entomol Res 97:211–215. https://doi.org/10.1017/S0007485307004920

Candolfi MP, Brown K, Grimm C, Reber B, Schmidli H (2004) A faunistic approach to assess potential side-effects of genetically modified Bt-Corn on non-target arthropods under field conditions. Biocontrol Sci Technol 14:129–170. https://doi.org/10.1080/0958315031000165570l

Cardinale BJ, Harvey CT, Gross K, Ives AR (2003) Biodiversity and biocontrol: emergent impacts of a multi-enemy assemblage on pest suppression and crop yield in an agroecosystem. Ecol Lett 6:857–865. https://doi.org/10.1046/j.1461-0248.2003.00508.x

Chang L, Liu X, Ge F (2011) Effect of elevated O3 associated with Bt cotton on the abundance, diversity and community structure of soil Collembola. Appl Soil Ecol 47:45–50. https://doi.org/10.1016/j.apsoil.2010.10.013

Chen M, Ye G, Lu X, Hu C, Peng Y, Shu Q, Altosaar I (2005) Biotransfer and bioaccumulation of Cry1Ab insecticidal protein in rice plant-brown planthopper-wolf spider food chain. Acta Entomol Sin 48:208–213

Chen M, Ye GY, Liu ZC, Yao HW, Chen XX, Shen ZC, Hu C, Datta SK (2006) Field Assessment of the Effects of Transgenic Rice Expressing a Fused Gene of cry1Ab and cry1Ac from Bacillus thuringiensis Berliner on Nontarget Planthopper and Leafhopper Populations. Environ Entomol 35:127–134. https://doi.org/10.1603/0046-225X-35.1.127

Chen M, Liu Z, Ye G, Shen Z, Hu C, Peng Y, Altosaar I, Shelton AM (2007) Impacts of transgenic cry1Ab rice on non-target planthoppers and their main predator Cyrtorhinus lividipennis

(Hemiptera: Miridae)—A case study of the compatibility of Bt rice with biological control. Biol Control 42:242–250. https://doi.org/10.1016/j.biocontrol.2007.05.005

Chen M, Ye G, Liu Z, Fang Q, Hu C, Peng Y, Shelton AM (2009) Analysis of Cry1Ab toxin bioaccumulation in a food chain of Bt rice, an herbivore and a predator. Ecotoxicol Lond Engl 18:230–238. https://doi.org/10.1007/s10646-008-0276-z

Chen Y, Tian J-C, Wang W, Fang Q, Akhtar ZR, Peng Y-F, Cui H, Guo Y-Y, Song Q-S, Ye G-Y (2012) Bt rice expressing Cry1Ab does not stimulate an outbreak of its non-target herbivore, Nilaparvata lugens. Transgenic Res 21:279–291. https://doi.org/10.1007/s11248-011-9530-x

Chen Y, Lai F, Sun Y, Hong L, Tian J, Zhang Z, Fu Q (2015) CrylAb rice does not impact biological characters and functional response of Cyrtorhinus lividipennis preying on Nilaparvata lugens eggs. J Integr Agric 14:2011–2018. https://doi.org/10.1016/S2095-3119(14)60978-3

Cooper AM, Silver K, Zhang J, Park Y, Zhu KY (2019) Molecular mechanisms influencing efficiency of RNA interference in insects. Pest Manag Sci 75:18–28. https://doi.org/10.1002/ps.5126

Couty A, Poppy GM (2001) Does host-feeding on GNA-intoxicated aphids by Aphelinus abdominalis affect their longevity and/or fecundity? Entomol Exp Appl 100:331–337. https://doi.org/10.1046/j.1570-7458.2001.00880.x

Couty A, de la Viña G, Clark SJ, Kaiser L, Pham-Delègue M-H, Poppy GM (2001) Direct and indirect sublethal effects of Galanthus nivalis agglutinin (GNA) on the development of a potato-aphid parasitoid, Aphelinus abdominalis (Hymenoptera: Aphelinidae). J Insect Physiol 47:553–561. https://doi.org/10.1016/S0022-1910(00)00148-7

Cowgill SE, Atkinson HJ (2003) A sequential approach to risk assessment of transgenic plants expressing protease inhibitors: effects on nontarget herbivorous insects. Transgenic Res 12:439–449. https://doi.org/10.1023/A:1024215922148

Cowgill SE, Wright C, Atkinson HJ (2002) Transgenic potatoes with enhanced levels of nematode resistance do not have altered susceptibility to nontarget aphids. Mol Ecol 11:821–827

Crall JD, Switzer CM, Oppenheimer RL, Versypt ANF, Dey B, Brown A, Eyster M, Guérin C, Pierce NE, Combes SA, de Bivort BL (2018) Neonicotinoid exposure disrupts bumblebee nest behavior, social networks, and thermoregulation. Science 362:683–686. https://doi.org/10.1126/science.aat1598

Dai P-L, Zhou W, Zhang J, Cui H-J, Wang Q, Jiang W-Y, Sun J-H, Wu Y-Y, Zhou T (2012) Field assessment of Bt cry1Ah corn pollen on the survival, development and behavior of Apis mellifera ligustica. Ecotoxicol Environ Saf 79:232–237. https://doi.org/10.1016/j.ecoenv.2012.01.005

Daly T, Buntin GD (2005) Effect of Bacillus thuringiensis transgenic corn for lepidopteran control on nontarget arthropods. Environ Entomol 34:1292–1302. https://doi.org/10.1603/0046-225X(2005)034[1292:EOBTTC]2.0.CO;2

Darvas B, Csoti A, Adel G, Peregovits L, Ronkay L, Lauber E, Polgar AL (2004) Some data on the risk analysis of BT-corn pollen and protected Lepidoptera species in Hungary. Novenyvedelem Hung 40:441–449

de Castro TR, Ausique JJS, Nunes DH, Ibanhes FH, Delalibera Júnior I (2013) Risk assessment of Cry toxins of Bacillus thuringiensis on the predatory mites Euseius concordis and Neoseiulus californicus (Acari: Phytoseiidae). Exp Appl Acarol 59:421–433. https://doi.org/10.1007/s10493-012-9620-3

Desneux N, Decourtye A, Delpuech J-M (2007) The sublethal effects of pesticides on beneficial arthropods. Annu Rev Entomol 52:81–106. https://doi.org/10.1146/annurev.ento.52.110405.091440

Desneux N, Ramírez-Romero R, Bokonon-Ganta AH, Bernal JS (2010) Attraction of the parasitoid Cotesia marginiventris to host (Spodoptera frugiperda) frass is affected by transgenic maize. Ecotoxicology 19:1183–1192. https://doi.org/10.1007/s10646-010-0502-3

Deutsch CA, Tewksbury JJ, Tigchelaar M, Battisti DS, Merrill SC, Huey RB, Naylor RL (2018) Increase in crop losses to insect pests in a warming climate. Science 361:916–919. https://doi.org/10.1126/science.aat3466

Devos Y, De Schrijver A, De Clercq P, Kiss J, Romeis J (2012) Bt-maize event MON 88017 expressing Cry3Bb1 does not cause harm to non-target organisms. Transgenic Res 21:1191–1214. https://doi.org/10.1007/s11248-012-9617-z

Dhillon MK, Sharma HC (2009) Effects of Bacillus thuringiensis δ-endotoxins Cry1Ab and Cry1Ac on the coccinellid beetle, Cheilomenes sexmaculatus (Coleoptera, Coccinellidae) under direct and indirect exposure conditions. Biocontrol Sci Technol 19:407–420. https://doi.org/10.1080/09583150902783801

Dicke M, Sabelis MW (1987) How plants obtain predatory mites as bodyguards. Neth J Zool 38:148–165. https://doi.org/10.1163/156854288X00111

Donegan KK, Seidler RJ, Fieland VJ, Schaller DL, Palm CJ, Ganio LM, Cardwell DM, Steinberger Y, Dynamac C (1997) Decomposition of genetically engineered tobacco under field conditions: persistence of the proteinase inhibitor I product and effects on soil microbial respiration and protozoa, nematode and microarthropod populations. J Appl Ecol UK

Down RE, Ford L, Woodhouse SD, Davison GM, Majerus MEN, Gatehouse JA, Gatehouse AMR (2003) Tritrophic interactions between transgenic potato expressing snowdrop lectin (GNA), an aphid pest (peach–potato aphid; Myzus persicae (Sulz.) and a beneficial predator (2-spot ladybird; Adalia bipunctata L.). Transgenic Res 12:229–241. https://doi.org/10.1023/A:1022904805028

Duan JJ, Head G, McKee MJ, Nickson TE, Martin JW, Sayegh FS (2002) Evaluation of dietary effects of transgenic corn pollen expressing Cry3Bb1 protein on a non-target ladybird beetle, Coleomegilla maculata. Entomol Exp Appl 104:271–280. https://doi.org/10.1046/j.1570-7458.2002.01013.x

Duan JJ, Marvier M, Huesing J, Dively G, Huang ZY (2008) A meta-analysis of effects of Bt crops on honey bees (Hymenoptera: Apidae). PLOS ONE 3:e1415. https://doi.org/10.1371/journal.pone.0001415

Dubelman S, Fischer J, Zapata F, Huizinga K, Jiang C, Uffman J, Levine S, Carson D (2014) Environmental fate of double-stranded RNA in agricultural soils. PLOS ONE 9:e93155. https://doi.org/10.1371/journal.pone.0093155

Dutton A, Klein H, Romeis J, Bigler F (2002) Uptake of Bt-toxin by herbivores feeding on transgenic maize and consequences for the predator Chrysoperla carnea. Ecol Entomol 27:441–447. https://doi.org/10.1046/j.1365-2311.2002.00436.x

Dutton A, Romeis J, Bigler F (2003) Assessing the risks of insect resistant transgenic plants on entomophagous arthropods Bt-maize expressing Cry1Ab as a case study. BioControl 48:611–636. https://doi.org/10.1023/A:1026313719424

Eliyahu D, McCall AC, Lauck M, Trakhtenbrot A, Bronstein JL (2015) Minute pollinators: the role of thrips (Thysanoptera) as pollinators of pointleaf manzanita, Arctostaphylos pungens (Ericaceae). J Pollinat Ecol 16:64–71

European Food Safety Authority (2015) Updating risk management recommendations to limit exposure of non-target Lepidoptera of conservation concern in protected habitats to Bt-maize pollen. EFSA J 13:4127. https://doi.org/10.2903/j.efsa.2015.4127

Felke M, Langenbruch G-A, Feiertag S, Kassa A (2010) Effect of Bt-176 maize pollen on first instar larvae of the Peacock butterfly (Inachis io) (Lepidoptera; Nymphalidae). Environ Biosafety Res 9:5–12. https://doi.org/10.1051/ebr/2010006

Fernandes FS, Ramalho FS, Nascimento JL, Malaquias JB, Nascimento ARB, Silva CAD, Zanuncio JC (2012) Within-plant distribution of cotton aphids, Aphis gossypii Glover (Hemiptera: Aphididae), in Bt and non-Bt cotton fields. Bull Entomol Res. https://doi.org/10.1017/S0007485311000381

Ferry N, Mulligan EA, Stewart CN, Tabashnik BE, Port GR, Gatehouse AMR (2006) Prey-mediated effects of transgenic canola on a beneficial, non-target, carabid beetle. Transgenic Res 15:501–514. https://doi.org/10.1007/s11248-006-0021-4

Flexner JL, Lighthart B, Croft BA (1986) The effects of microbial pesticides on non-target, beneficial arthropods. Agric Ecosyst Environ 16:203–254. https://doi.org/10.1016/0167-8809(86)90005-8

Food and Agriculture Organization of the United Nations (2015) FAO – News Article: Keeping plant pests and diseases at bay: experts focus on global measures. http://www.fao.org/news/story/en/item/280489/icode/. Accessed 1 Nov 2018

Fountain MT, Hopkin SP (2005) Folsomia candida (Collembola): a "standard" soil arthropod. Annu Rev Entomol 50:201–222. https://doi.org/10.1146/annurev.ento.50.071803.130331

Gao M-Q, Hou S-P, Pu D-Q, Shi M, Ye G-Y, Chen X-X (2010) Multi-generation effects of Bt rice on Anagrus nilaparvatae, a parasitoid of the nontarget pest Nilapavarta lugens. Environ Entomol 39:2039–2044. https://doi.org/10.1603/EN10035

García M, Ortego F, Castañera P, Farinós GP (2012) Assessment of prey-mediated effects of the coleopteran-specific toxin Cry3Bb1 on the generalist predator Atheta coriaria (Coleoptera: Staphylinidae). Bull Entomol Res 102:293–302. https://doi.org/10.1017/S0007485311000666

Goulson D, Nicholls E, Botías C, Rotheray EL (2015) Bee declines driven by combined stress from parasites, pesticides, and lack of flowers. Science 347:1255957. https://doi.org/10.1126/science.1255957

Gowda A, Rydel TJ, Wollacott AM, Brown RS, Akbar W, Clark TL, Flasinski S, Nageotte JR, Read AC, Shi X, Werner BJ, Pleau MJ, Baum JA (2016) A transgenic approach for controlling Lygus in cotton. Nat Commun 7:12213. https://doi.org/10.1038/ncomms12213

Gray S, Cilia M, Ghanim M (2014) Circulative, "nonpropagative" virus transmission: an orchestra of virus-, insect-, and plant-derived instruments. Adv Virus Res 89:141–199. https://doi.org/10.1016/B978-0-12-800172-1.00004-5

Green TR, Ryan CA (1972) Wound-induced proteinase inhibitor in plant leaves: a possible defense mechanism against insects. Science 175:776–777. https://doi.org/10.1126/science.175.4023.776

Gregorc A, Ellis JD (2011) Cell death localization in situ in laboratory reared honey bee (Apis mellifera L.) larvae treated with pesticides. Pestic Biochem Physiol 99:200–207. https://doi.org/10.1016/j.pestbp.2010.12.005

Griffiths BS, Caul S, Thompson J, Birch ANE, Scrimgeour C, Cortet J, Foggo A, Hackett CA, Krogh PH (2006) Soil microbial and faunal community responses to bt maize and insecticide in two soils. J Environ Qual 35:734–741. https://doi.org/10.2134/jeq2005.0344

Groot AT, Dicke M (2002) Insect-resistant transgenic plants in a multi-trophic context. Plant J 31:387–406. https://doi.org/10.1046/j.1365-313X.2002.01366.x

Gruber H, Paul V, Meyer HHD, Müller M (2012) Determination of insecticidal Cry1Ab protein in soil collected in the final growing seasons of a nine-year field trial of Bt-maize MON810. Transgenic Res 21:77–88. https://doi.org/10.1007/s11248-011-9509-7

Guo Y, Feng Y, Ge Y, Tetreau G, Chen X, Dong X, Shi W (2014) The cultivation of Bt corn producing Cry1Ac toxins does not adversely affect non-target arthropods. PLOS ONE 9:e114228. https://doi.org/10.1371/journal.pone.0114228

Gutiérrez-Campos R, Torres-Acosta JA, de Jesús Pérez-Martínez J, Gómez-Lim MA (2001) Pleiotropic effects in transgenic tobacco plants expressing the oryzacystatin I gene. HortScience 36:118–119

Hagenbucher S, Wäckers FL, Wettstein FE, Olson DM, Ruberson JR, Jörg R (2013) Pest trade-offs in technology: reduced damage by caterpillars in Bt cotton benefits aphids. Proc R Soc B Biol Sci 280:20130042. https://doi.org/10.1098/rspb.2013.0042

Hagenbucher S, Wäckers FL, Romeis J (2014a) Aphid honeydew quality as a food source for parasitoids is maintained in Bt cotton. PLOS ONE 9:e107806. https://doi.org/10.1371/journal.pone.0107806

Hagenbucher S, Wäckers FL, Romeis J (2014b) Indirect multi-trophic interactions mediated by induced plant resistance: impact of caterpillar feeding on aphid parasitoids. Biol Lett:10. https://doi.org/10.1098/rsbl.2013.0795

Haller S, Widmer F, Siegfried BD, Zhuo X, Romeis J (2019) Responses of two ladybird beetle species (Coleoptera: Coccinellidae) to dietary RNAi. Pest Manag Sci 0. https://doi.org/10.1002/ps.5370

Han P, Niu C-Y, Lei C-L, Cui J-J, Desneux N (2010) Use of an innovative T-tube maze assay and the proboscis extension response assay to assess sublethal effects of GM products and pesticides on learning capacity of the honey bee Apis mellifera L. Ecotoxicol Lond Engl 19:1612–1619. https://doi.org/10.1007/s10646-010-0546-4

Han Y, Meng J, Chen J, Cai W, Wang Y, Zhao J, He Y, Feng Y, Hua H (2014) Bt rice expressing Cry2Aa does not harm Cyrtorhinus lividipennis, a main predator of the nontarget herbivore Nilapavarta lugens. PLoS ONE 9. https://doi.org/10.1371/journal.pone.0112315

Hanley AV, Huang ZY, Pett WL (2003) Effects of dietary transgenic Bt corn pollen on larvae of Apis mellifera and Galleria mellonella. J Apic Res 42:77–81. https://doi.org/10.1080/00218839.2003.11101097

Hansen LS, Lövei GL, Székács A (2012) Survival and development of a stored-product pest, Sitophilus zeamais (Coleoptera: Curculionidae), and its natural enemy, the parasitoid Lariophagus distinguendus (Hymenoptera: Pteromalidae), on transgenic Bt maize. Pest Manag Sci 69:602–606. https://doi.org/10.1002/ps.3410

Harrison RL, Bonning BC (2010) Proteases as Insecticidal Agents. Toxins 2:935–953. https://doi.org/10.3390/toxins2050935

Head G, Brown CR, Groth ME, Duan JJ (2001) Cry1Ab protein levels in phytophagous insects feeding on transgenic corn: implications for secondary exposure risk assessment. Entomol Exp Appl 99:37–45. https://doi.org/10.1046/j.1570-7458.2001.00799.x

Hellmich RL, Siegfried BD, Sears MK, Stanley-Horn DE, Daniels MJ, Mattila HR, Spencer T, Bidne KG, Lewis LC (2001) Monarch larvae sensitivity to Bacillus thuringiensis- purified proteins and pollen. Proc Natl Acad Sci 98:11925–11930. https://doi.org/10.1073/pnas.211297698

Hellmich RL, Albajes R, Bergvinson D, Prasifka JR, Wang Z-Y, Weiss MJ (2008) The present and future role of insect-resistant genetically modified maize in IPM. In: Romeis J, Shelton AM, Kennedy GG (eds) Integration of insect-resistant genetically modified crops within IPM programs. Springer, Dordrecht, pp 119–158

Hendriksma HP, Härtel S, Steffan-Dewenter I (2011) Testing pollen of single and stacked insect-resistant Bt-maize on in vitro reared honey bee larvae. PloS One 6:e28174. https://doi.org/10.1371/journal.pone.0028174

Hendriksma HP, Härtel S, Babendreier D, von der Ohe W, Steffan-Dewenter I (2012) Effects of multiple Bt proteins and GNA lectin on in vitro-reared honey bee larvae. Apidologie 43:549–560. https://doi.org/10.1007/s13592-012-0123-3

Herbert LT, Vázquez DE, Arenas A, Farina WM (2014) Effects of field-realistic doses of glyphosate on honeybee appetitive behaviour. J Exp Biol 217:3457–3464. https://doi.org/10.1242/jeb.109520

Hilbeck A, Baumgartner M, Fried PM, Bigler F (1998a) Effects of transgenic Bacillus thuringiensis corn-fed prey on mortality and development time of immature Chrysoperla cornea (Neuroptera: Chrysopidae). Environ Entomol 27:480–487. https://doi.org/10.1093/ee/27.2.480

Hilbeck A, Moar WJ, Pusztai-Carey M, Filippini A, Bigler F (1998b) Toxicity of Bacillus thuringiensis Cry1Ab toxin to the predator Chrysoperla carnea (Neuroptera: Chrysopidae). Environ Entomol 27:1255–1263

Hopkins DW, Gregorich EG (2003) Detection and decay of the Bt endotoxin in soil from a field trial with genetically modified maize. Eur J Soil Sci 54:793–800. https://doi.org/10.1046/j.1351-0754.2003.0563.x

Howald R, Zwahlen C, Nentwig W (2003) Evaluation of Bt oilseed rape on the non-target herbivore Athalia rosae. Entomol Exp Appl 106:87–93. https://doi.org/10.1046/j.1570-7458.2003.00012.x

Huang ZY, Hanley AV, Pett WL, Langenberger M, Duan JJ (2004) Field and semifield evaluation of impacts of transgenic canola pollen on survival and development of worker honey bees. J Econ Entomol 97:1517–1523

ISAAA (2017) Global status of commercialized biotech/GM crops in 2017: biotech crop adoption surges as economic benefits accumulate in 22 years, ISAAA Brief No. 53. ISAAA, Ithaca

Jarosch A, Moritz RFA (2012) RNA interference in honeybees: off-target effects caused by dsRNA. Apidologie 43:128–138. https://doi.org/10.1007/s13592-011-0092-y

Jasinski JR, Eisley JB, Young CE, Kovach J, Willson H (2003) Select nontarget arthropod abundance in transgenic and nontransgenic Field crops in Ohio. Environ Entomol 32:407–414. https://doi.org/10.1603/0046-225X-32.2.407

Jesse LCH, Obrycki JJ (2002) Assessment of the non-target effects of transgenic Bt corn pollen and anthers on the milkweed tiger moth, Euchatias egle Drury (Lepidoptera: Arctiidae). J Kans Entomol Soc 75:55–58

Jia H-R, Dai P-L, Geng L-L, Jack CJ, Li Y-H, Wu Y-Y, Diao Q-Y, Ellis JD (2017) No effect of Bt Cry1Ie toxin on bacterial diversity in the midgut of the Chinese honey bees, Apis cerana cerana (Hymenoptera, Apidae). Sci Rep 7:41688. https://doi.org/10.1038/srep41688

Jurat-Fuentes JL, Gould FL, Adang MJ (2003) Dual resistance to Bacillus thuringiensis Cry1Ac and Cry2Aa toxins in Heliothis virescens suggests multiple mechanisms of resistance. Appl Env Microbiol 69:5898–5906. https://doi.org/10.1128/AEM.69.10.5898-5906.2003

Koch MS, Ward JM, Levine SL, Baum JA, Vicini JL, Hammond BG (2015) The food and environmental safety of Bt crops. Front Plant Sci 6. https://doi.org/10.3389/fpls.2015.00283

Konrad R, Ferry N, Gatehouse AMR, Babendreier D (2008) Potential effects of oilseed rape expressing oryzacystatin-1 (OC-1) and of purified insecticidal proteins on larvae of the solitary bee Osmia bicornis. PloS One 3:e2664. https://doi.org/10.1371/journal.pone.0002664

Lawo NC, Wäckers FL, Romeis J (2009) Indian Bt cotton varieties do not affect the performance of cotton aphids. PLOS ONE 4:e4804. https://doi.org/10.1371/journal.pone.0004804

Lawo NC, Wäckers FL, Romeis J (2010) Characterizing indirect prey-quality mediated effects of a Bt crop on predatory larvae of the green lacewing, Chrysoperla carnea. J Insect Physiol 56:1702–1710. https://doi.org/10.1016/j.jinsphys.2010.06.012

Li Y, Romeis J (2010) Bt maize expressing Cry3Bb1 does not harm the spider mite, Tetranychus urticae, or its ladybird beetle predator, Stethorus punctillum. Biol Control 53:337–344. https://doi.org/10.1016/j.biocontrol.2009.12.003

Li Y, Meissle M, Romeis J (2008) Consumption of Bt maize pollen expressing Cry1Ab or Cry3Bb1 does not harm adult green Lacewings, Chrysoperla carnea (Neuroptera: Chrysopidae). PloS One 3:e2909. https://doi.org/10.1371/journal.pone.0002909

Li G, Feng H, Chen P, Wu S, Liu B, Qiu F (2010) Effects of transgenic Bt cotton on the population density, oviposition behavior, development, and reproduction of a nontarget pest, Adelphocoris suturalis (Hemiptera: Miridae). Environ Entomol 39:1378–1387. https://doi.org/10.1603/EN09223

Li Y, Romeis J, Wang P, Peng Y, Shelton AM (2011) A comprehensive assessment of the effects of Bt cotton on Coleomegilla maculata demonstrates no detrimental effects by Cry1Ac and Cry2Ab. PLOS ONE 6:e22185. https://doi.org/10.1371/journal.pone.0022185

Lima MAP, Pires CSS, Guedes RNC, Campos LAO (2013) Lack of lethal and sublethal effects of Cry1Ac Bt-toxin on larvae of the stingless bee Trigona spinipes. Apidologie 44:21–28. https://doi.org/10.1007/s13592-012-0151-z

Lima M, Martins G, Oliveira E, Guedes R (2016) Agrochemical-induced stress in stingless bees: peculiarities, underlying basis, and challenges. J Comp Physiol Neuroethol Sens Neural Behav Physiol 202:733–747. 2016 Oct2029-10733-47. Epub 2016 Jul 11. https://doi.org/10.1007/s00359-016-1110-3

Liu XD, Zhai BP, Zhang XX, Zong JM (2005) Impact of transgenic cotton plants on a non-target pest, Aphis gossypii Glover. Ecol Entomol 30:307–315. https://doi.org/10.1111/j.0307-6946.2005.00690.x

Liu X, Chen M, Onstad D, Roush R, Shelton AM (2011) Effect of Bt broccoli and resistant genotype of Plutella xylostella (Lepidoptera: Plutellidae) on development and host acceptance of the parasitoid Diadegma insulare (Hymenoptera: Ichneumonidae). Transgenic Res 20:887–897. https://doi.org/10.1007/s11248-010-9471-9

Liu Y, Wang Y, Shu C, Lin K, Song F, Bravo A, Soberón M, Zhang J (2018) Cry64Ba and Cry64Ca, two ETX/MTX2-type Bacillus thuringiensis insecticidal proteins active against hemipteran pests. Appl Environ Microbiol 84. https://doi.org/10.1128/AEM.01996-17

Lorenc-Kukuła K, Amarowicz R, Oszmiański J, Doermann P, Starzycki M, Skała J, Zuk M, Kulma A, Szopa J (2005) Pleiotropic effect of phenolic compounds content increases in transgenic flax plant. J Agric Food Chem 53:3685–3692. https://doi.org/10.1021/jf047987z

Losey JE, Rayor LS, Carter ME (1999) Transgenic pollen harms monarch larvae. Nature 399:214. https://doi.org/10.1038/20338

Lövei GL, Andow DA, Arpaia S (2009) Transgenic insecticidal crops and natural enemies: a detailed review of laboratory studies. Environ Entomol 38:293–306

Lu YH, Qiu F, Feng HQ, Li HB, Yang ZC, Wyckhuys KAG, Wu KM (2008) Species composition and seasonal abundance of pestiferous plant bugs (Hemiptera: Miridae) on Bt Cotton in China. Crop Prot 27:465–472. https://doi.org/10.1016/j.cropro.2007.07.017

Lu Y, Wu K, Jiang Y, Xia B, Li P, Feng H, Wyckhuys KAG, Guo Y (2010) Mirid bug outbreaks in multiple crops correlated with wide-scale adoption of Bt cotton in China. Science 328:1151–1154. https://doi.org/10.1126/science.1187881

Lu Y, Wu K, Jiang Y, Guo Y, Desneux N (2012) Widespread adoption of Bt cotton and insecticide decrease promotes biocontrol services. Nature 487:362. https://doi.org/10.1038/nature11153

Lumbierres B, Albajes R, Pons X (2004) Transgenic Bt maize and Rhopalosiphum padi (Hom., Aphididae) performance. Ecol Entomol 29:309–317. https://doi.org/10.1111/j.0307-6946.2004.00597.x

Lundgren JG, Wiedenmann RN (2002) Coleopteran-specific Cry3Bb toxin from transgenic corn pollen does not affect the fitness of a nontarget species, Coleomegilla maculata DeGeer (Coleoptera: Coccinellidae). Environ Entomol 31:1213–1218. https://doi.org/10.1603/0046-225X-31.6.1213

Malone LA, Pham-Delègue M-H (2001) Effects of transgene products on honey bees (Apis mellifera) and bumblebees (Bombus sp.). Apidologie 32:287–304. https://doi.org/10.1051/apido:2001130

Malone LA, Burgess EPJ, Stefanovic D (1999) Effects of a Bacillus thuringiensis toxin, two Bacillus thuringiensis biopesticide formulations, and a soybean trypsin inhibitor on honey bee (Apis mellifera L.) survival and food consumption. Apidologie 30:465–473. https://doi.org/10.1051/apido:19990601

Malone LA, Burgess EPJ, Gatehouse HS, Voisey CR, Tregidga EL, Philip BA (2001) Effects of ingestion of a Bacillus thuringiensis toxin and a trypsin inhibitor on honey bee flight activity and longevity. Apidologie 32:57–68. https://doi.org/10.1051/apido:2001111

Malone LA, Todd JH, Burgess EPJ, Christeller JT (2004) Development of hypopharyngeal glands in adult honey bees fed with a Bt toxin, a biotin-binding protein and a protease inhibitor. Apidologie 35:655–664. https://doi.org/10.1051/apido:2004063

McPherson RM, Johnson WC, Mullinix BG, Mills WA, Peebles FS (2003) Influence of herbicide tolerant soybean production systems on insect pest populations and pest-induced crop damage. J Econ Entomol 96:690–698

Meissle M, Romeis J (2012) No accumulation of Bt protein in Phylloneta impressa (Araneae: Theridiidae) and prey arthropods in Bt maize. Environ Entomol 41:1037–1042. https://doi.org/10.1603/EN11321

Men X, Ge F, Edwards CA, Yardim EN (2004) Influence of pesticide applications on pest and predatory arthropods associated with transgenicBt cotton and nontransgenic cotton plants. Phytoparasitica 32:246–254. https://doi.org/10.1007/BF02979819

Milfont MO, Rocha EEM, Lima AON, Freitas BM (2013) Higher soybean production using honeybee and wild pollinators, a sustainable alternative to pesticides and autopollination. Environ Chem Lett 11:335–341. https://doi.org/10.1007/s10311-013-0412-8

Musser FR, Shelton AM (2003) Bt sweet corn and selective insecticides: impacts on pests and predators. J Econ Entomol 96:71–80. https://doi.org/10.1603/0022-0493-96.1.71

Nakasu EYT, Dias SC, Pires CSS, Andow DA, Paula DP, Togni PHB, Macedo TR, Sujii ER, de Sá MFG, Fontes EMG (2013) Bitrophic toxicity of Cry1Ac to Cycloneda sanguinea, a predator in Brazilian cotton. Entomol Exp Appl 148:105–115. https://doi.org/10.1111/eea.12082

Naranjo SE (2005) Long-term assessment of the effects of transgenic Bt cotton on the abundance of nontarget arthropod natural enemies. Environ Entomol 34:1193–1210. https://doi.org/10.1093/ee/34.5.1193

Naranjo SE (2011) Impacts of Bt transgenic cotton on integrated pest management. J Agric Food Chem 59:5842–5851. https://doi.org/10.1021/jf102939c

Ndolo DO, Songa JM, Lövei GL (2018) Tri-trophic impacts of Bt-transgenic maize on parasitoid size and fluctuating asymmetry in native vs. novel host-parasitoid interactions in East Africa. Insects 9. https://doi.org/10.3390/insects9020038

Niu L, Ma Y, Mannakkara A, Zhao Y, Ma W, Lei C, Chen L (2013) Impact of single and stacked insect-resistant Bt-cotton on the honey bee and silkworm. PloS One 8:e72988. https://doi.org/10.1371/journal.pone.0072988

Nunes FMF, Aleixo AC, Barchuk AR, Bomtorin AD, Grozinger CM, Simões ZLP (2013) Non-target effects of Green Fluorescent Protein (GFP)-Derived Double-Stranded RNA (dsRNA-GFP) used in honey bee RNA interference (RNAi) assays. Insects 4:90–103. https://doi.org/10.3390/insects4010090

Oberhauser KS, Prysby MD, Mattila HR, Stanley-Horn DE, Sears MK, Dively G, Olson E, Pleasants JM, Lam W-KF, Hellmich RL (2001) Temporal and spatial overlap between monarch larvae and corn pollen. Proc Natl Acad Sci 98:11913–11918. https://doi.org/10.1073/pnas.211234298

Oerke E-C (2006) Crop losses to pests. J Agric Sci 144:31–43. https://doi.org/10.1017/S0021859605005708

Pálinkás Z, Kiss J, Zalai M, Szénási Á, Dorner Z, North S, Woodward G, Balog A (2017) Effects of genetically modified maize events expressing Cry34Ab1, Cry35Ab1, Cry1F, and CP4 EPSPS proteins on arthropod complex food webs. Ecol Evol 7:2286–2293. https://doi.org/10.1002/ece3.2848

Palma L, Muñoz D, Berry C, Murillo J, Caballero P (2014) Bacillus thuringiensis toxins: an overview of their biocidal activity. Toxins 6:3296. https://doi.org/10.3390/toxins6123296

Paula DP, Andow DA (2016) Differential Cry toxin detection and effect on Brevicoryne brassicae and Myzus persicae feeding on artificial diet. Entomol Exp Appl 159:54–60. https://doi.org/10.1111/eea.12419

Paula DP, Andow DA, Timbó RV, Sujii ER, Pires CSS, Fontes EMG (2014) Uptake and transfer of a Bt toxin by a Lepidoptera to its eggs and effects on its offspring. PLOS ONE 9:e95422. https://doi.org/10.1371/journal.pone.0095422

Paula DP, Souza LM, Andow DA (2015) Sequestration and transfer of Cry entomotoxin to the eggs of a predaceous ladybird beetle. PLOS ONE 10:e0144895. https://doi.org/10.1371/journal.pone.0144895

Pérez-Hedo M, López C, Albajes R, Eizaguirre M (2012) Low susceptibility of non-target Lepidopteran maize pests to the Bt protein Cry1Ab. Bull Entomol Res 102:737–743. https://doi.org/10.1017/S0007485312000351

Perry JN, Devos Y, Arpaia S, Bartsch D, Ehlert C, Gathmann A, Hails RS, Hendriksen NB, Kiss J, Messéan A, Mestdagh S, Neemann G, Nuti M, Sweet JB, Tebbe CC (2012) Estimating the effects of Cry1F Bt-maize pollen on non-target Lepidoptera using a mathematical model of exposure. J Appl Ecol 49:29–37. https://doi.org/10.1111/j.1365-2664.2011.02083.x

Peterson JA, Lundgren JG, Harwood JD (2011) Interactions of transgenic Bacillus thuringiensis insecticidal crops with spiders (Araneae). J Arachnol 39:1–21. https://doi.org/10.1636/M10-98.1

Pilcher CD, Rice ME, Obrycki JJ, Rice ME (2005) Impact of transgenic Bacillus thuringiensis corn and crop phenology on five nontarget arthropods. Environ Entomol 34:1302–1317. https://doi.org/10.1603/0046-225X(2005)034[1302:IOTBTC]2.0.CO;2

Pinheiro PV, Quintela ED, Junqueira AMR, Aragao FJL, Faria JC (2014) Populational survey of arthropods on transgenic common bean expressing the rep gene from Bean golden mosaic virus. GM Crops Food 5:139–148. https://doi.org/10.4161/gmcr.29224

Pinheiro PV, Ghanim M, Alexander M, Rebelo AR, Santos RS, Orsburn BC, Gray S, Cilia M (2017) Host plants indirectly influence plant virus transmission by altering gut cysteine protease activity of aphid vectors. Mol Cell Proteomics 16:S230–S243. https://doi.org/10.1074/mcp.M116.063495

Pires VC, Silveira FA, Sujii ER, Torezani KRS, Rodrigues WA, Albuquerque FA, Rodrigues SMM, Salomão AN, Pires CSS (2014) Importance of bee pollination for cotton production in conventional and organic farms in Brazil. J Pollinat Ecol 13. https://doi.org/10.26786/1920-7603(2014)20

Pleasants JM, Hellmich RL, Dively GP, Sears MK, Stanley-Horn DE, Mattila HR, Foster JE, Clark P, Jones GD (2001) Corn pollen deposition on milkweeds in and near cornfields. Proc Natl Acad Sci 98:11919–11924. https://doi.org/10.1073/pnas.211287498

Pons X, Lumbierres B, Lopez C, Albajes R (2005) Abundance of non-target pests in transgenic Bt-maize: a farm scale study. Eur J Entomol Czech Repub 102:73–79

Potts SG, Biesmeijer JC, Kremen C, Neumann P, Schweiger O, Kunin WE (2010) Global pollinator declines: trends, impacts and drivers. Trends Ecol Evol 25:345–353. https://doi.org/10.1016/j.tree.2010.01.007

Price PW, Bouton CE, Gross P, McPheron BA, Thompson JN, Weis AE (1980) Interactions among three trophic levels: influence of plants on interactions between insect herbivores and natural enemies. Annu Rev Ecol Syst 11:41–65. https://doi.org/10.1146/annurev.es.11.110180.000353

Priestley AL, Brownbridge M (2009) Field trials to evaluate effects of Bt-transgenic silage corn expressing the Cry1Ab insecticidal toxin on non-target soil arthropods in northern New England, USA. Transgenic Res 18:425–443

Ramirez-Romero R, Chaufaux J, Pham-Delègue M-H (2005) Effects of Cry1Ab protoxin, deltamethrin and imidacloprid on the foraging activity and the learning performances of the honeybee Apis mellifera, a comparative approach. Apidologie 36:601–611. https://doi.org/10.1051/apido:2005039

Ramirez-Romero R, Bernal JS, Chaufaux J, Kaiser L (2007) Impact assessment of Bt-maize on a moth parasitoid, Cotesia marginiventris (Hymenoptera: Braconidae), via host exposure to purified Cry1Ab protein or Bt-plants. Crop Prot 26:953–962. https://doi.org/10.1016/j.cropro.2006.09.001

Raps A, Kehr J, Gugerli P, Moar WJ, Bigler F, Hilbeck A (2001) Immunological analysis of phloem sap of Bacillus thuringiensis corn and of the nontarget herbivore Rhopalosiphum padi (Homoptera: Aphididae) for the presence of Cry1Ab. Mol Ecol 10:525–533

Rauschen S, Schultheis E, Pagel-Wieder S, Schuphan I, Eber S (2009) Impact of Bt-corn MON88017 in comparison to three conventional lines on Trigonotylus caelestialium (Kirkaldy) (Heteroptera: Miridae) field densities. Transgenic Res 18:203–214. https://doi.org/10.1007/s11248-008-9207-2

Rawlings ND, Tolle DP, Barrett AJ (2004) MEROPS: the peptidase database. Nucleic Acids Res 32:D160–D164. https://doi.org/10.1093/nar/gkh071

Ray DK, Mueller ND, West PC, Foley JA (2013) Yield trends are insufficient to double global crop production by 2050. PLOS ONE 8:e66428. https://doi.org/10.1371/journal.pone.0066428

Raybould A, Vlachos D (2011) Non-target organism effects tests on Vip3A and their application to the ecological risk assessment for cultivation of MIR162 maize. Transgenic Res 20:599–611. https://doi.org/10.1007/s11248-010-9442-1

Raybould A, Burns A, Hamer M (2014) High concentrations of protein test substances may have non-toxic effects on Daphnia magna: implications for regulatory study designs and ecological risk assessments for GM crops. GM Crops Food 5:296–301. https://doi.org/10.4161/21645698.2014.950540

Riddick EW, Dively G, Barbosa P (1998) Effect of a seed-mix deployment of Cry3A-transgenic and nontransgenic potato on the abundance of Lebia grandis (Coleoptera: Carabidae) and Coleomegilla maculata (Coleoptera: Coccinellidae). Ann Entomol Soc Am 91:647–653. https://doi.org/10.1093/aesa/91.5.647

Roberts AF, Devos Y, Lemgo GNY, Zhou X (2015) Biosafety research for non-target organism risk assessment of RNAi-based GE plants. Front Plant Sci 6. https://doi.org/10.3389/fpls.2015.00958
Romeis J, Meissle M (2011) Non-target risk assessment of Bt crops – Cry protein uptake by aphids. J Appl Entomol 135:1–6. https://doi.org/10.1111/j.1439-0418.2010.01546.x
Romeis J, Babendreier D, Wäckers FL (2003) Consumption of snowdrop lectin (Galanthus nivalis-agglutinin) causes direct effects on adult parasitic wasps. Oecologia 134:528–536. https://doi.org/10.1007/s00442-002-1144-9
Romeis J, Dutton A, Bigler F (2004) Bacillus thuringiensis toxin (Cry1Ab) has no direct effect on larvae of the green lacewing Chrysoperla carnea (Stephens) (Neuroptera: Chrysopidae). J Insect Physiol 50:175–183. https://doi.org/10.1016/j.jinsphys.2003.11.004
Romeis J, Meissle M, Bigler F (2006) Transgenic crops expressing Bacillus thuringiensis toxins and biological control. Nat Biotechnol 24:63–71. https://doi.org/10.1038/nbt1180
Romeis J, Meissle M, Naranjo SE, Li Y, Bigler F (2014) The end of a myth—Bt (Cry1Ab) maize does not harm green lacewings. Front Plant Sci 5. https://doi.org/10.3389/fpls.2014.00391
Romeis J, Naranjo SE, Meissle M, Shelton AM (2019) Genetically engineered crops help support conservation biological control. Biol Control 130:136–154. https://doi.org/10.1016/j.biocontrol.2018.10.001
Rose R, Dively GP, Pettis J (2007) Effects of Bt corn pollen on honey bees: emphasis on protocol development. Apidologie 38:368–377. https://doi.org/10.1051/apido:2007022
Rovenská G, Zemek R, Schmidt J, Hilbeck A (2005) Altered host plant preference of Tetranychus urticae and prey preference of its predator Phytoseiulus persimilis (Acari: Tetranychidae, Phytoseiidae) on transgenic Cry3Bb-eggplants. Biol Control 33:293–300. https://doi.org/10.1016/j.biocontrol.2005.03.017
Saxena D, Stotzky G (2002) Bt toxin is not taken up from soil or hydroponic culture by corn, carrot, radish, or turnip. Plant Soil 239:165–172. https://doi.org/10.1023/A:1015057509357
Saxena D, Flores S, Stotzky G (2002) Bt toxin is released in root exudates from 12 transgenic corn hybrids representing three transformation events. Soil Biol Biochem 34:133–137. https://doi.org/10.1016/S0038-0717(01)00161-4
Saxena D, Stewart CN, Altosaar I, Shu Q, Stotzky G (2004) Larvicidal Cry proteins from Bacillus thuringiensis are released in root exudates of transgenic B. thuringiensis corn, potato, and rice but not of B. thuringiensis canola, cotton, and tobacco. Plant Physiol Biochem 42:383–387. https://doi.org/10.1016/j.plaphy.2004.03.004
Schmidt JEU, Braun CU, Whitehouse LP, Hilbeck A (2009) Effects of activated Bt transgene products (Cry1Ab, Cry3Bb) on immature stages of the ladybird Adalia bipunctata in laboratory ecotoxicity testing. Arch Environ Contam Toxicol 56:221–228. https://doi.org/10.1007/s00244-008-9191-9
Schuler TH, Potting RPJ, Denholm I, Clark SJ, Clark AJ, Stewart CN, Poppy GM (2003) Tritrophic choice experiments with Bt plants, the Diamondback Moth (Plutella xylostella) and the parasitoid Cotesia plutellae. Transgenic Res 12:351–361. https://doi.org/10.1023/A:1023342027192
Schuler TH, Denholm I, Clark SJ, Stewart CN, Poppy GM (2004) Effects of Bt plants on the development and survival of the parasitoid Cotesia plutellae (Hymenoptera: Braconidae) in susceptible and Bt-resistant larvae of the diamondback moth, Plutella xylostella (Lepidoptera: Plutellidae). J Insect Physiol 50:435–443. https://doi.org/10.1016/j.jinsphys.2004.03.001
Scott JG, Michel K, Bartholomay L, Siegfried BD, Hunter WB, Smagghe G, Zhu KY, Douglas AE (2013) Towards the elements of successful insect RNAi. J Insect Physiol 59. https://doi.org/10.1016/j.jinsphys.2013.08.014
Sears MK, Hellmich RL, Stanley-Horn DE, Oberhauser KS, Pleasants JM, Mattila HR, Siegfried BD, Dively GP (2001) Impact of Bt corn pollen on monarch butterfly populations: a risk assessment. Proc Natl Acad Sci 98:11937–11942. https://doi.org/10.1073/pnas.211329998
Seide VE, Bernardes RC, Pereira EJG, Lima MAP (2018) Glyphosate is lethal and Cry toxins alter the development of the stingless bee Melipona quadrifasciata. Environ Pollut 243:1854–1860. https://doi.org/10.1016/j.envpol.2018.10.020

Sétamou M, Bernal JS, Legaspi JC, Mirkov TE (2002) Parasitism and location of sugarcane borer (Lepidoptera: Pyralidae) by Cotesia flavipes (Hymenoptera: Braconidae) on transgenic and conventional sugarcane. Environ Entomol 31:1219–1225. https://doi.org/10.1603/0046-225X-31.6.1219

Singh IK, Singh S, Mogilicherla K, Shukla JN, Palli SR (2017) Comparative analysis of double-stranded RNA degradation and processing in insects. Sci Rep 7. https://doi.org/10.1038/s41598-017-17134-2

Sisterson MS, Carrière Y, Dennehy TJ, Tabashnik BE (2005) Evolution of resistance to transgenic crops: interactions between insect movement and field distribution. J Econ Entomol 98:1751–1762

Stanley-Horn DE, Dively GP, Hellmich RL, Mattila HR, Sears MK, Rose R, Jesse LCH, Losey JE, Obrycki JJ, Lewis L (2001) Assessing the impact of Cry1Ab-expressing corn pollen on monarch butterfly larvae in field studies. Proc Natl Acad Sci 98:11931–11936. https://doi.org/10.1073/pnas.211277798

Sujii ER, Togni PHB, de A Ribeiro P, de A Bernardes T, Milane PVGN, Paula DP, Pires CSS, Fontes EMG (2013) Field evaluation of Bt cotton crop impact on nontarget pests: cotton aphid and boll weevil. Neotrop Entomol 42:102–111. https://doi.org/10.1007/s13744-012-0094-0

Sun X, Yan M-J, Zhang A, Wang M-Q (2015) Transgenic cry1C(*) gene rough rice line T1C-19 does not change the host preferences of the non-target stored product pest, Rhyzopertha dominica (Fabricius) (Coleoptera: Bostrichidae), and its parasitoid wasp, Anisopteromalus calandrae (Howard) (Hymenoptera: Pteromalidae). Ecotoxicol Environ Saf 120:449–456. https://doi.org/10.1016/j.ecoenv.2015.06.034

Svobodová Z, Habuštová OS, Boháč J, Sehnal F (2016) Functional diversity of staphylinid beetles (Coleoptera: Staphylinidae) in maize fields: testing the possible effect of genetically modified, insect resistant maize. Bull Entomol Res 106:432–445. https://doi.org/10.1017/S000748531500111X

Svobodová Z, Shu Y, Habuštová OS, Romeis J, Meissle M (2017) Stacked Bt maize and arthropod predators: exposure to insecticidal Cry proteins and potential hazards. Proc R Soc B 284:20170440. https://doi.org/10.1098/rspb.2017.0440

Swevers L, Vanden Broeck J, Smagghe G (2013) The possible impact of persistent virus infection on the function of the RNAi machinery in insects: a hypothesis. Front Physiol 4. https://doi.org/10.3389/fphys.2013.00319

Tabashnik BE (2008) Delaying insect resistance to transgenic crops. Proc Natl Acad Sci 105:19029–19030. https://doi.org/10.1073/pnas.0810763106

Tabashnik BE, Carrière Y (2017) Surge in insect resistance to transgenic crops and prospects for sustainability. Nat Biotechnol 35:926–935. https://doi.org/10.1038/nbt.3974

Tabashnik BE, Van Rensburg JBJ, Carrière Y (2009) Field-evolved insect resistance to Bt crops: definition, theory, and data. J Econ Entomol 102:2011–2025

Tabashnik BE, Brévault T, Carrière Y (2013) Insect resistance to Bt crops: lessons from the first billion acres. Nat Biotechnol 31:510–521. https://doi.org/10.1038/nbt.2597

Tian JC, Liu ZC, Chen M, Chen Y, Chen XX, Peng YF, Hu C, Ye GY (2010) Laboratory and field assessments of prey-mediated effects of transgenic Bt rice on Ummeliata insecticeps (Araneida: Linyphiidae). Environ Entomol 39:1369–1377. https://doi.org/10.1603/EN10003

Tian J-C, Chen Y, Li Z-L, Li K, Chen M, Peng Y-F, Hu C, Shelton AM, Ye G-Y (2012) Transgenic Cry1Ab rice does not impact ecological Fitness and predation of a generalist spider. PLOS ONE 7:e35164. https://doi.org/10.1371/journal.pone.0035164

Tian Y, Zhou Y, Xiao K, Zhi W, Jinjun C, Lu X, Song Q (2013) Effect of Cry1Ab protein on hemocytes of the wolf spider Pardosa pseudoannulata. Biocontrol Sci Technol 23:423–432. https://doi.org/10.1080/09583157.2013.765940

Tian J-C, Yao J, Long L-P, Romeis J, Shelton AM (2015) Bt crops benefit natural enemies to control non-target pests. Sci Rep 5:16636. https://doi.org/10.1038/srep16636

Tomov BW, Bernal JS, Vinson SB (2003) Impacts of transgenic sugarcane expressing Gna Lectin on Parasitism of Mexican Rice Borer by Parallorhogas pyralophagus (Marsh) (Hymenoptera: Braconidae). Environ Entomol 32:866–872. https://doi.org/10.1603/0046-225X-32.4.866

Tompkins J-ML, Wratten SD, Wäckers FL (2010) Nectar to improve parasitoid fitness in biological control: Does the sucrose:hexose ratio matter? Basic Appl Ecol 11:264–271. https://doi.org/10.1016/j.baae.2009.12.010

Turlings TCJ, Tumlinson JH, Eller FJ, Lewis WJ (1991) Larval-damaged plants: source of volatile synomones that guide the parasitoid Cotesia marginiventris to the micro-habitat of its hosts. Entomol Exp Appl 58:75–82. https://doi.org/10.1111/j.1570-7458.1991.tb01454.x

US EPA O (2015) Pesticides industry sales and usage 2008–2012 market estimates. In: US EPA. https://www.epa.gov/pesticides/pesticides-industry-sales-and-usage-2008-2012-market-estimates. Accessed 1 Nov 2018

Vanbergen AJ, Initiative the IP (2013) Threats to an ecosystem service: pressures on pollinators. Front Ecol Environ 11:251–259. https://doi.org/10.1890/120126

Vélez AM, Jurzenski J, Matz N, Zhou X, Wang H, Ellis M, Siegfried BD (2016) Developing an in vivo toxicity assay for RNAi risk assessment in honey bees, Apis mellifera L. Chemosphere 144:1083–1090. https://doi.org/10.1016/j.chemosphere.2015.09.068

von Burg S, van Veen FJF, Álvarez-Alfageme F, Romeis J (2011) Aphid–parasitoid community structure on genetically modified wheat. Biol Lett 7:387–391. https://doi.org/10.1098/rsbl.2010.1147

Wakefield ME, Bell HA, Fitches EC, Edwards JP, Gatehouse AMR (2006) Effects of Galanthus nivalis agglutinin (GNA) expressed in tomato leaves on larvae of the tomato moth Lacanobia oleracea (Lepidoptera: Noctuidae) and the effect of GNA on the development of the endoparasitoid Meteorus gyrator (Hymenoptera: Braconidae). Bull Entomol Res 96:43–52

Wang Q, Zhu Y, Sun L, Li L, Jin S, Zhang X (2016) Transgenic Bt cotton driven by the green tissue-specific promoter shows strong toxicity to lepidopteran pests and lower Bt toxin accumulation in seeds. Sci China Life Sci 59:172–182. https://doi.org/10.1007/s11427-015-4920-6

Wang J, Peng Y, Xiao K, Wei B, Hu J, Wang Z, Song Q, Zhou X (2017) Transcriptomic response of wolf spider, Pardosa pseudoannulata, to transgenic rice expressing Bacillus thuringiensis Cry1Ab protein. BMC Biotechnol 17. https://doi.org/10.1186/s12896-016-0325-2

Whitehouse MEA, Wilson LJ, Fitt GP (2005) A Comparison of Arthropod Communities in Transgenic Bt and Conventional Cotton in Australia. Environ Entomol 34:1224–1242. https://doi.org/10.1603/0046-225X(2005)034[1224:ACOACI]2.0.CO;2

Wolfenbarger LL, Naranjo SE, Lundgren JG, Bitzer RJ, Watrud LS (2008) Bt crop effects on functional guilds of non-target arthropods: a meta-analysis. PLOS ONE 3:e2118. https://doi.org/10.1371/journal.pone.0002118

Wu K, Guo Y (2003) Influences of Bacillus thuringiensis Berliner cotton planting on population dynamics of the cotton aphid, Aphis gossypii Glover, in Northern China. Environ Entomol 32:312–318. https://doi.org/10.1603/0046-225X-32.2.312

Wu K, Li W, Feng H, Guo Y (2002) Seasonal abundance of the mirids, Lygus lucorum and Adelphocoris spp. (Hemiptera: Miridae) on Bt cotton in northern China. Crop Prot 21:997–1002. https://doi.org/10.1016/S0261-2194(02)00080-7

Yao Y-S, Han P, Niu C-Y, Dong Y-C, Gao X-W, Cui J-J, Desneux N (2016) Transgenic Bt cotton does not disrupt the top-down forces regulating the cotton aphid in Central China. PLOS ONE 11:e0166771. https://doi.org/10.1371/journal.pone.0166771

Yi D, Fang Z, Yang L (2018) Effects of Bt cabbage pollen on the honeybee Apis mellifera L. Sci Rep 8:482. https://doi.org/10.1038/s41598-017-18883-w

Yu L, Berry RE, Croft BA (1997) Effects of Bacillus thuringiensis toxins in transgenic Cotton and potato on Folsomia candida (Collembola: Isotomidae) and Oppia nitens (Acari: Orbatidae). J Econ Entomol 90:113–118. https://doi.org/10.1093/jee/90.1.113

Yu X, Liu T, Sun Z, Guan P, Zhu J, Wang S, Li S, Deng Q, Wang L, Zheng A, Li P (2012) Co-expression and synergism analysis of Vip3Aa29 and Cyt2Aa3 insecticidal proteins from Bacillus thuringiensis. Curr Microbiol 64:326–331. https://doi.org/10.1007/s00284-011-0070-7

Yuan Y, Ke X, Chen F, Krogh PH, Ge F (2011) Decrease in catalase activity of Folsomia candida fed a Bt rice diet. Environ Pollut Barking Essex 159:3714–3720. https://doi.org/10.1016/j.envpol.2011.07.015

Yuan Y, Xiao N, Krogh PH, Chen F, Ge F (2013) Laboratory assessment of the impacts of transgenic Bt rice on the ecological fitness of the soil non-target arthropod, Folsomia candida (Collembola: Isotomidae). Transgenic Res 22:791–803. https://doi.org/10.1007/s11248-013-9687-6

Zangerl AR, McKenna D, Wraight CL, Carroll M, Ficarello P, Warner R, Berenbaum MR (2001) Effects of exposure to event 176 Bacillus thuringiensis corn pollen on monarch and black swallowtail caterpillars under field conditions. Proc Natl Acad Sci 98:11908–11912. https://doi.org/10.1073/pnas.171315698

Zhang G, Wan F, Lövei GL, Liu W, Guo J (2006) Transmission of Bt toxin to the predator Propylaea japonica (Coleoptera: Coccinellidae) through its aphid prey feeding on transgenic Bt cotton. Environ Entomol 35:143–150. https://doi.org/10.1603/0046-225X-35.1.143

Zhang G-F, Wan F-H, Murphy ST, Guo J-Y, Liu W-X (2008) Reproductive biology of two non-target insect species, Aphis gossypii (Homoptera: Aphididae) and Orius sauteri (Hemiptera: Anthocoridae), on Bt and non-Bt cotton cultivars. Environ Entomol 37:1035–1042

Zhang B, Chen M, Zhang X, Luan H, Tian Y, Su X (2011) Expression of Bt-Cry3A in transgenic Populus alba × P. glandulosa and its effects on target and non-target pests and the arthropod community. Transgenic Res 20:523–532. https://doi.org/10.1007/s11248-010-9434-1

Zhao Y, Zhang S, Luo J-Y, Wang C-Y, Lv L-M, Wang X-P, Cui J-J, Lei C-L (2016) Bt proteins Cry1Ah and Cry2Ab do not affect cotton aphid Aphis gossypii and ladybeetle Propylea japonica. Sci Rep 6. https://doi.org/10.1038/srep20368

Zhou J, Xiao K, Wei B, Wang Z, Tian Y, Tian Y, Song Q (2014) Bioaccumulation of Cry1Ab protein from an herbivore reduces anti-oxidant enzyme activities in two spider species. PLOS ONE 9:e84724. https://doi.org/10.1371/journal.pone.0084724

Zwahlen C, Hilbeck A, Nentwig W (2007) Field decomposition of transgenic Bt maize residue and the impact on non-target soil invertebrates. Plant Soil 300:245–257. https://doi.org/10.1007/s11104-007-9410-6

# Impacts of Genetically Engineered Crops on the Soil Microbiome, Biological Processes, and Ecosystem Services

Robert J. Kremer

**Abstract** Information on the impacts of genetically engineered (GE) crops on the soil microbial community, biological processes, soil health, and ecosystem services is limited. Assessments to acquire this information are challenged further because the necessary comparisons of GE crops with non-GE crops are practically nonexistent. The objectives of this chapter are to provide a background on the impacts of genetically engineered (GE) crops on soil health, with a focus on the soil microbiome and biological processes and on ecosystem services. Genetic materials and pesticide chemicals are released from GE crops into soil and impose variable effects on the soil microbiome and other soil organisms that can cause changes in plant nutrient availability, soil properties, and ecosystem services such as water quality. However, some studies show little effect of GE crops on soil biological and ecosystem processes suggesting the need for balanced research approaches to assure the fair monitoring of impacts of GE crops on the environment. Soil health and ecosystem processes may be maintained or improved under GE cropping only when sustainable management is integrated into these systems. Several vital action approaches are suggested for conducting critical examination and assessment of the potential risks of GE crops.

**Keywords** Biotechnology · Bt crops · Genetically engineered crops · Glyphosate · Mycorrhizae · Rhizosphere · Soil health · Soil microbial communities · Soil microbial diversity · Transgenic crops

## Introduction

Current conventional cropping systems are typically dependent on synthetic agricultural chemicals (fertilizers and pesticides) and artificial irrigation that support production of food, feed, fiber, and bioenergy crops (Arriaga et al. 2017). These management systems may be considered under agricultural intensification whereby

---

R. J. Kremer (✉)
University of Missouri, Columbia, MO, USA
e-mail: kremerr@missouri.edu

 
A. Chaurasia et al. (eds.), *GMOs*, Topics in Biodiversity and Conservation 19,
https://doi.org/10.1007/978-3-030-53183-6_6

all practices are "optimized" with the goal of maximum output (yields) with little regard to environmental factors and long-term soil productivity or health. Numerous new crop varieties (cultivars) with enhanced disease resistance, greater drought, salt tolerance, and resistance to herbicides for simplified weed management have been developed through modern biotechnological techniques (Berg 2009). The vast majority of agroecosystems under agricultural intensification have included the cultivation of transgenic or genetically engineered (GE) crops first introduced in 1996. The global land area cultivated with GE crops had increased to 191.7 million hectares in 2018 (James 2019). Traits genetically engineered into these crops primarily include herbicide tolerance (HT) and insect resistance, as well as gene stacking in maize (*Zea mays*), cotton (*Gossypium hirsutum*), and soybean (*Glycine max*) for tolerance to two or more herbicides, resistance to multiple insect pests with several incorporated *Bacillus thuringiensis* (Bt) toxins (*cry* proteins), and combinations of HT and *cry* genes.

Development of GE crops has traditionally focused on the selection of plant phenotype characteristics, while important beneficial plant–microbe interactions that impact soil health and environmental services have been apparently largely overlooked (Berg 2009; Morrissey et al. 2004). The recent *Global Assessment on Biodiversity* finds that 23% of agricultural land is less productive than 5 years ago (IPBES 2019). Soil health degradation and ecosystem change or biodiversity loss is suspected to be due in part to agricultural intensification, deforestation, and other land-use changes. Despite the current interest in soil health for aiding agricultural management and land-use decisions, a standardized assessment for soil health is still in the developmental stages. Recent assessments primarily focus on broad comparisons of management systems that vary in general practices, including crop rotation, tillage, fertilizer or manure amendment, and soil conservation (Karlen et al. 2008, 2014; Stott et al. 2013). Understanding how more specific practices, such as inclusion of GE crops, affect soil biodiversity is very difficult because these types of comparisons are practically nonexistent. The objectives of this chapter are to provide a background on the impacts of genetically engineered (GE) crops on soil health, with a focus on the soil microbiome and biological processes and on ecosystem services.

## Soil Health

Soil health, as proposed by Doran (2002), is the capacity of a living soil to sustain plant and animal productivity, maintain or enhance water and air quality, and promote plant and animal health within ecosystem boundaries. Coleman et al. (1998) distinguished soil health from soil quality to emphasize that the health and balanced activity of all organism components within an ecosystem are implicit and specified this as integral parts of soil health. Lehman et al. (2015) affirmed the significance of microbial diversity and activity as the basis for soil function in describing the reliance of soil health on diverse soil biological communities that also support critical environmental services. Kibblewhite et al. (2008) defined soil health within the context of sustainable agriculture so that agricultural production does not have a priority status over the provision of ecosystem services. The working definition noted that a "healthy agricultural soil is capable of supporting the production of food and fiber to a level and with quality sufficient to meet human requirements, *and* deliver ecosystem services essential to maintain environmental quality, quality of life for humans, animals, plants and conservation of biodiversity" (Kibblewhite et al. 2008). Soil health processes are assigned for crop production (i.e., water infiltration capacity, storage/release of nutrients, disease suppression) but also provides for important ecosystem services (carbon (C) sequestration, water quality maintenance, biodiversity enhancement) (Stirling et al. 2016). Optimal soil health must be balanced between productivity, environmental quality, and plant and animal health, all of which are influenced by management and land-use decisions. Good management practices must consider all attributes of soil health rather than focus on single goals, such as crop productivity (Doran 2002). In summary, soil health focuses on the living, dynamic nature of soil that incorporates the biological attributes of biodiversity, soil food web structure, ecosystem processes, and the intimate relationships of soil microorganisms with plants and animals (Kremer 2016).

Soil health as an evolving concept is based on numerous indicators of chemical, physical, and biological properties for the assessment of management impacts on our soil resource. Many aspects of crop production and nearly all ecosystem services are mediated by biological processes (Kibblewhite et al. 2008); however, microbial diversity and microbial component groups are not widely used as standard soil health indicators in assessment models due to a lack of sufficient databases and the difficulty in devising on-site sampling methods that maintain in situ conditions (Lehman et al. 2015). Soil health assessment models are available to evaluate the effects of land management on the soil resource (Morrow et al. 2016; Stott et al. 2013); however, these models may be deficient in indicators representing important measures of soil properties such as soil loss (Morrow et al. 2016), disease suppression (van Bruggen and Semenov 2000), soil microbial diversity (Garbeva et al. 2004), and biological/biochemical activity (Stott et al. 2013).

## The Soil Microbiome

A realistic view of bacteria and fungi in soils is their existence within communities consisting of millions of individual species representing a vast array of taxonomic groups. Interactions among groups within communities mediate numerous important processes, including decomposition of natural and xenobiotic organic substances (including pesticides) and carbon cycling, nutrient mineralization and cycling, soil structure formation, plant growth promotion, and soil organic matter formation (Whalen and Sampedro 2010; Willey et al. 2008; Torsvik et al. 1990). When evaluating the impacts of the environment and management on soils, microbial diversity in these ecosystems must be considered from standpoints of structural diversity—the species' abundance and their distribution within a microbial community based on taxonomic characteristics—and diverse roles—the distribution and abundance of microbial groups based on metabolic processes (Brown 2014; Whalen and Sampedro 2010). Modern developments in molecular methods have advanced studies of soil microbial communities by expanding the concept of microbial diversity through detection of nonculturable microorganisms and functional genes that can be matched with soil biological processes and the microbial source of origin (Fierer et al. 2013). These studies have led to the recognition of the *microbiome*, defined as the multispecies community of microbes in a specific environment, including soil, derived using genome-enabled technology (Stulberg et al. 2016), which can be further characterized based on the total proteins of the community (proteomics) responsible for microbial activity and on the metabolites of the community (metabolomics) that identify the products of microbial activity (Bouchez et al. 2016). Soil biology encompassing the composition of microbial communities, or biodiversity, and biological functions is associated with soil health (Pankhurst 1997).

The soil microbiome includes groups of different organisms inhabiting particular niches within the soil, their "biomes," and is used interchangeably with "soil microbial community." Several approaches are available to comprehensively characterize the soil microbiome. Physiological profiling (phenotyping) of bacteria and fungi in soil is based on utilization patterns of an array of substrates (Zak et al. 1994). The phospholipid fatty acid (PLFA) technique detects lipids of microbial membranes as "biomarkers" for specific microbial groups producing a profile or "fingerprint" of the community structure. Biomarkers specific to ecological groups of microorganisms, including bacteria, actinobacteria, arbuscular mycorrhizal fungi (AMF), saprophytic or non-AM fungi, and protists, are reported. Thus, some rapid changes in the microbial community structure due to environmental or management stresses can be reflected in changes in PLFA patterns; also, total PLFA content indicates viable microbial biomass (Zelles 1999). A more thorough depiction of the soil microbiome is provided by the molecular technique of metagenomics, the whole-community sampling of the entire genetic information provided by the deoxyribonucleic acid (DNA) of all individual taxonomic units found within a soil sample (White et al. 2017). The soil metagenome, therefore, is the collection of all individual microorganisms comprising the entire community represented by the total genetic information encoded by DNA, which includes genes for taxonomic expression, physiological processes (metabolomics), and biochemical products (including proteomics).

## Soil Biological Processes

Activities of soil organisms and soil biological processes are influenced by interactions with soil chemical and physical properties under the influence of plant roots, the environment, and land management. For example, soil microbial communities and their activities are highly dependent on quantity and quality of above- and below-ground inputs of plant-derived organic matter (Meyers et al. 2001; Zak et al. 2003). Enzyme activities are promising indicators of soil health because of their rapid response to changes in soil management (Bandick and Dick 1999; Schloter et al. 2003). Soil enzymes drive key biochemical functions in many processes, including organic matter decomposition, nutrient mineralization and cycling, nutrient availability, biodegradation of synthetic compounds, and synthesis of plant-growth-regulating substances, thereby mediating critical roles in most biochemical and ecological processes in the soil ecosystem (Sinsabaugh et al. 1991; Bardgett and van der Putten 2014). Therefore, assessment of enzymatic activities in the ecosystem aids in quantifying and evaluating specific biological processes in the soil. Soil enzyme activities are sensitive soil health indicators for microbial activity used to differentiate various soil and crop management regimes and to quantify specific soil biological processes (Bandick and Dick 1999).

Soil biological processes and organismal groups proposed as bioindicators include soil C mineralization, active C (AC or permanganate-oxidizable C), water-extractable (soluble) C, soil enzymes, soil microbial community structure and biodiversity components, soil fauna (e.g., micro- and mesoarthropods, nematodes, earthworms, etc.), and plant disease criteria (Killham and Staddon 2002; Morrow et al. 2016; Stott et al. 2013; van Bruggen and Semenov 2000). Inclusion of reference soils with similar properties is necessary to document potential changes in soil health due to different management systems, suggesting that conventional cropping practices need to be included when assessing the impacts of GE crop production (Sparling 1997). A microbial community in a healthy soil depends on interactions with soil chemical and physical factors (Stirling et al. 2016); thus, these bioindicators will be considered in describing the impacts of GE crops on soil health and ecosystem services.

## Ecosystem Services

Ecosystems contain a diverse assemblage of living organisms that provide a range of essential services that contribute to various natural cycles and are fundamental in sustaining life (Stirling et al. 2016). Ecosystem services provided by a healthy soil microbiome include development of optimum soil structure through aggregation of soil particles and organic matter to provide adequate aeration and water infiltration; production, storage, and release of nutrients for plant growth; suppression of soil-borne pests and pathogens; plant growth promotion; and degradation of toxic compounds. For the soil biological community to wholly provide a full range of ecosystem services, a rich, diverse, and constant supply of organic matter is required to sustain the necessary diversity and abundance of soil microbes responsible for these services (Stirling et al. 2016). This is accomplished in agroecosystems through use of diverse crop rotations; cover cropping during the fallow season; maintenance or addition of organic substances such as crop residues, manure, compost, etc.; minimum or no tillage; and integration of livestock.

Key ecosystem services represent inputs to farming (e.g., soil fertility), mitigation of externalities associated with farming (e.g., energy-use efficiency), adaptation of farming to environmental change (e.g., resistance and resilience to extreme weather events), and outputs from farming (crop productivity) and the related services of soil quality maintenance, nutrient management, and water-holding capacity (Kremen and Miles 2012). These ecosystem services are negatively affected by both agricultural intensification (increased use of synthetic fertilizers and pesticides combined with reduced use of diversified farming system techniques) and landscape simplification on components of biodiversity. Biodiversity in the context of ecosystem services generally refers to the total of all species in any habitat. Studies on biodiversity of soils have primarily focused on the microbial components of the food web with less attention to individual communities of the soil ecosystem, including meso- and macrofauna represented by microarthropods, nematodes, earthworms, and many other larger organisms such as arthropods, crustaceans, and small mammals (Brussaard et al. 1997). Regardless, biodiversity is considered a critical ecosystem service provided by agroecosystems because it has a fundamental input role to other ecosystem services (Kremen and Miles 2012).

## GE Crops: Effects on the Soil Microbiome, Biological Activity and Soil Health

Despite numerous reports suggesting that transgenic cropping systems have no or little significant effect on soil and environmental biological processes (e.g., Bennett et al. 2013; Cerdeira and Duke 2010; Duke et al. 2012; Keppler et al. 2020), the literature is rife with reports of adverse effects of transgenic crop management on fundamental soil processes, many of which are considered critical indicators of soil health. Effects of GE crops on the soil microbiome, biological activity, and soil health are summarized in Table 1. A number of soil health and environmental services are influenced by environmental microbiomes, which are affected by GE crops due partly to nonspecific or indirect consequences of the genetic transformation, or *pleiotropy*, in addition to the presence of Bt toxins or herbicide within the plant (Turrin et al. 2015). For example, soluble carbohydrate and amino acid exudation from roots of GE soybean altered rhizosphere microbial community structure compared with non-GE cultivars (Kremer et al. 2005). Similarly, resistance of two Bt cotton lines to the *Fusarium oxysporum* root phytopathogen was inferior compared with the parental lines (Li et al. 2009). Several GE crops and trees negatively affected the soil fungal community even though expressed traits were not expected to affect fungi (Stefani and Hamelin 2010). This is of concern considering the important roles that fungi perform in soil structure formation, nutrient transfer (including to plants through mycorrhizae), and organic matter formation. Overall, potential impacts on the soil microbiome during development of transgenic crop cultivars were seemingly overlooked, thus emphasizing the need for monitoring key sensitive microbial groups that mediate important soil health processes in future environmental impact assessments of GE crops (Turrin et al. 2015).

**Table 1** Impacts of genetically engineered (GE) plants on the soil microbiome, biological processes, and soil health

| Soil biological property | Impact | GE plant | References |
|---|---|---|---|
| Carbon released from roots | Increased substrate concentrations alter rhizosphere microbiome | Soybean, maize | Kremer et al. (2005) |
| Resistance to root phytopathogens | Decreased | Bt cotton | Li et al. (2009) |
| Intact soil microbiome | Altered structure and abundance of microbial taxa due to glyphosate | Various | Kremer (2014), Kremer and Means (2009) |
| Intact soil microbiome | No change due to adaptation to glyphosate applied annually | HT maize, HT soybean | Dick et al. (2010) |
| Intact soil microbiome | No change in microbial structure due to glyphosate applications | HT maize, HT soybean | Keppler et al. (2020) |
| Soil fungal community | Altered taxonomic structure | GE crops, trees | Stefani and Hamelin (2010) |
| Soil microbial respiration | Decreased | Bt maize | Castaldini et al. (2005) |
| Soil metabolic activities | Decreased | Bt maize | Castaldini et al. (2005) |
| Carbon utilization patterns | Altered | Bt cotton | Zhang et al. (2019) |
| Soil microbiome functional diversity | Altered | Bt cotton | Zhang et al. (2019) |
| Rhizobacteria communities | Altered (GE vs non-GE) due to changed root exudates | Canola | Dunfield and Germida (2003) |
| Soil rhizobia community | Adapted to glyphosate, soluble aluminum due to GE cropping | HT soybean | Iturralde et al. (2019) |
| Rhizosphere colonization by PGPR | Decreased | Bt maize | Castaldini et al. (2005) |
| *Enterobacter cloacae* PGPR | Increased by frequently applied glyphosate to soils | Various | Kryuchkova et al. (2014) |
| Mycorrhizal root colonization | Altered | Bt maize | Cheeke et al. (2012) |
| Mycorrhizal symbiosis | Altered | Bt maize | Cheeke et al. (2012) |
| Mycorrhizal spore viability, abundance, germination | Reduced due to glyphosate | HT soybean, HT maize | Powell et al. (2009), Zaller et al. (2014) |
| Soil enzyme activity, general | Reduced | Bt cotton | Chen et al. (2011) |
| Soil protease activity | Suppressed | Bt cotton | Chen et al. (2012) |
| Soil enzyme activities – nitrogen cycling | Depressed | Bt *various* | Singh and Dubey (2016) |
| Decomposition | Depressed | Bt *various* | Singh and Dubey (2016) |

(continued)

**Table 1** (continued)

| Soil biological property | Impact | GE plant | References |
|---|---|---|---|
| Soil microbial biomass | Reduced | Bt cotton | Chen et al. (2012) |
| Nitrifying bacteria abundance | Reduced | Bt maize | Cotta et al. (2014) |
| Soil protist community | Reduced | Bt maize | Griffiths et al. (2005) |
| Soil nematode community | Reduced | Bt maize | Griffiths et al. (2005) |
| Root nodulation by bradyrhizobia | Reduced due to glyphosate | HT soybean | Kremer and Means (2009) |

*Bt Bacillus thuringiensis*, *PGPR* plant-growth-promoting rhizobacteria, *HT* herbicide-tolerant

Intact transgenic DNA released through root exudation may be incorporated by the soil microbial biomass through horizontal gene transfer and result in intercellular persistence and amplification of foreign DNA sequences within the microbial lateral gene transfer network and may lead to unforeseen environmental impacts (Levy-Booth et al. 2007). Transgene transfer from genetically modified (GM) plant roots and leaves to bacteria was demonstrated by Tepfer et al. (2003). Such studies confirmed that transgenic oilseed rape (*Brassica napus*), tobacco (*Nicotiana tabacum*), and alfalfa (*Medicago sativa*) could transfer genes released from roots and vegetative residues into rhizosphere soil and be incorporated into soil bacteria. Implications are that such extracellular DNA released into soil, although at very low concentrations, still poses a risk as it can be a source of the gene pool for surrounding microbial communities, especially bacteria and fungi having the competency of natural intake of DNA, as well as various phages (viruses) associated within soil bacteria that could incorporate DNA within their genomes and transfer from host to other organisms during lysogenic events.

Genetically engineered plants may indirectly influence the structure, processes, and diversity of soil and rhizosphere microbial communities. Rhizosphere microbial communities are vital components of the soil-root environment and contribute to essential ecosystem services such as decomposing crop residues, mediating biogeochemical cycles within the soil food web, and maintaining environmental quality and productivity. Maize, cotton, and rice have been engineered to incorporate *cry* genes (i.e., *Cry1Ab*, *Cry1Ab/Ac*) from *Bacillus thuringiensis* that codes for the synthesis of an insecticidal protein (*Bt cultivars*) that kills lepidopteran pests; however, these insecticides are also released into soil through root exudation (Saxena et al. 1999; Icoz and Stotzky 2008; Liu et al. 2018) and often cause adverse effects on nontarget soil microbial communities, leading to overall deterioration of soil health.

Reports of Bt crop cultivar effects on the soil microbiome are generally focused on broad taxonomic groups and functional processes and very few, if any, on specific microbial taxa. A greenhouse study with a field soil amended with vegetative residues of Bt maize cultivars and incubated for 4 months reduced soil respiration and mycorrhizal colonization of maize roots and altered rhizosphere bacterial community structure (Castaldini et al. 2005). Other Bt maize cultivars depressed the initial development of mycorrhizal symbiosis under field conditions (Cheeke et al.

2012; see also chapter "Environmental Analytical and Ecotoxicological Aspects of Bt Maize in the Pannonian Biogeographical Region of the European Union"). Shifts in soil microbial community composition; biological processes, including soil enzyme activities; and decomposition and nutrient cycling were reported by Singh and Dubey (2016). Carbon utilization patterns by the soil microbiome were altered in fields planted to Bt cotton, suggesting that the functional diversity of soil microbial communities was affected by Bt cotton (Zhang et al. 2019). Continuous cultivation of Bt cotton for more than 4 years resulted in the persistence of *Cry* proteins in soil, which contributed to significant reductions in soil microbial biomass carbon, microbial activities, and eight of ten soil enzyme activities (Chen et al. 2011). This study further described the apparent inhibition of soil protease activity by *Cry* proteins, which likely suppressed the hydrolysis of soil *Cry* proteins and thereby contributed to long-term persistence in soils (Chen et al. 2012). It was concluded that potential adverse effects on soil health biological components likely persist under continuous Bt cotton. Growth of Bt maize on tropical soils significantly reduced ammonia-oxidizing bacteria (nitrifying bacteria) abundance and was consistently responsive such that this microbial group was proposed as a sensitive bioindicator for assessing the impacts of GE crops on soil health (Cotta et al. 2014). Abundance of protist and nematode communities are reduced in fields cultivated with Bt maize (Griffiths et al. 2005). In contrast, some studies indicate no effect of Bt crop cultivars on soil microbial community structure and function, leading to suggestions that assessments of potential benefits and ecological and environmental risks of GE crops need to be conducted on a case-by-case basis as the most appropriate approach to assure the fair monitoring of impacts on soil health status (Singh and Dubey 2016).

Glyphosate-resistant crop cultivars were developed by the insertion of a transgene (cp4) from an *Agrobacterium* species to code for an insensitive version of 5-enolpyruvyl-shikimate-3-phosphatase synthase (EPSPS), a critical enzyme required for the synthesis of aromatic amino acids and phenolic compounds important in metabolic reactions but is blocked in glyphosate-sensitive plants and microorganisms (Franz et al. 1997). Soil microorganisms are indirectly affected because glyphosate applied to herbicide-resistant crops is released through roots into the rhizosphere environment (Kremer et al. 2005; Kremer and Means 2009). Impacts of GM crops modified for resistance to herbicides on soil health cannot be discussed without consideration of glyphosate because this herbicide is an integral component of the modern biotechnological cropping systems currently in practice. Consequently, glyphosate is the most widely used herbicide in the world, with a market predicted to exceed 1.2 million metric tons for use in weed management on a global scale by 2022 (Global Industry Analysts 2018). Thus, soil health impacts of glyphosate-resistant crops as major components of modern crop production systems are realistically a result of combined contributions of the GE crop and glyphosate necessarily applied during the growing season. Many recent reports document entry and persistence of glyphosate in soil and aquatic environments (Battaglin et al. 2014; Kremer 2017a; Primost et al. 2017; Lupi et al. 2019). Glyphosate interacts with soil biological components as it enters the soil via release from roots of both glyphosate-susceptible and -resistant plants where glyphosate may immediately contact microbial communities in the rhizosphere (Kremer 2014; Kremer et al. 2005). Glyphosate effects on microbial communities, the basis of important ecosystem services, including nutrient cycling, are critical to understand because of potential reduction in the functional sustainability of soils (Zabaloy et al. 2012). The symbioses between rhizobia and herbicide-tolerant (HT) soybean necessary for effective nitrogen fixation is negatively affected by glyphosate released through roots causing reduction in root nodulation (Kremer and Means 2009). One of the most detrimental effects associated with GE crops sprayed with glyphosate is the delivery of glyphosate into the soil ecosystem to result in reduction of spore abundance, spore viability and germination, and root colonization and hyphal extension of arbuscular mycorrhizae, the symbiotic root fungi that mediates plant uptake of water, phosphorus, and other essential nutrients (Powell et al. 2009; see also chapter "Impact of Genetically Modified Crops on the Biodiversity of Arbuscular Mycorrhizal Fungi"). However, research that detected no shifts in microbial community structure in soil from long-term glyphosate-resistant cropping system fields suggested that the soil microbiome adapted after prolonged exposure to annual and multiple glyphosate applications (Dick et al. 2010). Zaller et al. (2014) reported that glyphosate affected interactions between essential soil organisms such as earthworms and arbuscular mycorrhizal fungi whereby glyphosate significantly decreased root mycorrhization, soil mycorrhizae spore biomass, vesicles, and propagules. Despite inconsistencies in reports of glyphosate effects on microbial community function and structure in short-term studies, the changes in microbial metabolism upon exposure to repeated applications merits longer term field studies (Zabaloy et al. 2012).

Representative impacts of glyphosate on environmental microbial diversity are presented; however, more specific information on glyphosate effects may be found in several reviews, including Allegrini et al. (2015), Kremer (2017a), and Zaller et al. (2014). Glyphosate used in conjunction with glyphosate-tolerant GE crops alters the rhizosphere fungal community with increases in growth and virulence of potential root pathogenic *Fusarium* spp. and of the pathogens *Phytophthora, Pythium,* and *Gaeumannomyces* (Johal and Huber 2009; Kremer and Means 2009). Interestingly, most of these fungi are also increased in the rhizospheres of many weeds that develop resistance to glyphosate and other herbicides by mechanisms similar to transgenic crop (Kremer 2014). After five or more years of cropping to GE soybean with glyphosate applications, indigenous soybean rhizobia and heterotrophic bacteria (e.g., *Paenibacillus glycanilyticus*) became tolerant of soil residual glyphosate, acidity, soluble aluminum, and high temperatures (Iturralde et al. 2019). The tolerant indigenous rhizobia were potentially more competitive for soybean nodulation than rhizobia strains applied as inoculants, which also demonstrated likely impaired persistence in the soil. The rhizosphere of several plants able to grow normally in and tolerate high soil concentrations of glyphosate (>200 mg $kg^{-1}$) had greater proportion of the plant-growth-promoting rhizobacteria *Enterobacter cloacae*, which was also able to degrade glyphosate to sarcosine and glycine, suggesting that this bacterium enzymatically cleaved the C-P bond, thereby not accumulating the aminomethylphosphonic acid metabolite (Kryuchkova et al. 2014).

## GE Crops: Effects on Ecosystem Services

A survey of effects of GE crops on various ecosystem services is reported here (Table 2). The potential environmental risks implicated by GE crops include direct impacts of gene transfer to nontarget species, invasiveness, weediness, and genetic recombination of free DNA in the environment. On the contrary, indirect impacts include harmful and side effects of chemical control, that is, reduced efficiency of pest, disease and weed control, the effect on water and soil, and global decline of biodiversity (Tsatsakis et al. 2017). Indirect effects of GE cropping systems on ecosystem services include reduced soil erosion, an alleged benefit based on the widespread use of glyphosate that allows farmers to use no-tillage practices rather than rely on disruptive and intensive tillage (Cerdeira and Duke 2010). However, according to the Conservation Tillage Information Center (http://www.ctic.purdue.edu/), reduced tillage occurs on only 38% of cropped areas in the US, indicating that a majority of GE cropping systems still rely on some form of tillage. Furthermore,

**Table 2** Impacts of genetically engineered (GE) plants on some ecosystem services

| Ecosystem service | Impact | GE plant | References |
|---|---|---|---|
| Soil conservation | Reduced erosion | Various | Cerdeira and Duke (2010) |
| Soil conservation | Increased erosion | Various | Tsatsakis et al. (2017) |
| Ecosystem biodiversity | Reduced | Various | Tsatsakis et al. (2017) |
| Weed invasiveness | Increased | Various | Kremer (2014), Tsatsakis et al. (2017) |
| Pesticide inputs | Increased (herbicides for HR weed management) | Soybean | Beckie and Hall (2014) |
| Plant nutrient cycling | Immobilized due to glyphosate | Various | Johal and Huber (2009) |
| Plant nutrient cycling | Altered due to HR weeds | Various | Kremer (2014) |
| Beneficial arthropods | Decreased populations, altered symbiotic relationships | HT maize, HT canola | Schutte et al. (2017) |
| Crop and native plant pollination | Decrease | Various | Schutte et al. (2017) |
| Aquatic organisms – ecosystem health indicators | *Daphnia* mortality, disrupt aquatic food web by *Cry1Ab* toxins leached into streams | Bt cotton, Bt maize | Tank et al. (2010), Bohn et al. (2016) |
| Water quality of rice paddy | Release of Cry1Ab/Ac proteins from Bt rice roots | Bt rice | Liu et al. (2018) |
| Integrity of natural plant communities and soil microbiome | Altered – genes introduced through pollen movement and horizontal gene transfer | Bt/HT maize, Bt/HT cotton, Bt/HT canola | Tsatsakis et al. (2017), Powell and Dunfield (2007) |

*HR* herbicide resistant, *HT* herbicide tolerant, *Bt* *Bacillus thuringiensis*, *Bt/HT* *Bacillus thuringiensis*–herbicide tolerant "stacked traits"

because of the development of glyphosate-resistant weeds, the actual amount of herbicide active ingredient applied per hectare has increased, for example, by 25% from 1995 to 2008 in soybean alone (Beckie and Hall 2014).

Widespread herbicide-resistant (HR) weed infestations caused by overuse of glyphosate on GE HT crops may alter the soil microbiome and soil biological processes, effects that differ from weed-free GE crop monocultures. For example, depending on the HR weed biotype, changes in amount and type of labile carbon released from roots into soil, mycorrhizal viability and abundance, and plant nutrient release and uptake may occur in production fields (Kremer 2014). Thus, both soil health attributes and ecosystem services may be negatively impacted. The use of additional herbicides to manage HR weeds escaping control by glyphosate will seemingly further complicate mechanisms leading to negative environmental impacts. Understanding all complex interactions of HR weeds with soil microbes under extensive infestations occurring due to failed weed management with GE crops is ongoing as it is important in developing effective alternative weed management systems (Kremer 2014).

Potential threats to farmland and natural habitats are associated with the cultivation of herbicide-tolerant GM crops. Approximately 80% of transgenic crops under cultivation have transgenes expressing tolerance to glyphosate or glufosinate herbicides and/or stacked with transgenes for insect resistance. Apart from toxicity to plants themselves, the possibility of toxicity to other life forms also exist (Tsatsakis et al. 2017). Several studies have demonstrated how glyphosate weakened plant defense and increased root pathogen virulence in both glyphosate-resistant and -susceptible plants (Johal and Huber 2009; Kremer and Means 2009). Glyphosate inhibits the plant's defense and structural barriers and immobilizes micronutrients such as manganese (Mn), which play vital roles in disease resistance, and modifies plant nitrogen metabolism. Other indirect effects of herbicide tolerance include altered biodiversity of weeds, weed-inhabiting arthropods, pollinators, parasitoids, predators, and decomposers, which may lead to imbalanced symbiotic relationships, decreased beneficial insect populations, and rapid changes in ecosystem food chains (Schutte et al. 2017).

Residues of Bt maize and cotton often reach streams or water bodies where the *Cry* toxins may leach from plant tissues and contact aquatic organisms. A survey of 217 streams in Indiana found that 85% of the streams contained maize plant tissues, and the *Cry*1Ab toxin was detected in water from 23% of the sites (Tank et al. 2010). The toxins were detected 6 months after maize harvest in nearby fields, illustrating prolonged persistence in aquatic environments. Exposure of the freshwater crustacean *Daphnia magna* to *Cry* toxins caused high mortality, small body size, and low reproductive rates (Bohn et al. 2016). Because *D. magna* is very sensitive to changes in aquatic quality, the adverse response to *Cry* toxins suggests that these compounds may interrupt the aquatic food web with further potential disruption of ecosystem processes.

Advancements in molecular biology and plant genetics since the introduction of transgenic crop varieties have led to a new technique highly anticipated to yield varieties engineered for various traits that is simpler, more flexible and more accurate (Georges and Ray 2017). This technique, known as "clustered regularly interspaced short palindromic repeats" (CRISPR) and "CRISPR-associated proteins" (Cas) (CRISPR/Cas9), is a genome-editing approach whereby manipulation of the genome occurs at defined sites and provides more controlled introduction of specific alterations for gene silencing or expression in contrast to random mutations associated with older transgenic methods. However, recent reports of off-target mutations in base editing of the rice (*Oryza sativa*) genome using CRISPR/Cas9 suggests that the technique could result in unintended effects expressed in the engineered progeny (Jin et al. 2019). The subsequent impact of potentially newly released gene-edited crops on soil health and ecosystem services relative to transgenic crops is yet to be determined.

## Management Implications for Compatibility of GE Crops and Sustainability

Although GE crops have been cultivated for over 20 years, many studies suggest that GE cropping systems do not promote soil health and biodiversity unless sustainable practices such as no tillage, cover cropping or organic soil amendments, and/or mixed crop rotations are included in the management plan (Bedano and Dominguez 2016; Bedano et al. 2016; Scholes and Scholes 2013). Management for soil health and ecosystem sustainability with a GE cropping system is nearly completely comprised of strategies devised to lessen the detrimental impacts of this cropping system. Incorporation of traditional practices into these systems may improve soil and environmental health assessments. Manure applied to soils of various cropping systems, including GE maize–soybean rotations, in five diverse watersheds in northcentral US significantly increased soil health ratings using a soil health assessment model (Karlen et al. 2014). Integration of a variety of cover crops can maintain soil health and suppress weed infestations while minimizing inputs of herbicides in transgenic HT cropping systems (Davis et al. 2012). Inclusion of cover crops and organic amendments in an intercropping arrangement with Bt cotton overcame negative effects of Bt toxins on microbial activity, measured as soil enzymes, and increased most components of the soil microbiome (Singh et al. 2013).

The functional consequences of GE crops on the structure of soil microbial communities have not been adequately addressed (Powell and Dunfield 2007). Main areas of study concerning such effects may be the possibility of horizontal transfer of gene coding for herbicide resistance to the soil microbial community or direct effects on the community via contact within the weed rhizosphere. To date few, if any, attempts have been made to investigate these possibilities in the absence of herbicide treatment. A limited number of studies examining conventional and transgenic HT crop cultivars for effects on soil microbial communities serve as a reference for similar investigations of HT weeds. For example, Dunfield and Germida (2003) found differences between the bacterial communities associated with genetically modified *Brassica napus* and conventional varieties, which were presumably linked to differences in root exudate composition. This suggests that interactions of the biotypes with soilborne microorganisms may be similar as well. However, studies directly addressing these comparisons should be conducted to confirm the hypothesized effects on the soil microbial community (Hirsch and Mauchline 2012).

## Conclusions

Modern biotechnology involving the use of GE cropping systems has not been directly subjected to evaluations for comparing soil health with more sustainable cropping systems. Most studies reviewed for this chapter examined many soil health and ecosystem indicators and suggested that GE cropping systems do not promote soil health unless sustainable practices such as cover cropping or organic soil amendments are included in the management plan. The consistent negative effect of the continuous planting of Bt and herbicide-resistant crop varieties on soil fungi and mycorrhizae is of particular concern due to the critical functions that these microbial groups perform in soil structure improvement, soil organic matter formation, and nutrient cycling (see also chapter "Impact of Genetically Modified Crops on the Biodiversity of Arbuscular Mycorrhizal Fungi"). Application of strategies similar to those for improving agricultural biological processes by manipulating microbial consortia in the rhizosphere of GE crops to either degrade *Cry* proteins and herbicides or enhance mechanisms within the root environment to overcome detrimental factors associated with GE crop production should improve soil health under GE cropping systems (Kremer 2017b).

The global area cultivated to transgenic crops engineered with a single gene or stacked gene traits has increased greatly over the last three decades. Based on Tsatsakis et al. (2017), selected critical actions are summarized to emphasize the importance of critical examination and assessment of potential risks of GE crops:

1. To avoid harm to beneficial organisms, studies on the expression of transgenes should be conducted to understand the type of risk and its actual potential in broader agro- and natural ecosystems.
2. Large-scale investigations should be conducted to identify possible hosts for gene transfer in the environment.
3. Many of the transgenes already present in the environment need to be studied to investigate the comparative survival of identical genes in a GM crop.
4. Possible routes of gene (DNA) transferred into competent bacteria or viruses should be determined whether particular transgenes are able to flow through soil microorganisms and, if DNA will flow, assess the possible risk.
5. Impacts on complex food webs and microbial substrates should be considered before the release of new transgenes harboring various traits.
6. Monocrop GE cropping systems established over vast areas pose an increasing possible risk that may be suppressed with rotations, preferably with non-GE crops, implemented in cropping areas.
7. Effects on nontarget species should be assessed for several successive generations rather than single or two generations.

## References

Allegrini M, Zabaloy MC, Gómez EV (2015) Ecotoxicological assessment of soil microbial community tolerance to glyphosate. Sci Total Environ 533:60–68

Arriaga FJ, Guzman J, Lowery B (2017) Conventional agricultural production systems and soil functions. In: Al-Kaisi MM, Lowry B (eds) Soil health and intensification of agroecosystems. Academic, San Diego, pp 109–125

Bandick AK, Dick RP (1999) Field management effects on soil enzyme activities. Soil Biol Biochem 31:1471–1479

Bardgett RD, van der Putten WH (2014) Belowground biodiversity and ecosystem functions. Nature 515:505–511

Battaglin WA, Meyer MT, Kuivila KM, Dietze JE (2014) Glyphosate and its degradation product AMPA occur frequently and widely in U.S. soils, surface water, groundwater, and precipitation. J Am Water Resour Assoc 50:275e290

Beckie HJ, Hall LM (2014) Genetically-modified herbicide-resistant (GMHR) crops a two-edged sword? An Americas perspective on development and effect on weed management. Crop Protect 66:40–45

Bedano JC, Dominguez A (2016) Large-scale agricultural management and soil meso- and macrofauna conservation in the Argentine Pampas. Sustainability 8:653

Bedano JC, Dominguez A, Arolfo R, Wall LG (2016) Effect of good agricultural practices under no-till on litter and soil invertebrates in areas with different soil types. Soil Tillage Res 158:100–109

Bennett AB, Chi-Ham C, Barrows G, Sexton S, Zilberman D (2013) Agricultural biotechnology: economics, environment, ethics, and the future. Annu Rev Environ Resour 38:249–279

Berg G (2009) Plant-microbe interactions promoting plant growth and health: perspectives for controlled use of microorganisms in agriculture. Appl Microbiol Biotechnol 84:11–18

Bohn T, Rover CM, Semenchuk PR (2016) *Daphnia magna* negatively affected by chronic exposure to purified Cry-toxins. Food Chem Toxicol 91:130–140

Bouchez T, Blieus AL, Dequiedt S, Domaizon I, Dufresne A, Ferreira S, Godon JJ, Hellal J, Jouliam C, Quaizer A, Martin Laurent F, Mauffret A, Monier JM, Peyret P, Schmitt-Koplin R, Sibourg O, D'oiron E, Bispo A, Deprotes I, Grand C, Cuny P, Maron PA, Ranjard L (2016) Molecular microbiology methods for environmental diagnosis. Environ Chem Lett 14:423–441

Brown JW (2014) Principles of microbial diversity. American Society for Microbiology Press, Washington, DC

Brussaard L, Behan-Pelletierr VM, Bignell DE, Brown VK, Didden W, Folgarait P, Fragoso C, Freckman DW, Gupta VVSR, Hattori T, Hawksworth DL, Klopatek C, Lavelle P, Malloch DW, Rusek J, Söderström B, Tiedge JM, Virginia RA (1997) Biodiversity and ecosystem functioning in soil. Ambio 26:563–570

Castaldini M, Turrini A, Sbrana C, Benedetti A, Marchionni M, Mocali S, Fabiani A, Landi S, Santomassimo F, Pietrangeli B, Nuti MP, Miclaus N, Giovannetti M (2005) Impact of Bt corn on rhizospheric and soil eubacterial communities and on beneficial mycorrhizal symbiosis in experimental microcosms. Appl Environ Microbiol 71:6719–6729

Cerdeira AL, Duke SO (2010) Effects of glyphosate-resistant crop cultivation on soil and water quality. GM Crops 1:16–24

Cheeke TE, Rosenstiel TN, Cruzan MB (2012) Evidence of reduced arbuscular mycorrhizal fungal colonization in multiple lines of Bt maize. Am J Bot 99:700–707

Chen ZH, Chen LJ, Zhang YL, Wu ZJ (2011) Microbial properties, enzyme activities and the persistence of exogenous proteins in soil under consecutive cultivation of transgenic cotton (*Gossypium hirsutum* L.). Plant Soil Environ 57:67–74

Chen ZH, Chen LJ, Wu ZJ (2012) Relationships among persistence of Bacillus thuringiensis and cowpea trypsin inhibitor proteins, microbial properties and enzymatic activities in rhizosphere soil after repeated cultivation with transgenic cotton. Appl Soil Ecol 53:23–30

Coleman DC, Hendrix PF, Odum EP (1998) Ecosystem health: an overview. In: Huang PM (ed) Soil chemistry and ecosystem health, SSSA special publication 52. Soil Science Society of America, Madison, pp 1–20

Cotta SR, Dias ACF, Marriel IE, Andreote FD, Seldin L, van Elsas JD (2014) Different effects of transgenic maize and nontransgenic maize on nitrogen-transforming archaea and bacteria in tropical soils. Appl Environ Microbiol 80:6437–6445

Davis AS, Hill JD, Chase CD, Johanns AM, Liebman M (2012) Increasing cropping system diversity balances productivity, profitability and environmental health. PLoS One 7:e47149

Dick RP, Lorenz N, Wojno M, Lane M (2010) Microbial dynamics in soils under long-term glyphosate tolerant cropping systems. In: Gilkes RJ, Prakongkep N (eds) Proceedings of the 19th world congress of soil science. CSIRO Publishing, Brisbane, pp 153–156

Doran JW (2002) Soil health and global sustainability: translating science into practice. Agric Ecosyst Environ 88:119–127

Duke SO, Lydon J, Koskinen WC, Moorman TB, Chaney RL, Hammerschmidt R (2012) Glyphosate effects on plant mineral nutrition, crop rhizosphere microbiota, and plant disease in glyphosate-resistant crops. J Agric Food Chem 60:10375–10397

Dunfield KE, Germida JJ (2003) Seasonal changes in the rhizosphere microbial communities associated with field-grown genetically-modified canola (*Brassica napus*). Appl Environ Microbiol 69:7310–7318

Fierer N, Ladau J, Clemente JC, Leff JW, Owens SM, Pollard KS, Knight R, Gilber JA, McCulley RL (2013) Reconstructing the microbial diversity and function of preagricultural tallgrass prairie soils in the United States. Science 342(6158):621–624

Franz JE, Mao MK, Sikorski A (1997) Glyphosate: a unique global herbicide, ACS monograph 189. American Chemical Society, Washington, DC

Garbeva P, van Veen JA, van Elsas JD (2004) Microbial diversity in soil: selection of microbial populations by plant and soil type implications for disease suppressiveness. Annu Rev Phytopathol 42:243–270

Georges F, Ray H (2017) Genome editing of crops: a renewed opportunity for food security. GM Crops Food 8:1–12

Global Industry Analysts (2018) Glyphosate – market analysis, trends and forecasts. https://www.strategyr.com/market-report-glyphosate-forecasts-global-industry-analysts-inc.asp. Accessed 14 Oct 2019

Griffiths BS, Caul S, Thompson J, Birch ANE, Scrimgeour C, Anderson MN, Cortet J, Messean A, Sausse C, Lacroix B, Krogh PH (2005) A comparison of soil microbial community structure, protozoa and nematodes in field plots of conventional and genetically modified maize expressing the *Bacillus thuringiensis Cry1Ab* toxin. Plant Soil 275:135–146

Hirsch PR, Mauchline TH (2012) Mutualism: plant-microorganism interactions. In: Ogilvie LA, Hirsch PR (eds) Microbial ecological theory. Caister Academic Press, Norfolk, pp 43–55

Icoz I, Stotzky G (2008) Fate and effects of insect-resistant Bt crops in soil ecosystems. Soil Biol Biochem 40:559–586

Intergovernmental Science-Policy Platform on Biodiversity and Ecosystem Services (2019) Global assessment report on biodiversity and ecosystem services. IPBES secretariat, Bonn, Germany. https://www.ipbes.net/global-assessment-report-biodiversity-ecosystem-services

Iturralde ET, Covelli JM, Alvarez F, Pérez-Giménez J, Arrese-Igor C, Lodeiro AR (2019) Soybean-nodulating strains with low intrinsic competitiveness for nodulation, good symbiotic performance, and stress tolerance isolated from soybean-cropped soils in Argentina. Front Microbiol 10:1061. https://doi.org/10.3389/fmicb.2019.01061

James C (2019) Global status of commercialized biotech/GM crops: 2018, ISAAA brief 54, international service for the acquisition of agri-biotech applications. Cornell University, Ithaca

Jin S, Zong Y, Gao Q, Zhu Z, Wang Y, Qin P, Liang C, Wang D, Qiu JL, Zhang F, Gao C (2019) Cytosine, but not adenine, base editors induce genome-wide off-target mutations in rice. Science 364(6437):292–295

Johal GS, Huber DM (2009) Glyphosate effects on diseases of plants. Eur J Agron 31:144–152

Karlen DL, Andrews SS, Wienhold BJ, Zobeck TM (2008) Soil quality assessment: past, present and future. Electron J Integr Biosci 6:3–14

Karlen DL, Stott DE, Cambardella CA, Kremer RJ, King KW, McCarty GW (2014) Surface soil quality in five Midwestern cropland conservation effects assessment project watersheds. J Soil Water Conserv 69:393–401

Keppler RM, Schmidt DJE, Yarwook SA, Cavigelli MA, Reddy KN, Duke SO, Bradley CA, Williams MM, Buyer JS, Maul JE (2020) Soil microbial communities in diverse agroecosystems exposed to the herbicide glyphosate. Appl Environ Microbiol 86:e01744–e01719. https://doi.org/10.1128/AEM.01744-19

Kibblewhite MG, Ritz K, Swift MJ (2008) Soil health in agricultural systems. Philos Trans R Soc Lond B Biol Sci 363:685–701

Killham K, Staddon WJ (2002) Bioindicators and sensors of soil health and the application of geostatistics. In: Burns RG, Dick RP (eds) Enzymes in the environment: activity, ecology, and applications. Marcel Dekker, New York, pp 391–405

Kremen C, Miles A (2012) Ecosystem services in biologically diversified versus conventional farming systems: benefits, externalities, and trade-offs. Ecol Soc 17:40. https://doi.org/10.5751/ES-05035-170440

Kremer RJ (2014) Environmental implications of herbicide-resistant weeds: soil biology and ecology. Weed Sci 62:415–426

Kremer RJ (2016) Soil health, quality & microbial diversity. Acres U S A 46(12):38–43

Kremer RJ (2017a) Soil and environmental health after twenty years of intensive use of glyphosate. Adv Plants Agric Res 6:00224. https://doi.org/10.15406/apar.2017.06.00224

Kremer RJ (2017b) Biotechnology impacts on soil and environmental services. In: Al-Kaisi MM, Lowery B (eds) Soil health and intensification of agroecosystems. Academic Press, San Diego, pp 353–375

Kremer RJ, Means NE (2009) Glyphosate and glyphosate-resistant crop interactions with rhizosphere microorganisms. Eur J Agron 31:153–161

Kremer RJ, Means NE, Kim SJ (2005) Glyphosate affects soybean root exudation and rhizosphere microorganisms. Int J Environ Anal Chem 85:1165–1174

Kryuchkova YV, Burygin GL, Gogolev NE, Gogole YV, Chernyshova MP, Makarov OE, Fedorov EE, Turkovskay OV (2014) Isolation and characterization of a glyphosate-degrading rhizosphere strain, *Enterobacter cloacae* K7. Microbiol Res 169:99–105

Lehman RM, Acosta-Martinez V, Buyer JS, Cambardella CA, Collins HP, Ducey TF, Halvorson JJ, Jin VL, Johnson JMF, Kremer RJ, Lundgren JG, Manter DK, Maul JE, Smith JE, Stott DE (2015) Soil biology for resilient, healthy soil. J Soil Water Conserv 70:12A–18A

Levy-Booth DJ, Campbell RG, Gulden RH, Hart MM, Powell JR, Klironomos JN, Pauls KP, Swanton CJ, Trevors JT, Dunfield KE (2007) Cycling of extracellular DNA in the soil environment. Soil Biol Biochem 39:2977–2991

Li X, Liu B, Heia S, Liu D, Han Z, Zhou K, Cui J, Luo J, Zheng Y (2009) The effect of root exudates from two transgenic insect-resistant cotton lines on the growth of *Fusarium oxysporum*. Transgenic Res 18:757–767

Liu L, Wu L, Eickhorst T (2018) Accumulation of Cry1Ab/Ac proteins released from transgenic Bt-rice in the rhizosphere of a paddy soil. Rhizosphere 6:39–46

Lupi L, Bedmar F, Puricelli M, Marino D, Aparicio VC, Wunderlin D, Miglioranza KSB (2019) Glyphosate runoff and its occurrence in rainwater and subsurface soil in the nearby area of agricultural fields in Argentina. Chemosphere 225:906–914

Meyers RT, Zak DR, White DC, Peacock A (2001) Landscape-level patterns of microbial community composition and substrate use in upland forest ecosystems. Soil Sci Soc Am J 65:359–367

Morrissey JP, Dow JM, Mark GL, O'Gara F (2004) Are microbes at the root of a solution to world food production? EMBO Rep 5:922–926

Morrow JG, Huggins DR, Carpenter-Boggs LA, Reganold JP (2016) Evaluating measures to assess soil health in long-term agroecosystem trials. Soil Sci Soc Am J 80:450–462

Pankhurst CE (1997) Biodiversity of soil organisms as an indicator of soil health. In: Pankhurst CE, Doube BM, Gupta VVSR (eds) Biological indicators of soil health. CAB International, Oxford, pp 297–324

Powell JR, Dunfield KE (2007) Non-target impacts of genetically-modified, herbicide-resistant crops on soil microbial and faunal communities. In: Gulden RH, Swanton CJ (eds) The first decade of herbicide-resistant crops in Canada, Topics in Canadian weed science, vol 4. Canadian Weed Science Society, Sainte Anne de Bellevue, pp 127–137

Powell JR, Levy-Booth DJ, Gulden RH, Asbil WL, Campbell RG, Dunfield KE, Hamill AS, Hart MM, Lerat S, Nurse RE, Pauls KP, Sikema PH, Swanton CJ, Trevors JT, Klironomos JN (2009) Effects of genetically modified, herbicide-tolerant crops and their management on soil food web properties and crop litter decomposition. J Appl Ecol 46:388–396

Primost JE, Marino D, Aparicio VC, Costa JL, Carriquiriborde P (2017) Glyphosate and AMPA, "pseudo-persistent" pollutants under real-world agricultural management practices in the Mesopotamic Pampas agroecosystem, Argentina. Environ Pollut 229:771e779

Saxena D, Flores S, Stotzky G (1999) Insecticidal toxin in root exudates from Bt corn. Nature 402:480

Schloter M, Dilly O, Munch JC (2003) Indicators for evaluating soil quality. Agriculture, Ecosystems & Environment 98 (1-3):255-262

Scholes MC, Scholes RJ (2013) Dust to dust. Science 342:565–566

Schutte G, Eckerstorfer M, Rastelli V, Reichenbecher W, Restrepo-Vassalli S, Ruohonen-Lehto M, Wuest Saucy AG, Mertens M (2017) Herbicide resistance and biodiversity: agronomic and environmental aspects of genetically-modified herbicide-resistant plants. Environ Sci Eur 29:5. https://doi.org/10.1186/s12302-016-0100-y

Singh AK, Dubey SK (2016) Current trends in Bt crops and their fate on associated microbial community dynamics: a review. Protoplasma 253:663–681

Singh RJ, Ahlawat IPS, Singh S (2013) Effects of transgenic Bt cotton on soil fertility and biology under field conditions in a subtropical inceptisol. Environ Monit Assess 185:485–495

Sinsabaugh RL, Antibus RK, Linkins AE (1991) An enzymic approach to the analysis of microbial activity during plant litter decomposition. Agric Ecosyst Environ 34:43–54

Sparling GP (1997) Soil microbial biomass, activity and nutrient cycling as indicators of soil health. In: Pankhurst C, Doub BM, Gupta VVSR (eds) Biological indicators of soil health. CABI, Wallingford, pp 97–120

Stefani FOP, Hamelin RC (2010) Current state of genetically modified plant impact on target and non-target fungi. Environ Rev 18:441–475

Stirling G, Hayden H, Pattison T, Stirling M (2016) Soil health, soil biology, soilborne diseases and sustainable agriculture. CSIRO Publishing, Clayton South

Stott DE, Cambardella CA, Karlen DL, Harmel RD (2013) A soil quality and metabolic activity assessment after fifty-seven years of agricultural management. Soil Sci Soc Am J 77:903–913

Stulberg E, Fravel D, Proctor LM, Murray DM, LoTempio J, Chrisey L, Garland J, Goodwin K, Graber J, Harris MC, Jackson S, Mishkind M, Porterfield DM, Record A (2016) An assessment of US microbiome research. Nat Microbiol 1:15015

Tank JL, Rosi-Marshall EJ, Royer TV, Whiles MR, Griffiths NA, Frauendorf TC, Treering DJ (2010) Occurrence of maize detritus and a transgenic insecticidal protein (Cry1Ab) within the stream network of an agricultural landscape. Proc Natl Acad Sci U S A 107:17645–17650. www.pnas.org/cgi/doi/10.1073/pnas.1006925107

Tepfer D, Garcia-Gonzales R, Mansouri H, Seruga M, Message B, Leach F, Perica MC (2003) Homology-dependent DNA transfer from plants to a soil bacterium under laboratory conditions: implications in evolution and horizontal gene transfer. Transgenic Res 12:425–437

Torsvik V, Goksoyr J, Daae FL (1990) High diversity in DNA of soil bacteria. Appl Environ Microbiol 56:782–787

Tsatsakis AM, Nawaz MA, Kouretas D, Balias G, Savolainen K, Tutelyan VA, Golokhvast KS, Lee JD, Yang SH, Chung G (2017) Environmental impacts of genetically modified plants: a review. Environ Res 156:818–833

Turrin A, Sbrana C, Giovannetti M (2015) Belowground environmental effects of transgenic crops: a soil microbial perspective. Res Microbiol 166:121–131
van Bruggen AHC, Semenov AM (2000) In search of biological indicators for soil health and disease suppression. Appl Soil Ecol 15:13–24
Whalen JK, Sampedro L (2010) Soil ecology & management. CAB International, Oxfordshire
White RA III, Rivas-Ubach A, Borkum MI, Köberl M, Bilbao A, Colby SM, Hoy DW, Bingol K, Kim YM, Wendler JP, Hixson KK, Jansson C (2017) The state of rhizospheric science in the era of multi-omics: a practical guide to -omics technologies. Rhizosphere 3:212–221
Willey JM, Sherwood LM, Woolverton CJ (2008) Prescott, Harley, and Klein's microbiology, 7th edn. McGraw-Hill, Boston
Zabaloy MC, Gómez E, Garland JL, Gómez MA (2012) Assessment of microbial community function and structure in soil microcosms exposed to glyphosate. Appl Soil Ecol 61:333–339
Zak DR, Willig MR, Moorhead DL, Wildman HG (1994) Functional diversity of microbial communities: a quantitative approach. Soil Biol Biochem 26:1101–1108
Zak DR, Holmes WE, White DC, Peacock AD, Tilman D (2003) Plant diversity, soil microbial communities, and ecosystem function: are there any links? Ecology 84:2042–2050
Zaller JG, Heigl F, Ruess L, Grabmaier A (2014) Glyphosate herbicide affects belowground interactions between earthworms and symbiotic mycorrhizal fungi in a model ecosystem. Sci Rep 4:5634
Zelles L (1999) Fatty acid patterns of phospholipids and lipopolysaccharides in the characterization of microbial communities in soil: a review. Biol Fertil Soils 29:111–129
Zhang M, Feng M, Xiao L, Song X, Ding G, Yang W (2019) Persistence of Cry1Ac protein from transgenic *Bt* cotton cultivation and residue returning in fields and its effect on functional diversity of soil microbial communities. Pedosphere 29:114–122

# Environmental Analytical and Ecotoxicological Aspects of *Bt* Maize in the Pannonian Biogeographical Region of the European Union

**András Székács**

**Abstract** Insect-resistant transgenic crops expressing toxins originated from *Bacillus thuringiensis* (*Bt*) appear advantageous by not requiring field applications of *Bt* bioinsecticides, and by prevention of efficacy losses due to improper application timing, wash-off or inactivation. Through preventing insect damage potentially transmitting infection by toxinogenic fungi, *Bt* plants may indirectly reduce mycotoxin contamination. Strong disadvantages are, however, that Cry1Ab toxin-based *Bt* bioinsecticides and *Bt* plants differ in their active ingredients: *MON 810 Bt* maize expresses a single truncated (preactivated) CrylAb toxin, while the corresponding bioinsecticide contains a Cry1Ab protoxin (with other Cry1, Cry2 and Vip protoxins). This can facilitate rapid insect resistance development not only against Cry1Ab (see cross-resistance). Cry1Ab toxin protected from decomposition in plant tissues shows environmental persistence in the stubble. Protected butterflies (Lepidoptera) in Hungary, showing higher sensitivity to Cry1Ab than the target pest, are exposed to Cry1Ab toxin through the dispersal of *Bt* maize pollen. *Bt* maize showed moderate but statistically significant effects on parasitoid or predator beneficial insects in tritrophic studies. Finally, *Bt* plants produce *Cry* toxin during their entire vegetation period. Thus, toxin administration cannot be limited to the occurrence of the pest insect that contradicts the threshold-based treatment timing principle of integrated pest management.

**Keywords** Cry proteins · Protoxin · Preactivated toxin · Immunoassay · Pest resistance · Protected insects · Tritrophic assessment · *MON 810* · *DAS-59122-7*

A. Székács (✉)
Agro-Environmental Research Institute, National Agricultural Research and Innovation Centre, Budapest, Hungary
e-mail: szekacs.andras@akk.naik.hu



A. Chaurasia et al. (eds.), *GMOs*, Topics in Biodiversity and Conservation 19,
https://doi.org/10.1007/978-3-030-53183-6_7

## Abbreviations

| | |
|---|---|
| AM | arbuscular mycorrhizal |
| *Bt* | *Bacillus thuringiensis* |
| CR | the cross-reactivity |
| CRP | Co-operative Research Programmes |
| Cry | crystal *Bt* endotoxin |
| Cyt | cytolytic *Bt* endotoxin |
| EFSA | European Food Safety Authority |
| ELISA | enzyme-linked immunosorbent assay |
| EPA | Environmental Protection Agency |
| ERA | environmental risk assessment |
| EU | European Union |
| GM | genetically modified |
| GMO | genetically modified organism |
| HT | herbicide-tolerant |
| IPM | integrated pest management |
| IR | insect-resistant |
| IRM | insect resistance management |
| ITU | international toxic unit |
| OECD | Organisation for Economic Co-operation and Development |
| PPP | plant protection products |
| RNAi | ribonucleic acid interference |
| SAB | Scientific Advisory Body |
| UN | United Nations |
| US | United States (of America) |
| Vip | vegetative insecticidal proteins |

## Introduction

The European Union (EU) legislation specifies a genetically modified organism (GMO) as "*an organism in which the genetic material has been altered in a way that does not occur naturally by mating and/or natural recombination*" (EC 1990). The corresponding Hungarian law is even more specific defining a GMO (in the original text an organism modified by gene technology) as "*a natural organism in which the genetic material has been altered by genetic modification, including the progeny of such organisms carrying the properties appearing as a result of these modifications*", and a genetic modification as "*a method defined by the relevant law issued under the authorisation of this Act which extracts a gene or any part thereof from the cells and transplants it into another cell, or introduces synthetic genes or gene fragments into a natural organism to alter the genetic material of the recipient*" (Government of Hungary 1998). Recognising potential risks of unintended releases and reproduction of GMOs in the environment and possible irreversible consequences, commercial use of genetically modified (GM) crops is authorised only upon assessment of human health and environmental risks on the basis of the precautionary principle (EC 2001). Among currently registered GM plants, the vast majority (approximately 80%) are represented by crops that have been genetically modified for plant protection purposes (so-called first-generation GM plants). Of these first-generation GM plants, according to their acreage in 2017 (Clive 2017), 47% are herbicide-tolerant (HT) plant varieties, 12% are insect-resistant (IR) and 41% contain stacked events (HT and/or IR).

The only genetic event approved in the EU for cultivation for food and feed purposes is IR maize event *MON 810*, cultivated in two EU Member States, Spain and Portugal altogether on 131,535 hectares in 2017 representing a 4% decrease compared to the previous year (Clive 2017). This is a marginal level, representing 0.07%, 0.22% and approximately 2.2% of the global cultivation area of GM crops, GM maize and IR maize, respectively. It has to be also noted that the ratio of stacked events has been rapidly increasing lately. Reported global cultivation areas are, however, somewhat misleading: acreages of stacked event GM crops are biased as are considered as "trait hectares", i.e. actual crop acreage multiplied by the number of traits to "confer multiple benefits in a single biotech variety".

Insect resistance in GM crops is achieved by the incorporation of a transgene encoding an endotoxin protein (or its variety) from *Bacillus thuringiensis* (*Bt*), a well-known insect pathogenic, endospore-forming, soil-borne, Gram-positive bacterium. *B. thuringiensis*, first reported in 1901 in Japan, described in 1915, and proven to have numerous strains worldwide, forms characteristic parasporal bodies during sporulation containing crystal (Cry) and cytolytic (Cyt) endotoxins that are known to exert pore-forming effects in the insect midgut (Palma et al. 2014). Cry endotoxins produced by various *B. thuringiensis* strains are lectin-like proteins with a characteristic three-domain structure consisting of an α-helix subunit (domain 1) facilitating the incorporation of the toxin in membranes; as well as two β-sheets (domains 2 and 3) participating in binding to lectin receptors of the cell membranes in the midgut epithelium and upon oligomerisation-forming pores in the insect midgut (Schnepf et al. 1998). These pores disturb the ion channel functions in the cell membranes; the insect ceases feeding, its digestion stops, and subsequently dies of internal sepsis due to the microwounds created on the midgut wall. Commercial topical microbial *Bt*-based insecticides, containing Cry toxins as their active ingredients, have long been registered and applied in integrated pest management and in ecological farming, and have been found to be effective to control selected insect pests, more benign environmentally than broad-spectrum insecticides and safe for birds and mammals (Kaur 2000; Sanchis 2011; Sanahuja et al. 2011; Gatehouse et al. 2011; Székács and Darvas 2012a; Palma et al. 2014; Bravo et al. 2018). Factors limiting their applicability, however, include low field stability, narrow activity spectrum, and recently an assessment by the European Food Safety Authority (EFSA) raising concern regarding the ability to *Bt* strains to possibly infect humans via food (EFSA 2016) impugned later by *Bt* occurrence, epidemiological and phylogenic data (Raymond and Federici 2017).

Cry toxins have been classified by their primary protein structure (amino acid sequence) into 54 types (Cry1 to Cry54) and several subtypes (e.g. Cry1Aa, Cry1Ba). Different subtypes exert toxicity to different insect orders (Lepidoptera, Coleoptera, Diptera, Hemiptera, Hymenoptera), as well as to nematodes (Rhabditida) and snails (Gastropoda), or even to human cancer cells (Palma et al. 2014). They are typically produced in the microorganism as protoxins that require activation in the alkaline pH of the insect midgut and are stabilised by disulphide bonds in the parasporal protein crystals.

## *Bt* Crop Varieties

Since the cloning of *Bt* strains producing various Cry toxins and the introduction and expression of their genes responsible for Cry toxin production into other microorganisms and into plants, various *Bt* crops have emerged and are being cultivated worldwide (Clive 2017). Transgenic *Bt* potato (Cry3A) against Colorado potato beetle (*Leptinotarsa decemlineata*) and *Bt* cotton (Cry1Ac) against the American bollworm (*Helicoverpa armigera*), the spotted bollworm (*Earias vittella*) and the pink bollworm (*Pectinophora gossypiella*) became commercialised in the USA in 1995, followed by *Bt* maize (Cry1Ab) against the European corn borer (*Ostrinia nubilalis*) in 1996, and another *Bt* maize variety (Cry3) against the Western corn rootworm (*Diabrotica virgifera virgifera*) in 2003. The range of *Bt* plants rapidly increased worldwide in different additional crops including soybean and other field cultures (rice, alfalfa, canola), vegetables (tomato, chickpea), tobacco, sugarcane and poplar with transgenes encoding different Cry and secretable Vip (vegetative insecticidal proteins) *Bt* toxins (10 *Bt* toxins used in transgenic crops against 15 insect pests) in single and combined (stacked) genetic events (single and multiple traits) using the, in the case of "SmartStax" varieties using six stacked Cry genes, three toxins (Cry34Ab1/Cry35Ab1 and Cry3Bb) against coleopteran pests and three other (Cry1A.1.05, Cry2Ab and Cry1F) against Lepidopteran pests, in addition to two traits conferring HR. Although IR GM crops represent the lesser proportion of first-generation GM plants (see above), the overall worldwide cultivation area of *Bt* crops reached over 100 million hectares by 2017 (Clive 2017).

Technological, economic and social benefits of *Bt* crops have been reviewed extensively in the scientific literature (US National Research Council 2010; Hutchison et al. 2010; Royal Society 2016; Brookes and Barfoot 2017; Clive 2017; Carzoli et al. 2018; Dively et al. 2018; Zilberman et al. 2018; Brookes 2019; Catarino et al. 2019) and are reflected in their substantial utilisation in intensive agriculture. Thus, *Bt* crops certainly realise a considerable profit for the variety of owners and have been claimed to produce economic benefits for farmers. In addition, relying on environmentally favourable active ingredients, Cry toxin proteins, *Bt* crops represent environmental benefits relative to broad-spectrum insecticides (some even claim (Carzoli et al. 2018), practically no risks are associated with these crops), as well as communal benefits due to area-wide suppression of pests

(Tabashnik 2010), and increases in beneficial insect populations in around fields of *Bt* crops compared to conventional management (Gatehouse et al. 2011). Environmental risks potentially associated with Cry proteins may differ dramatically among various Cry proteins and by their various expression levels in different crops or different genetic events in the same crop (e.g. maize) (Clark et al. 2005; Székács and Darvas 2012a; Chátalová 2019). Yet, the amount of Cry toxin produced by these crops, increased in the case of stacked Bt events, is only rarely considered, when reduced insecticide applications due to GM technology are estimated (Benbrook 2012; Hilbeck and Otto 2015), and in the case of *Bt* crops the comparator plant protection products (PPPs) should not be the broad-spectrum insecticide agrochemicals, but bioinsecticide of the same type of active ingredient, Cry toxin-based bioinsecticides. This report does not intend to summarise agrotechnological, economical and societal aspects of *Bt* crops, but focuses on its environmental and ecotoxicological impacts as having been considered in their regulation in the EU and particularly in its Pannonian Biogeographical Region within the Natura 2000 protected area network.

In Hungary, as in other EU Member States, a single *Bt* maize variety registered in the EU has been filed for authorisation for public cultivation, genetic event *MON 810* (Monsanto Corp.), and therefore this variety has been assessed by environmental analysis and in ecotoxicity tests. It has to be emphasised that a safeguard clause moratorium on the cultivation of *MON 810* GM maize is effective in Hungary (Ministry of Agriculture, Hungary 2005) on the basis, to a great extent, of the studies summarised here. Therefore, just like in neighbouring Austria, deliberate releases of *MON 810* have been carried out in Hungary only for experimental purposes. In addition, a different *Bt* maize variety *DAS 59122–7* (Pioneer Hi-Bred International, Inc.) was assessed in analytical and tritrophic biological studies.

## Transgenic Cry Proteins Expressed, Methodological and Conceptual Problems in Their Analysis

To assess possible effects (main or side-effects) of *Bt* maize, the amount of *Bt* toxins (Cry or Vip) produced needs to be determined. This is of importance not only for the efficacy of the technology, but also for assessing unintended effects on non-target organisms depending on the level and distribution of the transgenic Cry or Vip toxin produced. It is a requirement for the registration of all PPPs that the active ingredient has to be quantitatively detectable, for which appropriate analytical method has to be available; moreover, analytical standards of the purified active substance of relevant metabolites have to be provided by the applicant or producer upon request (EC 2009, 2013). In accordance, analytical methods for detecting Cry toxin residues in commerce, as well as expression level data of the toxin (termed "plant-incorporated protectant") in various plant organs were requested by the US Environmental Protection Agency (EPA) for the reassessment of *Bt* crops (Mendelsohn et al. 2003).

Cry and Vip toxins are commonly analysed by enzyme-linked immunosorbent assays (ELISAs) in a so-called sandwich ELISA setup. In these immunoassays, Cry-specific antibodies are immobilised on the wall of 96-well microplates, and toxin protein captured from the sample is allowed to react with a second Cry-specific antibody labelled with a reporter enzyme, or this second antibody is further reacted with an immunoglobulin-specific antibody labelled with a reporter enzyme. Immunoassays are readily used in research and development, but are less frequently accepted for regulatory, surveillance or enforcement purposes by control authorities, particularly if chromatographic instrumental analytical methods are available. Exceptions to this include the use of immunoassay as a means of determination of the amount of a pesticidal *Bt* protein in GM crops and commodities, where ELISAs or lateral flow devices are the method of choice (Grothaus et al. 2006). These immunoassays appeared to be of good reproducibility, yet reported Cry toxin concentrations in GM plants highly vary among different laboratories, cultivation sites or even with the same GM variety at a given location. To test quantitative detectability of transgenic Cry toxins by the ELISA method approved by Monsanto Corp. for toxin determination, to corroborate its analytical features, and to follow Cry toxin production in the crop during vegetation, we tested the performance of Cry toxin-specific immunoassay method in detail. We have assessed the analytical performance of the immunoanalytical determination of Cry toxins, and have identified various sources of analytical variation and error (Takács et al. 2012a; Székács 2013). Such errors included discrepancy in the identity of the analyte, consequent inaccuracies in Cry toxin content reported, as well as tissue-specific and seasonal variabilities in Cry toxin levels produced in *Bt* maize (see below).

## *Protein Forms of Given Cry Toxins*

As mentioned above, the Cry toxin content in *B. thuringiensis* endotoxin crystals are mostly protoxins, from which the active toxin form is liberated by alkaline hydrolysis. In the case of Cry1Ab toxin, the molecular mass of the protoxin is 131 kDa, and the protein forms bipyramidal crystals stabilised by a maximum of 16 disulphide bonds per molecule. Upon reduction of the disulphide bonds and hydrolytic cleavage of the protoxin, an activated toxin with a molecular mass of approximately 63–65 kDa is formed. The transgene in *MON 810* encodes neither the protoxin, nor the activated toxin, but a protein form in between, a partially hydrolysed Cry1Ab protoxin of 91 kDa molecular mass; therefore, it produces this so-called preactivated toxin. As seen from the above, the active ingredients of the microorganism-based *Bt* bioinsecticide and of *MON 810* maize are different, being the Cry1Ab protoxin (131 kDa) and the preactivated Cry1Ab toxin (91 kDa), respectively, both hydrolysed in the insect to form the activated Cry1Ab toxin (63–65 kDa) responsible for insecticidal activity (Székács and Darvas 2012a, b).

An important analytical consequence of the above is that ELISA kits manufactured for the determination of bacterial Cry endotoxins (using Cry protoxin as an immunogen) will provide biased results when detecting the preactivated Cry toxin: antibodies directed to the protoxin are expected to show lower affinity to the truncated preactivated toxin protein, therefore, virtual signals sensed in quantitative immunoassays validated to detect protoxin molecules will correspond to higher concentrations of the preactivated toxin than those of the protoxin (as the antibody has lower immunoaffinity to the former than to the latter). The extent of the bias is described by the cross-reactivity (CR) between these toxin forms, defined as the percentage ratio of their $IC_{50}$ values in the ELISA test.

CR values determined for Cry1Ab protoxin/preactivated toxin ranged 0.41–0.56, indicating that the ELISA kits are suitable to detect Cry1Ab protoxin (in microbial samples), but require correction with the CR values determined when used on *MON 810* maize samples containing preactivated Cry1Ab toxin (Székács et al. 2010a). Actual preactivated Cry1Ab toxin concentrations in these *Bt* maize samples are 1.8–2.3-fold higher than detected by the Cry1Ab protoxin-specific ELISA kits. This applies to Cry1Ab values in *MON 810* maize reported to date in the scientific literature, including data by the variety owner, Monsanto Corp.

The other *Bt* maize studied was variety *DAS 59122-7* producing Cry34Ab1 (14 kDa) and Cry35Ab1 (44 kDa) binary toxins. Similar, but less substantial differences exist for these toxins between their microbial and plant-biosynthesised forms, where the maize-derived Cry34Ab1 and Cry35Ab1 proteins were found nearly identical to the microbe-expressed forms, with Cry34Ab1 having one amino acid missing at the N-terminal and exhibiting forms at 60, 50 and 42 kDa in addition to the expected 13.6 kDa protein (Latham et al. 2017). Therefore, significant differences in the CRs of these toxins originated from microbes and maize are not expected.

## *Matrix Effects in the Determination of Cry Toxin Levels*

Maize leaf material is a sample matrix commercial ELISA kits have been validated for. Therefore, Cry toxin measurement in foliage is unproblematic, other than toxin level fluctuation in the leaves, but that is not a question of tissue matrix. In addition to that, the ELISA kits were straight forward applicable on stem, root and seed samples as tissues of plant origin (Székács et al. 2010a, b; Takács et al. 2012a). More marked matrix effects were observed with pollen that required higher sample extract dilution due to its high fat, protein and mineral contents (Székács et al. 2010a). In addition, the Cry1Ab ELISA test was assessed and used on animal tissues as well (Takács et al. 2015).

## *Assay Validation*

Analytical characteristics and applicability of the ELISA kits were tested using Shewhart analytical control charts and quality control by internal standard reference samples to detect analytical goodness (precision, accuracy, stability) for the detection of both Cry1Ab (Takács et al. 2012a, 2015) and Cry34/35Ab1 toxins (Takács et al. 2010, 2012b). In addition, an inter-laboratory ring trial test has been carried out with the participation of specialised laboratories in Germany, Hungary, Norway and Switzerland to explore whether high variability in reported Cry toxin concentrations in the same *Bt* maize variety is due to the ELISA protocols, instrumentation, extraction methods, human error, sample reproducibility or plant variability (Székács et al. 2012). In turn, such ring tests have been proposed to be performed as a part of the standardised environmental risk assessment (ERA) of *Bt* maize effects on non-target insects as a means of external quality assurance (Lang et al. 2019). Reduction or elimination of sources of analytical variability allows feasible quality control of *Bt* plants and makes proper interpretation of differences or variability among published data from different laboratories possible, but the results underlined the importance of well-controlled reference materials, ELISA kits and protocol, particularly for reported concentrations of Cry toxins in pollen that render mathematical models for the environmental fate (Romeis et al. 2008) or biological effects (Perry et al. 2010) burdened with uncertainty.

## Estimated Production and Bioavailability of Cry Proteins

*Bt* maize varieties, *MON 810* and *DAS 59122-7* cultivars, were demonstrated to produce the corresponding Cry toxins (Cry1Ab and Cry34/35Ab1) in a tissue- and time-specific manner (see below). Cry1Ab in *MON 810* provides protection against Lepidopteran pests, particularly against larvae of the European corn borer feeding in the stem. This pest may damage in two or three generations in a season, therefore, the highest level of expression of the transgenic protein should preferably occur in the stem from the VT growth stage on. In contrast, Cry34/35Ab1 in *DAS 59122-7* provides protection against Coleopterans, e.g. the corn rootworm that damages at the larval stage of the root. Thus, the highest level of toxin production would be desirable in the root in the V12-R3 growth stages. To assess compliance of toxin production dynamics with these required protection times, actual toxin production was experimentally systematically monitored throughout the entire vegetation periods for these *Bt* maize varieties. When available, Cry toxin production was compared to the availability of the corresponding Cry toxin from *Bt*-based bioinsecticides.

## *Tissue and Temporal Variability of Cry Toxin Production in* Bt *Maize Varieties*

Levels of Cry1Ab toxin (corrected for active toxin content on the basis of cross-reactivities between the activated Cry1Ab toxin and the Cry1Ab protoxin) were found to fall between 9.6 ± 2.1 and 17.2 ± 1.7 µg/g in the leaves, 0.47 ± 0.03 and 5.0 ± 0.3 µg/g in the stem, 2.3 ± 0.3 and 5.3 ± 0.5 µg/g in the roots and 0.03 ± 0.01 and 0.5 ± 0.03 µg/g in the pollen of *MON 810* maize (plant material expressed in fresh weight) with characteristic patterns during the vegetation period tested in 3 different years within an 8-year period (Székács et al. 2005, 2010a). Since crop damage by the European corn borer (causing yield loss by decreased kernel number and weight due to disruption of plant growth, broken stalks and dropped ears) occurs mainly in the stem, it is a rather unfavourable feature of the *MON 810* maize variety (DK-440-BTY) that only 12–20% of the Cry1Ab toxin protein biosynthesised in the plant is produced in the stalk. This means the plant produces 7–8-times more toxin protein, than the amount being utilised in the crop protection mechanism.

Poor targeting of pesticide application is, however, not unique to *Bt* crops. Estimates of the efficacy of spray applications range from 1% (Pimentel 1995) to 30–40% (Matthews et al. 2014). The 12–20% accuracy of *Bt* crops regarding Cry toxin content in the target plant tissues also falls into this range. The accuracy of aerosol treatments, however, can be enhanced with targeted application and precision agricultural technologies (Pedersen and Lind 2017), while toxin production is determined by the genetic sequence of the GM crop. Therefore, GM crop development is strongly recommended to focus on varieties with target tissue-specific transgene expression.

The toxin content in pollen has been found strikingly different among different *MON 810* maize varieties provided by the variety owner, Monsanto Corp. Preactivated CryAb toxin quantity in the pollen of those varieties was determined to be 0.03 ± 0.01, 0.11 ± 0.02, and 0.47 ± 0.03 µg/g fresh weight, while pollen productivity was practically unchanged, 1.39 ± 0.33 g/plant among varieties and cultivation years. Pollen amount on the field was determined to be 3.5–5.5 × $10^{11}$ pollen/ha, which is only a fraction of the potentially produced pollen quantity (6.4–7.2 × $10^{11}$ pollen/ha).

Minor, but statistically significant variability was found in preactivated Cry1Ab toxin content in maize leaves diagonally, with approximately 20% higher levels (9.9 ± 0.9 µg/g fresh weight) near the leaf vein, than further towards at leaf edges. Longitudinal distribution of the preactivated toxin showed a much higher variability in the leaves, with the highest toxin concentration (8.9 ± 1.5 µg/g fresh weight) in the lamella middle between the base and the leaf tip, almost 5- and 2-fold higher than at the sheath and at the tip, respectively. Low levels at the sheath are explained by the leaf base being the most rapidly growing zone of the leaf, and at the tip with partial plant tissue necrotisation, as decreased toxin levels were seen only in slightly yellow leaf tips (Székács et al. 2010b). Necrotisation has been found a major cause of decrease in toxin concentrations among leaf levels (with outstandingly, 1.3- to

2.3-fold lower toxin concentrations 4.8 ± 1.0 μg/g fresh weight at the first leaf level than at all other leaf levels) and in the stem as well. Cry1Ab toxin in the plant tissue is protected from rapid decomposition, and can long remain in the stubble in maize roots (containing 7.7–9.7% of the overall toxin production of the plant) or as plant foliage biomass enters the soil unintentionally or intentionally during harvest. Results indicate that 1–8% of the toxin content in the stubble can be detected 1 year after harvest, indicating environmental persistence of the toxin protein in the stubble (Székács et al. 2005; Székács and Darvas 2012a).

Concentrations of Cry34Ab1 and Cry35Ab1 binary toxins in the leaves of *DAS 59122-7* maize were 81.1 ± 17.7 and 75.1 ± 11.9 μg/g dry weight, respectively. The longitudinal distribution of the toxin proteins showed a similar trend than seen for *MON 810* maize: Cry34Ab1 and Cry35Ab1 toxin levels were 3.1- and 2.7-fold higher, respectively, in the lamella middle of the leaf than in the leaf base. Because crop damage by the corn rootworm occurs in the root nodes, the efficacy of Cry34/35Ab1 toxin production of the *DAS 59122-7* maize variety is rather unfavourable as only 2–3% of the toxin protein biosynthesised in the plant is produced in the root, indicating that the plant produces 35–46-times more toxin protein, than the toxin amount providing the desired protective effect. The pollen contained 47.4 ± 12.3 and < 0.12 μg/g fresh weight (the latter being the limit of detection of the ELISA) of Cry34Ab1 and Cry35Ab1, respectively (Takács et al. 2011, 2012b).

Assessment of the production of preactivated Cry1Ab toxin in the tissues of *MON 810* and of Cry34Ab1 and Cry35Ab1 binary toxins in *DAS 59122-7* maize appeared to be proportional with chlorophyll content and therefore, photosynthetic activity in the tissue. This is of significance not only for plant physiology, but also for the exposure of herbivorous insects. Insects that prefer green plant tissues (unlike species that develop in the fruit (fructus), seed, root or saprophagous ones) will tend to become exposed to tissues with the highest toxin content. Such photosynthesis-related toxin production, however, is far not optimal for maize pest control: larvae of the European corn borer (*O. nubilalis*) feeds in the stem, where 3- to 17-fold lower preactivated Cry1Ab toxin production occurs than in the leaves of *MON 810* maize (Székács et al. 2010a), and larvae of the corn rootworm (*D. virgifera*) damages the root, where Cry34Ab1 and Cry35Ab1 toxin production is approximately 5- and 10-fold lower, respectively, than in the leaves of *DAS 59122-7* maize (Takács et al. 2011, 2012b).

## *Determination of Cry Toxin from* Bt*-Based Bioinsecticides*

Conventional insecticides are officially characterised by their net active ingredient content, while such specification is unfortunately no longer required for endotoxin-based bacterial preparations, as these bioinsecticides are assessed by their efficacy (ITU/g, ITU referring to international toxic units). However, it is not the bacterium, but its endotoxins that are responsible for the biological effect; therefore, efficacy should be attributed to the actual endotoxin content. To monitor active ingredient content, immunoanalytical (ELISA) determinations were applied to Cry1Ab and Cry4 endotoxin content in bioinsecticide preparations Dipel (Székács and Darvas 2012a) and Vectobac (Fejes et al. 2011), respectively. These studies with Dipel revealed that only a minor fraction of the toxin protein is immediately bioavailable (soluble) at neutral pH, the vast majority of the crystal mass is only bioaccessible (temporarily non-bioavailable), and a part of the entire endotoxin content is non-bioavailable due to decomposition during crystal digestion. Thus, the nominal concentration of a common formulation of Dipel, 3.2% (corresponding to 32 mg/g bacterial protein in the bioinsecticide) corresponded to average bioavailable Cry1Ab/Cry1Ac endotoxin content of 20.6 ± 2.6 μg/g and to bioaccessible Cry1Ab/Cry1Ac endotoxin content of 0.085–8.16 mg/g. In addition to a clarification of the active ingredient content, these measurements allowed to compare detected Cry1Ab production of *MON 810* maize to corresponding bioavailable and bioaccessible Cry1Ab/Cry1Ac endotoxin content in Dipel (Székács and Darvas 2012a). In the case of Vectobac, determination of the Cry4 endotoxin content by immunoassay has been correlated with efficacy measurement in dose–mortality bioassays (Fejes et al. 2011).

## Resistance Development and Non-target Effects

Beyond the technological advantages of the *Bt* maize varieties, their potential technological drawbacks also need to be assessed. These include the problems of emergence of pest resistance, potential non-target effects and applicability of these *Bt* corps in integrated pest management (IPM) practice. The first two issues are discussed in this section, and the third one is covered under "Legislatory measures" (see below).

Some of the general problems of pest control discussed in the context of the assessment of *Bt* crops are in fact not unique solely to *Bt* crops. This particularly applies to pest resistance development, where problems emerged, but achievements in their mitigation have also been accomplished (see below). Moreover, clear advantages of the environmentally benign characteristics of Cry proteins compared to broad-spectrum insecticides need to be emphasised.

### *Resistant Populations*

Pest resistance development is a practically inevitable natural response to intervention by agricultural technologies in the agro-ecosystem. Any method of pest control can potentially be overcome by the evolution of resistance in the pest population. Resistance to pesticides (synthetic and biological, including *Bt* sprays) is rife and is more a reflection of widespread and continual use of PPPs than the properties of those products themselves. Moreover, even crop rotations to suppress build-up of pest populations can be defeated by those pests: corn rootworm has evolved extended diapause in response to attempts to control it in growing soybeans between successive maize crops.

The occurrence of resistance against a single substance is easier to emerge if a simple mutation in the pest population can result in biochemical changes that inactivate the site of action of the compound abolishing susceptibility of the mutant individual to the agent. A common approach to prevent resistance is the parallel use of different agents acting by somewhat or completely different modes of action. Resistance development against *Bt* microorganisms or their preparations is slow, as their numerous, related, but different Cry toxins act in concert at different lectin receptors. This resistance to Dipel rapidly declined in the highly resistant populations of the target insect (the diamondback moth, *Plutella xylostella*) upon halting administration of Dipel, but returned upon resumed treatments (Tabashnik et al. 1994). The study not only indicated reduced and restored *in vitro* binding to the receptors in the midgut of the affected insects being associated with emerging and declining resistance, and not only revealed the importance of maintaining a susceptible sub-population of the insect pest that later became the fundamental aspect of insect resistance management (IRM) programmes, but also warned that continuous cultivation of *Bt* crops may also cause resistance problems by eliminating temporal refuges for susceptible insect sub-populations. Indeed, field-evolved resistance against single Cry toxins in *Bt* crops has later been reported in different insect pests in various regions from the United States to Australia, India, South Africa and China (Tabashnik et al. 2013), yet a more recent survey indicates that such occurrences are narrowly distributed (Tabashnik and Carrière 2017). Although the incidence of practical resistance (resistance occurring in at least 50% of the pest insect population resulting in an observable decrease in crop insecticidal efficacy) has been on the rise in the past two decades, pest susceptibility was somewhat more frequently sustained. Practical resistance to *Bt* crops occurred mostly in maize, followed by cotton. Various approaches including the "high dose/refuge" strategy using non-*Bt* plants in the cultivation area to allow limited reproduction of the susceptible insects and the "pyramid" strategy of parallel use of two or more toxins with affinity to different lectin receptors have been applied for IRM. This is commendable – *Bt* crops at least have mandated IRM programmes, as requested by the US EPA (Mendelsohn et al. 2003), unlike many other pest control products, and these programmes have been successful in their own terms; for example, IRM attempts to delay resistance evolution, not to prevent it altogether (which would be impossible). Combined action and synergism of several toxins, however, not only provide advantages against pest resistance, but may also result in combined side-effects on non-target organisms (Then 2010; Hilbeck and Otto 2015), although such side effects are expected to be additive, as synergism has been claimed quite rare among the Cry proteins used in *Bt* crops (Walters et al. 2018). Field-evolved resistance in the corn rootworm and the European corn borer occurs in 6–7 years of application, particularly when singe Cry toxins are applied, and in the case of extended chemical pressure applied by preactivated Cry1Ab toxin (produced by *MON 810* maize), resistance against Cry1Ab was found to be combined with cross-resistance to Cry2 toxins (Darvas 2011). In such cases, pest resistance triggered by *MON 810* maize renders the application of *Bt*-based bioinsecticides, such as Dipel, also ineffective.

## *Effects on Protected Insects (Lepidoptera)*

Effects of *Bt* crops on non-target organisms, e.g. non-target insects, have to be considered in their ERA (Wolfenbarger and Phifer 2000; Marvier 2001; Darvas et al. 2004; Andow and Hilbeck 2004; O'Callaghan et al. 2005; Andow and Zwahlen 2006; Romeis et al. 2006, 2008; Lang et al. 2007; Lang and Otto 2010; Hilbeck et al. 2011). Practically all methods of pest control will have effects on non-target organisms as well, the most obvious of which being non-target toxicity of insecticides. The more specific an anti-insect agent is, the more favourable it potentially is in terms of non-target effects. Due to the insect specificity of Cry1Ab toxin produced by *MON 810* maize, only Lepidopteran insect species are at hazard. These species are, however, not limited to herbivorous insects feeding on *Bt* maize, as air-drifting maize pollen may settle on other plants, and insects feeding on those plants may become thus exposed by ingesting *Bt* maize pollen along with their food. Sublethal physiological symptoms (decreased larval, pupal and adult weight, delay in development) heighten mortality of the affected individuals and possibly their population.

Three ruderal weed species, frequently emerging on the perimeters of maize fields, the stinging nettle (*Urtica dioica*), the European dewberry (*Rubus caesius*) and Jimsonweed (*Datura stramonium*) were proven to have substantial pollen capture capacity of 328 ± 200, 431 ± 334 and 339 ± 266 pollen grains/$cm^2$, respectively. Protected Lepidopteran species in the Pannonian Biogeographical Region, potentially exposed to the pollen of *MON 810* maize were identified by comparing their habitat preferences with the pollen shedding period of maize. There exist 213 protected butterfly species in Hungary (the Pannonian Biogeographical Region), 50 of which occur in the perimeters of maize fields (Darvas et al. 2004). Thus, during pollination, larvae of the comma butterfly (*Polygonia c-album*), the peacock butterfly (*Nymphalis io*, earlier *Inachis io*), the red admiral (*Vanessa atalanta*) and the small tortoiseshell (*Aglais urticae*) feeding on stinging nettle; larvae of the cardinal (*Pandoriana pandora*), the lesser marbled fritillary (*Brenthis ino*), the niobe fritillary (*Argynnis niobe*) and the red underwing skipper (*Spialia sertorius*) feeding on the European dewberry; as well as larvae of the death's-head hawkmoth (*Acherontia atropos*) feeding on Jimsonweed were specified as species that suffer the greatest level of exposure to the pollen of *Bt* maize (Lauber 2011).

Cry1Ab toxin content in the pollen of certain *MON-810-6* varieties (DK-440-BTY) (0.5 ± 0.03 μg Cry1Ab preactivated toxin/g pollen, see above) caused mortality on the larvae of protected butterflies, including the peacock butterfly (*N. io*). Sensitivities (assessed by $LC_{50}$ values against Dipel) of the larvae of the protected Lepidopteran species investigated to Cry1 toxin ranged between 1.9 and 15.1 μg/ml:

4.4 μg/ml of *N. io* in stage L1, significantly higher, 1.9 μg/ml in stage L2, 3.0–5.7 μg/ml in stages L3–L4 and slightly but significantly higher, 6.2 μg/ml in stage L5. Sensitivities of *N. c-album* and in *V. atalanta* in stage L1 were 1.7- and 3.5-fold lower than *N. io* in the same stage. Lepidopteran maize pest insects, the American bollworm (*H. armigera*) and the European corn borer (*O. nubilalis*) were 3.4–26.5-fold and 4.1–7.4-fold less sensitive in stages L1 and L2, respectively, than *N. io* in the same stages (Lauber 2011). The increased sensitivity of *N. io* was shown to be related to group behaviour of stage L1 larvae: mortality of lone larvae increase to 25–75% due to suppressed feeding in the absence of group stimuli (Lauber and Darvas 2009; Székács and Darvas 2012b), therefore, larval mortality due to consuming pollen containing Cry1Ab toxin triggers an avalanche-like effect that exaggerates mortality in larvae not lethally affected by Cry1Ab toxin but remaining solitary by the mortality of their groupmates. The exact extent of this effect could be ascertained by a detailed risk assessment as performed for the monarch butterfly (*Danaus plexippus*) in the USA (Sears et al. 2001). Major differences to the monarch butterfly case are, however, that the peacock butterfly, unlike the monarch butterfly, is a protected species, the habitat of which is safeguarded by law; and that the European corn borer is not a major pest in the Pannonian Biogeographical Region. On the basis of its outstanding sensitivity to Cry1Ab toxin, *N. io* was suggested as a model species for ERA of Cry1Ab (Lauber and Darvas 2009), which has later been implemented (Holst et al. 2013a, b; Fahse et al. 2018). As seen from the sensitivity data discussed, almost an order of magnitude difference in sensitivity to Cry1Ab occurs among larvae in various stages of protected butterflies, and larvae of pest insects are even less sensitive.

A strange sequel in light of the above has been that a mathematical model, authored by some of the members of the EFSA GMO Panel at that time (Perry et al. 2010), that analysed exposure of larvae of non-target species, e.g. *N. io* and *V. atalanta* to Cry toxins in four European countries, assumed larvae of *V. atalanta* and *N. io* equally susceptible to Cry1Ab. They cited Darvas et al. (2004) as a reference for such equitoxicity, even though the cited paper contains no data about species sensitivity. Lang et al. (2011) found that the incomplete and uncertain input data cause a higher uncertainty than indicated by Perry et al. (2010). In the mathematical model extended to non-target effects of Cry1F toxin in *Bt* maize pollen (Perry et al. 2012), the sensitivity of non-target insects has been considered purely on a theoretical basis, meanwhile the predictive power of a mathematical model rests on the certainty of its input data (species sensitivity in the current case), which cannot be speculative. Another flaw of the model is that it defines acceptable mortality thresholds, while no such thresholds apply to protected species in ERA. Pollen drifting from maize fields modifies habitat characteristics of protected species, which contradicts the Habitat Directive of the EU (EC 1992). The EFSA model (Perry et al. 2010, 2012; EFSA 2015) was later developed into the BtButTox model (Holst et al. 2013a, b) and the LepiX model (Fahse et al. 2018), but all these models, although lately became quite elaborated, rely on extrapolated data, while the only solid data measured on *N. io* are ours.

## *Effects on Soil-Borne Insects*

A collembollan species (*Folsomia candida*) showed avoidance of stubble residues of *MON 810* maize (DK-440 BTY, "YieldGard"), but adapted to it upon longer exposure, and no relationship was found between physiological parameters and feeding history, except that insects feeding on *MON 810* maize stubble had lower egg and faecal pellet production, demonstrating that food selection is a key factor in population dynamics (Bakonyi et al. 2006, 2011). The results indicate that long-term feeding on maize containing Cry1Ab toxin does not appear to be harmful to this collembolan, and therefore, avoidance of *MON 810* maize as a food source may have been a result of the modified composition of the maize variety. *Bt* maize appears to be a less preferred and therefore probably a less usable food source for *F. candida* than the corresponding isogenic maize variety (DK-440). The data also illustrate that effects on soil-forming, decomposing microorganisms have not yet been sufficiently explored.

## *Effects on Toxinogenic and Arbuscular Fungi*

Cry toxins may affect the production intensity of certain *Fusarium* mycotoxins by suppressing damage by insects serving as vectors for fungal infestation, with favourable health and economic consequences due to the hindrance of mycotoxin production (Wu 2006; Ostry et al. 2010; Folcher et al. 2010). The occurrence of *Fusarium* species, however, is only partially related to insect pest damage. Our corresponding studies also revealed that damage on *MON 810* maize cobs was caused predominantly by the cotton bollworm (*H. armigera*), where occasionally there occur insects surviving Cry1Ab toxin exposure, although *Fusarium* infestation is not transmitted in all cases (Darvas et al. 2011). By suppressing fungal infection by insect damage, the production of fumonisin B1 substantially decreased in *DAS-59122-7* maize (Bánáti et al. 2017).

The effect of Cry34/35Ab1 binary toxins produced by *DAS-59122-7* maize on the mycorrhizal colonisation on the roots by arbuscular mycorrhizal (AM) fungi was studied during the entire vegetation period (Seres et al. 2014). Statistically significantly (27–37%) reduced initial hyphal, arbuscule and arbuscular mycorrhizal colonisation was recorded on the root of the *DAS-59122-7* maize variety than in the control for up to 60 days after planting under field cropping conditions, but the effect vanished later (80–140 days), as the intensity of the arbuscular infection increased over time during plant maturation. In contrast, no reduction in vesicle colonisation was seen. The influence of GM crops on AM fungi is further discussed in chapter "Impact of Genetically Modified Crops on the Biodiversity of Arbuscular Mycorrhizal Fungi" of this book.

## Tritrophic Assessment of *Bt* Maize

To test non-target effects by Cry toxins exerted through the food chain, physiological parameters of a parasitoid and a predator insect preying on non-target herbivores were tested in tritrophic bioassays. The assay design allowed assessment of the effects of indirect exposure of the non-target parasitoid or predator to Cry toxins through prey. In the case of the tritrophic study with a predator insect, direct exposure though pollen could also be evaluated.

In a tritrophic assessment setup upon exposure to *MON 810* maize, survival and development parameters of a product storage pest, the maize weevil (*Sitophilus zeamais*) and its natural enemy the ectoparasitoid pteromalid wasp *Lariophagus distinguendus*, used in biological control of weevils, were assessed (Hansen et al. 2013). Preactivated Cry1Ab toxin content in the maize did not significantly affect emergence rates or development time of the maize weevil, but the body mass of the adult females that fed on *MON 810* maize was moderately (2–6%), but statistically significantly higher than the control (isogenic line) in the absence of the parasitoid. This can result in increasing reproduction rate of the weevil population through increased fecundity of the larger females. The presence of the parasitoid with a preference to larger females as hosts for oviposition can counterbalance this effect. No significant differences were observed in the development time, body size, sex ratio or wing length of the emerging adult parasitoids; however, significantly (approximately 40%) fewer female parasitoids emerged from the treatment with *MON 810* maize than the control. Thus, tritrophic effects of transgenic maize on this parasitoid were demonstrated.

In another study with *DAS-59122-7* maize, long-term effects on the fecundity and fertility of the seven-spotted ladybird (*Coccinella septempunctata*) preying on the bird cherry-oat aphid (*Rhopalosiphum padi*) was assessed (Takács et al. 2010, 2012b). No significant differences were observed in the sex ratio, fecundity or fertility parameters of the predator, but the average weight of adult *C. septempunctata* that developed and preyed on *R. padi* feeding on DAS-59122-7 maize was significantly (11–29%) lower than in the control (isogenic line). This has been seen separately for both females and males, females being uniformly 20–24% larger than males both in the treatment and the control groups. When, however, three other commercial maize hybrids were also considered (beyond the isogenic line) in the control, this significant difference disappeared in the standard deviation of the four controls (isogenic line + three commercial hybrids).

Similarly to direct non-target toxicity of insecticides, tritrophic effects are also inevitable outcomes of pest control. Tritrophic effects in the first study on the ectoparasitoid wasp *L. distinguendus* are not necessarily direct consequences of the composition or property of the *Bt* crop itself, but may be attributed to the effective pest control resulting in a decrease in the prey population. Nonetheless, in the second study on *C. septempunctata*, the effect appears to be more related to crop composition as the predator insect had access to ad libitum feeding.

## Legislatory Measures

On the basis of the early results of the above studies, a safeguard clause moratorium was announced in Hungary on the cultivation of *MON 810* GM maize (Ministry of Agriculture, Hungary 2005; Darvas and Székács 2011). This has met the criticism of EFSA, and the Hungarian environmental authority (along with corresponding authorities of Greece and Austria) was requested to a hearing by the GMO Panel of EFSA. Delegated by the Hungarian Ministry of the Environment and Water, three researchers, Prof. András Székács (author of this summary), Prof. Béla Darvas and Prof. Gábor Bakonyi presented results of their research groups on 11 June 2008 in Parma, Italy on environmental analysis, protected lepidopteran species and soil biology, respectively. No substantial rebuttal was expressed by the GMO Panel on the hearing and afterwards, yet no acceptance occurred, either. In contrast, EFSA maintained its position regarding ERA of *MON 810*, and Hungary renewed its moratorium on *MON 810*, and the number of European countries announcing such moratoria, joining GMO-free regions or opting out (at least regionally) of GM crop cultivation gradually rose to 19.

Legislation of *MON 810* maize gained recent actuality in the EU, where products containing this *Bt* maize variety have valid authorisation for food and feed purposes until 2027 (EC 2017), but re-registration for public cultivation of this genetic event is pending, while EFSA's position is supportive both in its scientific opinion statement (EFSA 2012) and assessment of its post-market environmental monitoring (EFSA 2019). In this context, ERA of *Bt* crops by EFSA has been criticised for underestimating exposure via pollen deposition (Maren Kruse-Plass et al. 2017) and for relying in some cases on experimental data of deficient or improper ecological relevance in impact assessment on honeybees and earthworms (Chátalová 2019).

A long-discussed issue in the scientific literature has been whether *Bt* crops comply with the principles of IPM. The use of crop cultivars tolerant or resistant to plant diseases, pests or stress factors has is definite preventive approach in IPM practices, and *Bt* plants as IR crop varieties have been argued on this basis to be compatible with IPM. *Bt* crops produce foreign substances that (or close derivatives of which) are registered insecticide active ingredients, therefore, their protection mechanism against the pest does not differ fundamentally from chemical pest control. Instead, these crops can be considered as "pesticides" formulated in the biological plant material. This has been reflected in the reassessment of *Bt* crops by US EPA, where the transgenic toxin was termed "plant-incorporated protectant" (Mendelsohn et al. 2003). There is, however, an essential element in IPM *Bt* crops cannot fulfil: the main ecological principle of IPM is that any protection measure should be initiated and timed only to periods, when pest damage exceeds a critical threshold, and *Bt* crops cannot comply with this requirement as they produce the toxin protein in their entire vegetation period, regardless of the pest population density. In addition, Cry toxin production and the corresponding (bio)chemical load on the environment is quite unfavourable in both *MON 810* and *DAS 59122–7* maize varieties, as the toxin proteins are produced in the highest concentration and amount in the foliage (leaves) of the plant, and not in the organs, where they produce crop protective effect (stem and root, respectively). Thus, *MON 810* and *DAS 59122–7* maize varieties produce their corresponding transgenic Cry toxin proteins in 7–8- and 35–46-fold higher amount, than the technologically utilised quantity, respectively.

One could argue that some (but not all) potential hazards or risks associated with *Bt* bioinsecticides or *Bt* crops are not specific to these technologies or regulated products, but are posed by all forms of pest control. This reasoning would be valid from the aspect that, indeed, all technologies affect their working environment, and the question is whether those effects would still allow sustainability. Such a notion could even lead to a number of philosophical questions. One of these is that ideally, regulation should be technology-neutral: equivalent safety regulation criteria would preferably be applied in different segments of industrial activities. This expectation is, however, currently unrealistic as perceptibly different safety requirements apply to various sectors, due to societal consensus, allowing certain technologies that would be considered hazardous operation in other industrial segments. Another fundamental question clearly reaching far beyond the scope of this report is how essentially the principles of agroecology should be considered in assessing sustainability of intensive agriculture. Yet another basic question could be whether the precautionary principle implemented in risk assessment in the EU is reasonable, as excessive precaution prevents the benefits of the technology (Zilberman et al. 2018). Beyond my conviction that it is reasonable, as established in its concept, implementation and normative standardisation (Myhr 2010), this is certainly not a point to be considered at the level discussed here. Hazard identification and risk assessment relates to given technologies, and should not depend on the safety of other technologies: decision-making on the basis of comparative analysis of various technologies is a part of risk management.

As seen from the above, although *Bt* toxins in insect control are environmentally more benign than broad-spectrum insecticides, and economic and social benefits of *Bt* crops have been highlighted (US National Research Council 2010; Dively et al. 2018), concerns regarding environmental effects of *Bt* crops have also been raised, and the lack of consensus on their safety has been published (Hilbeck et al. 2015) and has also been evidenced by the UN Cartagena Biosafety Protocol and the Guidelines of the *Codex Alimentarius*. To address the environmental and socio-economic risk assessment interface, a European Network for systematic GMO impact assessment (ENSyGMO) has been proposed to enhance ERA and post-market environmental monitoring of GM (including *Bt*) crops (Graef et al. 2011). Nonetheless, such concerns, accentuated by the precautionary principle of the EU, apply not only to transgenic GMOs, but also in case-by-case assessment to *Bt* technology applied in combination with RNA interference (RNAi) (Heinemann et al. 2013; Head et al. 2017) and to products of emerging biotechnologies including genome editing (Székács 2016), and it remains questionable whether currently dominant bioeconomy solutions do indeed represent a step towards the ultimate development goal of truly sustainable ecocycles (Székács 2017).

**Disclaimer** This work was supported by the Hungarian Ministry of the Environment and Water. The author is a member of the Management Board of the European Food Safety Authority (EFSA) and of the Scientific Advisory Body of the Co-operative Research Programmes (CRP) of the Organisation for Economic Co-operation and Development (OECD). This publication reflects the views of the author, and EFSA or OECD cannot be held responsible for any use which may be made of the information contained therein.

## References

Andow DA, Hilbeck A (2004) Science-based risk assessment for nontarget effects of transgenic crops. Bioscience 54(7):637–649. https://academic.oup.com/bioscience/article-abstract/54/7/637/223525

Andow DA, Zwahlen C (2006) Assessing environmental risks of transgenic plants. Ecol Lett 9(2):196–214. https://doi.org/10.1111/j.1461-0248.2005.00846.x

Bakonyi G, Szira F, Kiss I, Villányi I, Seres A, Székács A (2006) Preference tests with collembolas on isogenic and Bt-maize. Eur J Soil Biol 42:S132–S135. https://doi.org/10.1016/j.ejsobi.2006.06.005

Bakonyi G, Dolezsai A, Mátrai N, Székács A (2011) Long-term effects of Bt-maize (*MON810*) consumption on the collembolan *Folsomia candida*. Insects 2(2):243–252. https://doi.org/10.3390/insects2020243

Bánáti H, Székács A, Fehér-Tóth S, Czéh Á, Darvas B (2017) Determination of mycotoxin production of *Fusarium graminearum* and *Fusarium verticillioides* in genetically modified maize varieties by quantitative flow immunocytometry. Toxins 9(2):70. https://doi.org/10.3390/toxins9020070

Benbrook CM (2012) Impacts of genetically engineered crops on pesticide use in the U.S. – the first sixteen years. Environ Sci Eur 24:24. https://doi.org/10.1186/2190-4715-24-24

Bravo A, Soberón M, Gill SS (2018) *Bacillus thuringiensis*: mechanisms and use. Comp Mol Insect Sci 6:175–205. https://doi.org/10.1016/B978-0-12-809633-8.04071-1

Brookes G (2019) Twenty-one years of using insect resistant (GM) maize in Spain and Portugal: farm-level economic and environmental contributions. GM Crops Food 10(2):90–101. https://doi.org/10.1080/21645698.2019.1614393

Brookes G, Barfoot P (2017) Farm income and production impacts of using GM crop technology 1996–2015. GM Crops Food 8(3):156–193. https://doi.org/10.1080/21645698.2017.1317919

Carzoli AK, Aboobucker SI, Sandall LL, Lübberstedt TT, Suza WP (2018) Risks and opportunities of GM crops: Bt maize example. Glob Food Secur 19:84–91. https://doi.org/10.1016/j.gfs.2018.10.004

Catarino R, Areal F, Park J, Parisey N (2019) Spatially explicit economic effects of non-susceptible pests' invasion on Bt maize. Agric Syst 175:22–33. https://doi.org/10.1016/j.agsy.2019.05.008

Chátalová V (2019) A critical evaluation of EFSA's environmental risk assessment of genetically modified maize MON810 for honeybees and earthworms. Environ Sci Eur 31:52. https://doi.org/10.1186/s12302-019-0238-5

Clark BW, Phillips TA, Coats JR (2005) Environmental fate and effects of *Bacillus thuringiensis* (Bt) proteins from transgenic crops: a review. J Agric Food Chem 53:4643–4653. https://doi.org/10.1021/jf040442k

Clive J (2017) Global status of commercialized biotech/GM crops in 2017. ISAAA briefs no. 53. International Service for the Acquisition of Agri-biotech Applications (ISAAA), New York. https://www.isaaa.org/resources/publications/briefs/53/download/isaaa-brief-53-2017.pdf

Darvas B (2011) GM plants and resistance – resistance-management. In: Darvas B, Székács A (eds) Hungarian background on views of 1st generation genetically modified plants. Agricultural Committee of the Hungarian Parliament, Budapest, pp 140–141. http://www.kormany.hu/download/2/9d/20000/GenetEM.pdf

Darvas B, Székács A (eds) (2011) Hungarian background on views of 1st generation genetically modified plants. Agricultural Committee of the Hungarian Parliament, Budapest, pp 108–213. ISBN:978-963-9848-44-3. http://mek.oszk.hu/09900/09933/index.phtml

Darvas B, Csóti A, Gharib A, Peregovits L, Ronkay L, Lauber É, Polgár AL (2004) Some data to the risk analysis of Bt -corn pollen and protected Lepidoptera species in Hungary. Növényvédelem 40:441–449. (in Hungarian)

Darvas B, Bánáti H, Takács E, Lauber É, Szécsi Á, Székács A (2011) Relationships of *Helicoverpa armigera*, *Ostrinia nubilalis* and *Fusarium verticillioides* on *MON 810* maize. Insects 2(1):1–11. https://doi.org/10.3390/insects2010001

Dively GP, Venugopal PD, Bean D, Whalen J, Holmstrom K, Kuhar TP, Doughty HB, Patton T, Cissel W, Hutchison WD (2018) Regional pest suppression associated with widespread Bt maize adoption benefits vegetable growers. Proc Natl Acad Sci U S A 115(13):3320–3325. https://doi.org/10.1073/pnas.1720692115

EC (European Commission) (2013) Commission Regulation (EU) No 283/2013 of 1 March 2013 setting out the data requirements for active substances, in accordance with Regulation (EC) No 1107/2009 of the European Parliament and of the Council concerning the placing of plant protection products on the market Text with EEA relevance. Off J Eur Community L93(3.4.2013):1–84. https://eur-lex.europa.eu/legal-content/EN/ALL/?uri=CELEX%3A32013R0283

EC (European Commission) (2017) Commission implementing regulation (EU) 2017/1207 of 4 July 2017 of renewing the authorisation for the placing on the market of genetically modified maize MON 810 (MON-ØØ81Ø-6) products pursuant to Regulation (EC) No 1829/2003 of the European Parliament and of the Council. Off J Eur Community L173:18–22. https://eur-lex.europa.eu/legal-content/EN/TXT/?uri=CELEX%3A32017D1207

EC (European Council) (1990) Directive 90/220/EEC of 23 April 1990 on the deliberate release into the environment of genetically modified organisms. Off J Eur Community L117(8.5.1990):15–27. https://eur-lex.europa.eu/legal-content/EN/TXT/?uri=celex%3A31990L0220

EC (European Council) (1992) Directive 92/43/EEC of 21 May 1992 on the conservation of natural habitats and of wild fauna and flora. Off J Eur Community L206(22):1–7. https://eur-lex.europa.eu/legal-content/EN/ALL/?uri=celex%3A31992L0043

EC (European Council) (2001) Directive 2001/18/EC of the European Parliament and of the Council of 12 March 2001 on the deliberate release into the environment of genetically modified organisms and repealing Council Directive 90/220/EEC. Off J Eur Community L106(17.4.2001):1–39. https://eur-lex.europa.eu/legal-content/EN/TXT/?uri=celex%3A32001L0018

EC (European Council) (2009) Regulation (EC) No 1107/2009 of the European Parliament and of the Council of 21 October 2009 concerning the placing of plant protection products on the market and repealing Council Directives 79/117/EEC and 91/414/EEC. Off J Eur Community L309(24):11.2009):1–11.2009)50. https://eur-lex.europa.eu/legal-content/EN/TXT/?uri=CELEX%3A32009R1107

EFSA (European Food Safety Authority) (2012) Scientific opinion updating the risk assessment conclusions and risk management recommendations on the genetically modified insect resistant maize MON 810. EFSA J 10(12):3017. https://doi.org/10.2903/j.efsa.2012.3017

EFSA (European Food Safety Authority) (2015) Updating risk management recommendations to limit exposure of non-target Lepidoptera of conservation concern in protected habitats to Bt-maize pollen. EFSA J 13(7):4127. https://doi.org/10.2903/j.efsa.2015.4127

EFSA (European Food Safety Authority) (2016) Risks for public health related to the presence of *Bacillus cereus* and other *Bacillus* spp. including *Bacillus thuringiensis* in foodstuffs. EFSA J 14(7):4524. https://doi.org/10.2903/j.efsa.2016.4524

EFSA (European Food Safety Authority) (2019) Assessment of the 2017 post-market environmental monitoring report on the cultivation of genetically modified maize MON 810. EFSA J 17(6):5742. https://doi.org/10.2903/j.efsa.2019.5742

Fahse L, Papastefanou P, Otto M (2018) Estimating acute mortality of Lepidoptera caused by the cultivation of insect-resistant Bt maize – the LepiX model. Ecol Model 371:50–59. https://doi.org/10.1016/j.ecolmodel.2018.01.006

Fejes Á, Takács E, Fekete G, Darvas B, Ferguson BS, Saxena D, Székács A (2011) Aquatic effect study of Cry4 toxin with immunoassay and *Aedes aegypti* larval biotest. Aquat Insects 34(Suppl.1):207–222. https://doi.org/10.1080/01650424.2012.643062

Folcher L, Delos M, Marengue E, Jarry M, Weissenberger A, Eychenne N, Regnault-Roger C (2010) Lower mycotoxin levels in Bt maize grain. Agron Sustain Dev 30(4):711–719. https://doi.org/10.1051/agro/2010005

Gatehouse AMR, Ferry N, Edwards MG, Bell HA (2011) Insect-resistant biotech crops and their impacts on beneficial arthropods. Philos Trans R Soc Lond Ser B Biol Sci 366(1569):1438–1452. https://doi.org/10.1098/rstb.2010.0330

Government of Hungary (1998) Act XXVII of 1998 on gene technology activities. https://gmo.kormany.hu/download/6/36/70000/1998-27-gentorveny-2013-6-28-tol_EN.pdf

Graef F, Römbke J, Binimelis R, Myhr AI, Hilbeck A, Breckling B, Dalgaard T, Stachow U, Catacora-Vargas G, Bøhn T, Quist D, Darvas B, Dudel G, Oehen B, Meyer H, Henle K, Wynne B, Metzger MJ, Knäbe S, Settele J, Székács A, Wurbs A, Bernard J, Murphy-Bokern D, Buiatti M, Giovanetti M, Debeljak M, Andersen E, Paetz A, Dzeroski S, Tappeser B, van Gestel CAM, Wosniok W, Séralini G-E, Aslaksen J, Pesch R, Maly S, Doerpinghaus A, Werner A (2011) A framework for a European Network for a systematic environmental impact assessment of genetically modified organisms (GMO). BioRisk 7:73–97. https://doi.org/10.3897/biorisk.7.1969

Grothaus GD, Bandla M, Currier T, Giroux R, Jenkins GR, Lipp M, Shan G, Stave JW, Pantella V (2006) Immunoassay as an analytical tool in agricultural biotechnology. J AOAC Int 89(4):913–928. https://aoac.publisher.ingentaconnect.com/contentone/aoac/jaoac/2006/00000089/00000004/art00002

Hansen LS, Lövei GL, Székács A (2013) Survival and development of a stored product pest, *Sitophilus zeamais* (Coleoptera: Curculionidae) and its natural enemy, the parasitoid *Lariophagus distinguendus* (Hymenoptera: Pteromalidae) on transgenic Bt-maize. Pest Manag Sci 69(5):602–606. https://doi.org/10.1002/ps.3410

Head GP, Carroll MW, Evans SP, Rule DM, Willse AR, Clark TL, Storer NP, Flannagan RD, Samuel LW, Meinke LJ (2017) Evaluation of SmartStax and SmartStaxPRO maize against western corn rootworm and northern corn rootworm: efficacy and resistance management. Pest Manag Sci 73(9):1883–1899. https://doi.org/10.1002/ps.4554

Heinemann JA, Agapito-Tenfen SZ, Carman JA (2013) A comparative evaluation of the regulation of GM crops or products containing dsRNA and suggested improvements to risk assessments. Environ Int 55:43–55. https://doi.org/10.1016/j.envint.2013.02.010

Hilbeck A, Otto M (2015) Specificity and combinatorial effects of *Bacillus thuringiensis* cry toxins in the context of GMO environmental risk assessment. Front Environ Sci 3:71. https://doi.org/10.3389/fenvs.2015.00071

Hilbeck A, Meier M, Römbke J, Jänsch S, Teichmann H, Tappeser B (2011) Environmental risk assessment of genetically modified plants – concepts and controversies. Environ Sci Eur 23:13. https://doi.org/10.1186/2190-4715-23-13

Hilbeck A, Binimelis R, Defarge N, Steinbrecher R, Székács A, Wickson F, Antoniou M, Bereano PL, Clark EA, Hansen M, Novotny E, Heinemann J, Meyer H, Shiva V, Wynne B (2015) No scientific consensus on GMO safety. Environ Sci Eur 27:4. https://doi.org/10.1186/s12302-014-0034-1

Holst N, Lang A, Lövei GL, Otto M (2013a) Corrigendum to "Increased mortality is predicted of *Inachis io* larvae caused by Bt-maize pollen in European farmland". Ecol Model 265:126–133. 250. https://doi.org/10.1016/j.ecolmodel.2013.06.023

Holst N, Lang A, Lövei GL, Otto M (2013b) Increased mortality is predicted of *Inachis io* larvae caused by Bt-maize pollen in European farmland. Ecol Model 250:126–133. https://doi.org/10.1016/j.ecolmodel.2012.11.006

Hutchison WD, Burkness EC, Mitchell PD, Moon RD, Leslie TW, Fleischer SJ, Abrahamson M, Hamilton KL, Steffey KL, Gray ME, Hellmich RL, Kaster LV, Hunt TE, Wright RJ, Pecinovsky K, Rabaey TL, Flood BR, Raun ES (2010) Areawide suppression of European corn borer with Bt maize reaps savings to non-Bt maize growers. Science 330(6001):222–225. https://doi.org/10.1126/science.1190242

Kaur S (2000) Molecular approaches towards development of novel *Bacillus thuringiensis* biopesticides. World J Microbiol Biotechnol 16:781–793. https://doi.org/10.1023/A:1008931207374

Kruse-Plass M, Hofmann F, Kuhn U, Otto M, Schlechtriemen U, Schröder B, Vögel R, Wosniok W (2017) Reply to the EFSA (2016) on the relevance of recent publications (Hofmann et al. 2014, 2016) on environmental risk assessment and management of Bt-maize events (MON810, Bt11 and 1507). Environ Sci Eur 29:12. https://doi.org/10.1186/s12302-017-0106-0

Lang A, Otto M (2010) A synthesis of laboratory and field studies on the effects of transgenic *Bacillus thuringiensis* (Bt) maize on non-target Lepidoptera. Entomol Exp Appl 135:121–134. https://doi.org/10.1111/j.1570-7458.2010.00981.x

Lang A, Lauber É, Darvas B (2007) Early tier tests insufficient for GMO risk assessment. Nat Biotechnol 25:35–36. https://doi.org/10.1038/nbt0107-35

Lang A, Brunzel S, Dolek M, Otto M, Theißen B (2011) Modelling in the light of uncertainty of key parameters: a call to exercise caution in field predictions of Bt -maize effects. Proc R Soc B 278:980–981. https://doi.org/10.1098/rspb.2010.2085

Lang A, Lee M, Dolek M, Berchtold J, Otto M (2019) Laboratory tests with Lepidoptera to assess non-target effects of *Bt* maize pollen: analysis of current studies and recommendations for a standardised design. Environ Sci Eur 31:39. https://doi.org/10.1186/s12302-019-0220-2

Latham JR, Love M, Hilbeck A (2017) The distinct properties of natural and GM cry insecticidal proteins. Biotechnol Genet Eng Rev 33(1):62–96. https://doi.org/10.1080/02648725.2017.1357295

Lauber É (2011) Cry1 toxin content of *MON 810* Bt-corn and the effect of its pollen on protected butterfly species in Hungary. PhD thesis, Corvinus University, Budapest, Hungary. http://phd.lib.uni-corvinus.hu/598

Lauber É, Darvas B (2009) Increased mortality of isolated first instar larvae of *Inachis io* (Lepidoptera). Acta Phytopath Entomol Hung 44:111–117. https://doi.org/10.1556/APhyt.44.2009.1.11

Marvier M (2001) Ecology of transgenic crops. Am Sci 89(2):160–167. https://www.jstor.org/stable/27857439

Matthews GA, Bateman R, Miller P (2014) Pesticide application methods. Wiley, Chichester, p 24. ISBN:978-1-118-35130-7

Mendelsohn M, Kough J, Vaituzis Z, Matthews K (2003) Are Bt crops safe? Nat Biotechnol 21(9):1003–1009. https://doi.org/10.1038/nbt0903-1003

Ministry of Agriculture, Hungary (2005) The safeguard clause on the cultivation of *MON810* genetically modified maize. https://gmo.kormany.hu/safeguard-clauses-currently-in-effect-in-hungary

Myhr AI (2010) A precautionary approach to genetically modified organisms: challenges and implications for policy and science. J Agric Environ Ethics 23(6):501–525. https://doi.org/10.1007/s10806-010-9234-x

O'Callaghan M, Glare TR, Burgess EPJ, Malone LA (2005) Effects of plants genetically modified for insect resistance on nontarget organisms. Annu Rev Entomol 50:271–292. https://doi.org/10.1146/annurev.ento.50.071803.130352

Ostry V, Ovesna J, Skarkova J, Pouchova V, Ruprich J (2010) A review on comparative data concerning Fusarium mycotoxins in Bt maize and non-Bt isogenic maize. Mycotoxin Res 26:141–145. https://doi.org/10.1007/s12550-010-0056-5

Palma L, Muñoz D, Berry C, Murillo J, Caballero P (2014) *Bacillus thuringiensis* toxins: an overview of their biocidal activity. Toxins 6(12):3296–3325. https://doi.org/10.3390/toxins6123296

Pedersen SM, Lind KM (2017) Precision agriculture – from mapping to site-specific application. In: Pedersen SM, Lind KM (eds) Precision agriculture: technology and economic perspectives. Springer, Cham, pp 9–18. ISBN:978-3-319-68713-1

Perry JN, Devos Y, Arpaia S, Bartsch D, Gathmann A, Hails RS, Kiss J, Lheureux K, Manachini B, Mestdagh S, Neemann G, Ortego F, Schiemann J, Sweet JB (2010) A mathematical model of exposure of nontarget Lepidoptera to *Bt* -maize pollen expressing Cry1Ab within Europe. Proc R Soc B 277:1417–1425. https://doi.org/10.1098/rspb.2009.2091

Perry JN, Devos Y, Arpaia S, Bartsch D, Ehlert C, Gathmann A, Hails RS, Hendriksen NB, Kiss J, Messéan A, Mestdagh S, Neemann G, Nuti M, Sweet JB, Tebbe CC (2012) Estimating the effects of Cry1F *Bt* -maize pollen on non-target Lepidoptera using a mathematical model of exposure. J Appl Ecol 49:29–37. https://doi.org/10.1111/j.1365-2664.2011.02083.x

Pimentel D (1995) Amounts of pesticides reaching target pests: environmental impacts and ethics. J Agric Environ Ethics 8(1):17–29. https://doi.org/10.1007/BF02286399

Raymond B, Federici BA (2017) In defence of *Bacillus thuringiensis*, the safest and most successful microbial insecticide available to humanity – a response to EFSA. FEMS Microbiol Ecol 93(7):fix084. https://doi.org/10.1093/femsec/fix084

Romeis J, Meissle M, Bigler F (2006) Transgenic crops expressing *Bacillus thuringiensis* toxins and biological control. Nat Biotechnol 24(1):63–71. https://doi.org/10.1038/nbt1180

Romeis J, Bartsch D, Bigler F, Candolfi MP, Gielkens MMC, Hartley SE, Hellmich RL, Huesing JE, Jepson PC, Layton R, Quemada H, Raybould A, Rose RI, Schiemann J, Sears MK, Shelton AM, Sweet J, Vaituzis Z, Wolt JD (2008) Assessment of risk of insect-resistant transgenic crops to nontarget arthropods. Nat Biotechnol 26:203–208. https://doi.org/10.1038/nbt1381

Royal Society (2016) GM plants. Questions and answers. Royal Society, London. https://royalsociety.org/~/media/policy/projects/gm-plants/gm-plant-q-and-a.pdf

Sanahuja G, Banakar R, Twyman RM, Capell T, Christou P (2011) *Bacillus thuringiensis*: a century of research, development and commercial applications. Plant Biotechnol J 9:283–300. https://doi.org/10.1111/j.1467-7652.2011.00595.x

Sanchis V (2011) From microbial sprays to insect-resistant transgenic plants: history of the biospesticide *Bacillus thuringiensis*. A review. Agron Sustain Dev 31:217–231. https://doi.org/10.1051/agro/2010027

Schnepf E, Crickmore N, Van Rie J, Lereclus D, Baum J, Feitelson J, Zeigler DR, Dean HD (1998) *Bacillus thuringiensis* and its pesticidal crystal proteins. Microbiol Mol Biol Rev 62:775–806. https://www.ncbi.nlm.nih.gov/pmc/articles/PMC98934

Sears MK, Hellmich RL, Stanley-Horn DE, Oberhauser KS, Pleasants JM, Mattila HR, Siegfried BD, Dively GP (2001) Impact of *Bt* corn pollen on monarch butterfly populations: a risk assessment. Proc Natl Acad Sci U S A 98(21):11937–11942. https://doi.org/10.1073/pnas.211329998

Seres A, Kiss I, Nagy P, Sály P, Darvas B, Bakonyi G (2014) Arbuscular mycorrhizal fungi colonisation of *Cry3* toxin-producing *Bt* maize and near isogenic maize. Plant Soil Environ 60(12):569–573. https://doi.org/10.17221/674/2014-PSE

Székács A (2013) Cry1Ab toxin quantification in MON 810 maize. In: Breckling B, Verhoeven R (eds) GM-crop cultivation – Ecological effects on a landscape scale. Theorie in der Ökologie, vol 17. Peter Lang GmbH, Frankfurt, pp 123–124. ISBN:9783631628706

Székács A (2016) New technologies in agricultural biotechnology. Ecocycles 2(2):54–56. https://doi.org/10.19040/ecocycles.v2i2.65

Székács A (2017) Environmental and ecological aspects in the overall assessment of bioeconomy. J Agric Environ Ethics 30(1):153–170. https://doi.org/10.1007/s10806-017-9651-1

Székács A, Darvas B (2012a) Comparative aspects of Cry toxin usage in insect control. In: Ishaaya I, Palli SR, Horowitz R (eds) Advanced technologies for managing insect pests. Springer, Berlin, pp 195–230. https://doi.org/10.1007/978-94-007-4497-4_10

Székács A, Darvas B (2012b) Environmental assessment of *MON 810* maize in the Pannonian Biogeographical Region. Acta Phytopathol Entomol Hung 47(2):307–320. https://doi.org/10.1556/APhyt.47.2012.2.11

Székács A, Juracsek J, Polgar LA, Darvas B (2005) Levels of expressed Cry1Ab toxin in genetically modified corn DK-440-BTY (YIELDGARD) and stubble. FEBS J 272(Suppl.1):508

Székács A, Lauber É, Juracsek J, Darvas B (2010a) Cry1Ab toxin production of MON 810 transgenic maize. Environ Toxicol Chem 29(1):182–190. https://doi.org/10.1002/etc.5

Székács A, Lauber É, Takács E, Darvas B (2010b) Detection of Cry1Ab toxin in the leaves of *MON 810* transgenic maize. Anal Bioanal Chem 396(6):2203–2211. https://doi.org/10.1007/s00216-009-3384-6

Székács A, Weiss G, Quist D, Takács E, Darvas B, Meier M, Swain T, Hilbeck A (2012) Interlaboratory comparison of Cry1Ab toxin quantification in *MON 810* maize by enzyme-immunoassay. Food Agric Immunol 23(2):99–121. https://doi.org/10.1080/09540105.2011.604773

Tabashnik BE (2010) Communal benefits of transgenic corn. Science 330(6001):189–190. https://doi.org/10.1126/science.1196864

Tabashnik BE, Carrière Y (2017) Surge in insect resistance to transgenic crops and prospects for sustainability. Nat Biotechnol 35(10):926–935. https://doi.org/10.1038/nbt.3974

Tabashnik BE, Finson N, Groeters FR, Moar WJ, Johnson MW, Luo K, Adang MJ (1994) Reversal of resistance to *Bacillus thuringiensis* in *Plutella xylostella*. Proc Natl Acad Sci U S A 91(10):4120–4124. https://doi.org/10.1073/pnas.91.10.4120

Tabashnik BE, Brévault T, Carrière Y (2013) Insect resistance to Bt crops: lessons from the first billion acres. Nat Biotechnol 31(6):510–521. https://doi.org/10.1038/nbt.2597

Takács E, Bánáti H, Fónagy A, Darvas B (2010) Short term study of DAS-59122-7 maize on L1 and L2 larvae of seven-spotted ladybird (*Coccinella septempunctata*) feeding on the bird cherry-oat aphid (*Rhopalosiphum padi*). Cereal Res 59(Suppl):625–628. https://doi.org/10.1556/Novenyterm.59.2010.Suppl.6

Takács E, Juracsek J, Fónagy A, Székács A (2011) Monitoring Cry34Ab1 and Cry35Ab1 protein concentrations in transgenic maize (DAS-59122-7). In: Abstract book of the 5th EIGMO meeting. Ecological impact of genetically modified organisms – IOBC/wprs Working Group meeting, Česke Budějovice, Czech Republic, July 22–25, 2011. Biology Centre ASCR v.v.i., Prague, Czech Republic. ISBN:978-80-86668-14-7. https://www.iobc-wprs.org/expert_groups/Abstract-book-EIGMO-2011.pdf

Takács E, Darvas B, Székács A (2012a) Analytical difficulties and certain biological aspects of Cry1Ab toxin determination in *MON 810* genetically modified maize. Acta Phytopathol Entomol Hung 47(2):293–306. https://doi.org/10.1556/APhyt.47.2012.2.10

Takács E, Fónagy A, Juracsek J, Kugler N, Székács A (2012b) Characterisation of tritrophic effects of *DAS-59122-7* maize on seven-spotted ladybird (*Coccinella septempunctata*) feeding on the bird cherry-oat aphid (*Rhopalosiphum padi*). IOBC/wprs Bull 73:121–134. ISBN:978-92-9067-250-0

Takács E, Nagy A, Gelencsér É, Székács A (2015) Internal quality control of an enzyme-linked immunoassay for Cry1Ab toxin detection applied in animal tissues. Acta Aliment Hung 44(4):593–600. https://doi.org/10.1556/066.2015.44.0032

Then C (2010) Risk assessment of toxins derived from *Bacillus thuringiensis* – synergism, efficacy, and selectivity. Environ Sci Pollut Res 17(3):791–797. https://doi.org/10.1007/s11356-009-0208-3

US National Research Council (2010) The impact of genetically engineered crops on farm sustainability in the United States. National Academies Press, Washington, DC. ISBN:978-0-309-14708-8. https://www.nap.edu/catalog/12804/the-impact-of-genetically-engineered-crops-on-farm-sustainability-in-the-united-states

Walters FS, Graser G, Burns A, Raybould A (2018) When the whole is not greater than the sum of the parts: a critical review of laboratory bioassay effects testing for insecticidal protein interactions. Environ Entomol 47(2):484–497. https://doi.org/10.1093/ee/nvx207

Wolfenbarger LL, Phifer PR (2000) The ecological risks and benefits of genetically engineered plants. Science 290:2088–2093. https://doi.org/10.1126/science.290.5499.2088

Wu F (2006) Mycotoxin reduction in Bt corn: potential economic, health, and regulatory impacts. Transgenic Res 15:277–289. https://doi.org/10.1007/s11248-005-5237-1

Zilberman D, Holland TG, Trilnick I (2018) Agricultural GMOs – what we know and where scientists disagree. Sustainability 10(5):1514. https://doi.org/10.3390/su10051514

# Part III
# GM Insects

# GM Insect Biodiversity and Ecological Interactions

**Thomas A. Miller**

**Abstract** For many years we knew little about microbes because there was no reliable method to identify them. Advances in molecular genetics changed all that. With modern DNA methods, any life form can be identified. Recent studies have also begun to reveal how critically important microbes are to insect biology. The laboratory of Takema Fukatsu in Japan reported the astonishing result, that if you swap the gut microbes of two closely related stink bugs, the insects switch host plants. Genetic transformation has revolutionized plant breeding. It is now possible to apply that approach to insects in improving the ecologically friendly sterile insect technique. It is fair to say the world is having difficulty accepting these new genetic methods. Interdisciplinary studies are a powerful source of innovation.

**Keywords** Sterile insect technique (SIT) · Autocidal biological control (ABC) · Classical biological control (BC) · Chemical control · GM insects · Integrated pest management (IPM) · Symbiotic control · Paratransgenic insects · Wolbachia methods · GMO regulation · Invasive insects

## Introduction

By 2030 the world will need to produce 50% more food and energy and 30% more freshwater while at the same time dealing with climate change (Beddington 2009). As these demands arise, arable land is being encroached upon by urban development. Half of the global population is now city residents and this figure is expected to grow to 60% by 2030.

Beddington called for "crop improvement to increase yields and tolerance to stresses such as droughts; smarter use of water and fertilizers; new pesticides and their effective management to avoid resistance problems; introduction of novel

T. A. Miller (✉)
Jefferson Science Fellow, University of California, Riverside, CL, USA
e-mail: Thomas.miller@ucr.edu

 
A. Chaurasia et al. (eds.), *GMOs*, Topics in Biodiversity and Conservation 19,
https://doi.org/10.1007/978-3-030-53183-6_8

non-chemical approaches to crop protection; reduction of post-harvest losses; and more sustainable livestock and marine production."

Responses to these challenges require scientific research, which is going through a transformation. For example, microbes are the one constant on earth, and possibly the least understood (Yong 2016). We are just beginning to realize the key role played by microbes in insect ecology and physiology.

## Role of Microbes in Insect Ecology

As one example of the importance of microbes to insect ecology, a pest stink bug species, *M. punctatissima*, was successfully reared on soybeans. A closely related species, *M. cribraria* (not a pest insect), suffered a low egg hatch rate when reared on soybeans. Stink bug females typically defecate on a new batch of eggs and the resulting offspring consume the eggs and fecal deposit, presumably to acquire a characteristic gut microbiome.

The authors (Hosokawa et al. 2007) were able to swap fecal deposits between the two species of stink bugs. Thus, when the pest, *M. punctatissima*, with the transplanted gut microbiome attempted to feed on the crop legumes, they suffered a low hatch rate; whereas, the *M. cribraria* survived well on the crop legumes. Successful feeding on the crop legume depended entirely on the gut microbiome: the authors stated that:

> Our finding sheds new light on the evolutionary origin of insect pests, potentially leading to novel approaches to pest control and management.

A recent book expands on insect-microbe interactions (Douglas 2018). Douglas described that "All insects, including pest species, are colonized by microorganisms, variously located in the gut and within insect tissues (intracellular). Manipulation of these microbial partners can reduce the pest status of insects, either by modifying insect traits (e.g. altering the host range or tolerance of abiotic conditions, reducing insect competence to vector disease agents) or by reducing fitness" (Arora and Douglas 2017). This ushers in new approaches to pest control that are unfolding now.

## Historic Insect Control Methods

There are three historic types of insect control: insecticides (chemical), biological control (biological) and traps and fly swatters (physical). The latter category includes bed nets to keep mosquitoes from biting during sleep. Combinations are also possible, for example, bed nets are sometimes soaked with an insecticide to both act as a physical barrier and kill by chemical action.

## *Chemical Control*

Chemical control of insects is as old as flowering plants. Nicotine production was, presumably, naturally selected for in certain plants to protect against insect attack. Rotenone is another such plant-derived insecticidal chemical, still used today. It rapidly degrades in warm conditions and, being "natural," is accepted for use in insect control on "organic" crops.

One of the most interesting of these plant-based insecticides was discovered in the 1960s called the "paper factor" (Slama and Williams 1966). Karel Slama, from the Czechoslovakian Academy of Sciences in Prague, spent a research visit in the laboratory of Carroll Williams at Harvard University in the late 1950s. Karel brought along his own bugs from home. When he tried rearing them in cages with paper towels, they never developed to the adult stage. He traced the cause to a natural juvenile hormone analog present in the trees from which his paper towels were derived.

It became evident very quickly that chemical control as applied by broadcast sprays had negative effects. Broad-spectrum insecticides like DDT (toxic to all insects and few vertebrates) eliminated parasites and predators along with pests while new pests exploited the predator-free habitat. In addition to those side-effects, the development of resistance meant strategies were needed to avoid it.

Ironically, a broad-spectrum insecticide has a large market, while selective insecticides have narrow markets. This is ironic because the ideal chemical insecticide has a narrow spectrum, one with the least impact on the environment and greater maintenance of biodiversity including beneficial insects. Return on investment, thus, is greatest for a broad-spectrum chemical, and marginal for narrow-spectrum chemicals. Part of the development costs include required regulatory assessments.

## *(Classical) Biological Control*

The main biological method of controlling insects is called biological control (biocontrol or BC) (Hagen and Franz 1973). It stems partly from success in controlling cottony cushion scale, (*Icerya purchasi*) on citrus in California over a century ago. The vedalia beetle, *Rodolia cardinalis*, a predatory lady beetle, was introduced into California in 1888 and completely controlled cottony cushion scale insects (Debach 1973; Ebeling 1959).

From an initial position of prominence and promise, some pessimists are now warning of difficulty or decline of the biocontrol efforts (Duan et al. 2015; Mills 2017; Warner et al. 2011), while others remain optimistic (Hajek and Eilenberg 2018). Part of the conflict is due to excessive regulation, but this was based on spectacular failures such as imported cane toad in Australia, that became a major pest itself.

## *Sterile Insect Technique (SIT)*

When the sterile insect technique (SIT) was developed (Knipling 1955, 1959), it was considered a triumph of atomic energy. The method was originally designed to eradicate the screwworm fly, *Cochliomyia hominivorax*, from North America, but soon the effort spread through Central America and successfully eliminated the fly north of Colombia, South America with mass-rearing facilities in Panama preventing re-invasion of cleared areas (Scott et al. 2017a, b).

Scott et al. (2017a, b) provided an update of the screwworm operations in Panama. They reported improvements in all operations including trap designs and release mechanisms. They also reported using techniques to produce only males as one of the most important recent advances. This was done using a conditional female lethal gene. There have been surveys of the genetic makeup of field populations as a way of ensuring the mass-reared insects remain compatible for mating with wild types. Despite these advances, especially the genetic sexing breakthrough, the program still does not have permission for GM screwworm to replace radiation for sterility.

The SIT method is simplicity itself, a target pest insect, such as the Mediterranean fruit fly, *Ceratitis capitata*, is reared in large numbers, exposed to just enough radiation to cause sterility and then released in a target area to suppress (eradicate) an invasive wild population. The International Atomic Energy Agency (IAEA) in Vienna houses the global insect suppression efforts and remains active today.

There remain two serious impediments to radiation-based SIT. The level of radiation necessary to cause sterility has fitness costs. There has never been a way to get around this first impediment. Secondly, radiation levels necessary to induce sterility in certain insects make them uncompetitive for SIT. Those insects include the cotton boll weevil, *Anthonomus grandis*, and many of the most dangerous mosquitoes vectoring malaria and dengue fever, Anopheles and Aedes species. That is why for many years, classic radiation-based SIT was never used against these insects.

Indeed, the only genuine improvement in SIT as operated by IAEA is a method of producing only male medflies (Hendrichs et al. 1995; Scott et al. 2017a, b). This improves the method because males are the only necessary participants. If sterile females are released, they could only contribute by attracting males in the wild population, but SIT females are not the key element.

Jorge Hendrichs (pers. comm.) implied that IAEA had a statutory requirement to use radiation for sterilizing flies. He noted, however, that radiation creates multiple lethal genes, which makes it much more difficult for resistance to develop. He was arguing against using GM insects with one inserted lethal gene that yields a far more competitive insect than radiation. It is difficult to envision how resistance could occur in GM insects, as he suggests, without selection.

The boll weevil, a native of Central Mexico, was eradicated from most of the USA over a period of years but still re-invades parts of Texas. This eradication method used spot insecticide treatments based on extensive trap monitoring, not any SIT strategy. It may be that a GM insect SIT approach to suppressing (eradicating) boll weevil would work, particularly in its native range in central Mexico and certainly against invasive boll weevils in Egypt. Use depends on available financing.

## *SIT as the Ideal of Biocontrol*

SIT is the ideal form of biocontrol. It uses only the pest insect as an agent. It has no side effects, there are no residues, and no inherited genetic effects. The released insects are sterile so do not reproduce. They do not, therefore, represent an intrusion into the ecosystem. Adults do not feed on crops but obtain nectar from flowers. They are released, mate, and then expire without issue. Because they do not reproduce, there has not been any known instance of resistance developing in the five decades plus of SIT existence. For resistance to occur there must be a mechanism for selection. None exists here.

Because of the side-effects of radiation treatments, the use of lethal genes in insects in SIT is seen as closer to ideal. Mass-reared populations carrying lethal and marker genes can be outcrossed and back-crossed to improve viability; something that is routinely done in regular insect mass-rearing operations.

# GM Insects

## *Origins*

At the start of the GM insect era, there were few or no marker genes for selection of GM insects, no lethal genes known, and no transformation protocols. One by one these impediments were overcome. One of the first to fall was a demonstrable lethal gene. One was found in a strain of *Drosophila melanogaster* (Fryxell and Miller 1995). The Notch gene variety, $N^{60g11}$, is a sex-linked mutation that causes dominant, cold-sensitive lethality in heterozygous embryos. This strain of *D. melanogaster* could be mass-reared under permissive temperatures, released and the lethal gene effect would be expressed upon lethal males mating with feral females. This was a perfect, dominant lethal candidate for GM insect SIT, provided the lethal temperature was compatible with the climate in the target area.

Laboratory experiments showed the population collapse in three generations. Thus, the stage was set for SIT to be improved by using GM insects with inserted lethal genes instead of radiation.

A version of indirect GM insect control was born when Monsanto inserted the delta endotoxin, from *Bacillus thuringiensis* into the crop plants, cotton, and corn. By the late 1980s, the only insect that had been genetically transformed was *Drosophila melanogaster* and relatives. This was due to the discovery and use of the "P" element, a transposon. P expresses a transposase. This enzyme recognizes a short nucleotide sequence at both ends of the transposon, extracts the entire transposon and inserts it elsewhere in the genome at the same target sequence. In the popular press these were called "jumping genes" (Fedoroff 2012).

Some insects have very active transposons that move on a regular basis like those in the tobacco budworm and as a result have unstable genomes, and other insects like pink bollworm have intact transposons that are potentially active, but never move; their genomes are very stable (Miller 2012). It might be significant that tobacco budworm is a multi-host pest (a pest of cotton and tobacco) and pink bollworm has a narrow host range restricted to the *Malvaceae*. This group includes cotton, the vegetable, okra, and the ornamental flower, hibiscus. What we don't know is to what extent microbes might be playing a role in host selection of these two insects.

DNA from any source can be inserted into a transposon and the combination injected into a recipient egg along with a separate transposase. The transposase would then find its nucleotide target sequence, open it, insert the transposon, and close the insertion site of the target chromosome. This information appeared to be enough for the Cotton Pest Control Board in California who had been operating an SIT program based in Arizona to prevent pink bollworm from invading Central California cotton-growing areas for almost 20 years at the cost of millions of dollars a year. They were convinced by one advisor to launch a GM pink bollworm project; interestingly, the other three scientific advisors cautioned against developing the method. It was estimated that the 60:1 ratio of radiated SIT insects released compared to wild types for SIT suppression could be improved to 5:1 at enormous savings in operational costs.

The next advance was the discovery of a general-purpose transposable element. A *Drosophila* element, called Minos, was demonstrated to genetically modify the medfly by a Greek team (Loukeris et al. 1995). Next, a marker gene, *egfp*, was genetically inserted into pink bollworm, *Pectinophora gossypiella*, with another new transposon element, *piggyBac*, generously supplied by Mac Fraser at Notre Dame University (Peloquin et al. 2000; Thibault et al. 1998).

At the same time the pink bollworm was transformed, the Luke Alphey laboratory at Oxford University, UK had developed another lethal gene construct in *Drosophila* from scratch (Thomas et al. 2000). They set up a new company, Oxitec (Oxford Insect Technologies) to exploit their lethal gene. One of the first targets they sought to exploit was pink bollworm.

As it was, the introduction of GM insects for SIT was met with excessive precaution. Actionbioscience.org called for assessing the impact of GM insects on the environment, potential risks to human health and advantages and disadvantages in controlling crop pests (Miller 2004; Anonymous 2004). However, there was unmistakable interest perked by the advances (Alphey et al. 2009; Bruce 2012; Miller and Staten 2001; Miller 2007a, b; Baltzegar et al. 2018).

The precaution soon gave way to stagnation. Nearly 20 years later we are still waiting for GM insects to be broadly used in SIT. At the time GM pink bollworm was field-tested for SIT in Arizona, the cotton industry was moving toward final pink bollworm eradication (Antilla and Liesner 2008). Bob Staten was asked to direct the program since he had just retired from USDA APHIS in Phoenix, AZ. Staten and his team including Larry Antilla at Arizona Cotton Research and Protection Council in Phoenix and the National Cotton Council staff mounted the pink bollworm eradication effort. The first thing they did was ask for a waiver to grow 100% Bt cotton in the eradication area. This was met, at first, with opposition from Monsanto (owners of the Bt cotton patent) and federal regulators.

Planting non-Bt cotton with Bt cotton together was made a requirement for growers to provide pink bollworms carrying susceptible genes to dilute out any resistance that might occur in pink bollworms on Bt cotton. Resistance normally requires two copies of a given recessive trait for resistance to be expressed (homozygous). This strategy had successfully prevented resistance from occurring since the first use of Bt cotton in the western USA. At a review session, Staten's team laid out their reasons why the released SIT pink bollworms (being sterile) could substitute for the presence of non-Bt cotton as a resistance management strategy (Tabashnik et al. 2010).

After their presentation, the review team told the applicants that it was the best argument they had ever heard. The waiver was granted, and Monsanto also approved the strategy for eradication. Moreover, Tabashnik and colleagues eventually published an article (Wan et al. 2017) suggesting that hybridizing Bt and non-Bt cotton achieves resistance management as well. This was like another practical suggestion that Bt cotton and non-Bt cotton seeds merely be mixed at a desired ratio (say 4:1) before planting to achieve the same end (Tim Dennehey, personal communication).

At some point, a request was made to use the GM pink bollworm in the SIT eradication program. Several required field trials had, by then, been conducted (see Simmons et al. 2011). That permission was denied. Then the program asked for permission to use just the genetically marked pink bollworm. The reason for this was simple. Ordinary SIT pink bollworms are dusted with a dye to tell them apart from wild types in monitoring pheromone traps. The dye wears off making it difficult to tell older SIT insects from wild types with certainty. This meant the program continued releases with continuing costs beyond the point needed.

USDA approved this but then asked National Organic Program (NOP) for a ruling. NOP administrators said if one GM pink bollworm lands on one organic cotton plant in eradication operations, that field of organic cotton would lose certification for 1 year. Instead of suing the NOP, the national cotton industry decided to retreat and do nothing; stagnation continued. This sorry tale, described in Miller (2013), suggested that the organic grower industry is against the most sustainable methods of pest control. Perhaps this is yet another example of "Science Wars" (Wagner et al. 2018).

The pink bollworm eradication program started in 2005. The last pink bollworm adult males were caught in pheromone traps in May 2012. The eradiation area included California, Arizona, New Mexico, Texas, and northern Mexico (Miller 2013).

## GM Insects and IPM

Cooler heads are now a part of GM insect development. In 2019 the Fred Gould website stated: "Our project on "genetic pest management" reflects our belief that genetic engineering of insects can be used as a tool for reducing the impacts from pests of medical and agricultural importance.".

Collins et al. (2018) examined the effect of the removal of a vector mosquito species on the remaining environment. They specifically were comparing these genetic methods with after-effects of insecticide spraying or residues. They examined what the loss of the vector mosquito might mean in terms of species that prey on them and concluded the number of alternative hosts appeared sufficient to blunt the effect of suppressing one (pest) species.

The Oxitec GM *Aedes aegypti* mosquitoes were used in Brazil to suppress dengue outbreaks (de Campos et al. 2017). There was much attention paid to public acceptance. It is unclear if this effort will be maintained or copied elsewhere.

## GM Paratransgenic Insects

At the beginning of this chapter, microbes were mentioned as perhaps not being understood enough. In the early days of developing GM insect approaches, most of the workers were mosquito people beyond pink bollworm. One discontinuity was the late Frank Richards, then at Yale University medical school. Frank was working on a concept that David O'Brochta, subsequently coined, "paratransgenesis." Frank was trying to modify a gut microbial symbiont from the blood-sucking bug, *Rhodnius prolixus*, to deliver a gene product that would prevent *Rhodnius* from transmitting the pathogen causing Chagas' disease. He had a seemingly perfect candidate, *R. rhodnii*. This was an obligatory gut symbiont vital to survival (it supplied missing essential nutrients). It was routinely acquired by the adult bugs by coprophagic behavior, sampling conspecific feces. Frank even had a brilliant method to apply it. He made a concoction called "CRUZIGARD," an artificial fecal material spread in the habitat was a vehicle for a GM symbiotic bacterium to gain access to the hindgut and there release anti-pathogen products.

## Symbiotic Control

Frank's approach was so refreshing, new, and imaginative that it attracted a cult following and a growing field of research (Durvasula et al. 1997, 1999; Miller 2007b; 2011a, b, c, d). Some projects sought microbes that could prevent transmission of the bacterial pathogen, *Xylella fastidiosa*, to grapevines to prevent Pierce's disease (Bextine et al. 2004; Kuzina et al. 2006; Lacava et al. 2007; Ramirez et al. 2008;

Arora et al. 2015, 2018). Others sought paratransgenic solutions to citrus variegated chlorosis virus (Gai et al. 2009). Still others sought solutions to white leaf disease in sugarcane (Wangkeeree et al. 2012).

However, perhaps most interesting of all was that by Marcelo Jacobs-Lorena (Wang and Jacobs-Lorena, 2017) who sought to infest mosquitoes with a genetically modified symbiotic bacterium that could disrupt the acquisition of pathogens in the midgut. They called this a "population replacement" strategy for "interfering with pathogen development via genetic modification of symbiotic microbes to produce antipathogen effector molecules in the host." They also summarized other paratransgenic projects involving mosquitos.

The cycle of acquisition and loss of symbiotic microbes is still being revealed, as are different methods of delivery of paratransgenic microbes to mosquitos, both larvae and adults. Interestingly bacteria from the genus *Asaia* were reported to be stable symbionts in certain *Anopheles* species (Favia et al. 2007, 2008), but appear to be absent in others (Rani et al. 2009).

The laboratory of Raymond St. Leger in Maryland, USA, has been perfecting a creative way of improving classical biological control using entomopathogenic fungi, by genetically modifying fungi with a gene that produced a neurotoxin (Fang et al. 2015). The clear advantage is the classical fungi lethality took a very long time to develop, whereas the GM method is very rapid. Early results are very promising.

## *Wolbachia* Methods

Related to paratransgenesis is the use of *Wolbachia* to prevent mosquito vectored pathogens of human diseases. The similarity ends when it is revealed that *Wolbachia* infections, while they are symbiotic bacteria, maintain and spread infections via "cytoplasmic incompatibility." This driving mechanism appears to favor infected offspring with a sperm defect strategy (Wang and Jacobs-Lorena 2017). A cross of infected males with uninfected females yields no offspring, whereas infected males and females mating yield infected offspring of both sexes. Moreover, an uninfected male mating with an infected female yields viable infected offspring.

The huge difference between the *Wolbachia* approach and GM insects or paratransgenic insects was no genetic transformation was involved, only lengthy selection of a suitable strain of the intracellular symbiotic bacterium.

The differences didn't stop there, *Wolbachia* does not need to be released daily in mass numbers such as SIT does, the first attempt used half dozen releases of infected mosquitoes over about a month period (Hoffmann et al. 2011). From this modest introduction, the wild population eventually became 100% infected with *Wolbachia* and vector-incompetent.

This approaches a purer form of biocontrol. However, there is one huge difference compared to SIT. In the *Wolbachia* case, there are no side-effects. The target insect is an invasive population. By infecting it and adding it back to the field population, no other insects are involved. Outside of some extraordinary talent in the personnel involved, mass rearing is not needed, and the costs are accordingly drastically reduced.

By merely seeding a wild population of uninfected mosquitoes with *Wolbachia*-infected conspecifics, over time, the infection will spread naturally throughout the population in a displacement strategy. This operation is quite distinct from SIT, because daily mass releases are not necessary. *Wolbachia* strains are found or selected to render the infected mosquito vector-incompetent. While natural infections of the intracellular bacterium, *Wolbachia*, are enormous (40–69% of all arthropods species in germ cells to ensure sexual transmission; Wang and Jacobs-Lorena 2017). One possible drawback of such an approach is the unintended and natural loss of *Wolbachia* infections. However, that does not seem to be an insurmountable snag, given the simplicity of re-introducing infected individuals into a population.

One of the first practical demonstrations of this new *Wolbachia* method was conducted in Queensland, Australia to control dengue infections spread by *Aedes aegypti* mosquitos (Hoffmann et al. 2011). It is important that permission was obtained rather quickly for a full field trial. Further work (Fraser et al. 2017) has flushed out the characteristics of the agents. The methods have been adopted or suggested for other mosquitos (Marrelli et al. 2007; Hughes et al. 2011; Blood et al. 2018). *Wolbachia* has been extensively studied by now in *Aedes aegypti* (Moreira et al. 2009; Xi et al. 2005).

## *Gene Drive and CRISPR Methods*

Alphey et al. (2010) suggested a combination of SIT and gene drive for symbiotic control methods, the former called suppression, the latter called substitution. Several groups are trying to exploit this strategy (Austin et al. 2018; Carvalho et al. 2015).

Scott et al. (2017a, b) lay out a paradigm for population suppression in the future: "The traditional approach in agriculture has been to suppress insect pest populations using insecticides and other farming practices. Similarly, we suggest the main use of gene drives in agriculture will be for population suppression through targeting essential genes. We provide examples of gene drives that target specific genes including female-essential genes. Further, we discuss issues related to containment in the laboratory and eventual field testing of strains harboring a Cas9-mediated gene drive system."

They used examples of New World screwworm, spotted wing *Drosophila*, diamondback moth, whitefly, and red flour beetle. They described examples of gene drives and targets such as female-essential genes. The issue of containment and dealing with release outside of quarantine was addressed.

Austin et al. (2018) reported that: "Many different synthetic gene drive systems have been proposed to suppress the number of mosquitoes and/or reduce vector competence. As with any control measure, due attention should be paid to the possible evolution of resistance. No gene drive construct has yet been reported that is 'field-ready' for release, and when such constructs are developed, they should be assessed on a case-by-case basis. Gene drive approaches to vector control promise to have a number of key features that motivate their continued development, and scrutiny, by all concerned."

Note that, "No gene drive construct has yet been reported that is 'field ready.'" There is no question that this approach has attracted a following. While Burt and Crisanti (2018) caution that: "There are a relatively small number of species for which genetic control methods, including gene drive, may be appropriate." They still note that delivery by this means scores high for efficiency. The agents literally drive themselves through the target population with limited side-effects. If resistance issues develop, because spread is by sexual reproduction and selection is possible, counter-measures of follow up releases of gene drive agents could possibly counteract resistance development. No such follow up exists with any pest insect that becomes resistant to a chemical agent. Moreover, Burt (2014) suggested that genetic methods are self-limiting. A debate has emerged about whether CRISPR is GM (a form of genetic modification) or not.

Researchers at the Imperial College London have demonstrated the use of gene drives to completely eliminate populations of mosquitoes known to transmit malaria. Gene drives are a form of genetic engineering that involves spreading a gene or cluster of genes through a population:

> The research, published in *Nature Biotechnology*, reported the eradication of confined populations of *Anopheles gambiae* by blocking female reproduction using the gene editing technique known as CRISPR.

Dr. Crisanti, the corresponding author on the study, said, "2016 marked the first time in over two decades that malaria cases did not fall year-on-year despite huge efforts and resources, suggesting we need more tools in the fight."

The research group will now test the technique on larger populations of mosquitoes under more real-world conditions by bringing into play competition for food and other ecological factors.

"It will still be at least 5–10 years before we consider testing any mosquitoes with gene drive in the wild, but now we have some encouraging proof that we're on the right path," said Crisanti. "Gene drive solutions have the potential one day to expedite malaria eradication by overcoming the barriers of logistics in resource-poor countries."

## *pgSIT*

Using CRISPR methods in the SIT strategy was reported to have yielded all male *Drosophila melanogaster* populations. The authors (Kandul et al. 2018) call this precision-guided SIT or pgSIT. They suggest it for use with any target insect.

## *HEGAAs*

A new wrinkle in genetic modification was announced recently (Reeves et al. 2018). Called Horizontal Environmental Genetic Alteration Agents (HEGAAs) these are designed to be viruses carried by insects that are transported to plants. There the virus infects the plants (after feeding by the insects) and delivers a CRISPR cassette that then selectively edits the plant chromosome to achieve some function. Since this is a new program supported by DARPA (US Defense Advanced Research Projects Agency) the first greenhouse trials are not due for another 4 years.

While the HEGAA approach is not strictly a GM insect topic, the insects are roped into the project as carriers. It does sound a little like the *Wolbachia* approach, except the virus involved is transmitted to plants not another insect of the same species. The news focus announcing the new approach (Reeves et al. 2018) was careful to point out this method might violate existing agreements on containment of chemical and biological weapons. We can expect progress to be very slow and methodical.

## Law and GMO Regulation

It is not difficult to find arguments for and against the use of GM insects and related genetic methods. They are everywhere. "Two scientists explore the controversies over releasing genetically modified mosquitoes into the wild" was the headline introducing an interview of two opposing viewpoints.

Gabriela Steier (Steier 2018) describes the GM regulation in Europe as based on the "precautionary principle" and that in the USA as diametrically opposite to that approach. This principle is included in the "Cartagena Protocol on Biosafety." Others point out that the USA did not sign this international agreement. She claims market-driven GM crops contribute to monocultures that are incompatible with biodiversity. Further, she makes the point that food dependence "relinquishes control over how to obtain one's food in the broad sense" and goes on to suggest moving production back into cities by using vacant rooftop space to produce fresh fruits and vegetables.

When anti-GMO efforts fade in time, as is happening now with GM crops in Africa (Cohen 2005; Morse et al. 2004), it is due partly to lack of any side-effects found for application of new technologies. It is very difficult to maintain an argument against something new when evidence accumulates suggesting a lack of deleterious effect.

A similar situation has not occurred in the use of GM insects, particularly GM mosquitoes, in strategies aimed at controlling malaria, dengue, and now zika caused by pathogens carried by mosquitoes. Often push-back efforts occur during regulatory analyses. If a new technology, such as GM insects, has never been employed before, it is a challenge to know what to "regulate." (Morrison et al. 2010).

## GM Insects and Climate Change

Despite the success of GM crops, GM insects are not being applied to crop and health protection in a timely manner. There seems to be a parallel between lazy adoption of GM insects in SIT strategies and apparently stagnated efforts to address climate change. Climate change will bring greater yield losses due to crop pest insects (Deutsch et al. 2015; Riegler 2018), increased disease outbreaks caused by vector insects (Barrett 2018), and a coming transformation of ecosystems (Nolan et al. 2018).

Storm events appear to be increasing in severity and high-temperature records are being set (Francis and Vavrus 2012; Diffenbaugh et al. 2013). These things have been predicted for some time now (Schlenker and Roberts 2009; Dawson et al. 2011). Mainka and Howard (2010) pointed to climate change and invasive species as being the two key drivers of biodiversity loss. What is not changing are steps to reverse increasing carbon emissions. The predictions are put in dire terms: "… the world is on a path to have its temperature rise 1.5 °C by 2030 and 2 °C by 2050 if people do not take swift action. This increase would have catastrophic effects, …" (Johnson 2018). Johnson reported "Also needed are sweeping changes in agriculture, including implementing sustainable land-use practices, restoring ecosystems, and transitioning to less resource-intensive diets."

## Isolated Islands, Insects, and Ecological Interactions

The Palmyra Atoll is isolated 1000 miles south of the Hawaiian Islands. During World War II it was inhabited by US military personnel who built airstrips and other structures. They also brought rats and mosquitoes with them and left them there upon departing. There are no endemic mammals and only a few endemic insects. With abundant food and no predators, the rat population exploded, reaching some 40,000 by 2011 (Williams 2018).

A recent major rat poisoning effort (chemical) eliminated the rats entirely. When follow up visits were conducted to check on recovery of endemic species, it became evident (Lafferty et al. 2018; Williams 2018) that the previously introduced *Aedes albopictus* mosquitoes vanished along with the rats. Researchers speculated that absence of coconut half-shells that once littered the islets as remnants of rat feeding deprived the mosquitoes of freshwater breeding sites.

This last example of the interaction of insects with the environment, especially invasive insects, shows how delicate the balance of nature can be in places were food resources are limited. In this case the limitation was absence of humans and eradication of rats to provide blood meals for the Asian tiger mosquitoes.

## References

Alphey L, Bourtzis K, Miller T (2009) Genetically modified insects as a tool for biorational control. In: Ishaaya I, Horowitz AR (eds) Biorational control of arthropod pests: application and resistance management. Springer, Berlin/Heidelberg, pp 189–206

Alphey L, Benedict M, Bellini R, Clark GG, Dame DA, Service MW, Dobson SL (2010) Sterile-insect methods for control of mosquito-borne diseases: an analysis. Vector Borne Zoonotic Dis 10(3):295–311. https://doi.org/10.1089/vbz.2009.0014

Anonymous (2004) Bugs in the system? Issues in the science and regulation of genetically modified insects. The Pew Foundation Initiative on Food and Biotechnology, Washington, DC, www.pewagbiotech.org, p 109

Antilla L, Liesner L (2008) Program advances in the eradication of pink bollworm, Pectinophoragossypiella, in Arizona cotton. In: Beltwide cotton conferences, Nashville, TN

Arora AK, Douglas A (2017) Hype or opportunity? Using microbial symbionts in novel strategies for insect pest control. J Insect Physiol 103:10–17

Arora AK, Forshaw A, Miller TA, Durvasula R (2015) A delivery system for field application of Paratransgenic control. BMC Biotechnol 15:59. https://doi.org/10.1186/s12896-015-0175-3

Arora AK, Pesko N, Quintero-Hernandez V, Possani LD, Miller TA, Durvasula RV (2018) A paratransgenic strategy to block transmission of *Xylella fastidiosa* from the glassy-winged sharpshooter *Homalodisca vitripennis*. BMC Biotechnol 18:50. https://doi.org/10.1186/s12896-018-0460-z

Austin B, Coulibaly M, Crisanti A, Diabate A, Kayondo JK (2018) Gene drive to reduce malaria transmission in sub-Saharan Africa. J Respons Innov 5(sup1):S66–S80. https://doi.org/10.1080/23299460.2017.1419410

Baltzegar JF, Elsensohn JE, Gutzmann NE, Webster SH (2018) What is the single best strategy for decreasing dengue fever virus (breakbone fever) incidence worldwide? Genetically engineered mosquitoes with lethal genes. Am Entomol 64(3):168–169

Barrett ADT (2018) The reemergence of yellow fever, since 2016, yellow fever outbreaks have become a major public health concern. Science 361(6505):847–848. https://doi.org/10.1126/science.aau8225

Beddington J (2009) Food, energy, water and the climate: a perfect storm of global events. In: Lecture to sustainable development UK 09 conference, vol 19

Bextine B, Lauzon C, Potter S, Lampe D, Miller TA (2004) Delivery of a genetically marked *Alcaligenes* sp. to the glassy-winged sharpshooter for use in a paratransgenic control strategy. Curr Microbiol 48:327–331

Blood BL, Eason J, Hughes GP, Stack SC, Stewart TJ (2018) What is the sing best strategy for decreasing dengue fever virus (breakbone fever) incidence worldwide? Biological control using Wolbachia. Am Entomol 64(3):169–171

Bruce TJA (2012) GM as a route for delivery of sustainable crop protection. J Exp Bot 63(1):537–541. https://doi.org/10.1093/jxb/err281

Burt A (2014) Heritable strategies for controlling insect vectors of disease. Phil Trans R Soc B 369:20130432. https://doi.org/10.1098/rstb.2013.0432

Burt A, Crisanti A (2018) Gene drive: evolved and synthetic. ACS Chem Biol 13(2):343–346. https://doi.org/10.1021/acschembio.7b01031

Carvalho DO, McKemey AR, Garziera L, Lacroix R, Donnelly CA, Alphey L et al (2015) Suppression of a field population of Aedes aegypti in Brazil by sustained release of transgenic male mosquitoes. PLoS Negl Trop Dis 9(7):e0003864. https://doi.org/10.1371/journal.pntd.0003864

Cohen JI (2005) Poorer nations turn to publicly developed GM crops. Nat Biotech 23:27–33

Collins CM, Bonds JAS, Quinlan MM, Mumford JD (2018) Effects of the removal or reduction in density of the malaria mosquito, Anopheles gambiaes.L., on interacting predators and competitors in local ecosystems. Med Vet Entomol. https://doi.org/10.1111/mve.12327

Dawson TP, Jackson ST, House JI, Prentice IC, Mace GM (2011) Beyond predictions: biodiversity conservation in a changing climate. Science 332(6025):53–58. https://doi.org/10.1126/science.1200303
de Campos AS, Hartley S, de Konis C, Lezaun J, Velho L (2017) Responsible innovation and political accountability: genetically modified mosquitoes in Brazil. J Respons Innov 4(1):5–23. https://doi.org/10.1080/23299460.2017.1326257
Debach P (1973) Biological control of insect pests and weeds. Chapman and Hall, London, p 844
Deutsch CA, Tewksbury JJ, Tigehelaar M, Battisti DS, Merrill SC, Huey RB, Naylor RL (2015) Increase in crop losses to insect pests in a warming climate. Science 361:916–919
Diffenbaugh NS, Scherer M, Trapp RJ (2013) Robust increases in severe thunderstorm environments in response to greenhouse forcing. Proc Natl Acad Sci 110:16361–16366. https://doi.org/10.1073/pnas.1307758110
Douglas A (2018) Fundamentals of microbiome science, how microbes shape animal biology. Princeton University Press, p 249
Duan JJ, Bauer LS, Abell KJ, Ulyshen MD, Van Driesche RG (2015) Population dynamics of an invasive forest insect and associated natural enemies in the aftermath of invasion: implications for biological control. J Appl Ecol 52(5). https://doi.org/10.1111/1365-2664.12485
Durvasula RV, Gumbs A, Panackal A, Kruglov O, Aksoy S, Merrifield RB, Richards FF, Beard CB (1997) Prevention of insect-borne disease: an approach using transgenic symbiotic bacteria. Proc Natl Acad Sci U S A 94(7):3274–3278
Durvasula RV, Kroger A, Goodwin M, Panackal A, Kruglov O, Taneja J, Gumbs A, Richards FF, Beard CB, Cordon-Rosales C (1999) Strategy for introduction of foreign genes into field populations of Chagas disease vectors. Ann Entomol Soc Amer 92(6):937–943. https://doi.org/10.1093/aesa/92.6.937
Ebeling (1959) Subtropical Fruit Pests. University of California Press, Los Angeles, p 436
Fang W, Lu HL, King GF, St J, Leger R (2015) Construction of a hypervirulent and specific mycoinsecticide for locust control. Sci Rep 4:7345. https://doi.org/10.1038/srep07345
Favia G, Ricci I, Damiani C et al (2007) Bacteria of the genus *Asaia* stably associate with *Anopheles stephensi*, an Asian malarial mosquito vector. Proc Natl Acad Sci U S A 104(21):9047–9051
Favia G, Ricci I, Marzorati M et al (2008) Bacteria of the genus *Asaia*: a potential paratransgenic weapon against malaria. Transgenesis and the Management of Vector-Borne Disease. Adv Exp Med Biol 627:49–59
Fedoroff NV (2012) Transposable elements, epigenetics, and genome evolution. AAAS presidential address. Science 338:758–767
Francis JA, Vavrus SJ (2012) Evidence linking Arctic amplification to extreme weather in mid-latitudes. Geophys Res Lett 39(6):L06801. https://doi.org/10.1029/2012GL051000
Fraser JE, De Bruyne JT, Iturbe-Ormaetxe I, Stepnell J, Burns RL, Flores HA, O'Neill SL (2017) Novel Wolbachia-transinfected Aedes aegypti mosquitoes possess diverse fitness and vector competence phenotypes. PLoS Pathog 13(12):e1006751. https://doi.org/10.1371/journal.ppat.1006751
Fryxell KJ, Miller TA (1995) Autocidal biological control: a general strategy for insect control based on genetic transformation with a highly conserved gene. J Econ Entomol 88(5):1221–1232. https://doi.org/10.1093/jee/88.5.1221
Gai CS, Lacava PT, Quecine MC, Auriac M-C, JRS L, Araújo WL, Miller TA, Azevedo JL (2009) Transmission of *Methylobacterium mesophilicum* by Bucephalogoniaxanthophis for paratransgenic control strategy of citrus variegated chlorosis. J Microbiol 47:448–454
Hagen KS, Franz JM (1973) A history of biological control, pp 433–476, in history of entomology, ed by R. F. Smith. T. E. Mittler and C. N. Smith. Annual Reviews, Palo Alto, CA, p 517
Hajek A, Eilenberg J (2018) Biological control: concerns, changes, and challenges. In: Natural enemies: an introduction to biological control. Cambridge University Press, Cambridge, pp 325–388

Hendrichs J, Franz G, Rendon P (1995) Technique through male-only releases for control of Mediterranean fruit flies during fruiting seasons. J Appl Entomol 119(1–5):371–377. https://doi.org/10.1111/j.1439-0418.1995.tb01303.x

Hoffmann AA, Montgomery BL, Popovici J, Iturbe-Ormaetxe I, Johnson PH, Muzzi F, Greenfield M, Durkan M, Leong YS, Dong Y, Cook H, Axford J, Callahan AG, Kenny N, Omodei C, McGraw EA, Ryan PA, Ritchie SA, Turelli M, O'Neill SL (2011) Successful establishment of Wolbachia in Aedes populations to suppress dengue transmission. Nature 476(7361):454–457. https://doi.org/10.1038/nature10356

Hosokawa T, Kikuchi Y, Shimada M, Fukatsu T (2007) Obligate symbiont involved in pest status of host insect. Proc R Soc B 274:1979–1984. https://doi.org/10.1098/rspb.2007.0620

Hughes GL, Koga R, Xue P, Fukatsu T, Rasgon JL (2011) Wolbachia infections are virulent and inhibit the human malaria parasite *Plasmodium falciparum* in *Anopheles gambiae*. PLoS Pathog 7(5):e1002043

Johnson J (2018) Last chance to curb greenhouse gas emissions, climate change. Chem & Eng News, p 18. A summary of a report by the Intergovernmental Panel of Climate Change (IPCC) http://www.ipcc.ch/report/sr15/

Kandul NP, Liu J, Sanchez HM, Wu SL, Marshall JM, Akbari OS (2018) Transforming insect population control with precision guided sterile males. bioRxiv Preprint Server for Biology. https://www.biorxiv.org/content/early/2018/07/26/377721

Knipling EF (1955) Possibilities of insect control or eradication through the use of sexually sterile males. J Econ Entomol 48:459–469

Knipling EF (1959) Sterile-male method of population control. Science 130(3380):902–904. https://doi.org/10.1126/science.130.3380.902

Kuzina LV, Miller TA, Cooksey DA (2006) In vitro activities of antibiotics and antimicrobial peptides against the plant pathogenic bacterium Xylellafastidiosa. Lett Appl Microbiol 42:514–520

Lacava PT, Parker J, Andreote FD, Dini-Andreote F, Ramirez JL, Miller TA (2007) Analysis of the bacterial community in glassy-winged sharpshooter heads. Entomol Res 37:261–266

Lafferty KD, McLaughlin JP, Gruner DS et al (2018) Local extinction of the Asian tiger mosquito (*Aedes albopictus*) following rat eradication on Palmyra atoll. Biol Lett 14:20170743. https://doi.org/10.1098/rsbl.2017.0743

Loukeris TG, Livadaras I, Arcà B, Zabalou S, Savakis C (1995) Gene transfer into the medfly, *Ceratitis capitata*, with a *Drosophila hydei* transposable element. Science 270(5244):2002–2005

Mainka SA, Howard GW (2010) Climate change and invasive species: double jeopardy. Integ Zool 5(2):102–111. https://doi.org/10.1111/j.1749-4877.2010.00193.x

Marrelli MT, Li CY, Rasgon JL, Jacobs-Lorena M (2007) Transgenic malaria-resistant mosquitoes have a fitness advantage when feeding on Plasmodium-infected blood. Proc Natl Acad Sci U S A 104(13):5580–5583

Miller TA (2004) Designing insects. http://www.actionbioscience.org/biotech/miller.html

Miller TA (2007a) Applying insect transgenic technology: scientific and regulatory experiences. Entomol Res 37:34–40

Miller TA (2007b) Symbiotic control in agriculture and medicine. Symbiosis 44:67–73

Miller TA (2012) Random thoughts on recombinant insects. Nova Science Publishers, New York, p 166

Miller TA (2013) Delivery. Pest Manag Sci 69:1199–1204

Miller TA, Staten R (2001) The pink bollworm, tale of a transgenic tool. Agrichem Environ News 182

Mills NJ (2017) Rapid evolution of resistance to parasitism in biological control. PNAS 114(15):3792–3794. https://doi.org/10.1073/pnas.1702753114

Moreira LA, Iturbe-Ormaetxe I, Jeffery JA et al (2009) A Wolbachia symbiont in Aedes aegypti limits infection with dengue, Chikungunya, and Plasmodium. Cell 139(7):1268–1278

Morrison NI, Franz G, Koukidou M, Miller TA, Saccone G, Alphey LS, Beech CJ, Nagaraju J, Simmons GS, Polito LC (2010) Genetic improvements to the sterile insect technique for agricultural pests. AsPac J Mol Biol Biotechnol 18(2):275–295

Morse S, Bennett R, Ismael Y (2004) Why Bt cotton pays for small scale producers in South Africa. Nat Biotech 22:379–380

Nolan C, Overpeck JT, Allen JRM et al (2018) Past and future global transformation of terrestrial ecosystems under climate change. Science 361(6405):920–923. https://doi.org/10.1126/science.aan5360

Peloquin JJ, Thibault ST, Staten R, Miler TA (2000) Germ-line transformation of pink bollworm (Lepidoptera: Gelechiidae) mediated by the piggyBac transposable element. Insect Molec Bio 9:323–333

Ramirez JL, Perring TM, Miller TA (2008) Fate of a genetically modified bacterium in the foregut of glassy-winged sharpshooter. J Econ Entomol 101:1519–1525

Rani A, Sharma A, Rajagopal R, Adak T, Bhatnagar RK (2009) Bacterial diversity analysis of larvae and adult midgut microflora using culture-dependent and culture-independent methods in lab-reared and field-collected *Anopheles stephensi*-an Asian malarial vector. BMC Microbiol 9:96

Reeves RG, Voeneky S, Caetano-Anolles D, Beck F, Boete C (2018) Agricultural research, or a new bioweapon system? Science 362(6410):35–37. https://doi.org/10.1126/science.aat7664

Riegler M (2018) Insect threats to food security. Science 361(6405):846. https://doi.org/10.1126/science.aau7311

Schlenker W, Roberts MJ (2009) Nonlinear temperature effects indicate severe damages to U.S. crop yields under climate change. PNAS 106(37):15594–15598. https://doi.org/10.1073/pnas.0906865106

Scott MJ, Concha C, Welch JB, Phillips PL, Skoda SR (2017a) Review of research advances in the screwworm eradication program over the past 25 years. Entomol Exp et Applicata 164:226–236. https://doi.org/10.1111/eea.12607

Scott MJ, Gould F, Lorenzen MD, Grubbs N, Edwards OR, O'Brochta DA (2017b) Agricultural production: assessment of the potential use of Cas9-mediated gene drive systems for agricultural pest control. J Respons Innov 5:S98–S120. https://doi.org/10.1080/23299460.2017.1410343

Simmons GS, McKemey AR, Morrison NI, O'Connell S, Tabashnik BE, Claus J, Fu G, Tang G, Sledge M, Walker AS, Phillips CE, Miller ED, Rose RI, Staten RT, Donnelly CA, Alphey L (2011) Field performance of a genetically engineered strain of pink bollworm. PLoS One 6(9):e24110. https://doi.org/10.1371/journal.pone.0024110

Slama K, Williams CM (1966) 'Paper factor' as an inhibitor of the embryonic development of the European bug, *Pyrrhocoris apterus*. Nature 210:329–330

Steier G (2018) Advancing food integrity. GMO regulation, agroecology, and urban agriculture. CRC Press, Taylor & Francis Group, p 229. https://doi.org/10.1201/b22381

Tabashnik BE, Sisterson MS, Ellsworth PC, Dennehy TJ, Antilla L et al (2010) Suppressing resistance to Bt cotton with sterile insect releases. Nat Biotechnol 28(12):1304–1307

Thibault ST, Luu HT, Vann N, Miller TA (1998) Precise excision and transposition of piggyBac in pink bollworm embryos. Insect Mol Biol 8:119–123

Thomas DD, Donnelly CA, Wood RJ, Alphey LS (2000) Insect population control using a dominant, repressible, lethal genetic system. Science 287:2474–2476

Wagner W, Fisher E, Pascual P (2018) Whose science? A new era in regulatory "science wars". Science 362(6415):636–639. https://doi.org/10.1126/science.aau3205

Wan P, Xu D, Cong S, Jiang Y, Huang Y, Wang J, Wu H, Wang L, Wu K, Carriere Y, Mathias A, Li X, Tabashnik BE (2017) Hybridizing transgenic Bt cotton with non-Bt cotton counters resistance in pink bollworm. PNAS 114(21):5413–5418. www.pnas.org/cgi/doi/10.1073/pnas.1700396114

Wang S, Jacobs-Lorena M (2017) Chapter 13: Paratransgenesis applications: fighting malaria with engineered mosquito symbiotic bacteria. In: Arthropod vector: controller of disease transmission, volume 1, Vector microbiome and innate immunity of arthropods, pp 219–234. https://doi.org/10.1016/B978-0-12-805350-8.00013-1

Wangkeeree J, Miller TA, Hanboonsong Y (2012) Candidates for symbiotic control of sugarcane white leaf disease. Appl Environ Microbiol 78(19):6804–6811. https://doi.org/10.1128/AEM.01439-12

Warner KD, Daane KM, Getz CM, Maurano SP, Calderon S, Powers KA (2011) The decline of public interest agricultural science and the dubious future of crop biological control in California. Agr Human Values 28(4):483–496. https://doi.org/10.1007/s10460-010-9288-4

Williams R (2018) Paradise regained: how the Palmyra Atoll got rid of invasive mosquitoes. The Scientist. https://www.the-scientist.com/news-opinion/paradise-regained-how-the-palmyra-atoll-got-rid-of-invasive-mosquitoes-30063

Xi Z, Khoo CCH, Dobson SL (2005) *Wolbachia* establishment and invasion in an *Aedes aegypti* laboratory population. Science 310(5746):326–328. https://doi.org/10.1126/science.1117607

Yong E (2016) I contain multitudes. The microbes within us and a grander view of life. Harper Collins Publishers, p 355

# Invasive Species Control and Resolution of Wildlife Damage Conflicts: A Framework for Chemical and Genetically Based Management Methods

**Larry Clark, John Eisemann, John Godwin, Katherine E. Horak, Kevin Oh, Jeanette O'Hare, Antoinette Piaggio, Kimberley Pepin, and Emily Ruell**

**Abstract** Vertebrate wildlife damage management relates to developing and employing methods to mitigate against damage caused by wildlife in the areas of food production, property damage, and animal or human health and safety. Of the many management tools available, chemical methods (e.g., toxicants) draw the most attention owing to issues related to environmental burden, species specificity, and humaneness. Research and development focusing on RNA interference and gene drives may be able to address the technical aspects of performance goals. However, there remain many questions about regulation, environmental risk, and societal acceptance for these emerging biological technologies. Here we focus on the development and use of these biological technologies for use in vertebrate pest management and conservation (e.g., management of wildlife diseases). We then discuss the regulatory framework and challenges these technologies present and conclude with a discussion on factors to consider for enabling these technologies for pest management and conservation applications under a commercially applied framework.

**Keywords** Gene drive · Gene silencing · RNAi · Wildlife · Regulation · Product development · Pest control

L. Clark · J. Eisemann (✉) · K. E. Horak · K. Oh · J. O'Hare · A. Piaggio · K. Pepin · E. Ruell
National Wildlife Research Center, Wildlife Services, Animal and Plant Health Inspection Service, United States Department of Agriculture, Fort Collins, CO, USA
e-mail: john.d.eisemann@usda.gov

J. Godwin
Department of Biological Sciences, North Carolina State University, Raleigh, NC, USA



A. Chaurasia et al. (eds.), *GMOs*, Topics in Biodiversity and Conservation 19,
https://doi.org/10.1007/978-3-030-53183-6_9

## Introduction

The field of vertebrate wildlife damage management relates to developing and employing methods to mitigate damage caused by wildlife in the areas of food production, property damage, and animal or human health and safety (Conover 2001). The methods used to resolve such conflicts include the use of physical devices (e.g., traps, sonic or visual scaring methods), chemical methods (e.g., reproductive inhibitors, repellents, toxicants), direct lethal control (e.g., culling), and alterations to the landscape (e.g., habitat manipulation). Unlike other areas of wildlife management that involve natural resource protection for conservation or sustainable consumptive and nonconsumptive recreational use, wildlife damage management seeks to alter animal behavior or circumstances to prevent damage to human activity or interests that may be caused by wildlife and hence involve more direct and interactive approaches to management.

For the most part, the technologies and methods used by wildlife damage managers have long histories (Reidinger Jr and Miller 2013). Among the many management tools, chemical toxicants have proven to be both effective and controversial (Eason et al. 2010). For the most part, criticism has focused on five areas: effectiveness, need and/or alternatives, humaneness, nontarget effects, and environmental burden. Thus, imperative for managers in the development of new technologies is to consider developing methods that provide for sustainable agricultural production and natural resource stewardship within a social license framework.

If the use of vertebrate pesticides is required, then it is desirable for them to be species-specific, thus eliminating direct nontarget risks. The pesticide should also reduce environmental and ecological burden. That is to say, once the target effect is achieved, diffusion into the environment and amplification through the food chain should not occur, thus eliminating indirect nontarget effects. The mode of action for the pesticide should reduce animal awareness to pain and have a short time to death, both traits that contribute to current standards for humane death (Underwood et al. 2013). As the science evolves and the options for mode of action and nature of the pesticide broaden, there should be regulatory clarity from authorizing agencies and guidance to users in the form of approved labeling and licensing of products. Because the research, development, and regulatory processes are complex and not necessarily familiar to wildlife damage managers, it is vital that communication among the developing partners is well founded and follows an orderly process so that formulation, effectiveness testing, technology transfer, production, delivery, and scalability result in a cost-effective product that is easily applied on a landscape scale. Finally, and perhaps most importantly, it is imperative that a communication and engagement plan is developed for users and the community. It does little good to develop an expensive tool only to have it fail because of unanticipated or unaddressed concerns within the human social and political framework.

Traditional strategies to control pest species include biological (use of predators, crop rotation), mechanical (physical removal), and chemical (toxicants, repellents). Biological and mechanical methods are often chosen to limit the use of chemicals as a means to reduce impacts to the environment and nontarget species (Damalas and Eleftherohorinos 2011; Kogan 1998; Witmer 2007). Yet chemical methods may be favored because pest population control often occurs more quickly and with less monetary investment than when other techniques are used. More recently, research has focused on development of biological pesticides that combine target specificity, humaneness, and low environmental burden with the response and cost efficiencies of traditional chemical applications, such as gene drive technology and RNA interference. These technologies hold the promise to address the objectives identified above and perhaps decrease the reliance on use of chemical pesticides.

Here we focus on the development and use of two technologies that hold promise as tools for vertebrate wildlife management and conservation: gene silencing through RNA interference and gene drives.

## Gene Silencing

Interest in RNA interference (RNAi) as a tool for basic research and for the treatment of diseases started with its discovery in the 1990s. Since then, the utility of RNAi as a means to control pest species has become increasingly popular. Successes in the development of RNAi toxicants to control insects and novel methods using plants as a means to deliver RNAi to feeding insects have expanded initial expectations of the applications of this technology. In 2017, the US Environmental Protection Agency (EPA) registered the first RNAi-based insecticide for the control of Western corn rootworm (*Diabrotica virgifera virgifera*). This EPA registration ushers in a new era in toxicant development with the ability to design toxicants that are species-specific alleviating concerns of risks to nontarget species, and following the promise shown for insect control, there is increasing interest for use of this technology for applications in vertebrate pest management.

RNA interference (RNAi) is a fundamental cellular process that controls what genes are turned on and off by determining what proteins are synthesized from the messenger RNA (mRNA) messages in the cytoplasm. RNAi shows promise as a class of species-specific toxicants because RNAi molecules, by design, bind specifically to a single mRNA triggering its destruction. The instructions for protein synthesis travel from DNA in the nucleus to the cytoplasm by messenger RNA (mRNA) (Crick 1970). These single-stranded mRNA molecules are composed of unique sequences of nucleotides that code for specific proteins.

The first indication that translation from mRNA to protein may be influenced by exogenous RNA sequences came when Jorgensen (1998) attempted to increase the purple pigment in petunia flowers by injecting extra copies of the pigment gene into the plant. The result of these injections was not flowers that were more purple, but surprisingly yielded either two-toned or totally white flowers (Fire et al. 1998; Jorgensen et al. 1998; Napoli et al. 1990). Jorgensen termed this phenomenon "co-suppression." This was the first experimental demonstration of the process now called RNA interference for which Andrew Z. Fire and Craig C. Mello were awarded a Nobel Prize in 2006.

RNAi is dependent on the inherent protective pathway within cells that degrades mRNA. Double-stranded RNA (dsRNA) binds to a protein called Dicer that cleaves the dsRNA into smaller fragments that are then integrated into the RNA-induced silencing complex, RISC (Liu et al. 2004; Tang 2005; Zhang et al. 2004). The RISC complex then separates the short dsRNA fragments into single-stranded segments, one of which is shuttled to and binds the matching mRNA sequencing that was transcribed from DNA. The complex formed by the RISC (with fragment of the initial dsRNA) and the mRNA then cleaves and degrades the mRNA guided by properties of the fragment of the initial dsRNA (Khvorova et al. 2003; Schwarz et al. 2003). The gene for which the mRNA is coded is, therefore, not synthesized, and the gene is "silenced." Since its discovery, this sequence-specific gene silencing has shown great promise as both a research tool and in the treatment of diseases (Lares et al. 2010). More recently, interest in using RNAi as a means to control pest species has become increasingly popular; however, to date, this interest has been focused on insect pests (Baum and Roberts 2014; Burand and Hunter 2013; Mamta and Rajam 2017; Niu et al. 2018; Zotti et al. 2018).

## *Gene Silencing: Pest Control and Risk*

The success of RNAi as a "chemical" means to control pest species depends on multiple factors. Though the RNAi pathways are inherently present in cells, introduction of exogenous RNAi is still recognized as nonself and therefore can elicit an immune response and subsequent release of inflammatory cytokines (Alexopoulou et al. 2001; Heil et al. 2004). Responses such as these are called off-target effects as they are not related to the desired physiological response from the RNAi-induced gene silencing. Significant effort is made during the design of RNAi molecules to reduce off-target effects and minimize immune response (Schwarz et al. 2003). Off-target effects are a concern in nontarget species as they can occur even if the target sequence of the RNAi is not present in the nontarget animal. Nontarget species will not be affected by the directed silencing of specific genes, but the RNAi must not elicit an immune response in nontarget species. If successful, one of the benefits of RNAi over traditional chemical toxicants is its species specificity. Comparisons between gene sequences in target and nontarget species are done to determine regions of the genes that are the most different. RNAi molecules are then designed to those regions. These sequence differences result in a mismatch between the RNA incorporated in the RISC complex and the mRNA in the cytoplasm, blocking degradation of the mRNA (Amarzguioui et al. 2003). It is possible to design RNAi molecules that differ from target mRNA by a single nucleotide and therefore do not bind; purine/purine mismatches offer the highest level of discrimination (Schwarz et al. 2006). Nucleotide mismatches in different regions of the RNAi also affect specificity, with the last two nucleotides of the same types of RNAi molecules not contributing to binding to the mRNA (Elbashir et al. 2001), so mismatches in those regions do not affect binding and subsequent mRNA degradation. The requirement of exact matching of RNAi and target mRNA sequences places a burden on the development of RNAi-based chemical toxicants, as genes of wildlife species are often not sequenced. In addition, ensuring the target sequence in conserved diverse animal populations that are geographically separated needs to be done before sequences are registered for use as toxicants.

Using RNAi as a means to control pest populations is contingent on delivery of the nucleotide sequence to both the target species and the desired location in the organism. Toxicant baits containing RNAi face the same requirements as traditional chemical toxicant baits of attractiveness to target species and stability in harsh environmental conditions. This is a challenge as RNAi is, by its nature, unstable. Modifications to the RNAi sequences, such as the addition of 2'-O-methylpurines or 2-fluoropyrimidines, have been shown to increase stability without decreasing effectiveness (Czauderna et al. 2003). Once the pest species consumes the RNAi bait, getting the RNAi to the target tissue in appropriate concentration becomes the next hurdle. RNAi baits will be formulated for oral delivery meaning the RNAi will have to survive the extreme conditions of the gastrointestinal tract before absorption into the systemic circulation. Lipid nanoparticle carriers have shown promise in protecting the RNAi from degradation at low digestive pH (Ball et al. 2018). Once in systemic circulation, the RNAi must get to the target location. This can be achieved through the addition of cell-type-specific ligands, antibodies, or receptors on the carrier molecule. Uptake of the RNAi by target cells can be facilitated by the use of cell-penetrating peptides, nanoparticles, and polymer-based delivery systems (Ahmadzada et al. 2018; Avila et al. 2016; Singh et al. 2018a, b). Once inside the cell, the RNAi will then silence the target gene by directed destruction of the mRNA.

There are numerous obstacles for delivery of an RNAi toxicant for use in vertebrate pests, from the development of the oral bait to shuttling the RNAi to the target tissue. However, research from human drug development and successes in the fields of insect control lay the foundation for vertebrate RNAi toxicant development and wildlife disease treatments in the field.

## Gene Drives

Pest control technologies using RNAi have the potential to come to use sooner and under a clear regulatory framework (see below). Such technologies will also most likely be applied similar to traditional chemical pesticides. However, gene drive technologies hold the promise of being self-sustaining once released, thus eliminating the need for constant reapplication. It is this feature which is both appealing from an economic logistical perspective and warrants caution from an environmental risk perspective.

It is not difficult to find popular news stories about the promise and potential catastrophe that could be realized through the application of clustered regularly interspaced short palindromic repeats (CRISPR) and CRISPR-associated proteins for genome editing and DNA. CRISPR-Cas systems have been identified from bacteria and archaea and provide immunity to ward off bacteriophages (Barrangou 2015). When harnessed for genome editing, the system has been shown to provide targeted, sequence-specific cleavage of double-stranded DNA (Mali et al. 2013; Jiang et al. 2013). This is accomplished through the precise delivery of endonuclease enzymes by synthetic single-guide RNA (sgRNA) that are engineered to bind only to specific target sequences within the genome. These DNA cleavages will then be repaired by the targeted genome through either homology-directed repair (HDR) or nonhomologous end joining (NHEJ). HDR repairs with a homologue piece of DNA, whereas NHEJ directly ligates the cleaved pieces of DNA and thus can lead to a loss of nucleotides and other potential errors. With an engineered template, the HDR process will repair the nicked DNA with a synthetic portion of DNA on both DNA strands (Mali et al. 2013; Alphey 2014; Doudna and Charpentier 2014). Thus, through the CRISPR-Cas system, double-stranded DNA can be cleaved at a targeted site and repaired with a synthetic piece of DNA, which can then be copied in the organism through regular genomic copying. Application of this technology has been coined "synthetic biology," and its utility has been realized in the pharmaceutical and agricultural production already (Carlson 2010; Church and Regis 2012; Doudna and Charpentier 2014; Gantz et al. 2015; Lander 2016; Hussain et al. 2018).

Because gene drive applications would not require constant application in the sense of traditional chemical pesticides or even RNAi, there is inherent appeal to managers that the management tool will be self-sustaining, whether for pest management or for control of disease susceptibility for wildlife conservation purposes, e.g., avian malaria for protection of birds or plague resistance for protection of endangered species.

## *Gene Drives: Uses for Disease Mitigation*

Disease control is typically accomplished by reductions in host or vector abundance, reduction of contact between hosts and pathogens, or increase in the refractoriness or resistance of hosts or vectors to infections (Sokolow et al. 2019). Traditional techniques include host culling, pesticides for vector control, physical barriers between hosts and vectors, vaccination to reduce transmission, or treatment to reduce severity and transmission. All of these interventions can be expensive and time-consuming and have variable levels of effectiveness depending on ecological conditions. Pesticide application can present health risks to humans and domestic animals through contamination of soil or water supplies. Genetic modification (GM) techniques can be developed to target genes that could affect any of these processes and have the potential to be cheaper and more effective and have lower host or environmental burden. For example, GM approaches can be designed to target specific isolates or strains of a bacterium, which is advantageous over antibiotics because the specificity allows "good" bacteria to be unaffected (Barrangou and

Doudna 2016). Strain-specific editing of bacterial populations is particularly useful in food biotechnology where CRISPR systems have been used to vaccinate industrial bacterial cultures against viruses or to engineer antibiotic resistance uptake or probiotic cultures (Selle and Barrangou 2015a, b). Similarly, it was recently shown that a CRISPR-Cas13a system could be used to engineer potato plants to be resistant to potato virus Y while having no effect on related viruses such as potato virus A (Zhan et al. 2019). The ability to efficiently protect crops or livestock against specific agricultural diseases could dramatically improve food security while reducing the ecological footprint of agriculture (Herman et al. 2019; Van Eenennaam and Young 2014).

One of the most active areas of disease control research has been to create gene drives that repress vector populations or make them refractory to pathogens that cause human diseases such as dengue or malaria (Ferguson 2018; Shaw and Catteruccia 2019). For example, *Anopheles stephensi*, a vector of the malaria parasite *Plasmodium* spp., has been successfully engineered to have much reduced vector competence relative to wild-type vector individuals (Gantz et al. 2015; Ito et al. 2002). More recently, it has been demonstrated that a CRISPR-Cas9 system for delivery of a female sterility trait could spread to 100% prevalence in 7–11 generations in caged *Anopheles* mosquitoes (Kyrou et al. 2018). Similar results have been obtained using other gene targets (Hammond et al. 2016), suggesting that there is ample opportunity to choose a target that will be successful in a particular ecological context (Champer et al. 2016). While not yet realized, the ability to protect livestock against specific diseases efficiently using CRISPR technology is on the horizon (Conklin 2019; Lamas-Toranzo et al. 2017).

## *Gene Drives: Uses for Conservation*

Invasive mammalian predators represent a major threat to biodiversity worldwide. Doherty et al. (2016) estimated mammalian predators are responsible for the extinction of at least 142 vertebrate species since AD 1500 (58% of the total including 87 bird, 45 mammal, and 10 reptile species) and threaten another 596 species. The key invasive vertebrate predator threats involve species from 13 mammalian families including rodents, felids, canids, and mustelids with seven species/groups in particular accounting for the bulk of the documented impacts on birds, mammals, and reptiles: cats, rodents, dogs, pigs, small Indian mongoose, red fox, and stoats. Of these, cats and rodents including three species of rats (*Rattus* spp.) and house mice (*Mus musculus*) have proven particularly damaging with island faunas being especially hard hit for several reasons. First, islands harbor a disproportionate share of biodiversity. Despite representing approximately 5% of the Earth's land area, islands are home to 20% of described vertebrate species and approximately 40% of threatened and endangered species (Tershy et al. 2015; Spatz et al. 2017). Second, the populations of island endemic species are often much smaller and are therefore typically more vulnerable. Lastly, these island species have often evolved with few or no predators and consequently lack adaptive antipredator defenses (Adler and Levins 1994; Cuthbert et al. 2016).

Because islands are hot spots for both biodiversity and threatened and endangered species, managing threats in island ecosystems has also been a central focus for both conservationists and managers. In addition, for logistical reasons, islands represent isolated contained systems of limited geographic scale which will be critical for effective early phase testing and evaluation.

A major focus of efforts on islands has been eradication of invasive mammalian predators, and the outcomes have been tremendously positive (Cuthbert et al. 2011; Jones et al. 2016). Despite the significant conservation benefits realized through traditional eradication approaches, primarily aerial broadcast of rodenticides when targeting rats and mice, these methods have both significant drawbacks and some fundamental limitations (Campbell et al. 2015, 2019). The drawbacks include high fixed costs for operations (often in the millions of US dollars), nontarget species exposure to toxicants, and animal welfare concerns for both target and nontarget species (Mackay et al. 2007; Holmes et al. 2015). Failures and other rationales can often lead to significant social and political opposition. Lastly and critically, applying toxicant-based methods is extremely challenging on islands with human inhabitants, which represent the majority of islands where invasive mammals threaten biodiversity. Indeed, Campbell et al. (2019) estimate that fewer than 15% of islands worldwide where invasive rodents threaten critically threatened or endangered species are amenable to rodenticide-based eradication approaches. New approaches are clearly needed.

Genetic methods of pest control potentially offer a useful alternative to these established approaches. Although engineered gene drives harnessing either natural or synthetic drive mechanisms are still in the development phase, population modeling supports their potential effectiveness in reducing invasive mammal populations (Backus and Gross 2016; Prowse et al. 2017, 2019; Sudweeks et al. 2019). As detailed above, harnessing natural or synthetic selfish genetic elements in the form of gene drives could provide options not burdened by many of the drawbacks of rodenticide-based approaches. Specifically, genetic approaches could provide flexibility in financial models relative to the high fixed costs of rodenticide-based eradications, where operations typically must be conducted within short time windows. Genetic approaches should also be species-specific, eliminating at least direct effects on nontarget organisms. A gene drive approach would likely affect target pest populations through either biasing offspring sex ratios or inducing infertility in drive carriers, thereby leading to population reduction through natural attrition. This could alleviate animal welfare concerns that arise from the mechanism of action of toxicants currently. The species specificity of genetic approaches could facilitate use on inhabited islands due to the reduced threat to humans, pets, and livestock. However, use on inhabited islands could raise other concerns, including increased potential for movement of gene drive carriers from the targeted island and introduction of resistant individuals onto an island from nontarget populations. These are also new, unfamiliar, and as yet untested and unproven technologies, so a great deal of effort will need to be dedicated to engaging stakeholders and relevant publics in order to decide if, when, and how these approaches should be employed.

If a gene drive approach is employed for biodiversity conservation on islands, how is such an effort likely to proceed? While rats and cats represent more significant overall threats to biodiversity and have been discussed as potential target organisms for gene drive approaches (Moro et al. 2018), efforts to date have been focused on establishing feasibility in the house mouse. The feasibility of synthetic gene drive approaches has thus far only been demonstrated in insects and yeast (Gantz and Bier 2015; Gantz et al. 2015; Hammond et al. 2016; DiCarlo et al. 2015). The house mouse is the preeminent, most manipulable, and intensively studied mammalian genetic model organism with a rich knowledge base to support efforts to affect sex determination and fertility in this species (Campbell et al. 2019). No functional synthetic gene drive has yet been described for a vertebrate, and a recent report suggests generating one could prove challenging (Grunwald et al. 2019). Therefore, focusing on the most genetically tractable mammal is likely the best approach as well as advancing the general knowledge base necessary to advance efforts in other mammalian pest species should efforts in mice prove successful.

A number of factors including invasive species threaten terrestrial vertebrates worldwide (Allan et al. 2019). The foregoing discussion focused on use of genetic technologies on islands, which are favorable due to physical containment provided by geographic isolation. There is discussion of employing these technologies in a continental context and a recently launched effort to explore the potential of genetic approaches for controlling cats in Australia. Feral cats and red foxes represent the major current threat to Australia's terrestrial mammal fauna (Woinarski et al. 2015), and existing control strategies are not equal to this challenge. This has prompted consideration of gene drive approaches for control of feral cats (Moro et al. 2018; Kinnear 2018). Key knowledge gaps remain, and advances in understanding the potential of gene drive approaches for cats and other species will likely depend on progress in implementing these approaches in rodents first.

## *Gene Drives: Risk*

There are several types of risk that must be considered in evaluating the potential utility of different pest control methods, such as ecological, evolutionary, economic, and ethical (Gould 2008). All methods of pest control share some risks in common (Herman et al. 2019). For example, reduction of pest populations can have ecological consequences in terms of ecosystem maintenance or community ecology (Gould 2008). Likewise, pesticide or repellant application or GM organisms can have economic risks due to evolution of resistance by the pest or ecological risks through negative impacts on nontarget species. Any pest control method can pose ethical risks through negative effects on human health or belief systems. All types of risks should be thoroughly evaluated before a new technology can be applied.

For innovative pest control approaches, the first type of risk assessment pursuits typically are: Will the method work to reduce pest populations? How can we make it work most efficiently? How will it affect nontarget species or the ecosystem? Research to answer these questions then ensues. However, for GM organisms, there is an additional layer of complexity: what research directions are safe and ethical (Collins 2018; Courtier-Orgogozo et al. 2017)? The concept of GMOs, even for low-risk improvements to crop production, is not supported broadly in the human population (Linnhoff et al. 2017). It is important that differences in belief systems are thoroughly evaluated while defining research and technology directions for GM organisms (Shinwari et al. 2017). Questions of safety and ethics also have been hotly debated in related fields such as studying gain-of-function mutations for potentially pandemic viruses (National Research Council 2015; Lipsitch 2018). The rationale is that, without better safeguards in place, the risk of releasing a synthetically engineered strain that could overcome natural barriers in the fitness landscape and cause widespread devastation is much greater than the potential benefits that could be gained by understanding this strain's pathology and epidemiological dynamics. Similar concerns have arisen for GM pest control approaches, especially those with gene drive delivery (Abbasi 2016), because with super-Mendelian inheritance, these GM methods present a high risk of uncontrolled spread of genes (Akbari et al. 2015; Backus and Gross 2016; Dhole et al. 2018; Esvelt et al. 2014) that could have devastating consequences on nontarget populations. Thus, the perceived ecological risk of GM methods is so high that stringent containment conditions must be used even to study these methods in the laboratory (Benedict et al. 2018), and effects of these methods must be very well understood before they can proceed to contained field trials. This poses a severe but necessary limitation on the rate at which the technology should advance (Abbasi 2016).

Most of the literature applicable to wildlife focuses on the use of CRISPR-Cas systems to produce a "gene drive" or to push a trait introduced into a wildlife population to fixation or near fixation by avoiding Mendelian inheritance through inheritance by all offspring (Champer et al. 2016). This approach has been applied to some mosquito species that carry malaria that infects humans or Hawaiian birds and is being tested in laboratory and field experiments (Alphey 2014; Gantz et al. 2015; Hammond et al. 2016; Kyrou et al. 2018). In wildlife management, the application of this approach would need to be demonstrated in vertebrate species. This is a large technological leap from cells, insects, and even plants. However, the work of investigating the feasibility of CRISPR-Cas-mediated gene drives has begun in house mice (Piaggio et al. 2017; Grunwald et al. 2019) and thus holds promise for the control of invasive species and agricultural pests, which has been a promise of genetic engineering for over a decade (Burt 2003; Gould 2008).

Given the potential for synthetic gene drives to propagate rapidly within populations, the development of safeguards to spatially and temporally limit spread to nontarget organisms is a key technological challenge (Noble et al. 2018). Unlike most chemical-based management methods, RNA-guided gene drives are vertically transmitted, and thus, species-specificity is largely ensured by normal assortative mating among conspecifics. However, in many cases, the potential ecological impacts of uncontrolled spread into wildlife populations outside of the treatment area may present an unacceptable risk (Gould 2008).

A second major risk factor that is currently poorly understood is potential evolution of a gene drive system. History has shown over and over that strong selection can repeatedly and predictably produce resistance in pest species – antimicrobial resistance is a primary example. For traits that strongly influence fitness, mutations that occur during propagation of a deleterious gene can rapidly predominate, especially if they confer a fitness advantage. This has recently been observed in laboratory experiments with fruit flies (*Drosophila melanogaster*) to examine the efficiency of a CRISPR-Cas9 homing element for driving inheritance of a "Killer-Y chromosome" that results in all male offspring (KaramiNejadRanjbar et al. 2018). Here, gene-drive-resistant mutations readily arose in the mothers by in-frame indel mutations in the recognition site of the guide RNA, and these drive-resistant alleles strongly impacted efficiency of the drive system (KaramiNejadRanjbar et al. 2018), posing a potential economic risk for use in the wild. However, evolution of resistance could also pose ecological risks, and these mechanisms remain largely unexplored. For example, the mechanism by which a drive evolves could reduce target population specificity or even improve pest reproductive performance, thus increasing the risk to nontargets or other ecosystem impacts. Exploring the evolutionary landscape of candidate gene drive mechanisms using experimental evolution and loss-of-function mutational analyses in high containment settings are critical risk assessment steps to take. Importantly, these experiments need to occur using the specific target species and gene drive mechanisms of interest because specific evolutionary mechanisms can differ across systems.

In addition to experiments, risk assessment based on expert opinion (Beech et al. 2009) can help to prioritize risk factors to be investigated in more detail, thus improving efficiency of risk assessment. Simulation models are another efficient and safe approach to risk assessment. Simulation models can help guide the design of experiments efficiently (Restif et al. 2012) and improve our understanding of how different ecological and evolutionary processes interact to determine risk to nontarget individuals (Edgington and Alphey 2018). Models can be especially useful in high-dimensional systems where it is infeasible to test all potential factors empirically. While models are not a substitute for experimental data, they can predict which characteristics of a GM system might be safest and most efficient in a given ecological context (Dhole et al. 2018; Gemmell et al. 2013), improving efficiency of experimental design for risk assessment studies. Recent modeling work on examining the spread rates of gene drive systems in vertebrate pests shows that homing rates are an important characteristic of low-risk gene drive systems that confer high eradication probabilities of the pest species (Prowse et al. 2017). This work emphasizes that understanding factors that affect successful homing is a critical avenue for empirical research (Prowse et al. 2017). To date, models of gene drive systems have focused primarily on combined population genetic-dynamic models of a two-deme or island-mainland system (Dhole et al. 2018; Edgington and Alphey 2018; Sudweeks et al. 2019). However, individual-level spatial processes due to social structure or movement behavior, and mating structure, can have important consequences for structuring genetic variation in space, suggesting that these are important future directions for models and experiments to explore.

## Gene Drives: Risk Mitigation

Several molecular strategies have been proposed to limit gene drive spread including physical separation of gene drive components ("split drive"; DiCarlo et al. 2015) or gene drives that only function above a certain population frequency threshold (Buchman et al. 2018; Leftwich et al. 2018). Engineering the system such that there is a marker gene could also be useful for monitoring containment (Beech et al. 2009). Less efficient drive systems will be easier to contain than highly efficient systems (Dhole et al. 2018). Thus, a clear understanding of the potential spread rates for particular drive mechanisms is crucial for evaluating containment risk (Dhole et al. 2018). Containment risk in a target area will depend on the demographic and spatial dynamics of the pest species within the target zone (Edgington and Alphey 2018; Wilkins et al. 2018) and its connectivity to surrounding nontarget populations. Because measuring this risk in the field is in itself risky, important preliminary steps are to understand the ecological risk landscape in silico using simulations and experimental data collected in a virtual environment under stringent confinement conditions (Abbasi 2016; Akbari et al. 2015).

Another promising approach capitalizes on the precise genome editing afforded by CRISPR-Cas systems to target locally unique sequences that are fixed in the population of interest (i.e., locally fixed alleles, LFA) but absent (or at low frequency) in nontarget populations (Campbell et al. 2019). Evidence suggests that a single nucleotide change in the proto-spacer adjacent motif (PAM) associated with a sgRNA target site can be sufficient to preclude endonuclease binding (Hsu et al. 2013). Thus, population specificity might be accomplished through designing sgRNA that bind genomic regions harboring polymorphisms that form a functional PAM site in the target populations, but not in nontarget populations. Recent modeling efforts (Sudweeks et al. 2019) demonstrate that such an approach can effectively achieve localized population suppression under a variety of scenarios. Interestingly, this work suggests that escape and interbreeding of gene-drive-bearing individuals out of the treatment area are likely to result in only transient suppression of nontarget populations, even when the "susceptible" (i.e., target) allele is present at high frequencies. This phenomenon is explained by the presence of "resistance" alleles (i.e., naturally occurring genetic variants that preclude gene drive homing) in nontarget populations that will be rapidly driven to high frequencies as a result of selection against drive-bearing individuals, subject to the assumptions of the model. This finding also emphasizes the critical importance of thorough population genetic evaluation of the target population prior to sgRNA design to identify sequences that are locally invariant, as even a low level of polymorphism would reduce effectiveness of gene-drive-mediated population suppression. Likewise, both recent theoretical (Unckless et al. 2017) and empirical studies (Champer et al. 2017) suggest that resistant alleles will inevitably arise spontaneously within populations from de novo mutations in the target site or by the gene drive itself as a consequence of errors in the cleavage repair process (e.g., NHEJ). One proposed solution to the evolution of resistance to gene drives is the design of drive systems with multiplexed sgRNA (Cong et al. 2013; Champer et al. 2018), that is, multiple sgRNA that each targets adjacent locally fixed alleles, wherein there is a low likelihood of resistance arising simultaneously in all targets. Indeed, evidence from modeling efforts suggests that multiplexed sgRNA is likely to be necessary for population suppression, even under low levels of NHEJ (Prowse et al. 2017).

The feasibility of the LFA strategy for gene drive containment in the context of vertebrate pest management will depend critically on several aspects of the population structure and ecological setting. As gene drive effectiveness will be diminished by the influx of resistant individuals, relatively isolated populations with low levels of gene flow to nontarget populations, such as oceanic islands, would provide ideal settings. Small populations of introduced species, which are often founded by a small number of individuals, are also expected to harbor reduced allelic diversity, thereby providing a greater number of potential locally fixed allele targets (Morgan et al. 2018). Overall, it is clear that the success of the LFA approach will depend on rigorous population genetic survey of allelic variation within the target population prior to any action.

## Impacts and Effectiveness

The most straightforward measure of effectiveness of a pest control technique is to evaluate how rapidly it reduces the target pest population. However, determining the effects of control on the resources being protected relative to the effort invested is critical for choosing a technique with optimal effectiveness (Hone et al. 2017). Yet this "effort-outcomes relationship" is rarely reported or understood in vertebrate pest management (Hone et al. 2018). The effort-outcomes relationship not only highlights how much effort is needed for a desired outcome but also can reveal which techniques are optimal. It provides additional information over effectiveness alone for choosing between techniques because there may be multiple techniques with the same outcome but with drastically different effort levels. For GM organisms, once a technology is ready for application, there is potential for a much lower application cost than traditional methods (Herman et al. 2019), leading to a more favorable effort-outcome level relative to other techniques. Appropriately defining the dimensions of "effort" and "outcomes" in these comparisons is not straightforward, however, and is worthy of much consideration. For example, over what time scale do we measure outcomes? There may be downstream effects such as increased crop yield per hectare such that less land must be farmed to protect food security (Herman et al. 2019). This type of higher efficiency could also lead to lower farming effort over time, thus affecting a component of management effort (Herman et al. 2019). Lastly, the measure of effectiveness needs to be defined based on the management objective. For example, if we define our objective as maximum agricultural sustainability with the smallest ecological footprint, then the optimal technique, or measure of effectiveness, might be different than if we define our objective simply as maximum crop yield.

Another important component of outcomes is "side effects" or "impacts" of a control method. Impacts of GM organisms could be beneficial such as reduced carbon emissions (Herman et al. 2019) or reduced health hazards from chemical exposure. Alternatively, GM methods for pest control could alter the population genetics of an entire species, potentially causing unforeseen impacts on ecosystem function and stability. Defining and quantifying impacts are as complicated as defining effectiveness and require a systematic decision framework for risk assessment involving multiple stakeholders (Sanvido et al. 2012).

## Regulatory Framework in the United States

A new era of genetic modification began in the 1970s with the generation of a new plasmid from DNA segments of two distinct plasmid species that was inserted into a bacterium (*Escherichia coli*) (Cohen et al. 1973). Questions soon arose regarding whether products derived from the new genetically modified organisms (GMOs) would pose greater risks than those products achieved through traditional techniques and whether the regulatory mechanisms were sufficient evaluate safety.

The distribution and use of almost all chemical and biological wildlife damage control products used to protect agriculture or to control invasive species are regulated under a set of US federal statutes under the jurisdiction of one or more federal regulatory agencies. The three agencies that have regulatory authority over biotechnology products in the United States are the Food and Drug Administration (FDA), the Environmental Protection Agency (EPA), and the United States Department of Agriculture (USDA). These regulatory agencies have oversight of a broad spectrum of products, including GMOs, and subscribe to risk-based assessments to ensure human and environmental health. In some instances, the agency responsible for the regulation of a specific product is uncertain and requires a multiagency conference and decision as to the most appropriate agency or agencies best suited to provide regulatory oversight. Regulatory agencies sometimes have overlapping jurisdictions for a single product. Both of these situations may cause confusion and frustration for both the public and the regulated community and lead to a lack of confidence in the processes.

In an effort to stem these concerns, the US White House Office of Science and Technology Policy (OSTP) formed a workgroup in 1984 to assess the existing regulatory mechanisms for their capability to ensure safety while also fostering a supportive environment for technological development of new biotechnology products. The OSTP released the "Coordinated Framework for Regulation of Biotechnology" in 1986 (OSTP 1986). The Coordinated Framework concluded that the existing laws address most health and safety regulatory processes for biotechnology products. Further, the existing laws provide immediate regulatory oversight for the biotechnology industry, and the implementation of new laws would create uncertainty disruptive to the advancement of new technologies and thus be counterproductive.

The Coordinated Framework was updated in 1992 to explain the federal regulatory agencies' oversight roles and responsibilities as provided by statute, describe a science-based risk assessment process for oversight of biotechnology products released into the environment, and reaffirm that the regulatory oversight will focus "on the characteristics of the biotechnology product and the environment into which it is being introduced" and not whether the process employed to create the product is safe (Bromley 1992).

Federal regulatory agencies were directed by the Executive Office of the President in 2015 to further clarify their roles and responsibilities and to develop a strategy to ensure the regulatory processes are adaptable to scientific advances leading to new types of products. The Emerging Technologies Interagency Policy Coordination Committee Biotechnology Working Group was formed to develop the strategy and update the Coordinated Framework. The result was the National Strategy for Modernizing the Regulatory System for Biotechnology Products was released in 2016 (OSTP 2019). The Strategy's priorities are to increase transparency, increase predictability and efficiency, and support the science that underpins the regulatory system. The goals of the updated Coordinated Framework, issued in 2017, were to increase public understanding and confidence in the regulatory system and "to prevent unnecessary barriers to innovation and competitiveness." The statutory authorities and roles of the FDA, EPA, and USDA were again reaffirmed (OSTP 2016, 2017).

The FDA has broad statutory authority under the Federal Food, Drug, and Cosmetic Act (FDCA) over human food and animal feed including pesticide residues, drugs, cosmetics, and biological products. Genetically engineered animals are regulated under the drug provisions with exceptions including GMO mosquitoes intended only for mosquito population control. Drugs are defined in statue as "articles (other than food) intended to affect the structure or any function of the body of man or other animals" (21 CFR § 321).

Early research in vertebrate species focused on the genetic modification of domestic animals for the production of human drugs. This precedent formed a natural fit for FDA to regulate all genetically engineered animals irrespective of the genetic alteration. Other genetically engineered organisms regulated under the FDCA are human and animal foods derived from non-pesticidal GM plants and human drugs, biological products, and medical devices derived from GE sources. It is important to note that FDA regulates the genetic construct and not the animal itself.

All pesticides registered, distributed, and used in the United States are regulated by EPA under the Federal Insecticide, Fungicide, and Rodenticide Act (FIFRA). The EPA definition of a pesticide is "any substance (or group of structurally similar substances if specified by the Agency) that will prevent, destroy, repel or mitigate any pest, or that functions as a plant regulator, desiccant, or defoliant..." with specific exceptions provided in 40 CFR § 174.3. Under this authority, EPA regulates three major classes of pesticides: conventional, microbial, and biopesticides. Historically, all pesticides were regulated under a single framework. In time, microbial pesticides and biopesticides were split from conventional pesticides, and each is now regulated under their own respective requirements more suited to their formulations and uses. GMO pesticides include microbes modified to produce a pesticide, crops tolerant to specific herbicides, plant-incorporated protectants (PIP) that express pesticidal properties, and most recently mosquitoes genetically engineered to control a pest. In addition to FIFRA, EPA has authority to regulate the allowable level of pesticide residues allowed in food and feed under the FDCA.

Several agencies within the USDA exercise authority over certain plants and animals, food and feed, and products intended to mitigate plant and animal disease, all of which may include biotechnology products. The USDA's statutory authority is provided by the Plant Protection Act (PPA, Public Law 106–224), Animal Health Protection Act (AHPA, Pub. L. 107–171, 116 Stat. 494, 7 U.S.C. 8301), and the Virus-Serum-Toxin Act (VSTA, 21 USC 151–159). The USDA Animal and Plant Health Inspection Service (APHIS) has regulatory oversight under PPA and AHPA over biotechnology products that are considered plant pests or noxious weeds and livestock pests including but not limited to those that cause disease. The USDA Center for Veterinary Biologics has regulatory authority under VSTA over veterinary biological products to prevent, diagnose, and treat disease in animals. Several other statutes are administered by the USDA Food Safety and Inspection Service (FSIS).

In summary, primary regulatory authority is determined by how the various federal laws define the different classes of products, which is based on their intended use, composition, route of administration, and mechanism of action rather than the technology used to create them (Ruell et al. 2016; OSTP 2017; Wozniak 2018). These regulatory laws also encompass completely novel biotechnologies or novel uses of existing technologies that alter the structure or function of organisms, unless they have already been specifically exempted by law or agency rulemaking (OSTP 2017; Wozniak 2018). All three of the federal agencies have shared roles and responsibilities that require coordination (Table 1). In particular, the USDA and FDA often coordinate due to overlapping jurisdiction, but development of these newer technologies emphasizes the challenges for regulatory agencies to adapt existing regulatory frameworks to this changing technological landscape.

The different agencies charged with regulating these products have developed different sets of standards and requirements based on the particularities of the underlying laws they are charged with implementing (Ruell et al. 2016; OSTP 2017). Therefore, early knowledge of a potential product's regulatory situation is crucial for researchers when determining the feasibility of gaining authorization for the product and in order to comply with and meet the relevant regulatory requirements during the research and development phase. Any field testing of experimental products usually requires some form of authorization from the agency with primary regulatory jurisdiction. In addition, notification and authorization are also often required prior to importing, exporting, or interstate transporting of experimental

**Table 1** Summary of pesticide category, claims, characteristics, and application method

| Product category | Product claim | Product characteristics | Application method |
|---|---|---|---|
| Vertebrate animal with intentionally altered genomic DNA | Population control/eradication of a target vertebrate pest through reproduction of the edited animal with wild individuals to produce predominantly male, infertile, inviable, or reduced fitness offspring | Derived from vertebrate pest species<br>Edited to sire predominantly male, infertile, or inviable offspring<br>With or without gene drive<br>Vertebrate pest is native or invasive<br>Not a *pest* under the Livestock Health Protection Act (LHPA)[a] | Environmental release of live individuals to breed with target vertebrate pest |
| Invertebrate animal with intentionally altered genomic DNA | Repel/kill/contracept a target vertebrate pest | Edited to produce the product (a substance)<br>Product is not harvested, but rather is applied as a release of the live invertebrate<br>Not a *pest* under the LHPA[a]<br>Not a *plant pest* under the Plant Protection Act (PPA)[a] | Environmental release of the live invertebrates to be consumed by target vertebrate pest (oral application)<br>Oral application of dead invertebrate |
| | | Edited to produce the product (a substance)<br>Product is harvested from and applied apart from the invertebrate itself<br>Not a *pest* under the LHPA[a]<br>Not a *plant pest* under the PPA[a] | Oral, dermal, inhalation application to target vertebrate pest |
| Modified bacterium or fungus | Kill a target vertebrate pest | Designed to cause disease (target-specific) in a vertebrate pest<br>Not a *pest* under the LHPA[a]<br>Not a *plant pest* under the PPA[a] | Oral, dermal, inhalation application to target vertebrate pest |

(continued)

**Table 1** (continued)

| Product category | Product claim | Product characteristics | Application method |
|---|---|---|---|
| Substance causing RNA interference | Kill or contracept a target vertebrate pest | Designed to cause degradation of mRNA and prevent translation of a specific protein (target-specific) within the target vertebrate pest | Oral, dermal, inhalation application to target vertebrate pest |
| Nanomachines with pesticides | Kill or contracept a target vertebrate pest | Toxicant or contraceptive combined with nanorobotics (delivery device) | Oral or inhalation application to target vertebrate pest<br>Nanorobotics used to deliver pesticide to target organ |
| Substance causing change in sensory organ perception or ill feeling | Repel a target vertebrate pest from the food, forage, or prey through aversive conditioning | Causes unpleasant sensory feedback or function, or causes sickness when combined with normally attractive foods, forage, or prey | Oral, dermal, inhalation application to target vertebrate pest |
| Device combined with a substance that causes infertility | Contracept a target vertebrate pest | Contraceptive | Insertion into a female reproductive body cavity |

[a]Would not fall under USDA APHIS regulatory authority

products. Conventional wildlife damage products with clear regulatory precedents are relatively easy to classify under the different regulatory jurisdictions.

## *Regulatory Framework: Gene Silencing Using RNAi*

The regulatory framework for the new generation of RNAi-based pesticides would likely follow the general guidelines for biopesticides, which are modified from conventional pesticides (OSTP 1986). If applied to growing crops, the FDA ensures the food from crops containing RNAi is as safe as its conventional counterpart. The USDA ensures there is no risk to agriculture from the use of RNAi. The EPA ensures that the product can perform its intended function with a reasonable certainty of no harm to people from dietary and residential exposure and no unreasonable risks to the environment.

## *Regulatory Framework: Gene Editing*

Regulatory jurisdiction of novel GM products for wildlife damage control can be more difficult to classify and may require prolonged consultation between the regulatory agencies themselves before a final designation is made. As an example, products consisting of microorganisms or invertebrates, whether unmodified or modified (including GM), which are used to change the structure or function of animals, have been on the market for some time, and, for the most part, are now distributed among the regulatory agencies (Wozniak et al. 2012; Ruell et al. 2016; FDA CVM 2017a, b; OSTP 2017; EPA OPP 2018; Wozniak 2018). They are regulated by USDA based on whether they qualify as a pest to livestock as defined by the AHPA or as a veterinary biologic as defined by VSTA. They are regulated by EPA if they or their byproducts are used as pesticides against pests as defined by FIFRA. Any other product that uses microorganisms with the intent to directly change the structure or function of animals is regulated by FDA.

In contrast, products consisting of vertebrate animals fall into much different regulatory classifications that are not as clear cut. Non-GM vertebrate animals released to control a pest species, aka biological control agents (e.g., the release of mongooses to control rats on islands), do not qualify as products requiring authorization under the FDCA and have been exempt from the registration requirements of FIFRA by EPA (40 C.F.R. § 152.20(a)(3)). However, EPA has left the door open to revoke this exemption for any biological control agent it considers to be inadequately regulated by other federal agencies (40 C.F.R. § 152.20(a)(2)). In contrast, GM vertebrate animals do fall under the provisions of the FDCA, or more accurately, the intentionally altered genomic DNA within the GM animal becomes the regulated article because it changes the structure of function of the animal for an intended purpose (FDA CVM 2015, 2017a; OSTP 2017). This intentionally altered genomic DNA in a GM vertebrate animal is not currently classified as a regulated article under the AHPA, VSTA, or FIFRA, although a substance produced by a GM vertebrate animal could be regulated separately by another agency if it meets their definition of a regulated article (FDA CVM 2015; Ruell et al. 2016; FDA CVM 2017a). Several GM vertebrate animal products have made their way through part of or the whole approval process with FDA to date (e.g., GE salmon, chickens, goats), although none so far were designed for use in wildlife damage control or pest management (EPA OPP 2018).

It is possible that the Congress or FDA and EPA may one day determine that certain GM vertebrate animals intended for population control of vertebrate pests will be classified as pesticides and regulated under FIFRA instead of under the provisions of the FDCA, similar to GM mosquitos intended solely for mosquito population control (FDA CVM 2017b). The two agencies have started working together to determine statutory authority over these types of GM animal products for pest management on the horizon, but it is unclear whether this will change their regulatory jurisdiction in the near future (EPA OPP 2018). Until alternative guidance is issued by FDA and EPA, the primary regulatory authority over all GM vertebrate animals will remain FDA (FDA CVM 2017a; OSTP 2017). FDA will work closely with EPA and other federal regulatory agencies charged with implementing environmental laws before authorizing any experimental field use or eventual approval of these novel products (FDA CVM 2017a). FDA recommends that product developers contact them early in the development phase of product development of GM vertebrates (FDA CVM 2017a).

There can be significant disadvantages for developers submitting the first product of its kind to a regulatory agency that result in unforeseen delays and costs prior to authorization. Many agencies, including EPA, will convene Scientific Advisory Panels at the expense of the applicant to help them determine how to appropriately evaluate the risk of the novel product type and the appropriate set of regulatory data they will require from the registrant prior to authorization (EPA OPP 2018). Once the agency's risk assessment process has been determined, study guidelines must then be developed and finalized for each data requirement. However, they will not usually initiate these processes until after they receive an application or information on an actual product, because these processes take considerable time and resources.

There is also the potential that a novel product could change regulatory hands midstream in the regulatory authorization process or even after a product has been authorized by an agency, like what happened with GE mosquitos used for mosquito population control, and this can pose additional difficulties and added costs for the applicant. For example, data that were collected or contracted for one authorization process may not be directly applicable or adequate for the next. The manufacturing and National Environmental Policy Act requirements for the product also differ depending on the agency and regulatory statute. Some regulatory statutes allow government agencies to take products through the full authorization process, and some require a private or nonprofit entity to be the applicant. Thus, last minute changes to the regulatory jurisdiction of a product can result in considerable challenges for the product developer in predicting the practicality and use timeline for novel products such as GM vertebrate animals used in animal damage control.

## Translational Product Development

Translational product development can be thought of as a pipeline or continuum as an idea moves from research and development to fully implemented product or technique. Figure 1 graphically illustrates that pipeline for a regulated product. Early in the process, pilot and confirmatory development steps are undertaken to provide the information needed to assess the product development costs, market potential, and ultimately the viability of the product. Early in the product development pipeline, decisions are normally made regarding protecting intellectual property and the need to form development and commercialization partnerships. In the case of regulated products such as a pesticide, biologic, vaccine, or drug, a realistic assessment must be made as to the cost of regulatory studies and the time it will take

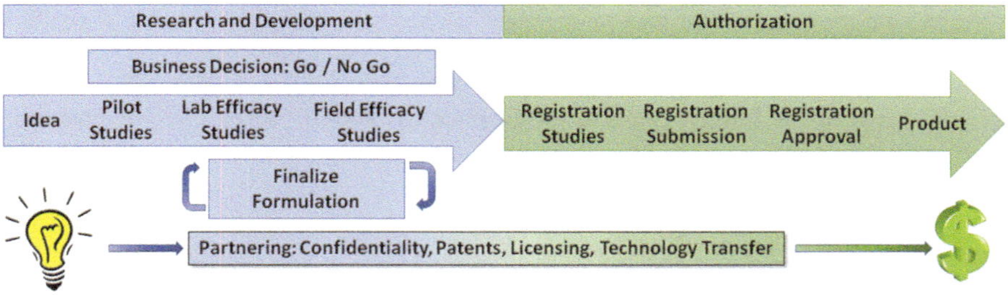


**Fig. 1** Conceptual product development pipeline for regulated products developed in a federal laboratory

to obtain a product authorization. All of these aspects inform decisions in advancing development efforts.

As this chapter outlines, developments in our ability to manipulate or target the genetic code of an organism have opened up an entirely new world for product development. The application of genetic tools appears to be boundless in advancing medical, agricultural, and conservation goals. At the heart of biotechnology product development is the decision and ability to protect the intellectual property. To illustrate this, two potential biotech techniques or tools that will prove useful in agricultural production and conservation will be briefly presented. Both of these technologies, gene silencing and gene editing, share the same general product development pipeline but have significant variations along that path.

## Intellectual Property Protection

At the heart of translational product development is technology transfer and, if appropriate, protection of intellectual property. There are many definitions of technology transfer, but in essence, it is the process by which technology or knowledge developed in one place or for one purpose is applied in and used in another place (FLC 2006). Two conditions are responsible for initiating or accomplishing technology transfer, a technology is created and pushed into practice, or some identified need creates the opportunity for a new technology. Regardless of what initiated the product development effort, the successful transfer from development to practice is the goal. An idea, technique, or tool is underutilized until it is put into practice.

Patent protection from the US Patent and Trademark Office or other international patent authorities can be a major driver of technology transfer. Patent protection allows the patent owner to exclusively pursue the development of the patented tool or technique or to license the rights to practice the patent to another entity, typically for a fee. As advances in genetic technologies allowed us to sequence genomes and identify specific genome sequences that were responsible for specific physiological functions, scientists and businesses sought patent protection of gene sequences they had identified in the hopes of capitalizing on those discoveries and future applications of that knowledge. The idea of patenting naturally occurring genetic code quickly became a hotly debated topic. Many found the idea of "owning" the knowledge of a person's genetic code unethical. Despite the debate, 4300 patents were issued for the human genome alone prior to 2013. In 2012, the US Supreme Court ruled that DNA in its natural form cannot be patented; however, DNA manipulated in a lab is eligible to be patented because DNA sequences altered by humans are not found in nature (NIH 2017, 2019; SCOTUS 2013). While this ruling closed the door on capitalizing on a person's genetic code, it defined what was allowed to be patented and provides the guidance for those seeking intellectual property protection for intentionally modified genetic tools.

## *Intellectual Property: Gene Silencing Using RNAi*

As previously described, the ability to intentionally prevent or inhibit normal RNA translation has profound impacts on an organism's physiological function and has enormous potential to be utilized in human medicine, agricultural production, and many other beneficial areas. An example of a gene-silencing tool for agricultural protection is the development of new pest management tools, such as active ingredients used in pesticides. Typical pesticides function by disrupting a biochemical mechanism critical to life. Common biochemical mechanisms include, but are not limited to, cholinesterase inhibitors, which function by inhibiting nerve impulse transmission across synaptic junctions, and other pesticides disrupt the production of ATP by impacting enzyme production critical to normal Krebs cycle functions. Gene silencing could target similar critical functions but at the genetic level, by introducing engineered strands of RNA into the cell, which disrupt translation or transcription processes.

Taking advantage of an organism's unique genetic code could potentially lead to the development of species-specific, humane pesticides with reduced risk of unintended environmental or human health consequences. Such an approach could revolutionize pest management, leading to dramatic advances in agricultural and human health protection, and associated economic benefits to protected resources in addition to significant economic benefits to companies advancing these technologies. For these reasons, there will be a great desire to protect the intellectual property behind the technologies. Patenting is available for gene-silencing tools because the patented technology would not be based on the native genetic code or an organism, but rather on the man-made genetic sequences introduced into cells which disrupt normal RNA translation or transcription processes. In the case of pest management, most of these types of products would be regulated as pesticides by the US EPA. Consequently, there is clear understanding of the regulatory path these products would face. Even though this line of product development is high risk, having regulatory clarity paves the way for clear product development efforts.

## *Intellectual Property: Gene Editing*

Directly editing the genetic sequence of an organism has the potential for significant contributions to agriculture, human health, and natural resource protection. As described above, the discovery of native CRISPR-Cas systems and the knowledge of how to utilize these systems to make specific gene edits have revolutionized our ability to produce single-generation designer organisms. Current lines of research are focused on developing gene-edited organisms that pass the desired trait to a high proportion of their offspring in a way that it is active in subsequent generations, known as "gene drive" systems. Driving a trait through a population has enormous social implications and elicits very controversial discussions, especially when it could be applied to natural resource conservation issues. Despite those important discussions, work will continue to develop applications of gene-edited and gene drive organisms because associated economic benefits are presumed to be high. If the resulting products include unnatural genetic sequences in an organism, there is a potential for seeking patent protection.

The decision to protect intellectual property is only one step of the technology transfer pipeline. The ultimate proof that product development efforts were successful is seeing that technology adopted somewhere in society. However, an important consideration is that the ultimate landing place of a technology may not be what the original researchers and product developers intended.

## Future Directions

Wildlife damage conflicts continue to increase as the world's population increases. Generally speaking, mitigation of a conflict implies an active management or control program. Typically, conflicts manifest themselves as invasive species and disease impacts on native ecosystems and damage caused by wildlife to commodities of value to humans (e.g., agriculture, property, health, safety, property). Managers must weigh the options available to them to mitigate the conflict in terms of economics, effectiveness, environmental effects, and humaneness, all within a social framework. A question becomes whether today's technologies will be well suited to constraints and problems we may face into the future. For example, will today's reliance on chemical pesticides continue into the future, and if not, what mitigation tools and strategies will replace these methods? Like many new technologies that were introduced in the past, there is great promise and apprehension about these new genetically based methods in terms of development and use. Herein, we reviewed the context, opportunities, and challenges of genetically based biopesticides and provided some social, technical, and regulatory practicalities in research and development for technologies such as gene silencing and gene drives. The encouraging news is that the scope and depth of discussions are vigorous and inclusive (e.g., scientists, ethicists, managers, regulators, the public, and policy makers), all of which bodes well for informed decision processes.

**Acknowledgments** This work was supported in part by the US Department of Agriculture's Animal and Plant Health Inspection Service. The findings and conclusions in this publication are those of the authors and should not be construed to represent any official USDA or US Government determination or policy.

## References

Abbasi J (2016) National Academies hit the brakes on gene drive–modified organisms. JAMA 316:482–483. https://doi.org/10.1001/jama.2016.8830

Adler GH, Levins R (1994) The island syndrome in rodent populations. Q Rev Bio 69:473–490

Ahmadzada T, Reid G, McKenzie DR (2018) Fundamentals of siRNA and miRNA therapeutics and a review of targeted nanoparticle delivery systems in breast cancer. Biophys Rev 10:69–86

Akbari OS, Bellen HJ, Bier E, Bullock SL, Burt A, Church GM, Cook KR, Duchek P, Edwards OR, Esvelt KM, Gantz VM (2015) Safeguarding gene drive experiments in the laboratory. Science 349:927–929. https://doi.org/10.1126/science.aac7932

Alexopoulou L, Holt AC, Medzhitov R, Flavell RA (2001) Recognition of double-stranded RNA and activation of NF-κB by toll-like receptor 3. Nature 413:732

Allan JR, Watson JE, Di Marco M, O'Bryan CJ, Possingham HP, Atkinson SC, Venter O (2019) Hotspots of human impact on threatened terrestrial vertebrates. PLoS Biol. https://journals.plos.org/plosbiology/article?id=10.1371/journal.pbio.3000158

Alphey L (2014) Genetic control of mosquitoes. Annu Rev Entomol 59:205–224

Amarzguioui M, Holen T, Babaie E, Prydz H (2003) Tolerance for mutations and chemical modifications in a siRNA. Nucleic Acids Res 31:589–595

Avila LA, Aps LR, Ploscariu N, Sukthankar P, Guo R, Wilkinson KE, Games P, Szoszkiewicz R, Alves RPS, Diniz MO, Fang Y (2016) Gene delivery and immunomodulatory effects of plasmid DNA associated with branched Amphiphilic peptide capsules. J Control Release 241:5–24

Backus GA, Gross K (2016) Genetic engineering to eradicate invasive mice on islands: modeling the efficiency and ecological impacts. Ecosphere 7. https://doi.org/10.1002/ecs2.1589

Ball RL, Bajaj P, Whitehead KA (2018) Oral delivery of siRNA lipid nanoparticles: fate in the GI tract. Sci Rep 8:2178

Barrangou R (2015) The roles of CRISPR–Cas systems in adaptive immunity and beyond. Curr Opin Immunol 32:36–41

Barrangou R, Doudna JA (2016) Applications of CRISPR technologies in research and beyond. Nat Biotechnol 34:933

Baum JA, Roberts JK (2014) Progress towards RNAi-mediated insect Pest management. In: Dhadialla TS, Gill SS (eds) Advances in insect physiology. Academic Press, New York, pp 249–295

Beech CJ, Nagaraju J, Vasan SS, Rose RI, Othman RY, Pillai V, Saraswathy TS (2009) Risk analysis of a hypothetical open field release of a self-limiting transgenic *Aedes aegypti* mosquito strain to combat dengue. APJMBB 17:97–108

Benedict MQ, Burt A, Capurro ML, De Barro P, Handler AM, Hayes KR, Marshall JM, Tabachnick WJ, Adelman ZN (2018) Recommendations for laboratory containment and management of gene drive systems in arthropods. Vector-Borne Zoonotic Dis 18:2–13. https://doi.org/10.1089/vbz.2017.2121

Bromley A (1992) Exercise of federal oversight within scope of statutory authority: planned introductions of biotechnology products into the environment. White house, Office of Science and technology policy (OSTP). Fed Regist 57:6753–6762

Buchman AB, Ivy T, Marshall JM, Akbari OS, Hay BA (2018) Engineered reciprocal chromosome translocations drive high threshold, reversible population replacement in Drosophila. ACS Synthc Bio 7:1359–1370
Burand JP, Hunter WB (2013) RNAi: future in insect management. J Invertebr Path 112:S68–S74
Burt A (2003) Site-specific selfish genes as tools for the control and genetic engineering of natural populations. Proc R Soc Lond B Biol Sci 270:921–928
Campbell KJ, Beek J, Eason CT, Glen AS, Godwin J, Gould F, Holmes ND, Howald GR, Madden FM, Ponder JB, Threadgill DW (2015) The next generation of rodent eradications: innovative technologies and tools to improve species specificity and increase their feasibility on islands. Biol Conserv 185:47–58
Campbell KJ, Saah JR, Brown PR, Godwin J, Gould F, Howald GR, Piaggio A, Thomas P, Tompkins DM, Threadgill D, Delborne J, Kanavy DM, Kuiken T, Packard H, Serr M, Shiels A (2019) A potential new tool for the toolbox: assessing gene drives for eradicating invasive rodent populations. In: Veitch CR, Clout MN, Martin AR, Russell JC, West CJ (eds) Island invasives: scaling up to meet the challenge, Occasional Paper SSC no. 62. IUCN, Gland, pp 6–14
Carlson RH (2010) Biology is technology. Harvard University Press, Cambridge, p 288
Champer J, Buchman A, Akbari OS (2016) Cheating evolution: engineering gene drives to manipulate the fate of wild populations. Nat Rev Genet 17:146–159
Champer J, Reeves R, Oh SY, Liu C, Liu J, Clark AG, Messer PW (2017) Novel CRISPR/Cas9 gene drive constructs reveal insights into mechanisms of resistance allele formation and drive efficiency in genetically diverse populations. PLoS Genet 13(7). https://doi.org/10.1371/journal.pgen.1006796
Champer J, Liu J, Oh SY, Reeves R, Luthra A, Oakes N, Clark AG, Messer PW (2018) Reducing resistance allele formation in CRISPR gene drive. PNAS 115:5522–5527
Church GM, Regis E (2012) Regenesis: how synthetic biology will reinvent nature and ourselves. Basic Books, New York, p 284
Cohen SN, Chang AC, Boyer HW, Helling RB (1973) Construction of biologically functional bacterial plasmids in vitro. PNAS 70:3240–3244. https://doi.org/10.1073/pnas.70.11.3240
Collins JP (2018) Gene drives in our future: challenges of and opportunities for using a self-sustaining technology in pest and vector management. BMC Proc 12:9. https://doi.org/10.1186/s12919-018-0110-4
Cong L, Ran FA, Cox D, Lin S, Barretto R, Habib N, Hsu PD, Wu X, Jiang W, Marraffini LA, Zhang F (2013) Multiplex genome engineering using CRISPR/Cas systems. Science 339:819–823
Conklin BR (2019) On the road to a gene drive in mammals. Nature 566:43–45. https://doi.org/10.1038/d41586-019-00185-y
Conover MR (2001) Resolving human-wildlife conflicts: the science of wildlife damage management. CRC press, New York
Courtier-Orgogozo V, Morizot B, Boëte C (2017) Agricultural pest control with CRISPR-based gene drive: time for public debate. EMBO Rep 18:878–880. https://doi.org/10.15252/embr.201744205
Crick F (1970) Central dogma of molecular biology. Nature 227:561
Cuthbert RJ, Visser P, Louw H, Rexer-Huber K, Parker G, Ryan PG (2011) Preparations for the eradication of mice from Gough Island: results of bait acceptance trials above ground and around cave systems. Wildl Res 38:196–203
Cuthbert RJ, Wanless RM, Angel A, Burle MH, Hilton GM, Louw H, Visser P, Wilson JW, Ryan PG (2016) Drivers of predatory behavior and extreme size in house mice *Mus musculus* on Gough Island. J Mammal 97:533–544
Czauderna F, Fechtner M, Dames S, AyguÈn H, Klippel A, Pronk GJ, Giese K, Kaufmann J (2003) Structural variations and stabilising modifications of synthetic siRNAs in mammalian cells. Nucleic Acids Res 31:2705–2716
Damalas CA, Eleftherohorinos IG (2011) Pesticide exposure, safety issues, and risk assessment indicators. Int J Environ Res Public Health 8:1402–1419

Dhole S, Vella MR, Lloyd AL, Gould F (2018) Invasion and migration of spatially self-limiting gene drives: a comparative analysis. Evol Appl 11:794–808. https://doi.org/10.1111/eva.12583
DiCarlo JE, Chavez A, Dietz SL, Esvelt KM, Church GM (2015) Safeguarding CRISPR-Cas9 gene drives in yeast. Nat Biotechnol 33:1250–1255
Doherty TS, Glen AS, Nimmo DG, Ritchie EG, Dickman CR (2016) Invasive predators and global biodiversity loss. PNAS 113:11261–11265
Doudna JA, Charpentier E (2014) The new frontier of genome engineering with CRISPR-Cas9. Science 346:1258096
Eason CT, Fagerstone KA, Eisemann JD, Humphrys S, O'Hare JR, Lapidge SJ (2010) A review of existing and potential New World and Australasian vertebrate pesticides with a rationale for linking use patterns to registration requirements. Int J Pest Manag 56:109–125
Edgington MP, Alphey LS (2018) Population dynamics of engineered underdominance and killer-rescue gene drives in the control of disease vectors. PLOS Comput Bio 14. https://doi.org/10.1371/journal.pcbi.1006059
Elbashir SM, Martinez J, Patkaniowska A, Lendeckel W, Tuschl T (2001) Functional anatomy of siRNAs for mediating efficient RNAi in Drosophila melanogaster embryo lysate. EMBO J 20:6877–6888
EPA OPP (2018) U.S. Environmental Protection Agency, Office of Pesticide Programs, November 1. Biotechnology/Emerging Technologies Seminar. https://register.gotowebinar.com/register/6494848564946662914
Esvelt KM, Smidler AL, Catteruccia F, Church GM (2014) Emerging technology: concerning RNA-guided gene drives for the alteration of wild populations. elife 3. https://doi.org/10.7554/eLife.03401
FDA CVM (2015) U.S. Food and Drug Administration, Center for Veterinary Medicine, June. #187 Guidance for industry on regulation of genetically engineered animals containing heritable recombinant DNA constructs. https://wayback.archive-it.org/7993/20170111005939/http://www.fda.gov/downloads/AnimalVeterinary/GuidanceComplianceEnforcement/GuidanceforIndustry/UCM113903.pdf
FDA CVM (2017a) U.S. Food and Drug Administration, Center for Veterinary Medicine, January. Draft Guidance: #187 Guidance for industry on regulation of intentionally altered genomic DNA in animals. http://www.fda.gov/downloads/AnimalVeterinary/GuidanceComplianceEnforcement/GuidanceforIndustry/UCM113903.pdf
FDA CVM (2017b) U.S. Food and Drug Administration, Center for Veterinary Medicine, October. #236 Clarification of FDA and EPA Jurisdiction Over Mosquito-Related Products. https://www.fda.gov/downloads/AnimalVeterinary/GuidanceComplianceEnforcement/GuidanceforIndustry/UCM533600.pdf
Ferguson NM (2018) Challenges and opportunities in controlling mosquito-borne infections. Nature 559:490. https://doi.org/10.1038/s41586-018-0318-5
Fire A, Xu S, Montgomery MK, Kostas SA, Driver SE, Mello CC (1998) Potent and specific genetic interference by double-stranded RNA in *Caenorhabditis elegans*. Nature 391:806
FLC (2006) FLC technology transfer desk reference: a comprehensive introduction to technology transfer. Federal Laboratory Consortium for Technology Transfer, Cherry Hill, NJ. http://ictt.by/Rus/Portals/0/T2_Desk_Reference_2006.pdf
Gantz VM, Bier E (2015) The mutagenic chain reaction: a method for converting heterozygous to homozygous mutations. Science 348:442–444
Gantz VM, Jasinskiene N, Tatarenkova O, Fazekas A, Macias VM, Bier E, James AA (2015) Highly efficient Cas9-mediated gene drive for population modification of the malaria vector mosquito *Anopheles stephensi*. PNAS 112:E6736–E6743
Gemmell NJ, Jalilzadeh A, Didham RK, Soboleva T, Tompkins DM (2013) The Trojan female technique: a novel, effective and humane approach for pest population control. Proc R Soc B: Bio Sci 280:20132549. https://doi.org/10.1098/rspb.2013.2549
Gould F (2008) Broadening the application of evolutionarily based genetic pest management. Evolution 62:500–510

Grunwald HA, Gantz VM, Poplawski G, Xu XRS, Bier E, Cooper KL (2019) Super-Mendelian inheritance mediated by CRISPR–Cas9 in the female mouse germline. Nature 566:105
Hammond A, Galizi R, Kyrou K, Simoni A, Siniscalchi C, Katsanos D, Gribble M, Baker D, Marois E, Russell S, Burt A (2016) A CRISPR-Cas9 gene drive system targeting female reproduction in the malaria mosquito vector Anopheles gambiae. Nature Biotechnol 34:78–83
Heil F, Hemmi H, Hochrein H, Ampenberger F, Kirschning C, Akira S, Lipford G, Wagner H, Bauer S (2004) Species-specific recognition of single-stranded RNA via toll-like receptor 7 and 8. Science 303:1526–1529
Herman RA, Zhuang M, Storer NP, Cnudde F, Delaney B (2019) Risk-only assessment of genetically engineered crops is risky. Trends Plant Sci 24:58–68. https://doi.org/10.1016/j.tplants.2018.10.001
Holmes ND, Griffiths R, Pott M, Alifano A, Will D, Wegmann AS, Russell JC (2015) Factors associated with rodent eradication failure. Bio Conserv 185:18–16
Hone J, Drake VA, Krebs CJ (2017) The effort–outcomes relationship in applied ecology: evaluation and implications. Bio Science 67:845–852
Hone J, Drake VA, Krebs CJ (2018) Evaluating wildlife management by using principles of applied ecology: case studies and implications. Wildlife Res 45:436–445
Hsu PD, Scott DA, Weinstein JA, Ran FA, Konermann S, Agarwala V, Li Y, Fine EJ, Wu X, Shalem O, Cradick TJ (2013) DNA targeting specificity of RNA-guided Cas9 nucleases. Nature Biotechnol 31:827
Hussain W, Mahmood T, Hussain J, Ali N, Shah T, Qayyum S, Khan I (2018) CRISPR/Cas system: a game changing genome editing technology, to treat human genetic diseases. Gene 685:70
Ito J, Ghosh A, Moreira LA, Wimmer EA, Jacobs-Lorena M (2002) Transgenic anopheline mosquitoes impaired in transmission of a malaria parasite. Nature 417:452–455. https://doi.org/10.1038/417452a
Jiang W, Bikard D, Cox D, Zhang F, Marraffini LA (2013) RNA-guided editing of bacterial genomes using CRISPR-Cas systems. Nature Biotechnol 31:233
Jones HP, Holmes ND, Butchart SH, Tershy BR, Kappes PJ, Corkery I, Aguirre-Muñoz A, Armstrong DP, Bonnaud E, Burbidge AA, Campbell K (2016) Invasive mammal eradication on islands results in substantial conservation gains. PNAS 113:4033–4038
Jorgensen RA, Atkinson RG, Forster RL, Lucas WJ (1998) An RNA-based information superhighway in plants. Science 279:1486–1487
KaramiNejadRanjbar M, Eckermann KN, Ahmed HM, Dippel S, Marshall JM, Wimmer EA (2018) Consequences of resistance evolution in a Cas9-based sex conversion-suppression gene drive for insect pest management. PNAS 115:6189–6194. https://doi.org/10.1073/pnas.1713825115
Khvorova A, Reynolds A, Jayasena SD (2003) Functional siRNAs and miRNAs exhibit strand bias. Cell 115:209–216
Kinnear JE (2018) Mammal conservation and invasive species control in Australia: harnessing a potential extinction machine. Aust Mammal 40:131–135
Kogan M (1998) Integrated Pest management: historical perspectives and contemporary developments. Annl Rev Entomol 43:243–270
Kyrou K, Hammond AM, Galizi R, Kranjc N, Burt A, Beaghton AK, Nolan T, Crisanti A (2018) A CRISPR–Cas9 gene drive targeting doublesex causes complete population suppression in caged Anopheles gambiae mosquitoes. Nature Biotechnol 36:1062–1066
Lamas-Toranzo I, Guerrero-Sánchez J, Miralles-Bover H, Alegre-Cid G, Pericuesta E, Bermejo-Álvarez P (2017) CRISPR is knocking on barn door. Reprod Domest Anim 52:39–47. https://doi.org/10.1111/rda.13047
Lander ES (2016) The heroes of CRISPR. Cell 164:18–28
Lares MR, Rossi JJ, Ouellet DL (2010) RNAi and small interfering RNAs in human disease therapeutic applications. Trends Biotechnol 28:570–579
Leftwich PT, Edgington MP, Harvey-Samuel T, Paladino LZC, Norman VC, Alphey L (2018) Recent advances in threshold-dependent gene drives for mosquitoes. Biochem Soc Trans 46:1203–1212

Linnhoff S, Volovich E, Martin H, Smith M (2017) An examination of millennials' attitudes toward genetically modified organism (GMO) foods: is it Franken-food or super-food? IJARGE 13:371–390. https://doi.org/10.1504/IJARGE.2017.088403
Lipsitch M (2018) Why do exceptionally dangerous gain-of-function experiments in influenza? In: Influenza virus. Humana Press, New York, pp 589–608. https://doi.org/10.1007/978-1-4939-8678-1_29
Liu J, Carmel MA, Rivas FV, Marsden CG, Thomson JM, Song JJ, Hammond SM, Joshua-Tor L, Hannon GJ (2004) Argonaute2 is the catalytic engine of mammalian RNAi. Science 305:1437–1441
MacKay JWB, Russell JC, Murphy EC (2007) Eradicating house mice from islands: successes, failures and the way forward. In: Witmer GW, Pitt WC, Fagerstone KA (eds) Managing vertebrate invasive species. Proceedings of an international symposium, Fort Collins, pp 294–304
Mali P, Yang L, Esvelt KM, Aach J, Guell M, DiCarlo JE, Norville JE, Church GM (2013) RNA-guided human genome engineering via Cas9. Science 339:823–826
Mamta B, Rajam MV (2017) RNAi technology: a new platform for crop pest control. Physiol Mol Bio Plants 23:487–501
Morgan AP, Didion JP, Hughes JJ, Searle JB, Jolley WJ, Campbell KJ, Threadgill DW, Manuel de Villena FP (2018) Genetic characterization of invasive house mouse populations on small islands. BioRxiv. https://doi.org/10.1101/332064
Moro D, Byrne M, Kennedy M, Campbell S, Tizard M (2018) Identifying knowledge gaps for gene drive research to control invasive animal species: the next CRISPR step. Glob Ecol Conserv 13. https://doi.org/10.1016/j.gecco.2017.e00363
Napoli C, Lemieux C, Jorgensen R (1990) Introduction of a chimeric chalcone synthase gene into petunia results in reversible co-suppression of homologous genes in trans. Plant Cell 2:279–289
National Research Council (2015) Potential risks and benefits of gain-of-function research: summary of a workshop. National Academies Press. https://doi.org/10.17226/21666
NIH (2017) Intellectual property and genomics. National Institute of Health, Bethesda, MD. https://www.genome.gov/19016590/
NIH (2019) Help Me Understand Genetic Testing. Lister Hill National Center for Biomedical Communications. National Institute of Health. Bethesda MD, March 19, 2019. https://ghr.nlm.nih.gov/primer/testing.pdf
Niu J, Taning CNT, Christiaens O, Smagghe G, Wang JJ (2018) Rethink RNAi in insect Pest control: challenges and perspectives. In: Smagghe G (ed) Advances in insect physiology, vol 2018. Academic Press, pp 1–17. https://doi.org/10.1016/bs.aiip.2018.07.003
Noble C, Adlam B, Church GM, Esvelt KM, Nowak MA (2018) Current CRISPR gene drive systems are likely to be highly invasive in wild populations. elife 7. https://doi.org/10.7554/eLife.33423
OSTP (1986) Executive Office of the President, Office of Science and Technology Policy, Coordinated Framework for Regulation of Biotechnology. 122 p. 51 FR 23302. https://www.aphis.usda.gov/brs/fedregister/coordinated_framework.pdf
OSTP (2016) Executive Office of the President, National Strategy for Modernizing the Regulatory System for Biotechnology Products, p 14 https://obamawhitehouse.archives.gov/sites/default/files/microsites/ostp/biotech_national_strategy_final.pdf
OSTP (2017) Executive Office of the President, Office of Science and Technology Policy, Modernizing the Regulatory System for Biotechnology Products: Final Version of the 2017 Update to the Coordinated Framework for Regulation of Biotechnology. 51 F.R. 23302-93. https://obamawhitehouse.archives.gov/sites/default/files/microsites/ostp/2017_coordinated_framework_update.pdf
OSTP (2019) Executive order on modernizing the regulatory framework for agricultural biotechnology products. https://www.whitehouse.gov/presidentialactions/executive-order-modernizing-regulatory-framework-agricultural-biotechnology-products/

Piaggio AJ, Segelbacher G, Seddon PJ, Alphey L, Bennett EL, Carlson RH, Friedman RM, Kanavy D, Phelan R, Redford KH, Rosales M (2017) Is it time for synthetic biodiversity conservation? Trends Ecol Evol 32:97–107

Prowse TA, Cassey P, Ross JV, Pfitzner C, Wittmann TA, Thomas P (2017) Dodging silver bullets: good CRISPR gene-drive design is critical for eradicating exotic vertebrates. Proc R Soc B Bio ScI 284:20170799

Prowse TA, Adikusuma F, Cassey P, Thomas P, Ross JV (2019) A Y-chromosome shredding gene drive for controlling pest vertebrate populations. eLife 8:e41873. https://doi.org/10.7554/eLife.41873

Reidinger RF Jr, Miller JE (2013) Wildlife damage management: prevention, problem solving, and conflict resolution. JHU Press, Baltimore

Restif O, Hayman DTS, Pulliam JRC, Plowright RK, George DB, Luis AD, Cunningham AA, Bowen RA, Fooks AR, O'Shea TJ, Wood JLN, Webb CT (2012) Model-guided fieldwork: practical guidelines for multidisciplinary research on wildlife ecological and epidemiological dynamics. Ecol Lett 15:1083–1094

Ruell EW, Gatewood DM, O'Hare JR, Eisemann JE (2016) A decision support tool for determining federal regulatory authority over products for vertebrate animals. Proc Vertebr Pest Conf 27:422–431

Sanvido O, Romeis J, Gathmann A, Gielkens M, Raybould A, Bigler F (2012) Evaluating environmental risks of genetically modified crops: ecological harm criteria for regulatory decision-making. Environ Sci Pol 15:82–91. https://doi.org/10.1016/j.envsci.2011.08.006

Schwarz DS, Hutvágner G, Du T, Xu Z, Aronin N, Zamore PD (2003) Asymmetry in the assembly of the RNAi enzyme complex. Cell 115:199–208

Schwarz DS, Ding H, Kennington L, Moore JT, Schelter J, Burchard J, Linsley PS, Aronin N, Xu Z, Zamore PD (2006) Designing siRNA that distinguish between genes that differ by a single nucleotide. PLoS Genet 2:e140. https://doi.org/10.1371/journal.pgen.0020140

SCOTUS (2013) Association for Molecular Pathology et al. v. Myriad Genetics, Inc. et al. Supreme Court of the United States. Decided June 13, 2013. https://www.supremecourt.gov/opinions/12pdf/12-398_1b7d.pdf

Selle K, Barrangou R (2015a) CRISPR-based technologies and the future of food science. J Food Sci 80:R2367–R2372

Selle K, Barrangou R (2015b) Harnessing CRISPR–Cas systems for bacterial genome editing. Trends Microbio 23:225–232

Shaw WR, Catteruccia F (2019) Vector biology meets disease control: using basic research to fight vector-borne diseases. Nature Microbio 4:20–34. https://doi.org/10.1038/s41564-018-0214-7

Shinwari ZK, Tanveer F, Khalil AT (2017) Ethical issues regarding CRISPR-mediated genome editing. Cris Syst Emerg Technol Appl 26:103–110. https://doi.org/10.21775/cimb.026.103

Singh A, Trivedi P, Jain NK (2018a) Advances in siRNA delivery in cancer therapy. Artif Cells Nanomed Biotechnol 46:274–283

Singh T, Murthy AS, Yang HJ, Im J (2018b) Versatility of cell-penetrating peptides for intracellular delivery of siRNA. Drug Deliv 25:1996–2006

Sokolow SH, Nova N, Pepin KM, Peel AJ, Pulliam JRC, Manlove K, Cross PC, Becker DJ, Plowright RK, McCallum H, De Leo Giulio A (2019) Ecological interventions to prevent and manage zoonotic pathogen spillover. Phil Trans R Soc B. https://doi.org/10.1098/rstb.2018.0342

Spatz DR, Zilliacus KM, Holmes ND, Butchart SH, Genovesi P, Ceballos G, Tershy BR, Croll DA (2017) Globally threatened vertebrates on islands with invasive species. Sci Adv 3(10). https://doi.org/10.1126/sciadv.1603080

Sudweeks J, Hollingsworth B, Blondel DV, Campbell KJ, Dhole S, Eisemann JD, Edwards O, Godwin J, Howald GR, Oh K, Piaggio AJ, ProwseTAA RJV, Saah JR, Shiels AB, Thomas P, Threadgill DW, Vella MR, Gould F, Lloyd A (2019) Locally fixed alleles: a method to localize gene drive to island populations. Sci Rep 9:15821. https://doi.org/10.1038/s41598-019-51994-0

Tang G (2005) siRNA and miRNA: an insight into RISCs. Trends Biochem Sci 30:106–114

Tershy BR, Shen KW, Newton KM, Holmes ND, Croll DA (2015) The importance of islands for the protection of biological and linguistic diversity. Bioscience 65:592–597

Unckless RL, Clark AG, Messer PW (2017) Evolution of resistance against CRISPR/Cas9 gene drive. Genetics 205:827–841

Underwood W, Anthony R, Cartner S, Corey D, Grandin T, Greenacre CB, Gwaltney-Bran S, McCrackin MA, Meyer R, Miller D (2013) AVMA guidelines for the euthanasia of animals: 2013 edition. American Veterinary Medical Association, Schaumburg, IL. ISBN 978-1-882691-21-0

Van Eenennaam AL, Young AE (2014) Prevalence and impacts of genetically engineered feedstuffs on livestock populations. J Anim Sci 92:4255–4278. https://doi.org/10.2527/jas2014-8124

Wilkins KE, Prowse TA, Cassey P, Thomas PQ, Ross JV (2018) Pest demography critically determines the viability of synthetic gene drives for population control. Math Biosci 305:160–169. https://doi.org/10.1016/j.mbs.2018.09.005

Witmer G (2007) The ecology of vertebrate pests and integrated pest management (IPM). In: Kogan M, Jepson P (eds) Perspectives in ecological theory and integrated pest management. Cambridge University Press, Cambridge, UK, pp 393–341

Woinarski JC, Burbidge AA, Harrison PL (2015) Ongoing unraveling of a continental fauna: decline and extinction of Australian mammals since European settlement. PNAS 112:4531–4540

Wozniak CA (2018) Regulatory avenues at US EPA for products of novel breeding technologies. U.S. Environmental Protection Agency, Office of Pesticide Programs. http://cropbioengineering.iastate.edu/wp-content/uploads/2018/06/2-Day2_Wozniak.pdf

Wozniak CA, McClung G, Gagliardi J, Segal M, Matthews K (2012) Regulation of genetically engineered microorganisms under FIFRA, FFDCA and TSCA. In: Regulation of agricultural biotechnology: the United States and Canada. Springer, Dordrecht, pp 57–94. https://doi.org/10.1007/978-94-007-2156-2_4

Zhan X, Zhang F, Zhong Z, Chen R, Wang Y, Chang L, Bock R, Nie B, Zhang J (2019) Generation of virus-resistant potato plants by RNA genome targeting. Plant Biotechnol J 17:1814–1822

Zhang H, Kolb FA, Jaskiewicz L, Westhof E, Filipowicz W (2004) Single processing center models for human Dicer and bacterial RNase III. Cell 118:57–68

Zotti M, dos Santos EA, Cagliari D, Christiaens O, Taning CNT, Smagghe G (2018) RNA interference technology in crop protection against arthropod pests, pathogens and nematodes. Pest Managt Sci 74:1239–1250

# Risk Assessment of Transgenic Silkworms

**Natuo Kômoto and Shuichiro Tomita**

**Abstract** The domesticated silkworm, *Bombyx mori*, cannot survive without human assistance. Commercial use of transgenic silkworms has recently begun, producing recombinant proteins and fluorescent silks. If the silkworms are reared in controlled facilities, such as factories, it is easy to implement containment measures to avoid adverse effects on the environment because domesticated silkworms do not escape, even when starved. On the other hand, in the case of silkworms reared in conventional sericulture farms, risks to biodiversity potentially exist and have been assessed. As *B. mori* larvae are unable to climb mulberry trees to feed themselves, they do not threaten host plants or compete with other insects. The only possible risk is an adverse effect on a wild relative *Bombyx mandarina*, which is widespread throughout East Asia. Although male moths of the wild species can mate with female moths of the domesticated species, no interspecific hybrids have been found in natural populations in Japan based on nucleotide sequence of the mitochondrial *cox1* of more than 20,000 moths. Hence, there will be no risks caused by hybridization, even if transgenic silkworms are reared using conventional methods on sericulture farms.

**Keywords** Fluorescent silk · Sericulture farm · Containment measure · *Bombyx mandarina* · Domesticated animals · Hybridization · Introgression · Cartagena Act

## Introduction

The silkworm, *Bombyx mori*, has been reared for 5000 years since it was bred from a wild relative, the wild mulberry silkmoth *B. mandarina*. It is arguably one of the most highly domesticated animals; it cannot survive without human involvement and hence would likely become extinct if humans stop raising silkworms. The major

---

N. Kômoto · S. Tomita (✉)
National Agriculture and Food Research Organization, Tsukuba, Ibaraki, Japan
e-mail: tomita@affrc.go.jp



A. Chaurasia et al. (eds.), *GMOs*, Topics in Biodiversity and Conservation 19,
https://doi.org/10.1007/978-3-030-53183-6_10

product of *B. mori* as a crop is silk, which is reeled from its cocoons. Silk is a natural protein fiber known as the most beautiful of all textile fibers, which makes it one of the most expensive and popular fibers. *Bombyx mori* has been under human protection for a long time owing to its commercial value in silk production. Due to its economic importance, *B. mori* has been the subject of considerable biological research since medieval times to improve its productivity and the quality of its cocoon silk and to understand its basic characteristics as an organism.

In modern times, these research activities were extended to many basic biological fields, and *B. mori* became one of the most highly studied experimental animals in genetics, physiology, and developmental biology. As an experimental animal, *B. mori* has several advantages: (1) large numbers of strains harboring genetic mutations and/or unique commercial traits with considerable genomic variation are stocked as a genetic resource (Goldsmith et al. 2005; Banno et al. 2010), (2) precise developmental staging enables detailed physiological analyses of embryonic and postembryonic development (Takami and Kitazawa 1960; Goldsmith et al. 2005), and (3) its relatively massive body makes it suitable starting material for the isolation and purification of biochemical compounds (Tazima 1978). Advancements such as the establishment of a baculovirus expression system (Maeda et al. 1985; Palhan et al. 1995; Sumathy et al. 1996) and a high-quality draft genome (Mita et al. 2004; Xia et al. 2004; The International Silkworm Genome Consortium 2008) along with germline transgenesis (Tamura et al. 2000) and the availability of genome editing (Daimon et al. 2014) have further extended its contributions to basic biology.

In recent years, biological research has come to require modern model organisms suitable for analyses of gene functions. To this end, misexpression, gene knockout, and gene knockdown are the most frequently used procedures, and germline transformation, RNAi, and gene editing, respectively, are the key technologies for realizing these genetic manipulations.

## Advent of Transgenic Silkworms

The transgenesis technique to produce genetically modified *Bombyx mori* reproducibly and commercially employs the *piggyBac* transposon. The first successful germline transformation of *B. mori* using the *piggyBac* vector was reported in the year 2000 (Tamura et al. 2000), in which the vector plasmid along with the helper plasmid that expresses transposase is injected into the eggs of syncytial blastoderm stage. Whereas *B. mori* transgenesis had been reported several times previously (Nikolaev et al. 1993; Nagaraju et al. 1996; Yamao et al. 1999), transformation of the silkworm was not widely used in biological studies until after the introduction of the technique employing the *piggyBac* vector.

Although the major motivation for establishing a transformation system is primarily rooted in its utility in biological studies and further enhancement of a particular species as a model organism, the production of recombinant proteins using transgenic organisms has often been used as rationale for research projects that aim to establish genetic manipulation technology. Likewise, in the case of the silkworm, researchers have used this reasoning for acquiring funding to advance fundamental research projects. However, soon after the advent of successful silkworm transgenesis, researchers realized that in fact, transgenesis may also be used for commercially feasible applications. To demonstrate the utility of *B. mori* as an experimental model animal along with the feasibility of the transgenic technology, the system used with *B. mori* was designed to express green fluorescent protein (GFP) fused with a silk protein, fibroin heavy chain, thus producing "fluorescent cocoons" (Kojima et al. 2007). These cocoons emit green fluorescence under blue light for excitation. News of this result spread more quickly and widely than expected, and inquiries assuming that "fluorescent silk" was ready for commercial use began arriving. In reality, it took more than 5 years to establish the technologies that collectively made it possible to produce fluorescent textiles from cocoons containing GFP which would otherwise be inactivated during the conventional reeling and degumming process in which cocoons, silks, and textiles are treated with hot water, by applying low-temperature treatments (Iizuka et al. 2013). Sericulture using transgenic silkworms producing GFP-containing cocoons began in Japan in 2017 under the regulation of genetically modified (GM) organisms.

Recent advancements in the availability of this transgenic system in *B. mori* have led to its rapid commercialization as a system for recombinant protein expression. Transgenic silkworms, as well as the baculovirus expression system, have been shown to be capable of producing a wide variety of biologically active proteins, including antibodies and membrane proteins for pharmaceutical and medical applications (Sato et al. 2012; Matsumoto et al. 2014; Tada et al. 2015). Several recombinant proteins produced by transgenic silkworms are now commercially sold in Japan. Recombinant proteins expressed by the silkworm are, in general, biologically active and achieve useful levels of important posttranslational modifications.

In terms of risk assessment for GM silkworms, only their use in sericultural farms is currently relevant because all other manufacturers that produce recombinant proteins for pharmaceutical and medical applications must implement effective containment measures by law.

## Risk Assessment of Transgenic Silkworms

### *Rearing of Transgenic Silkworms and Containment Measures*

Environmental risk assessment of transgenic silkworms depends on how they are reared. One way to avoid adverse effects to wildlife and the environment is to contain the silkworms in closed facilities, such as laboratories and factories. Because silkworms are reared indoors, using containment measures is a realistic option. Moreover, larvae of domesticated silkworm do not move around even when they are starved, and moths do not fly at all (Kômoto 2017; Kômoto et al. 2014). Thus, it is easy to keep them under control by closing doors and windows and covering openings such as for fans and drains. Nevertheless, silkworms could potentially escape from a facility if staff exiting a rearing room failed to notice larvae attached to their bodies or hidden in clothing, such as a pocket. However, this circumstance can be avoided simply by leaving coats in a separate room. It should be noted that transgenic silkworms must be sacrificed before they are brought out of such a closed system. Even waste left in rearing containers must be inactivated because it may hold transgenic silkworms. Freezing or autoclaving is a simple and effective way to kill silkworms at any developmental stage, such as egg, larva, pupa, or moth.

In contrast to the situation in laboratories and factories, rearing transgenic silkworms in conventional sericulture farms requires careful assessment of risks to the environment because it is difficult to completely close off the rearing houses or to inactivate large quantities of waste (Kômoto et al. 2014).

## *Silkworms in Sericulture Farms*

Silkworms are reared in sericulture farms mainly to obtain cocoons, from which silk thread is spun. In Japan and other major sericultural countries, including China and India, hybrid strains of silkworm are popular because of their uniform growth and high productivity, as with many other crops. Thus, sericulture farmers do not produce eggs by themselves but rather buy them from companies that produce eggs. Typically, the eggs are brought into rearing centers where newly hatched larvae are kept until the middle of larval development, the third or fourth instar. The centers rear larvae for multiple farms under clean conditions to avoid infectious diseases. The grown larvae are then brought to sericulture farms, where the silkworms are fed mulberry leaves. Although silkworms are kept in rearing houses, windows and doors are sometimes opened to maintain optimal temperature and humidity. When the larvae become mature, they are brought into separate rooms used only for cocooning. The cocoons collected are soon transported to silk mills, where they are heat-dried to kill the pupae and thus maintain the cocoons in good condition without emergence of any moths. Waste left in the rearing containers is brought outdoors for composting.

## *Biology of* **Bombyx mandarina***, a Wild Silkmoth*

*Bombyx mandarina* is a wild mulberry silkmoth inhabiting a wide area of East Asia, including China, Korea, Taiwan, Japan, and the far eastern part of Russia (Nakamura et al. 1999). It is presumed that *B. mori* was domesticated from *B. mandarina* in China around 5000 years ago and was subsequently introduced to various regions of the world, including East Asia, Southeast Asia, South Asia, and even Europe (Yoshitake 1970; Goldsmith et al. 2005).

Based on the mitochondrial *nad5* gene, the Japanese population of *B. mandarina* is estimated to have separated from the Chinese population, which is the direct ancestor of *B. mori*, seven million years ago, long before domestication in China (Yukuhiro et al. 2002). Molecular phylogeny shows a clear grouping of *B. mori* and Chinese *B. mandarina*, which are almost indistinguishable based on genomic sequences (Yukuhiro et al. 2012a). Chromosome numbers also support the close relationship between *B. mori* and Chinese *B. mandarina*, as both are $2n = 56$, whereas Japanese *B. mandarina* is $2n = 54$ (Nakamura et al. 1999; Goldsmith et al. 2005).

The wild mulberry silkmoth overwinters as diapause eggs in an early stage of embryonic development (Kobayashi 1989). The larvae of the first generation hatch in spring concurrently with the budding of mulberry leaves. In contrast to larvae of the domesticated silkworm, *B. mandarina* larvae can crawl up and down mulberry trees and feed themselves (Sasaki et al. 1984), and late-stage larvae extend their thoraxes to mimic twigs in the daytime (Sasaki et al. 1984). After three or four larval molts, they begin spinning yellowish cocoons attached between leaves or rolled in leaves. Moths do not eat or drink because they lack mouthparts. The moths usually emerge from their cocoons in the morning (Sasaki et al. 1984). Female moths then emit sex pheromone and wait near their cocoons. Male moths are attracted to the sex pheromone, bombykol, and find and mate with females (Kuwahara et al. 1984; Daimon et al. 2012). After they separate from males, female moths start flying around from place to place to lay eggs during the night. First-generation moths lay non-diapause eggs, from which larvae hatch in 10–15 days. *Bombyx mandarina* in Japan is bivoltine or trivoltine, meaning moth appearance peaks two or three times a year. The last generation of moths each year lays diapause eggs that overwinter.

## *Risk Assessment of Transgenic Silkworms Reared in Farms*

Despite the long history of sericulture, there have been no reports to date of the domesticated silkworm becoming a wild insect as active as *Bombyx mandarina* for survival and reproduction in nature without human assistance. It is also unlikely that genetic modification will cause such reversion of the silkworm, because functions of many genes responsible for domesticated behaviors, about which there is no information at present, will not be changed inadvertently by introduction of several genes. Thus, although risk assessment should be conducted according to the traits of each transgenic strain, we provide a general discussion of the environmental risk of rearing transgenic silkworms based on the premise that their behavioral traits will not differ greatly from those of non-transgenic silkworms or at least that they will not become wild.

First, transgenic silkworms will not cause any harm to mulberry trees because the larvae are unable to reach or climb mulberry trees to obtain leaves (Kômoto et al. 2014). They do not walk to find food even if they are starved (Kômoto 2017). This is widely observed in sericulture farms, where silkworms are reared in containers without lids. Moreover, the larvae can hardly adhere to branches, because their appendages are too weak to hold their heavy bodies.

Second, transgenic silkworms will not compete with other insects because they are unable to survive and reproduce in nature without human assistance. As described above, the larvae cannot feed themselves. Moreover, the larvae can easily be found by predators because their bodies are white, and they continuously move and wave them without mimicking branches, unlike wild silkworms, which extend their bodies and remain motionless during the day (Kômoto 2017). Indeed, when domesticated silkworm larvae or moths are placed on the ground, they are eaten or killed by birds, ants, and wasps because they have no means of escape (Kômoto et al. 2014). Moreover, the moths cannot find mating partners because they are flightless. Even if silkworms were engineered to be resistant to viruses (see below), there would be no environmental risks of competition with other insects, because such transgenic silkworms would still be unable to survive and reproduce in the field.

## *Hybridization Between the Two* **Bombyx** *Species*

*Bombyx mori* and *B. mandarina* can mate and produce fertile hybrids, a process which has been used in silkworm breeding to introgress characteristics of the wild species into the domesticated one (Goldsmith et al. 2005). Although male *B. mandarina* moths are attracted to and mate with female *B. mori* moths that are left outdoors because the two species share the same sex pheromone (Kuwahara et al. 1984; Daimon et al. 2012), male *B. mori* moths are unable to reach female *B. mandarina* moths waiting on trees because the domesticated silkmoths cannot fly. Mating experiments in laboratories have also shown that the combination of *B. mori* females with *B. mandarina* males leads to successful mating to produce fertilized eggs, whereas the opposite combination of *B. mandarina* females with *B. mori* males ends in failure even if the moths are kept in small cages (Nakamura et al. 1997). The likely reason for this is not sterility of the hybrid embryos, but this mating never occurs because domesticated male silkmoths cannot fly and are too inactive to catch wild females.

To confirm whether interspecific *Bombyx* hybrids have occurred naturally, male *Bombyx* moths were collected from all around Japan and applied in genetic analyses. The moths were collected using pheromone traps baited with female *B. mori* moths or synthesized bombykol that was inoculated on rubber sleeves (Yukuhiro et al. 2017a). The nucleotide sequences of the mitochondrial *cox1* gene of more than 20,000 moths showed that the haplotypes of *B. mori* and *B. mandarina* in Japan are clearly separated, with no traces of hybridization (Yukuhiro et al. 2012a, 2017b). Because interspecific mating occurs only in the combination of *B. mori* females and *B. mandarina* males, mitochondrial DNA, which is contributed only by the female, is a good index of hybridization (Fig. 1). Genotyping of a nuclear gene, *CAD*, also showed that there has been no introduction of the domesticated species into the wild population (Yukuhiro et al. 2012b). Moreover, the incidence of $F_1$ hybrids was

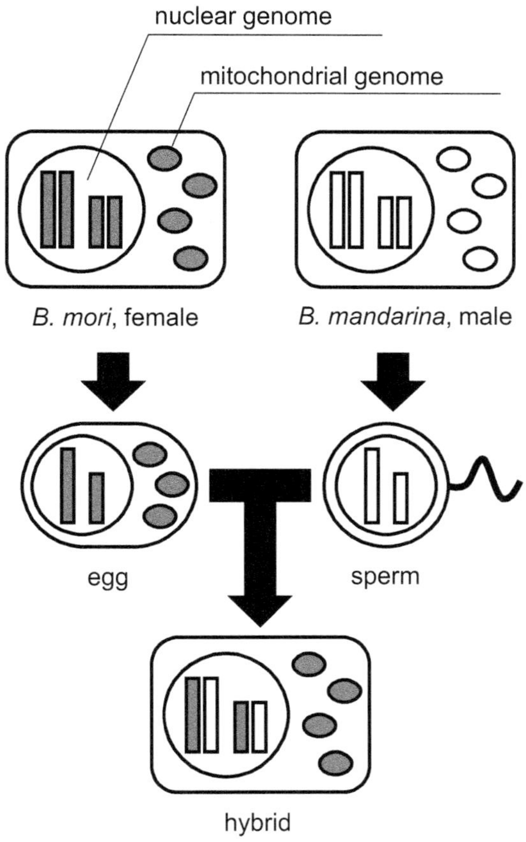


**Fig. 1** Diagram of inheritance of mitochondrial and nuclear genomes in hybrids between *Bombyx mori* and *B. mandarina*. Since interspecific mating occurs only between female moths of *B. mori* and male moths of *B. mandarina*, their hybrids inherit the *B. mandarina* mitochondrial genome, while their nuclear genome is heterozygous. Therefore, the haplotype of the mitochondrial genome distinguishes hybrids in wild populations

studied by genotyping the *cox1* gene of moths caught by pheromone traps around sericulture farms where silkworms are reared conventionally without specific care to avoid mating between the two *Bombyx* species. As a result, no $F_1$ moths carrying the *B. mori*-typed *cox1* gene were found in 3750 moths even in such a situation where interspecific mating was expected to occur with the highest frequency (Kômoto et al. 2016). These data showing only the presence of *B. mandarina* genotypes indicate that there is no or very little chance of interspecific hybridization in nature.

There are several more reasons why hybrids between the domesticated and wild silkmoths in nature are so unlikely. One is the absence of *B. mori* moths in sericulture farms. Because farmers buy eggs rather than produce them, it is unnecessary for them to obtain or produce moths. Moreover, farmers prevent moths from emerging from cocoons because the emergence would diminish cocoon quality by creating a hole and allowing the moth to stain cocoons by spraying them with meconium (pupal waste products).

It is possible that mature larvae, which actively walk around to find suitable places for cocooning, could escape from farmers and produce cocoons somewhere in the room, such as in corners or on the ceiling, where moths could emerge unintentionally. However, even if wild male *B. mandarina* moths attracted by the sex pheromone and somehow entering the cocooning room were to mate with female *B. mori* moths, the females would lay eggs around themselves inside the room because they cannot fly (Kômoto et al. 2014). The newly hatched hybrid larvae would be unable to reach mulberry trees growing outside. This scenario was confirmed by experiments in which around 3000 $F_1$ hatchlings in total were placed on the ground 2 m away from a mulberry tree and none of them grew up to spin cocoons probably because crawling on the ground is a significant obstacle for 3-mm larvae (Kômoto et al. 2016). It has also been shown that $F_1$ hatchlings walk less actively than *B. mandarina* (Shimoda and Kanekatsu 2016). On the other hand, *B. mandarina* larvae don't have to walk on the ground because their mother moths lay eggs on mulberry trees. It is also possible that cocoons hidden in waste could be brought out of rearing houses for composting and that moths could emerge there. Nonetheless, even if female *B. mori* moths were to mate outdoors with wild male *B. mandarina* moths, eggs would be laid in the compost, from which newly hatched larvae could hardly reach mulberry trees to survive.

Since the absence of hybrids mainly depends on the biological characteristics of *B. mori* and the way of sericulture, it is supposed that interspecific hybridization occurs rarely, if ever, in other silk-producing countries, such as China.

In the unlikely event that hybrids between transgenic *B. mori* and wild *B. mandarina* could occur, the influence of the introduced genes on fitness should be assessed individually depending on the gene functions. For example, the addition of fluorescence to silk has no evident advantage for survival or reproduction, whereas resistance to viruses falls into a different category and will require comprehensive assessment of various aspects, including resistance to nontarget diseases and adverse effects on growth and/or reproduction.

In conclusion, the risks associated with hybridization can be classified as follows, depending on the presence of the wild relative and the traits added by transgenes. (1) There are no risks in areas uninhabited by *B. mandarina*, such as India and Vietnam. (2) If introduced genes do not increase fitness, risks are not probable, even if transgenic silkworms are reared in conventional sericulture farms in areas where *B. mandarina* is present. (3) Careful risk assessment will be required if introduced genes are expected to increase fitness and the transgenic silkworms are reared within the natural habitat of the wild silkmoth.

## *Horizontal Transfer of Transgenes from* **Bombyx mori** *to Other Species*

Horizontal gene transfer between *Bombyx mori* and different insect species or microorganisms has supposedly occurred during their evolution according to various studies mainly based on molecular phylogenetic analyses, although mechanisms of horizontal transfer are unknown (Daimon et al. 2003; Zhang et al. 2013; Schneider and Thomas 2014; Wang et al. 2019). The possibility of horizontal transfer, however, is too small to estimate its risk on biodiversity by introgression of transgenes into wild organisms.

## Examples of Transgenic Silkworm Rearing in Farms in Japan

In 2017, transgenic silkworms were reared in a sericulture farm in Japan to produce green fluorescent silk. It was the first case of introducing transgenic silkworms into a farm for commercial silk production. In this section, we describe the situation in Japan to show how risk evaluation is applied to the actual process.

### *Risk Assessment Process Under Regulations in Japan*

In Japan, the cultivation or rearing of all genetically modified organisms is regulated under the "Cartagena Act," which has the central purpose of avoiding adverse effects on biodiversity. Under the law enforcement of the act, the rearing of transgenic silkworms in farms is overseen jointly by the Ministry of Agriculture, Forestry and Fisheries and Ministry of the Environment. After an application is filed, the legal process begins by consulting with the regulatory authorities. Then, the application describing the transgenic strains and the methods to control them is examined by a committee of scientists who provide advice if necessary. After admission by the committee and consideration of public comments, the two ministers approve the application to rear the transgenic silkworm.

For the risk assessment, each application is appended with biological data on the transgenic strain to show that there are no significant differences in traits that may be involved in adverse effects compared with those of non-transgenic silkworms. Data are collected in laboratories as follows: copy number, stability, and expression of the inserted genes; hatching rate; days from hatch to cocooning; weight of larvae, pupae, and cocoons; cocooning rate; egg number; walking distance of larvae; oviposition area; and influence on plant germination and microorganisms. The characteristics of transgenic silkworms must be reconfirmed several times in different seasons in isolated rearing areas under the equivalent rearing conditions as in farms, in the same way as for other transgenic crops.

## Management Measures and Monitoring

Although almost no environmental risk was indicated, some specific management measures were implemented in the process of approving the rearing of transgenic silkworms producing green fluorescent silk, probably because it was the first case. It was also a rare case because of the intention to grow a transgenic organism in Japanese farms. Although many transgenic crops have been approved for cultivation in Japan, almost none of these are actually cultivated in Japan but imported from foreign countries.

One measure for controlling the transgenic silkworm is to restrict farm rearing to only the latter half of the larval stage until cocooning, excluding eggs, the early larval stages, and moths. The reason for excluding eggs and early larvae is that they are too small to avoid unintentionally bringing them out of rearing houses, although it is clear that eggs and small larvae will not cause any harm to the environment because they will not survive in nature. Another example of a special management measure is to close windows and doors of rearing houses to prevent wild male silkmoths from flying in and mating with transgenic female silkmoths. If windows are opened to control the temperature and humidity, they must be covered with screens. However, it is clear that the larvae in farms are too immature to mate with wild silkmoths. Even if accidentally emerged transgenic silkmoths were to mate with wild moths, it would be impossible for their hybrids to survive in the rearing house, as previously discussed. Proper waste disposal is also required to avoid mating between transgenic and wild moths. The waste should be chipped into small pieces to kill hidden pupae or kept covered with netting for 30 days to insure no stranded silkmoths survive. Such preventive measures are expected to become more balanced through accumulation of additional data about hybridization between transgenic and wild silkmoths.

To confirm the absence of hybrids, male *Bombyx* moths are collected with pheromone traps to monitor them. Because domesticated silkmoths usually lay diapause eggs that hatch the following spring, pheromone traps are placed around sericulture farms during early summer of the next year of rearing. In mark-recapture experiments, $F_1$ male moths were caught by pheromone traps at a higher rate than *B. mandarina* (unpublished data). Field observations suggest that $F_1$ male moths were attracted to sex pheromone more effectively than *B. mandarina* male moths although there was no apparent difference in flight activity between the two groups.

## The First Transgenic Silkworms Reared in Sericulture Farms

After ministerial approval in September 2017, the first transgenic silkworms of a strain producing green fluorescent silk were reared in a conventional sericulture farm in Gunma Prefecture in October 2017 (Fig. 2). Newly hatched larvae were grown to third instar in the closed facilities of a prefectural experimental station and

**Fig. 2** Rearing of transgenic silkworms producing green fluorescent silk in a conventional sericulture farm. (**a**) Feeding mulberry leaves to the larvae. (**b**) Transgenic silkworm larvae. (**c**) Larvae transferred to the apparatus to spin cocoons. (**d**) Cocoons. (**e**) Cocoons emitting green fluorescence under blue excitation light. (**f**) Collected cocoons with greenish color from the fluorescent protein

were then transported to the farm. Although the rearing house was renovated to prevent wild insects from entering by filling gaps in the walls and placing netting on the windows, silkworms were fed with mulberry leaves as usual, and cocoons were spun on conventional apparatuses. In 2018, another farm participated in the production of fluorescent cocoons. Then, from May to July 2018, the emergence of hybrids between the transgenic silkmoths and wild mulberry silkmoths was monitored using pheromone traps set around the farm where transgenic silkworms were reared in 2017. No hybrids were found. This supports the premise of the approval that no hybrids will occur naturally or under the preventive measures.

So far in Japan, there has been no intense opposing movement against transgenic silkworms. For example, in annual briefings of test rearing in isolated rearing areas, participants often ask how they can rear transgenic silkworms, while few of them express anxieties about risks to the environment.

## Application of Transgenic Silkworms in Other Countries

The application of transgenic technology, like other biological technologies, has been intensively explored in *Bombyx mori*. Because the silkworm is used primarily for silk production in sericulture farms, the most important aspect for implementing this technology is the breeding of new strains with improved agricultural traits such as cocoon yield, silk quality, robustness against the environment, and resistance to infectious diseases. Among these, studies aiming to enhance resistance against the nuclear polyhedrosis virus (NPV) were started early (Isobe et al. 2004). *Bombyx mori* NPV (BmNPV) is one of the most serious silkworm pathogens and causes grasserie disease, which may be the largest contributor to all damage to cocoon crops by silkworm diseases. Researchers have targeted BmNPV genes *lef-1* (Isobe et al. 2004) and *ie-1* (Kanginakudru et al. 2007), in combination with *lef-3*, *p74*, *helicase*, *gp64*, and *vp39* (Jiang et al. 2013; Subbaiah et al. 2013), in searching for a good combination of targets for an efficient RNAi effect. A transgenic line that expressed hairpin double-stranded RNA for *lef-1*, *ie-1*, *lef-3*, and *p74*, which showed a significantly improved survival rate upon viral infection (Subbaiah et al. 2013), was further used to breed new practical varieties by crossing with strains already used commercially in India. Four new transgenic varieties that are resistant to BmNPV were developed and test-reared in sericultural institutes at various locations in India from 2015 to 2017 (Central Sericultural Research & Training Institute 2017) in hopes of receiving approval from a committee on genetically modified organisms for trials on farms in India.

Production of spider silks is another target to apply to transgenic silkworms. Although spider silks are strong fibers and highly anticipated as industrial materials, mass rearing of spiders is unrealistic. Several groups have reported that expression of spider silk proteins in transgenic silkworm cocoons improves the mechanical properties of silk fibers (Teulé et al. 2012; Kuwana et al. 2014; Xu et al. 2018; You et al. 2018). Recently, a US-based biotechnology company, Kraig Biocraft Laboratories, announced that it started rearing of recombinant spider silk silkworms in Vietnam (Kraig Biocraft Laboratories 2019). The process of risk assessment and legal approval of the transgenic silkworms there has not been disclosed, at least to our knowledge.

## Conclusion and Perspectives

Recombinant protein production by transgenic silkworms has precedents and is expected to be applied for various proteins, especially for pharmaceutical drugs, which have a huge market. The containment measures for such transgenic silkworms will not be a bottleneck because they will be encompassed by the good manufacturing practices required for drug safety.

Transgenic silkworm rearing in sericulture farms has passed legal barriers, as described in detail above. Other strains can follow the same route to obtain approval. Some strains have been through test rearing in isolated areas and are on track to be introduced into sericulture farms.

Nevertheless, two major issues still remain to be solved. One is the tightness of restrictions introduced as preventive measures, such as the closure of the rearing house and stringent waste disposal. It is necessary, of course, to maintain measures to avoid adverse environmental effects, but restrictions that are not meaningful practically or scientifically speaking should be removed. Applicants for the approval of transgenic silkworm rearing are expected to present scientific data that will result in changing these restrictions. The other issue is to stimulate demand for genetically modified silks. Because sericulture in Japan has long faced the serious problem of cocoon prices being lower than rearing costs, which have been made highly efficient with almost no room for cost cutting, the success of rearing transgenic silkworms in sericulture farms depends on the development of even more attractive silks with greater market appeal.

## References

Banno Y, Shimada T, Kajiura Z et al (2010) The silkworm – an attractive BioResource supplied by Japan. Exp Anim 59:139–146

Central Sericultural Research & Training Institute (2017) Annual report 2016–2017. Mysuru, India. http://www.csrtimys.res.in/annual-report-0

Daimon T, Hamada K, Mita K et al (2003) A *Bombyx mori* gene, *BmChi-h*, encodes a protein homologous to bacterial and baculovirus chitinases. Insect Biochem Mol Biol 33:749–759

Daimon T, Fujii T, Fujii T et al (2012) Reinvestigation of the sex pheromone of the wild silkmoth *Bombyx mandarina*: the effects of bombykal and bombykyl acetate. J Chem Ecol 38:1031–1035

Daimon T, Kiuchi T, Takasu Y (2014) Recent progress in genome engineering techniques in the silkworm, *Bombyx mori*. Develop Growth Differ 56:14–25

Goldsmith MR, Shimada T, Abe H (2005) The genetics and genomics of the silkworm, *Bombyx mori*. Annu Rev Entomol 50:71–100

Iizuka T, Sezutsu H, Tatematsu K et al (2013) Colored fluorescent silk made by transgenic silkworms. Adv Funct Mater 23:5232–5239

Isobe R, Kojima K, Matsuyama T et al (2004) Use of RNAi technology to confer enhanced resistance to BmNPV on transgenic silkworms. Arch Virol 149:1931–1940

Jiang L, Zhao P, Cheng T et al (2013) A transgenic animal with antiviral properties that might inhibit multiple stages of infection. Antivir Res 98:171–173

Kanginakudru S, Royer C, Edupalli SV (2007) Targeting *ie-1* gene by RNAi induces baculoviral resistance in lepidopteran cell lines and in transgenic silkworms. Insect Mol Biol 16:635–644

Kobayashi J (1989) Effects of photoperiod on the induction of egg diapause of tropical races of the domestic silkworm, *Bombyx mori*, and the wild silkworm, *B mandarina*. JARQ 23:202–205

Kojima K, Kuwana Y, Sezutsu H et al (2007) A new method for the modification of fibroin heavy chain protein in the transgenic silkworm. Biosci Biotechnol Biochem 71:2943–2951

Kômoto N (2017) Behavior of the larvae of wild mulberry silkworm *Bombyx mandarina*, domesticated silkworm *B. mori* and their hybrid. J Insect Biotechnol Sericol 86:17–20

Kômoto N, Tsuda M, Okada E et al (2014) Development of methods for risk assessment of transgenic silkworms rearing on biodiversity. Sanshi Konchu Biotec 83:171–179. (in Japanese with an English abstract)

Kômoto N, Kuwabara N, Yukuhiro K (2016) Absence of hybrids between the domesticated silkmoth, *Bombyx mori*, and the wild mulberry silkmoth, *B. mandarina*, in natural populations around sericulture farms. J Insect Biotechnol Sericol 85:67–71

Kraig Biocraft Laboratories (2019) Kraig Biocraft Laboratories' spider silk technology successfully delivered to Vietnam. Press release, January 7, 2019

Kuwahara Y, Mori N, Yamada S et al (1984) Evaluation of bombykol as the sex pheromone of *Bombyx mandarina* (Lepidoptera: Bombycidae). Appl Entomol Zool 19:265–267

Kuwana Y, Sezutsu H, Nakajima K et al (2014) High-toughness silk produced by a transgenic silkworm expressing spider (*Araneus ventricosus*) dragline silk protein. PLoS One 9:e105325

Maeda S, Kawai T, Obinata M et al (1985) Production of human alpha-interferon in silkworm using a baculovirus vector. Nature 315:592–594

Matsumoto Y, Ishii M, Ishii K et al (2014) Transgenic silkworms expressing human insulin receptors for evaluation of therapeutically active insulin receptor agonists. Biochem Biophys Res Commun 455:159–164

Mita K, Kasahara M, Sasaki S et al (2004) The genome sequence of silkworm, *Bombyx mori*. DNA Res 11:27–35

Nagaraju J, Kanda T, Yukuhiro K et al (1996) Attempt of transgenesis of the silkworm (*Bombyx mori* L.) by egg-injection of foreign DNA. Appl Entomol Zool 31:589–598

Nakamura T, Banno Y, Kawaguchi Y et al (1997) Degree of reproductive isolation between *Bombyx mori* and *B. mandarina*. Kyushu Sanshi 28:30. (in Japanese)

Nakamura T, Banno Y, Nakada T et al (1999) Geographic dimorphism of the wild silkworm, *Bombyx mandarina*, in the chromosome number and the occurrence of a retroposon-like insertion in the arylphorin gene. Genome 42:1117–1120

Nikolaev AI, Tchkinia TT, Kafiani-Eristavi CA et al (1993) Preferential extrachromosomal localization of exogenous DNA in transgenic silkworm *Bombyx mori* L. Mol Gen Genet 236:326–330

Palhan VB, Sumathy S, Gopinathan KP (1995) Baculovirus mediated high-level expression of luciferase in silkworm cells and larvae. BioTechniques 19:102–104

Sasaki M, Jjbiki F, Hirobe T (1984) Comparison of the daily rhythmic behavior between wild and domestic silkmoths, *Bombyx mandarina* and *B. mori*. Bull Fac Agric Tamagawa Univ 24:26–42

Sato M, Kojima K, Sakuma C et al (2012) Production of scFv-conjugated affinity silk powder by transgenic silkworm technology. PLoS One 7:e34632

Schneider SE, Thomas JH (2014) Accidental genetic engineers: horizontal sequence transfer from parasitoid wasps to their lepidoptera hosts. PLoS One 9:e109446

Shimoda M, Kanekatsu R (2016) Walking activity of silkworm larvae and the possibility of survival in natural conditions. Sanshi Konchu Biotec 85:145–151. (in Japanese with an English abstract)

Subbaiah EV, Royer C, Kanginakudru S et al (2013) Engineering silkworms for resistance to baculovirus through multigene RNA interference. Genetics 193:63–75

Sumathy S, Palhan VB, Gopinathan KP (1996) Expression of human growth hormone in silkworm larvae through recombinant *Bombyx mori* nuclear polyhedrosis virus. Protein Expr Purif 7:262–268

Tada M, Tatematsu K, Ishii-Watabe A et al (2015) Characterization of anti-CD20 monoclonal antibody produced by transgenic silkworms (*Bombyx mori*). MAbs 7:1138–1150

Takami T, Kitazawa T (1960) Normal stages of the embryonic development of in the silkworm, *Bombyx mori*. Tech Bull Sericult Exp Sta 75:1–31. (in Japanese)

Tamura T, Thibert C, Royer C et al (2000) Germline transformation of the silkworm *Bombyx mori* L. using a *piggyBac* transposon-derived vector. Nat Biotechnol 18:81–84

Tazima Y (ed) (1978) The silkworm: an important laboratory tool. Kodansha, Tokyo

Teulé F, Miao YG, Sohn BH et al (2012) Silkworms transformed with chimeric silkworm/spider silk genes spin composite silk fibers with improved mechanical properties. Proc Natl Acad Sci U S A 109:923–928

The International Silkworm Genome Consortium (2008) The genome of a lepidopteran model insect, the silkworm *Bombyx mori*. Insect Biochem Mol Biol 38:1036–1045

Wang CF, Sun W, Zhang Z (2019) Functional characterization of the horizontally transferred 4,5-DOPA extradiol dioxygenase gene in the domestic silkworm, *Bombyx mori*. Insect Mol Biol. https://doi.org/10.1111/imb.12558

Xia Q, Zhou Z, Lu C et al (2004) A draft sequence of the genome of the domesticated silkworm (*Bombyx mori*). Science 306:1937–1940

Xu J, Dong Q, Yu Y et al (2018) Mass spider silk production through targeted gene replacement in *Bombyx mori*. Proc Natl Acad Sci U S A 115:8757–8762

Yamao M, Katayama N, Nakazawa H et al (1999) Gene targeting in the silkworm by use of a baculovirus. Genes Dev 13:511–516

Yoshitake N (1970) Origin and differentiation of the silkworm races. Jpn Agric Res Quart 5:38–43

You Z, Ye X, Qian Q et al (2018) Extraordinary mechanical properties of composite silk through hereditable transgenic silkworm expressing recombinant major ampullate spidroin. Sci Rep 8:15956

Yukuhiro K, Sezutsu H, Itoh M et al (2002) Significant levels of sequence divergence and gene rearrangements have occurred between the mitochondrial genomes of the wild mulberry silkmoth, *Bombyx mandarina*, and its close relative, the domesticated silkmoth, *Bombyx mori*. Mol Biol Evol 19:1385–1389

Yukuhiro K, Iwata K, Kômoto N et al (2012a) Nucleotide sequences of mitochondrial cytochrome C oxidase subunit I (COI) gene show clear differences between the domesticated silkmoth *Bombyx mori* and the wild mulberry silkmoth *Bombyx mandarina* from Japan. J Insect Biotechnol Sericol 81:29–35

Yukuhiro K, Sezutsu H, Tamura T et al (2012b) Little gene flow between domestic silkmoth *Bombyx mori* and its wild relative *Bombyx mandarina* in Japan, and possible artificial selection on the *CAD* gene of *B. mori*. Genes Genet Syst 87:331–340

Yukuhiro K, Kuwabara N, Kômoto N (2017a) Pheromone dose and set height of pheromone traps for efficient collection of wild mulberry silkmoths, *Bombyx mandarina*. J Insect Biotechnol Sericol 86:55–57

Yukuhiro K, Sakaguchi H, Kômoto N (2017b) Three single nucleotide polymorphisms indicate four distinctive distributions of Japanese *Bombyx mandarina* populations. J Insect Biotechnol Sericol 86:77–84

Zhang HH, Shen YH, Xu HE et al (2013) A novel *hAT* element in *Bombyx mori* and *Rhodnius prolixus*: its relationship with miniature inverted repeat transposable elements (MITEs) and horizontal transfer. Insect Mol Biol 22:584–596

# Part IV
# GM Vertebrates

# Genetically Engineered Fish: Potential Impacts on Aquaculture, Biodiversity, and the Environment

**Rex A. Dunham and Baofeng Su**

**Abstract** Studies on transgenic fish for the aquaculture industry have focused on improving growth rates, enhancing disease resistance, altering body composition, acting as biological factories for medical proteins, and even altering temperature tolerance and coloration. The future impact of transgenesis will likely be quite large. Growth hormone-transgenic salmon has been approved for human consumption and has been introduced to the market in Canada and soon to the USA. This is the first human consumption of approved transgenic meat. Transgene insertion has many pleiotropic effects. Several studies have projected the fitness of transgenic fish to be low, in general, compared to non-transgenic and wild fish; thus, their environmental risk is likely low and they would have minimal, if any, long-term impact on ecosystems or biodiversity. However, there have been no actual escapements; thus, only projections of risk are available based on small-scale experiments and the characteristics of transgenic fish compared to controls. An active area of research is repressible transgenic sterilization and sterilization using gene editing, both of which would allow application of transgenic fish with only short-term consequences for ecosystems in the worst-case scenario. Transgenic technology could also be potentially used to reduce or eliminate populations of nuisance species.

**Keywords** Genetic engineering · Biodiversity · Aquaculture · Environmental risk · Transgenic · Fitness · Transgenic sterilization · Genetic enhancement

## Aquaculture and Selective Breeding

Most agricultural crops are genetically modified (traditional approaches) products that have been bred for hundreds or even thousands of years. Natural selection and conventional breeding drove the phenotypic and genetic changes in food organisms,

R. A. Dunham (✉) · B. Su
School of Fisheries, Aquaculture and Aquatic Sciences, Auburn University, Auburn, AL, USA
e-mail: dunhara@auburn.edu; subfeng@auburn.edu



A. Chaurasia et al. (eds.), *GMOs*, Topics in Biodiversity and Conservation 19,
https://doi.org/10.1007/978-3-030-53183-6_11

such as corn, watermelon, and peach as well as animals, including fish. For example, corn was revolutionized from small cobs (19 mm in length), hard and a meager kernel (5–10 in number), into a larger (1000-fold larger) and tastier modern version (rich in carbohydrates, oil, and proteins), easy to peel and cook through domestication and breeding activities. The compositions changed with 2% less juice and 2.5-fold more sugars (Beadle 1980; VOX 2014) in modern corn. Although these changes utilizing selective breeding were dramatic, the genetic enhancement and phenotypic change is usually slow. In some cases, phenotypic variance or lack of additive genetic variation prevents genetic enhancement from selective breeding.

Selective breeding in aquatic organisms has yielded similar results. Selection has been used to double growth rates and increase disease resistance as well as improve other production traits (Dunham 2011) over multiple generations.

## Transgenic Technology

With the development of molecular biology theory and technology, transgenic technology emerged (Palmiter et al. 1982), providing a new method for breeding and genetic enhancement. By 1985, fish scientists were adapting transgenic technology with the goal of improving productivity of aquaculture. After a large initial thrust in this area, research slowed as it became clear public concern and regulation would make application difficult and slow. In addition, genomics emerged as an alternative area to practice molecular skills as well as generate resources for future genetic enhancement. A tremendous amount of research in aquatic genomics has been conducted during the last 25 years (Abdelrahman et al. 2017), although very little application in selective breeding has yet to occur. However, the Japanese flounder (*Paralichthys olivaceus*) industry has been transformed by marker-assisted selection (Fuji et al. 2007), while research on quantitative trait loci and genome-wide associated study has greatly expanded (Abdelrahman et al. 2017; Liu et al. 2018; Wang et al. 2019).

### *Growth*

The greatest amount of transgenic fish work, especially early work, focused on transfer of growth hormone (GH) genes. This was especially true in the 1980s and 1990s, but a greater variety of transgenes are explored today. Maclean and Talwar (1984) working at the University of Southampton, UK, were the first to inject cloned genes into fish eggs (rainbow trout, *Oncorhynchus mykiss*). However, Zhu et al. (1985) and his team at the Institute of Hydrobiology in China were the first to report production of a transgenic fish, goldfish (*Carassius auratus*), using the hGH gene and resulting in a 4.6× increase in growth, although they did not report any integration data.

Growth (size and rate) improvement has ranged from 0% to, in some cases, an amazing 3000%. Several species, including loach (*Misgurnus mizolepis*), common carp (*Cyprinus carpio*), crucian carp (*Carassius carassius*), Atlantic salmon (*Salmo salar*), channel catfish (*Ictalurus punctatus*), Nile tilapia (*Oreochromis niloticus*), medaka (*Oryzias latipes*), and northern pike (*Esox lucius*), containing either human, bovine, or salmonid GH genes, grew 10–80% faster than non-transgenic fish. Du et al. (1992) used an "all-fish" GH gene construct to make transgenic Atlantic salmon that grew 2–6× faster than non-transgenic controls.

Zhu followed up the goldfish work by making transgenic GH Yellow River common carp. An "all-fish" growth hormone (GH) chimeric gene construct, pCAgGH, using a promoter $\beta$-actin gene from Yellow River common carp linked to the growth hormone gene from the grass carp (*Ctenopharyngodon idellus*) was developed and transferred to fertilized embryos of Yellow River common carp to produce the "all-fish" growth-transgenic Yellow River carp (Zhu 1992). The growth rate of transgenic carp was 42–114.92% faster than the control. The control Yellow River carp needed 2 years to reach market size, while the transgenic carp needed only 1 year. Since the growth hormone gene had an inhibitory function on reproductive development, the weight of the gonads was reduced, and the edible portion of the fish was correspondingly increased.

Mori and Devlin (1999) examined the expression of the sockeye salmon (*Oncorhynchus nerka*) metallothionein-B (MTB)-sockeye GH1 gene in transgenic coho salmon (*Oncorhynchus kisutch*), resulting in 40× elevated circulating GH levels, and, in some cases, inducing 5–11-fold increases in weight after 1 year of growth. GH expression was greater in younger, smaller, transgenic coho salmon (20–21 g) compared with older, larger, transgenic salmon (400–500 g).

Response to transgene insertion can vary based on transgene, promoter, position effect, copy number, epigenetics, species, family, and genetic background. Heterozygous $F_1$ and $F_2$ lines of transgenic Nile tilapia possessing one copy of an eel (ocean pout) promoter-chinook salmon GH fusion grew 2.5–4-fold faster and converted feed 20% better than control siblings (Rahman et al. 1998, 2001; Rahman and Maclean 1999). However, $F_1$ fish transgenic for the sockeye salmon MT promoter-sockeye salmon GH gene exhibited no growth enhancement (Rahman et al. 1998), although salmon transgenic for this construct had greatly accelerated growth.

Similarly, sockeye salmon-sockeye GH cDNA1 introduced into coho salmon increased growth from 11-fold to 37-fold (Devlin et al. 2001). Results with Atlantic salmon are not as dramatic as with coho salmon. Transgenic Atlantic salmon containing the opAFP-chinook salmon GH cDNA1 gene construct had a three- to six-fold increased growth rate compared to non-transgenic salmon (Du et al. 1992; Cook et al. 2000), and insertion of sockeye MTB-sockeye GH cDNA1 (Devlin 1997) produced a similar result, fivefold growth enhancement.

## *Dramatic Growth of Transgenic Fish: Explanations and Limitations*

As indicated above, growth enhancement varies greatly among different transgenic fish systems when GH transgenes are integrated (Devlin 1997; Dunham and Devlin 1998). Family effects, position effects, and others have been presented as potential explanations of this variable response. Family effect was observed for growth hormone-transgenic coho salmon produced from a wild strain. Promoters of sockeye salmon metallothionein-B or histone 3 were fused to a growth hormone-1 coding region from the same species (OnMTGH1 and OnH3GH1 constructs, respectively) and were used for evaluation of the growth rate. Salmon transgenic for the OnMTGH1 construct had consistently higher weight than those containing the OnH3GH1 construct, and both transgenic groups had greatly enhanced growth over non-transgenic fish. However, strong family effects were observed as some OnH3GH1 families had similar weight to OnMTGH1 families while others did not (Leggatt et al. 2012).

Domestication effects, innate growth rate, and life-history traits have been put forward as potential factors responsible for the tremendous range of GH transgene responses, with one of the first hypotheses revolving around domestication. Domestication in vertebrates may use the same genetic and physiological pathways in GH endocrine axis to regulate growth rate. Microarray analysis confirmed that transgene insertion and domestication affect the gene expression in concordant ways and implied that the two genetic processes modified the same regulatory pathway for growth (Devlin et al. 2009, 2013). Strains or species that have been selected to near maximum growth rates may have many of their metabolic and physiological processes optimized, and further growth enhancement might be more difficult to obtain by insertion of GH or other growth-related genes.

Insertion of salmon metallothionein growth hormone (OnMTGH1) transgenes increased growth 17× in slow-growing, wild rainbow trout strains (with naturally low growth rates), while the transgene did not stimulate growth (4.4% increase) in fast-growing, non-transgenic, domestic rainbow trout (Devlin et al. 2001). However, these $P_1$ domestic rainbow trout were mosaic, and very few families were evaluated.

However, additional data on transgenic rainbow trout (Devlin et al. 2001) are not consistent with the hypothesis that wild fish when made GH-transgenic immediately reach a growth plateau already existing for selected domestic lines. When OnMTGH1 was transferred to another wild rainbow trout strain, F77, growth was enhanced sevenfold, which exceeded by fourfold the growth exhibited by a non-transgenic domestic rainbow trout. In this case, the wild transgenic rainbow trout is actually superior to the domestic selected strain, indicating that genetic engineering can have a greater effect than, rather than an equivalent effect to, domestication and selection. When F77 was crossbred with the domestic strain, growth of the crossbreed was intermediate to the parent strains (Devlin et al. 2001). However, the transgenic wild X domestic crossbreed was the largest genotype, 18 times larger than the non-transgenic wild parent, 13 times larger than the non-transgenic wild X domestic crossbreed, 9 times larger than the non-transgenic domestic parent and more than 2.5 times larger than the wild F77-transgenic parent (Devlin et al. 2001). The combined effects of transgenesis and crossbreeding had a much greater growth enhancement than crossbreeding or transgenesis alone (Dunham 2011). A transgenic rainbow with 50% of its heritage from a domestic genome was much larger than a transgenic with a wild genome. Strain effects, in general, epistasis, and genetic background may be more significant in regard to affecting transgene response, rather than the domestic or wild nature of the fish (Dunham 2011).

An alternative explanation for the hyper growth response of GH-transgenic salmonids is that growth of non-transgenic salmonids is normally relatively slow prior to sexual maturity, and is extremely low when water temperatures are low and food resources in nature are scarce (Dunham 2011; Leggatt et al. 2017b), and transgenic individuals are less affected by these factors. The other amazingly dramatic example of growth enhancement is the 30× size attained by GH-transgenic mud loach, *Misgurnus mizolepis*, above that of the non-transgenic biological maximum (Nam et al. 2001). A pattern appears to be emerging that GH transgenesis has the most profound effects on slow-growing species. However, the very small model species, medaka and zebrafish, only grew 75% (Howard et al. 2004) and three times (Silva et al. 2015) faster than controls, respectively, when GH constructs were introduced; thus, the hypothesis that GH transgenesis is more effective in slow-growing species, strains, and lines is not universal.

Salmon may represent a unique case with their life-history and physiology making them especially amenable for growth enhancement via GH transgenesis. Genetic advantages could lead to further magnification of differences due to environmental advantages (Moav and Wohlfarth 1974). Even prior to first feeding, transgenic progeny were 21.2% heavier and 11.9% longer than their non-transgenic full-siblings, suggesting that the expression of GH in early development affected the rate and/or efficiency of conversion of yolk energy reserves (Devlin et al. 1995a, b). GH expression increased by 40-fold in cold temperatures, when GH expression is normally low (Mori and Devlin 1999). Parr-smolt transformation occurred 6 months early in the transgenic fish compared to the control fish. This becomes another advantage that can be further magnified genetically and environmentally as smolts are naturally in a faster-growing life stage.

## Other Growth Genes

Overexpressing a growth hormone gene is not the only strategy to increase growth through transgenesis in fish. Jiang et al. (2017) introduced the grass carp follistatin gene into blunt nose bream (*Megalobrama amblycephala*). $F_2$ fish exhibited double muscling, increased size, body depth, and body width. The follistatin expression resulted in hypertrophic muscle growth.

## *Disease Resistance*

Disease resistance is one of the most important aquaculture traits. Genetic gain is possible through traditional selective breeding, but the rate of genetic improvement and the likelihood of attaining genetic improvement for disease resistance will probably be better via transgenesis. Several successful examples of significant disease improvement using genetic engineering have been accomplished.

Bacterial disease resistance may be improved up to fourfold through gene transfer of antibacterial peptide genes. Cytomegalovirus (CMV)-cecropin-transgenic channel catfish had higher survival rate (100%) than non-transgenic channel catfish (27.3%) during an epizootic of *Flavobacterium columnare* in an earthen pond (Dunham et al. 2002d). Transfer of cecropin genes to Japanese rice fish (*Oryzias latipes*) resulted in an increased resistance to *Pseudomonas fluorescens* and *Vibrio anguillarum*, which killed about 40% of the control fish in both cases (Sarmasik et al. 2002), while only 0–10% of the $F_2$-cecropin-transgenic medaka were killed by *P. fluorescens* and about 10–30% killed by *V. anguillarum*. Cecropin-transgenic rainbow trout exhibited not only increased bacterial, but also increased viral disease resistance (Chiou et al. 2014).

Grass carp transfected with a carp $\beta$-actin/human lactoferrin gene resulted in $P_1$ individuals that were more resistant to *Aeromonas hybrophila* and showed enhanced phagocytosis and more viral resistance than controls (Mao et al. 2004). $F_2$-transgenic zebrafish containing the Japanese flounder keratin promoter linked to the hen egg white (HEW) lysozyme gene exhibited 1.75× higher lytic activity from liver protein extracts than that in the wild-type zebrafish (Yazawa et al. 2006). This translated to increased disease resistance as 65% of the $F_2$-transgenic fish survived an infection of *Flavobacterium columnare* and 60% survived an infection of *Edwardsiella tarda* (likely *Pfiesteria pisicida*), whereas 100% of the control fish were killed by both pathogens. Gao et al. (2012) confirmed the lytic activity of tilapia C-3 lysozyme against Gram-positive bacteria *Streptococcus agalactiae* along with other Gram-positive bacteria and Gram-negative bacteria by comparing the activities of recombinant lysozymes in a bacterial challenge test. The same research team reported that expression of Hsp70 in the liver of Nile tilapia was induced after *S. agalactiae* infection and Hsp70 could drive expression of GFP in zebrafish (Zhang et al. 2014). They further fused tilapia Hsp70 promoter with tilapia lysozyme-C3 and a reporter gene, GFP, using the goldfish Tgf2 transposon system to produce transgenic zebrafish (Sun et al. 2017). The turbidimetric assay of extracted protein from liver of transgenic zebrafish were 1.6 times higher than that of wild-type zebrafish, indicating that tilapia C-type lysozyme 3 gives transgenic zebrafish more resistance to *S. agalactiae* infection than wild-type zebrafish.

## Cold Tolerance

Antifreeze proteins (AFPs) were first found in Arctic (Scholander et al. 1957) and Antarctic (DeVries and Wohlschlag 1969) fishes. These discoveries were key to understanding how these species survive in water colder than the freezing point of their blood, which gave early fish genetic engineers ideas on how to produce transgenic fish that could be farmed under Arctic conditions. Early transgenic research in this area involved the transfer of the antifreeze protein gene of the winter flounder (Fletcher et al. 1988; Shears et al. 1991; Hobbs and Fletcher 2008) into Atlantic salmon, but expression levels obtained have been inadequate for increasing the cold tolerance of salmon. However, preliminary results with goldfish showed some promise for increasing survival within the normal cold temperature range. Goldfish transgenic for ocean pout (*Macrozoarces americanus*) type III antifreeze protein (AFP) gene had significantly higher survival at the lower end of their normal temperature tolerance than controls (Wang et al. 1995). Similar to results observed with the goldfish, ovarian and testicular tissues of $F_3$ generation mice transgenic for ocean pout type III antifreeze protein gene driven by chicken $\beta$-actin promoter maintained normal morphology at 4 °C as compared to non-transgenic control tissues (Bagis et al. 2006). These studies indicate that the antifreeze protein has functional roles slightly above freezing point temperature.

## Body Composition

Transgenic technology can be used to improve body composition and nutrient content and produce bioactive substances in fish. Zebrafish transfected with $\beta$-actin salmon desaturase genes had enhanced levels of omega-3 fatty acids, docosahexaenoic acid (DHA), and eicosapentaenoic acid (EPA) in their flesh (Alimuddin et al. 2007). Cheng et al. (2014) inserted a masu salmon (*Oncorhynchus masou*) $\Delta$5-desaturase gene driven by the $\beta$-actin gene of carp into common carp, which increased unsaturated fatty acids. However, genotype–environment interactions occurred, and feed needed to contain sufficient precursors to allow the transgenic common carp to produce elevated levels of *n*-3 fatty acids. Pang et al. (2014) performed a codon optimization of the $\Delta$12-desaturase gene (fat1) and the $\Delta$15-desaturase gene (fat2) from the nematode, and then produced a transgenic fat1 and fat2 zebrafish. These fish had a doubling to tripling of *n*-3 fatty acids compared to controls. Kabeya et al. (2016) produced transgenic Nibe croaker (*Nibea mitsukurii*), a marine species, containing elongation of very long-chain fatty acids protein 2 (Elovl2) gene isolated from masu salmon, and the transgenic fry were able to produce increased omega-3 fatty acids. Thus, de novo synthesis of long-chain polyunsaturated fatty acids in fish has been achieved using transgenesis.

## Flavor

Flavor and texture are high priority traits in all, but especially Chinese cultures. Transgenic technology has greater potential to genetically alter these traits than traditional selective breeding. Although large quantities of carp are consumed, the Chinese would like to have the flavor and umami taste of these fish enhanced and even prefer the flavor of slow-growing goldfish to that of traditional carp. Yan et al. (2011) successfully microinjected total DNA extracted from Chinese shrimp (*Fenneropenaeus chinensis*) to the zygote of common carp with the goal of enhancing flavor. Analysis of AFLP markers indicated that prawn DNA was integrated into the carp and transmitted to the $F_2$ generation. Muscle-nutrient analysis of the transgenic carp showed that the protein content and total amino acids were higher in the transgenics than the control group, including the four kinds of umami taste from aspartic acid, glutamic acid, glycine, and alanine.

## Transgenic Fish as Bioreactors

Transgenic mammals have been used as biological factories to produce biomedical proteins such as clotting factors. Such technology is especially important in the current millennium since human extracted products have the potential to be contaminated with HIV, hepatitis viruses, Ebola, prions, and other human pathogens (Dunham 2019). Transgenically produced biomedical compounds should be free of human pathogens, be less expensive, and more widely available than those extracted from people. Several examples are now available demonstrating the potential of fish as bioreactors for medical products as well as compounds that can be used in fish spawning (Dunham 2011). However, quality control and regulation have as yet prevented commercialization of this technology.

CMV-human coagulation factor VII was microinjected into fertilized eggs of zebrafish, African walking catfish (*Clarias gariepinus*), and Nile tilapia (Hwang et al. 2004). Clotting activity was detected, indicating proper post-translational modifications. Proteins could be collected in eggs, serum, or possibly different proteins in different tissues for other types of genes, which demonstrated the possiblity of application of trangenic fish as bioractors. Transgenic Nile tilapia secreted human insulin in Brockmann Bodies (Pohajdak et al. 2004). Goldfish follicle-stimulating hormone (FSH) and luteinizing hormone (LH) gene were respectively constructed, driven by medaka $\beta$-actin promoter/enhancer, and microinjected into rainbow trout eggs (Morita et al. 2003). At 4 days, goldfish LH and FSH were isolated from the transgenic rainbow trout embryos. In vitro bioassay showed that single chain gold fish (scgf) FSH and scgfLH were expressed in rainbow trout embryos, and significantly elevated testosterone levels about three times compared to negative controls in testis. Hu et al. (2011) used the oocyte-specific promoter zp3 to initiate expression of tilapia-like, insulin-like growth factors (IGFs [plasmid: ZP: tIGFs: hrGFP]) in fertilized zebrafish eggs to produce recombinant proteins.

## Pleiotropy

### *Growth Hormone Gene*

Pleiotropic effects are common and numerous in transgenic fish. This is not surprising as insertion of transgenes usually affect the expression of a multitude of other genes. The pleiotropic effects can be positive, negative, or neutral.

Lo et al. (2014) observed 478 differentially expressed genes in cecropin-transgenic rainbow trout. Roberts et al. (2004) found that insertion of GH transgenes in salmon altered expression of some, but not all, myostatin-related genes. Hepatic gene expression was also altered in transgenic coho salmon and transgenic amago salmon (*Oncorhynchus masou*; Rise et al. 2006; Mori et al. 2007). Gene expressions of appetite-regulating, gastric-regulating, muscle immune function genes were altered in GH-transgenic coho salmon (Kim et al. 2015, 2018; Alzaid et al. 2018), common carp (Zhong et al. 2013), and zebrafish (Dalmolin et al. 2015). However, the alteration in gene expression for GH-transgenics varied from one species to another.

The transfer of GH genes has pleiotropic effects on body composition, body shape, feed-conversion efficiency, disease resistance, reproduction, tolerance of low oxygen, carcass yield, swimming ability, and predator avoidance. How the transgene affects overall phenotype and performance dictates whether or not a transgenic genotype has commercial potential as well as its fitness and potential impact on the environment. Delayed reproductive development was reported in GH-transgenic common carp (Chen et al. 2018). Silva et al. (2015) found that GH-transgenic zebrafish had greatly impaired reproduction that was corrected by making a double transgenic with both GH and GH receptor genes.

Improved feed-conversion efficiency is usually a component of fast-growing, transgenic GH fish including common carp, channel catfish, Nile tilapia, and salmon (Chatakondi 1995; Stevens and Devlin 2000a; Rahman et al. 2001; Dunham and Liu 2002). In most of these cases, the feed conversion was improved by approximately 20%. Transgenic tilapia expressing the hCMV-tiGH cDNA had an amazing feed-conversion efficiency that was 290% better for the transgenic tilapia (Martínez et al. 2000), which was similar to results obtained with mud loach (Nam et al. 2001).

The surface area of the intestine of GH-transgenic coho salmon, *Oncorhynchus kisutch*, was 2.2 times that of control salmon and the growth rate was about twice that of controls (Stevens and Devlin 2000a). The relative intestinal length was the same in transgenic and control salmon, but the surface area was greater for transgenics as a result of an increased number of folds. These differences could be related to the level of food consumption or GH may have a direct effect on intestinal growth (Stevens and Devlin 2000a). This phenomenon occurred in both GH Atlantic and GH coho salmon. This change in intestinal surface area could be a partial explanation for the increased feed-conversion efficiency of transgenic salmon. Regarding a related efficiency issue, GH-transgenic salmon grew better on high plant protein diets (Ganga et al. 2015) and high carbohydrate diets (Higgs et al. 2009; Leggatt et al. 2009) than controls, which could result in more profitable and more environmentally friendly aquaculture.

Integration of the rtGH gene alters the survival of common carp (Chatakondi 1995). Transgenic individuals had higher survival than controls when subjected to a series of stressors and pathogens, such as low oxygen, anchor worms, *Lernaea*, *Aeromonas*, and dropsy. GH-transgenic common carp had higher lysozyme activity in the serum compared to age-matched, non-transgenic control fish (Wang et al. 2006). The serum bactericidal activity in the transgenics was 20% higher than in the controls. Values for leukocrit and phagocytic percent of macrophages in head kidney were higher in transgenics than controls, but the phagocytic indices and relative spleen weights in the transgenics and the controls were not different. GH transgene expression apparently not only stimulated growth, but also the non-specific immune functions of common carp.

GH gene transfer does not always confer increased disease resistance as GH-transgenic salmon were more sensitive to *Vibrio* compared to controls (Jhingan et al. 2003). Again, family effects were important as survival among GH salmon families sometimes improved, sometimes decreased, and sometimes remained unchanged relative to controls. Alzaid et al. (2018) found that mimicked viral infection suppressed muscle immune response in GH salmon.

Pleiotropic effects of GH gene insertion on oxygen tolerance characteristics vary from one species to another (Dunham 2011). GH tilapia have a 58% higher metabolism than controls, compensate for oxygen consumption, and have the same maximum swim speed as non-transgenics. GH tilapia tolerate hypoxia equally well as controls despite higher demand for oxygen. GH-transgenic salmon have an increased need for dissolved oxygen; however, after 4 days of starvation, GH individuals had the same oxygen uptake as controls. After feeding, GH-transgenics had 40–70% increased $O_2$ demand even when controls consumed equivalent amounts of feed. Adult GH-transgenic salmon had higher oxygen demand, poorer swimming ability, and longer recovery time compared to ocean ranched salmon (Lee et al. 2003; Dunham 2011; Leggatt et al. 2017a).

The definition of survival and how survival traits are measured can alter the outcome and conclusions. When exposed to low dissolved oxygen, survival was the same for GH-transgenic and control common carp. However, when mean survival time was calculated for all fish, dead or alive, the transgenic individuals had longer mean survival time than the non-transgenic full-siblings (Dunham et al. 2002a). Family effects were important as transgenic common carp in some families had higher and longer survival than control common carp when subjected to low oxygen, but in some cases control full-siblings were more tolerant. Transgenic channel catfish with the same rtGH construct as the common carp have a lower ventilation rate when subjected to low dissolved oxygen, compared with controls.

GH transgenesis has a dramatic effect on body composition in mammals (Ebert et al. 1988), with a drastic reduction in fat deposition in transgenic mice (Pomp et al. 1992; Knapp et al. 1994), pigs (Ebert et al. 1988; Pursel et al. 1990; Wieghart et al. 1990), and lamb (Nancarrow et al. 1991). Transgenic mammals possessing recombinant GH genes also show elevated levels of protein. GH-transgenic fish also exhibit similar changes in the fat-to-protein ratios in the muscle, but they are not as dramatic as those observed in mammals. $F_1$ and $F_2$ rtGH-transgenic common carp had more protein, less fat, and less moisture than non-transgenic full-siblings (about a 10% change; Dunham et al. 2002b). Transgenic channel catfish with the same rtGH cDNA also had more protein, less fat, and less moisture in their edible muscle than non-transgenic full-siblings (about a 10% change). Amino acid ratios can be altered. For example, hCMV-tiGH cDNA-transgenic tilapia, *Oreochromis urolepis hornorum*, had lower levels of cholesterol, free alanine, and aspartic acid in the muscle compared with controls (Martínez et al. 2000).

GH transgenesis also affects muscle-cell characteristics and activity. GH-transgenic channel catfish had increased numbers of mitochondria in the cell, increased numbers of glycogen globules, and increased numbers of muscle fibers, but a reduced number of fat globules (Dunham 2011). Muscle-fiber size was unchanged. Perhaps due to these changes in amino acid levels and ratio, changes in fat and ultrastructure of the muscle, the flavor and texture of transgenic catfish flesh were slightly better than those of non-transgenic controls. Heterozygous GH-transgenic coho salmon also had higher numbers of small-diameter fibers in somite muscles (Hill et al. 2000) or a doubling of muscle-fiber recruitment (Johnston et al. 2014).

Morphological changes from GH gene transfer is common in transgenic fish. GH-transgenic common carp were slightly more truncated than full-sibling controls (Chatakondi et al. 1994; Chatakondi 1995; Dunham et al. 2002b). Similar changes are seen in GH-transgenic Nile tilapia, as the head:total length ratio was higher in transgenic fish relative to controls (Rahman et al. 2001). However, GH-transgenic rainbow trout derived from a wild strain had a slender body shape similar to that of controls (Devlin et al. 2001). Family or strain effects had a role in pleiotropy for body shape as domestic transgenic rainbow trout derived from a deep-bodied strain had an even deeper body depth than the controls caused by either increased muscle or tremendous visceral fat deposits or both.

Some families in the $P_1$ generation of transgenic Pacific salmon containing chinook salmon GH gene had excessive levels of GH resulting in morphological abnormalities in head, fin, jaw, and operculum as a result of excessive cartilage and bone growth of the fastest-growing transgenic fish (Devlin et al. 1995a). Insertion of an *Oncorhynchus* metallothionein GH1 plasmid (pOnMTGH1) gene construct into coho salmon altered centroid size, and the dorsal caudal peduncle and abdominal regions were also distinctly enhanced in transgenic fish when compared with controls (Ostenfeld et al. 1998). Morphological changes in both whole body and syncranium were prominent. The endocrine stimulation had been elevated to pathological levels in these GH-transgenic salmon, and excessive, deleterious deposition of cartilage was observed (Devlin et al. 1995a, b) analogous to the mammalian acromegaly syndrome. This effect can be sufficiently severe for impaired feeding and respiration to result in reduced growth and poor viability. Consequently, salmon that ultimately display the greatest growth enhancement as adults are those that have been only moderately stimulated (Devlin et al. 1995a, b).

Despite their minimal growth enhancement, domestic transgenic rainbow trout exhibited cranial deformities (Devlin et al. 2001), and Devlin et al. (2001) suggested that this is evidence that transgenesis affects growth pathways outside the range supported by the homoeostatic processes that maintain normal morphology and viability. This hypothesis is supported by the data of Maclean et al. (1987), as domestic rainbow trout receiving exogenous GH not only showed modest increases in growth (9%), but also had cranial abnormalities and silver body coloration, whereas controls did not have these characteristics. However, no abnormalities were observed in rapidly growing, GH-transgenic Nile tilapia, although minor changes to skull shape were observed in some fish (Rahman et al. 1998).

The altered body shape of rtGHcDNA-transgenic common carp resulted in improved dressout percentage in the $F_2$ generation, and a similar result was obtained for transgenic channel catfish containing the same GH construct. Dressout percentage was higher in almost all transgenic families. pCAgGH Yellow River common carp had higher dressout percentage than controls (Zhu 1992) because the weight of the gonads was reduced due to the inhibitory effect from the transgene, and the edible flesh of the fish was correspondingly increased.

GH gene transgenesis can also affect gill morphology in salmonids. Transgenic Atlantic salmon (Stevens and Sutterlin 1999) and Pacific salmon (Stevens and Devlin 2000b) had altered gill morphology compared to controls, but the difference was expressed in different ways in the two species. Pacific transgenic salmon had gill filaments similar to those of controls in length but had smaller lamellar spacing. Atlantic transgenics had longer gill filaments than controls, but with similar lamellar spacing to controls. Pleiotropic effects can be dissimilar for even closely related species.

Phenotypic variation is a key and underappreciated aquaculture trait with both beneficial and harmful effects dependent upon the situation. GH-transgenic channel catfish had more uniform growth than controls (Dunham et al. 1992a). When the mean body weight of an $F_1$-transgenic GH common carp family was greater than that of the control full-siblings, the coefficient of variation for body weight was smaller for transgenic fish than for non-transgenic fish (Zhang et al. 1990).

Growth hormone gene insertion can affect tolerance of various physical parameters. GH-transgenic channel catfish can survive water temperatures of −0.5 °C dependent upon the salinity, which was lethal to non-transgenic full-siblings (Abass et al. 2016). Additionally, these transgenic catfish are more tolerant of high salinity as fry (Youssef 2017), which is not surprising since GH has a role in osmoregulation (Tang et al. 1993). Similarly, GH-transgenic coho salmon also had slightly expanded temperature tolerance as Arrhenius breakpoint temperature was higher than that for controls (Chen et al. 2015).

## *Cecropin*

A vast amount of information has been generated on pleiotropic effects of GH gene, but little for other transgenes. Cecropin-transgenic channel catfish and cecropin-transgenic rainbow trout, although, theoretically, healthier, had identical growth rate compared to controls.

## *GFP*

Reporter genes are commonly used in some exploratory transgenic research as markers coupled with the transgene of interest. However, results obtained from such experiments may have limited applicability to predict aquaculture performance or fitness in natural environments as expression of green fluorescent protein reduces cardiac function and aerobic performance in transgenic zebrafish (Avey et al. 2018).

## Food Safety

Food safety of transgenic aquatic organisms is beyond the scope of this chapter, but is a critical issue to address for transgenic fish to be commercialized. WHO/FAO, the U.S. National Academy of Sciences, the Royal Society of London, and the European Food Safety Authority have all reviewed the science of transgenic technology and concluded that in the vast majority of cases, except for potential allergenicity and perhaps specific cases, transgenic meat should be a safe product (FAO/WHO 2004).

## Environmental Risk and Evaluation

Ecological impacts are a significant concern of the public regarding transgenic aquatic organisms. The ecological impact evaluation of genetically modified fish includes two main aspects, perturbances to ecological balance through altering the food web and habitat destruction, and population genetic changes by matings with wild conspecifics, potentially altering biodiversity and genetic biodiversity. Intentional or accidental stocking of transgenic fish has not occurred in the natural environment, and, of course, such an event is being tried to be prevented. Potential risk is being evaluated with various confined, laboratory-scale experiments.

The published scientific literature generally considers the escape and establishment of transgenic aquatic organisms as a negative event. Establishment of a transgenic fish could have negative, neutral, or positive impacts on ecosystems. Societal values as well as science dictate what is desirable and what is undesirable. Both science and society define what is a natural environment and a natural ecosystem. For example, the State of Alabama has more miles of stream per land area than any other state in the USA except for Tennessee. However, dams have been erected that impact all but 11 miles of Alabama rivers and many miles of river are now reservoirs. All over the world, habitat has been altered and destroyed. These altered environments are now considered "natural." Prior to erection of the dams, ictalurid catfish were 60% of fish biomass in Alabama. Now centrarchids are the dominant species with catfish constituting only 11% of the fish biomass. One could argue that establishment of a genetically altered catfish resulting in an increase in catfish biomass would also result in a more "natural state" based on the population distributions 70–80 years ago.

This is likely an unpopular line of thought. Even if it were correct, thorough and careful scientific evaluation of environmental risk and potential impact on ecosystems should be, and is being, conducted prior to use of transgenic fish in potentially inappropriate environments.

The key factor determining the environmental risk of a transgenic fish is fitness. If the fitness of a transgenic fish is less than that of a wild conspecific, the transgenic genotype should be selected against in the natural environment. The key components of fitness are reproduction, predator avoidance, foraging ability, and swimming ability, which affects the first three traits.

Fitness is not easy to measure in a small research environment or mesocosm trying to mimic a river, lake, or ocean. Other difficulties include identifying whether or not effects are caused by age, size, culture, or transgene. Also, results and conclusions can vary depending upon the length of the experiment.

## *Transgenic Exotic*

Exotic species are the most likely type of fish to cause disturbance to ecosystems once they are established (Dunham 2011). Introducing transgenes that can alter the geographic range of a species would be of high environmental risk. The escaped transgenics could enter an area where there were no wild conspecifics to compete against since it would have a larger geographic range than the wild conspecific. A hypothetical example would have been the successful decrease in lower lethal temperature in salmon by introducing antifreeze constructs. At the time this would have allowed expansion of salmon aquaculture, but at great environmental risk.

## *Domestication and Incumbent Wild Types*

In general, but not always, it is difficult for even stocked wild conspecifics to have a genetic impact when moved from one watershed to another based on data generated on freshwater sportfish in the Southeastern USA (Norgren et al. 1986; Dunham et al. 1992b, 2002c; Dunham 2011). In most of these examples, strains from other watersheds were stocked repeatedly in a body of water with the purpose of changing allele frequencies, but this often failed. The resident genetic type appears to have some reproductive advantage, which may or may not persist with climate change, and should also translate to a transgenic type also having difficulty in becoming established. Factors dictating whether or not the allele frequencies could be altered were the number of stockings, the stocking densities, the years of stocking, and water quality parameters.

In general, domesticated fish appear to have reduced fitness compared to wild conspecifics (although this is very controversial in the case of salmon) and, since most transgenic fish would be produced from domestic founders, this should add to the reduced fitness of transgenic fish, further decreasing their risk. In the case of channel catfish, DNA analysis indicated no detectable genetic impact of domestic populations on wild populations after three decades of likely escapements (Simmons et al. 2006).

Weir and Grant (2005) examined multiple studies on the mixing of domestic and wild Atlantic salmon. Seven studies indicated the fitness of the wild fish was higher in regard to survival and reproduction, 13 documented significant phenotypic differences, and 10 found distinct genetic differences between the wild and domestic populations, but the authors indicated that no conclusion could be made on population impact, so caution should be practiced in regard to the domestic escapes.

Erkinaro et al. (2010) found repeat domestic spawners of Atlantic salmon in the River Teno, but introgression determination was complicated as the microsatellite markers indicated that escapees came from a multitude of locations. Their review of the literature indicated that domestic Atlantic salmon have inferior reproduction in the wild, but can mate with wild-type Atlantic salmon. However, Karlsson et al. (2016) and Glover et al. (2017) found widespread introgression of domestic Atlantic salmon into wild populations in Norway with the lowest levels found in national protected areas. In Newfoundland, Wringe et al. (2018) found extensive crossing of domestic and wild Atlantic salmon. Over time the percentage of $F_1$ and feral individuals decreased, but introgression persisted as domestic alleles in backcross individuals in both directions.

Data to date indicate lower fitness of domestic fish. However, under the right conditions such as massive or repeat escapes or stockings, domestic alleles can be established at least in the short term (Glover et al. 2017).

## *Fitness of Transgenic Fish*

### Reproduction

Female GH-transgenic Nile tilapia had a lower gonadosomatic index than non-transgenic siblings in both mixed and separate culture conditions (Rahman et al. 1998, 2001). The transgenic male gonadosomatic index was higher in mixed culture and lower in separate culture than that of their non-transgenic siblings. However, GH-transgenic male tilapia had reduced sperm production (Rahman et al. 1998).

Fecundity was not affected by insertion of rtGH in common carp, and precocious sexual development was not observed in these carp (Dunham 2011). Similarly, transgenic channel catfish harboring rainbow trout or coho growth hormone gene had normal reproductive ability under artificial conditions (Dunham et al. 1992a).

Liu et al. (2011) studied the effects of starvation on the growth and gonadal development of *Oncorhynchus keta* growth hormone-transgenic common carp. When fed sufficiently or short fed, the salmon growth hormone-transgenic common carp had a faster growth rate than the control and a slightly better gonad development than the control. Guan and Liang (2013) compared the microstructure and gonadal development of testis between *O. keta* growth hormone-transgenic common carp and non-transgenic common carp. The structures of the testis in transgenic and non-transgenic males were similar, developmental degree of testis had no significant difference, and both genetic groups were able to reach sexual maturity. However, in another study, fast-growing, GH-transgenic common carp exhibited delayed sexual maturation and decreased gonad size (Cao et al. 2014). In this case, overexpression of GH depressed reproduction by directly inhibiting luteinizing hormone (LH) production and release through GH receptors in the pituitary gonadotrophs. Chen et al. (2018) conducted follow-up gene expression analysis with these fish and further found that pituitary gonadotropin inhibitory hormone (*gnih*), dopamine receptor D1 (*drd1*), dopamine receptor D3 (*drd3*), and dopamine receptor D4 (*drd4*) had increased expression in the fast-growing, GH common carp, and this expression profile was associated with the retarded reproductive development. Additional alterations in neuroendocrine factor gene expression and reduced hepatic leptin signaling to the pituitary were likely part of the response cascade to overexpression of GH, resulting in delayed sexual maturation.

Several examples of reduced reproduction in GH-transgenic fish exist. Bessey et al. (2004) and Fitzpatrick et al. (2011) observed reduced courtship and spawning in GH-transgenic coho salmon, and wild salmon outcompeted them reproductively in a semi-natural environment. GH-transgenic salmon put more energy into somatic growth and had reduced gonad size (Bessey et al. 2004) as did GH Nile tilapia (Rahman et al. 1998). Male GH-transgenic salmon displayed reduced nest loyalty, quivering frequency, and spawning participation (Moreau et al. 2011).

**Predator Avoidance and Foraging Ability**

Fitness of $F_2$- and $F_3$-transgenic, RSVLTR-rtGH1, and RSVLTR-csDNA (salmonid growth hormone genes) channel catfish and non-transgenic channel catfish fingerlings (BW: 3.45~4.31 g) and fry (BW: 0.37~0.85 g) were evaluated under natural conditions without supplemental feeding in confined earthen ponds (Dunham et al. 1999). Transgenic fish were more vulnerable to predators, largemouth bass *Micropterus salmoides* and green sunfish *Lepomis cyanellus*, than non-transgenic channel catfish (Dunham et al. 1999). An alternative explanation could be starvation due to higher metabolism of the transgenic fry. GH-transgenic coho salmon lost weight faster than non-transgenic coho salmon when starved (Abernathy et al. 2015). When supplemental feed was applied, the same groups of transgenic channel catfish grew 33–50% faster than non-transgenic individuals; while with natural food, no difference in growth was found between the two genetic groups (Dunham et al. 1999).

Marnis et al. (2016) conducted similar experiments with GH-transgenic African walking catfish fry, which had the potential to grow 19% faster than non-transgenic controls, but evaluated in aquaria. They conducted a unique experiment that addressed foraging ability and predator avoidance and behavior in a single experiment in which older, larger GH-transgenic and control African walking catfish were used as the predators. The transgenic fry were ten times more vulnerable to being cannibalized than the non-transgenic controls. All of the cannibalism came from the non-transgenic predators as the GH-transgenic African walking catfish exhibited minimal or no cannibalism. Under the restricted feeding regime, there was no growth difference between the transgenic and control fry. GH transgenesis totally altered the normal cannibalistic behavior. The opposite phenomenon occurred for GH coho salmon as under low food conditions some transgenic individuals turned cannibalistic (Devlin et al. 2004).

GH salmon are more vulnerable to predators than controls. GH-transgenic Atlantic salmon do not show the appropriate fear response in the presence of predators (Abrahams and Sutterlin 1999), and GH coho salmon fry are more easily preyed upon than controls (Sundström et al. 2004).

Several papers indicate that GH-transgenic fish have greater foraging ability. In most of these studies, appetite and foraging ability are being confused as pellets were the food. Foraging is the ability to obtain natural food items as one would not find pellets in the natural environment, and this is essentially spoon feeding and requires no food searching behavior. Foraging ability can be different at different life stages and the type of food or prey presented. Zhu et al. (2017) conducted two experiments providing GH-transgenic common carp with gastropods as prey, one providing a single species of gastropod and second providing four species of gastropods. In the first experiment with a single species of gastropod, transgenic and control common carp had the same minimal consumption, and gastropod biomass was not affected as apparently this species was not a preferred food. However, this prey item was mostly shunned in the second experiment, and the other three gastropod populations were strongly affected with the foraging of the transgenic genotype almost three times that of the control.

## Models

Ecological and mathematical models have been explored to predict the risks of bioengineered fish. The ability of a transgene to establish and spread will depend on the fitness of the transgenic fish. Muir and Howard (1999, 2001, 2002) and Howard et al. (2004) used a GH transgenic medaka to model the potential effects of growth hormone gene transgenesis for environmental risk, and observed the "Trojan gene effect." Control females preferentially chose GH-transgenic males as mates because of their larger size. However, the GH-transgenic progeny had lower survival than controls, and the model indicates that this genetic load could, in some cases, lead to extinction of the population.

However, there are many flaws in the model. The fish were fed artificial feed, which would not occur in the natural environment; thus, foraging ability was not considered and genotype–environment interactions were not in the model. There was no predation or habitat present. Most species of fish do not preferentially choose large males for mates. Medaka have very low fecundity. It is doubtful that this experiment and model yielded realistic results.

Realistic models are important for identifying additional factors that impact behavior and genotype–environment interactions. When predators were present in the aquatic environment, female fish of various species decreased selectiveness with regard to male size (Forsgren 1992; Bierbach et al. 2011; Pennington and Kapuscinski 2011) and coloration (Godin and Dugatkin 1996). Furthermore, Atlantic molly (*Poecilia mexicana*) females (lab-reared) chose small males in the presence of the cichlid *Cichlasoma salvini*, a natural predator of *P. mexicana* in dichotomous choice tests. Wild-caught females did not respond to the same extent to the presence of a predator, and could alter their mate choice only when a natural predator was present, most likely due to a learned ability to evaluate their predators' motivation to prey (Bierbach et al. 2011). Genetic background (wild or domesticated) of fish impacted their mate choice, and wild fish have evolved visual predator recognition mechanisms; thus, risk evaluation should take this point into consideration.

A likely invasion case model using growth hormone-transgenic Atlantic salmon showed reduced breeding performance, such as fertilization success, compared to controls. However, the transgenic genotype had the capability to mate with their non-transgenic rivals, leading to gene flow to wild salmon populations (Moreau et al. 2011).

## Semi-Natural Risk Experiments

Semi-natural environments, where conditions are kept as close as possible to natural in a closed system, are under evaluation to estimate potential ecological risks of transgenic fish. Interaction among strengths of promoter and transgene, strain and environment should gain attention during determining risk assessments (Leggatt et al. 2012, 2017b). Performance of wild-type coho salmon (*Oncorhynchus kisutch*) fry was evaluated for a fast-growing, GH-transgenic strain containing a sockeye salmon metallothionein promoter (MT, OnMTGH1), and three coho salmon lines/strains containing a reportedly weaker sockeye salmon histone-3 promoter (H3, OnH3GH1, H3-A,B,C) in hatchery conditions and semi-natural stream tanks, and they had varied growth and survival demonstrating the importance of these interactions and genotype–environment interactions in environmental risk evaluation.

Growth hormone-transgenic rainbow trout having the same OnMTGH1 were assessed in naturalized stream mesocosms in the presence of predators and without predators for two life stages, first-feeding fry and 60-day post-first-feeding (Crossin et al. 2015). For the first experiment with first-feeding fry, they found that in the late summer, the transgenic rainbow trout had lower survival rate either in the presence of predator, potentially due to the additive effect of the transgene that negatively decreased their foraging ability and risk of being predated, or in the absence of predator, potentially due to the transgene requiring a great metabolic demand in a food-limited environment. In regard to the growth rates, transgenic rainbow trout had lower growth rates than their sibling control in these two environments.

Results in the second experiment were much different. When the rainbow trout were 60 days old and past the critical mortality bottleneck period (2–3 weeks after emergence) in winter, effects of transgene on wild and domestic X wild genetic background were evaluated. Two genetic backgrounds, transgenic with wild genetic and transgenic with wild X domestic genetic background, had similar survival rate to their control siblings that had the same genetic backgrounds. Transgenic types grew faster than non-transgenic control siblings in the predator and predator-free environments. Thus, risk results can be life-stage dependent and vary from one life stage to another. Once again, genotype–environment interactions were important in environmental risk evaluation of transgenic fish.

The largest-scale and most complex "mesocosm" experiment was conducted in China. Hu et al. (2007b) constructed a 6.7-ha artificial lake containing mollusks, shrimps, rye grass, and other fish species (12 families and 23 genera with majority carp for approximately 65.2% of the total species), mimicking the carp habitat of the Yangtze River in China to evaluate the ecological risk of "all-fish" transgenic carp. A follow-up study (Lian et al. 2013) showed that in a natural aquatic environment, "all-fish" transgenic common carp (carp $\beta$-actin gene driving grass carp GH gene) had identical mating competitiveness to wild common carp, large fish did not have advantage in fertility, and juvenile viability of transgenic common carp was low. Additionally, the swimming speed of the GH-transgenic common carp was slower (Li et al. 2009). The authors concluded that fitness of the "all-fish" growth-transgenic carp was lower than the control carp.

Behavior could affect the key fitness traits. GH-transgenic salmon were more active and had a higher average swimming speed in a simulated ocean mesocosm (Hollo et al. 2017). Genotype–environment interactions related to abnormal behavior under laboratory conditions are key for environmental risk assessment and can confound risk evaluation. GH-transgenic coho salmon exhibited similar predator avoidance behavior regardless of environment as wild coho salmon raised in natural settings (Sundström et al. 2016). However, when wild coho salmon were reared in hatchery conditions, they exhibited extreme adverse predator avoidance behavior.

Taken in total, the results of environmental risk evaluation are quite variable. Often similar transgenes had different effects on different species. Although there are exceptions for the results for each fitness trait, there are some trends. In general, transgenic fish are less fit for reproductive traits and foraging for natural food, which usually eliminates the growth advantage of most GH-transgenic types. Early fry survival of GH-transgenics is greatly reduced compared to controls, GH-transgenics are more vulnerable to predators, swimming speed is reduced, requirements for oxygen are increased, and major behavior changes occur. In combination, it appears that GH-transgenic fish are less fit than non-transgenic and wild fish, and transgenes would likely be selected against in the natural environment. However, at this time confinement should be our major goal for prevention of transgenes making their way to the environment, even though these genotypes would likely, but not certainly, be eliminated.

## Containment and Confinement

Risk-based frameworks or platforms need to be developed to evaluate individual transgenes and transgenic fish species under the actual (best simulation of) ecosystem on a case-by-case basis. Regardless of risk, several options exist for containment of transgenic fish. Simultaneous application of multiple containment strategies was recommended by the National Research Council (NRC 2004). Wong and Van Eenennaam (2008) reviewed and compared different containment strategies, and grouped them into different categories including physical, biological, and genetic containment. Physical containment and biological containment have various advantages and disadvantages. Physical containment and most other forms of containment cannot be 100% effective.

Genetic confinement is one strategy. One option for genetic containment is triploid induction, and that is one of the confinement aspects of the *AquAdvantage* salmon. However, these fish are 99% + triploid, so risk has not been totally eliminated. Triploid induction rates vary between 10% and 100%, depending on the species, shock conditions, and egg batch quality. Triploidy can decrease performance in fish (Dunham 2011), and is not feasible on a commercial scale in the case of catfish, tilapia, and many other species of fish. Triploid induction reduced performance of GH-transgenic fish, negating about half of the enhancement from the transgenesis in salmon and loach (Dunham 2011; Nam et al. 2004). AquaBounty reported that family variation exists for this phenomenon, and the problem can be solved with family selection (Xu et al. 2013). Triploidy also has the disadvantage that it requires fertile diploid brood stock; so again, the possibility of escape and risk can be reduced, but not eliminated. Repressible transgenic sterilization and gene editing have great promise as effective confinement strategies.

## Transgenic Sterilization

The ultimate means of preventing environmental or ecological impact of transgenic, domestic hybrid, or exotic fish are repressible transgenic sterilization or gene editing coupled with hormone therapy. In this case, escaped fish are incapable of breeding or their progeny are incapable of breeding, resulting in absolute reproductive confinement and the prevention of introgression of transgenes into wild populations and any potential associated impacts. Several potential transgenic approaches have been evaluated including antisense, shRNAi, and overexpression of cDNAs. Alternative systems have shown promise, but they require fertile individuals at some point in the process. In this case, long-term environmental risk cannot be eliminated. If repressible systems are used and all individuals in the population are homozygous for the sterilization construct at all life stages, environmental risk can only be short-term in the worst-case scenario.

One option is the disruption of translation of reproductive genes with antisense followed by hormone therapy when reproduction is needed. Uzbekova et al. (2000) used antisense Atlantic salmon sGnRH cDNA driven by the sGnRH Pab promoter in production of transgenic rainbow trout. They had positive results in their initial studies, but did not follow-up. Hu et al. (2007a) reported that a recombinant construct using carp $\beta$-actin driving antisense Atlantic salmon GnRH gene microinjected into fertilized common carp eggs resulted in 30% of the founders having no gonads. They also reported that the fertility could be restored by exogenous hormone administration.

Wu et al. (2010) also modified a similar system to control reproduction of transgenic zebrafish. The construct was driven by an ovary-specific and a testis-specific promoter. The transgene of interest was a suicide gene, consisting of a reductase and a photosensitizer, while the reductase gene was linked to a reporter gene. The novelty of this concept is that infertility is induced if transgenic fish expressing the reductase are treated with an effective amount of reductase-activated cytotoxic prodrug or if the transgenic fish expressing photosensitizer are treated with light irradiation. They reported 100% reliable infertility in zebrafish. However, to effectively produce a transgenic male with three copies of fusion transgene is unclear and doubtful, because the need for high copy number of the transgene prevents high transfection rate of the cell, and integration sites are random events. A piece of Simian virus 40 sequence, a polyomavirus, was used for design of the construct, which would be disadvantage when considering commercialization. The system is too complicated for practical use, and fertile transgenic fish are part of the system, thus, risk cannot be eliminated.

One form of transgenic sterilization, Sterile Feral (SF) technology, has shown promise for achieving transgenic reproductive confinement. Components of these constructs are fused so that a specific promoter is coupled to a repressible element that in turn drives expression of a blocker gene, antisense RNA, dsRNA, sense RNA, or ribozyme to an early key developmental gene (Dunham 2004; Thresher et al. 2005, 2009).

This strategy has two parts: one is to suppress the expression of critical gene related to embryonic development, gonad development, or sexual maturity via a knockdown construct in the absence of a repressible element. The second step is to reverse the sterility of transgenic fish by administrating exogenous compound to shut off expression of the blocker gene allowing rescue of the embryos and/or to produce brood stock. This knockdown strategy could be used to generate reversibly sterile transgenic fish. This is essentially a Tet-off or "Tet-off-like" system. Proof of principle of the sterile feral system has been demonstrated with the repression of the knockout function demonstrated in zebrafish, oysters, channel catfish, and common carp for disruption of embryonic development (Templeton 2005; Thresher et al. 2005; Chaimongkol 2009; Thresher et al. 2009). "Sterile feral technology" based on the Tet-off system was successfully evaluated in zebrafish and channel catfish (Thresher et al. 2009; Chaimongkol 2009). Mortality rates of integrated transgenic lines of channel catfish overexpressing a dorsolateral gene (BMP2) were less than 50% and were not significantly different from the control in the presence of 100 ppm of doxycycline in the hatching water. Without doxycycline, 95.6% of embryos died that carried BMP2, which was being up-regulated (Chaimongkol 2009).

The SF approach could be utilized to disrupt gamete production by preventing primordial germ cell (PGC) migration during embryogenesis. PGCs migrate far from the site of developing gonads to the genital ridge where they differentiate into gametes. A number of genetic markers are associated with and expressed in PGCs, such as *vasa*, *nanos*, *dead end*, *cxcr4b*, and *dazl* (Raz 2004). Tet-off systems for transgenic sterilization for knocking out PGCs have shown great promise as repressible systems. The primary problem with these systems is that the transgene contains small viral sequences. This does not pose any food safety or biological risk; however, public perception and negative advertising could prevent the marketing of such aquatic organisms.

Wong and Collodi (2013) used HSP70 promoter to initiate expression of stromal-derived factor 1a (SDF 1a) and control zebrafish fertility by controlling migration of primordial stem cells. Expression of the transgenic SDF 1a requires induction of fertilized eggs at 34.5 °C for 18 hours, thereby producing 100% sterile male zebrafish. Such a strategy has its potential risks. For commercialization, it will require large-scale, high-temperature incubation equipment. Many species cannot be incubated at this temperature. Fertile brood stock are required, so risk remains.

Zhang et al. (2015) used the GAL4/UAS transgenic technology system to inhibit endogenous dead end expression by transcription of the antisense dead end (primary germ stem cell marker). The advantages of this system are that fertile brood stock are effectively produced, and the sterile offspring are easily and simply produced by crossbreeding. However, the use of CMV sequences in plasmids as promoters limits the application of commercialization in this system. Additionally, the creation and maintenance of two fully fertile brood stock lines does not eliminate risk. Another potential problem with this system is that GAL4 is temperature sensitive (Fortier and Belote 2000).

"Tet-off-like" systems have been recently developed, which contain no viral sequences (Su 2012; Li 2016). Modified Tet-off approaches were applied with a number of different combinations of promoters, target genes, and repressible elements in channel catfish and common carp. Su et al. (2014, 2015a, 2015b) and Li et al. (2017, 2018) modified the Tet-off system to induce expression of shRNAi and overexpression of cDNA knockdown constructs for nanos and dead end. Chemicals such as copper sulfate and sodium chloride used at high dosage turned off the construct in developing embryos and restored the gonadal development to allow production of brood stock that can produce sterile progeny.

One potential drawback of transgenic sterilization is potential negative pleiotropic effects, as knocking out reproduction can affect other traits similar to what might be observed with triploidy. For example, knockout of some primordial germ cell genes resulted in channel catfish with no gonads, and these fish had a 25% reduction in growth rate and survival (Li et al. 2018). Gene editing of reproductive genes followed by hormone therapy to restore fertility (Qin et al. 2016) may be one option to overcome this problem.

## Transgenic Technology to Control Invasive Species

Invasive, exotic, or feral species/populations commonly damage biodiversity once established in their new habitat or geographic range. In Australia, 90% of the freshwater fish biomass is feral common carp (Zhang 2016). However, one option to reduce or eliminate these populations requires a willingness to intentionally stock transgenic conspecifics of the invaders into the environment.

Autocidal technologies have been proposed as a mechanism and strategy to reduce or eliminate invasive species populations. Eight autocidal approaches have been explored to control invasive species (Thresher et al. 2013). Autocidal refers to controlling or eradicating populations of noxious organisms (such as the screwworm) by reducing their capacity to produce viable or fertile offspring. This concept covers a wide array of strategies, including "lethal construct," "sex or stage-specific lethality/sterility," "inducible mortality," "Trojan gene," "mutual incompatibility," "engineered under dominance," and "Daughterless" (Thresher 2008; Thresher et al. 2009; Thresher et al. 2013). The main weaknesses in the strategy are the requirements for stocking large numbers of carriers and the high numbers of independently segregating copies in each carrier. Other pros and cons were also compared and discussed (Table 2 in Thresher et al. 2013).

"Trojan Y Chromosome," a non-transgenic genetic strategy, is the most promising technique that would likely have public acceptance (Teem et al. 2014; Teem and Gutierrez 2014). YY females are stocked to shift sex-ratio of a target population toward males. The advantages are: (1) it is relatively inexpensive to develop; (2) it is likely to be publicly acceptable as any organism having YY is acceptable; (3) it may not require legislative changes; and (4) as female carriers outperform wild-type females that are decreasing over time, relative stocking rates increase. However, the internal potential disadvantage within itself is limitation to use in only fish and amphibians (Thresher et al. 2013). In reality, sex determination of fish could be plastic, and two typical sex-determination mechanisms exist, that is, genetic-sex determination and temperature-sex determination. Other environmental factors, such light, pH, dissolved oxygen, and water pressure also influence the sex determination and differentiation (Yan et al. 2017). This strategy requires ten generations under optimal conditions and could take longer. This strategy remains to be tested and documented (Thresher et al. 2013). Additionally, not all genetic-sex determination is XY, and YY females are not viable in all species of fish (Dunham 2011).

Teem and Gutierrez (2014) proposed that combining the Trojan Y chromosome and daughterless carp (a transgenic approach) eradication strategies could be more effective than both strategies working independently, modeling a rapid decline of females in the population and a shorter time to extinction. Theoretically, these strategies, the preliminary data, and the modeling of the results appear promising. However, none of these approaches have been fully developed or field tested.

## Status of Commercialization

Eighteen years after the first transgenic fish was produced, this technology began to impact commercial aquaculture, not as a food organism, but as ornamental fish. In the spring of 2003, the first transgenic fish, green fluorescent protein medaka, TK-1, was sold in Taiwan (Wikipedia 2019). In December of 2003, the US Food and Drug Administration (FDA) approved the sale of red fluorescent protein zebrafish, GloFish, in the USA, but they continued to be banned in California. Even California decided there were no food safety or environmental risks and legalized GloFish in 2015. Now a multitude of transgenic aquarium species fluorescing with a wide array of colors from several source species are sold in the aquaria trade in the USA, Canada, and other countries. These fish are still banned in the European Union, but Dutch authorities found 1400 transgenic fluorescent fish that originated from several aquaria shops in 2006.

In 2013, 28 years after the first transgenic fish was produced, a transgenic food fish was approved by Canada making the first impact of transgenic fish food technology. In November of 2013, the Canadian government became the first to approve a transgenic food fish, growth hormone-transgenic Atlantic salmon, for commercial sale, however, with the stipulation that they should be triploid embryos for export only to countries with both containment and approval of transgenic flesh (DFO 2013). History was made when the flesh of these triploid salmon produced by AquaBounty was approved by the US Food and Drug Agency for consumption in the USA (US Food and Drug Administration 2015). This was the first known approval of the consumption of transgenic animal meat in the world, but approval was not granted for the culture of transgenic salmon in the USA. Then on November 19, 2015, Canada also approved the consumption of GH-transgenic salmon, *AquAdvantage Salmon* (Ledford 2015). On May 19, 2016, the Health Canada and the Canadian Food Inspection Agency approved the sale of *AquAdvantage Salmon* into the market. The first 4.5 tons were placed in the Canadian market on August 4, 2017, and were all quickly sold (Waltz 2017). This is the first government-approved consumption of any transgenic animal. Now, more than 18 tons have been sold in Canada.

After first approval in the USA, an import ban was triggered preventing importation and sale of transgenic salmon in any form because of a lack of labeling laws in the USA. Anticipating that this would be resolved, AquaBounty purchased an indoor recirculating aquaculture system (RAS) in Indiana, and this is where the first US grown genetically engineered fish were to be produced (IntraFish Media 2017). On December 20, 2018, the National Bioengineered Food Disclosure Standard was announced, which requires food manufacturers, importers, and certain retailers to label foods containing genetically modified or bioengineered ingredients (Federal Register 2018). This lifted the import ban and AquaBounty is now raising transgenic salmon in Indiana. First sale of GH-transgenic salmon in the USA is anticipated in the next few months. Production of GH-transgenic salmon has also now been approved in Canada (CBC 2019).

## Summary and Conclusions

Transgenic fish have been produced that have traits that can impact aquaculture in major ways allowing increased production, efficiency, profitability, and, if used properly, increased environmental friendliness of aquaculture. A concern is how these fish could impact the natural environment, ecosystems, biodiversity, and genetic biodiversity if these were to escape from an aquaculture facility. Evaluation of fitness traits indicates that transgenic genotypes are, in general, less fit than non-transgenic and wild fish when evaluated for fitness traits in simulated natural systems. However, in some cases, there is great variability for some fitness traits. It is likely that if transgenic fish were to escape into the natural environment, the transgene would be selected against, and there would be no long-term effect on ecosystems or biodiversity. However, it is better to be cautious, and provide strong confinement to prevent transgenic fish from entering the natural environment. Repressible transgenic sterilization and related genetic strategies are likely the best strategy to provide absolute biological confinement. In this case, the fish can only reproduce with the intervention of man. Any environmental impacts would be short term. There are no physical containment systems that can guarantee 100% confinement.

In the future, there may be strategies employing transgenic technology to control feral populations, which could actually have a positive impact on biodiversity by controlling these damaging organisms. Much of our world has been altered, and genetic biodiversity has been lost. As of today, it could be considered heresy, but there may be times and locations in the future where transgenesis might be used to create positive and increased biodiversity.

The generation of a multitude of species and varieties of GloFish has certainly added biodiversity by strict definition to the world. However, there is no ecological value or detriment from these fish. Their value is primarily to aquaria enthusiasts who enjoy the unusual, strange, and colorful. In actuality, from a pure ecological point of view, these are analogous to the large number of "varieties" of ornamental goldfish developed by selective breeding of mutants over hundreds and thousands of years, but are of no environmental consequence because of their extremely low fitness. One segment of society deems these as beautiful useful pets, and another segment of society deems them ugly, bizarre, and useless.

Application of transgenic fish in the future will likely be common in aquaculture and be of great benefit. Used properly, it should reduce pressure on the natural environment and natural resources, and in that way more efficiently use our food-producing footprint. The relief for natural ecosystems should impact biodiversity in a positive manner. More traditional genetic approaches have already started this process, as in the case of catfish for which production per surface area has increased 5× during the last 30 years, and transgenesis will assist in making food production increasingly efficient.

## References


Abass NY, Elwakil HE, Hemeida AA, Abdelsalam NR, Ye Z, Su B, Alsaqufi AS, Weng CC, Trudeau VL, Dunham RA (2016) Genotype-environment interactions for survival at low and sub-zero temperatures at varying salinity for channel catfish, hybrid catfish and transgenic channel catfish. Aquaculture 458:140–148

Abdelrahman H, ElHady M, Alcivar-Warren A et al (2017) Aquaculture genomics, genetics and breeding in the United States: current status, challenges, and priorities for future research. BMC Genom 18:191–214. https://doi.org/10.1186/s12864-017-3557-1

Abernathy J, Panserat S, Welker T, Plagne-Juan E, Sakhrani D, Higgs DA, Audouin F, Devlin RH, Overturf K (2015) Food shortage causes differential effects on body composition and tissue-specific gene expression in salmon modified for increased growth hormone production. Mar Biotechnol (NY) 17:753–767

Abrahams MV, Sutterlin A (1999) The foraging and antipredator behaviour of growth-enhanced transgenic Atlantic salmon. Anim Behav 58(5):933–942

Alimuddin YG, Kiron V, Satoh S, Takeuchi T (2007) Expression of masu salmon delta5-desaturase-like gene elevated EPA and DHA biosynthesis in zebrafish. Mar Biotechnol (NY) 9:92–100

Alzaid A, Kim JH, Devlin RH, Martin SAM, Macqueen DJ (2018) Growth hormone transgenesis in coho salmon disrupts muscle immune function impacting cross-talk with growth systems. J Exp Biol 221(Pt 13), pii: jeb173146

Avey SR, Ojehomon M, Dawson JF, Gillis TE (2018) How the expression of green fluorescent protein and human cardiac actin in the heart influences cardiac function and aerobic performance in zebrafish *Danio rerio*. J Fish Biol 92:177–189

Bagis H, Aktoprakligil D, Mercan HO, Yurdusev N, Turgut G, Sekmen S, Arat S, Cetin S (2006) Stable transmission and transcription of newfoundland ocean pout type III fish antifreeze protein (AFP) gene in transgenic mice and hypothermic storage of transgenic ovary and testis. Mol Reprod Dev 73:1404–1411

Beadle GW (1980) The ancestry of corn. Sci Am 242:112–119. Accessed online at https://www.jstor.org/stable/pdf/24966237.pdf

Bessey C, Devlin RH, Liley NR, Biagi CA (2004) Reproductive performance of growth-enhanced transgenic coho salmon. Trans Am Fish Soc 133:1205–1220

Bierbach D, Girndt A, Hamfler S, Klein M, Mücksch F, Penshorn M, Schwinn M, Zimmer C, Schlupp I, Streit B, Plath M (2011) Male fish use prior knowledge about rivals to adjust their mate choice. Biol Lett 7:349–351

Cao M, Chen J, Peng W, Wang Y, Liao L, Li Y, Trudeau VL, Zhu Z, Hu W (2014) Effects of growth hormone over-expression on reproduction in the common carp *Cyprinus carpio* L. Gen Comp Endocrinol 195:47–57

CBC (2019) GM salmon approved for commercial production in P.E.I, AquaBounty announces. Online resources accessed at https://www.cbc.ca/news/canada/prince-edward-island/pei-aquabounty-salmon-commercial-production-1.5080914

Chaimongkol A (2009) Disruption of embryonic development in common carp, *Cyprinus Carpio*, and channel catfish, *Ictalurus Punctatus*, via knock down of BMP2 gene for repressible transgenic sterilization. Ph.D. Dissertation, Auburn University, Auburn, Alabama, USA

Chatakondi NG (1995) Evaluation of transgenic common carp, *Cyprinus carpio*, containing rainbow trout growth hormone in ponds. Ph.D. Dissertation, Auburn University, Auburn, AL, USA

Chatakondi N, Ramboux AC, Nichols A, Hayat M, Duncan PL, Chen TT, Powers DA, Dunham RA (1994) The effect of rainbow trout growth hormone gene on the morphology, dressing percentage and condition factor in the common carp, *Cyprinus carpio*. Proceedings V World Congress of Genetics and Applied Livestock Production 17:481–484

Chen Z, Devlin RH, Farrell AP (2015) Upper thermal tolerance of wild-type, domesticated and growth hormone-transgenic coho salmon *Oncorhynchus kisutch*. J Fish Biol 87:763–773

Chen J, Cao M, Zhang A, Shi M, Tao B, Li Y, Wang Y, Zhu Z, Trudeau VL, Hu W (2018) Growth hormone overexpression disrupts reproductive status through actions on leptin. Front Endocrinol (Lausanne) 9:131

Cheng Q, Su B, Qin Z, Weng CC, Yin F, Zhou Y, Fobes M, Perera DA, Shang M, Soller F, Shi Z, Davis A, Dunham RA (2014) Interaction of diet and the masou salmon Δ5-desaturase transgene on Δ6-desaturase and stearoyl-CoA desaturase gene expression and N-3 fatty acid level in common carp (*Cyprinus carpio*). Transgenic Res 23:729–742

Chiou PP, Chen MJ, Lin CM, Khoo J, Larson J, Holt R, Leong JA, Thorgarrd G, Chen TT (2014) Production of homozygous transgenic rainbow trout with enhanced disease resistance. Mar Biotechnol (NY) 16:299–308

Cook JT, McNiven MA, Richardson GF, Sutterlin AM (2000) Growth rate, body composition and feed digestibility/conversion of growth-enhanced transgenic Atlantic salmon (*Salmo salar*). Aquaculture 188:15–32

Crossin GT, Sundström LF, Vandersteen WE, Devlin RH (2015) Early life-history consequences of growth-hormone transgenesis in rainbow trout reared in stream ecosystem mesocosms. PLoS One 10:e0120173

Dalmolin C, Almeida DV, Figueiredo MA, Marins LF (2015) Food intake and appetite control in a GH-transgenic zebrafish. Fish Physiol Biochem 41(5):1131–1141

Devlin RH (1997) Transgenic Salmonids. In: Houdebine LM (ed) Transgenic animals: generation and use. Harwood Academic Publishers, Amsterdam, pp 105–117

Devlin RH, Yesaki TY, Donaldson EM, Du SJ, Hew CL (1995a) Production of germline transgenic Pacific salmonids with dramatically increased growth performance. Can J Fish Aquat Sci 52:1376–1384

Devlin RH, Yesaki TY, Donaldson EM, Hew CL (1995b) Transmission and phenotypic effects of an antifreeze/GH gene construct in coho salmon (*Oncorhynchus kisutch*). Aquaculture 137:161–169

Devlin RH, Biagi CA, Yesaki TY, Smailus DE, Byatt JC (2001) Growth of domesticated transgenic fish. Nature 409:781–782

Devlin RH, D'Andrade M, Uh M, Biagi CA (2004) Population effects of growth hormone transgenic coho salmon depend on food availability and genotype by environment interactions. Proc Natl Acad Sci U S A 101(25):9303–9308

Devlin RH, Sakhrani D, Tymchuk WE, Rise ML, Goh B (2009) Domestication and growth hormone transgenesis cause similar changes in gene expression in coho salmon (*Oncorhynchus kisutch*). Proc Natl Acad Sci U S A 106:3047–3052

Devlin RH, Sakhrani D, White S, Overturf K (2013) Effects of domestication and growth hormone transgenesis on mRNA profiles in rainbow trout (*Oncorhynchus mykiss*). J Anim Sci 91(11):5247–5258

DeVries AL, Wohlschlag DE (1969) Freezing resistance in some Antarctic fishes. Science 163:1073–1075

DFO (2013) Summary of the environmental and indirect human health risk assessment of AquAdvantage® Salmon. DFO Can Sci Advis Sec Sci Resp 2013/023. Accessed online at https://waves-vagues.dfo-mpo.gc.ca/Library/361091.pdf

Du SJ, Gong ZY, Fletcher GL, Shears MA, King MJ, Idler DR, Hew CL (1992) Growth enhancement in transgenic Atlantic salmon by the use of an "all fish" chimeric growth hormone gene construct. Biotechnology (N Y) 10:176–181

Dunham RA (2004) Aquaculture and fisheries biotechnology: genetic approaches. CABI Publishing, Cambridge, MA, p 372

Dunham RA (2011) Aquaculture and fisheries biotechnology: genetics approaches, 2nd edn. CABI, Cambridge, MA, p 504

Dunham RA (2019) Chapter 7: Genetics. In: Lucas JS, Southgate PC, Tucker CS (eds) Aquaculture: farming aquatic animals and plants, 3rd edn. Wiley-Blackwell/Wiley, pp 127–156

Dunham RA, Devlin R (1998) Comparison of traditional breeding and transgenesis in farmed fish with implications for growth enhancement and fitness. In: Murray JD, Anderson GB, Oberbauer AM, McGloughlin MN (eds) Transgenic Animals in Agriculture. CAB International, Wallingford, pp 209–229

Dunham RA, Liu Z (2002) Gene mapping, isolation and genetic improvement in catfish. In: Shimizu N, Aoki T, Hirono I, Takashima F (eds) Aquatic genomics: steps toward a great future. Springer, Tokyo, pp 45–60

Dunham RA, Ramboux AC, Duncan PL, Hayat M, Chen TT, Lin CM, Kight K, Gonzalez-Villasenor I, Powers DA (1992a) Transfer, expression, and inheritance of salmonid growth hormone genes in channel catfish, *Ictalurus punctatus*, and effects on performance traits. Mol Mar Biol Biotechnol 1:380–389

Dunham RA, Turner CJ, Reeves WC (1992b) Introgression of Florida largemouth bass genome into native populations in Alabama Public Lakes. N Am J Fish Manag 12:494–498

Dunham RA, Chitmanat C, Nichols A, Argue B, Powers DA, Chen TT (1999) Predator avoidance of transgenic channel catfish containing salmonid growth hormone genes. Mar Biotechnol (NY) 1:545–551

Dunham RA, Chatakondi N, Nichols A, Chen TT, Powers DA, Kucuktas H (2002a) Survival of $F_2$ transgenic common carp (*Cyprinus carpio*) containing pRSVrtGH1 complementary DNA when subjected to low dissolved oxygen. Mar Biotechnol (NY) 4:323–327

Dunham RA, Chatakondi N, Nichols AJ, Kucuktas H, Chen TT, Powers DA, Weete JD, Cummins K, Lovell RT (2002b) Effect of rainbow trout growth hormone complementary DNA on body shape, carcass yield, and carcass composition of $F_1$ and $F_2$ transgenic common carp (*Cyprinus carpio*). Mar Biotechnol (NY) 4:604–611

Dunham RA, Kucuktas H, Liu Z, Keefer L, Spencer M, Robinson S (2002c) Biochemical genetics of brook trout in Georgia: management implications. Southeastern Association of Fish and Wildlife Agencies 55:63–80

Dunham RA, Warr GW, Nichols A, Duncan PL, Argue B, Middleton D, Kucuktas H (2002d) Enhanced bacterial disease resistance of transgenic channel catfish *Ictalurus punctatus* possessing cecropin genes. Mar Biotechnol (NY) 4:338–344

Ebert KM, Low MJ, Overstrom EW, Buonomo FC, Baile CA, Roberts TM, Lee A, Mandel G, Goodman RH (1988) A Moloney MLV-rat somatotropin fusion gene produces biologically active somatotropin in a transgenic pig. Mol Endocrinol 2:277–283

Erkinaro J, Niemelä E, Vähä J, Primmer CR, Brørs S, Hassinena E (2010) Distribution and biological characteristics of escaped farmed salmon in a major subarctic wild salmon river: implications for monitoring. Can J Fish Aquat Sci 67:130–142

FAO/WHO (2004) Safety assessment of foods derived from genetically modified animals, including Fish. FAO Food and Nutrition Paper 79. Report of the FAO/WHO Expert Consultation Rome, 17–21 November 2003.World Health Organization, Food and Agriculture Organization of the United Nations, Rome

Federal Register (2018) National Bioengineered Food Disclosure Standard, 83: 65814–65876

Fitzpatrick JL, Akbarashandiz H, Sakhrani D, Biagi CA, Pitcher TE, Devlin RH (2011) Cultured growth hormone transgenic salmon are reproductively out-competed by wild-reared salmon in semi-natural mating arenas. Aquaculture 312:185–191

Fletcher GL, Shears MA, King MJ, Davies PL, Hew CL (1988) Evidence for antifreeze protein gene transfer in Atlantic salmon (*Salmo salar*). Can J Fish Aquat Sci 45:352–357

Forsgren E (1992) Predation risk affects mate choice in a gobiid fish. Am Nat 140:1041–1049

Fortier E, Belote JM (2000) Temperature-dependent gene silencing by an expressed inverted repeat in *Drosophila*. Genesis 26(4):240–244

Fuji K, Hasegawa O, Honda K, Kumasaka K, Sakamoto T, Okamoto N (2007) Marker-assisted breeding of a lymphocystis disease-resistant Japanese flounder (*Paralichthys olivaceus*). Aquaculture 272:291–295

Ganga R, Tibbetts SM, Wall CL, Plouffe DA, Bryenton MD, Peters AR, Runighan CD, Buchanan JT, Lall SP (2015) Influence of feeding a high plant protein diet on growth and nutrient utilization to combined 'all-fish' growth-hormone transgenic diploid and triploid Atlantic salmon (*Salmo salar* L.). Aquaculture 446:272–282

Gao FY, Qu L, Yu SG, Ye X, Tian YY, Zhang LL, Bai JJ, Lu M (2012) Identification and expression analysis of three c-type lysozymes in *Oreochromis aureus*. Fish Shellfish Immunol 32:779–788

Glover KA, Solberg MF, McGinnity P, Hindar K, Verspoor E, Coulson MW, Hansen MM, Araki H, Skaala Ø, Svåsand T (2017) Half a century of genetic interaction between farmed and wild Atlantic salmon: status of knowledge and unanswered questions. Fish Fish 18:890–927

Godin JG, Dugatkin LA (1996) Female mating preference for bold males in the guppy, *Poecilia reticulata*. Proc Natl Acad Sci U S A 93:10262–10267

Guan H, Liang L (2013) Comparison of testis tissue construction and development of the genetically modified carp and the carp. J Northeast Agric Univ 44:95–100. (in Chinese with English abstract)

Higgs DA, Sutton JN, Kim H, Oakes JD, Smith J, Biagi C, Rowshandeli M, Devlin RH (2009) Influence of dietary concentrations of protein, lipid and carbohydrate on growth, protein and energy utilization, body composition, and plasma titres of growth hormone and insulin-like growth factor-1 in non-transgenic and growth hormone transgenic coho salmon *Oncorhynchus kisutch* Walbaum. Aquaculture 286:127–137

Hill JA, Kiessling A, Devlin RH (2000) Coho salmon (*Oncorhynchus kisutch*) transgenic for a growth hormone gene construct exhibit increased rates of muscle hyperplasia and detectable levels of differential gene expression. Can J Fish Aquat Sci 57:939–950

Hobbs RS, Fletcher GL (2008) Tissue specific expression of antifreeze protein and growth hormone transgenes driven by the ocean pout (*Macrozoarces americanus*) antifreeze protein OP5a gene promoter in Atlantic salmon (*Salmo salar*). Transgenic Res 17:33–45

Hollo T, Watson BM, Johnston SV, Devlin RH (2017) Behaviour of growth hormone transgenic coho salmon *Oncorhynchus kisutch* in marine mesocosms assessed by acoustic tag telemetry. J Fish Biol 90:1660–1667

Howard RD, DeWoody JA, Muir WM (2004) Transgenic male mating advantage provides opportunity for Trojan gene effect in a fish. Proc Natl Acad Sci U S A 101:2934–2938

Hu W, Li S, Tang B, Wang Y, Lin H, Liu X, Zou J, Zhu Z (2007a) Antisense for gonadotropin-releasing hormone reduces gonadotropin synthesis and gonadal development in transgenic common carp (*Cyprinus carpio*). Aquaculture 271:498–506

Hu W, Wang Y, Zhu Z (2007b) Progress in the evaluation of transgenic fish for possible ecological risk and its containment strategies. Sci China C Life Sci 50(5):573–579

Hu SY, Liao CH, Lin YP, Li YH, Gong HY, Lin GH, Kawakami K, Yang TH, Wu JL (2011) Zebrafish eggs used as bioreactors for the production of bioactive tilapia insulin-like growth factors. Transgenic Res 20:73–83

Hwang G, Müller F, Rahman MA, Williams DW, Murdock PJ, Pasi KJ, Goldspink G, Farahmand H, Maclean N (2004) Fish as bioreactors: transgene expression of human coagulation factor VII in fish embryos. Mar Biotechnol (NY) 6:485–492

IntraFish Media (2017) Accessed online at https://www.intrafish.com/finance/1281270/aquabounty-acquires-us-land-based-farmer-bell-fish

Jhingan E, Devlin RH, Iwama GK (2003) Disease resistance, stress response and effects of triploidy in growth hormone transgenic coho salmon. J Fish Biol 63:806–823

Jiang XY, Huang CX, Zhong SS, Sun CF, Zou SM (2017) Transgenic overexpression of *follistatin 2* in blunt snout bream results in increased muscle mass caused by hypertrophy. Aquaculture 468:442–450

Johnston IA, de la Serrana DG, Devlin RH (2014) Muscle fibre size optimisation provides flexibility for energy budgeting in calorie-restricted coho salmon transgenic for growth hormone. J Exp Biol 217:3392–3395

Kabeya N, Takeuchi Y, Yazawa R, Haga Y, Satoh S, Yoshizaki G (2016) Transgenic modification of the n-3 HUFA biosynthetic pathway in nibe croaker larvae: improved DPA (docosapentaenoic acid; 22:5n-3) production. Aquac Nutr 22:472–478

Karlsson S, Diserud OH, Fiske P, Hindar K (2016) Widespread genetic introgression of escaped farmed Atlantic salmon in wild salmon populations. ICES J Mar Sci 73:2488–2498

Kim JH, Leggatt RA, Chan M, Volkoff H, Devlin RH (2015) Effects of chronic growth hormone overexpression on appetite-regulating brain gene expression in coho salmon2. Mol Cell Endocrinol 413:178–188

Kim JH, Chatchaiphan S, Crown MT, White SL, Devlin RH (2018) Effect of growth hormone overexpression on gastric evacuation rate in coho salmon. Fish Physiol Biochem 44:119–135

Knapp JR, Chen WY, Turner ND, Byers FM, Kopchick JJ (1994) Growth patterns and body composition of transgenic mice expressing mutated bovine somatotropin genes. J Anim Sci 72:2812–2819

Ledford H (2015) Salmon approval heralds rethink of transgenic animals. Nature 527:417–418

Lee CG, Farrell AP, Lotto A, MacNutt MJ, Hinch SG, Healey MC (2003) The effect of temperature on swimming performance and oxygen consumption in adult sockeye (*Oncorhynchus nerka*) and coho (*O. kisutch*) salmon stocks. J Exp Biol 206(Pt 18):3239–3251

Leggatt RA, Raven PA, Mommsen TP, Sakhrani D, Higgs D, Devlin RH (2009) Growth hormone transgenesis influences carbohydrate, lipid and protein metabolism capacity for energy production in coho salmon (*Oncorhynchus kisutch*). Comp Biochem Physiol B Biochem Mol Biol 154(1):121–133

Leggatt RA, Biagi CA, Smith JL, Devlin RH (2012) Growth of growth hormone transgenic coho salmon *Oncorhynchus kisutch* is influenced by construct promoter type and family line. Aquaculture 356–357:193–199

Leggatt RA, Biagi CA, Sakhrani D, Dominelli D, Eliason EJ, Farrell AP, Devlin RH (2017a) Fitness component assessments of wild-type and growth hormone transgenic coho salmon reared in seawater mesocosms. Aquaculture 473:31–42

Leggatt RA, Sundström LF, Woodward K, Devlin RH (2017b) Growth-enhanced transgenic coho salmon (*Oncorhynchus kisutch*) strains have varied success in simulated streams: implications for risk assessment. PLoS One 12(1):e0169991

Li H (2016) Repressible transgenic sterilization in channel catfish, *Ictalurus punctatus*, by knockdown of primordial germ cell genes. Ph.D. Dissertation, Auburn University, Auburn, AL, USA
Li D, Hu W, Wang Y, Zhu Z, Fu C (2009) Reduced swimming abilities in fast-growing transgenic common carp *Cyprinus carpio* associated with their morphological variations. J Fish Biol 74:186–197
Li H, Su B, Qin G, Ye Z, Alsaqufi A, Perera DA, Shang M, Odin R, Vo K, Drescher D, Robinson D, Zhang D, Abass N, Dunham RA (2017) Salt sensitive Tet-off-like systems to knockdown primordial germ cell genes for repressible transgenic sterilization in channel catfish, *Ictalurus punctatus*. Mar Drugs 15(6):155
Li H, Su B, Qin G, Ye Z, Elaswad A, Alsaqufi A, Perera DA, Qin Z, Odin R, Vo K, Drescher D, Robinson D, Dong S, Zhang D, Shang M, Abass N, Das SK, Bangs M, Dunham RA (2018) Repressible transgenic sterilization in channel catfish, *Ictalurus punctatus*, by knockdown of primordial germ cell genes with copper-sensitive constructs. Mar Biotechnol (NY) 20:324–342
Lian H, Hu W, Huang R, Du F, Liao L, Zhu Z, Wang Y (2013) Transgenic common carp do not have the ability to expand populations. PLoS One 8:e65506
Liu C, Xu L, Liang L, Liu J, Chang Y (2011) Effects of semi-starvation on the growth and gonad development of transgenic carps. J Jilin Agric Univ 33:450–454, 463 (in Chinese with English abstract)
Liu S, Vallejo RL, Evenhuis JP, Martin KE, Hamilton A, Gao G et al (2018) Retrospective evaluation of marker-assisted selection for resistance to bacterial cold water disease in three generations of a commercial rainbow trout breeding population. Front Genet 9:286
Lo JH, Lin CM, Chen MJ, Chen TT (2014) Altered gene expression patterns of innate and adaptive immunity pathways in transgenic rainbow trout harboring Cecropin P1 transgene. BMC Genomics 15:887
Maclean N, Talwar S (1984) Injection of cloned genes into rainbow trout eggs. J Embroyl Exp Morphol 82(Suppl):187
Maclean N, Penman D, Talwar S (1987) Introduction of novel genes into rainbow trout. In: Tiews K (ed) EIFAC/FAO symposium on selection, Hybridization and Genetic Engineering in Aquaculture of Fish and Shellfish for Consumption and Stocking, vol II. Heenemann Verlagsgesellschaft mbH, Berlin, pp 325–334
Mao W, Wang Y, Wang W, Wu B, Feng J, Zhu Z (2004) Enhanced resistance to *Aeromonas hydrophila* infection and enhanced phagocytic activities in human lactoferrin-transgenic grass carp (*Ctenopharyngodon idellus*). Aquaculture 242:93–103
Marnis H, Iswanto B, Imron FS, Dewi RRSPS (2016) The ability of fast-growing transgenic African catfish (*Clarias gariepinus*) on predator avoidance. Indones Aquac J 11:9–13
Martínez R, Juncal J, Zaldívar C, Arenal A, Guillén I, Morera V, Carrillo O, Estrada M, Morales A, Estrada MP (2000) Growth efficiency in transgenic tilapia (*Oreochromis* sp.) carrying a single copy of an homologous cDNA growth hormone. Biochem Biophys Res Commun 267:466–472
Moav R, Wohlfarth GW (1974) Magnification through competition of genetic differences in yield capacity in carp. Heredity (Edinb) 33:181–202
Moreau DT, Conway C, Fleming IA (2011) Reproductive performance of alternative male phenotypes of growth hormone transgenic Atlantic salmon (*Salmo salar*). Evol Appl 4:736–748
Mori T, Devlin RH (1999) Transgene and host growth hormone gene expression in pituitary and nonpituitary tissues of normal and growth hormone transgenic salmon. Mol Cell Endocrinol 149:129–139
Mori T, Hiraka I, Kurata Y, Kawachi H, Mano N, Devlin RH, Nagoya H, Araki K (2007) Changes in hepatic gene expression related to innate immunity, growth and iron metabolism in GH-transgenic amago salmon (*Oncorhynchus masou*) by cDNA subtraction and microarray analysis, and serum lysozyme activity. Gen Comp Endocrinol 151:42–54
Morita T, Yoshizaki G, Kobayashi M, Takeuchi T (2003) Production of biologically-active recombinant goldfish gonadotropins in transgenic rainbow trout. Fish Physiol Biochem 28:473–474

Muir WM, Howard RD (1999) Possible ecological risks of transgenic organism release when transgenes affect mating success: sexual selection and the Trojan gene hypothesis. Proc Natl Acad Sci U S A 96:13853–13856

Muir WM, Howard RD (2001) Fitness components and ecological risk of transgenic release: a model using Japanese medaka (*Oryzias latipes*). Am Nat 158:1–16

Muir WM, Howard RD (2002) Assessment of possible ecological risks and hazards of transgenic fish with implications for other sexually reproducing organisms. Transgenic Res 11:101–114

Nam YK, Noh JK, Cho YS, Cho HJ, Cho KN, Kim CG, Kim DS (2001) Dramatically accelerated growth and extraordinary gigantism of transgenic mud loach *Misgurnus mizolepis*. Transgenic Res10:353–362

Nam YK, Park I, Kim DS (2004) Triploid hybridization of fast-growing transgenic mud loach *Misgurnus mizolepis* male to cyprinid loach *Misgurnus anguillicaudatus* female: the first performance study on growth and reproduction of transgenic polyploid hybrid fish. Aquaculture 231:559–572

Nancarrow CD, Marshall JT, Clarkson JL, Murray JD, Millard RM, Shanahan CM, Wynn PC, Ward KA (1991) Expression and physiology of performance regulating genes in transgenic sheep. J Reprod Fertil Suppl 43:277–291

Norgren KG, Dunham RA, Smitherman RO, Reeves WR (1986) Biochemical genetics of largemouth bass populations in Alabama. Proc Annu Conf Southeast Assoc Fish Wildl Agencies 40:194–205

NRC (National Research Council) (2004) Biological confinement of genetically engineered organisms. National Academies Press, Washington, DC, p 255

Ostenfeld TH, McLean E, Devlin RH (1998) Transgenesis changes body and head shape in Pacific salmon. J Fish Biol 52:850–854

Palmiter RD, Brinster RL, Hammer RE, Trumbauer ME, Rosenfeld MG, Birnberg NC, Evans RM (1982) Dramatic growth of mice that develop from eggs microinjected with metallothionein-growth hormone fusion genes. Nature 300:611–615

Pang SC, Wang HP, Li KY, Zhu ZY, Kang JX, Sun YH (2014) Double transgenesis of humanized *fat1* and *fat2* genes promotes omega-3 polyunsaturated fatty acids synthesis in a zebrafish model. Mar Biotechnol (NY) 16:580–593

Pennington KM, Kapuscinski AR (2011) Predation and food limitation influence fitness traits of growth-enhanced transgenic and wild-type fish. Trans Am Fish Soc 140:221–234

Pohajdak B, Mansour M, Hrytsenko O, Conlon JM, Dymond LC, Wright JR Jr (2004) Production of transgenic tilapia with Brockmann bodies secreting [desThrB30] human insulin. Transgenic Res 13:313–323

Pomp D, Nancarrow CD, Ward KA, Murray JD (1992) Growth, feed-efficiency and body-composition of transgenic mice expressing a sheep metallothionein 1a-sheep growth hormone fusion gene. Livest Prod Sci 31:335–350

Pursel VG, Hammer RE, Bolt DJ, Palmiter RD, Brinster RL (1990) Integration, expression and germ-line transmission of growth-related genes in pigs. J Reprod Fertil Suppl 41:77–87

Qin Z, Li Y, Su B, Cheng Q, Ye Z, Perera DA, Fobes M, Shang M, Dunham RA (2016) Editing of the luteinizing hormone gene to sterilize channel catfish, *Ictalurus punctatus*, using a modified zinc finger nuclease technology with electroporation. Mar Biotechnol (NY) 18:255–263

Rahman MA, Maclean N (1999) Growth performance of transgenic tilapia containing an exogenous piscine growth hormone gene. Aquaculture 173:333–346

Rahman MA, Mak R, Ayad H, Smith A, Maclean N (1998) Expression of a novel piscine growth hormone gene results in growth enhancement in transgenic tilapia (*Oreochromis niloticus*). Transgenic Res 7:357–369

Rahman MA, Ronyai A, Engidaw BZ, Jauncey K, Hwang GL, Smith A, Roderick E, Penman D, Varadi L, Maclean N (2001) Growth and nutritional trials on transgenic Nile tilapia containing an exogenous fish growth hormone gene. J Fish Biol 59:62–78

Raz E (2004) Guidance of primordial germ cell migration. Curr Opin Cell Biol 16:169–173

Rise ML, Douglas SE, Sakhrani D, Williams J, Ewart KV, Rise M, Davidson WS, Koop BF, Devlin RH (2006) Multiple microarray platforms utilized for hepatic gene expression profiling of GH transgenic coho salmon with and without ration restriction. J Mol Endocrinol 37:259–282

Roberts SB, McCauley LA, Devlin RH, Goetz FW (2004) Transgenic salmon overexpressing growth hormone exhibit decreased myostatin transcript and protein expression. J Exp Biol 207:3741–3748

Sarmasik A, Warr G, Chen TT (2002) Production of transgenic medaka with increased resistance to bacterial pathogens. Mar Biotechnol (NY) 4:310–322

Scholander PF, van Dam L, Kanwisher JW, Hammel HT, Gordon MS (1957) Supercooling and osmoregulation in arctic fish. J Cell Comp Physiol 49:5–24

Shears MA, Fletcher GL, Hew CL, Gauthier S, Davies PL (1991) Transfer, expression and stable inheritance of antifreeze protein genes in Atlantic salmon (*Salmo salar*). Marine Mol Biol Biotech 1:58–63

Silva AC, Almeida DV, Nornberg BF, Figueiredo MA, Romano LA, Marins LF (2015) Effects of double transgenesis of somatotrophic axis (GH/GHR) on skeletal muscle growth of zebrafish (*Danio rerio*). Zebrafish 12:408–413

Simmons M, Mickett K, Kucuktas H, Li P, Dunham R, Liu Z (2006) Comparison of domestic and wild channel catfish (*Ictalurus punctatus*) populations provides no evidence for genetic impact. Aquaculture 252:133–146

Stevens ED, Devlin RH (2000a) Intestinal morphology in growth hormone transgenic coho salmon. J Fish Biol 56:191–195

Stevens ED, Devlin RH (2000b) Gill morphometry in growth hormone transgenic Pacific coho salmon, *Oncorhynchus kisutch*, differs markedly from that in GH transgenic Atlantic salmon. Environ Biol Fish 58:113–117

Stevens ED, Sutterlin A (1999) Gill morphometry in growth hormone transgenic Atlantic salmon. Environ Biol Fish 54:405–411

Su B (2012) Reproductive confinement of common carp, *Cyprinus carpio*, and channel catfish, *Ictalurus punctatus*, via transgenic sterilization. Ph.D. Dissertation, Auburn University, Auburn, AL, USA

Su B, Peatman E, Shang M, Thresher R, Grewe P, Patil JG, Pinkert CA, Irwin MH, Li C, Perera DA, Duncan PL, Fobes M, Dunham RA (2014) Expression and knockdown of primordial germ cell genes, *vasa*, *nanos* and *dead end* in common carp (*Cyprinus carpio*) embryos for transgenic sterilization and reduced sexual maturity. Aquaculture 420-421(Suppl 1):S72–S84

Su B, Shang M, Grewe PM, Patil JG, Peatman E, Perera DA, Cheng Q, Li C, Weng CC, Li P, Liu Z, Dunham RA (2015a) Suppression and restoration of primordial germ cell marker gene expression in channel catfish, *Ictalurus punctatus*, using knockdown constructs regulated by copper transport protein gene promoters: potential for reversible transgenic sterilization. Theriogenology 84:1499–1512

Su B, Shang M, Li C, Perera DA, Pinkert CA, Irwin MH, Peatman E, Grewe P, Patil JG, Dunham RA (2015b) Effects of transgenic sterilization constructs and their repressor compounds on hatch, developmental rate and early survival of electroporated channel catfish embryos and fry. Transgenic Res 24:333–352

Sun C, Qu L, Ye X, Dong J, Tian Y, Lu M (2017) Establishing a zebrafish transgenic line expressing tilapia lysozyme with enhanced antibacterial activity. Aquaculture Res 48:760–766

Sundström LF, Lõhmus M, Johnsson JI, Devlin RH (2004) Growth hormone transgenic salmon pay for growth potential with increased predation mortality. Proc Biol Sci 271(Suppl 5):S350–S352

Sundström LF, Löhmus M, Devlin RH (2016) Gene–environment interactions influence feeding and anti-predator behavior in wild and transgenic coho salmon. Ecol Appl 26:67–76

Tang Y, Lin CM, Chen TT, Kawauchi H, Dunham RA, Powers DA (1993) Structure of the channel catfish (*Ictalurus punctatus*) growth hormone gene and its evolutionary implications. Mol Mar Biol Biotechnol 2:198–206

Teem JL, Gutierrez JB (2014) Combining the Trojan Y chromosome and daughterless carp eradication strategies. Biol Invasions 16:1231–1240

Teem JL, Gutierrez JB, Parshad RD (2014) A comparison of the Trojan Y chromosome and daughterless carp eradication strategies. Biol Invasions 16:1217–1230

Templeton CS (2005) Disruption of embryonic development in channel catfish, *Ictalurus punctatus*, using "sterile feral" gene constructs. Ms. Thesis, Auburn University, Auburn, AL, USA

Thresher RE (2008) Autocidal technology for the control of invasive fish. Fisheries 33:114–121

Thresher RE, Grewe P, Patil J, Hinds L (2005) Genetic control of sex ratio in animal populations. Australian Patent Number 2001291520, Commonwealth Scientific and Industrial Research Organisation

Thresher R, Grewe P, Patil JG, Whyard S, Templeton CM, Chaimongol A, Hardy CM, Hinds LA, Dunham R (2009) Development of repressible sterility to prevent the establishment of feral populations of exotic and genetically modified animals. Aquaculture 290:104–109

Thresher RE, Hayes K, Bax NJ, Teem J, Benfey TJ, Gould F (2013) Genetic control of invasive fish: technological options and its role in integrated pest management. Biol Invasions 16:1201–1216

U.S. Food and Drug Administration (2015) FDA had determined that the AquAdvantage Salmon is as safe to eat as non-GE salmon. Accessed online at https://web.archive.org/web/20151121055636/http://www.fda.gov/downloads/ForConsumers/ConsumerUpdates/UCM473578.pdf

Uzbekova S, Chyb J, Ferrière F, Bailhache T, Prunet P, Alestrom P, Breton B (2000) Transgenic rainbow trout expressed sGnRH-antisense RNA under the control of sGnRH promoter of Atlantic salmon. J Mol Endocrinol 25:337–350

VOX (2014) Here's what 9,000 years of breeding has done to corn, peaches, and other crops. Accessed online at https://www.vox.com/2014/10/15/6982053/selective-breeding-farming-evolution-corn-watermelon-peaches

Waltz E (2017) First genetically engineered salmon sold in Canada. Nature 548:148

Wang R, Zhang P, Gong Z, Hew CL (1995) Expression of the antifreeze protein gene in transgenic goldfish (*Carassius auratus*) and its implication in cold adaptation. Mol Mar Biol Biotechnol 4:20–26

Wang WB, Wang YP, Hu W, Li AH, Cai TZ, Zhu ZY, Wang JG (2006) Effects of the "all-fish" growth hormone transgene expression on non-specific immune functions of common carp, *Cyprinus carpio* L. Aquaculture 259:81–87

Wang W, Tan S, Luo J, Shi H, Zhou T, Yang Y, Jin Y, Wang X, Niu D, Yuan Z, Gao D, Dunham R, Liu Z (2019) GWAS analysis indicated importance of NF-κB signaling pathway in host resistance against motile aeromonas septicemia disease in catfish. Mar Biotechnol (NY) 21(3):335–347

Weir LK, Grant JWA (2005) Effects of aquaculture on wild fish populations: a synthesis of data. Environ Rev 13:145–168

Wieghart M, Hoover JL, McGrane MM, Hanson RW, Rottman FM, Holtzman SH, Wagner TE, Pinkert CA (1990) Production of transgenic pigs harbouring a rat phosphoenolpyruvate carboxykinase-bovine growth hormone fusion gene. J Reprod Fertil Suppl 41:89–96

Wikipedia (2019) GloFish. https://en.wikipedia.org/wiki/GloFish

Wong TT, Collodi P (2013) Inducible sterilization of zebrafish by disruption of primordial germ cell migration. PLoS One 8:e68455

Wong AC, Van Eenennaam AL (2008) Transgenic approaches for the reproductive containment of genetically engineered fish. Aquaculture 275:1–12

Wringe BF, Jeffery NW, Stanley RRE, Hamilton LC, Anderson EC, Fleming IA, Grant C, Dempson JB, Veinott G, Duffy SJ, Bradbury IR (2018) Extensive hybridization following a large escape of domesticated Atlantic salmon in the Northwest Atlantic. Commun Biol 1:108

Wu JL, Hu SY, Her GM (2010). Infertility control of genetically modified fish. United States Patent Number US 8383880 B2

Xu Q, Feng CY, Hori TS, Plouffe DA, Buchanan JT, Rise ML (2013) Family-specific differences in growth rate and hepatic gene expression in juvenile triploid growth hormone (GH) transgenic Atlantic salmon (*Salmo salar*). Comp Biochem Physiol Part D Genomics Proteomics 8(4):317–333

Yan X, Zhang X, Liu C, Sun X, Liang L, Cao D, Li C, Ge Y (2011) Introgressive hybridization of exogenous gene by micro-injection in carp (*Cyprinus carpio* L.) breeding. J Fish Sci China 18:275–282. (In Chinese with English Abstract)

Yan H, Liang L, Chang Y, Sun B, Su B (2017) Effects of inheritance and temperature on sex determination and differentiation related genes and sex ratio in fish: a review. J Dalian Ocean Univ 32:111–118. (In Chinese with English Abstract)

Yazawa R, Hirono I, Aoki T (2006) Transgenic zebrafish expressing chicken lysozyme show resistance against bacterial diseases. Transgenic Res 15:385–391

Youssef N (2017) Comparisons among channel catfish *Ictalurus punctatus*, blue catfish, channel ♀ × *I. furcatus* blue ♂ hybrid catfish and Transgenic Channel catfish for growth, cold and salinity tolerance. Ph.D. dissertation, Auburn University, Auburn, AL, USA

Zhang D (2016) Transgenic disruption of aromatase using the daughterless construct to Alter sex ratio in common carp, *Cyprinus carpio*. M.S. Thesis, Auburn University, Auburn, AL, USA

Zhang PJ, Hayat M, Joyce C, Gonzalez-Villaseñor LI, Lin CM, Dunham RA, Chen TT, Powers DA (1990) Gene transfer, expression and inheritance of pRSV-rainbow trout-GH cDNA in the common carp, *Cyprinus carpio* (Linnaeus). Mol Reprod Dev 25:3–13

Zhang L, Sun C, Ye X, Zou S, Lu M, Liu Z, Tian Y (2014) Characterization of four heat-shock protein genes from Nile tilapia (*Oreochromis niloticus*) and demonstration of the inducible transcriptional activity of Hsp70 promoter. Fish Physiol Biochem 40:221–233

Zhang Y, Chen J, Cui X, Luo D, Xia H, Dai J, Zhu Z, Hu W (2015) A controllable on-off strategy for the reproductive containment of fish. Sci Rep 5:7614

Zhong C, Song Y, Wang Y, Zhang T, Duan M, Li Y, Liao L, Zhu Z, Hu W (2013) Increased food intake in growth hormone-transgenic common carp (*Cyprinus carpio* L.) may be mediated by upregulating agouti-related protein (AgRP). Gen Comp Endocrinol 192:81–88

Zhu Z (1992) Chapter 6: Generation of fast growing transgenic fish: methods and mechanisms. In: Hew CL, Fletcher GL (eds) Transgenic fish. World Scientific Publishing, Singapore, pp 92–119. Accessed online at https://www.worldscientific.com/doi/abs/10.1142/9789814503600_0006

Zhu Z, Li G, He L, Chen S (1985) Novel gene transfer into the fertilised eggs of goldfish (*Carossius auratus* L. 1758). J Appl Ichthyol 1:32–34

Zhu TB, Zhang LH, Zhang TL, Wang YP, Hu W, Ringø E, Zhu ZY (2017) Effects of sustained predation by fast-growing transgenic common carp (*Cyprinus carpio* Linnaeus, 1758) on gastropods in artificial environments. J Appl Ichthyol 33:22–28

# GM Farm Animals: Potential Impact on Biodiversity Including Ethical Concerns

**George E. Seidel Jr**

**Abstract** For most traits in farm animals, the environment and chance influence phenotype much more than genetics. However, genetic tools including genetic modification (GM) are extremely powerful approaches to changing average phenotypes of populations. In nearly all cases, adding a new genotype to a population via conventional or GM approaches will increase biodiversity initially. The ultimate effect of the new genotype on the biodiversity of the general population will depend on how the new genotype is used in breeding programs. A special case, often termed gene editing, is using GM approaches to move genotypes within a species, e.g., from beef to dairy cattle. The end product will be identical whether produced conventionally or via GM, so the only real difference is that the GM approach will be more rapid, in some cases producing the change a decade sooner. To date, there are no GM farm animals outside of laboratory settings, university farms, etc., so at present, there has been no practical effect on biodiversity or ecosystems. How GM farm animals are used in the future will determine whether resulting biodiversity will increase, decrease, or stay about the same.

**Keywords** Ethics · Biodiversity · Genetics · Genetic modification · Farm animals · Selective breeding · Cost/benefit ratios · Heritability

## Introduction

Genetic modification (GM) is generally defined as changing DNA using tools of molecular biology, in most cases by incorporating the change into germ-line cells so that this change is inherited by succeeding generations. There usually is an additional implication that the change was derived or copied from a different species or was simply modified from nature for one reason or another. Such changes, however,

G. E. Seidel Jr (✉)
Colorado State University, Fort Collins, CO, USA
e-mail: George.Seidel@colostate.edu



A. Chaurasia et al. (eds.), *GMOs*, Topics in Biodiversity and Conservation 19,
https://doi.org/10.1007/978-3-030-53183-6_12

are both minor and more rational than selective breeding, which relies on pretty much random mutations that arise naturally.

So far, there has been no impact of what most consider to be GM farm animals on biodiversity or ecosystems because the very few GM farm animals produced to date have been confined to laboratory settings, university farms, etc. There appear to be no GM farm animals on commercial farms, and they are not providing any food or fiber to be sold, bought, or consumed.

The first GM farm animals, pigs and sheep, were produced in the US Department of Agriculture (USDA) laboratories in Beltsville, Maryland, in the early 1980s (Hammer et al. 1985). In those days, knowledge about regulation of genes was in its infancy, and procedures were very crude compared to those available today. Those experiments were very informative scientifically but did not lead to commercially viable agricultural products.

To date, the only animal GM food product approved by the US Food and Drug Administration is a strain of salmon (Aquabounty) that to date is not being sold for food in the USA, despite its proven safety and the ability to produce fish to eat much more efficiently than non-GM salmon in terms of feed requirements, months to harvest, and many potential effects on ecosystems such as decreased use of water and energy and fewer greenhouse gasses.

The one area in which GM farm animals currently have commercial value is to produce pharmaceuticals and biologicals such as antibodies for human medical applications. These applications are not agricultural and involve only a few confined animals. Another upcoming application is to produce organs and tissues for transplantation to people; pigs are especially suitable due to similarities in size and physiology but do require some GM to prevent graft rejection.

Over 90% of corn and soybeans and over 70% of cotton grown in North America and over 90% of papaya produced worldwide are now GM. We and our farm animals have been eating those products (including cotton seed oil) for decades with no convincing scientific evidence of detrimental effects to human or animal health and many ecological benefits such as less land needed per unit of food/feed and less erosion because no-till systems become feasible (Van Eenennaam and Young 2014, 2017)

Although there are concerns about GM crops, discussed elsewhere in this volume, there is a much greater pushback from the public concerning GM farm animals. Even the US government evaluates food produced by GM animals differently (evaluated as a drug by the Food and Drug Administration) from food from GM plants which is evaluated more conventionally (by the US Department of Agriculture). Surely, GM procedures should be considered as just one more tool, and we should put effort into assessing appropriate or inappropriate uses.

Another interesting perspective is that "Nature" has provided essentially all of the GM tools that we use (e.g., Konforti 2000) and has applied essentially all of them through the process of evolution, resulting in all the life forms on earth. An illustrative example is sweet potatoes, which in the course of domestication about 5 millennia ago incorporated the Bt gene to fight off insect damage (Kyndt et al. 2015). This is the very same gene that scientists have added to the GM crops used in North America to decrease insecticide use. I consider it unethical not to use GM tools for agricultural and medical applications, since there are so many opportunities to benefit mankind, other life forms, and our planet, although testing needs to continue to be done on a case by case basis as with any new product.

## Definition(s) of Genetic Modification

These days, many definitions are stipulative, that is, definitions are interpreted as something a user wants rather than what they really mean. Marketing forces are especially creative in this regard and are exacerbated by agendas of the many constituencies, including consumers, farmers, business, religious leaders, and scientists – including those who write review papers. The agendas often, even usually, include presenting the "truth" as understood by the interest groups involved. As "Nature" has been the ultimate genetic engineer and nearly every GM tool has existed naturally (see above), I, among others, contend that the most powerful genetic engineering tool used to date is selective breeding (Capper et al. 2009). An obvious visual example is the myriad breeds of dogs, horses, cattle, sheep, goats, etc., which range in mature size within a species by a factor of more than 10, just to mention 1 trait. We can cite numerous other traits with similar variation, for example, hair length and milk production. Many single gene traits have also been selected for in populations, traits such as coat color, polled (absence of horns), specific milk proteins, sex, and, of course, the absence of deleterious mutations.

The currency for selective breeding is natural variation in the population that continues to be added with each generation due to changes in germ-line DNA from radiation, peroxidation, errors in DNA synthesis (Seidel Jr 2015), genetic drift, and errors in chromosome segregation, in addition to the new mixing of the DNA combinations at meiosis with each succeeding generation. Despite the huge genetic modifications occurring naturally, these approaches are not considered GM by most people.

There is a special case of GM that is often defined as non-GM, because the result is the exact DNA modification that could be produced with selective breeding, even though the same GM tools are used as those to add genes from a different species. This is illustrated by the "polled" gene (Carlson et al. 2016), but there are many others such as "slick" for heat tolerance (Dikmen et al. 2014). The Angus beef breed of cattle is polled (no horns), and the Holstein dairy breed of cattle is not (horns). Horns are removed from dairy calves when they are young via methods that are moderately painful for a short time because horns can cause injuries to cattle and people and exacerbate dominance tendencies within herds. The polled gene can be introgressed by crossing Angus and Holstein breeds, so as to get half Holstein/half Angus crosses (not particularly good for milk production), and then by crossing again to get 3/4, 7/8, 15/16, etc. Holsteins that are great for milk production and have no horns. The problem is that this approach is both very expensive, because many suboptimal animals are produced *en route*, and time-consuming due to the long generation interval, which is over 2 years in cattle (Mueller et al. 2019). Cows first give birth at about 2 years of age and have 9-month gestations. Thus, it would take over 8 years to get to the 15/16 Holstein. A further complication is that only half of the offspring from any mating after the first step will inherit the polled gene. Moreover, this would need to be repeated with multiple lines because to exploit this by only producing one or two lines would result in high rates of inbreeding to get the trait into the general population, especially in the homozygous configuration.

With GM methodology, such a change can be effected in one step and, in 2 years rather than 8, gives 100% Holstein milk genes plus homozygous polled. So, while the polled gene was taken from Angus and added to a Holstein embryo, the final product is identical for the polled DNA that would have occurred with introgression via selective breeding. For some traits, simply deleting some DNA accomplishes the change. A rational analysis of this GM approach is that it simply uses a different procedure to get to an identical endpoint. The simplest solution to consumer acceptance is not to refer animals produced in this way as GM animals, and "gene edited" appears to be increasing in usage for these animals.

## Genetic Variability

An often mentioned objection to GMOs is that the result will be less genetic variation and thus less biodiversity in populations. However, the exact opposite occurs with GM, as it is adding new genetic variations. Interestingly, even cloning can be used to increase biodiversity, for example, by making a steer into a bull. Again, this depends on how the tool is used to determine its eventual effects on biodiversity. It is important to define biodiversity and provide the context. First, it is necessary to distinguish between genetic and phenotypic biodiversity. Phenotypic biodiversity is determined in part by genetics, but for many traits, the environment has a much greater effect on phenotype than genetics, and chance has an even greater effect than environment for some traits. The average percentage effect genetics has on a trait is termed heritability and ranges from near 0 to 100%, depending on the trait and species. Traits of low heritability include resistance to disease and fertility. Traits of moderate heritability, i.e., 20–40%, include milk production and growth rates. Mature size and certain carcass characteristics are in the 40–60% range, and there of course are traits with 100% heritability such as sex, coat color, and polled/horns. This discussion of heritability and effects of environment and chance on biodiversity may seem like a "red herring" to some readers. Among the reasons such discussion is relevant are genetic x environment interactions, for example, the slick gene currently may be added to certain dairy breeds via GM methods from other breeds to provide heat tolerance. This may be quite inappropriate for dairy breeds in colder climates but clearly increases biodiversity; however, its value depends on the environment.

Biodiversity also differs when considering the concept on a within herd basis, a within local population basis, or a global basis. Within cattle herds, biodiversity has increased greatly due to artificial insemination, even within herds that do not practice artificial insemination, since those farmers almost always buy bulls conceived by artificial insemination or whose parents were conceived by artificial insemination. Historically, most herds used a single bull for breeding at any given time, and bulls on neighboring farms frequently were closely related to each other or even shared among farms (Baker and Seidel Jr 1984). Thus, biodiversity within a herd was more limited than today, and herd is the functional unit when considering animal agriculture. Global biodiversity is irrelevant unless there is gene flow among regions, or put differently, biodiversity needs to be put to use to be of value. Considering the functional unit, there is an optimum phenotype for any given flock or herd, in part due to different global environments, and many farmers/ranchers aim for homogeneity that approaches that optimum phenotype for their situation; they do this not only via genetic tools but also by making the environment homogeneous, such as by what and how much they feed. A second reason for striving for intra-herd uniformity is marketing; a homogeneous set of cattle or hogs or poultry is much more attractive to most buyers than a heterogeneous product. To illustrate the complexity of the situation, in my own beef cattle herd, I strive for phenotypic homogeneity via selective breeding by using a dozen or more bulls via artificial insemination to complement the characteristics of the cows; e.g., small cows are mated to bulls that sire larger offspring, or cows that are inefficient for growth are mated to bulls whose offspring grow more efficiently. Also note that for most traits there is an optimum, not a minimum or maximum. In my beef cattle herd, I want to maximize the growth of calves but do it in the context of an optimum amount of milk production from their mothers; with too little milk, the calf grows too slowly (though, in some circumstances, this can be compensated by feeding the calf grain, but not in our extensive grazing environments), and with too much milk, the mothers lose body weight and reproduction declines, so she may be culled due to non-pregnancy. Thus, instead of using a single bull (or two or three), which would decrease genetic diversity of the herd, I use many bulls, which increases genetic diversity while increasing phenotypic homogeneity. With this approach, I could eventually be indirectly selecting for alleles, and for some traits, this probably occurs. However, it is becoming clear that there are multiple genetic ways of making a tall cow or a high-milk-producing cow. For example, one approach might be to select for more growth hormone secretion and another to select for more efficient growth hormone receptors on cells.

Another consideration is the huge value of heterosis for many health and fertility traits, accomplished, for example, by crossing different breeds or strains. One then has genetic diversity within the animal, one set of alleles from the father and another from the mother. The problem is that these crosses will not breed true; for the next generation, one can cross with one of the original breeds to get ¼, ¾ genetics, cross with a third breed and get ¼, ¼, ½ genetics of the three breeds, or cross males and females of this original cross, which results in considerable variation in genotypes and phenotypes.

There also are potential effects on biodiversity from selection at the cellular level. One example, often practiced with human embryos, is preimplantation genetic diagnosis, which involves biopsying embryos, determining their genetic makeup using molecular biology methodology, and then selecting those with desired genotypes. The net effect is selective culling of certain genotypes, which would appear to decrease genetic diversity. However, this is not necessarily the case, because one usually creates many more embryos than can be gestated, and at least theoretically, there can be selection for diversity. Sexing sperm is another example of cellular selection, and since sperm are interrogated individually (Seidel Jr. and Garner 2002), traits other than sex can be selected simultaneously. From a purist perspective, selecting for sex decreases biodiversity more than most any other practice; many dairy farms have no bulls other than young calves; roosters are absent in most egg production systems; killing males at younger ages than females has been practiced in many ancient human societies for millennia. Castration clearly increases phenotypic diversity, in some ways resulting in a third sex. These are not necessarily trivial considerations; for example, GM methodology is being used to produce pigs that do not require castration to eliminate boar taint in meat.

Superimposing GM on all of the above simply provides more options to optimize genetics and resulting phenotypes and especially to make changes faster than when using conventional approaches. As is continually being demonstrated in GM plant breeding, conventional genetic tools continue to be used along with field trials to optimize new GM traits, a process that can take years for most species of plants. Similarly, for many production traits in animals, the important point is how the single or few GM traits interact with the whole genome. For example, rapid growth is of little value if it results in infertility.

## The Main Question

Will GM farm animals result in decreased biodiversity within the functional unit, the herd or flock, or in global populations? This will depend on how GM is used. GM is a tool, and it will likely speed up improving animals within flocks and herds. Various constituencies have different definitions of improve, and one can also speed up the pace of making "mistakes" using GM tools, although conversely one can also speed up correcting mistakes! I consider that there is more potential to increase biodiversity than decrease it with GM, simply because GM increases, not decreases, genetic variability (even if the GM procedure is to delete genes, because the deleted version is a diversification).

For perspective, it is already clear that GM plants have had a profound effect on animal agriculture, at least in North America, via less expensive and more nutritious feed; the feed is grown with fewer inputs, especially amount of land and water per unit of product. It cannot be overemphasized how much less land is needed for agricultural production due to GM crops, saving it for other purposes such as the millions of hectares in the Conservation Reserve Program of the USDA. It appears that in the long run GM farm animals will further change animal agriculture profoundly, such as having fewer, healthier animals to produce the same amount of products but probably with only minor effects on genetic biodiversity.

## Ethical Considerations

As only minor effects of GM on biological diversity of farm animals are to be expected, there are not likely to be substantive resulting ethical issues. Ethical issues are much more likely to arise in other contexts. Most of the traits of farm animals currently being considered for GM concern improving animal welfare with comparatively few concerning production levels. Examples of traits being considered (Carlson et al. 2016; Dikmen et al. 2014) include disease resistance, removing horns genetically rather than by painful methods, incorporating the slick gene to improve heat tolerance, and eliminating the need to castrate male pigs. These modifications concern welfare and convenience traits. One contemplated GM modification, more efficient use of phosphorous by pigs, results in less phosphorous in urine and thus less phosphorous pollution (Zhang et al. 2018). It is clearly unethical not at least to consider these GM.

Perhaps the main ethical issue for GM in any context including effects on biodiversity is that there are plusses and minuses to almost any change in practices (including making no change) and the extent to which things need to be tested before recommending or permitting widespread use and furthermore whether GM products require more rigorous testing than non-GM products. There were huge concerns about this by scientists when new tools to modify DNA first became available in the 1970s, resulting in the Asimolar Conference on Recombinant DNA in 1975 and a self-imposed moratorium by scientists for certain kinds of experiments until more was known. It turned out that in those days many scientists were relatively naïve about what occurs every day naturally, which became increasingly clear from doing GM experiments.

There continue to be concerns over some kinds of research such as using GM (or any other method) to weaponize biology, for example, to make potent, easily transmissible lethal viruses. Another example is making a genetic modification that spreads rapidly in the population by making it to be inherited by more than the conventional 50:50 transmission. An example application is to introduce a lethal gene that then rapidly makes a species of mosquito that transmits disease to go extinct (Kyrou et al. 2018), an approach that has been used several times in the past with non-GM approaches to wipe out pests locally, such as gypsy moths (Leonhardt et al. 1966). Eliminating pest species will sometimes have unexpected consequences, but perhaps not always, as we have eliminated the small pox virus with no obvious negative consequences and have almost achieved that with polio; measles is currently another candidate for annihilation.

The scientific consensus is that, in retrospect, there is nothing special about using GM approaches to improve plant or animal agriculture but that the usual safety and efficacy testing that evolved for non-GM approaches should be practiced.

Much early disagreement resulted from a lack of familiarity and understanding. Once scientists gained knowledge of GM, their concerns decreased, both from scientific evidence and increased familiarity. A telling example for the general population is vaccinations. Most people recognize that the benefit/cost ratio is extremely positive but that there are negatives for a few individuals. Independent of all that, if there was no knowledge of immunology and efficacy of vaccination these days and one proposed to do vaccination experiments, one can imagine a huge outcry from many quarters, especially the lay public who would be alarmed by the prospect of taking a virus or bacterium that causes a disease, inactivating it in some way, actually injecting it into helpless laboratory animals, and, if efficacious, actually injecting it into children. It would take decades to get such experiments approved by review boards today and perhaps even longer until used by the general public. Vitamins are another example, the prospect of taking chemicals orally, even daily, and possibly adding them to food.

One huge problem that still needs to be understood is that it is impossible to prove a negative, from logical, scientific or philosophical perspectives. One can *never* be 100% sure that nothing happens, for example, that taking a vitamin pill daily never is detrimental or that eating a particular plant product, whether GM or not, never will cause some problem for someone. The opposite can be proven that a particular product causes harm or has some effect. One can prove a "yes," but not a "no," scientifically. Thus, one always tests things for probable untoward effects harmful in some way, and if one never or rarely finds problems, it is the absence of yeses that gets the product approved; it is impossible to prove that there never will be a problem.

Pragmatism is essential; there usually will be some provable good and bad characteristics of a new product, and virtually, any product used to excess may cause a problem. For example, we need calories in our food to survive, but too many result in obesity and associated issues.

GM approaches and products have the same issues as non-GM approaches and products: one can never prove a negative, and almost anything can be proven to be detrimental in excess.

## Conclusions

The special case of developing an identical product, whether via GM or conventional genetic approaches, by definition means the GM-produced product is no better or worse than that produced conventionally. Similarly, if a new product is produced via GM, it needs to be checked for safety and efficacy, just as for non-GM products. These arguments also apply to potential effects of GM on biodiversity. At least initially, GM farm animals not previously present naturally, by definition, will increase biodiversity. How GM animals are used in breeding programs will ultimately determine if the biodiversity in the general population will increase or decrease and if the ecological and public health consequences will be negative or positive.

## References

Baker AEM, Seidel GE Jr (1984) Why do cows mount other cows? Appl Anim Behav Sci 13:237–241
Capper JL, Cady RA, Bauman DE (2009) The environmental impact of dairy production: 1944 compared with 2007. J Anim Sci 87:2160–2167
Carlson DF, Lancto CA, Zang B, Kim ES et al (2016) Production of hornless dairy cattle from genome-edited cell lines. Nat Biotechnol 34:479–481
Dikmen S, Khan FA, Huson HJ, Sonstegard TS et al (2014) The SLICK hair locus derived from Senepol cattle confers thermotolerance to intensively managed lactating Holstein cows. J Dairy Sci 97:5508–5528
Hammer RE, Purcel VG, Rexroad CE Jr, Wall RJ, Bolt DJ, Ebert KM, Palmiter RD, Brinster RL (1985) Production of transgenic rabbits, sheep, and pigs by microinjection. Nature 315:680–683
Konforti B (2000) The servant with the scissors. Nat Struct Biol 7:99–100
Kyndt T, Quispe-Huamanquispe D, Zhai H et al (2015) The genome of cultivated sweet potato contains agrobacterium T-DNAs with expressed genes: an example of a naturally transgenic food crop. Proc Natl Acad Sci U S A 112:5844–5849
Kyrou K, Hammond AM, Galizi RJ, Kranjc N et al (2018) A CRISPR-Cas9 gene drive targeting doublesex causes complete population suppression in caged Anopheles gambiae mosquitoes. Nat Biotechnol 36:1062–1066
Leonhardt BA, Mastro VC, Leonard DS, McLane W, Reardon RC, Thorpe KW (1966) Control of low density gypsy moth (Lepidoptera: Lymantriidae) populations by mating disruption with pheromone. J Chem Ecol 22:1255–1272
Mueller ML, Cole JB, Sonstegard TS, Van Eenennam AL (2019) Comparison of gene editing versus conventional breeding to introgress the POLLED allele into the US dairy population. J Dairy Sci 102:4215–4226
Seidel GE Jr (2015) Maintaining integrity of germline DNA: individuals age, species do not. Reprod Fertil Dev 28:125–129
Seidel GE Jr, Garner DL (2002) Current status of sexing mammalian sperm. Reproduction 124:733–743
Van Eenennaam AL, Young AE (2014) Prevalence and impacts of genetically engineered feedstuffs on livestock populations. J Anim Sci 92:4255–4427
Van Eenennaam AL, Young AE (2017) Detection of dietary DNA, protein, and glyphosate in meat, milk, and eggs. J Anim Sci 95:3247–3269
Zhang X, Li Z, Yang H, Liu D et al (2018) Novel transgenic pigs with enhanced growth and reduced environmental impact. elife 7:34286

# GM Animals: Biodiversity and Bioethical Concerns and Analysis

**Diana Caballero-Hernandez, Alejandra E. Arreola-Triana, and Cristina Rodríguez-Padilla**

**Abstract** The development of GM animals has been hampered by public concerns, and the discussion of their risks and ethical implications is limited when compared to that of GM crops. Here we discuss the ethical implications of GM animals on biodiversity. In this chapter, the ethical issue of GM animals is framed as a component of a larger issue: the impact of domestication in the face of the Anthropocene extinction. We provide a background of modern biotechnology, its history and progress in developing genetically engineered animals. Next, we challenge the purported antagonism between genetic engineering, a form of domestication, and biodiversity. For this, we examine the risks that transgenic and cloning technologies pose on biodiversity, understood as species richness. We then address the ethical issues around genetic engineering of animals and its potential impact on biodiversity, drawing upon the Aristotelian concept of telos and the modern approach of functional traits as proxies for animal welfare and biodiversity.

**Keywords** Domestication · Genetic engineering · GMOs · Living modified organisms · Conservations ethics · Animal welfare · Telos

## Introduction

The modification of species through genetic engineering (GE) remains controversial, despite that genetic modification of plants and animals has been a constant in human society: This is known as artificial selection, and it has played a crucial part in agriculture and the rise of civilization. But unlike artificial selection, the modern tools for genetic modification bypass sexual reproduction and species barriers and are thus distrusted because they are seen as artificial and unnatural, and consequently deserving of a closer scrutiny than artificial selection by selective breeding.

D. Caballero-Hernandez (✉) · A. E. Arreola-Triana · C. Rodríguez-Padilla
Universidad Autónoma de Nuevo León (UANL), Facultad de Ciencias Biológicas, Laboratorio de Inmunología y Virología, San Nicolás de los Garza, Nuevo León, México
e-mail: diana.caballerohr@uanl.edu.mx



A. Chaurasia et al. (eds.), *GMOs*, Topics in Biodiversity and Conservation 19,
https://doi.org/10.1007/978-3-030-53183-6_13

Both artificial selection and genetic engineering modify the genetic makeup of organisms with a different degree of precision and control. Whereas "genetically modified organisms" (GMOs) is the preferred term in most policies enacted to regulate those organisms whose genome has been modified using the tools of molecular biology, we prefer to use the term "genetically engineered" (GE) to distinguish them from those organisms that have been modified through artificial selection.

In comparison to the debates about GMO crops, conversations about policies and ethical analysis of the creation and use of genetically engineered animals have been slow to start, in part due to the sporadic success in developing new races or strains of animals, and in part due to the low acceptance of these products by the public, driven by concerns for human health and wildlife contamination.

According to the GM Contamination Register, incidents of contamination must meet at least one of four criteria, primarily developed for GMO plants; two are pertinent here: (1) illegal plantings of GMOs or unauthorized releases to the environment or food/feed chain and (2) establishment of feral population(s) of a GM crop or presence of the genetic insert within wild or feral populations, including wild or weedy relatives (Price and Cotter 2014). In the case of GE animals, only four incidents have been recorded, all involving experimental GE pigs that accidentally or intentionally entered into the food or feed supply. No incidents of contamination of GE insects or livestock have been recorded. In the case of GE fish, several incidents of the transgenic zebrafish *Danio rerio* have been reported; however, we could not find official records for these incidents. Taken together, these reports represent only 2.3% of the total contamination incidents recorded between 1997 and 2013 (Price and Cotter 2014).

Currently, there is not enough evidence to support the claim that GE animals represent a significant risk for biodiversity, and yet, the subject remains open if we consider the potential of GE technologies to have a positive impact on biodiversity or, alternatively, if we consider a broader meaning of biodiversity beyond species richness, one that considers the welfare of animals. Here we explore the ethical issues raised by the creation of genetically engineered animals, from mice and rats bred and contained in laboratories to disease-fighting mosquitoes deployed into the wild, focusing the analysis on their potential impact on biodiversity.

## Domestication Involves Genetic Modification

Climate change and loss of biodiversity are characteristic events of the Anthropocene (Corlett 2015). Whereas the climate crisis is mainly attributed to dependence on fossil fuels, the causes for the loss of biodiversity are more diverse, including loss of ecosystems to agriculture and urban and industrial development, hunting of endangered species (poaching), and domestication, particularly the intensive farming of domesticated species.

Domestication started between 11,000 and 9000 BC and involves the repurposing of nonhuman living beings to supply the needs of human society; this

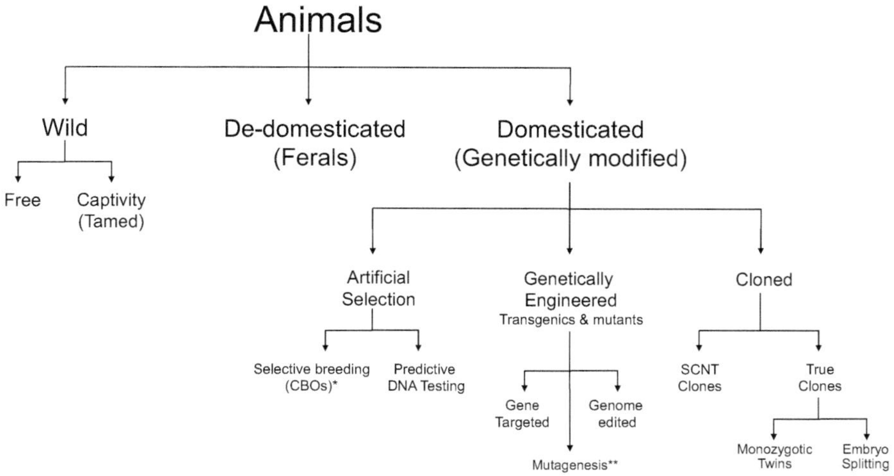


**Fig. 1** Earth is now populated by an array of animals that go from wild to domesticated, including populations of de-domesticated animals, and of tamed animals that are kept and bred in captivity for their economic and educational value. The domestication of animals has involved the harnessing of their reproduction, in the past through conventional breeding, and in modern times with genetic engineering, a technique that allows to overcome not only sexual reproduction but also species barriers. *Conventionally Bred Organisms **Recently, the Court of Justice of the European Union ruled that organisms obtained by mutagenesis, modified without the insertion of DNA, must be considered, and regulated, as GMOs. This ruling is directed at newer techniques for mutagenesis, specifically genome editing. (Grand Chamber of Court of Justice of the European Union 2018)

was achieved empirically at first, and later with increasing expertise and ability to control the outcomes, thus reshaping the biosphere and enabling human societies to thrive (Bull and Maron 2016). The Earth is now populated by an array of animals that go from wild to domesticated, including populations of de-domesticated animals, ferals, and tamed organisms that are kept and bred in captivity for their economic and educational value. Among these animal populations we include those whose genetic makeup has been modified by selective breeding through directional selection and those that are product of genetic engineering, including recombinant DNA techniques, cloning, and genome editing (Fig. 1). These methods have become a source of much controversy.

We consider that, notwithstanding legal definitions, all domesticated animals are genetically modified living beings produced through manipulations with several degrees of control and precision, be it selective breeding, conventional, or informed by predictive DNA testing or genetic engineering, including gene targeting and genome editing, and reproductive cloning (Fig. 1). Domestication is not synonymous with taming (Driscoll et al. 2009), but rather describes "a permanent genetic modification of a bred lineage that leads, to, among other things, a heritable predisposition toward human association" and defines a domestic animal as "one whose mate choice is influenced by humans and whose tameness and tolerance of humans is genetically determined." The domestication of plants and animals has been an integral part of the development of human societies and has progressed relentlessly, supported by an accumulation of scientific knowledge and technological advance. Genetic engineering has expanded the potential of domestication to improve human lives and modify the biosphere, but it also has created conflicts that reflect the concerns of the scientific community regarding the ethics of their goals, the uneasiness of the general public toward these endeavors, and the complexity of the tools chosen to these ends.

Perhaps because it has been integral to the development of human society, historical domestication has remained mostly unchallenged from an ethical standpoint. In the past decades, however, there has been a growing consensus between scientists and laypeople that the use and manipulation of living beings is worthy of ethical reflection, but the motivations for these concerns may come from different places. In the particular case of domestication achieved by genetic engineering, besides the recurring concerns of animal welfare and biosafety, concerns regarding the genetic integrity of animals and their potential impact on the environment have been growing, compounded with distrust for the apparent power these emerging technologies vest on a select few.

Both genetic engineering and selective breeding rely on genetic modification, which can have unintended effects on the modified individuals and can become a threat to the survival of the species.

## Progress in GE Animal Science

Genetic engineering techniques such as microinjection, vector systems, cell fusion, and, lately, CRISPR/Cas9 result in animals with traits that could be difficult or impossible to obtain through traditional methods (Bulfield et al. 2001; Bolland et al. 2010). Here we describe GE animals that have been successfully developed, according to the purpose of their modification. It must be noted that most of these modifications are only used in biomedical research and are not meant to enter the food/feed chain.

### *Growth Enhancement*

Growth enhancement is perhaps the modification that has received most attention; recently, the AquAdvantage salmon was approved for human consumption in the USA (2015) and Canada (2016) (Waltz 2017). Atlantic, Chinook, and Coho salmon, tilapia, rainbow trout, Northern pike, rohu, loach, carp, and channel catfish have been modified with growth hormone (GH) genes or their promoters sourced from fish or other species (Maclean and Laight 2000; Devlin et al. 2006). This modification results in fish that can reach commercial sizes in about half the usual time (Menozzi et al. 2012). GH constructs have also been added to pigs, which showed improved weight gain (Vize et al. 1988).

## *Metabolic Modifications*

Metabolic modification is another application of genetic engineering that leads to better assimilation of nutrients and thus greater savings for the industry, as well as improved, more healthful products for consumers. Rainbow trout and other carnivorous fishes have been modified with a transgene for a glucose transporter and hexokinase to improve carbohydrate metabolism and reduce dependence on "trash" fish that are being overfished. Other modifications seek to improve the metabolism of fats, vitamins, and minerals (Devlin et al. 2006). Chickens have been inserted the beta-galactosidase gene, which allows them to use lactose as an energy source (Forabosco et al. 2013). Pigs have been modified with the fat-1 gene, an n-3 fatty acid desaturase, to make "heart-healthy pork" with higher levels of n-3 fatty acids than wild-type pigs (Lai et al. 2006). A line of GE pigs, the Enviropig (TM), was developed in Canada; these animals were able to secrete phytase and other active enzymes in their saliva, making more efficient use of the phosphates in their food and reducing their phosphate excretion up to 60% (Golovan et al. 2001). Other modifications in pigs include the insertion of insulin transgenes to produce more loin mass and the knockout of the myostatin gene in pigs and cattle to produce leaner muscle mass (Telugu et al. 2017). Although not precisely a metabolic modification, some GE animals are modified to produce more wool, milk, or eggs.

## *Disease Resistance and Control*

Livestock production requires conditions that make animals susceptible to many diseases; resistance to these diseases would not only be beneficial to the animal but also for those that raise them because treatment and replacement of the animals is expensive (Bulfield et al. 2001).

Fish have been modified to resist several pathogens such as *Aeromonas*, grass carp hemorrhage virus, and infectious hematopoietic necrosis virus. These modifications increase survival rate and allow fish to be kept in higher densities, but may provide fitness advantages if fish were to escape into the environment (Mozdziak et al. 2003).

Other GE animals modified for disease resistance are sheep, goats, and cattle resistant to prion diseases, as well as cows resistant to mastitis (Telugu et al. 2017). Chickens and pigs are repositories of influenza and can transmit it to humans; these animals have had myxovirus-resistance genes added as protection against this disease (Forabosco et al. 2013; Looi et al. 2018).

Animals not only are repositories of human diseases; they also act as vectors, and GE technology can be used to control the populations of animals that spread diseases to humans. The Oxitec mosquito (*Aedes aegypti* OX513A) contains a self-limiting gene that hinders their survival into adulthood; its aim is to reduce mosquito populations and, with that, the spread of dengue virus (Nading 2015). There are also initiatives to create GE-Anopheles mosquitoes unable to carry the parasites responsible for malaria (Beisel and Boëte 2013). When released, these GE mosquitoes would greatly reduce wild-type populations; several trials have already taken place, such as the Cayman Island trial of the Oxitec mosquito in 2010.

## *Population Control*

Gene drive systems are being proposed as an alternative for the population control of invasive species, pests, and disease vectors. In a gene drive, transgenic organisms that carry a desirable gene are released into a wild population, and the genes, whether or not they provide a fitness advantage, spread through the population. Gene drives can be used to introduce advantageous traits into a population (modification drives), such as disease resistance, or they may be used to introduce genes that reduce fitness (suppression drives), resulting in the disappearance of traits, or the reduction or elimination of entire populations (Champer et al. 2016).

## *Bioreactors*

Poultry and livestock already possess the ability to synthesize large amounts of molecules in their milk and eggs. Genetic engineering techniques are used in animals such as fish, chickens, cows, sheep, and goats so that they can function as bioreactors in the production of biopharmaceuticals at a lower cost than traditional methods (Bolland et al. 2010; Herron et al. 2018). Some of the biopharmaceuticals produced in GE animals are antithrombin III in goat milk (Echelard et al. 2006), human blood-clotting factor VII in zebrafish (Bolland et al. 2010), human C1 esterase inhibitor produced in rabbits, active alpha-1-antitrypsin in sheep (Telugu et al. 2017), and human cytokines in chickens (Herron et al. 2018).

## *Biomedical Research*

Thousands of GE animals, transgenics, and mutants are used in biomedical research; these animals spend all their lives inside research facilities and there is little risk of escape from confinement and interaction with wild species—they are considered risk class 1 (Maclean and Laight 2000; Chaible et al. 2010). Recombinant DNA-based genetic engineering technologies (knockouts, knock-ins, transgenesis, etc.) allow for the study of the mechanisms of development or disease and the function of specific genes, which would be otherwise impossible. GE animals are also used in pharmacological and toxicology research (Bulfield et al. 2001; Chaible et al. 2010; Lee et al. 2015).

Mice and fruit flies are the GE organism most commonly used because their upkeep is relatively simple and inexpensive, and the methods used to produce them are well-understood (Jennings 2011; Perleberg et al. 2018). There is a mouse model for every human condition imaginable—there are already strains of knockout mice for more than 10% of human genes, and there are plans to create many more (Chaible et al. 2010). There are more than 20,000 transgenic fly lines that are used for studying animal development, regenerative biology, drug discovery, bioengineering, and medicine—many genes for human diseases have their equivalent in the *Drosophila* genome (Jennings 2011).

Mice and fruit flies, however, are much smaller than humans, and their lives are many times shorter, so GE technology is being used to create animal models that resemble humans in size and life span (Garas et al. 2014). The pig is a popular model organism because it shares many anatomical and physiological characteristics with humans, and its husbandry is easily adapted to laboratory conditions. Currently, there are pig models for the study of colorectal cancer, osteosarcoma, diabetes, cystic fibrosis, cardiovascular disease, Duchenne's muscular dystrophy, and Huntington's and Alzheimer's disease (Perleberg et al. 2018). GE pigs are also being used in xenotransplantation—the modifications are aimed at preventing immune rejection of the transplanted tissues (Bulfield et al. 2001; Garas et al. 2014). Other GE organisms used for research are nematodes, frogs, zebrafish, guinea pigs, and rats. Goats and sheep are also used as models, for example, for Huntington's disease and atrial fibrosis (Jacobsen et al. 2010; Polejaeva et al. 2016).

Genome editing has expanded the possibilities of what can be done with living beings, including their domestication and biodiversity (Fig. 2). Some even envision

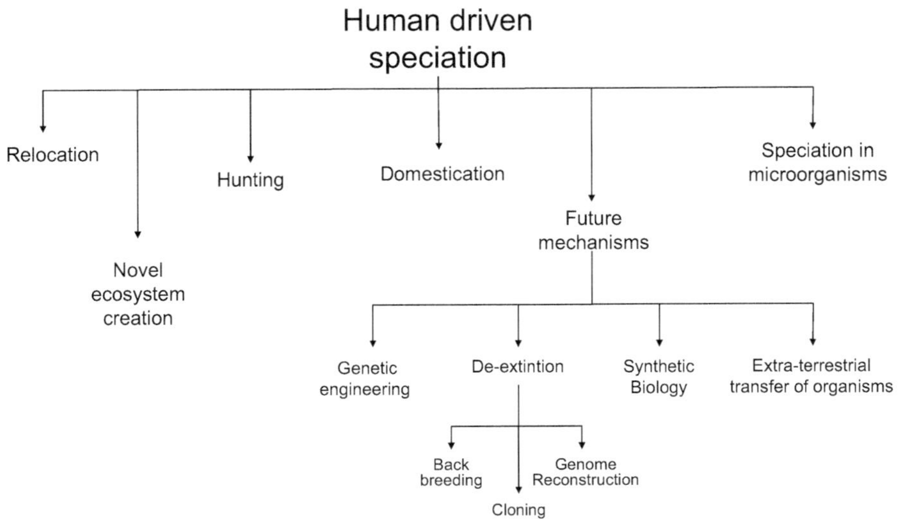


**Fig. 2** Human ingenuity has reshaped the biosphere, and now with the tools of genetic engineering, genome editing, synthetic biology, and cloning, it has the potential for further transformation of the animal landscape. (Adapted from Bull and Maron 2016)

a future populated by novel organisms created in the laboratory and by long-gone species brought back from extinction (Charo and Greely 2015). This de-extinction of species, such as the mammoth, the thylacine, and the passenger pigeon, will make use of genomic tools (Novak 2018) and the most sophisticated of the breeding technologies: reproductive cloning.

## Animal Cloning as Genetic Modification

Reproductive cloning is a breeding technology that aims at the replication of an adult individual. The birth of Dolly the sheep in 1996 was a breakthrough (Campbell et al. 1996), significant not only for reproductive science, but for science in general. Dolly's birth also prompted the scientific community to reflect on the nature and purpose of the research behind this achievement (García-Sancho 2015). Part of the discussion focused on biosafety and animal welfare concerns, but for others the prospect of human cloning has been at the forefront. The premature death of Dolly at 6 years old fueled fears regarding the long-term safety of the technique, since cloned animals appeared to suffer from shortened life spans, an observation that did not bode well for the successful implementation of the technique. However, as recently as 2016, Dolly's sisters, all clones themselves bred using the same procedure, were shown to be healthy and thriving animals at 9 years old, the equivalent of 70 years old for humans (Sinclair et al. 2016). Although not completely overcome, fears around cloning have been receding and, despite the ethical issues, the reproductive cloning of animals is becoming a profitable business. Livestock, horses, dogs, and cats are being cloned in laboratories around the world; it has been estimated by the hundreds.

Reproductive cloning is not considered part of genetic engineering, but a breeding technology. An argument can be made, however, in favor of reproductive cloning as a type of manipulation of the genome, because the most commonly performed artificial cloning technique is the somatic cell nuclear transfer (SCNT). In this technique, the nucleus of a somatic cell from an adult individual is transferred to an enucleated egg to be implanted in a gestational surrogate. The embryo then possesses the nuclear genome from the parent cell and the mitochondrial genome of the receiving cell (National Research Council 2002). This mismatching of genomes has practical implications, and it could be involved in the low rate success of cloning (Takeda 2013). SCNT clones are not considered true clones, however, because the mitochondrial genome is much smaller than the nuclear, and most of the genes it contains deal with cellular respiration, not performance traits of interest for breeders. In a sense, SCNT cloning is a technique that allows the simultaneous transfer of not one but many genes—a whole nuclear genome.

## Domestication as a Threat to Biodiversity

One misconception regarding the process of domestication is that it is antagonistic to biodiversity—in its simplest definition, the variety of extant species. The underlying view is that domestication implies selecting some species over others, if and when they display traits of interest to humans (performance traits). This process gives priority to their capacity to serve as means of transportation, labor, food, or companionship, so their usefulness for human society acts as a safety net, insurance for survival. A different approach to the subject, however, suggests that domestication is a form of conservation: animals are brought to a human-controlled environment, and those individuals that display traits of interest are bred and reared, their utility for humans working as a safeguard for the species (Uchola 2016). Domestication has been suggested as an approach for the conservation of endangered species, an idea thoroughly analyzed by Teletchea (2017).

When it comes to genetic engineering technologies for animal conservation, transgenesis and cloning are commonly thought as opposites, but Wall et al. (2009) offer a more nuanced understanding of these technologies and their role in diversity. Although at first impression, cloning, by definition, seems to reduce genetic variability because it bypasses sexual reproduction, it has a positive potential for diversity. For instance, cloning can maintain genetic diversity when used to rescue endangered species, and it can also result in gains for diversity if used to obtain copies of reproductively compromised animals whose genetic makeup otherwise would be lost. Transgenesis, the insertion of exogenous genetic material, implies a gain of genetic diversity simply by adding genetic information otherwise inaccessible to an individual, bypassing cross-species barriers. In transgenic technology, the DNA code is a source of solutions that can be transferred between organisms.

## An Ethical Reflection on Animal Modification Through Genetic Engineering

Although at first glance, when considering the technical details, neither transgenic nor cloning technologies seem to represent a significant threat to biodiversity, in order to discuss the impact of the genetic engineering of animals on biodiversity, it is also necessary to acknowledge the question of the inherent value of species. This has also become an urgent issue, because of recent developments in genetic engineering that are aimed at the eradication of species, specifically the gene drive technology for population control of invasive species and disease vectors (Pugh 2016).

Historically and in practice, there has not been a need for ethical or moral guidelines regarding the rightful existence of species other than humans, a question directly relevant to the conservation of biodiversity. The scrutiny of the relationship between humans and animals, wild and domesticated, has been mostly concerned with how they exist and how well they fare, so animal welfare has come to be the main issue regarding the domestication of animals. In the Anthropocene, the way we act toward animals may include invasive procedures for the manipulation of their genomes.

We have different systems to judge the morality of our acts, for instance, the four ethical principles: beneficence, non-maleficence (*Primum non nocere*), justice, and autonomy (Lawrence 2007). These are, however, more oriented to judge the way we act toward other human beings and more suited to medicine and biomedical research. Through the years, with the increasing concern regarding how we treat animals, some other concepts have been developed for guidance. In the 3 Rs guideline proposed by Russell and Burch (1959), the concepts of replacement, reduction, and refinement are concerned with the ethical use of animals in biomedical research and have been widely adopted in laboratories around the world. The concept of the Five Freedoms (Brambell 1965) considers that animals must be free from: (1) hunger or thirst; (2) discomfort; (3) pain, injury, or disease; (4) inability to express normal behavior; and (5) fear and distress. Because these systems focus on the welfare of animals and not their inherent value, they are useful to interrogate the ethics of genetic manipulation and how this intervention affects the quality of life of animals—the individuals and not the species. This is an important distinction that according to each case could reflect adherence to a particular bioethical theory (Verhoog 2000).

As for animal welfare, although genetic engineering of animals has been focused on increasing their value for humans, for example, through the production of therapeutic drugs, it also has attempted the modification of animals with the aim of reducing their suffering under intensive farming, such as in the genetic dehorning of livestock (see chapter "Governance of Emerging Technologies/Applications in the Bio/Life Sciences: Genome Editing and Synthetic Biology"). Some, however, question even this approach, contending that this kind of interventions violates the genetic integrity of the animal (Warkentin 2009).

When considering the ethical limits of genetic engineering, and its potential role in the biodiversity crisis, a concept that can be helpful to this discussion is the animal *telos*, first proposed by Aristotle in an attempt to describe the nature and purpose of living beings, especially human beings (Hauskeller 2005). The telos has been extensively discussed and through time has assimilated modern concepts such as genetics and has come to be used to discuss the ethical nature of genetic engineering (Rollin 2015). According to Rollin (2015) the telos is "the sum total of how an organism does" elaborating that the telos, "in modern terminology, is roughly what is encoded in an animal's genetics, as expressed in its normal environment"; he exemplifies this idea, "We can explain the sharp edge of a knife by reference to what a knife does, namely cut, without assuming consciousness on the part of the knife. In a similar manner, we can explain the building of dams by beavers in terms of such dams increasing the likelihood of catching fish without assuming either that beavers have a conscious purpose in mind when they build, or that they were consciously designed to do so; evolution by natural selection is perfectly adequate as an explanation." From this we can conclude that it is not necessary to attribute consciousness to an animal to recognize that it has a natural way of being, of existing, of pursuing what it is good for them, and that this is largely, but not exclusively, encoded in its DNA.

Regarding animal domestication, Rollins adds: "We could never have domesticated them [animals] if we failed to understand at least the basics of their telos, and as we domesticated them, we changed their telos to suit domestication, making them more docile and tractable, and more dependent on us." This is an important assertion, because independence from humans can be used to determine the intrinsic value of animals, and by creating or increasing their dependence on us, we are creating moral obligations toward them, and this dependence becomes part of their telos. It follows that some interventions, which are bound to increase an animal's dependence on us, violates the telos. When genetic engineering is used to create transgenic animals to model diseases, for example those that limit mobility like amyotrophic lateral sclerosis, we are not only increasing animal dependence on their keepers but also their suffering. There are limits to the animal telos, which leads into concerns of animal welfare; here we suggest that the welfare of an animal is a component of its identity, its telos.

## Biodiversity, the Animal Telos and Traits

In discussing the impact of the genetic engineering of animals on biodiversity, a fundamental question to address is the inherent value of a species to exist, a question that lies at the heart of conservation biology (Hare et al. 2018). The main aim of conservation science is the management of biodiversity, be it the number of species in an ecosystem or the number of variants of a domesticated species, oriented toward the preservation of all species, following the ideal that all species are intrinsically valuable or have potential value for humans. The conservation of species requires knowledge of species and ecosystems in order to implement conservation practices. In this context, genetic engineering, genomics, synthetic biology, and cloning have been suggested as tools for conservation, inserting modifications to reduce the fitness of invasive species that are a threat to native flora and fauna, creating microorganisms capable of repairing polluted soils, and inserting genes for resistance to extreme environments into the genome of endangered species, etc. (Piaggio et al. 2017).

In conservation biology, a relatively new concept of functional-trait ecology has developed (Cernansky 2017). Functional traits "are morphological, biochemical, physiological, structural, phenological, or behavioral characteristics that are expressed in phenotypes of individual organisms and are considered relevant to the response of such organisms to the environment and/or their effects on ecosystem properties" (Garnier et al. 2007).

New traits obtained through genetic modification may confer fitness advantages such as higher fecundity, increased growth rate, and greater tolerance to environmental conditions (Devlin et al. 2006). For domestic species such as dairy cattle, these traits include milk production, longevity, udder health, and fertility, among others. Some, such as udder health, may not be of particular concern for wild cattle, but would certainly be for a species whose telos—or purpose—as a domesticated animal is to produce milk year-round. Interestingly, in some cases the variation in these traits can be attributed only in a small extent to genetics because they depend more on the environment: feeding, housing, breeding management, welfare, and freedom; such traits might become a useful measurable proxy for the telos of wild and domesticated animals.

## Final Thoughts

It has been estimated that 44 species of land animals, 230 species of aquatic animals, and 269 of plants (land and marine) have been domesticated over the last 4000 years (Duarte et al. 2007); this is a just a fraction of the approximately 8.7 million of species that have been estimated for the Earth, even if around 86% have yet to be described. However, if we consider this in terms of biomass, humans and livestock outweigh all other wild vertebrates (Bar-On et al. 2018)—clear evidence of the impact of domestication on the biosphere.

The extraordinary success of animal domestication has been crucial in enabling human expansion, so it is interesting that the genetic engineering of wild and domesticated species is viewed as a threat to biodiversity, when it is rather the intensive farming of animals and plants that remains one of the main threats to the ecological balance in the biosphere, regardless of the modifying technology involved. Genetic engineering may spearhead the domestication of wild species of interest that prove difficult to tame, can help in rescuing endangered species, and can also be used to reduce animal suffering and diminish biological risks to humans—the possibilities are endless. The ethical issue of genetic engineering of animals may be approached as a question about the ethical boundaries of domestication, both for the individual (animal welfare) and for the species (biodiversity).

## References

Bar-On YM, Phillips R, Milo R (2018) The biomass distribution on earth. Proc Natl Acad Sci 115:6506–6511. https://doi.org/10.1073/pnas.1711842115

Beisel U, Boëte C (2013) The flying public health tool: genetically modified mosquitoes and malaria control. Sci Cult (Lond) 22:38–60. https://doi.org/10.1080/09505431.2013.776364

Bolland D, Kerin G, Stein J et al (2010) Defining environmental risk assessment criteria for genetically modified fishes to be placed on the EU market. https://doi.org/10.2903/sp.efsa.2010.EN-69

Brambell F (1965) Report of the technical committee to enquire into the welfare of animals kept under intensive livestock husbandry systems Cmd. (Great Britain. Parliament), Cmnd 2836 Volume 2836 of Command papers, H.M. Stationery Office

Bulfield G, Fisher E, Goodfellow P, et al (2001) The use of genetically modified animals. Science Advice Section, The Royal Society, 6 Carlton House Terrace, London SW1Y 5AG. https://royalsociety.org/-/media/Royal_Society_Content/policy/publications/2001/10026.pdf

Bull JW, Maron M (2016) How humans drive speciation as well as extinction. Proc R Soc B Biol Sci 283:20160600. https://doi.org/10.1098/rspb.2016.0600

Campbell KHS, McWhir J, RItchie WA, Wilmut I (1996) Sheep cloned by nuclear transfer from a cultured cell line. Nature 380(6569):64–66

Cernansky R (2017) Biodiversity moves beyond counting species. Nature 546:22–24. https://doi.org/10.1038/546022a

Chaible LM, Corat MA, Abdelhay E, Dagli ML (2010) Genetically modified animals for use in research and biotechnology. Genet Mol Res 9(3):1469–1482. https://doi.org/10.4238/vol9-3gmr867

Champer J, Buchman A, Akbari OS (2016) Cheating evolution: engineering gene drives to manipulate the fate of wild populations. Nat Rev Genet 17:146–159. https://doi.org/10.1038/nrg.2015.34

Charo RA, Greely HT (2015) CRISPR critters and CRISPR cracks. Am J Bioeth 15:11–17. https://doi.org/10.1080/15265161.2015.1104138

Corlett RT (2015) The Anthropocene concept in ecology and conservation. Trends Ecol Evol 30:36–41. https://doi.org/10.1016/j.tree.2014.10.007

Devlin RH, Sundström LF, Muir WM (2006) Interface of biotechnology and ecology for environmental risk assessments of transgenic fish. Trends Biotechnol 24:89–97. https://doi.org/10.1016/j.tibtech.2005.12.008

Driscoll CA, Macdonald DW, O'Brien SJ (2009) From wild animals to domestic pets, an evolutionary view of domestication. Proc Natl Acad Sci 106:9971–9978. https://doi.org/10.1073/pnas.0901586106

Duarte CM, Marba N, Holmer M (2007) ECOLOGY: rapid domestication of marine species. Science 316:382–383

Echelard Y, Ziomek CA, Meade HM (2006) Production of recombinant therapeutic proteins in the milk of transgenic animals. BioPharm Int 19:36–46

Forabosco F, Löhmus M, Rydhmer L, Sundström LF (2013) Genetically modified farm animals and fish in agriculture: a review. Livest Sci 153:1–9. https://doi.org/10.1016/j.livsci.2013.01.002

Garas LC, Murray JD, Maga EA (2014) Genetically engineered livestock: ethical use for food and medical models. Annu Rev Anim Biosci 3:559–575. https://doi.org/10.1146/annurev-animal-022114-110739

García-Sancho M (2015) Animal breeding in the age of biotechnology: the investigative pathway behind the cloning of Dolly the sheep. Hist Philos Life Sci 37:282–304. https://doi.org/10.1007/s40656-015-0078-6

Garnier E, Navas M-L, Fortunel C et al (2007) Let the concept of trait be functional! Oikos 116:882–892. https://doi.org/10.1111/j.2007.0030-1299.15559.x

Golovan SP, Meidinger RG, Ajakaiye A et al (2001) Pigs expressing salivary phytase produce low-phosphorus manure. Nat Biotechnol 19:741–745. https://doi.org/10.1038/90788

Hare D, Blossey B, Reeve HK (2018) Value of species and the evolution of conservation ethics. R Soc Open Sci 5:181038. https://doi.org/10.1098/rsos.181038

Hauskeller M (2005) Telos: the revival of an Aristotelian concept in present day ethics. Inquiry 48:62–75. https://doi.org/10.1080/00201740510015356

Herron LR, Pridans C, Turnbull ML et al (2018) A chicken bioreactor for efficient production of functional cytokines. BMC Biotechnol 18:82. https://doi.org/10.1186/s12896-018-0495-1

Jacobsen JC, Bawden CS, Rudiger SR et al (2010) An ovine transgenic Huntington's disease model. Hum Mol Genet 19:1873–1882. https://doi.org/10.1093/hmg/ddq063

Jennings BH (2011) Drosophila-a versatile model in biology & medicine. Mater Today 14:190–195. https://doi.org/10.1016/S1369-7021(11)70113-4

Lai L, Kang JX, Li R et al (2006) Generation of cloned transgenic pigs rich in omega-3 fatty acids. Nat Biotechnol 24:435–436. https://doi.org/10.1038/nbt1198

Lawrence DJ (2007) The four principles of biomedical ethics: a foundation for current bioethical debate. J Chiropr Humanit 14:34–40. https://doi.org/10.1016/S1556-3499(13)60161-8
Lee O, Green JM, Tyler CR (2015) Transgenic fish systems and their application in ecotoxicology. Crit Rev Toxicol 45:124–141. https://doi.org/10.3109/10408444.2014.965805
Looi FY, Baker ML, Townson T et al (2018) Creating disease resistant chickens: a viable solution to avian influenza? Viruses 10:1–10. https://doi.org/10.1016/j.pediatrneurol.2017.04.020
Maclean N, Laight RJ (2000) Transgenic fish: an evaluation of benefits and risks. Fish Fish 1:146–172. https://doi.org/10.1046/j.1467-2979.2000.00014.x
Menozzi D, Mora C, Merigo A (2012) Genetically modified salmon for dinner? Transgenic salmon marketing scenarios. AgBioforum 15:276–293. https://doi.org/10.1002/pat
Mozdziak PE, Pophal S, Borwornpinyo S, Petitte JN (2003) Transgenic chickens expressing β-Galactosidase hydrolyze lactose in the intestine. J Nutr 133:3076–3079. https://doi.org/10.1093/jn/133.10.3076
Nading AM (2015) The lively ethics of global health GMOs: the case of the Oxitec mosquito. BioSocieties 10:24–47. https://doi.org/10.1057/biosoc.2014.16
National Research Council (2002) Scientific and medical aspects of human reproductive cloning. The. National Academies Press, Washington, DC
Novak BJ (2018) De-extinction. Genes (Basel) 9(11):548. https://doi.org/10.3390/genes9110548
Perleberg C, Kind A, Schnieke A (2018) Genetically engineered pigs as models for human disease. Dis Model Mech 11:dmm030783. https://doi.org/10.1242/dmm.030783
Piaggio AJ, Segelbacher G, Seddon PJ et al (2017) Is it time for synthetic biodiversity conservation? Trends Ecol Evol 32:97–107. https://doi.org/10.1016/j.tree.2016.10.016
Polejaeva IA, Ranjan R, Davies CJ et al (2016) Increased susceptibility to atrial fibrillation secondary to atrial fibrosis in transgenic goats expressing transforming growth factor-β1. J Cardiovasc Electrophysiol 27:1220–1229. https://doi.org/10.1111/jce.13049
Price B, Cotter J (2014) The GM contamination register: a review of recorded contamination incidents associated with genetically modified organisms (GMOs), 1997-2013. Int J Food Contam 1:5. https://doi.org/10.1186/s40550-014-0005-8
Pugh J (2016) Driven to extinction? The ethics of eradicating mosquitoes with gene-drive technologies. J Med Ethics 42:578–581. https://doi.org/10.1136/medethics-2016-103462
Rollin B (2015) On Telos and genetic manipulation. Between Species An Online J Study Philos Anim 2:88–89. https://doi.org/10.15368/bts.1986v2n2.9
Russell WMS, Burch RL (1959) The principles of humane experimental technique. Methuen https://caat.jhsph.edu/principles/the-principles-of-humane-experimental-technique
Sinclair KD, Corr SA, Gutierrez CG et al (2016) Healthy ageing of cloned sheep. Nat Commun 7:12359
Takeda K (2013) Mitochondrial DNA transmission and confounding mitochondrial influences in cloned cattle and pigs. Reprod Med Biol 12:47–55. https://doi.org/10.1007/s12522-012-0142-9
Teletchea F (2017) Wildlife conservation: is domestication a solution? In: Global exposition of wildlife management. InTech. https://doi.org/10.5772/65660
Telugu BP, Park KE, Park CH (2017) Genome editing and genetic engineering in livestock for advancing agricultural and biomedical applications. Mamm Genome 28:338–347. https://doi.org/10.1007/s00335-017-9709-4
Uchola BE (2016) Resource domestication: an introduction to biodiversity and wildlife in agriculture. Am J Agric For 4:23. https://doi.org/10.11648/j.ajaf.20160402.13
Verhoog H (2000) Defining positive welfare and animal integrity. In: Hovi M, García-Trujillo R (eds) Proceedings of the second NAHWOA workshop. Cordoba, pp 1–12
Vize PD, Michalska AE, Ashman R et al (1988) Introduction of a porcine growth hormone fusion gene into transgenic pigs promotes growth. J Cell Sci 90. (Pt 2:295–300

Wall R, Laible G, Maga E et al (2009) Animal productivity and genetic diversity: cloned and transgenic animals - animal agriculture's future through biotechnology, part 8. Issue Pap - Counc Agric Sci Technol

Waltz E (2017) First genetically engineered salmon sold in Canada. Nature 548:148–148. https://doi.org/10.1038/nature.2017.22116

Warkentin T (2009) Dis/integrating animals: ethical dimensions of the genetic engineering of animals for human consumption. In: Gigliotti C (ed) Leonardo's choice: genetic technologies and animals, pp 1–256

# Part V
# Risk Assessment

# Hypothesis-Led Ecological Risk Assessment of GM Crops to Support Decision-Making About Product Use

**Alan Raybould**

**Abstract** Ecological risk assessment assesses the probability that harmful effects to biodiversity will arise from using a particular genetically modified (GM) crop and the severity of those effects should they occur. Potential harmful effects to biodiversity include reductions in the abundance of species that are valued for their own sake or for the ecological services that they provide. Mechanisms by which harm could occur include exposure of species to toxic substances produced by the crop or displacement of species by the crop because it is able to colonise new habitats. Ecological risk assessments should contribute to decisions about whether to approve certain uses of the GM crop. They do this most effectively by being hypothesis-led, that is, testing hypotheses that the crop does not possess properties that indicate that ecological risk is unacceptable. These properties are defined after consideration of the aims of the decision-maker, which in most cases is a regulatory authority enacting government policy. Ecological risk assessment is ineffective when it tests null hypothesis that the GM crop is no different from a similar non-GM crop without first defining what differences should be regarded as important. Such data-led risk assessments offer no clear support for decision-making and tend to create controversy as decisions appear to be made arbitrarily. Current regulatory risk assessments for GM crops follow a mixture of hypothesis-led and data-led approaches.

**Keywords** Problem formulation · Risk analysis · Acceptable risk · Decision-making criteria · Gene flow · Non-target organisms · Weediness potential

A. Raybould (✉)
Science, Technology and Innovation Studies, The University of Edinburgh, Edinburgh, UK

Global Academy of Agriculture and Food Security, The University of Edinburgh, Midlothian, UK
e-mail: alan.raybould@ed.ac.uk



A. Chaurasia et al. (eds.), *GMOs*, Topics in Biodiversity and Conservation 19,
https://doi.org/10.1007/978-3-030-53183-6_14

## Introduction

The use of genetically modified (GM) crops is strictly regulated worldwide. Before these products may be cultivated in a particular country, specific uses must be approved by a regulatory authority with the power to make these decisions (Jaffe 2004). An important consideration for decision-makers is the likelihood and severity of the harm to biodiversity that may be caused by cultivation of the GM crop in question, that is, the amount of ecological risk posed by cultivating the crop.

GM crops seem to pose no greater ecological risk than do crops developed using other breeding techniques (Carpenter 2011; Romeis et al. 2019). However, this does not mean that genetic modification could never produce a crop that is harmful to biodiversity. Hence, it is important that the ecological risks from growing a given GM crop are assessed and used in decision-making about its proposed cultivation.

Regulation of GM crops imposes significant costs on product developers and on society generally (Raybould and Poppy 2012; Smyth 2017). It is important, therefore, that the cost of regulation does not outweigh its benefits. Ensuring that ecological risk assessment (ERA) to support regulatory decision-making about GM crops is efficient and effective is crucial component in developing such proportionate regulation.

In this article, I argue that efficient and effective ERA is led by developing and testing hypotheses that support decision-making. The method may be summarised briefly:

- Determine clear objectives of regulatory policy.
- Use the objectives to set criteria for acceptable ecological risk.
- Test hypotheses that specific uses of the GM crop meet these acceptability criteria.
- Use the results of the tests to support decision-making about whether to approve those uses.

The method emphasises the importance of policy in developing hypotheses for testing in ERA. Risk is not a completely objective property, like, say, the charge on the electron. Instead, it is a combination of objective and subjective properties: the likelihood of an event happening and the seriousness of the harm represented by the event's occurrence. Decision-makers may agree completely on the likelihood of an event but disagree profoundly about the amount of risk because their divergent policy objectives lead to different views on what represents serious harm.

The method also emphasises that the priority for hypotheses is to support decision-making, not necessarily to maximise the production of knowledge. In philosophical terms, hypotheses in risk assessment should value practical utility not epistemic utility and hence favour accuracy and relevance and seek only precision that is sufficient for decision-making (Niiniluoto 1993, 1994; Hendry 2001; Hill and Sendashonga 2003; Calvert 2004).

The method does not imply that scientific rigour is sacrificed. Hypotheses in risk assessment should be tested just as rigorously as those in basic research

(Fenner-Crisp and Dellarco 2016; Kaltenhäuser et al. 2017), but they try to answer different questions. In essence, basic research seeks answers to "What will happen if we cultivate this GM crop?" whereas risk assessment seeks answers to "What is the likelihood of specified harmful events happening if we cultivate this GM crop?" (Raybould 2010a).

A final consideration is that the method discourages the testing of null hypotheses of the form that a GM crop and non-GM progenitor are not phenotypically different or do not cause different ecological effects when cultivated. While testing null hypotheses often produces little knowledge (Stephens et al. 2007) and hinders decision-making (Raybould and Macdonald 2018), our increasing ability to make multiple measurements of organisms or ecosystems very quickly using various "omics" methods is driving calls for even greater use of null-hypothesis-led approaches in evaluating GM crops (Christ et al. 2018). For omics approaches to be useful in risk assessment, they should test a hypothesis about the likelihood of a specified harmful effect, not simply accumulate data on differences between a GM crop and a comparator.

The sections below develop the arguments behind the hypothesis-led risk assessment scheme outlined above.

## What Is Risk Assessment?

### *What Is Risk?*

In everyday life, we make risk assessments all the time. Almost all our decisions involve intuitive or explicit consideration of what could go wrong if we choose one course of action over another. Some decisions, such as whether to permit the use of a pesticide or the cultivation of a GM crop, are out of our hands; they are taken by government officials after considering advice from scientific experts. In these circumstances, risk assessment is often seen as a highly technical activity with minimal or no non-scientific input. For example, risk to human health from using chemicals is often defined as hazard multiplied by exposure, with hazard being the toxicity of the chemical and exposure the amount of the chemical with which people come into contact. Armed with an estimate of toxicity from animal studies and a prediction of exposure derived from the chemical's physical and chemical properties and its proposed pattern of use, a risk assessor may feel confident in assigning a level of risk to human health were the chemicals to be used (Dorne and Fink-Gremmels 2013).

Treating risk as a function of hazard and exposure means that risk assessments may not be explicit about "what could go wrong" if a particular decision is made. Experts will immediately understand that hazard and exposure help to predict human mortality and morbidity resulting from exposure to a chemical. And as mortality and morbidity are almost always regarded as harmful effects, determination of hazard and exposure may appear to be all that is needed to conclude how much risk the use of the chemical poses. However, this overlooks a crucial point: morality and ethics, not science, define human mortality and morbidity as harmful effects. Similarly, the value we place on biodiversity and hence the size of the risk we assign to a predicted reduction in biodiversity must come from cultural, religious or economic considerations. Indeed, the idea that health and environmental risks are a separate category from socio-economic risks (e.g. Binimelis and Myhr 2016) seems wrong: all risks are socio-economic.

Failure to define harm explicitly in terms of human values can have serious consequences. Scientific experts may make incorrect assumptions about what people regard as the ultimate harmful effects of an activity and how much they fear them – not only "what could go wrong" but also "how much would we care if things were to go wrong" – leading to loss of trust in decisions founded on "science-based" risk assessment (Frewer et al. 2003).

Further problems arise from the treatment of risk assessment as purely scientific. Decision-makers may fail to define any harmful effects and instead expect risk assessors to infer them from scientific analysis (Evans et al. 2006). ERA is particularly prone to this problem because definitions of ecological harm are more ambiguous than those for human health (Sanvido et al. 2012). Failure to define harm explicitly at the start leads to inefficient and ineffective risk assessments, and often decision-making based on such assessments appears arbitrary (Raybould and Macdonald 2018).

Also, in some aspects of risk assessment for GM crops, it may not be obvious how to classify an event, and trying to force the properties of GM crops into the categories "hazard" and "exposure" may lead to some risks being ignored. If a GM crop hybridised with a wild species, many scientific experts would regard the presence of transgenes outside agriculture as exposure (Poppy 2004). The presence of the transgene would pose a risk only if it had hazardous properties, perhaps producing a toxin that has adverse effects on wildlife or causing the population of the wild species to expand leading to ecological damage. Many non-experts, however, would regard the presence of the transgene in a wild species as harm, the presence of the gene being a form of "genetic pollution" violating the purity of the species' gene pool (Bruce 2003). Similar considerations apply to GM crops themselves spreading into non-agricultural habitats: is their presence harmful per se, or must they have some hazardous property to pose a risk, and who is to make this decision?

The problems caused by highly technical definitions of risk – failure to define harm unambiguously or at all, failure to consider some scenarios by which harm may arise, loss of trust in decision-making – can be alleviated, at least partially, by defining the risk of an activity as a combination of the likelihood that it will cause harm and the severity of the harm if it were to occur. This formulation has two major advantages. First, it defines risk in terms of the harm the activity may cause rather than as properties of components of that activity, such as the toxicity and persistence of a chemical. If harm is defined in a manner acceptable to people likely to be affected by the activity, which may not be straightforward (Jasanoff and Hurlbut

2018), then decisions based on the assessment of the likelihood and severity of harm may engender greater trust than those based on estimates of hazard and exposure.

The second big advantage of defining risk as likelihood and severity of harm is that it focuses risk assessment on mechanisms by which harm may arise from the activity under assessment. It thereby discourages an open-ended search for hazards, which in the absence of definitions of harm becomes an attempt to catalogue all possible effects (beneficial, harmful and neither) that the activity may cause (Raybould 2010a). Defining mechanisms, or pathways, by which the specific activity under assessment may lead to clearly defined harmful effects enables risk to be characterised by the testing of hypotheses with clear relevance to decision-making. Testing such risk hypotheses with existing data and requiring further data only if the testing is insufficient for decision-making produce efficient and effective risk assessments. We now turn to the formulation of risk hypotheses.

## *Risk Assessment as Hypothesis Testing*

Defining risk as the likelihood and severity of harm means that for a proposed activity, such as cultivation of a particular GM crop, to pose non-negligible risk, we must be able to describe at least one plausible scenario, or series of events, that leads from the activity to a harmful effect. As we have seen, harm must be defined by the decision-maker, who may be an individual person or a public or private institution that sets its own criteria for defining harm. In the case of GM crops, the decision-maker is often an official body that has the task of enforcing regulations that implement public policy. Relevant policy may have fairly broad objectives, such as protecting biodiversity, and the first task in hypothesis-led ERA must therefore be to translate the broad objectives into specific targets that are amenable to scientific analysis; these specific targets are sometimes called operational protection goals (Sanvido et al. 2012; Garcia-Alonso and Raybould 2014; Devos et al. 2015).

The next stage is to formulate risk hypotheses. All ERAs for GM crops can be regarded as having risk hypotheses with the same basic form: growing GM crop A does not lead to unacceptable risks X, Y, Z, etc., where A is the crop being assessed and X, Y, Z, etc. are the probability and severity of different harmful effects that plausibly result from growing crop A. Harmful effects are adverse effects on the operational protection goals. A harmful effect is implausible if no pathway to harm (Fig. 1) can be constructed that is not "obviously" highly unlikely based on current knowledge (Raybould 2010b). Criteria for judging acceptability are usefully summarised by Sanvido et al. (2012) and Dolezel et al. (2018). Here, it is sufficient to note that acceptability may be judged relative to the opportunities (the probability and value of beneficial effects) from growing crop A or against a threshold; in other words, risks above the threshold are unacceptable regardless of the size of the opportunities (Rozell 2018).

Accepting or rejecting the overall risk hypothesis usually involves a test of a subsidiary hypothesis. A useful technique for formulating these hypotheses is to

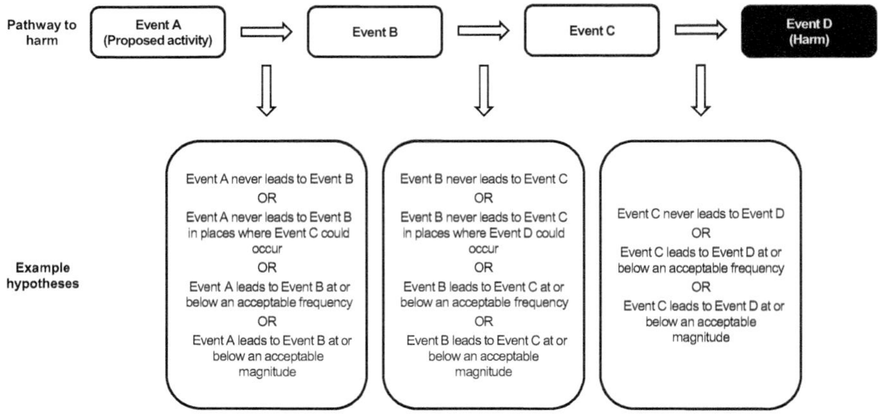


**Fig. 1** A generic pathway to harm with examples of the types of hypothesis that may be tested to characterise risk

specify events that must occur for harm to be realised; these events may be called a pathway to harm (Fig. 1). A hypothetical example may help to demonstrate the use of such pathways and illustrate the essential features of good risk assessment based on hypothesis testing.

Suppose an ERA is required for decision-making about cultivation of GM maize A in country C. Crop A was produced by genetic modification of a non-GM maize to make it produce an insecticidal protein (IP) for control of a soil-dwelling beetle pest. Based on the environmental and agricultural policy objectives of country C, one harmful potential effect of growing GM maize A is reduction in the population size of a valued beetle V. This species provides biological control of weeds by consuming their seeds. Beetle V may also consume seeds of GM crop A and thereby be adversely affected by the toxicity of IP. These adverse effects could lead to abundance of beetle V being lower in fields where GM maize A has been cultivated compared with fields where non-GM maize varieties have been cultivated.

These considerations lead to a simple pathway to harm: cultivation of GM maize A (proposed activity) → production of seeds containing IP → consumption by beetle V of spilled seeds containing IP → adverse effects of IP on beetle V → reduction of population size of beetle V (the harmful effect). This pathway could be extended to include increases in weed abundance caused by loss of beetle V and the economic and environmental costs of that increase in abundance. However, we will take decrease in beetle V numbers to be the effect of ultimate concern.

Unacceptable risk might be defined in terms of the probability and size of the reduction in the population size of beetle V in fields following cultivation of GM maize A relative to cultivation of non-GM maize, for example, a 50% probability of at least a 50% reduction for at least 6 months after harvest of GM crop A.

Direct testing of the hypothesis "cultivating GM maize A in country C does not reduce the population size of beetle V by 50% or more for longer than 6 months after harvest" would require a long and expensive programme of field trials that may have limited statistical power to detect the effect of interest (Perry et al. 2003). Also, conducting such a programme would require a decision to allow crop A to be grown in country C – the very event the ERA is intended to inform.

However, it is not necessary to test this hypothesis directly to demonstrate that an unacceptable decrease in population size is unlikely to occur and thereby show that the risk is acceptable. Rather than testing the cumulative probability of all the steps in the pathway using field studies, acceptable risk can be demonstrated by showing that a single step in the pathway has an acceptably low probability. Hence, one option is to demonstrate that if beetles were to be exposed to IP via seeds of GM maize, they would suffer no adverse effect, or more precisely, any adverse effect would be insufficient to lead to unacceptable reduction in population size.

The first step in this approach would be to define a criterion for agreeing that the risk is acceptable. A simple measure of risk is a toxicity/exposure ratio or TER, which compares a measure of toxicity obtained in a laboratory study under standard conditions with an estimate of exposure (Damalas and Eleftherohorinos 2011). A typical TER is $LC_{50}$/PEC, where $LC_{50}$ is the concentration of a substance that kills 50% of a test population of an organism in the laboratory and the PEC, or predicted environmental concentration, is a worst-case estimate of the exposure of the organism to the substance following its proposed use. Acceptability of risk is set as a trigger value; if the TER exceeds the trigger, risk is acceptable (Hamer 2000).[1]

A high trigger value is conservative not only because most activities that would cause serious harm are identified as posing unacceptable risk but also because many activities that would cause little or no harm and which may provide great benefit would also be identified as posing unacceptable risk. Conversely, a low trigger value reduces the chances of wrongly deeming beneficial activities to pose unacceptable risk but may fail to identify many activities that would be seriously harmful. Thus, setting the trigger value is partly a matter for expert scientific judgement about how accurately the trigger value predicts effects of the activity under real conditions (Forbes and Calow 2002). It is also partly a matter for policy about how opportunities and risks should be balanced (Chapman et al. 1998).

Let us suppose that we set the criterion for acceptability of the risk to beetle V from cultivating GM maize A in country C as $LC_{50}/PEC > 5$. $LC_{50}$ is the concentration of IP that kills 50% of test population of beetle V in a laboratory bioassay, and we may set PEC to be the concentration of IP in the seeds of GM maize A; this PEC is conservative because it assumes that the diet of beetle V comprises only seeds of the GM crop. We could start from scratch and acquire new data to estimate a TER and test the hypothesis that it exceeds 5. We might need to develop a laboratory bioassay to measure the effects of IP on beetle V and estimate the $LC_{50}$. We could also grow GM maize A in confined field trials in country C to estimate the concentration of IP in the seeds and use this value as the PEC.

[1] In some risk assessment schemes, exposure is divided by toxicity to give a risk quotient (RQ); risk is acceptable if the RQ is below a specified level of concern (Hamer 2000).

However, it is good risk assessment practice to test hypotheses with existing data and conduct new experiments only if the test is insufficient for decision-making. Using existing data saves time and money and may reduce overall risk because new product uses that pose lower risk than current practices are not delayed unnecessarily (Cross 1996). In our hypothetical example, developing a bioassay for beetle V in a timely manner may prove difficult or impossible (see Romeis et al. (2011) for criteria for designing reliable bioassays), and field trials of GM maize A would require regulatory approvals, again delaying decision-making perhaps unnecessarily.

A good test of the hypothesis TER > 5 can be made with toxicity data on IP from species taxonomically related to beetle V, especially if we know that the spectrum of activity of IP against pest species is limited to species in one or a few closely related taxonomic families (Romeis et al. 2013). On the exposure side, we could use data on the concentration of IP in seeds of GM maize grown in a different country, N, especially if environmental conditions and agronomic practices are similar to those in country C (Garcia-Alonso et al. 2014).

Suppose we have toxicity data on IP from beetle W, which is closely related to beetle V, and data on the concentration of IP in seeds of GM maize A grown in country N, which neighbours country C. Calculations based on these data give $LC_{50}$/PEC = 10. A number of scenarios could develop:

- We accept that data from species W and country N are sufficiently representative of species V and country C and conclude that the hypothesis TER > 5 for beetle V in country C has been corroborated satisfactorily; hence, we determine the risk to beetle V from cultivating GM maize in country C to be acceptable and require no further work to characterise this risk (at least via this pathway).
- We have serious concerns that data from species W and country N are not sufficiently representative of species V and country C and hence conclude that TER = 10 for W and N is insufficient corroboration of the hypothesis TER > 5 for V in C; hence, we cannot conclude that the risk to beetle V from cultivating GM maize in country C is acceptable from the available toxicity and protein expression data, and further work to characterise the risk is required. One option is to look at published research on the diet of beetle V. If several studies show that the proportion of maize seeds in the diet of beetle V in country N never exceeds 50%, we could halve the PEC and hence double the TER to 20. We may decide that the TER for W and N based on this "refined" exposure assessment is sufficient corroboration of the hypothesis TER > 5 for V and C and deem the risk to beetle V acceptable via this pathway.
- We have minor concerns that data from species W and country N are insufficiently representative, but they are not strong enough to require further risk characterisation. Instead, if cultivation of GM maize A in country C is acceptable in all other respects, we might recommend monitoring beetle V abundance to check that our conclusions are sound. In taking this decision, we transfer further work from the realm of risk assessment to that of risk management (Fig. 2).

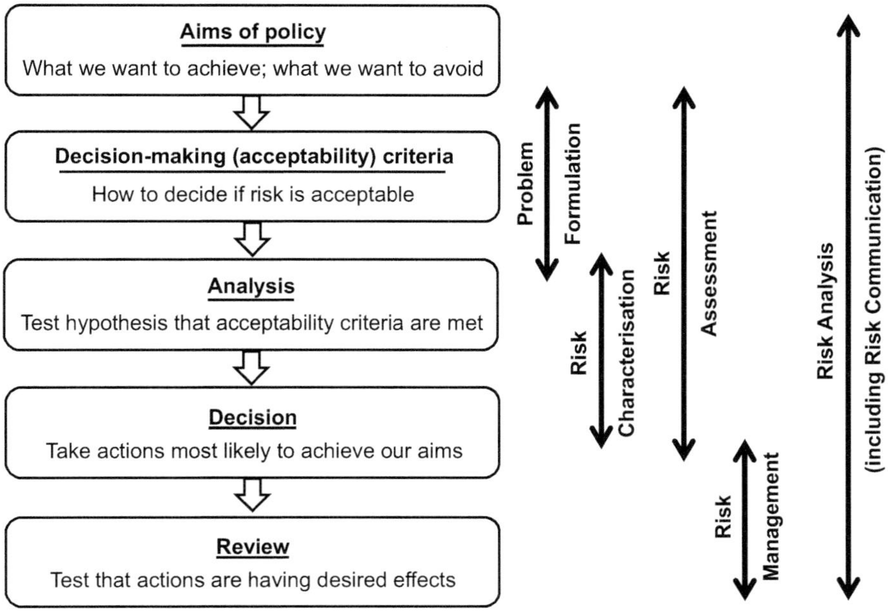


**Fig. 2** A conceptual model showing the position of risk assessment in risk analysis. Risk assessment comprises a planning phase (problem formulation) and an analysis phase that feeds into decision-making (risk characterisation)

- We categorically reject the idea of using data from species W and country N and insist that data from species V and country C are needed to make decisions about the acceptability of risk. In these circumstances, it is important to be clear why this decision is reached; for example, is the decision based on science (e.g. we have good reason to believe that species W is far less sensitive to IP than is species V), is it based on a regulatory requirement (e.g. that expression data must be obtained from country C) or is it for reasons of risk communication (e.g. low public acceptance of data from country N). Any or all of these reasons may be valid, but identifying the real underlying reason is important for designing experiments to acquire the necessary data.

This hypothetical example illustrates some key points for risk assessment. Setting acceptability criteria and testing that the proposed use of the product meets those criteria are effective ways to identify and arrange scientific knowledge to help decision-making. By having an explicit objective – ensuring that the population of beetle V does not fall by more than 50% for more than 6 months after cultivation of GM maize A in country C – and an explicit decision-making criterion, $LC_{50}/PEC > 5$, risk assessors are able to organise existing data on protein toxicity and its concentration in seeds most effectively for decision-making. Risk assessors can also concentrate on accuracy over precision: high confidence that the ratio of $LC_{50}/PEC$

is greater or less than 5 is far more important than high precision in the estimates of the $LC_{50}$ and the PEC.

Setting acceptability criteria does not remove judgement from decision-making. First, setting the criteria to deliver policy objectives requires judgement. Second, even with clear acceptability criteria, decision-makers may legitimately come to different conclusions based on the same data; in the example above, approve cultivation of GM maize A in country C, approve after a refined exposure assessment, conditionally approve and require monitoring or not approve and ask for the production of new data. However, because the difficult work of setting policy objectives and acceptability criteria has been done, discussion can focus on technical questions such as the reliability of predicting the toxicity of IP to beetle V based on data from beetle W. The risk assessors do not have to become de facto policymakers.

Another key point is that risk assessment should be flexible. If a hypothesis can be tested sufficiently with existing data, new studies should not be required. When asking for new data, we need to be sure that the data to be collected will provide a test of the hypothesis that acceptability criteria for the proposed activity are met that is at least as rigorous as that provided by existing data (Raybould 2006). If the data provide a weaker test, or no test at all, then they should not be required for the purposes of risk assessment. If the data are required for regulatory or risk communication reasons, that should be explicit when designing the study or other means of acquiring them. If data are continually required for regulatory and risk communication reasons, but not for risk assessment, then a review of the regulations or risk communication strategy may be sensible, particularly if the data are time-consuming and expensive to collect or involve experiments on animals.

In summary, risk assessment is a policy-led activity. Acceptability criteria are derived from policy objectives, not from scientific analysis. And while good science – that is, science that rigorously tests the hypothesis that acceptability criteria are met – is crucial for risk assessment, ultimately, decisions must be judgements. Science cannot prove the hypothesis that use of a product will never have a particular effect; therefore, all decisions reflect opinions that certain criteria have, or have not, been demonstrated with sufficient certainty. Hence, policy is important not only for setting questions at the beginning of the risk assessment but also for deciding when the risk assessment should be concluded.

## *Risk Assessment in Decision-Making: Risk Analysis*

Risk assessment is a tool that helps decision-makers achieve their policy aims. The setting of policy aims and methods for decision-making is part of risk analysis, a broader activity that includes and guides risk assessment (Johnson et al. 2007). Regular demands are made for improvements to risk assessment for GM crops (e.g. Pott et al. 2018). These demands are rarely, if ever, associated with examples of where current risk assessment practice has failed and hence why it is in need of improvement. Probably what the authors mean is that new scientific methods are available that could be used to measure certain attributes of GM crops that older methods cannot. However, unless these measurements improve risk analysis, they offer no improvement to risk assessment.

If risk assessment is viewed primarily as a scientific activity (McHughen 2007), it is not surprising that using new scientific knowledge and technical advances in acquiring data are seen as self-evident improvements to risk assessment. Greater accuracy and precision in the predictions of ecological models (Houlahan et al. 2017), conceptual advances in describing desired attributes of ecosystems (Vighi and Rico 2018) and greater speed and sensitivity in performing ecological surveys (Macel et al. 2010; Bohan et al. 2017) are all valuable results of ecological research. However, achieving greater accuracy and precision of scientific predictions, using new descriptors of ecosystem function or collecting larger amounts of data with more modern equipment are not necessarily improvements to ERA.

An improvement in risk assessment ought to mean that decisions based on the new assessments are better than those based on the ones they replace, where "better" means that the decision-makers' aims are more likely to be achieved and the improvement in the decisions is worth any extra effort. Greater precision may not help if acceptability criteria are not similarly precise, new descriptors of ecosystem function may not represent policy aims, more data do not help if they do not test a hypothesis that acceptability criteria are met and greater accuracy may even be detrimental if the increase in the probability of selecting the best option is outweighed by increases in the cost and time needed to make the decision.

Goldstein (2011) succinctly described the problems of trying to shoehorn scientific advances into risk assessment:

> My particular concern is that many of the repeated efforts to reform risk assessment for chemical risks impair the ability to use what has been a reasonably effective tool for risk management and *lose sight of public health and environmental objectives* [emphasis added].

Viewing risk assessment as a decision-support tool helps to distinguish between real improvements (achieving objectives; practical utility), scientific advances (basic research; epistemic utility) and analytical advances (ability to make measurements; possibly no or even negative utility) and to analyse the effectiveness of various aspects of ERA for GM crops (see section "Major areas of risk assessment of GM crops").

The first step in risk analysis is to determine the objectives of the activity we are proposing to undertake, which in our case is the cultivation of a given GM crop. The term risk analysis suggests an examination of unintended harmful effects; however, risk can also be regarded as the probability of failing to realise an objective. In addition, deciding the amount of risk that is acceptable requires consideration of opportunities; hence, risk analysis will usually require identification of potential benefits ("what we want to achieve") as well as potential harms ("what we want to avoid") of the proposed activity.

Aims are set by policy. Most ERAs for GM crops are shaped by regulatory policy formulated by governmental bodies; however, personal preferences or the aims of public or private institutions may also be regarded as "policy" depending on who is making the decision and why. Regulatory policy for GM crops could be drawn narrowly, perhaps focusing exclusively on the safety of products to people and the environment or, more widely, bringing in relevant aspects of, among others things, agricultural, health, industrial and rural policies. A narrow focus would aid the identification of clear aims, whereas a broader approach would probably require trade-offs between the objectives of various policies. Regulatory policy that has explicitly considered trade-offs and explained how those trade-offs have been made will probably work better than policy focused on a single objective, such as environmental protection. A narrow focus risks having contradictory aims across a range of policy areas (Barling et al. 2003; Masip et al. 2013) and may lead people to think that their concerns have been ignored. Explaining that trade-offs are inevitable, admitting that there is no perfect solution, giving a clear explanation of the risks and opportunities that have been considered and acknowledging, but not deferring to, those who would make different compromises may reduce controversy, at least among people motivated by interests not ideology (Tait 2001).

The next step is to convert the broad aims of policy into acceptability criteria. In part, this stage comprises translating the broad aims of policy into unambiguous targets (sometimes called operational protection goals) that are amenable to scientific analysis; for example, we might interpret the aim of protecting biodiversity as maintaining the abundance of a particular species. The choice of species may be because we value it for cultural reasons, because we think that it is a good indicator of the status of many other species or because it provides an economically valuable service, such as pollination. The concept of ecosystem services is potentially useful in selecting species or processes for protection; it acknowledges that species may be valued for different reasons and, in principle, enables comparison and prioritisation of the different types of value (Forbes and Calow 2012; Schäfer 2012).

Next, we should try to define what change to this species or process we would regard as unacceptable. This is often the most difficult step in ecological risk analysis because the definition may need to take into account many factors and reconcile many different opinions. Even defining change in abundance could be tricky given natural fluctuations in population size (Turchin and Taylor 1992). We would also need to consider what type of change would be regarded as harm. For most species that we are interested in protecting, a reduction in their population size would be regarded as a harmful effect, but this may depend on exactly where and when the reduction occurs and the reason that we value the species concerned; it is conceivable that an increase in abundance of an otherwise valued species could be regarded as harmful if it occurred in the wrong place, at the wrong time or above a certain threshold.

Finally, we would need to define acceptable risk: a combination of the probability and severity of harm that we would find unacceptable if exceeded. A key decision is whether the risk posed by the proposed activity will be considered against its potential benefits (utilitarian decision-making) or against a threshold regardless of any potential benefit (ethical or deontological decision-making) (Sanvido et al. 2012; Rozell 2018).

Ideally, these considerations would lead us to a formulation of the following type: effect X of magnitude of Y or less with probability of Z or less is acceptable. It is unlikely that many disparate and potentially conflicting aims can be operationalised to this extent. However, the process of trying to reach such a definition is valuable because it makes us consider what is important in our decision-making. Definitions of acceptable risk can be left qualitative or categorical (e.g. no worse than now) provided they usefully guide scientific analysis and help the setting of acceptability criteria.

Even if we can agree on precise quantification of X, Y and Z in our definition of acceptable risk, we are unlikely to have the scientific wherewithal to make precise and accurate predictions about the values of X, Y and Z if we were to approve the cultivation of a certain GM crop. However, the process of thinking about the combination of values of X, Y and Z that indicate acceptable ecological risk should enable us to identify acceptability criteria in our GM crop.

Regarding ERA as similar to the breeding of a new plant variety is helpful for showing how acceptability criteria may be developed (Raybould 2019). At some high level, breeders will have aims set by their organisation's policy, for example, making money, maintaining a good reputation and being recognised as innovative by their peers. To meet these aims, breeders may decide to develop a variety of maize that resists a new strain of a virus. The breeders will have some idea of what a successful new variety needs to achieve, perhaps a level of yield protection under heavy virus pressure necessary to gain a target market share and profitability. Yield protection is the breeders' equivalent of a risk assessor's operational protection goal. Breeders will also have an idea of a minimum acceptable probability that a variety will meet these objectives under real conditions of use. Adding the probability element gives the equivalent of acceptable risk, that is, acceptable opportunity.

Breeders are unlikely to be able to judge the probability that the progeny of a particular cross can be developed into a commercial variety that will fulfil his aims under actual conditions of use without lengthy evaluation. And the breeders may have the progeny of many crosses to assess. Hence, they will set up a quick test to decide which plants are worth further evaluation, perhaps inoculating seedlings with the virus and selecting those that stay green for two weeks and discarding those that turn yellow. Staying green for two weeks is an acceptability criterion for the breeders opportunity assessment and corresponds to an acceptably criterion in risk assessment, except that risk assessment tends to focus on traits that we want to avoid not those that we want to encourage (Raybould 2019). The pathway to harm technique described below can help to identify these traits which are also discussed further later.

Failure to meet the acceptability criteria in an ERA does not necessarily mean that the risk is unacceptable, only that our quick screening test has failed to demonstrate that the risk is acceptable. Acceptability criteria should favour minimising false negatives and accept that false positives may occur; thus, GM crops posing acceptable risk can be quickly and confidently identified. In our example above, meeting an acceptability criterion of TER > 5 may prompt a decision that no further work is needed to characterise toxicity and exposure. Failure to meet the criterion could trigger further work to refine the toxicity or exposure estimates or lead to a decision to stop development of the particular crop as a potential product. The decision to do more work or discontinue development would depend, among other things, on by how much the acceptability criterion was missed and the size of the potential opportunity. In contrast to risk assessors, breeders usually wish to minimise false positives to avoid wasting resources on developing a variety that fails to provide acceptable opportunity. However, the concept is the same: identifying plants that warrant further analysis using a quick and simple screening test.

A plan to test the hypothesis that specific uses of the GM crop meet the acceptability criteria is the next stage in risk assessment. The translation of policy aims into operational protection goals and acceptability criteria, and creation of a plan to test the hypothesis that those criteria are met, is called problem formulation (Fig. 2). This is the planning stage of the risk assessment (Norton et al. 1992). As we have seen, problem formulation may have to make difficult choices to distil complicated policy objectives into simple acceptability criteria. However, this effort will be worthwhile for the clarity and predictability it brings to decision-making and its focus on identifying relevant data. Working in the opposite direction, collecting lots of data and then trying to work out what is relevant are inefficient and ineffective and risk setting policy on the basis of potentially spurious statistical significance (Raybould and Macdonald 2018).

With hypothesis testing, risk assessment moves from problem formulation to risk characterisation. As described in section "Risk assessment as hypothesis testing", hypotheses can be tested with existing data. The adequacy of existing data to complete a risk assessment depends on their reliability and relevance. Reliability is based on the methods by which the data were obtained, in particular the suitability of the experimental methods to provide findings that are clear and plausible. Klimisch et al. (1997) propose criteria by which reliability can be judged and many regulatory authorities and intergovernmental bodies have detailed guidance on judging the reliability of data (Fenner-Crisp and Dellarco 2016). The most reliable data are those obtained using internationally recognised test guidelines, and peer-reviewed papers may also be a source of reliable data.

Relevance relates to the ability of the data to test the hypothesis that the acceptability criteria are met. Powerful tests, in other words those that are likely to show the hypothesis to be false if it really is false, are more relevant than weak tests that have limited ability to reject a false hypothesis. If a study is incapable of rejecting a hypothesis in any circumstances, then it is irrelevant. This situation is often found in phenotypic or compositional analysis trials in which a GM crop is compared with a genetically similar non-GM crop for numerous characteristics (see section "Weediness and invasiveness potential of GM crops"). If the study measures leaf length and we can envisage no difference in leaf length between the GM and non-GM line that would indicate unacceptable ecological risk from growing the GM crop, then a study of leaf length is irrelevant for ERA. This conclusion does not mean that measurement of leaf length never has value; for example, changes to leaf length may make a crop commercially unacceptable, perhaps because it is difficult to harvest. It does, however, mean that data on leaf length should be required for

ERA only if there is a plausible means by which a change to leaf length could result in ecological harm.

If sufficient relevant and reliable data are available and they corroborate the hypothesis that acceptability criteria are met, the risk assessor may conclude that there are adequate data to complete the risk assessment. If further data are required, because existing data either inadequately corroborate the hypothesis of acceptable risk or reject it, then the same criteria used to evaluate existing data may be used to design new studies. A fruitful approach to designing studies for GM crop ERA has been to adapt ecotoxicology study guidelines used for pesticide ERA; common changes include extending the duration of the tests to account for slower-acting toxins and using dietary instead of contact exposure to the toxins (Romeis et al. 2011; Karstens et al. 2012).

If problem formulation is done well, risk characterisation will focus on testing hypotheses that acceptability criteria are met. Testing hypotheses that do not relate directly to acceptability criteria may impair decision-making, either by diverting resources away from studies that would improve other decisions or by confusing the present decision by obscuring what may be important. In particular, risk characterisation should avoid testing null hypotheses that the product or product use under evaluation does not differ from another product or product use. Such testing will always reveal differences at some level, and large profiling experiments may create huge burdens on risk assessors who have to interpret them. Relevant differences (i.e. those that violate acceptability criteria) should be derived from policy objectives during problem formulation. Risk characterisation can then conduct the most suitable experiments to test whether those differences occur. Risk characterisation should not try to derive definitions of relevant differences from profiling data; this would be inefficient and ineffective and likely lead to poor and controversial decisions (Evans et al. 2006; Raybould and Macdonald 2018). This topic is discussed further in section "Major areas of risk assessment of GM crops".

Characterisation of a particular risk, or set of risks (e.g. ecological risks), contributes to decision-making. As we have seen, in theory, it should be possible to set acceptability criteria such that a decision can flow more or less directly from risk characterisation. However, in practice, when we begin the risk assessment, we may not have all the information required to set unequivocal acceptability criteria; for example, if we are using risk – (potential) benefit analysis to make decisions, we may not have the completed benefit assessment for use in problem formulation. Hence, we will not be able to decide how much risk is acceptable. A solution to this might be to set a level of risk that is unacceptable no matter how large the opportunity (the probability and value of a benefit) and use this to screen out potential product uses that are too risky. Those uses that pass will be evaluated further against refined criteria as we gain more knowledge of the opportunities they present.

Decision-making is not necessarily "yes" or "no" to a proposed activity. For GM crops, the decision-maker may set conditions on the approval of a product use to reduce risk from an unacceptable to an acceptable level, perhaps specifying certain times or places where the use pattern should be modified from that originally envisaged (see section "Major areas of risk assessment of GM crops" for examples).

These conditions are risk management measures (Fig. 2). The design of risk management can use the pathway to harm method outlined in Fig. 1. In essence, a risk management measure is an event inserted into a pathway with the effect that the combination of the probability and severity of harm is reduced to an acceptable amount.

Evaluating the likely effectiveness of a particular risk management method can be based on tests of hypotheses similar to those outlined in Fig. 1. If risk management aims to, say, reduce the probability of A leading to B, then we would test a hypothesis of the kind "given A from the proposed use, intervention X will reduce the probability of B such that this pathway no longer leads to unacceptable risk." Testing such hypotheses would usually be completed using existing knowledge, although it is perfectly feasible to design experiments to evaluate whether the proposed risk management is likely to be effective.

Risk analysis does not end with a decision to go ahead with or prohibit a proposed activity. No decision can be perfect, that is, the chosen option cannot guarantee to deliver the aims of the decision-maker. Second, conditions may change; the decision-maker may change his or her mind about policy aims; and even if the aims remain unchanged, the decision may have unforeseen consequences, or external factors, such environmental conditions, may change such that the acceptability criteria no longer align with the aims.

For these reasons, Miller (1994) suggested that in many circumstances, decision-makers should concentrate on reviewing the consequences of a decision and intervene if the decision is not delivering the desired aims:

> … in complex circumstances where there is a limited quantity of scientific knowledge, the aim of the rational agent is not really to make the right decision (there may be no such thing), it is to make the decision right … making a decision is rarely the end of the affair, each decision has to be followed by innumerably many more, correcting and refining the initial one.

In risk analysis, the "innumerably many more" decisions that correct and refine the initial one are a second component of risk management – the initial decision being the first. Monitoring would measure specified indicators that the decision is having the desired effects and is not having undesired effects. The undesired effects should be plausible based on the risk assessment; in other words, monitoring should be hypothesis-led and test that specified effects that we would regard as harmful have not resulted from the decision. Depending on the confidence in the initial decision and the likely seriousness of any harm, monitoring could be more or less continuous and widespread immediately after the initial decision or a limited check after several years that no unacceptable changes have occurred. Whatever method is used, monitoring should not be a "fishing expedition" for unintended changes that are justified as problematic based on statistical significance rather than because they indicate damage to something we value (Chapman 2012).

Risk management, and monitoring in particular, is relevant to risk assessment because having the option to impose conditions on or review impacts of product uses can take some of the load off risk assessment. Risk assessment of GM crops can sometimes be bogged down in trying to eliminate scientific uncertainty that perhaps cannot be eliminated by laboratory experiments or confined field trials. Early claims that growing GM crops would "let the genie out of the bottle" and conflation of these concerns with those about cloning animals and people (Nerlich et al. 2001) have perhaps led to a situation where we ask too much of risk assessment for GM crops because decision-making is regarded as irreversible and potentially catastrophic. As we shall see in section "Gene flow from GM crops", risk management offers another way to deal with that uncertainty by creating options to ensure that decisions to allow cultivation of GM crops are not irrevocable and hence move risk analysis out of risk assessment and into decision-making (Nickson and Head 1999).

Risk communication is the remaining and crucial component of risk analysis. Risk communication is not simply the announcement of a risk management decision; it should be conducted at every stage of risk analysis and is a two-way flow of knowledge and opinion (Hope 2006):

> Ideally it [risk communication] is a focal point for communication and cooperation between decision-makers and affected or regulated parties (often termed "stakeholders"), thus enabling decision-makers and stakeholders to make more informed decisions.

Problem formulation can make a great contribution to risk communication. It should be consultative because it has to understand the aims of policymakers to be able to set acceptability criteria. Using pathways to harm to identify and explain acceptability criteria should mean that regulatory decisions do not appear arbitrary, even though they may be controversial (Raybould and Macdonald 2018). Finally, studies may be done for purposes of risk communication when they may add little or nothing to the risk assessment. As we will see in section "Weediness and invasiveness potential of GM crops", sometimes demonstrating that an event does not happen is more effective than overwhelming corroboration of the hypothesis that it will not happen.

Having reviewed the position of risk assessment within risk analysis, we can turn to the major sections of ERA that are required by most regulatory authorities. The intention is not to provide a comprehensive review, but to highlight areas where assessments have been done well and where they may be improved.

## Major Areas of Risk Assessment of GM Crops

### *Gene Flow from GM Crops*

Movement of transgenes from crops to wild species via sexual hybridisation was one of the earliest concerns raised about releasing GM crops into the environment. This stemmed from the need to conduct field trials to evaluate the performance of the first GM crops that were potential products. There was strict risk management of these early trials, with the aim of preventing transgene spread and persistence outside the confines of the field trial. There was no reason to suppose that genetic modification would change the sexual compatibility of the crop with wild species (Raybould and Gray 1994). Hence, risk management measures, such as locating GM field trials away from areas with compatible wild relatives or destroying the trials before flowing if there were likely to be flowering wild relatives nearby, were based on existing knowledge of sexual compatibility between crops and wild species (de Vries et al. 1992; Raybould and Gray 1993). Knowledge of the crop's biology also allowed the design of programmes to monitor the trial site for persistence of the GM crop itself.

With the impending commercial cultivation of GM crops, the policy aim could no longer be the confinement of transgenes within field trials. For some crops in some countries, the policy aim was to prevent the movement of transgenes into wild species or local crop varieties that have huge cultural importance. Of particular concern was the potential for transgenes to move from GM maize to teosinte – the wild ancestor of maize – or local landraces of maize in Mexico and other countries in Central America, either because the presence per se of the gene was unacceptable or because its presence could reduce the diversity of the land races, making them less valuable to local farmers (Bellon and Berthaud 2004; Ortiz-García et al. 2005).

Another situation where the policy aim was to prevent gene flow concerned GM crops producing pharmaceuticals – so-called pharming. Here the concern was that genes that coding for the production of a pharmaceutical, such as an antibody, might cross with crops intended for food production (Gepts and Papa 2003).

Surveys of public opinion in many countries have shown that producing pharmaceuticals in food crops growing in open fields is unacceptable, even if food crops were located many miles away (Pardo et al. 2008). If assuaging this concern is the overriding aim of regulatory policy, a very low or even zero frequency of the transgene in food crops would be the limit of acceptability. If existing data indicate that the probability of hybridisation is not negligible, then the risk from cultivating the GM crop is unacceptable and risk management decisions can be made on that basis: either ban cultivation of the GM crop in the area concerned, as was the case with maize in Mexico (Ortiz-García et al. 2005), or require that the crops be confined in glasshouses or other secure site (Pardo et al. 2008).

In the above examples, the risk assessment was driven entirely by exposure: the presence of any transgene would be unacceptable regardless of its effect on the environment or human health. In other situations, the presence per se of the transgene in a wild relative was not the concern; rather, it was that the transgene would confer properties that led the wild relative to spread and cause ecological harm. A general pathway to harm is given in Fig. 3.

The first step in the pathway is cross-fertilisation of a wild relative by pollen of the GM crop to produce hybrids. Although predicting the number of hybrids formed during a given time is feasible (Wilkinson et al. 2003), it is doubtful that it is worth the effort for risk assessment. First, the predictions will have large uncertainty (de Jong and Rong 2013). Second, it is hard to envisage acceptability criteria based simply on the number of first-generation hybrids between a GM crop and a wild relative. Transgenes can spread through subsequent generations, so the initial

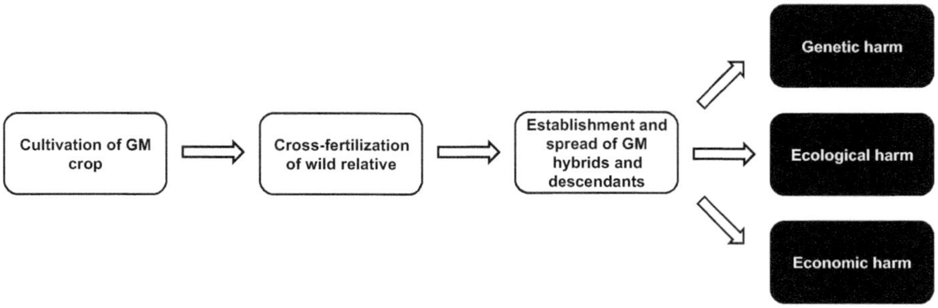


**Fig. 3** Pathways to harm from hybridisation between a GM crop and a sexually compatible wild relative. Harm may be ecological, such as reduced biodiversity; economic, such as increased costs of weed control; or genetic, where the spread of the transgene leads to reduced genetic resources for crop breeding

number of hybrids may be a poor predictor of the severity of harm, although it may suggest the speed with which it may arise. Hence, once risk assessment has decided that the probability of one or more hybrids forming is not negligible, the risk assessment should concentrate on assessing the consequences of transgene spread not on trying to quantify it precisely.

General approaches to assessing the ecological consequences of transgene spread are available (Raybould and Cooper 2005; Hokanson et al. 2010). They concentrate on determining whether the transgene will lead to "ecological release" (Keane and Crawley 2002), that is, whether the wild relative can overcome a constraint, such as a disease or herbivory, that was keeping its population size in check. If the transgene allows the wild relative to increase in abundance, then, in theory, it could cause ecological damage similar to that caused by invasive plant species (Pimentel 2009). Also, genetic variation within the wild relative may be lost if there is strong selection for the transgene, resulting a so-called selective sweep at loci closely linked to the insertion site (Hokanson et al. 2010, 2016).

GM crops have rarely been cultivated in regions where they could hybridise with wild species in non-agricultural habitats. In part, this is because risk managers have restrained the cultivation of GM crops in such regions; for example, the United States Environmental Protection Agency restricts the sale and distribution of insect-resistant GM cotton in states and territories where wild or feral cotton may occur (Mendelsohn et al. 2003).

An exception are GM varieties of squash, *Cucurbita pepo*, that have been genetically modified to resist infection by cucumber mosaic virus (CMV), zucchini yellow mosaic virus (ZYMV) and watermelon mosaic virus-2 (WMV-2). The GM squash can hybridise with undomesticated *C. pepo* in the southern United States. Studies have shown that CMV, ZYMV and WMV-2 occur in the wild *C. pepo* populations. Also, these viruses reduce the fecundity of wild *C. pepo* in controlled trials when mechanically inoculated or spread naturally by aphids. Crosses between wild *C. pepo* and the cultivated GM varieties are protected against the virus in these controlled trials (Fuchs et al. 2004; Laughlin et al. 2009). Hence, it is possible that hybridisation with GM squash could protect wild *C. pepo* from virus infections in non-agricultural habitats and increase its fecundity over that of non-GM populations.

Despite the possible increase in fecundity of wild *C. pepo*, the United States Department of Agriculture's Animal and Plant Health Inspection Service (USDA/APHIS) decided that the ecological risk from cultivating GM virus-resistant squash was acceptable (Acord 1996). A key element in the decision was the existence of non-GM squash varieties with resistance to CMV, ZYMV and WMV-2 (Sitterley 1972; Pink and Walkey 1984). USDA/APHIS stated that the GM squash was "no more likely to become a weed than virus resistant squash developed by traditional breeding techniques … [and therefore] is unlikely to increase the weediness potential for any other cultivated or wild species with which it can interbreed." The low incidence of virus infections in wild *C. pepo* is another reason why the ecological risk may have been considered acceptable by USDA/APHIS (Laughlin et al. 2009). Later work has also suggested that virus-resistant wild *C. pepo* may be more susceptible to wilt owing to their greater attractiveness to cucumber beetles, which spread the bacteria that cause the disease (Sasu et al. 2009).

Gene flow from GM crops to weeds of agriculture has probably been a concern for regulatory risk assessment more often than has gene flow to wild species in non-agricultural habitats. The risks posed by a weed becoming more difficult to control are often considered as part of ecological (or environmental) risk assessment for GM crops, even though the ultimate harm is economic in the form of increased costs of control or lower crop yield or quality. Of particular concern is the transfer of transgenes conferring herbicide tolerance, which would negate or at least complicate the effectiveness and simplicity of GM herbicide tolerance for weed management. Gene flow occurs easily between rice and weedy red rice (Gealy et al. 2003), and concerns about creating herbicide-resistant red rice may be a reason why herbicide-tolerant rice has yet to be commercialised in China (Lu et al. 2016). Nevertheless, some authors (e.g. Gealy et al. 2003) consider that suitable risk management measures can reduce the risk to acceptable levels. Such measures should include crop rotation and herbicide rotation and have been recommended to manage the risk of creating multiple-herbicide-tolerant weeds of *Brassica napus* in Canada (Hall et al. 2000) and herbicide-tolerant weeds of introduced teosinte in Europe (Devos et al. 2018).

One area where gene flow continues to pose difficulties is the uncontained use of GM trees. As with arable crops, experimental field trials can be implemented with suitable risk management measures to prevent cross-fertilisation with non-GM trees (Häggman et al. 2013). However, despite successful demonstration in field trials of the potential benefits of GM trees, their commercial use has been extremely limited owing to regulatory uncertainty and forest certification schemes that prohibit use of GM trees and thereby discourage investment (Chang et al. 2018). Fears about long-distance gene flow and the potential long periods over which its effects could occur contribute to these problems (Farnum et al. 2007). However, progress in genetic containment technologies may reduce the probability of gene flow to acceptable levels and, when combined with continuing demonstration of the potential benefits, may overcome the regulatory and social barriers to the acceptance of large-scale cultivation of GM trees (Strauss et al. 2016).

Overall, regulatory risk assessment of gene flow from GM crops to wild species has worked well by implementing some of the principles described in section "What is risk assessment?". Risk management measures based on the biology of the crop allowed field trials to proceed without requiring huge amounts of data that would have been difficult or impossible to attain from laboratory studies. For commercial cultivation, non-zero gene flow from the GM crop to a wild species has been used as a decision-making criterion, either to trigger evaluation of information about the likely effects of transgene spread, as with virus-resistant *Cucurbita*, or to implement risk management measures, as with insect-resistant cotton. This criterion has largely prevented the need to quantify gene flow for regulatory ERA even though in the early 2000s research seemed to be heading that way (Wilkinson et al. 2003).

## *Weediness and Invasiveness Potential of GM Crops*

Inadvertently creating crops that have increased potential to invade non-agricultural habitats or to worsen weed problems in agriculture is another long-standing concern about the use of genetic modification in plant breeding (Keeler 1989). Pathways to harm can begin not only with cultivation of the crop but also with the spillage of seed imported for food and feed use (Fig. 4). Although spillage of seed may result in only small, sporadic roadside populations that are unlikely to persist or to cause any ecological harm if they do (Crawley and Brown 1995; Saji et al. 2005), some regulatory authorities require ERAs for import of GM seeds that are similar to ERAs for cultivation elsewhere, and decision-making over import of GM seeds in some jurisdictions can take longer than that for cultivation in others (Smyth 2017).

While the pathways are similar to those for gene flow (Fig. 3), ERAs for weediness and invasiveness potential tend to be far more data intensive for two reasons, one reasonable but the other less so. First, the environmental exposure to GM plants, at least via the cultivation pathway, will be higher than for many gene flow pathways where hybridisation between a GM crop and a wild species is restricted by limited sexual compatibility or risk management measures. Second, and much less reasonably, considerations of weediness and invasiveness potential for ERA have become

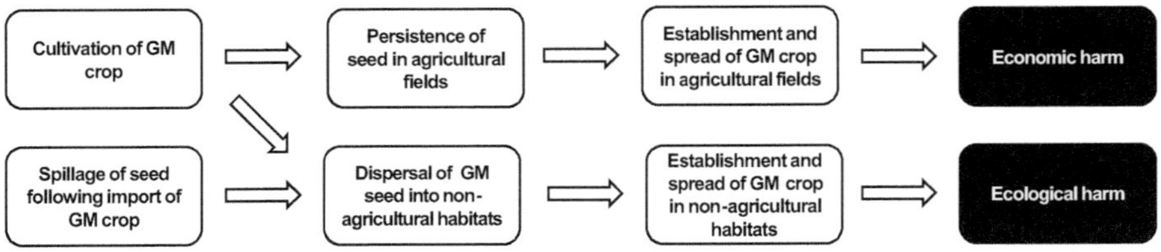


**Fig. 4** Outlines of pathways to harm that result from increased weediness or invasiveness potential of a GM crop

confounded with studies designed to assess product performance or even evaluate genetic modification generally.

As we saw in section "Gene flow from GM crops", in situations where formation of GM crop × wild species hybrids is plausible, the risk assessment focuses on whether the transgene may lead to ecological release of the wild species. Hence, the focus of the risk assessment is on side effects of the intended trait. This is also the case in ERA for increased weediness and invasiveness potential of the GM crop itself. Perhaps the clearest case is herbicide-tolerant crops that could persist as volunteer weeds in a following crop. If the next crop is tolerant of the same herbicide and that herbicide is used as the main weed management tool, the volunteers may cause more severe problems than a similar non-GM (or at least a non-herbicide tolerant) crop (Simard et al. 2002).

ERAs also consider whether the intended trait will increase the invasiveness potential of the GM crop, that is, will the GM crop be more likely than its non-GM counterpart to colonise non-agricultural habitats and, if so, would their appearance or increased abundance in these habitats represent environmental harm. Invasive plants can dominate communities and exclude other species (Dogra et al. 2010), and if the excluded species are considered to be more valuable than the species that has replaced them, harm has occurred (Boltovskoy et al. 2018).

Some arable crops can form self-sustaining "feral" populations outside agriculture (Sukopp and Sukopp 1993), although it is not clear whether they ever cause ecological harm. Nevertheless, ERAs for the cultivation, and increasingly the import, of GM crops test hypotheses about the acceptability of risk via the pathways in Fig. 4. The protection goals are often not particularly well defined – a notable exception being endangered and threatened species and their habitats in the United States – but a screening assessment will implicitly test the hypothesis that the GM crop in question is no more likely to establish and spread in non-agricultural habitats than are non-GM varieties of that crop. If this hypothesis were falsified, further hypotheses that the likely increase in abundance would not be harmful could be tested.

For side effects of the intended trait, the ERA usually concentrates on existing knowledge of the crop's ecology. For example, the abundance of feral populations of maize is known to be limited by sensitivity to frost and inability to compete with perennial vegetation (Raybould et al. 2012a). Hence, introducing a transgene for resistance to insects is unlikely to lead to ecological release. That the intended trait is not intended to confer frost tolerance or perenniality might be sufficient corroboration of the "no more likely to establish and spread" hypothesis to complete this part of the ERA. If not, tests that frost tolerance has not increased and that the maize does not regenerate after harvest could be conducted.

The weediness and invasiveness part of regulatory ERAs for GM crops would be simple and quick to complete if they targeted side effects of the introduced trait and factors known to limit the establishment of (potentially harmful) feral populations. However, most ERAs also include "phenotypic characterisation" studies that compare the GM crop with a non-GM crop for numerous traits related to crop performance. It is doubtful that such studies are useful for ERA.

Phenotypic characterisation studies for regulatory ERAs require substantial investments of time, money and materials. They must meet a long list of criteria for acceptability set out in guidance documents (e.g. EFSA 2015). Criteria include the number and representativeness of the trial sites, the materials used, the characters that must be measured and the experimental design and statistical analysis.

Phenotypic characterisation studies are an example of profiling (see section "Risk assessment in decision-making: risk analysis"): they are designed to test the null hypothesis of no difference between the GM crop and whatever material it is being compared with, usually one or more non-GM varieties of the same crop. Horak et al. (2007, 2015) describe typical examples. Sometimes, the hypothesis under test is an equivalence test of the form "there is a difference between the GMO and its reference [comparator] of a certain minimum size" (van der Voet et al. 2011). In both cases, the idea is that the ERA must assess any "unintended effects" caused by transformation. As "unintended" is not synonymous with "potentially harmful," there are problems with relevance of this type of study for risk assessment.

To understand the problems with current phenotypic characterisation studies, we must remember that risk assessment is part of risk analysis. Policy aims are decided and used to set decision-making or acceptability criteria. The scientific parts of risk assessment test whether the criteria are met. The hypotheses tested in phenotypic characterisation studies do not flow from policy aims and acceptability criteria; they are simply an example of the "fishing expeditions" described by Chapman (2012) and discussed in section "Risk assessment in decision-making: risk analysis".

When faced with a phenotypic characterisation study as currently performed, the risk assessor's job is to determine whether any statistically significant differences detected (whether a rejection of a null hypothesis of no difference or acceptance of an equivalence hypothesis of a minimum size) are "biologically relevant." The definition of biological relevance is lacking in guidance documents, but we could define it as a difference that shows that an acceptability criterion has not been met (Raybould et al. 2019).

Depending on the power of the study, risk assessors may have to evaluate numerous differences and work out whether they have any relevance. This data-led or "unbiased" approach is enormously wasteful and also raises a danger of setting policy based on statistical significance rather than careful consideration of what the aims of relevant policy ought to be (Fig. 5; Raybould and Macdonald 2018).

A more effective option is to work from policy aims to set acceptability criteria (a hypothesis-led approach; Fig. 5). If our objective is to prevent GM maize from harming biodiversity by colonising non-agricultural habitats, we have seen that an effective approach is to look at what currently prevents maize from colonising non-agricultural habitats and test the hypothesis that the GM maize in question has not acquired those attributes (e.g. frost tolerance or perenniality). Whether we actually run studies to do this, or use existing knowledge, will depend on the genetic modification: testing insect-resistant maize for frost tolerance may seem unnecessary, while we may choose to run a test if the maize has been modified for some kind of abiotic-stress tolerance.

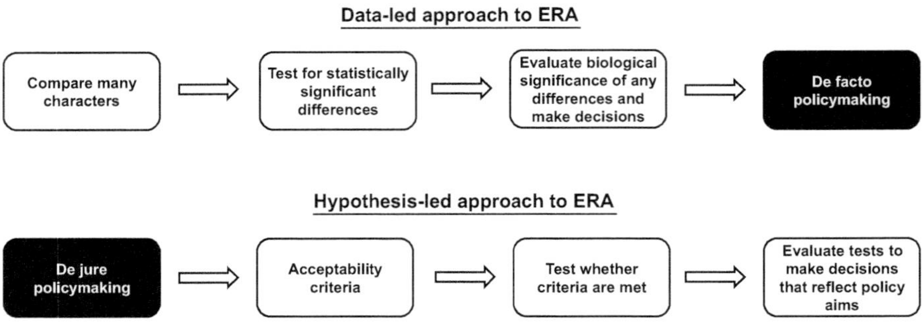


**Fig. 5** A comparison of data-led and hypothesis-led ERA for GM crops. Phenotypic analysis as currently conducted is data-led ERA

While it may seem strange to test whether a GM maize modified for insect resistance has become frost tolerant, it makes more sense than running a phenotypic characterisation study for ERA. Starting with our policy objective of projecting biodiversity enables us to derive an acceptability criterion: the GM maize should not be frost tolerant (the degree of frost tolerance – e.g. the minimum temperature and maximum duration survived – would need to be defined based on the conservatism we wish to build into our decision-making). If the maize meets this criterion and we can identify no other attributes of maize that would lead it to colonise non-agricultural habitats, our ERA for this pathway to harm can be completed; we do not need to trawl through data that compares the GM maize with a comparator, look for statistically significant differences and work out if they are biologically relevant. If we can identify an attribute, say plant height, we should set an acceptability criterion, perhaps no more than 25% taller than a comparator, and test the hypothesis that the GM maize will not exceed this height. Whether we need to run a test or use existing knowledge will depend on the intended GM trait and the conservatism of the decision-maker. If we cannot specify a minimum difference in any trait that would cause us concern, then no phenotypic characterisation study is needed for ERA: it would have no practical utility.

Hypothesis-led ERA leads us to specify characteristics that are unacceptable and test the hypothesis that the GM crop under evaluation does not possess them. We can design rigorous tests that are most likely to reveal whether the GM crop has unacceptable traits if indeed it does have them. The unbiased approach may appear to provide reassurance by measuring many things "just in case we have missed something important." However, "important" is determined by policy aims not statistical significance. Running a huge study may actually make it more likely that we miss something important. If we measure 100 phenotypic characters that have no relevance for determining whether a GM maize will colonise non-agricultural habitats (i.e. we cannot specify a priori what quantity or quality of the character is unacceptable), we may become overconfident and not consider frost tolerance. If an observation that a GM crop fails to meet an acceptability criterion is a needle and the data we collect is a haystack, if we want reassurance that there is no needle, time is more profitably spent defining what the needle looks like and designing methods to find it, than by making the haystack bigger.

In questioning the value of current phenotypic analysis studies for ERA, I am not suggesting they have no value to anyone. First, they have practical utility for breeders because they measure characters that breeders have defined as being important for their products. They were also useful to breeders in comparing genetic modification with other plant breeding techniques; for example, a breeder may be interested in whether genetic modification tends to produce more or fewer "off-types" than do other methods of plant breeding. In such circumstances, phenotypic characterisation may have practical utility in technology assessment.

A second practical utility of phenotypic characterisation studies may be risk communication. Showing that something does not happen may be more convincing than a powerful argument that it will not happen. Raybould et al. (2012b) conducted a field trial of non-GM and GM insect-resistant maize in Texas. The maize was initially cultivated to ensure that it established. Once established, cultivation was stopped to simulate conditions in non-agricultural habitats. A few months after maturity, both non-GM and GM maize had become overwhelmed by native vegetation, and after 12 months, there were no living maize plants at the trial site. The results of the experiment were easy to predict because maize is well known as a poor competitor in uncultivated land. However, photographs showing that GM insect-resistant maize had no advantage over non-GM maize and that both were highly unlikely to become feral in nearby Mexico were powerful tools for risk communication. Phenotypic characterisation may have served a similar purpose when GM crops were first commercialised, but the value of such studies for risk communication ought to decrease as we become more familiar with GM crops. Finally, phenotypic characterisation studies may have epistemic utility for molecular biologists testing theories about how genetic modification generally, or specific modifications to metabolic pathways, affect plant development. Value to breeders, risk communicators or molecular biologists should not be a reason for continuing to require untargeted phenotypic characterisation of GM crops for regulatory ERA.

In summary, assessments of weediness and invasiveness potential of GM crops are a mixed bag. When assessing side effects of the intended modification, ERA largely follows the problem formulation and risk analysis framework outlined in Fig. 2. Weediness and invasiveness potential assessments are made unnecessarily long and complicated by requirements for untargeted phenotypic characterisation studies. These studies add little or nothing to decision-making and may cause harm by diverting resources away from the identification of products that pose high risk and by delaying the introduction of products that are potentially beneficial (Cross 1996).

## *Risks to Non-target Organisms*

The term non-target organism (NTO) is used widely in ERAs for GM crops. A broad definition of NTO is a species that is not a target pest of the GM crop being assessed and is roughly synonymous with "biodiversity." This definition is problematic because not all GM crops have target organisms (e.g. a GM crop with improved water-use efficiency does not). A working definition of NTO for regulatory risk assessment of GM crops is roughly synonymous with "non-pest," but even this needs clarification. First, not all non-pest species are necessarily valued or valued to the extent that greater abundance is always regarded as beneficial and lower abundance as harmful; given the finite energy flows and space in ecosystems, we must value some organisms more than others or define acceptable limits to abundance. Second, assessment of the risks to biodiversity, particularly plants, via gene flow from GM crops (section "Gene flow from GM crops") or GM crops colonising non-agricultural habitats (section "Weediness and invasiveness potential of GM crops") tends not to be considered as NTO risk assessment, even though many of the acceptability criteria relate to potential effects on non-pest species. Finally, harm arising through effects on microorganisms, such as disruption of nutrient cycling, tends to be considered outside NTO risk assessments.

By the exclusion of plants, microorganisms and animals potentially adversely affected by gene flow or colonisation of non-agricultural habitats, NTO risk assessment has come to mean characterisation of the risks arising from exposure of non-pest animal species to toxic substances in GM crops. These substances may be produced intentionally by the genetic modification (e.g. new proteins or RNAs) or be unintended changes in endogenous toxins. Using this formulation, all GM crops may be the subject of NTO risk assessments. Some NTO risk assessment schemes also consider the risks posed by herbicides that are applied to the GM crop; others examine these risks separately under pesticide regulations.

The core of NTO risk assessment focuses on substances intentionally produced by the GM crop and is hypothesis-led, following schemes similar to that outlined in the hypothetical example in section "Risk assessment as hypothesis testing". Broadly, the policy aim is to prevent unacceptable reductions in the abundance of organisms that are valued in their own right (e.g. the monarch butterfly) or for the ecological services that they provide (e.g. pollination) (Sanvido et al. 2012).

Organisms are assumed to be exposed to toxic substances in the GM crop orally by eating the GM crop itself (e.g. non-pest species may eat pollen or nectar), eating organisms that have eaten the crop or eating crop residues (Head et al. 2001; Mendelsohn et al. 2003). Studies that measure the concentration of the substance, say an insecticidal protein from *Bacillus thuringiensis* (*Bt*), along with knowledge of dietary habits allow predictions of the amount of substance to which various types of organism will be exposed (the PEC, section "Risk assessment as hypothesis testing"; also called the estimated environmental concentration (EEC); Bascietto et al. (1990)). Among the organisms usually considered in NTO assessments are wild birds and mammals, freshwater fish, foliar and soil-dwelling arthropods (particularly those that are biological control organisms), earthworms and freshwater invertebrates (Mendelsohn et al. 2003).

The next stage is to test the hypothesis that a measure of toxicity of the substance divided by the PEC is above a trigger value (TER > trigger value; see section "Risk assessment as hypothesis testing"). This is done separately for each group of organisms at risk (valued organisms that are potentially exposed) and usually uses a purified preparation of the substance. The measure of toxicity could be the $LC_{50}$ or the NOEC – the no observed effect concentration, the highest concentration of the substance that has no effect on the organism concerned. The $LC_{50}$ or NOEC is obtained in laboratory studies following internationally accepted guidelines, which may be modified from their original design to account for the oral route of exposure. Details of study design, choice of test organisms and preparation of the test substance and testing that it is a suitable surrogate for the substance produced in the crop are given by Romeis et al. (2011), Romeis et al. (2013) and Raybould et al. (2012b), respectively. If toxicity data were obtained for the same toxin for an NTO assessment of a different crop, they may be suitable for subsequent assessments of other crops, so reducing or removing the need for testing (Romeis et al. 2009). Also, no testing may be needed at all if the substance is known not to be toxic to NTOs at concentrations greatly in excess of those found in the plant; such arguments could be based on knowledge of the function of the substance and prior exposure rather than toxicity testing (CERA 2011).

The TER for each group of organisms comprises the estimate of the PEC (or EEC) for that group and the suitable measure of toxicity to a representative member of that group (called a surrogate species). If all of the TERs are above the relevant trigger value for the organisms concerned, the NTO risk assessment for the intended substance is usually considered complete (Romeis et al. 2008). For further details of the concepts behind NTO risk assessments, see Garcia-Alonso et al. (2006) and Romeis et al. (2008); for the scope of testing, choice of surrogate organisms, setting of TERs and overall risk conclusions for regulatory NTO risk assessments for products producing *Bt* proteins or insecticidal double-stranded RNA, see Raybould et al. (2007), Raybould and Vlachos (2011), Burns and Raybould (2014) and Bachman et al. (2016).

If the hypothesis that the TER is above the relevant trigger is shown to be false, several options are available. First, the exposure could be refined; PECs often make the worse-case assumption that the diet of an organism is composed entirely of items containing the toxin at the highest plausible concentration, usually the highest concentration in the relevant tissue of the crop. Allowing for dilution of the substance through the exposed organism having a mixed diet often brings the PEC down and raises the TER above the trigger value (see Raybould et al. (2007) for refinement methods). Second, it may make sense to test the surrogate organisms under more realistic conditions. For example, if an organism is potentially exposed via pollen, the initial study may have exposed the surrogate organism to 10X the highest concentration of the toxin measured in the pollen. A follow-up study might expose the surrogate to pollen itself to obtain a new measure of toxicity which can be used to calculate a new TER and be compared with a relevant trigger. Semi-field and field studies are also options if laboratory studies and exposure calculations cannot establish acceptable risk (Romeis et al. 2006, 2008; Duan et al. 2009).

As with the weediness and invasiveness potential ERA, the clear, hypothesis-led NTO assessment often has a data-led component bolted onto it. A common form of data-led ERA is the requirement for a field study to compare the abundance of NTOs in the GM crop and a non-GM comparator even though TERs based on laboratory ecotoxicology studies and worst-case exposure estimates revealed acceptable risk, or even when laboratory studies reveal no adverse effect (Arpaia et al. 2014). A field study in these circumstances is analogous to the phenotypic characterisation studies described in section "Weediness and invasiveness potential of GM crops"; it is simply a test for unintended effects, or reassurance that something important has not been missed, and is unwise for the same reasons: it diverts resources from defining and looking for potentially harmful effects (it fails to describe the needle and simply makes the haystack bigger) and risks making policy ad hoc.

Perhaps the best example of a data-led approach to field studies is the Farm-Scale Evaluations (FSEs), a huge field experiment that compared plant and invertebrate biodiversity in crop fields managed using conventional herbicide regimes or using GM herbicide-tolerant (GMHT) crops (Firbank et al. 1999). The FSEs tested the null hypothesis that "[GMHT] crops had no effect on farmland biodiversity compared with a conventional cropping system" (Squire et al. 2003). This is a classic example of unbiased profiling: the objective was not to test for potentially harmful differences in biodiversity, but merely to test for "effects." Numerous statistically significant differences in plant and invertebrate abundance were found between the GMHT and non-GM fields, and the UK government made decisions about whether to permit the cultivation of certain GMHT crops based on these differences and their predicted effects on populations of farmland birds (Street 2007).

The FSEs lay bare perfectly the data-led approach to risk assessment and decision-making. The UK government did not decide that the policy priority in crop management ought to be the conservation of farmland birds, set acceptability criteria for changes in bird abundance and then design experiments to test whether GMHT crop management met these criteria. Instead, it conducted a huge profiling exercise and based its decision-making on those variables that were statistically significantly different between the treatments; thus, its policy of favouring farmland birds was determined by the experimental results and appeared ad hoc. It is moot whether this policy would have emerged had the results of the FSEs have been different.

The FSEs produced valuable ecological data that are still the basis of groundbreaking research over 15 years later (Ma et al. 2019). Nevertheless, the value of the £4.4 million spent on them has been questioned (Qi et al. 2008; Champion 2011). Concerns centre on the cost-effectiveness of measuring particular organisms: more data could have been collected if different organisms had been assessed (or the same amount of data could have been collected more cheaply). However, the data-led approach to risk assessment and decision-making seems not to have been criticised. There is no suggestion that the experiments should have been designed to test the hypothesis that GMHT crops did not pose unacceptable risks to farmland birds rather than test a hypothesis of no effect. If we use the needle in a haystack analogy again, the criticisms are that we could have built an even bigger haystack for the same money, not that we might have made better decisions (and maybe saved money) by defining what the needles look like and designing experiments to search for them.

The pressure for profiling approaches to NTO risk assessment is likely to increase with the advent of "omics" approaches that allow profiling of GM crops at the molecular level (Heinemann et al. 2011; Christ et al. 2018). The justification for molecular profiling is explicitly data-led: profiling identifies hazards to NTOs in GM crop tissues that would otherwise be missed (Heinemann et al. 2011). However, this justification is predicated on the false idea that an unintended effect is a hazard. A hazard is something that can lead to harm, and as we have seen, harm is defined by policy not by statistical significance. Hence, just because profiling uncovers more unintended differences between a GMO and a comparator does not mean that more hazards have been discovered. Neither does risk assessment that does not use profiling becomes less adequate as profiling methods become more sensitive and hence detect more unintended differences at a finer scale.

Advocates of profiling argue that it will discover more potential hazards than do other methods. This is trivially true if one regards a comparator as free from hazards – all hazards will be statistically significant differences from the comparator; however, not all statistically significant differences will be hazards. Profiling is likely to discover vastly more false negatives than real positives, increasing the probability of missing something important for assessing risk and making decisions. Second, and more fundamentally, profiling sells lack of bias as its major advantage. This promotes the pernicious idea that risk assessment is not a tool for making decisions that achieve societal objectives, but a way of discovering in data what societal objectives ought to be.

In everyday language, "bias" has negative connotations associated with unfair prejudice against a person or group. In risk assessment, however, we should see bias as something positive. Policy aims lead us to set acceptability criteria and then test hypotheses that the acceptability criteria have been met. Rigorous testing that the criteria are met requires us to choose to measure particular variables that could falsify these hypotheses and not measure others that could never falsify them. In the language of profiling, this choice represents bias, when in reality it means increasing the practical utility of risk assessment by measuring things that help us make decisions. If omics methods are the best test of hypothesis that acceptability criteria are met, then they have a valuable role in risk assessment. If they are used only for profiling, they are unlikely to be valuable. Our scientific methods should improve the rates of detecting needles, not make the haystacks bigger.

## Conclusions

This article argues for a hypothesis-led approach to ERA for the use of GM crops. As far as possible, our hypotheses should aim for practical utility by being of the form that use of the GM crop will not pose unacceptable risk, and we should define clear criteria for accepting or rejecting such hypotheses. Hypotheses that aim for epistemic utility by improving our knowledge of the processes that underlie the creation of GM crops may be valuable, but should not form part of risk assessment for product use.

Some aspects of GM crop ERA follow these principles closely. For example, assessment of risks to NTOs from exposure to toxins uses predetermined trigger values based on the ratio of a measure of toxicity and a predicted exposure. If the ratio is above the trigger, the risk is acceptable; if it is below the trigger, acceptable risk has not been shown. Acceptability or unacceptability in these screening tests leads to various decisions about how to proceed.

Some elements of ERA, however, are data-led. Instead of using policy aims to design acceptability criteria and test that they are met, data-led ERA uses profiling. The term profiling is often associated with omics methods but applies equally to untargeted phenotypic characterisation and studies of biodiversity. In essence, profiling searches for statistically significant differences between the GM crop and a comparator and regards all of these differences as hazards, or at least potential hazards. The ERA tries to determine which, if any, of these differences are important. Decisions are then based on these determinations.

The advantage of hypothesis-led ERA is that it contributes to effective decision-making. The decisions may be controversial, as not everyone will agree with the policy aims, but at least the reasons why certain results lead to certain decisions are clear. Data-led ERA, on the other hand, leads to capricious decision-making as the policy aims become clear only after the results are known (Raybould and Macdonald 2018).

Technology for plant breeding is developing rapidly (Schaart et al. 2016) leading to discussion of how risks from products of the new breeding methods should be assessed (Casacuberta et al. 2015). The lesson from ERA of GM crops is that we should concentrate on trying to define societal objectives for the use of products of these new methods (Lyall and Tait 2019) and thereby set acceptability criteria that can be tested in risk assessments, should they be required. This approach should be described as policy-led rather than biased.

We should avoid "unbiased" profiling approaches that seek a comprehensive characterisation of unintended effects of the method that was used to breed the crop under assessment. Every method of plant breeding, indeed everything we do, has unintended effects. Perfect knowledge of unintended effects does not help decision-making; it merely overwhelms us with data of unknown relevance. Even if we could perfectly predict the consequences of each unintended effect, our decision-making would be paralysed by our inability to weigh up the trade-offs and synergies between the consequences.

To repeat an analogy, risk assessment is like searching for needles in a haystack, where the needles are indicators of unacceptable risk and the haystack is data that have no bearing on risk. Hypothesis-led risk assessment seeks to define the properties of the needles and thereby design methods to detect them most effectively and minimise the size of the haystack. Data-led risk assessment seems to want to build the largest possible haystack and search for needles without first defining their properties.

Building a huge haystack of data seems to offer reassurance that risks have been assessed thoroughly and decision-making is sound. However, as we have seen, risks are defined by policy, and sound decision-making requires judgement. The misguided idea that the "right decision" becomes obvious once we have "all the facts" actually opens the door for "stealth advocates" to unduly influence decision-making. Stealth advocates are scientists or other experts who claim to offer disinterested advice but really are arguing for their own opinions (Pielke 2007). The more data that are collected, the greater the opportunity for stealth advocates to find results that support their view (Sarewitz 2004; Carolan 2008). To continue the analogy, larger haystacks increase the chances of stealth advocates being able to find a peculiar piece of straw (a statistically significant difference of no consequence) that they can claim is a needle (an indicator of unacceptable risk).

Society could derive great benefits from crops bred by new techniques but only if public policymaking focuses on defining the properties of products that society wishes to encourage and scientific assessments focus on rigorously testing that products have such characteristics. Building giant haystacks of data on unintended effects will lead to futile attempts to make decisions about products based on scientific certainty and increase the chances of public policy being determined by stealth advocates with narrow sectional interests in promoting or suppressing all products of certain technology regardless of their properties.

## References

Acord BR (1996) Asgrow Seed Co.; availability of determination of nonregulated status for squash line genetically engineered for virus resistance. Fed Regist 61:33484–33485

Arpaia S, Messéan A, Birch NA, Hokannen H, Härtel S, van Loon J, Lovei G, Park J, Spreafico H, Squire GR et al (2014) Assessing and monitoring impacts of genetically modified plants on agro-ecosystems: the approach of AMIGA project. Entomologia 2:154. https://doi.org/10.4081/entomologia.2014.154

Bachman PM, Huizinga KM, Jensen PD, Mueller G, Tan J, Uffman JP, Levine SL (2016) Ecological risk assessment for DvSnf7 RNA: a plant-incorporated protectant with targeted activity against western corn rootworm. Regul Toxicol Pharmacol 81:77–88. https://doi.org/10.1016/j.yrtph.2016.08.001

Barling D, Lang T, Caraher M (2003) Joined-up food policy? The trials of governance, public policy and the food system. Soc Policy Adm 36:556–574. https://doi.org/10.1111/1467-9515.t01-1-00304

Bascietto J, Hinkley D, Plafkin J, Slimak M (1990) Ecotoxicity and ecological risk assessment. Regulatory applications at the EPA. Environ Sci Technol 24:10–15. https://doi.org/10.1021/es00071a001
Bellon MR, Berthaud J (2004) Transgenic maize and the evolution of landrace diversity in Mexico. The importance of farmers' behavior. Plant Physiol 134:883–888. http://www.plantphysiol.org/content/plantphysiol/134/3/883.full.pdf
Binimelis R, Myhr AI (2016) Inclusion and implementation of socio-economic considerations in GMO regulations: needs and recommendations. Sustainability 8:62. https://doi.org/10.3390/su8010062
Bohan DA, Vacher C, Tamaddoni-Nezhad A, Raybould A, Dumbrell AJ, Woodward G (2017) Next-generation global biomonitoring – large-scale, automated reconstruction of ecological networks. Trends Ecol Evol 32:477–487. https://doi.org/10.1016/j.tree.2017.03.001
Boltovskoy D, Sylvester F, Paolucci EM (2018) Invasive species denialism: sorting out facts, beliefs, and definitions. Ecol Evol 8:11190–11198. https://doi.org/10.1002/ece3.4588
Bruce D (2003) Contamination, crop trials, and compatibility. J Agric Environ Ethics 16:595–604. https://doi.org/10.1023/B:JAGE.0000004961.84390.ee
Burns A, Raybould A (2014) Nontarget organism effects tests on eCry3.1Ab and their application to the ecological risk assessment for cultivation of Event 5307 maize. Transgenic Res 23:985–994. https://doi.org/10.1007/s11248-013-9778-4
Calvert J (2004) The idea of "basic research" in language and practice. Minerva 42:251–268. https://doi.org/10.1023/B:MINE.0000038307.58765.b4
Carolan MS (2008) The bright- and blind-spots of science: why objective knowledge is not enough to resolve environmental controversies. Crit Sociol 35:725–740. https://doi.org/10.1177/0896920508093365
Carpenter JE (2011) Impact of GM crops on biodiversity. GM Crops 2:7–23. https://doi.org/10.4161/gmcr.2.1.15086
Casacuberta JM, Devos Y, du Jardin P, Ramon M, Vaucheret H, Nogué F (2015) Biotechnological uses of RNAi in plants: risk assessment considerations. Trends Biotechnol 33:145–147. https://doi.org/10.1016/j.tibtech.2014.12.003
CERA (Center for Environmental Risk Assessment, ILSI Research Foundation) (2011) A review of the environmental safety of the CP4 EPSPS protein. Environ Biosaf Res 10:5–25. https://doi.org/10.1051/ebr/2012001
Champion G (2011) Lessons learned from the farm scale evaluation of GMHT crops. J Verbrauch Lebens 6(suppl 1):73–78. https://doi.org/10.1007/s00003-011-0678-3
Chang S, Mahon EL, MacKay HA, Rottmann WH, Strauss SH, Pijut PM, Powell WA, Coffey V, Lu H, Mansfield SD et al (2018) Genetic engineering of trees: progress and new horizons. In Vitro Cell Dev Biol Plant 54:341–376. https://doi.org/10.1007/s11627-018-9914-1
Chapman PM (2012) Adaptive monitoring based on ecosystem services. Sci Total Environ 415:56–60. https://doi.org/10.1016/j.scitotenv.2011.03.036
Chapman PM, Fairbrother A, Brown D (1998) A critical evaluation of safety (uncertainty) factors for ecological risk assessment. Environ Toxicol Chem 17:99–108. https://doi.org/10.1002/etc.5620170112
Christ B, Pluskal T, Aubury S, Weng J-K (2018) Contribution of untargeted metabolomics for future assessment of biotech crops. Trends Plant Sci 23:1047–1056. https://doi.org/10.1016/j.tplants.2018.09.011
Crawley MJ, Brown SL (1995) Seed limitation and the dynamics of feral oilseed rape on the M25 motorway. Proc R Soc Lond B 259:49–54. https://doi.org/10.1098/rspb.1995.0008
Cross FB (1996) Paradoxical perils of the precautionary principle. Wash Lee L Rev 53:851–925
Damalas CA, Eleftherohorinos IG (2011) Pesticide exposure, safety issues, and risk assessment indicators. Int J Environ Res Publ Health 8:1402–1419. https://doi.org/10.3390/ijerph8051402
De Jong TJ, Rong J (2013) Crop to wild gene flow: does more sophisticated research provide better risk assessment? Environ Sci Policy 27:135–140. https://doi.org/10.1016/j.envsci.2012.12.002

De Vries FT, van der Meijden R, Brandenburg WA (1992) Botanical files. A study of the real chances for spontaneous gene flow from cultivated plants to the wild flora of the Netherlands. Gorteria Suppl 1:1–100. http://www.persistent-identifier.nl/urn:nbn:nl:ui:19-528195
Devos Y, Romeis J, Luttik R, Maggiore A, Perry JN, Schoonjans R, Streissl F, Tarazona JV, Brock TCM (2015) Optimising environmental risk assessments. EMBO Rep 16:1060–1063. https://doi.org/10.15252/embr.201540874
Devos Y, Ortiz-García S, Hokanson KE, Raybould A (2018) Teosinte and maize × teosinte hybrid plants in Europe – environmental risk assessment and management implications for genetically modified maize. Agric Ecosyst Environ 259:19–27. https://doi.org/10.1016/j.agee.2018.02.032
Dogra KS, Sood SK, Dobhal PK, Sharma S (2010) Alien plant invasion and their impact on indigenous species diversity at global scale: a review. J Ecol Nat Environ 2:175–186
Dolezel M, Miklau M, Helssenberger A, Reichenbecher W (2018) Limits of concern: suggestions for the operationalisation of a concept to determine the relevance of adverse effects in the ERA of GMOs. Environ Sci Eur 30:39. https://doi.org/10.1186/s12302-018-0169-6
Dorne JLCM, Fink-Gremmels J (2013) Human and animal health risk assessments of chemicals in the food chain: comparative aspects and future perspectives. Toxicol Appl Pharmacol 270:187–195. https://doi.org/10.1016/j.taap.2012.03.013
Duan JJ, Lundgren JG, Naranjo S, Marvier M (2009) Extrapolating non-target risk of *Bt* crops from laboratory to field. Biol Lett 6:74–77. https://doi.org/10.1098/rsbl.2009.0612
EFSA (European Food Safety Authority) (2015) Guidance on the agronomic and phenotypic characterisation of genetically modified plants. EFSA J 13:4128. https://doi.org/10.2903/j.efsa.2015.4128
Evans J, Wood G, Miller A (2006) The risk assessment – policy gap: an example from the UK contaminated land regime. Environ Int 32:1066–1071. https://doi.org/10.1016/j.envint.2006.06.002
Farnum P, Lucier A, Meilan R (2007) Ecological and population genetics research imperatives for transgenic trees. Tree Genet Genomes 3:119–133. https://doi.org/10.1007/s11295-006-0063-z
Fenner-Crisp PA, Dellarco VL (2016) Key elements for judging the quality of a risk assessment. Environ Health Perspect 124:1127–1135. https://doi.org/10.1289/ehp.1510483
Firbank LG, Dewar AM, Hill MO, May MJ, Perry JN, Rothery P, Squire GR, Woiwod IP (1999) Farm-scale evaluations of GM crops explained. Nature 339:727–728. https://doi.org/10.1038/21516
Forbes VE, Calow P (2002) Extrapolation in ecological risk assessment: balancing pragmatism and precaution in chemical controls legislation. BioScience 52:249–257. https://doi.org/10.1641/0006-3568(2002)052[0249:EIERAB]2.0.CO;2
Forbes VE, Calow P (2012) Use of ecosystem services concept in ecological risk assessment. Integr Environ Assess Manag 9:269–275. https://doi.org/10.1002/ieam.1368
Frewer LJ, Scholderer J, Bredahl L (2003) Communicating about the risks and benefits of genetically modified foods: the mediating role of trust. Risk Anal 23:1117–1133. https://doi.org/10.1111/j.0272-4332.2003.00385.x
Fuchs M, Chirco EM, McFerson JR, Gonsalves D (2004) Comparative fitness of a wild squash species and three generations of hybrids between wild × virus-resistant transgenic squash. Environ Biosaf Res 3:17–28. https://www.ebr-journal.org/articles/ebr/pdf/2004/01/E4105.pdf
Garcia-Alonso M, Raybould A (2014) Protection goals in environmental risk assessment: a practical approach. Transgenic Res 23:945–956. https://doi.org/10.1007/s11248-013-9760-1
Garcia-Alonso M, Jacobs E, Raybould A, Nickson TE, Sowig P, Willekens H, van der Kouwe P, Layton R, Amijee F, Fuentes AM et al (2006) A tiered system for assessing the risk of genetically modified plants to non-target organisms. Environ Biosaf Res 5:57–65. https://doi.org/10.1051/ebr:2006018
Garcia-Alonso M, Hendley P, Bigler F, Mayeregger E, Parker R, Rubinstein C, Satoore E, Solari F, McLean M (2014) Transportability of confined field trial data for environmental risk assessment of genetically engineered plants: a conceptual framework. Transgenic Res 23:1025–1041. https://doi.org/10.1007/s11248-014-9785-0

Gealy DR, Mitten DH, Rutger JN (2003) Gene flow between red rice (*Oryza sativa*) and herbicide-resistant rice (*O. sativa*): implications for weed management. Weed Technol 17:627–645. https://doi.org/10.1614/WT02-100

Gepts P, Papa R (2003) Possible effects of (trans)gene flow from crops on the genetic diversity from landraces and wild relatives. Environ Biosaf Res 2:89–103. https://doi.org/10.1051/ebr:2003009

Goldstein BD (2011) Risk assessment of environmental chemicals: if it ain't broke …. Risk Anal 31:1356–1362. https://doi.org/10.1111/j.1539-6924.2010.01447.x

Häggman H, Raybould A, Borem A, Fox T, Handley L, Hertzberg M, Lu M-Z, Macdonald P, Oguchi T, Pasquali G, Pearson L et al (2013) Genetically engineered trees for plantation forests: key considerations for environmental risk assessment. Plant Biotechnol J 11:785–798. https://doi.org/10.1111/pbi.12100

Hall L, Topinka K, Huffman J, Davis L, Good A (2000) Pollen flow between herbicide-resistant *Brassica napus* is the cause of multiple-resistant *B. napus* volunteers. Weed Sci 48:688–694. https://doi.org/10.1614/0043-1745(2000)048[0688:PFBHRB]2.0.CO;2

Hamer M (2000) Ecological risk assessment for agricultural pesticides. J Environ Monit 2:104N–109N

Head G, Brown CR, Groth ME, Duan JJ (2001) Cry1Ab protein levels in phytophagous insects feeding on transgenic corn: implications for secondary exposure risk assessment. Entomol Exp Appl 99:37–45. https://doi.org/10.1046/j.1570-7458.2001.00799.x

Heinemann JA, Kurenbach B, Quist D (2011) Molecular profiling – a tool for addressing emerging gaps in the comparative risk assessment of GMOs. Environ Int 37:1285–1293. https://doi.org/10.1016/j.envint.2011.05.006

Hendry RF (2001) Are realism and instrumentalism methodologically indifferent? Philos Sci 68(S3):S25–S37. https://doi.org/10.1086/392895

Hill RA, Sendashonga C (2003) General principles for risk assessment of living modified organisms: lessons from chemical risk assessment. Environ Biosaf Res 2:81–88. https://doi.org/10.1051/ebr:2003004

Hokanson KE, Ellstrand NC, Ouedraogo JT, Olweny PA, Schaal BA, Raybould AF (2010) Biofortified sorghum in Africa: using problem formulation to inform risk assessment. Nat Biotechnol 28:900–903. https://doi.org/10.1038/nbt0910-900

Hokanson KE, Ellstrand NC, Dixon AGO, Kulembeka HP, Olsen KM, Raybould A (2016) Risk assessment of gene flow from genetically engineered virus resistant cassava to wild relatives in Africa: an expert panel report. Transgenic Res 25:71–81. https://doi.org/10.1007/s11248-015-9923-3

Hope BK (2006) A n examination of ecological risk assessment and management practices. Environ Int 32:983–995. https://doi.org/10.1016/j.envint.2006.06.005

Horak MJ, Rosenbaum EW, Woodrum CL, Martens AB, Mery RF, Cothren JT, Burns JA, Mickson TE, Pester TA, Jiang C et al (2007) Characterization of Rondup Ready Flex cotton, "MON 88913", for use in ecological risk assessment. Crop Sci 47:268–277. https://doi.org/10.2135/cropsci2006.02.0063

Horak MJ, Rosenbaum EW, Kendrick DL, Sammons B, Phillips SL, Nickson TE, Dobert RC, Perez T (2015) Plant characterization of Roundup Ready 2 Yield® soybean, MON 89788, for use in ecological risk assessment. Transgenic Res 24:213–225. https://doi.org/10.1007/s11248-014-9839-3

Houlahan JE, McKinney ST, Anderson M, McGill BJ (2017) The priority of prediction in ecological understanding. Oikos 126:1–7. https://doi.org/10.1111/oik.03726

Jaffe G (2004) Regulating transgenic crops: a comparative analysis of different regulatory processes. Transgenic Res 13:5–9

Jasanoff S, Hurlbut JB (2018) A global observatory for gene editing. Nature 555:435–437. https://doi.org/10.1038/d41586-018-03270-w

Johnson KL, Raybould AF, Hudson MD, Poppy GM (2007) How does scientific risk assessment of GM crops fit within the wider risk analysis? Trends Plant Sci 12:1–5. https://doi.org/10.1016/j.tplants.2006.11.004

Kaltenhäuser J, Kneuer C, Marx-Stoelting P, Niemann L, Schubert J, Stein B, Solecki R (2017) Relevance and reliability of experimental data in human health risk assessment of pesticides. Regul Toxicol Pharmacol 88:227–237. https://doi.org/10.1016/j.yrtph.2017.06.010

Karstens K, Anderson J, Bachman P, de Schriver A, Dively G, Federici B, Hamer M, Gielkens M, Jensen P, Lamp W, Rauschen S, Ridley G, Romeis J, Waggoner A (2012) Genetically modified crops and aquatic ecosystems: considerations for environmental risk assessment and non-target organism testing. Transgenic Res 21:813–842. https://doi.org/10.1007/s11248-011-9569-8

Keane RM, Crawley MJ (2002) Exotic plant invasions and the enemy release hypothesis. Trends Ecol Evol 17:164–170

Keeler KH (1989) Can genetically engineered crops become weeds? Nat Biotechnol 7:1134–1139. https://doi.org/10.1038/nbt1189-1134

Klimisch K, Andreae M, Tillman U (1997) A systematic approach for evaluating the quality of experimental toxicological and ecotoxicological data. Regul Toxicol Pharmacol 25:1–5. https://doi.org/10.1006/rtph.1996.1076

Laughlin KD, Power AG, Snow AA, Spenser LJ (2009) Risk assessment of genetically engineered crops: fitness effects of virus-resistance transgenes in wild *Cucurbita pepo*. Ecol Appl 19:1091–1101. https://doi.org/10.1890/08-0105.1

Lu BR, Yang X, Ellstrand NC (2016) Fitness correlates of crop transgene flow into weedy populations: a case study of weedy rice in China and other examples. Evol Appl 9:857–870. https://doi.org/10.1111/eva.12377

Lyall C, Tait J (2019) Beyond the limits to governance: new rules of engagement for the tentative governance of the life sciences. Res Policy. https://doi.org/10.1016/j.respol.2019.01.009

Ma A, Lu X, Gray C, Raybould A, Tamaddoni-Nezhad A, Woodward G, Bohan D (2019) Ecological networks reveal resilience of agro-ecosystems to changes in farming management. Nat Ecol Evol 3:260–264. https://doi.org/10.1038/s41559-018-0757-2

Macel M, van Dam NM, Keurentjes JJB (2010) Metabolomics: the chemistry between ecology and genetics. Mol Ecol Resour 10:583–593. https://doi.org/10.1111/j.1755-0998.2010.02854.x

Masip G, Sabalza M, Pérez-Massot E, Banakar R, Cebrian D, Twymann RM, Capell T, Albajes R, Christou P (2013) Paradoxical EU agricultural policies on genetically engineered crops. Trends Plant Sci 18:312–324. https://doi.org/10.1016/j.tplants.2013.03.004

McHughen A (2007) Fatal flaws in agbiotech regulatory policies. Nat Biotechnol 25:725–727. https://doi.org/10.1038/nbt0707-725

Mendelsohn M, Kough J, Vaitusis Z, Matthews K (2003) Are *Bt* crops safe? Nat Biotechnol 21:1003–1009. https://doi.org/10.1038/nbt0903-1003

Miller DW (1994) Critical rationalism: a restatement and defence. Open Court, Chicago and La Salle. Br J Philos Sci 46:610–616

Nerlich B, Clarke DD, Dingwall R (2001) Fictions, fantasies, and fears: the literary foundations of the cloning debate. J Lit Semant 30:37–52. https://doi.org/10.1515/jlse.30.1.37

Nickson TE, Head GP (1999) Environmental monitoring of genetically modified crops. J Environ Monit 1:101N–105N. https://doi.org/10.1039/A908763J

Niiniluoto I (1993) The aim and structure of applied research. Erkenntnis 38:1–21. https://doi.org/10.1007/BF01129020

Niiniluoto I (1994) Approximation in applied science. In: Kuokkanen M (ed) Idealization VII: structuralism, idealization and approximation. Rodopi, Amsterdam, pp 127–139

Norton SB, Rodier DJ, Gentile JH, van der Schale WH, Wood WP, Slimak MW (1992) A framework for ecological risk assessment at the EPA. Environ Toxicol Chem 11:1663–1672. https://doi.org/10.1002/etc.5620111202

Ortiz-García S, Ezcurra E, Schoel B, Acevedo F, Soberón J, Snow AA (2005) Absence of detectable transgenes in local landraces of maize in Oaxaca, Mexico (2003–2004). Proc Natl Acad Sci U S A 102:12338–12343. https://doi.org/10.1073/pnas.0503356102

Pardo R, Engelhard M, Hagen K, Jørgensen RB, Rehbinder E, Schnieke A, Szmulewicz M, Thiele F (2008) The role of means and goals in technology acceptance. A differentiated landscape of public perceptions of pharming. EMBO Rep 10:1069–1075. https://doi.org/10.1038/embor.2009.208
Perry JN, Rothery P, Clark SJ, Heard MS, Hawes C (2003) Design, analysis and statistical power of the Farm-Scale Evaluations of genetically modified herbicide-tolerant crops. J Appl Ecol 40:17–31. https://doi.org/10.1046/j.1365-2664.2003.00786.x
Pielke R (2007) The honest broker. Cambridge University Press, Cambridge, p 188. https://doi.org/10.1017/CBO9780511818110
Pimentel D (2009) Invasive plants: their role in species extinctions and economic losses to agriculture in the USA. In: Inderjit S (ed) Management of invasive weeds. Springer, Dordrecht, pp 1–7. https://doi.org/10.1007/978-1-4020-9202-2_1
Pink DAC, Walkey DGA (1984) Resistance in marrow (*Cucurbita pepo* L.) to different strains of cucumber mosaic virus. J Agric Sci 103:519–521. https://doi.org/10.1017/S0021859600043616
Poppy GM (2004) Geneflow from GM plants – towards a more quantitative risk assessment. Trends Biotechnol 22:436–438. https://doi.org/10.1016/j.tibtech.2004.07.005
Pott A, Otto M, Schlz R (2018) Impact of genetically modified organisms on aquatic environments: review of available data for the risk assessment. Sci Total Environ 635:687–698. https://doi.org/10.1016/j.scitotenv.2018.04.013
Qi A, Perry JN, Pidgeon JD, Haylock LA, Brook DR (2008) Cost-efficacy in measuring farmland biodiversity – lessons from the Farm Scale Evaluations of genetically modified herbicide-tolerant crops. Ann Appl Biol 152:93–101. https://doi.org/10.1111/j.1744-7348.2007.00193.x
Raybould A (2006) Problem formulation and hypothesis testing for environmental risk assessments of genetically modified crops. Environ Biosaf Res 5:119–125. https://doi.org/10.1051/ebr:2007004
Raybould A (2010a) Reducing uncertainty in regulatory decision-making for transgenic crops: more ecological research or clearer environmental risk assessment? GM Crops 1:25–31. https://doi.org/10.4161/gmcr.1.1.9776
Raybould A (2010b) The bucket and the searchlight: formulating and testing risk hypotheses about the weediness and invasiveness potential of transgenic crops. Environ Biosaf Res 9:123–133. https://doi.org/10.1051/ebr/2011101
Raybould A (2019) Problem formulation and phenotypic characterisation for the development of novel crops. Transgenic Res 28:135–145
Raybould A, Cooper JI (2005) Tiered tests to assess the environmental risk of fitness changes in hybrids between transgenic crops and wild relatives: the example of virus resistant *Brassica napus*. Environ Biosaf Res 4:127–140. https://doi.org/10.1051/ebr:2005018
Raybould A, Gray AJ (1993) Genetically modified crops and hybridization with wild relatives: a UK perspective. J Appl Ecol 30:199–219. https://doi.org/10.2307/2404623
Raybould A, Gray AJ (1994) Will hybrids of genetically modified crops invade natural communities? Trends Ecol Evol 9:85–89. https://doi.org/10.1016/0169-5347(94)90201-1
Raybould A, Macdonald P (2018) Policy-led comparative environmental risk assessment of genetically modified crops: testing for increased risk rather than profiling phenotypes leads to predictable and transparent decision-making. Front Bioeng Biotechnol 6:43. https://doi.org/10.3389/fbioe.2018.00043
Raybould A, Poppy GM (2012) Commercialising GM crops under EU regulations: objectives and barriers. GM Crops Food 3:9–20. https://doi.org/10.4161/gmcr.18961
Raybould A, Vlachos D (2011) Non-target organism effects tests on Vip3A and their application to the ecological risk assessment for cultivation of MIR162 maize. Transgenic Res 20:599–611. https://doi.org/10.1007/s11248-010-9442-1
Raybould A, Stacey D, Vlachos D, Graser G, Li X, Joseph R (2007) Non-target organism risk assessment of MIR604 maize expressing mCry3A for control of corn rootworm. J Appl Entomol 131:391–399. https://doi.org/10.1111/j.1439-0418.2007.01200.x

Raybould A, Higgins LS, Horak MJ, Layton RJ, Storer NP, De La Fuente JM, Herman RA (2012a) Assessing the ecological risks from the persistence and spread of feral populations of insect-resistant transgenic maize. Transgenic Res 21:655–664. https://doi.org/10.1007/s11248-011-9560-4

Raybould A, Kilby P, Graser G (2012b) Characterising microbial protein test substances and establishing their equivalence with plant-produced proteins for use in risk assessments of transgenic crops. Transgenic Res 22:445–460. https://doi.org/10.1007/s11248-012-9658-3

Raybould A, Holt K, Kimber I (2019) Using problem formulation to clarify the meaning of weight of evidence and biological relevance in environmental risk assessments for genetically modified crops. GM Crops Food 10:63–76. https://doi.org/10.1080/21645698.2019.1621615

Romeis J, Meissle M, Bigler F (2006) Transgenic crops expressing *Bacillus thuringiensis* toxins and biological control. Nat Biotechnol 24:63–71. https://doi.org/10.1038/nbt1180

Romeis J, Lawo NC, Raybould A (2009) Making effective use of existing data for case-by-case risk assessments of genetically engineered crops. J Appl Entomol 133:571–583. https://doi.org/10.1111/j.1439-0418.2009.01423.x

Romeis J, Hellmich RL, Candolfi M, Carstens K, De Schrijver A, Gatehouse AMR, Herman RA, Huesing JE, McLean M, Raybould A, Shelton AM, Waggoner A (2011) Recommendations for the design of laboratory studies on non-target arthropods for risk assessment of genetically engineered plants. Transgenic Res 20:1–22. https://doi.org/10.1007/s11248-010-9446-x

Romeis J, Raybould A, Bigler F, Candolfi MP, Hellmich RL, Huesing J, Shelton A (2013) Deriving criteria to select arthropod species for laboratory tests to assess the ecological risks from cultivating arthropod-resistant transgenic crops. Chemosphere 90:901–909. https://doi.org/10.1016/j.chemosphere.2012.09.035

Romeis J, Naranjo SE, Meissle M, Shelton AM (2019) Genetically engineered crops help support conservation biological control. Biol Control 130:136–154. https://doi.org/10.1016/j.biocontrol.2018.10.001

Rozell DJ (2018) The ethical foundations of risk analysis. Risk Anal 38:1529–1533. https://doi.org/10.1111/risa.12971

Saji H, Nakajima N, Aono M, Tamaoki M, Kubo A, Wakiyama S, Hatase Y, Nagatsu M (2005) Monitoring the escape of transgenic oilseed rape around Japanese ports and roadsides. Environ Biosaf Res 4:217–222. https://doi.org/10.1051/ebr:2006003

Sanvido O, Romeis J, Gathmann A, Gielkens M, Raybould A, Bigler F (2012) Evaluating environmental risks of genetically modified crops – ecological harm criteria for regulatory decision-making. Environ Sci Policy 15:82–91. https://doi.org/10.1016/j.envsci.2011.08.006

Sarewitz D (2004) How science makes environmental controversies worse. Environ Sci Policy 7:385–403. https://doi.org/10.1016/j.envsci.2004.06.001

Sasu MA, Ferrari MJ, Du D, Winsor JA, Stephenson AG (2009) Indirect costs of a nontarget pathogen mitigate the direct benefits of a virus-resistant transgene in wild *Cucurbita*. Proc Natl Acad Sci U S A 106:19067–19071. https://doi.org/10.1073/pnas.0905106106

Schaart JG, van der Wiel CCM, Lotz LAP, Smulders MJM (2016) Opportunities for products of new breeding techniques. Trends Plant Sci 21:438–449. https://doi.org/10.1016/j.tplants.2015.11.006

Schäfer RB (2012) Biodiversity, ecosystem functions and services in environmental risk assessment: introduction to the special issue. Sci Total Environ 415:1–2. https://doi.org/10.1016/j.scitotenv.2011.08.012

Simard SJ, Légère A, Pageau D, Lajeunesse J, Warwick S (2002) The frequency and persistence of volunteer canola (*Brassica napus*) in Québec cropping systems. Weed Technol 16:433–439. https://doi.org/10.1614/0890-037X(2002)016[0433:TFAPOV]2.0.CO;2

Sitterley WR (1972) Breeding for disease resistance in cucurbits. Annu Rev Phytopathol 10:471–490

Smyth SJ (2017) Genetically modified crops, regulatory delays, and international trade. Food Energy Secur 6:78–86. https://doi.org/10.1002/fes3.100

Squire GR, Brooks DR, Bohan DA, Champion GT, Daniels RE, Haughton AJ, Hawes C, Heard MS, Hill MO, May MJ et al (2003) On the rationale and interpretation of the farm scale evaluations of genetically modified herbicide-tolerant crops. Philos Trans R Soc Lond B 358:1779–1799. https://doi.org/10.1098/rstb.2003.1403

Stephens PA, Buskirk SW, del Rio CM (2007) Inference in ecology and evolution. Trends Ecol Evol 22:192–197. https://doi.org/10.1016/j.tree.2006.12.003

Strauss SH, Ma C, Ault K, Klocko AL (2016) Lessons from two decades of field trails with genetically modified trees in the USA: biology and regulatory compliance. In: Vettori C, Gallardo F, Häggman H, Kazana V, Migliacci F, Pilate G, Fladung M (eds) Biosafety of forest transgenic trees. Springer, Dordrecht, pp 101–124. https://doi.org/10.1007/978-94-017-7531-1_5

Street P (2007) Constructing risks: GMOs, biosafety and environmental decision-making. In: Somsen H (ed) The regulatory challenge of biotechnology: human genetics, food, and patents. Edward Elgar, Cheltenham/Northampton, pp 95–117

Sukopp H, Sukopp U (1993) Ecological long-term effects of cultigens becoming feral and of naturalization of non-native species. Experientia:210–218. https://doi.org/10.1007/BF01923528

Tait J (2001) More Faust than Frankenstein: the European debate about the precautionary principle and risk regulation for genetically modified crops. J Risk Res 4:175–189. https://doi.org/10.1080/13669870010027640

Turchin P, Taylor AD (1992) Complex dynamics in ecological time series. Ecology 73:289–305. https://doi.org/10.2307/1938740

Van der Voet H, Perry JN, Amzal B, Paoletti C (2011) A statistical assessment of differences and equivalences between genetically modified and reference plant varieties. BMC Biotechnol 11:15. https://doi.org/10.1186/1472-6750-11-15

Vighi M, Rico A (2018) The concept of resilience in ecological risk assessment: scientific and regulatory issues. Integr Environ Assess Manag 14:581–585. https://doi.org/10.1002/ieam.4080

Wilkinson MJ, Elliott LJ, Allainguillaume J, Shaw MW, Norris C, Welters R, Alexander M, Sweet J, Mason DC (2003) Hybridization between *Brassica napus* and *B. rapa* on a national scale in the United Kingdom. Science 302:457–459. https://doi.org/10.1126/science.1088200

# Risk Assessment of Insect-Resistant Genetically Modified Crops on Non-target Arthropods and Benefits to Associated Biodiversity of Agro-Ecosystems

**Patricia Gladys Gadaleta**

**Abstract** Biodiversity is threatened by anthropic activities. Among them, agriculture is a key driver of environmental change because agro-ecosystems are optimized for production. Agricultural biodiversity is the subset of biodiversity that contributes to food, feed, fibre and biofuel production. It also encompasses what is known as 'associated biodiversity', the vast range of organisms that live in and around agro-ecosystems, sustaining them and contributing to their output. Agricultural biodiversity also is seen as the outcome of interactions among genetic resources, environment and management systems and practices used by farmers. In this context, introduction of crops derived from modern biotechnology such as insect-resistant genetically modified (IRGM) crops and resulting interactions in agro-ecosystems are currently creating a novel scenario for associated biodiversity. Two main questions arise with respect to these interactions. First, do IRGM crops have non-target effects on other organisms, particularly those that enhance crop production; if so, how are these effects assessed? Second, to what extent does the use of IRGM crops reduce application of broad-spectrum pesticides that can in turn have impacts on the associated biodiversity? This chapter examines the process for assessing environmental impacts of IRGM crops, with a focus on non-target organisms, and also reviews substantial evidence suggesting that there is a potential value of this technology in reducing pesticide use and protecting beneficial insects in agro-ecosystems.

**Keywords** Insect-resistant genetically modified crops · Environmental risk assessment · Non-target arthropods · Biodiversity · Natural enemies · Pesticide reduction · Integrated pest management

P. G. Gadaleta (✉)
Ministerio de Agricultura, Ganadería y Pesca, Buenos Aires, Argentina
e-mail: pgadal@magyp.gob.ar



A. Chaurasia et al. (eds.), *GMOs*, Topics in Biodiversity and Conservation 19,
https://doi.org/10.1007/978-3-030-53183-6_15

## Introduction

Agricultural biodiversity is a broad concept that includes all components of biological diversity related to food production and agriculture, and all components of biological diversity that constitute the agricultural ecosystems, also named agro-ecosystems. It comprises the variety and variability of animals, plants and prokaryotes, at the genetic, species and ecosystem levels, which are necessary to sustain key functions of agro-ecosystems and their structure and processes (Convention on Biological Diversity, CDB COP 5 decision V/5, annex The Scope of Agricultural Biodiversity). According to the Commission on Genetic Resources for Food and Agriculture of the Food and Agriculture Organization of the United Nations (FAO), biodiversity for food and agriculture is the subset of biodiversity that contributes in one way or another to agriculture and food production. It includes domesticated plants and animals that are part of crop, livestock, forest or aquaculture systems, harvested forest and aquatic species, the wild relatives of domesticated species and other wild species harvested for food and other products. It also encompasses what is known as 'associated biodiversity', the vast range of organisms that live in and around food and agricultural production systems, sustaining them and contributing to their output (FAO 2019). Associated biodiversity includes beneficial organisms, such as pollinators, natural enemies of pests and organisms necessary for plant and soil health, litter decomposition and nutrient cycling such as earthworms, collembolans, nematodes, fungi and microorganisms. Biodiversity, in general, is threatened by numerous anthropic activities. Among them, agriculture is a key driver of environmental change because agro-ecosystems are optimized for food, feed, fibre and biofuel production.

The most direct negative impact of agriculture on biodiversity results from the transformation of natural ecosystems into land devoted to agriculture. In response, the sustainable use of biodiversity and the development of technologies to improve yield and quality of foods on less land area are key components of strategic optimization of the natural resources intended for agriculture.

One of the most commonly used tools for agriculture improvement is plant breeding, which originated independently in many parts of the world 5000–13,000 years ago (Balter 2007). Through artificial selection and selective breeding, genes of plants were modified and adapted to make crops more nutritious, higher yielding and more resilient to biotic and abiotic stresses (such as to drought and pestilence). Since then, advancements in agriculture, science and technology have brought about the modern agricultural biotechnology revolution. Although these new agricultural technologies are revolutionary, agriculture is still highly disruptive of the environment, presenting significant threats to sustainability. Additionally, agro-ecosystems are influenced by major global trends, such as changes in climate, international markets and demography. All these factors contribute to other challenges, such as land-use change, pollution, overuse, overharvesting and the proliferation of invasive species, which in turn further directly or indirectly affect biodiversity and the ecosystem services it provides (FAO 2019).

The importance of biodiversity in food production, environmental conservation and human well-being is a timely and important topic for discussion. The international community has observed and is starting to understand that there is a need to integrate biodiversity into most human activities. However, this requires careful consideration about how to accomplish this efficiently, safely and sustainably into multiple areas where humans and biodiversity intersect. At the 13th Conference of Parties of Convention on Biological Diversity in 2016, there was extensive discussion about how conservation and sustainable use of biodiversity could be integrated into sectoral and cross-sectoral plans, resulting in the Cancun Declaration on Mainstreaming the Conservation and Sustainable Use of Biodiversity for Well-being. In this declaration, governments and organizations would commit to work at all levels and across all sectors to maintain biodiversity through multiple actions. These actions would include structured and coherent actions for conservation, sustainable use, management and restoration of biological diversity and ecosystems, all of which would require the development of plans, programmes and policies, as well as legal and administrative measures and budgets (CBD COP13 2016). Such a commitment is critical to the implementation of the 2030 Agenda for Sustainable Development and its Sustainable Development Goals (SDG) and is necessary to achieve SDG 2 (end hunger, achieve food security and improved nutrition and promote sustainable agriculture) and SDG 15 (protect, restore and promote sustainable use of terrestrial ecosystems, sustainably manage forests, combat desertification, and halt and reverse land degradation and halt biodiversity loss).

## Genetically Modified (GM) Crops

Plant genetic resources derived from biotechnology also are part of today's agricultural biodiversity. From this perspective, biotechnology is an important tool for enhancing genetic diversity of crop species. New traits are developed by introducing novel genes or other kinds of genetic modifications, which help in dealing with increasing human demands for food, fibre and biofuels and also with conservation of agro-ecosystems. Genetically modified (GM) plants, by the use of genetic engineering, were introduced into commercial agriculture in the mid-1990s. Since then, the total area cultivated with GM crops has progressively increased worldwide. In 2018, the worldwide area of GM crops reached 191.7 million ha. In this context, 54% of the global GM crop area was located within developing countries, compared with 46% within industrial countries (ISAAA 2018). From 1992 to 2018, there have been 4349 approvals of GM plants granted by regulatory authorities in 70 countries, with approval of 387 GM events from 27 GM crops. Additionally, 44 countries formally approved importation of GM crop products for consumption purposes, including food, feed and further processing.

## Insect-Resistant Genetically Modified (IRGM) Crops

The two major traits developed in GM crops are herbicide tolerance and resistance to insects. This chapter focuses on insect-resistant genetically modified (IRGM) crops and the assessment of non-target impacts and the high potential value of this technology to protecting beneficial insects in agro-ecosystems. Herbicide-tolerant GM crops developed by introducing new genes (some of them derived from soil bacteria or plants) confer tolerance to herbicides externally applied on these crops, but the products of these genes do not have specific resulting toxicity for any organism. Nevertheless, the use of more herbicides in GM crops might indirectly affect the abundance and diversity of associated biodiversity by reducing weeds that are sources of habitat for many beneficial organisms, such as pollinators and natural enemies of pests. Interactions between associated biodiversity and weeds are very complex and, in general, the presence or absence of different species of beneficial organisms is related to crop management practices, including crop rotation, intercropping, cover cropping and crop-field border vegetation manipulation. More studies on intriguing aspects of associated biodiversity–herbicide–weed interactions are needed (Sharma et al. 2018).

The majority of IRGM plants were modified by the introduction of genes from *Bacillus thuringiensis*. This bacterium produces crystalline delta endotoxins called Cry and Cyt proteins during the sporulation growth phase (Agaisse and Lereclus 1995; Guerchicoff et al. 2001) and other insecticidal proteins called Vip proteins during the vegetative growth phase (Estruch et al. 1996). These proteins are active in the gut of target species of insects and have different modes of action (Lee et al. 2003; Bravo et al. 2007), but all have a high level of specificity in their spectrum of insect toxicity. Since the first insecticidal crystalline protein gene was cloned and sequenced in 1981, 993 toxin-encoding genes have been cloned and classified, including 801 Cry genes, 40 Cyt genes and 152 Vip genes (Sanchis 2011; Crickmore et al. 2018; Xiao and Wu 2019). Current GM crops varieties could have two or more insecticidal genes, either obtained by genetic engineering or by the conventional breeding techniques. When two or more genes targeting the same pest are combined, this is referred to as a pyramid event. And when two or more genes having different targets (i.e. lepidopteran and coleopteran pests) or different functions (i.e. herbicide tolerance and insect resistance) are combined, this is called a stack event. Both pyramid and staked events can be even more effective in controlling target pests if they are used in accordance with good agriculture practices, such as crop rotation and use of refuge areas; the combined approach can help slow down the evolution of insect resistances to insecticidal proteins (Carrière et al. 2016; Gressel et al. 2017).

Despite the large number of pests that impact all crops, IRGM crops are currently limited to maize, cotton, soybean, rice, eggplant, potato, sugarcane, cowpea and poplar trees (ISAAA's GM Approval Database 2020). They were mainly developed against lepidopteran and coleopteran pests. In maize, IRGM varieties have been developed against Lepidoptera that are pests globally, such as European corn borer (*Ostrinia nubilalis*), corn earworm (*Helicoverpa zea*) and fall armyworm (*Spodoptera frugiperda*) and Coleoptera that are principally maize pests in North America, such as corn rootworm (*Diabrotica virgifera*). In soybean, IRGM events target Lepidopteran pests, such as soybean looper (*Pseudoplusia includens*) and velvetbean caterpillar (*Anticarsia gemmatalis*), which present significant problems occurring largely in South America. IRGM cotton was developed to combat Lepidopteran pests, tobacco budworm (*Heliothis virescens*), cotton bollworm (*H. gelotopoeon, H. zea* and *H. armigera*, which together are often referred to as the bollworm complex) and against cotton leafworm (*Alabama argillacea*) and pink bollworm (*Pectinophora gossypiella*). A cotton GM event recently has been approved only for processing in the United States, Canada, Australia, New Zealand, Japan and Taiwan against Hemiptera (*Lygus hesperus* and *L. lineolaris*) (ISAAA's GM Approval Database 2020) but to date has not been approved for cultivation. For rice, two lepidopteran-resistant GM events against leaf-folders and yellow stem borer were developed by Huazhong Agricultural University in Wuhan (Tu et al. 2000). Chinese authorities granted a safety certificate of these events in 2009. However, these events have not been approved for commercial production even though rice is the most important food staple in this country. In 2018, one of these events obtained regulatory approval for food use or processing in the United States (ISAAA's GM Approval Database 2020). China is the only country worldwide with significant commercial GM poplar plantations. GM poplar trees contain the *cry1a* gene and a proteinase inhibitor from the weed *Sagittaria sagittifolia* to control beetle species (such as *Lymantria dispar*, *Clostera anachoreta* and *Micromelalopha troglodyte*). To date, nearly 22 insect-resistant poplar varieties have been developed and approved for small-scale field testing, environmental release or pilot-scale production (Wang et al. 2018). Potato cultivars expressing the Cry3A toxin for resistance to Colorado potato beetle (*Leptinotarsa decemlineata*) was the first GM crop type approved for human consumption and commercially produced in the United States. This occurred in 1995, but because of consumer concern and the introduction of more effective insecticides, GM potatoes were taken off the market in 2000 (Grafius and Douches 2008). Eggplant (also known as brinjal in India and Bangladesh and talong in the Philippines) is one of the most important, inexpensive and popular vegetables grown and consumed in Asia. IRGM eggplant for protection against the eggplant fruit and shoot borer, such as *Leucinodes orbonalis*, was approved in Bangladesh in 2013 (Hautea et al. 2016). Since then, the adoption of IRGM eggplant by farmers has increased significantly (Shelton et al. 2018). The cane borer is a widespread insect that causes an estimated 10% production loss for Brazilian farmers and results in about 4600 million dollars per year in pesticide expenses (Oliveira et al. 2014). Two events of IRGM sugarcane against cane borer were approved by Brazil in 2017, and Canada and United States granted approvals of GM sugarcane for food consumption and processing in 2018 (Cristofoletti et al. 2018; Kennedy et al. 2018). Lastly, IRGM cowpea was developed against Lepidoptera pest *Maruca vitrata*. The event was recently approved for cultivation and consumption in Nigeria (ISAAA's GM Approval Database 2020).

Another approach to developing crops with insect resistance is the use of genes that encode for natural defensive compounds of plants, such as proteinase and alpha amylase inhibitors that target Lepidoptera, Coleoptera and Hemiptera pests (Malone et al. 2008). GM crops expressing these products enhance resistance against pests by reducing the digestibility and nutritional quality of leaves. In addition to poplar trees mentioned above, IRGM cotton cultivars expressing Cry toxin together with a modified cowpea trypsin inhibitor were developed and approved commercially for cultivation in China in 2000 (Zhang et al. 2000). IRGM maize carrying a combination of *cry* gene and proteinase inhibitor gene of *Solanum tuberosum* was approved and commercialized in the United States and Canada in 1997 (ISAAA's GM Approval Database 2020).

A more recently developed technology is the use of RNA interference (RNAi) that leads to gene silencing (i.e. blocking the expression of target genes) by eliminating the corresponding messenger RNA (mRNA). RNAi is an ancient mechanism present in most eukaryotic cells that regulates gene expression via mRNA degradation, repression of translation and chromatin remodelling (Valencia-Sanchez et al. 2006; Hannon 2002). The RNA interference mechanism was discovered in the nematode *Caenorhabditis elegans* (Grishok 2005), but it was first reported in plants in 1928 as an immune response to viral infection (Baulcombe 2004). Currently, RNAi technology comprises a suite of tools with a vast range of potential applications in agriculture, including resistance management and control of pests and pathogens in a wide range of crops. Because of its highly specific mode of action compared to other pest control strategies, such as chemical pesticides, double-stranded RNA can be specifically designed against a pathway for regulation of gene expression in a single target species or a group of related species, without affecting other species (Zotti et al. 2018; Zhang et al. 2017). Recently, IRGM expressing RNAi was approved for commercial cultivation in the United States and Canada (ISAAA's GM Approval Database 2020). An inverted repeat sequence of a 240-base pair fragment of the western corn rootworm (*D. virgifera virgifera*) *Snf7* gene was introduced in maize to cause down-regulation of the targeted *DvSnf7* gene, a component of the endosomal sorting complex required for transport (Bachman et al. 2013). When western corn rootworm feeds on the plant-produced RNA, a rapid decrease results in DvSnf7 mRNA and protein levels produced by RNAi machinery, leading to growth inhibition followed by mortality (Bolognesi et al. 2012). In addition, a new generation of pyramided events that combine RNAi with Cry toxin provides two modes of action against corn rootworm (Levine et al. 2015).

## Interactions Between IRGM Crops and Associated Organisms

Two main questions arise with respect to the interactions between associated organisms and IRGM crops. First, do IRGM crops have non-target effects on other organisms, particularly those that enhance crop production; if so, how are these effects assessed? And second, to what extent does the use of IRGM crops reduce the application of broad-spectrum pesticides that can in turn have impacts on the associated biodiversity?

## *Non-target Effect on Other Organisms*

To address the first question, it is important to highlight that for almost 40 years, general principles have been developed and used for the assessment of human health and environmental risks from chemicals and other stressors. These principles are the basis of environmental risk assessment (ERA) and can be summarized as: problem formulation methodology, assessment of case-specific exposure to a stressor, assessment of the relationships between the level of exposure and the magnitude of associated effects and overall characterization of risks in terms of the likelihood and magnitude of effects. In a similar way, these principles are applied in ERA of GM plants to facilitate regulatory decisions related to the approval of this technology (US EPA 1998; Hill and Sendashonga 2003; Hill 2005).

Problem formulation is the first step in ERA where policy goals, scope, assessment end points and methodology are defined to an explicitly stated problem and approach for later analysis. Problem definition shapes ERA into a manageable form for analysis through consideration of the case-specific attributes of the GM crop in question, identification of logically relevant concerns and description of cause–effect relationships (Wolt et al. 2010). Information considered in problem formulation could include the published scientific literature, expert opinions, research data and relevant data derived from molecular, compositional and agronomic analyses performed during GM plants development. One of the key steps in problem formulation is to recognize and consider which policy protection goals are in play. In most countries, policy protection goals are set by legislation, but they usually are defined in very broad terms and are too ambiguous to be directly applicable in ERA, for example, 'avoid (or minimize) harm to the environment, humans and animals caused by human activities'. Therefore, policy protection goals should be translated into unambiguous operational protection goals to establish relevant assessment end points and to test risk hypotheses that can be used in ERA (Garcia-Alonso and Raybould 2014). A risk hypothesis is an assumption regarding the cause–effect relationships among changed attributes, sources, exposure routes, end points, responses and measures relevant to ERA (Wolt et al. 2010). In the case of an insecticidal protein of an IRGM crop, the risk hypothesis that emerges from problem formulation is usually that the stressor does not harm non-target organisms at the concentration expressed in the field.

In the analysis phase of the ERA, the risk hypothesis is used to develop one or more experimental hypotheses that are used for testing and corroboration. Specific tiered toxicological testing systems for IRGM plants are used based on the problem formulation. This approach provides a logical road map for assessing potential risks of insecticidal products (purified protein or plant material) on non-target arthropods (Dutton et al. 2003; Garcia-Alonso et al. 2006; Romeis et al. 2008). Early-tier testing involves exposing test organisms to elevated doses, often using laboratory procedures with purified protein in artificial diets (Rose 2007; Romeis et al. 2008). Recommendations for: (1) the design of laboratory studies on non-target arthropods with quality standards to support submissions to regulatory authorities (Romeis et al. 2011) and (2) deriving criteria for the selection of arthropod species for laboratory tests, which are well established, particularly early-tier studies in arthropods (Romeis et al. 2013). Based on the risk hypotheses, early-tier laboratory experiments are conducted on representative species of non-target organisms present in the receiving environment that are likely to be exposed to the insecticidal protein. These representative non-target organisms, referred as surrogates, are exposed to concentrations of the purified protein that are higher (often >10X) than concentrations they would encounter in the field. This high-dose testing increases the likelihood of detecting any adverse effects. The selection process of surrogate species poses numerous challenges for developers of GM crops as well as for regulators who will interpret test results (Carstens et al. 2014). This process can be informed by existing databases of arthropod communities associated with major field crops (Meissle et al. 2012; Romeis et al. 2014; Riedel et al. 2016; Li et al. 2017). Given the volume of data on non-target organism effects generated through the use of surrogate species, the depth of analysis to which these data have been subjected, and the similarities between the agricultural environment where GM crops are released, results from one region could be used to inform ERA of a particular IRGM crop in another region (Wach et al. 2016). Based on the degree to which agricultural ecosystems are exposed to crop residues, such as soil ecosystems (Icoz and Stotzky 2008) or aquatic ecosystems, these too should be analysed with the problem formulation approach in order to define the necessary laboratory studies to test risk hypothesis (Carstens et al. 2012).

In all these assessments, if potential hazards are detected in early tests or if unacceptable uncertainties about possible hazards remain, the sequence of testing continues using increasingly realistic scenarios outside the laboratory. In such cases, higher tier tests, which include more complex semi-field (e.g. employing greenhouse under containment conditions) or open-field tests, can serve to confirm whether an effect is detectable in these scenarios. In cases where uncertainty about the risk still remains after such higher tier studies, it is possible to then return to lower tiers to conduct additional studies in an iterative way, for example, by including additional surrogate test species (Romeis et al. 2008). The conceptual pathway of the tiered approach leads from relatively simple and controllable lower tier assessments to increasingly complex higher tier assessments. In 2010, a study based on meta-analyses was performed to test whether laboratory studies of non-target effects of Cry proteins are consistent with results from field studies that compared the abundance of non-target arthropods in GM crops versus non-GM counterparts (Duan et al. 2010). This study showed that laboratory studies of insecticidal proteins derived from IRGM crops predicted effects that were on average either more conservative than or consistent with effects of IRGM crops measured in the field. Others studies have used this approach of evaluating several independent trials in an improved statistical analysis to study the non-target effects of different IRGM crops

(Marvier et al. 2007; Duan et al. 2008; Wolfenbarger et al. 2008; Naranjo 2009; Comas et al. 2013; Pellegrino et al. 2018). The findings of these meta-analysis studies also reinforce the conclusions of individual studies that IRGM crops have little to no effect on the most common non-target organisms in the agro-ecosystem.

## *Do IRGM Crops Reduce Use of Broad-Spectrum Pesticides?*

In order to assess the extent the use of IRGM crops reduces the application of broad-spectrum pesticides, that in turn have impacts on associated biodiversity, it is important to highlight that expressed proteins or RNAi in IRGM crops have high specificity to control targeted pests. A careful consideration of all available pest control practices and the subsequent integration of appropriate measures to avoid the development of pest populations is recognized as a desirable international standard for crop protection (FAO 2002). In view of that, IRGM crops contribute to sustainable crop protection systems, such as integrated pest management (IPM). FAO outlines IPM as an ecosystem approach to crop production and protection that combines different management strategies and practices to grow healthy crops and minimize the use of pesticides. Two important features in this definition are the use of less pesticides and the use of a diverse range of pest control tactics. In this sense, a recent paper by Brookes and Barfoot (2018) updated previous assessments of the environmental impacts associated with changes in pesticide use and greenhouse gas emissions arising from the use of GM crops. They estimated that since 1996 to 2016 the use of pesticides on GM crops has decreased by 671.4 million kg of active ingredient, relative to the amount expected if the same crop area had been planted to conventional cultivars, resulting in a reduced environmental impact. In particular, cumulatively gains have included a 92.1 million kg reduction in maize pesticide active ingredients and a 288 million kg reduction in cotton pesticide active ingredients (especially from the adoption of IRGM cotton in China and India). IRGM soybean has only 8 years of commercial use in South America (mostly Brazil and Argentina); during the period of 2013 from 2016, the pesticide use (active ingredient) reduction relative to the amount reasonably expected to be used if this crop area had been planted with conventional soybeans was 7.4 million kg (6% of the total soybean pesticide use).

Carpenter (2010) summarizes some farmer surveys in different countries that have shown decreases in the amounts of pesticide and the number of pesticide applications used on IRGM crops compared to conventional crops. Reductions range from 14% to 75% in terms of the amount of active ingredient and similarly from 14% to 76% for the number of applications (Carpenter 2010). A report published in 2014 by the United States Department of Agriculture examined relevant topics related to three major stakeholders in agricultural biotechnology (GM seed suppliers and technology providers, farmers and consumers). This report indicates that farmers generally use less pesticide when they plant IRGM maize and GM cotton, and in particular the pesticide use on maize farms has declined by an even greater percentage than earlier observed by Carpenter: from 0.21 pound per planted acre in

1995 (before IRGM were introduced) to just 0.02 pound in 2010 (Fernandez-Cornejo et al. 2014). Meta-analysis of 147 original studies published on the most important GM crops, including herbicide-tolerant soybean, maize and cotton, as well as insect-resistant maize and cotton, Klümper and Qaim (2014) showed that on average GM technology adoption has reduced the chemical pesticide use by 36.9%.

Reduction in pesticide use has obvious direct benefit in terms of reduced adverse impact on soil, drinking water and general human health, all of which is critically important but beyond the scope of this chapter. As to pest control: Reductions in pesticide use allows different communities of natural enemies to increase in agro-ecosystems. In this context, biological pest control by natural enemies is a key ecological service associated with agro-ecosystem biodiversity. Predators and parasitoids that biologically control herbivorous pests diversify pest control tactics and play an important role in IPM. There is a need to consider both natural and introduced biological control and the inner plant defence of IRGM crops to tackle insect pests in IPM (Poppy and Sutherland 2004); however, there have been few attempts to combine these approaches. Presumably, this is because the economic impact of biological control in IPM is difficult to assess (for review see Naranjo et al. 2015). Nevertheless, in the United States, it was estimated in 2006 that the annual value of insect ecological services was almost $60 billion. In this estimation, insects were responsible for controlling 33% of pests. The estimated annual value of this ecological service averages around $4.5 billion (Losey and Vaughan 2006). Adoption of IRGM crops controls targeted pests, and because fewer insecticides are used, these fields also indirectly provide biocontrol services that can spill onto crops in neighbouring fields. In long-term landscape-level study of IRGM cotton, on the basis of data from 1990 to 2010 performed at 36 sites in six provinces of northern China, Lu et al. (2012) evaluated three types of generalist arthropod predators (ladybirds, lacewings and spiders). They found a marked increase in the abundance of these natural enemies and a corresponding decrease in the abundance of aphid pests due to the widespread adoption of IRGM cotton. In a recent review on IPM and GM crops, Anderson et al. (2019) give some examples of the use of IRGM in the application of IPM and analyse the challengers of develop a successful implementation of IPM plan. The authors highlight that sustainable and eco-rational IPM strategies rely on a diversified portfolio of tactics, of which GM crops represent a valuable tool.

In sum, there is substantial evidence that IRGM crops directly reduce target pest populations over broad scales, and there is also evidence these crops reduced pest populations, leading to the reduction of pesticide use on non-GM crops (Carrière et al. 2003; Dively et al. 2018; Hutchison et al. 2010; Wan et al. 2012; Wu et al. 2008; Zhang et al. 2018). That, in turn, leads to increased pest control by natural enemies. A recent review (Romeis et al. 2019) showed how the change in pesticide use by the introduction of IRGM crops has benefitted non-target species, including insects that provide biological control services. The same review summarizes evidence and the literature that demonstrates current IRGM crops have negligible or no impact on non-target arthropods.

## Future of Agricultural Biodiversity

In agro-ecosystems, there is a trade-off between different ecosystem services. If provisioning services such as food, feed, fibre or fuel are maximized, this often comes at the expense of others. However, agricultural biotechnology could reduce the adverse impacts of this trade-off. Advances in agricultural research, genomics and precision breeding, such as the application of genome editing tools like the use of clustered regularly interspaced short palindromic repeats (CRISPR) technology, will allow scientists to introduce precise and accurate modifications to genomes in order to produce crops better adapted to biotic and abiotic stresses (Eş et al. 2019; Chen et al. 2019). These technologies have the potential to greatly facilitate the development of new GM crops, including new types of GM crops and even minor crops of importance in the regional economies of countries. Improvement of these crops is important not only for productivity and quality of agricultural products, but also for reduction in usage of environmentally stressing pesticides and resulting conservation of beneficial associated biodiversity and their ecological services.

## Conclusions

Introduction of crops derived from modern biotechnology such as IRGM crops is currently creating a novel scenario for agricultural associated biodiversity. IRGM crops have great potential for enhancing agricultural production, and the environmental risk assessment of IRGM crops had shown to have little to no adverse direct effect on the most common non-target organisms in the agro-ecosystem, particularly beneficial arthropods. However, the potential impacts of IRGM crops on soil invertebrates, fungi and bacteria are only partially understood and information on their long-term impact on soil biota is limited. In terms of indirect impact, the introduction of IRGM crops has benefitted non-target species that provide biological control services by reducing the use of pesticides and helping to conserve the biodiversity of the overall agro-ecosystem. New arising technologies, such as the use of CRISPR-Cas9 genome editing in crop plants, may further revolutionize food production not only in terms of yield and the increasing number of crop species of crop that can be improved, but also in terms of sustainability of agro-ecosystems by the development of newly modified crops that are resistant to specific pests or pathogens with no adverse impact on agricultural biodiversity.

**Acknowledgements** The author would like to express her gratitude to Mr. James B. Isaacs for his kind assistance with English grammar and wording and thank to the two anonymous reviewers for their comments and suggestions.

**Disclaimer** The opinions expressed in this article are the author's own and do not reflect the view of the Ministry of Agriculture, Livestock and Fisheries or any other branch of the government of Argentina.

## References

Agaisse H, Lereclus D (1995) How does *Bacillus thuringiensis* produce so much insecticidal crystal protein? J Bacteriol 177:6027–6032
Anderson JA, Ellsworth PC, Faria JC et al (2019) Genetically engineered crops: importance of diversified integrated pest management for agricultural sustainability. Front Bioeng Biotechnol 7:24
Bachman PM, Bolognesi R, Moar WJ et al (2013) Characterization of the spectrum of insecticidal activity of a double-stranded RNA with targeted activity against Western Corn Rootworm *(Diabrotica virgifera virgifera* LeConte). Transgenic Res 22(6):1207–1222
Balter M (2007) Seeking agriculture's ancient roots. Science 316(5833):1830–1835
Baulcombe D (2004) RNA silencing in plants. Nature 431(7006):356
Bolognesi R, Ramaseshadri P, Anderson J et al (2012) Characterizing the mechanism of action of double-stranded RNA activity against western corn rootworm (*Diabrotica virgifera virgifera* LeConte). PLoS One 7(10):e47534
Bravo A, Gill SS, Soberon M (2007) Mode of action of *Bacillus thuringiensis* cry and Cyt toxins and their potential for insect control. Toxicon 49(4):423–435
Brookes G, Barfoot P (2018) Environmental impacts of genetically modified (GM) crop use 1996-2016: impacts on pesticide use and carbon emissions. GM Crops Food 9(3):109–139
Carpenter JE (2010) Peer-reviewed surveys indicate positive impact of commercialized GM crops. Nat Biotechnol 28(4):319–321
Carrière Y, Ellers-Kirk C, Sisterson M et al (2003) Long-term regional suppression of pink bollworm by *Bacillus thuringiensis* cotton. Proc Natl Acad Sci U S A 100(4):1519–1523
Carrière Y, Fabrick JA, Tabashnik BE (2016) Can pyramids and seed mixtures delay resistance to Bt crops? Trends Biotechnol 34(4):291–302
Carstens K, Anderson J, Bachman P et al (2012) Genetically modified crops and aquatic ecosystems: considerations for environmental risk assessment and non-target organism testing. Transgenic Res 21(4):813–842
Carstens K, Cayabyab B, De Schrijver A et al (2014) Surrogate species selection for assessing potential adverse environmental impacts of genetically engineered insect-resistant plants on non-target organisms. GM Crops Food 5(1):11–15
Chen K, Wang Y, Zhang R et al (2019) CRISPR/Cas genome editing and precision plant breeding in agriculture. Annu Rev Plant Biol 70:667–697
Comas C, Lumbierres B, Pons X et al (2013) No effects of *Bacillus thuringiensis* maize on nontarget organisms in the field in southern Europe: a meta-analysis of 26 arthropod taxa. Transgenic Res 23(1):135–143
Convention on Biological Diversity (CBD) (2016) Cancun declaration on mainstreaming the conservation and Sustainable Use of Biodiversity for Well-Being. https://www.cbd.int/cop/cop-13/hls/in-session/cancun-declaration-draft-dec-03-2016-pm-en.pdf
COP 5 Decision V/5: Agricultural biological diversity: review of phase I of the programme of work and adoption of a multi-year work programme. Annex, The Scope of Agricultural Biodiversity. Paragraph 1. https://www.cbd.int/decision/cop/default.shtml?id=7147
Crickmore N, Baum J, Bravo A et al (2018) *Bacillus thuringiensis* toxin nomenclature. http://www.btnomenclature.info/
Cristofoletti PT, Kemper EL, Capella AN et al (2018) Development of transgenic sugarcane resistant to sugarcane borer. Trop Plant Biol 1(11):17–30
Dively GP, Venugopal PD, Bean D et al (2018) Regional pest suppression associated with widespread Bt maize adoption benefits vegetable growers. Proc Natl Acad Sci U S A 115:3320–3325
Duan JJ, Marvier M, Huesing J et al (2008) A meta-analysis of effects of Bt crops on honey bees (Hymenoptera: Apidae). PLoS One 3(1):e1415
Duan JJ, Lundgren JG, Naranjo S et al (2010) Extrapolating non-target risk of Bt crops from laboratory to field. Biol Lett 6(1):74–77

Dutton A, Romeis J, Bigler F (2003) Assessing the risks of insect resistant transgenic plants on entomophagous arthropods: Bt-maize expressing Cryl 1Ab as a case study. BioControl 48:611–636

Eş I, Gavahian M, Marti-Quijal FJ et al (2019) The application of the CRISPR-Cas9 genome editing machinery in food and agricultural science: current status, future perspectives, and associated challenges. Biotechnol Adv 37(3):410–421

Estruch JJ, Warren GW, Mullins MA et al (1996) Vip3A, a novel *Bacillus thuringiensis* vegetative insecticidal protein with a wide spectrum of activities against lepidopteran insects. Proc Natl Acad Sci 93(11):5389–5394

FAO (2002) International code of conduct on the distribution and use of pesticides –revised version: June 2013. Rome, Food and Agriculture Organization of the United Nations. http://www.fao.org/agriculture/crops/thematic-sitemap/theme/pests/code/en/

FAO (2019) The state of the world's biodiversity for food and agriculture. In: Bélanger J, Pilling D (eds) FAO commission on genetic resources for Food and Agriculture Assessments, Rome, p 572. http://www.fao.org/3/CA3129EN/CA3129EN.pdf

Fernandez-Cornejo J, Wechsler S, Livingston M et al (2014) Genetically engineered crops in the United States, ERR-162 U.S. Department of Agriculture, Economic Research Service. https://www.ers.usda.gov/webdocs/publications/45179/43668_err162.pdf?v=41690

Garcia-Alonso M, Raybould A (2014) Protection goals in environmental risk assessment: a practical approach. Transgenic Res 23(6):945–956

Garcia-Alonso M, Jacobs E, Raybould A et al (2006) A tiered system for assessing the risk of genetically modified plants to non-target organisms. Environ Biosaf Res 5(2):57–65

Grafius EJ, Douches DS (2008) The present and future role of insect-resistant genetically modified potato cultivars in IPM. In: Romeis J, Shelton AM, Kennedy GG (eds) Integration of insect-resistant genetically modified crops within IPM programs. Springer, Dordrecht, pp 195–221

Gressel J, Gassmann AJ, Owen MD (2017) How well will stacked transgenic pest/herbicide resistances delay pests from evolving resistance? Pest Manag Sci 73(1):22–34

Grishok A (2005) RNAi mechanisms in *Caenorhabditis elegans*. FEBS Lett 579(26):5932–5939

Guerchicoff A, Delécluse A, Rubinstein CP (2001) The *Bacillus thuringiensis* cyt genes for hemolytic endotoxins constitute a gene family. Appl Environ Microbiol 67(3):1090–1096

Hannon GJ (2002) RNA interference. Nature 418(6894):244–251

Hautea DM, Taylo LD, Masanga APL et al (2016) Field performance of Bt eggplants (*Solanum melongena L.*) in the Philippines: Cry1Ac expression and control of the eggplant fruit and shoot borer (*Leucinodes orbonalis Guenee*). PLoS One 11:e0157498

Hill RA (2005) Conceptualizing risk assessment methodology for genetically modified organisms. Environ Biosaf Res 4(2):67–70

Hill RA, Sendashonga C (2003) General principles for risk assessment of living modified organisms: lessons from chemical risk assessment. Environ Biosaf Res 2(2):81–88

Hutchison WD, Burkness EC, Mitchell PD et al (2010) Areawide suppression of European corn borer with Bt maize reaps savings to non-Bt maize growers. Science 330:222–225

Icoz I, Stotzky (2008) Fate and effects of insect-resistant Bt crops in soil ecosystems. Soil Biol Biochem 40(3):559–586

ISAAA (Int Serv Acquis Agri-biotech Appl) (2018) Global status of commercialized biotech/GM crops in 2018: biotech crops continue to help meet the challenges of increased population and climate change, Brief 54. Int Serv Acquis Agribiotech Appl, Ithaca

ISAAA's GM Approval Database (2020). http://www.isaaa.org/gmapprovaldatabase/

Kennedy RD, Cheavegatti-Gianotto A, de Oliveira WS et al (2018) A general safety assessment for purified food ingredients derived from biotechnology crops: case study of Brazilian sugar and beverages produced from insect-protected sugarcane. Front Bioeng Biotechnol 6:45. https://doi.org/10.3389/fbioe.2018.00045

Klümper W, Qaim M (2014) A meta-analysis of the impacts of genetically modified crops. PLoS One 9:e111629

Lee MK, Walters FS, Hart H et al (2003) The mode of action of the *Bacillus thuringiensis* vegetative insecticidal protein Vip3A differs from that of Cry1Ab δ-endotoxin. Appl Environ Microbiol 69(8):4648–4657
Levine SL, Tan J, Mueller GM et al (2015) Independent action between DvSnf7 RNA and Cry3Bb1 protein in southern corn rootworm, *Diabrotica undecimpunctata howardi* and Colorado potato beetle, *Leptinotarsa decemlineata*. PLoS One 10(3):e0118622
Li Y, Zhang Q, Liu Q et al (2017) Bt rice in China—focusing the non-target risk assessment. Plant Biotechnol J 15:1340–1345
Losey JE, Vaughan M (2006) The economic value of ecological services provided by insects. Bioscience 56(4):311–323
Lu Y, Wu K, Jiang Y et al (2012) Widespread adoption of Bt cotton and insecticide decrease promotes biocontrol services. Nature 487(7407):362
Malone LA, Gatehouse AM, Barratt BI (2008) Beyond Bt: alternative strategies for insect-resistant genetically modified crops. In: Romeis J, Shelton AM, Kennedy GG (eds) Integration of insect-resistant genetically modified crops within IPM programs. Springer, Dordrecht, pp 357–417
Marvier M, McCreedy C, Regetz J et al (2007) A meta-analysis of effects of Bt cotton and maize on nontarget invertebrates. Science 316:1475–1477
Meissle M, Álvarez-Alfageme F, Malone LA et al (2012) Establishing a database of bio-ecological information on non-target arthropod species to support the environmental risk assessment of genetically modified crops in the EU. EFSA Supporting Publications 9(9):334E
Naranjo SE (2009) Impacts of Bt crops on non-target invertebrates and insecticide use patterns. CAB Rev: Perspect Agric Vet Sci Nutr Nat Resour 4:1–23
Naranjo SE, Ellsworth PC, Frisvold GB (2015) Economic value of biological control in integrated Pest management of managed plant systems. Annu Rev Entomol 60:621–645
Oliveira CM, Auad AM, Mendes SM et al (2014) Crop losses and the economic impact of insect pests on Brazilian agriculture. Crop Prot 56:50–54
Pellegrino E, Bedini S, Nuti M et al (2018) Impact of genetically engineered maize on agronomic, environmental and toxicological traits: a meta-analysis of 21 years of field data. Sci Rep 8(1):3113
Poppy GM, Sutherland JP (2004) Can biological control benefit from genetically-modified crops? Tritrophic interactions on insect-resistant transgenic plants. Physiol Entomol 29(3):257–268
Riedel J, Romeis J, Meissle M (2016) Update and expansion of the database of bioecological information on non-target arthropod species established to support the environmental risk assessment of genetically modified crops in the EU. Supporting Publication 2016: EN–956. European Food Safety Authority (EFSA), Parma, Italy. http://www.efsa.europa.eu/en/supporting/pub/956e
Romeis J, Bartsch D, Bigler F et al (2008) Assessment of risk of insect-resistant transgenic crops to nontarget arthropods. Nat Biotechnol 26(2):203–208
Romeis J, Hellmich RL, Candolfi MP et al (2011) Recommendations for the design of laboratory studies on non-target arthropods for risk assessment of genetically engineered plants. Transgenic Res 20(1):1–22
Romeis J, Raybould A, Bigler F et al (2013) Deriving criteria to select arthropod species for laboratory tests to assess the ecological risks from cultivating arthropod-resistant genetically engineered crops. Chemosphere 90(3):901–909
Romeis J, Meissle M, Alvarez-Alfageme F et al (2014) Potential use of an arthropod database to support the non-target risk assessment and monitoring of transgenic plants. Transgenic Res 23(6):995–1013
Romeis J, Naranjo SE, Meissle M et al (2019) Genetically engineered crops help support conservation biological control. Biol Control 130:136–154
Rose R (2007) White paper on tier-based testing for the effects of proteinaceous insecticidal plant-incorporated protectants on non-target arthropods for regulatory risk assessments. United States Environmental Protection Agency, Washington, DC

Sanchis V (2011) From microbial sprays to insect-resistant transgenic plants: history of the biospesticide *Bacillus thuringiensis*. A review. Agron Sustain Dev 31:217–231
Sharma A, Jha P, Reddy GVP (2018) Multidimensional relationships of herbicides with insect-crop food webs. Sci Total Environ 643:1522–1532
Shelton AM, Hossain MJ, Paranjape V et al (2018) Bt eggplant project in Bangladesh: history, present status, and future direction. Front Bioeng Biotechnol 6:106. https://doi.org/10.3389/fbioe.2018.00106
Tu J, Zhang G, Datta K et al (2000) Field performance of transgenic elite commercial hybrid rice expressing *Bacillus thuringiensis* d-endotoxin. Nat Biotechnol 18:1101–1104
US EPA (1998) Guidelines for ecological risk assessment. Washington, DC: United States Environmental Protection Agency
Valencia-Sanchez MA, Liu J, Hannon GJ et al (2006) Control of translation and mRNA degradation by miRNAs and siRNAs. Genes Dev 20(5):515–524
Wach M, Hellmich RL, Layton R et al (2016) Dynamic role and importance of surrogate species for assessing potential adverse environmental impacts of genetically engineered insect-resistant plants on non-target organisms. Transgenic Res 25(4):499–505
Wan P, Huang Y, Tabashnik BE et al (2012) The halo effect: suppression of pink bollworm on non-Bt cotton by Bt cotton in China. PLOS One:e42004
Wang G, Dong Y, Liu X et al (2018) The current status and development of insect-resistant genetically engineered poplar in China. Front Plant Sci 9:1408–1408
Wolfenbarger LL, Naranjo SE, Lundgren JG et al (2008) Bt crops effects on functional guilds of non-target arthropods: a meta-analysis. PLoS One 3(5):e2118
Wolt JD, Keese P, Raybould A et al (2010) Problem formulation in the environmental risk assessment for genetically modified plants. Transgenic Res 19(3):425–436
Wu KM, Lu YH, Feng HQ et al (2008) Suppression of cotton bollworm in multiple crops in China in areas with Bt toxin–containing cotton. Science 321(5896):1676–1678
Xiao Y, Wu K (2019) Recent progress on the interaction between insects and *Bacillus thuringiensis* crops. Philos Trans R Soc B 374(1767):20180316
Zhang BH, Liu F, Yao CB et al (2000) Recent progress in cotton biotechnology and genetic engineering in China. Curr Sci 79(1):37–44
Zhang J, Khan SA, Heckel DG et al (2017) Next-generation insect-resistant plants: RNAi-mediated crop protection. Trends Biotechnol 35:871–882
Zhang W, Lu Y, van der Werf W et al (2018) Multidecadal, county-level analysis of the effects of land use, Bt cotton, and weather on cotton pests in China. Proc Natl Acad Sci 115(33):E7700–E7709
Zotti M, dos Santos EA, Cagliari D et al (2018) RNA interference technology in crop protection against arthropod pests, pathogens and nematodes. Pest Manag Sci 74(6):1239–1250

# Ecological Risk Assessment for Soil Invertebrate Biodiversity and Ecosystem Services

**Paul Henning Krogh**

**Abstract** Ecological risk assessment (ERA) of genetically modified crops for the soil environment is supported by guidelines already available for pesticides. They have very much in common and here we point out the salient features of empirical ERA methodology for soil invertebrates and the need to address ecosystem services through assessment of ecological processes contributed by soil invertebrate functional traits. The soil quality, bait-lamina test, and earthworm burrowing activity are current candidates underpinning an ecosystem services assessment approach. Establishing functional trait databases, baseline data and making links between them will enable future assessment of soil ecosystem services.

**Keywords** Soil invertebrates · Soil ecosystem services · Functional traits · Ecological risk assessment · Earthworm burrows · Water infiltration · Tiered approach · Baseline · Databases · Mesocosm

## Introduction

Since the creation of the ecosystem service (ES) concept (Costanza et al. 1997), neither soil biodiversity nor the single populations are the sole focus of ecological risk assessment (ERA) but also their role in sustainable agricultural cropping systems (EFSA 2010a, b; EFSA PPR Panel et al. 2017). While species abundances can be assessed, it is more challenging to assess their joint contribution to ecosystem processes. Moreover, the translation of the ecosystem processes into ecosystem services (ESs) is not straightforward. Economists perform the final quantification of a soil ES in terms of the monetary value. Here, we will treat only some aspects of the ERA and ES.

P. H. Krogh (✉)
Department of Bioscience, Science and Technology, Aarhus University, Aarhus, Denmark
e-mail: phk@bios.au.dk



A. Chaurasia et al. (eds.), *GMOs*, Topics in Biodiversity and Conservation 19,
https://doi.org/10.1007/978-3-030-53183-6_16

ERA of genetically modified crops needs to be considered in the context of the farming practice and at the farming system level. It pertains to the first phase of the GMO (genetically modified organism) ERA (EFSA 2010b), that is, the problem formulation, to identify the pesticide use pattern and the tillage practice, both being integral parts of the cropping system with significant impact on the soil ecosystem. The particular crop rotations, including catch crops and cover crops, that is, crops growing between successive main crops and crops protecting the soil against erosion and loss of fertility, also need to be taken into consideration. In spite of the fact that farming practice has such a dominant influence on the soil life, current ERA practices rarely deal with this and still focus on the protection of species and ecosystem processes through ecotoxicological studies of direct and indirect effects. This chapter will present these two ways of assessing the influence of genetically modified organisms (GMOs) on the soil ecosystem, with special attention to genetically modified plants used for agricultural production. Key important background materials for ERA of the soil environment in this respect are the EFSA (European Food Safety Agency) GMO ERA Guideline (EFSA 2010b) and the scientific opinion on pesticide ERA for soil organisms (EFSA PPR Panel et al. 2017). The EU legislation referring to the GMO ERA guidance document is laid down in the European Union directives 2001/18/EC and 2018/350 and the pesticide opinion is the first public EU document addressing soil health and soil ecosystem services.

## The Tiered Approach

Risk assessments of GMOs are based on practical testing tools at three levels of complexity (Römbke et al. 2009; EFSA PPR Panel et al. 2017; Fig. 1). Before turning to costly long-term field experiments, simple Tier 1 laboratory tests are

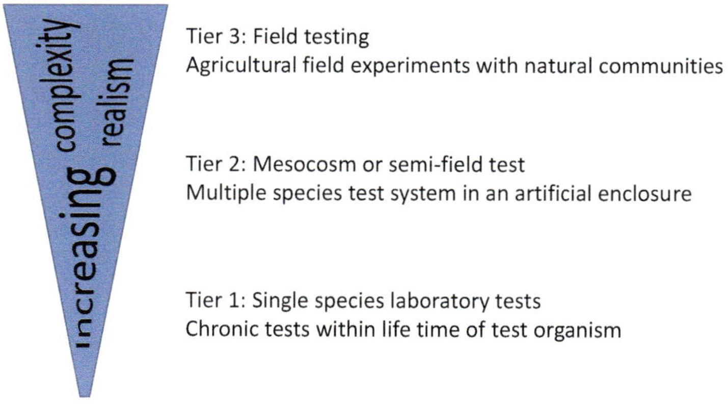


**Fig. 1** Testing methodology addressing increasing levels of ecosystem complexity from the single species test of chronic life history parameters to the long-term field test under natural conditions in an agricultural practice setting

performed on a complement of tests with soil invertebrates to reveal off-target effects (Römbke et al. 2009). Depending on the outcome, the second and third levels could be triggered; however, unlike for pesticides (European Commission 2002), those trigger values have not been defined yet for GMO studies, but because protection goals are the same for pesticides and GMOs, the trigger values from pesticide ERA should be adopted for GMOs. This tiered approach is employed both for pesticides and for GMO (EFSA 2010b; EFSA PPR Panel et al. 2017). It aims for only going to the next tier if an effect level threshold has been surpassed.

## *Tier 1*

Tier 1 is the simplest type of testing system supported by a range of ISO (International Organization for Standardization) and OECD (Organisation for Economic Co-operation and Development) standards (Römbke et al. 2009; EFSA 2010b). The older mortality tests such as OECD (1984) and ISO (1993) should be abandoned, as they will not reflect the population performance, which need more than just the mortality to become predictable at the population level. The standard chronic tests still provide the mortality end point along with sublethal life history related end points, such as growth and reproduction. Extrapolation to real communities of a suite of tests is sometimes done (Boeckman and Layton 2017). The population performance of non-target soil invertebrates (NTO) will reflect the food quality of plant material from genetically modified plants, so the mode of action is not only straightforward direct toxicity, but conspicuous nutrient effects either directly from the plant material or indirectly through its input into the decomposer food web mediated through bacteria, fungi, and invertebrate decomposers.

## *Tier 2*

This level of testing strives to mimic realistic conditions concerning exposure scenarios, timescale and ecosystem and community characteristics. Currently, two types of multispecies test systems have been suggested, where one contains an artificially composed soil invertebrate community and the other uses native species living in a soil monolith and neither of them has yet been employed commonly in soil ERA.

A gnotobiotic[1] type of mesocosm was introduced for studies of processes and species interactions in an assembled mesofauna community by Filser and Krogh (2002) and was hereafter employed on several occasions in the assessment of effects of chemicals on this artificially composed mesofauna community (Sechi et al. 2014;

[1] A biotic system where the composition of species is known.

Cortet et al. 2006, 2003; Schnug et al. 2014; Scott-Fordsmand et al. 2008). The sources of species added to this artificial system are laboratory cultures and field collected species. The standard mite test predator *Geolaelaps aculeifer* (OECD 2008), *Enchytraeus crypticus* (OECD 2004) and *Folsomia fimetaria* (OECD 2009), as well as a few other collembolans representing the three life forms hemiedaphic, euedaphic, and epedaphic, were included in the assemblage. In some cases, it was improved by field-collected species (Larsen et al. 2007; Pernin et al. 2006). This test system is labelled "SMS" (soil multi-species) (Jensen and Scott-Fordsmand 2012; Scott-Fordsmand et al. 2008). This gnotobiotic type of mesocosm has been used in only one case for assessment of GMO influence on soil invertebrates (D'Annibale et al. 2015).

The terrestrial model ecosystem (TME) was proposed for ecotoxicity testing (Weyers et al. 2004; EFSA PPR Panel et al. 2017; Römbke et al. 2009). It contains the innate community of soil organisms of isolated field collected monoliths.

### *Tier 3*

Field studies of soil invertebrates are supported by the series of ISO standards 23611:1 to 6 (e.g. ISO 1999, 2006a, b). Post-market monitoring strategies of GMOs are described in general terms by Ruf et al. (2013). Testing of GMO crops is very similar to pesticide testing concerning field test design and dimensioning. As for any testing of substances and materials used in agriculture, a positive control and a reference substance are paramount for quality assurance of the field test.

## Ecosystem Services

### *Functional Traits and Ecosystem Processes*

Two terms are key to the delivery of ESs: the functional trait and the ecosystem process. A function of an organism contributes to an ecosystem process. The mechanistic relationship between at the one end species populations of a community and the other end an ES is illustrated diagrammatically in Fig. 2. The properties of a

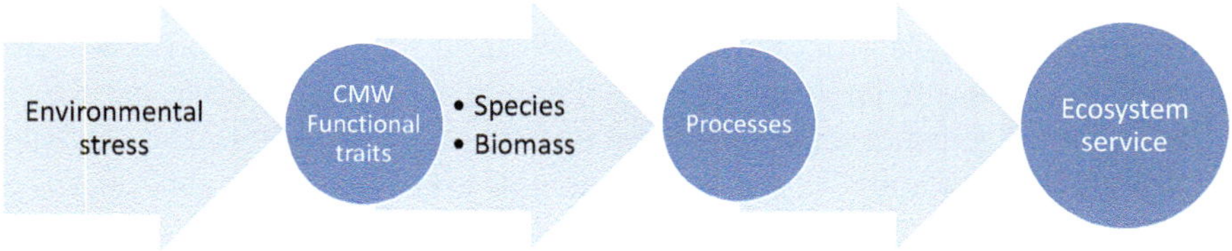


**Fig. 2** Conceptual presentation of the basic link between anthropogenic impact and environmental stress on functional traits and ESs. *CWM* Community weighted mean

species that affect ESs are reflected in its trait profile through the contribution of functional traits to ecosystem processes and derived ecosystem structures (Garnier et al. 2016; Hevia et al. 2017; Wong et al. 2019). This is the first level where a functional trait of an organism could affect nature, that is, the ecosystem. In terms of functional trait analyses, this type of function is an effect functional trait or just an effect trait, e.g. Díaz et al. (2013) and Nock et al. (2016). So certain properties of an organism will affect the surrounding environment and the tool to do so is the functional trait. When operating in the environment, the processes effected by invertebrate species will result in translocation and transformation of matter, that is, perform a process. A broad range of taxonomically unrelated species will often contribute to the resulting process, that is, interacting species forming a community. Traits analyses calculate a community weighted mean (CMW) across species traits and ultimately disregard species composition over the composition of traits.

In an agricultural setting, non-target organisms (NTOs) may be inhibited by the agricultural practice, as will be revealed through the tiered approach or even at the initial problem formulation step. The level of impact will be included in the assessment of ESs and identifying a potential loss of valuable services may trigger improvements or preventive actions.

## *Soil Quality*

The concept of ESs includes soil quality and soil health as deliveries from the soil ecosystem (Bünemann et al. 2018), which even allow to quantify its economic impact. A proper level of soil quality will require that the soil invertebrates will provide their contribution to ESs. Soil ESs are based on an assessment of functional traits of soil organisms and subsequently the ecosystem processes to which they contribute, but currently assessment and quantification of functional traits have no standardized methodology for the soil ecosystem (Lima et al. 2013; Bünemann et al. 2018) except for the bait-lamina test and the litter-bag test (OECD 2006; ISO 2016).

Soil quality has been subject to economic valuation. Soil quality is one of the ESs contributed by earthworms; in fact, they are considered the dominant taxon contributing to this service (Keith and Robinson 2012; Alam et al. 2014; Plaas et al. 2019). However, it must be stressed that soil quality depends on a range of organisms, including bacteria, fungi, and plants, so the interactions of their processes creates soil and determines soil quality.

## *Soil Macroporosity Contributed by Earthworms*

With the crucial impact of earthworms on ESs (Blouin et al. 2013), the earthworm contribution to soil macroporosity is used to demonstrate a quantifiable ES performed by a soil invertebrate. Large earthworm burrows of more than 3 mm diameter are found down to a depth of 2 m in loamy soils. They have a capacity to drain soil together with the system of biopores formed by roots. Measurement of their hydrological activity is done by pouring a dye trace solution, Brilliant Blue, on the soil and identifying the blue colored burrows at, for example, 0.5 m depth or at deeper horizons (Fig. 3). Thus, the contribution of earthworm burrows to the infiltration capacity can be quantified in this manner (van Schaik et al. 2014). The infiltration capacity of the soil can be measured by the double ring infiltrometer (DIN 19682). The infiltration capacity will contribute to the soil drainage capacity influencing crop productivity, that is, a crucial ES. If the relationship between the earthworm community and the hydrologically active biopores can be established, the community will be a proxy for the water infiltration process, eventually being translated into crop productivity. As roots also form biopores, their contribution must be included in the infiltration assessment.

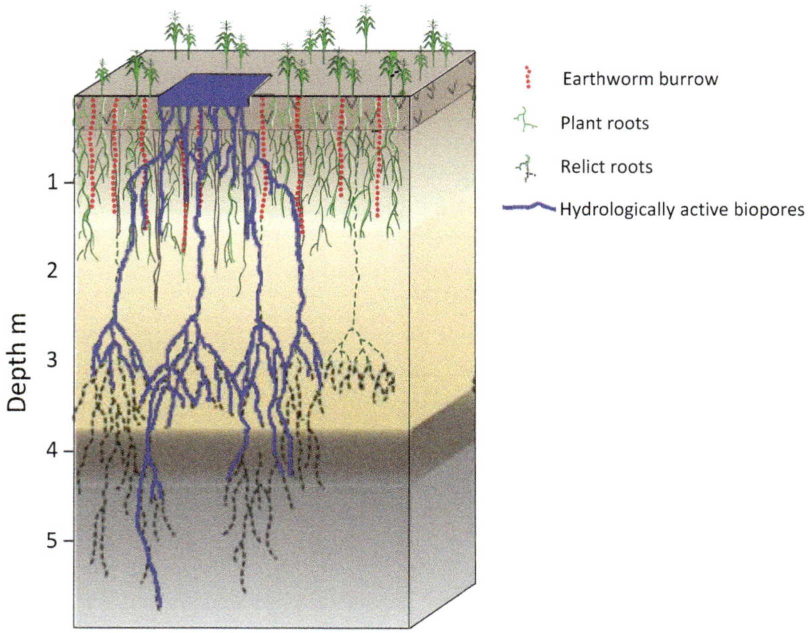


**Fig. 3** The permanent system of biopores in an agricultural clay till running from the soil surface to 5 m depth. Biopores were colored by Brilliant Blue to reveal hydrological activity during the infiltration of water (Jørgensen et al. 2017)

## Baseline Data

To enable an assessment of soil ES through traits (Fig. 2), proper baseline data are needed. These data should include the normal operation range or baseline of soil organisms in representative habitats and a trait database for conversion of the abundance and biomass data into quantitative measures of the processes. Such databases are *edaphobase* (edaphobase.org) and *BETSI* (betsi.cesab.org). While edaphobase holds abundance data, *BETSI* hosts soil invertebrate trait data. When joining these two forces, the assessment of soil ESs will have reached an important milestone.

## Future Directions

Assessment of soil ecosystem services depends on further accumulation of knowledge of functional traits across taxa contributing to the same ecosystem processes. The case study of biopores illustrates that we should develop ecohydrology modelling of the joint activity of plants and earthworms to optimize the capacity of water infiltration important to avoid waterlogging. As earthworm communities respond to land-use and soil management, so do their functional diversity, and so land use is the key metadata for predicting their burrowing activity resulting in spatial macroporosity. We still need to assess the contribution of soil invertebrate diversity to many ecosystem services, including nutrient cycling, decease suppression and soil aggregates. Current endeavors of mapping soil biodiversity from existing data combined with species trait information will be the first important stepping stone for future large-scale assessment of soil ecosystem services.

## References

Alam M, Olivier A, Paquette A, Dupras J, Revéret J-P, Messier C (2014) A general framework for the quantification and valuation of ecosystem services of tree-based intercropping systems. Agrofor Syst 88(4):679–691. https://doi.org/10.1007/s10457-014-9681-x

Blouin M, Hodson ME, Delgado EA, Baker G, Brussaard L, Butt KR, Dai J, Dendooven L, Pérès G, Tondoh JE, Cluzeau D, Brun JJ (2013) A review of earthworm impact on soil function and ecosystem services. Eur J Soil Sci 64(2):161–182. https://doi.org/10.1111/ejss.12025

Boeckman CJ, Layton R (2017) Use of species sensitivity distributions to characterize hazard for insecticidal traits. J Invertebr Pathol 142:68–70. https://doi.org/10.1016/j.jip.2016.08.006

Bünemann EK, Bongiorno G, Bai Z, Creamer RE, De Deyn G, de Goede R, Fleskens L, Geissen V, Kuyper TW, Mäder P, Pulleman M, Sukkel W, van Groenigen JW, Brussaard L (2018) Soil quality – a critical review. Soil Biol Biochem 120:105–125. https://doi.org/10.1016/j.soilbio.2018.01.030

Cortet J, Joffre R, Elmholt S, Krogh PH (2003) Increasing species and trophic diversity of mesofauna affects fungal biomass, mesofauna structure community and organic matter decomposition processes. Biol Fertil Soils 37(5):302–312

Cortet J, Joffre R, Elmholt S, Coeurdassier M, Scheifler R, Krogh PH (2006) Interspecific relationships among soil invertebrates influence pollutant effects of phenanthrene. Environ Toxicol Chem 25(1):120–127

Costanza R, dArge R, de Groot R, Farber S, Grasso M, Hannon B, Limburg K, Naeem S, Oneill RV, Paruelo J, Raskin RG, Sutton P, van den Belt M (1997) The value of the world's ecosystem services and natural capital. Nature 387(6630):253–260

D'Annibale A, Larsen T, Sechi V, Cortet J, Strandberg B, Vincze É, Filser J, Audisio PA, Krogh PH (2015) Influence of elevated $CO_2$ and GM barley on a soil mesofauna community in a mesocosm test system. Soil Biol Biochem 82:127–136. https://doi.org/10.1016/j.soilbio.2015.02.009

Díaz S, Purvis A, Cornelissen JHC, Mace GM, Donoghue MJ, Ewers RM, Jordano P, Pearse WD (2013) Functional traits, the phylogeny of function, and ecosystem service vulnerability. Ecol Evol 3(9):2958–2975. https://doi.org/10.1002/ece3.601

EFSA Panel on Genetically Modified Organisms (GMO) (2010a) Guidance on the environmental risk assessment of genetically modified plants. EFSA J 8(11):1879. https://doi.org/10.2903/j.efsa.2010.1879

EFSA Panel on Genetically Modified Organisms (GMO) (2010b) Scientific opinion on the assessment of potential impacts of genetically modified plants on non-target organisms. EFSA J 8(11):1877. https://doi.org/10.2903/j.efsa.2010.1877

EFSA PPR Panel (EFSA Panel on Plant Protection Products and their Residues), Ockleford C, Adriaanse P, Berny P, Brock T, Duquesne S, Grilli S, Hernandez-Jerez AF, Bennekou SH, Klein M, Kuhl T, Laskowski R, Machera K, Pelkonen O, Pieper S, Stemmer M, Sundh I, Teodorovic I, Tiktak A, Topping CJ, Wolterink G, Craig P, de Jong F, Manachini B, Sousa P, Swarowsky K, Auteri D, Arena M, Rob S (2017) Scientific opinion addressing the state of the science on risk assessment of plant protection products for in-soil organisms. EFSA J 15(2):e04690. https://doi.org/10.2903/j.efsa.2017.4690

European Commission (2002) guidance document on terrestrial ecotoxicology under council directive 91/414/EEC. DRAFT, Working document (SANCO/10329/2002 rev 2 final)

Filser J, Krogh PH (2002) Interactions between *Enchytraeus crypticus*, collembolans, gamasid mites and barley plants – a greenhouse experiment. In: Christensen B, Standen V (eds) Proceedings of the 4th International Symposium on Enchytraeidae, Mols Laboratory, Denmark, 2–4 June 2000. Natura Jutlandica, Occasional papers no. 2. Natura Jutlandica, pp 32–42

Garnier E, Navas M, Grigulis K (2016) Trait-based ecology: definitions, methods, and a conceptual framework. In: Garnier E, Navas M-L, Grigulis K (eds) Plant functional diversity: organism traits, community structure, and ecosystem properties. Oxford University Press, pp 9–25. https://doi.org/10.1093/acprof:oso/9780198757368.003.0002

Hevia V, Martín-López B, Palomo S, García-Llorente M, de Bello F, González JA (2017) Trait-based approaches to analyze links between the drivers of change and ecosystem services: synthesizing existing evidence and future challenges. Ecol Evol 7:831–844. https://doi.org/10.1002/ece3.2692

ISO (1993) International Organization for Standardization. ISO 11268-1:1993 (E) Soil quality – Effects of pollutants on earthworms (*Eisenia fetida*) - Part 1: Determination of acute toxicity using artificial soil substrate. Geneva (CH)

ISO (1999) International Organization for Standardization. ISO No. 11268-3 Soil quality – Effects of pollutants on earthworms. Part 3: Guidance on the determination of effects in field situations. Geneva (CH)

ISO (2006a) International Organization for Standardization. ISO 23611-1:2006 – Soil quality – Sampling of soil invertebrates. Part 1: Hand-sorting and formalin extraction of earthworms. Geneva (CH)

ISO (2006b) International Organization for Standardization. ISO 23611-2:2006 – Soil quality – Sampling of soil invertebrates. Part 2: Sampling and extraction of micro-arthropods (Collembola and Acarina). Geneva (CH)

ISO (2016) International Organization for Standardization. ISO 18311:2016 – Soil quality – method for testing effects of soil contaminants on the feeding activity of soil dwelling organisms – bait-lamina test. Geneva (CH)

Jensen J, Scott-Fordsmand JJ (2012) Ecotoxicity of the veterinary pharmaceutical ivermectin tested in a soil multi-species (SMS) system. Environ Pollut 171(0):133–139. https://doi.org/10.1016/j.envpol.2012.07.014

Jørgensen PR, Krogh PH, Hansen S, Petersen CT, Nielsen MH, Rasmussen SB, Heinrichson K, Spliid NH (2017) Occurence and importance of deep biopores for pesticide leaching in clay till [Dybe bioporers forekomst og betydning for pesticidudvaskning i moræneler] in Danish with English summary. Bekæmpelsesmiddelforskning, vol BEKF nr 171

Keith AM, Robinson DA (2012) Earthworms as natural capital: ecosystem service providers in agricultural soils. Economol J 2:91–99

Larsen T, Luxhøi J, Magid J, Jensen LS, Krogh PH (2007) Properties of anaerobically digested and composted municipal solid waste assessed by linking soil mesofauna dynamics and nitrogen modelling. Biol Fertil Soils 44(1):59–68

Lima A, Brussaard L, Totola M, Hoogmoed W, De Goede R (2013) A functional evaluation of three indicator sets for assessing soil quality. Appl Soil Ecol 64:194–200

Nock CA, Vogt RJ, Beisner BE (2016) Functional traits. eLS/Wiley, Chichester, pp 1–8. https://doi.org/10.1002/9780470015902.a0026282

OECD (1984) OECD Guidelines for testing of chemicals no. 207. Earthworm acute toxicity tests. Adopted 4

OECD (2004) Test No. 220: enchytraeid reproduction test. OECD guidelines for testing chemicals. OECD. doi:https://doi.org/10.1787/9789264070301-en

OECD (2006) Guidance document on the breakdown of organic matter in litter bags. ENV/JM/MONO(2006)23 OECD series on testing and assessment Number 56

OECD (2008) Test No. 226: predatory mite reproduction test in soil (*Hypoaspis* (*Geolaelaps*) *aculeifer*). OECD guidelines for the testing of chemicals. https://doi.org/10.1787/9789264067455-en

OECD (2009) Test No. 232: collembolan reproduction test in soil. OECD guidelines for the testing of chemicals. OECD. https://doi.org/10.1787/9789264076273-en

Pernin C, Ambrosi JP, Cortet J, Joffre R, Le Petit J, Tabone E, Torre F, Krogh PH (2006) Effects of sewage sludge and copper enrichment on both soil mesofauna community and decomposition of oak leaves (Quercus suber) in a mesocosm. Biol Fertil Soils 43(1):39–50

Plaas E, Meyer-Wolfarth F, Banse M, Bengtsson J, Bergmann H, Faber J, Potthoff M, Runge T, Schrader S, Taylor A (2019) Towards valuation of biodiversity in agricultural soils: a case for earthworms. Ecol Econ 159:291–300. https://doi.org/10.1016/j.ecolecon.2019.02.003

Römbke J, Jänsch S, Meier M, Hilbeck A, Teichmann H, Tappeser B (2009) General recommendations for soil ecotoxicological tests suitable for the environmental risk assessment (ERA) of genetically modified plants (GMPs). Integr Environ Assess Manag 6(2):287–300

Ruf A, Beylich A, Blick T, Buechs W, Glante F, Hoess S, Roß-Nikoll M, Ruess L, Russell DJ, Römke J (2013) Soil organisms as an essential element of a monitoring plan to identify the effects of GMO cultivation. Requirements–methodology–standardisation. BioRisk 8:73

Schnug L, Jensen J, Scott-Fordsmand JJ, Leinaas HP (2014) Toxicity of three biocides to springtails and earthworms in a soil multi-species (SMS) test system. Soil Biol Biochem 74:115–126. https://doi.org/10.1016/j.soilbio.2014.03.007

Scott-Fordsmand JJ, Maraldo K, van den Brink PJ (2008) The toxicity of copper contaminated soil using a gnotobiotic Soil Multi-species Test System (SMS). Environ Int 34(4):524–530. https://doi.org/10.1016/j.envint.2007.11.008

Sechi V, D'Annibale A, Maraldo K, Johansen A, Bossi R, Jensen J, Krogh PH (2014) Species composition of a soil invertebrate multi-species test system determines the level of ecotoxicity. Environ Pollut 184(0):586–596. https://doi.org/10.1016/j.envpol.2013.10.008

van Schaik L, Palm J, Klaus J, Zehe E, Schröder B (2014) Linking spatial earthworm distribution to macropore numbers and hydrological effectiveness. Ecohydrology 7(2):401–408. https://doi.org/10.1002/eco.1358

Weyers A, Sokull-Klüttgen B, Knacker T, Martin S, Van Gestel CAM (2004) Use of terrestrial model ecosystem data in environmental risk assessment for industrial chemicals, biocides and plant protection products in the EU. Ecotoxicology 13(1–2):163–176

Wong MKL, Guénard B, Lewis OT (2019) Trait-based ecology of terrestrial arthropods. Biol Rev 94:999–1022. https://doi.org/10.1111/brv.12488

# Part VI
# Gene Drive Approaches

# Engineered Gene Drives: Ecological, Environmental, and Societal Concerns

**Jennifer Kuzma**

**Abstract** This chapter overviews the types, purposes, and potential impacts of gene drive organisms (GDOs) and discusses challenges with foreseeing and assessing these impacts prior to their environmental release. It concludes with a few examples of risk analysis methods and governance systems that scholars have proposed to cope with the novelty of GDOs and uncertainties associated with their use. With GDOs poised for release in the near future, it is urgent that technologists, ecologists, social scientists, ethicists, stakeholders, and publics work together to grapple with the immense challenges associated with assessments of GDOs and the design of governance systems to ensure their responsible development and potential use.

**Keywords** Risk analysis · GMO · Biotechnology · Gene drive · Regulation · Governance

## Introduction

Advances in technology are usually incremental, building on previous discoveries and inventions. Yet, there are moments when something special happens, and new technological capabilities seem to emerge in a leap. Gene editing is one such example. No longer do genetic engineers have to blindly launch a novel gene into a host cell, hoping it lands in a good spot and works in the new environment. Now they can precisely cut and delete particular sites of DNA; replace portions of genes; or add entirely new genes in specific places. Gene editing is akin to our abilities to take pen to paper to correct typos, delete words or phrases, rearrange sentences, or add new ones.

Gene drives rely on gene editing but take it a step further in order to spread genes through wild populations. Usually, an introduced gene is carried on one of a pair of

J. Kuzma (✉)
Genetic Engineering & Society Center, North Carolina State University, Raleigh, NC, USA
e-mail: jkuzma@ncsu.edu



A. Chaurasia et al. (eds.), *GMOs*, Topics in Biodiversity and Conservation 19,
https://doi.org/10.1007/978-3-030-53183-6_17

chromosomes and is thus inherited by about half of the offspring in the first generation. Eventually the gene will get diluted in the natural population if there is no selective advantage to it. However, "gene drive" systems allow for an edited gene on one chromosome to copy itself into its partner chromosome. The result is that nearly all offspring will inherit the engineered gene. The idea is that even if just a few organisms with gene drives are released into the wild, the whole population could end up with the edited gene.

Several reasons to use gene drives to engineer populations in the wild have been proposed. For example, they could spread killer-genes to destroy unwanted pest populations, invasive species, or disease-carrying organisms. The release of just a few individuals with gene drive systems that are designed to kill organisms could theoretically cause the whole population to collapse. This could come in handy for eradicating mosquitos carrying dengue, malaria, or Zika virus or for eliminating invasive species like mice that threaten endangered birds on islands. In contrast, gene drives could also be used to add beneficial genes to populations. Editing systems like CRISPR-Cas9 could carry cargo genes with them to immunize an endangered species against disease or protect it from the effects of climate change.

However, ecosystems are complex and sensitive. Unintended effects could accompany engineering species in the wild. For example, a more dangerous pest could fill a niche left vacant by a gene drive organism (GDO), or beneficial predators could be harmed from reductions in prey. Although researchers are working on systems to recall gene drives, certain effects could be irreversible, and others unpredictable. Furthermore, there are social and cultural impacts that may be positive or negative. Species that are nuisances in one region, may be important for religious or economic purposes in another, making the deployment of gene drives to decrease populations a potentially contentious multi-national issue.

With engineered gene drive organisms poised for release in the near future, it is urgent that developers, stakeholders, and publics work together to grapple with challenges for assessing and governing GDOs. This chapter lays some groundwork for this purpose. First, it overviews the types, purposes, and potential ecological and societal impacts of gene drives. Then it discusses challenges with foreseeing and assessing these impacts prior to the release of GDOs. It concludes with examples of risk governance practices and systems that scholars have proposed for GDOs in light of their novelty, associated uncertainties, and potential wide-scale spread.

## *Types of Gene Drives*

Gene drive is a generic term for a variety of processes that in sexually reproducing organisms cause genes to be transmitted to successive generations at ratios greater than the classical Mendelian ratio (Fig. 1). Natural gene drives, such as homing endonuclease genes (HEGs), have been proposed as ways to suppress or modify populations that carry disease for several decades (Curtis 1968; Burt 2003; Sinkins and Gould 2006; Deredec et al. 2008). Engineered gene drives utilize gene editing

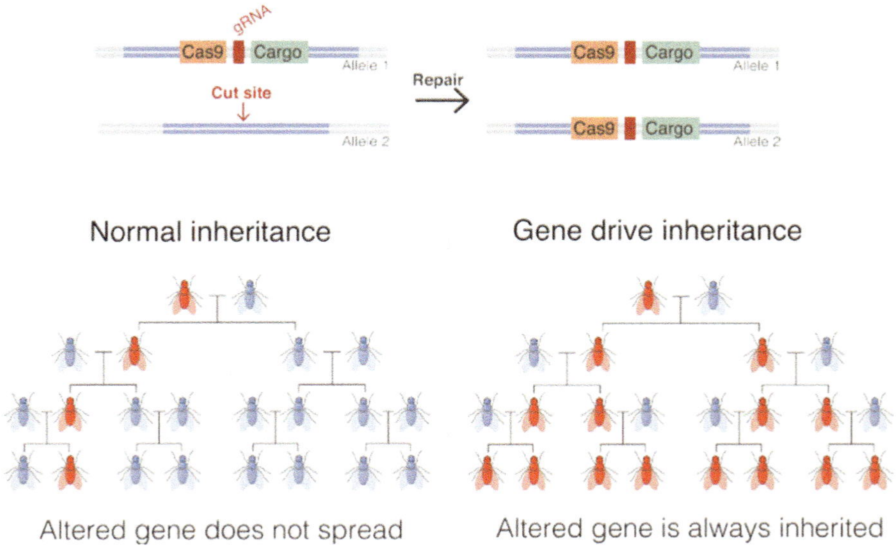


**Fig. 1** How Gene Drives Bias Inheritance (from Mauriwalter https://commons.wikimedia.org/wiki/File:Gene_Drive.png#filehistory)

tools at their core, with site-directed nucleases as perhaps the most common, and they allow biotechnologists to mutate, swap, or add multiple genes at specific sites in a genome. Nucleases are naturally occurring proteins that cut DNA, and they have been redesigned by genetic engineers for gene editing. Zinc finger nucleases and TALENs were early site-directed nucleases used for gene editing in the mid-2000s; however, in 2012, CRISPR-Cas 9 was discovered (Jinek et al. 2012), and soon thereafter, it was proposed as an easier way to edit genes and drive them through populations (Esvelt et al. 2014).

Cas proteins are nucleases that cleave DNA at "clustered regularly interspaced short palindromic repeats" (CRISPR), which are present in multiple places in genomes. CRISPR Cas9 can be guided more specifically to any site in the DNA by its accompanying RNA sequences (called "guide RNA" or gRNA). After the CRISPR-Cas 9 system (with the gRNA) cuts the target DNA site, a double-strand break results which can either be successfully repaired by the cell or result in a mutation. However, if engineers provide a DNA template sequence with homology to either side of the break at its ends, it can be used for repair instead and copied into the break site, causing a larger edit or deletion in that gene, or the introduction of a new gene depending on how the template is designed. Furthermore, if the repair templates also include DNA sequences of CRISPR-Cas and the gRNA (also known as the CRISPR-Cas 9 system), then CRISPR-Cas9 system can copy itself into cleavage sites via homology directed repair. If these constructs are incorporated into germ-line cells the system will be inherited at a super-Mendelian rate and is a gene drive (Esvelt et al. 2014).

To achieve the desired effect on populations, the gene drive system can be engineered to cut a sex-linked gene (e.g., lethal to females at the larval stage) so that the drive causes the population to decline (only males survive). Alternatively, a drive system can be engineered to carry extra "cargo" genes into populations to confer desirable traits, like disease resistance. If the gene drive is linked to an engineered genetic allele of interest, it can result in that allele being inherited by almost 100% of the offspring. This drives the gene into each successive generation until the entire population contains it (Fig. 1). Theoretically, cargo genes can come from any species and be introduced into any host.

More broadly speaking, specific modes of action and intended purposes of gene drives are limited only by the traits that can be inactivated, replaced, or introduced. As described by Champer et al. (2016), the technological choices associated with gene drives include: (1) whether the gene drive is designed to suppress the target population or to replace it with a genetically modified population; (2) the rate of its spread; (3) whether it is locally confined or not; (4) whether it has a fitness cost; (5) the rate of DNA sequences resistant to the gene drive with each generation; (6) whether it is reversible; and (7) whether it can be reversed to the original wild-type sequence. Some gene drives are designed to act globally with no limitations on spread if the drive is neutral toward fitness, if the population exhibits random mating and sexual reproduction, and if all members in the target area have no physical or other barriers toward interacting to mate. These are termed "global drives," and theoretically the release of one individual can drive the genes through the target population to achieve fixation. Other gene drives can be engineered to be "limited" in theory (e.g., spread to reduce only 20% of the population given the type of genes affected or introduced "self-limited," or to be limited to certain genetic variants of the organism in a particular geographic region "local"), or to require a number of individuals to be released that exceed a certain threshold in order to drive the gene into the population (e.g., 1000 individuals per 10,000 wild population need to be released to achieve full spread—a "threshold drive").

The ecological, health, and societal impacts of gene drive organisms (GDOs) depend on whether the drive is global, self-limited, local, or threshold, as well as on other choices technology developers make about modes of action, types of cargo genes, intended purposes, and gene drive reversibility or removal. This variation makes discussions about the implications of GDOs as a whole category difficult, and specific cases are usually considered in workshops to identify policy, risks, and societal issues (e.g., Kuzma et al. 2018).

## *Purposes of Gene Drives and Example Projects*

Gene drives have been considered and are being developed for multiple purposes. General applications for gene drives introduced into populations in the environment include eradicating vector-borne human disease, enhancing agricultural safety and sustainability, protecting threatened species, and controlling invasive species (Esvelt et al. 2014; Kuzma and Rawls 2016). In addition to the purposes, the goals of gene drives can be achieved by having different effects on species. Populations can be suppressed (or eliminated, e.g., GDOs with global, self-sustaining gene drives that prevent reproduction), enhanced (e.g., with cargo genes that confer an advantage to the GDO), immunized (e.g., with cargo genes that protect GDOs from disease), or sensitized (e.g., with cargo genes that make an invasive species susceptible to pesticides). Ethical, societal, regulatory, and ecological issues depend at least in part on the purpose and mode of action of gene drives. For example, if no other options exist for saving an endangered species, the risks caused by population suppression of its prey may be tolerable to the communities surrounding GDO deployment. Arguments can be made that there is an ethical obligation to deploy GDOs in cases where no alternatives exist for saving human lives or endangered species (e.g., Kuzma and Rawls 2016).

Much work is on suppression and eradication drives and has been focused on human health applications, especially gene drives to control human disease vectors like mosquitoes transmitting Zika, dengue, and malaria. A London-based research team recently reported on the development of a CRISPR gene drive that achieved a total population collapse in *Anopheles gambiae* mosquitoes, the carrier of the malaria parasite (Kyrou et al. 2018). They achieved this by designing the gene drive to insert itself into (and thus disrupt) a sex determination gene (*doublesex*). The females were less fit and not able to reproduce, leading to eventual crash of the population within 7–11 generations in laboratory cage trials. This gene drive is reported as the most effective drive to date. However, a theoretical modeling study found that the impacts of GDO mosquitos for malaria control in West Africa are likely to vary, from population suppression to complete elimination, depending on sub-regional environmental and physical characteristics (North et al. 2019).

To harness and promote gene drives for controlling mosquito borne diseases, research consortia have arisen. For example, Target Malaria is a group of scientists and stakeholders joining together in a non-profit consortium, funded by the Gates and Open Philanthropy foundations, to develop *Anopheles* malaria-fighting GDOs for use in sub-Saharan Africa (targetmalaria.org).

Also using suppression drives, another research consortium is tackling the conservation application of controlling mice on islands. The Genetic Biocontrol of Invasive Rodents (GBIRd) project, led by the non-governmental organization (NGO), Island Conservation, aims to reduce rodent populations on islands where endangered birds and other species are being destroyed by invasive mice (geneticbiocontrol.org). Currently, rodenticides like Brodifacoum are dispensed through bait stations or aerial methods, and they not only cause a painful death of internal bleeding in the mice but also harm non-target species that may be endangered. The GBIRd group is harnessing a natural gene drive that works during meiosis, called the t-haplotype, and inserting into it a male-determining gene called Sry (Leitschuh et al. 2018). Its work on this GDO is partially supported by the Defense Advanced Research Projects Administration (DARPA) which has been investing heavily in gene drive research through its "Safe Genes" program. Another group has also pursued gene drives in mice and recently reported on the use of a split CRISPR-Cas 9 system with female mice carrying the gene Cas9 and males carrying the gene for the gRNA and a gene that modifies the mouse's coat color (Grunwald et al. 2019). They were able to increase inheritance of the gene from the predicted 50% for Mendelian inheritance to 72%. In the United States on Nantucket Island, Lyme disease is being considered for eradication by a gene drive with an immunization mode-of-action, in which antibodies are spread through Lyme's disease reservoir species, such as white-footed mice. The "Mouse Against Ticks" project lead is consulting with the local citizens before and during technology development in order to make the decision whether to pursue a gene drive eradication strategy (Davies and Esvelt 2018).

Population suppression drives are being considered and developed for agricultural pests. Target genes and traits for CRISPR-Cas 9 gene drives have been proposed for New World screwworm (a pest of livestock still a problem in tropical South America and on some Caribbean islands), the fruit fly *Drosophila suzukii*, Diamond Back Moth, and the red flour beetle (Scott et al. 2018). With funding from the California Cherry Board, scientists at the University of California in San Diego set up a pair of companies last year to commercialize gene drives, and recently reported on the successful construction of a gene drive for population suppression in the fruit pest, Spotted Winged Drosophila (SWD) (*D. suzukii*). The gene drive system uses synthetic *Medea* drive with a maternal toxin and an antidote in the zygote (fertilized egg), to kill females with each generation (Buchman et al. 2018). They demonstrated that this drive system can bias inheritance up to 100% in the laboratory, but modeling studies suggest that in the field, a relative high numbers of *Medea* drive SWD will need to be released to reduce populations (functioning like a threshold drive), likely due to the development of resistance to the engineered toxin. Other groups have proposed gene drives for control of agricultural weeds, for example, by driving genes that confer susceptibility to herbicides into weeds that are resistant to the herbicides (NASEM 2016). There are significant challenges to plant GDOs; however, in that gene drives rely on sexual reproduction and many plants self-cross, reproduce asexually, have perennial life cycles, or produce seed banks that are dormant and can survive a long time before seedlings emerge (Neve 2018). Other places where gene drives are being considered include in coral reefs so that they can withstand rising sea temperatures, and in invasive-species efforts of countries, like Predator-free New Zealand (Dearden et al. 2018; Rode et al. 2019).

## Types of Gene Drive Impacts

### *Efficacy, Resistance, and Stability*

Before considering the impacts of gene drives, the question of whether they work as intended to address the problem for which they are designed should be addressed. Financial losses may stem from reduced GDO efficacy and stability so that other measures, like chemical pesticides in the case of pest control, are still required.

Unintended consequences to the environment or health may arise from a lack of stability and efficacy as well. Scientists have been studying the effectiveness of gene drives in laboratories to reduce populations of insects, such as those that carry diseases like *A. gambiae* and *Aedes aegypti* mosquitos. Results from these studies have been mixed depending on the CRISPR-Casx system, target sites for gene editing, cargo genes, and DNA templates used. No engineered gene drives have been released into the wild yet as of July 2020.

With introduced genes for population suppression and pest eradication (e.g., female killing systems), resistance to the gene drive may develop over time depending on the mutation rate of the target site for the gene drive nuclease and fitness costs of the introduced genes like the CRISPR-Casx system or DNA template for insertion. The design of gene drives to include target sites that have lower genetic variation in the population (low polymorphisms) can increase the probability that the drive will be propagated through the intended population and decrease the chance that it will work in non-target species in the ecosystem. This design can reduce risk of resistance and loss of other species important to ecosystem functioning.

Theoretical studies using mathematical modeling have investigated the influence of different factors on the efficacy of gene drives. These include fitness cost compared to the wild-type gene, ratio of number organisms released to total target population, initial population size, degree of dominance of the gene, mating characteristics, and spatial features of the population. Gene drives require sexual reproduction to work and short generation times to fixate into the population within a reasonable time frame. With ideal assumptions like complete population mixing and mating, models have predicted it would take 10–20 generations to fix gene drives into wild populations when the initial frequency of GDO individuals released to the wild population was 0.001 (Unckless et al. 2015).

In experimental studies, one of the main reasons that gene drives fail, or disappear from the population, is due to emergence of mutations causing resistance to cutting at the DNA recognition or target site (Unckless et al. 2017). The rapid evolution of resistance could present an important risk for eradication and suppression drives, as the released GDOs would not lead to a population decrease but instead would increase the population size (from released GDOs adding to the wild population) and thus potentially increase the chance of disease transmission. Mutations can arise from non-homologous end-joining (NHEJ) repair that could occur after the nuclease cuts the DNA target site, even in the absence of template DNA introduction. This has the potential to make the wild-type chromosomes resistant to further cleavage by the Cas9 endonuclease to cease the spread of the gene drive. A study with *Drosophila melanogaster* showed that the probability of these "NHEJ-induced indel" mutations in the germ line could be several orders of magnitude higher in drive/wild-type heterozygotes compared to wild-type homozygous (Champer et al. 2017). To combat resistance, proposals have suggested using several gRNAs that target multiple sites (Noble et al. 2017), much like using multiple antibiotics to combat bacteria resistant to disease treatment. Experimental studies have found that targeting multiple sites does indeed decrease resistance (Champer et al. 2018). However, multi-site targeting might also lead to greater unintended effects by increasing the potential for the gene drive to cut and mutate off-target sites (discussed below).

## *Unintended Molecular Malfunctions*

Site directed nucleases, including CRISPR-Casx, bind and cut at specific sites in the genome. However, this process is not 100% specific, and there is always potential for off-target binding, cutting, and edits or deletions in DNA regions with some homology to the target site. Furthermore, the gRNA used to target sites with CRISPR-Casx gene drives could also mutate causing additional off-target effects (Scharenberg et al. 2016). These off-target edits could have a variety of impacts including fitness costs. CRISPR-Casx gene drives are designed to be active over many generations, and with every generation, the chance of mutation at off-target sites increases. With each generation in the gene drive heritance chain, mutations could therefore accumulate.

Cutting at "off-target" sites could disrupt genes that are important for survival. If the gene drive is meant to immunize a valuable or endangered species, for example against a disease, an off-target mutation *that is detrimental to the organism* could spread and lead to a substantial risk to the health and survival of the species instead of achieving the intended benefit of increased survival. Furthermore, the gene drive could be transferred to another species or subspecies that is important to the ecosystem either through mating (if sexually compatible with the GDO) or horizontal gene transfer (albeit there is a low probability for the latter). Off-target mutations in the recipient species could then accumulate and cause a reduction in fitness. On the flip side, with gene drives intended to suppress or eradicate a population, off-target mutations could instead counteract this goal and *make the organisms more fit or a bigger threat* to the ecosystem. The unexpected survival of the population, despite the suppression drive, could lead to increased pestilence, disease transmission, or predation of other important species.

Unintended off-target mutations could lead to ecological or human health risks which are outlined in the sections below. Generally, off-target mutation rates will depend on the specificity of guide RNA sequences used and the uniqueness of the target site in the GDO species. Some studies have shown no off-target mutations after careful selection of unique target sequences and optimization of both the gRNA and Cas nuclease (Cho et al. 2014). Under experimental conditions, a metanalysis of mouse studies using CRISPR-Cas9 for gene editing found off-target edits in 23% of the experiments, as defined by *at least one animal with at least one allele with Cas mutations* in *at least one* off-target loci (Anderson et al. 2018).

## *Ecological Impacts*

### Potential Ecological Risks from Intended Effects of Gene Drives

Changes to populations of important species in the ecosystem may have wide ranging effects on biodiversity, food webs, and ecosystem services. For population suppression or eradication drives, where the goal is species decline, the demise of that target population could lead to concomitant decreases in their predators or increases in species on which they prey. If their predator is an important or endangered species, significant ecological risk would occur if that species does not have alternative food sources. Alternatively, if the GDO with the suppression drive is a predator that keeps another pest population in check, increases in pestilence or disease may result from the overabundance of the prey. Understanding the ecological role of the candidate organism for a gene drive is therefore crucial (Kuzma et al. 2018).

Learning from the history of classical biological control can also provide insights into the kinds of ecological risks to consider (Webber et al. 2015). If the GDO is an invasive species to an ecosystem and it is eradicated through gene drive technology, another more harmful alien invader could take its place, potentially causing more damage to the ecosystem. For example, the eradication of feral goats and pigs on the Sarigan islands in the Western Pacific led to the proliferation of a new invasive vine in the region (Kessler 2002). Furthermore, the GDO even as an invasive species could have a long history of presence in an ecosystem in which it has come to take on important roles as a predator or food resource. Removing a species (whether native or invasive) with gene drive technology "could produce unintended cascades that may represent a greater net threat than that of the target species" (Webber et al. 2015).

Risk could arise from unintended, lower-probability events instead of the intended disappearance of the target species. These pathways to ecological risk are discussed below.

### Unintended Genetic Transfer Events and Potential Ecological Risks

Unanticipated ecological impacts may arise from the spread of a gene drive to a non-target population of the same or a different species, which is referred to as a "spillover." Spillover effects can be beneficial, neutral, or detrimental to the ecological health of the recipient species, predators or symbionts that depend on the species, or ecosystem services to which the species contribute. Risk has two components, likelihood of exposure to a hazard and the severity of adverse effects stemming from that exposure. The ecological risk from gene drive spillover depends on both. In other words, the mere presence of gene drives in non-target species, or mutations from them at off-target sites, does not necessarily lead to harm. Rather, it is the effect of those events that matter.

Classical biological control can provide insights for dealing with the novelty of gene drive modified organisms and anticipating adverse effects, with some exceptions (Webber et al. 2015). Three risk pathways leading to unanticipated gene drive movement and exposure to non-target populations or species are discussed below—migration, hybridization, and horizontal gene transfer—with some examples of possible adverse effects.

A species that is invasive or considered a pest in one geographic region may be native or desired in another region. If a GDO containing an eradication or suppression drive unexpectedly moves outside of the target area, it could cause beneficial populations in those areas to crash. Controllable features of gene drives introduced into a wild population include gene drive phenotype and efficiency rates of the drive mechanism. However, migration patterns and changing ecological circumstances cannot be controlled and for many species are not well understood. Models have typically taken into consideration the fitness costs of the gene drive, degree of dominance, and the life-stage of where selection for the gene drive takes place; but given uncertainty, migration patterns and other ecological or weather-related variables are difficult to model and predict (Greenbaum et al. 2019). In addition, human travel patterns and commodity trading in a global market could lead to the movement of a GDO far beyond the expected range. For example, *A. aegypti* mosquito species first appeared in Florida in the United States but is believed to have moved from there to California via passive transport, such as with automobiles (Gloria-Soria et al. 2014). Rats and mice arrived on islands likely through shipwrecks and travel that occurred centuries ago. Now a target for population suppression gene drives, as they prey upon endangered bird species, a concern is that gene drive modified mice or rats deployed on islands could reach mainland via similar human-caused events and interbreed with populations in their habitat of origin resulting in species decline (Leitschuh et al. 2018).

Hybridization of GDOs with sexually compatible species could also be problematic. There is precedent for transgenes from genetically engineered (GE) plantfield trials contaminating native populations. For example, glyphosate-resistance genes arising from contained field trials of genetically engineered bent grass have been found in native grass populations on National Parklands and in intergeneric crosses with other grass species (Zapiola and Mallory-Smith 2012). Invasive species can have a wide geographic range and often occur in proximity to closely related and sexually compatible native species (e.g., Zuber et al. 2012; Lack et al. 2012). In these cases, population suppression drives introduced into the invasive species may be transferred to the beneficial native species in that area through sexual hybridization. If the target DNA site is conserved between the two species, the gene drive would be active in the native population, and if it is a suppression drive, the desired population could also decline. Even if the target site for the gene drive is carefully selected to be unique to the invader, the transfer of the gene drive may lead to off-target mutations in the native species and potentially cause harm if essential genes are inactivated.

In addition to hybridization, gene drives could be transferred from one species to another through horizontal gene transfers (HGT). HGT can occur via symbiotic or parasitic viruses, bacteria, fungi, and insects which can act as vehicles to transfer DNA between species. Transfer from prokaryotes to eukaryotes seems to be more common than the reverse (Keeling and Palmer 2008). The impacts of HGT of the gene drive system would be a low probability event, but potentially have high consequences. These "fault tree" events are largely unpredictable in risk analysis although with better understanding of genomic regions with propensity for HGT may assist with prediction in the future (e.g., Clasen et al. 2018).

### Ecological Benefits

Despite the potential ecological risks discussed above, genetic approaches to pest and disease management may be superior for protecting ecosystems over harmful chemical approaches which often indiscriminately kill many species. The benefits of GDOs to ecosystems or human health will outweigh the potential adverse and unintended effects in some cases. A gene drive may be the last hope for preserving an endangered species for future generation (Kuzma and Rawls 2016). For example, Hawaiian bird species are seriously threatened by avian malaria disease, which is carried by *Culex* mosquitos. Using a suppression drive to eradicate the mosquito carrier may be the only option to save these birds (NASEM 2016). Another example is the suppression of human malaria. *A. gambiae* mosquitos, which carry the parasite for malaria, are a target for gene drive eradication in Africa. Here, the species does not seem to play an important ecological role, as it is not native to Africa and does not provide nonredundant ecosystem services. Furthermore, biodiversity impacts have not been detected with conventional eradication programs for *Culex* mosquitos (Roberts et al. 2017). Despite the availability of medicine to treat malaria, GDOs might be a superior option as the benefits to human health could be great and come with little ecological risk. Careful risk-benefit analyses will be important for making responsible decisions about gene drive deployment.

## *Human Health Impacts*

GE insects for population suppression without gene drives have been deployed for disease control. The company Oxitec (purchased by Intrexon) has field tested GE *A. aegypti* for dengue control in several lesser-developed nations and territories, such as the Cayman Islands, Panama, Malaysia, and Brazil. Some success in population reductions of the target wild-type mosquito has been achieved through the use of GE larval killing genes (Nimmo and Beech 2016). Although field trials of these mosquitos were also proposed in the Florida Keys, USA, the city slated for deployment, Key Haven, voted to reject the release of the GE mosquito. At the same time, the Food and Drug Administration (FDA) and United States Environmental Protection Agency (EPA) were deciding which agency would have authority for disease—or pesticide—treatment with GE mosquitos. Oxitec initially applied to

FDA for release, and the FDA granted approval for the release, but that has since been withdrawn, and now the Oxitec mosquito has been reviewed and cleared for release by the EPA. The risk assessments that Oxitec performed for the regulatory package can be instructive for the types of human health risks that should be considered for GDOs with population suppression gene drives (Meghani and Kuzma 2018; Kuzma 2019). Another instructive example of risk assessment for GDOs comes from work done in Australia's Commonwealth Scientific and Industrial Research Organisation (CSIRO) for the release of mosquitos modified with Wolbachia bacteria for population suppression (Murphy et al. 2010). Some of these risks are discussed below.

Human-health concerns that could be associated with gene drives for mosquito-borne disease control include: (1) increased probability of exposure to the target disease (e.g., from higher prevalence of mosquitos); (2) a more severe disease replacing the target disease and carrier (e.g., due to a different carrier species moving into the newly vacant niche); and (3) higher severity of disease (e.g., from mutations in the virus due to presence of the gene editing machinery). The first concern is probably the most likely on a short-term horizon given that thousands of gene drive mosquitos might need to be released for self-limited and threshold gene drive approaches. For example, for the Oxitec mosquito, continual releases each week over the course of 2 years were proposed and would result in millions of additional mosquitos in the ecosystem (Kuzma 2019). With threshold and self-limited GDOs these levels of releases might also be required to achieve population reductions over time. The second category of risk is tied to the ecological risks. Population eradication of one mosquito species in an area could result in a different species moving into the area and that one might carry a more severe or less treatable disease. Finally, the mutation rate of diseases carried in the mosquito may be increased by the gene drive system, depending on the specificity for the target site. This could increase the transmissibility of the disease agent or the potency. Thus, the ecological and molecular events are connected to the human health risks.

Other human health impacts could be intertwined with social, behavioral, and economic variables. For example, with a transient or temporary increase in mosquitos upon the first release of threshold GDOs, people might perceive a greater risk and use more chemicals to control the pests. This could lead to greater toxicity to humans from these chemicals, as well as greater environmental and ecological impacts from the use of those pesticides. Systems approach to study the ecological, social, and economic impacts and their interconnectedness to human behavior will be needed for a full understanding of the risks and benefits of using GDOs (Kuzma et al. 2018).

Less likely are the toxic or allergenic effects of the gene drive system itself to humans. The likelihood of consuming the gene drive or coming in contact with it is low, and the adverse effects of such small exposures are likely to be close to zero. There is an extremely low probability that the drive system could be integrated into human cells and operate to cause mutations. However, the released GD mosquitos could include a certain percentage of females that bite humans, and people are concerned about increased bites even if they do not come with additional risks.

Risks can arise from decisions that are made without full knowledge of a situation and its dynamics over time. When the decision maker does not bear the risk, this can lead to a "moral hazard" (e.g., Lin 2013). For example, if we know that a GDO can help to mitigate human diseases or ecological risks in the future, we could be less likely to invest in prevention or control methods today, as future generations will bear the risk. Without comprehensive cost–benefit analyses of GDOs deployment that account for a range of health and environmental externalities into the future (Kuzma and Rawls 2016), we might naively forgo investing in safer, better known, and more effective control methods like bed nets or vaccine development.

## *Social, Cultural, and Economic Impacts*

Early literature on the societal implications of gene drives focused on ecological risk (e.g., Oye and Esvelt 2014, Oye et al. 2014; NASEM 2016), governance (e.g., Kuzma and Rawls 2016; NASEM 2016; Carter and Friedman 2016), population modeling (e.g., Gould and Schliekelman 2004), and laboratory biosafety issues (e.g., Akbari et al. 2015; Esvelt et al. 2014; NASEM 2016). Analyses in these areas, as well as research for gene drive technology development, have exploded in the past 5 years. However, scant attention has been paid to the political, social, economic, and cultural impacts of GDOs (societal impacts) and how they interact with natural-world variables like health and ecological risk. These societal impacts will vary based on the type of GDO, geographical setting, governance system, social and cultural setting, and ownership and power structures. Furthermore, societal impacts of GDOs are intertwined with each other and the socio-ecological systems into which they are deployed. This complexity precludes a comprehensive discussion of types of GDOs and societal impacts. Instead, a few examples are provided below, focusing on their integration in complex systems.

In the absence of broader socio-economic and cultural assessments, political conflicts between groups or nations might ensue from GDO deployment. For example, pigs were brought to Hawaii by the Polynesians, and later the Europeans when settling the Hawaiian Islands. The pigs soon established themselves in the wild, and in doing so, disrupted native ecosystems and allowed for other invasive species to move into the area, which ultimately impacted the health of native birds and forests (Maguire 2004). The eradication of wild pigs in Hawaii using population suppression by conventional techniques (traps and shooting) is seen as desirable from an ecosystem damage perspective, but Native Hawaiian communities, relying on the feral pigs for cultural events and food, value the pigs for cultural preservation (Maguire 2004). Wild pig eradication remains a contentious issue. GDOs may face similar situations where cultural and ecological values conflict.

Differences in the value of species may also be specific to adjacent geographic regions. GDOs transcend national boundaries, yet the country on the receiving end might not have approved or been informed of the release. If negative consequences arise from the movement of GDOs across national boundaries, political conflicts might result. Neighboring countries have different regulations and laws regarding genetically modified organisms. There is some harmonization under the United Nation's Convention on Biological Diversity's Cartagena Protocol on Biosafety (UN CBD-BSP), but not all countries, and notably not the United States, are parties this protocol. Under the protocol, countries must notify each other if a "living modified organism" (also known as GMO) is exported to another country. If any damage results from a LMO, there is an additional article under the protocol for liability and compensation. Without all countries being parties to the protocol, it remains unclear if damage due to a GDO crossing borders would be compensated (Brown 2017). This could lead to retribution like in other areas of trade or political conflict.

Eradication of an important species could cause direct or indirect economic damage. Direct economic damage could result if the target species for the GDO has economic value itself (e.g., for food, fiber, timber, or fuel). Indirect economic damage may arise from broader ecological consequences. For example, if the target species plays an important role in maintaining ecosystem services or keeping human diseases under control, its decline can then cascade into economic costs, such as lost revenues from natural products or increased expenses in health care. It is important to take not only the health and ecological risks but also the broader socio-economic impacts into consideration in decisions about whether to release a GDO organism. Integration of social, cultural, and economic values has been recommended for decision modeling of biocontrol of invasive species in the past (Maguire 2004).

Non-use values of species are also important to consider in deploying GDOs. For example, if GDOs become pervasive and persist in the environment, such as with population replacement or immunization to protect endangered species, people may view the natural world as tainted. Public rejection of current GMOs often relates to lack of "naturalness" (Lull and Scheufele 2017). Even if the species is preserved and can provide ecosystem services through the use of GDOs, current and subsequent generations may obtain less enjoyment from their natural-world surroundings (Kuzma and Rawls 2016). The tendency to be inspired by or healed by the natural world may be reduced, analogous to feelings we get when in a national park preserve versus an urban park.

Animal welfare is another important consideration of GDOs. In some cases, a gene drive approach may be a superior choice to the suffering caused by chemical or other eradication measures. For example, the anticoagulant Brodifacoum has been used to eradicate invasive rodents on islands to protect endangered birds. This chemical kills the animals over a period of days and can cause great suffering to them. Gene drive options that affect rodent fertility may be a superior approach and are currently being developed (Leitschuh et al. 2018).

It will be important for the impacts of gene drives to be fairly and equitably distributed. Environmental justice includes making sure that marginalized or underrepresented communities do not bear the risks of GDOs disproportionately (distributive justice) as well as have voice in decision making affecting them (procedural justice). Another is economic justice, for example, if GDOs are deployed in agriculture for pest control, organic farmers may suffer lost sales and revenue due to contamination by GMOs. Target genes and CRISPR-based gene drives are under consideration for controlling the fruit fly *D. suzukii* on soft fruits, such as cherries, blueberries, and raspberries (Scott et al. 2018). It is currently not clear if the presence of GDO insect parts in organic berries will impact organic certification and associated product premiums (Baltzegar et al. 2018).

The political economy of gene drive R&D seems distinct from the first generation of genetically modified organisms which were developed by multi-national corporations that could sell seeds or companion chemicals each year. In contrast, the key attraction of gene drives is the possibility of solving the problem, like pest presence, over wide geographic areas on the first try. However, this benefit also poses challenges for commercialization, as opportunities to repeatedly sell a GDO will be limited if they are effective the first time that they are released. In contrast, current approaches like sterile insect technology (SIT) or Oxitec's GE larval killing systems require the repeated release of thousands to millions of insects (Kuzma 2019). Furthermore, the economic model for GDOs resembles a public good in that benefits will be widely shared in the area of release, and they are non-excludable (Brown 2017; Kuzma et al. 2018). This feature makes it unlikely for private sector investment in global gene drives, and perhaps even self-limited or threshold drives depending on the need for repeated releases. SIT and biocontrol are historical cases where the public sector has taken leadership in development and deployment for public goods. For example, the USDA and government agencies from Mexico and Central America invested heavily in screwworm eradication from livestock in North and Central America in the 1960s (Brown 2017). Large-scale projects to develop gene drives are currently led by groups outside of private companies, like foundations, non-government organizations, government agencies, and academe (see section "Purposes of gene drives and example projects"). Without profit-seeking (although there are other financial and non-financial motivations both personal and professional), multi-sector public leadership for gene drives might proceed at a slower pace. However, more socially robust decisions might be made with this pace, and there is early indication that non-private groups involved in developing GDOs are more likely to take a step-wise, cautious approach that engages local and stakeholder communities in anticipation of potential impacts of gene drives (Kuzma et al. 2018).

## Uncertainty and Irreversibility

Information and analysis of potential downstream impacts prior to release is important in staged models of GDO testing, which go incrementally from lab to field tests, to deployment. The National Academies of Science, Engineering and Medicine suggests the need for great care and precaution in moving from almost 100% contained laboratory studies to mostly contained field cage experiments, to confined but geographically limited releases, and finally to wholly unconfined releases (NASEM 2016; Kaebnick et al. 2016). The juncture to move from lab to the first field test is a key analytical point for decision making, especially for self-sustaining drives, as one escapee from a field cage or limited trial could theoretically turn it into a full-scale release. Although there are guidelines for the steps a developer of a GDO mosquito should move through (James et al. 2018)—from laboratory to cage trials, small-scale open release, large-scale open release, and finally to post-surveillance—they do not explicate the types of risk studies, non-target endpoints to be assessed, or tolerable risk levels that would be used in decision models to move from lab or cage to the first open-release field trials. The uncertainties associated with GDOs are immense, and more specific decision protocols are needed to help determine when the first open release does not present an unreasonable risk. A field trial turn into a full-scale intervention if the risks are not fully anticipated.

### *Recall and Reversibility*

Given that GDOs are designed to spread (whether locally or globally) and that we cannot foresee all the potential adverse impacts of their release, it is very important to have the ability to stop, recall, or reverse a GDO. Various methods based on molecular biology have been proposed for doing so (Vella et al. 2017). A straightforward method to stop an ongoing gene drive is to release drive-resistant individuals who carry a synthetic yet functional copy of the targeted gene without the Cas9 (or general nuclease) recognition sequence (called a synthetic resistant SR drive) (Vella et al. 2017). This approach is likely effective for eradication drives, which impose significant fitness costs, but not for rescue drives or suppression drives that have mild, neutral, or advantageous fitness costs. A second way to achieve this same result is to release a GDO with a different guide RNA to alter the recognition site of the original gene drive so that it is no longer recognized by the original Cas9 nuclease (called a reversal drive, RD) (Esvelt et al. 2014). Immunizing reversal drives (IRD) replace both the wild-type allele with the first drive and then replace the first drive with a second immunization drive by introducing multiple guide RNAs and Cas9 at the same time (Esvelt et al. 2014). In all strategies, the goal is to remove the target site for the gene drive in the population.

Theoretical modeling studies have shown that SRs and RDs are not guaranteed to eliminate an unwanted gene drive from a population and instead result in a mixture of organisms containing the unwanted gene drive, wild type, and RD or SR allele in the species (Vella et al. 2017). Other schemes for gene drives that limit themselves have been proposed. One is a CRISPR-based "daisy-chain drive" that contains genetic elements that are serially dependent and arranged to work in a chain, but are unlinked (e.g., on different chromosomes). Each element drives the next, but their ability to spread is limited by the successive loss of the elements from the end of the chain via natural selection (Noble et al. 2019). These could theoretically drive a useful genetic element to local fixation in a population, while limiting the geographic range and making the changes transient. However, modeling studies have suggested that there is only a narrow range of conditions under which daisy chain drives would work as intended. Although daisy drives could theoretically be efficient and effective, they might present a higher chance of unintended spread than older approaches to self-limiting drives, such as those based on engineered underdominance (e.g., promoter-toxin/suppressor systems on unlinked chromosomes, so that one copy of each is needed for survival) (Dhole et al. 2018).

Regardless, there is significant worry that molecular approaches based on gene-drive technology to counteract gene drive technology would not only fail in the field given the ecological complexities, but also lead to additional, unintended adverse effects. Publics may also be uncomfortable with using a technological fix to prevent future technological failures. For example, with reversal and immunizing drives, the "wild" population would continue to carry engineered genes for Cas9 and guide RNA. This could perpetuate off-target mutations in the species, leading to potential ecological, health, or societal impacts as described in section "Types of gene drive impacts". Thus, relying on technological fixes or "silver bullets" to recall gene drives after release could be reckless. Anticipating adverse effects before any release is still of paramount importance. To do this, we must carefully consider the sources of uncertainties that accompany GDO and fill in knowledge and data gaps to address them.

## *Natural-World Factors Affecting Risk*

Uncertainty due to natural-world variables stems from several dimensions, many of which were discussed in sections "Unintended molecular malfunctions", "Ecological impacts" and "Human health impacts". To summarize, ecological sources include, but are not limited to: (1) the low, but non-negligible, probability of horizontal gene transfer of a population suppression drive to a desirable or beneficial species resulting in its demise; (2) the ramifications of population reductions of the target species on other species like predators; (3) the possibility that another, more harmful species could fill the ecological niche of the eradicated population; and (4) potential impacts on ecosystem services from reductions in the target population. To assess these impacts, the role and behavior of the target species in the surrounding ecosystem need to be understood (Kuzma et al. 2018). Roles of species in the environment and their population dynamics are important baseline data for environmental risk assessment prior to release of GDOs. Unfortunately, this area of research is not often well-funded in comparison to biomedical research or research on technological development of GEOs.

Generally, there is great uncertainty associated with the role of pests in the environment, yet this role must be understood to anticipate the potential impacts of GDOs designed to target pests. Although some might be non-native to an ecosystem, they might have established themselves for decades and other species may have come to rely on them. Potential effects of GDOs on off-target species either through horizontal gene flow or food webs are difficult to predict, and it is even more difficult to quantify the severity of potential adverse events. A significant challenge is that field trials are the best way to study such interactions and gather data, yet gene drives are not likely to be confined to the field trial area (especially for self-sustaining drives), so essentially one must assume "release" without field trial information. This problem might mean that sufficient information on behavior of the GDO under real-world circumstances may never exist prior to release. Risk assessments will need to rely on laboratory or field cage data.

Not only do population, ecological, mating, and genetic characteristics matter for the impacts of gene drives, but so do biophysical attributes of weather and climate, and geographic features of habitats, such as barriers. Sporadic and severe weather and climate events make prediction of risk difficult. These events will affect the spread of GDOs and its distribution for mating with other subpopulations. Even if a field trial can be confined, it is unlikely to capture the range of physical conditions under which gene drives will be deployed. These conditions will impact interactions with and potential risks to other species, such as predators and prey. There is a need for better ecosystem and population models of GDOs that account for variability in biophysical parameters across temporal and geographic scales.

## *Social and Political Factors Affecting Risk*

Attributes of social systems will also influence the effectiveness and spread of gene drives in passive and active ways. For example, patterns and behaviors of human movement may carry GDOs into unwanted areas, even across national borders through trade or travel. As mentioned, passive transport of *A. aegypti* on humans via motorized travel is thought to have resulted in the distribution of the mosquito in Florida and California in the United States (Gloria-Soria et al. 2014). Unfortunately, the unintended movement of species via humans or goods can be sporadic, causing great uncertainty in the probability of occurrence. To minimize risk from stochastic events, principles for gene drive deployment are emerging. For example, it has been suggested that the first open releases of GDOs should be on isolated islands with no-to-low human traffic, good border control, and large physical distances from the shore (Webber et al. 2015). Others have recommended that self-sustaining and global gene drives should only be used on target species for which global eradication of the species would not be seen as a problem, such as for *Anopheles* mosquitos and malaria prevention (Noble et al. 2018).

Policy systems related to gene drives will also affect how risk is managed. Gene drive governance has parallels to the governance of other common pool resources (Ostrom 2011; Kuzma et al. 2018). In these areas, behavioral and value systems of communities are important for managing risk through shared governance and collective action (Ostrom 2009). Gene drive release will require ongoing cooperation between different sectors and geographic regions to plan for, execute, and monitor gene drive releases and their impacts. Shared goals are important for collective-action settings, and in limited geographic areas, goals are more likely shared. As self-sustaining gene drives are designed for greater geographic areas and even for crossing national borders, the potential for shared values and norms is lower (Kuzma et al. 2018). Risk governance for gene drives will be a greater challenge across national or cultural boundaries, than for local, self-limited gene drives unlikely to travel outside of a defined area within a nation. However, local drives might fail and spread globally, so this risk must also be considered with them.

As previously mentioned, gene drives also share features with public goods, in that their impacts—both positive and negative—are likely to be non-excludable. Parties without direct control over deployment are likely to experience benefits or harm from the technology as it spreads across landscapes. Because the deployers of gene drives might not bear the adverse impacts, they might make riskier decisions to release a gene drive than socially desirable (Mitchell et al. 2018).

Policies and regulations may limit the types of impacts considered. In current U.S. regulatory decision making about GEOs, direct harms, such as toxicity to humans or non-target organisms, are a primary (and often sole) focus of decision making (Thompson 2007; Meghani and Kuzma 2011). For certain GDOs, the types of risks considered in regulatory decision making may be further limited depending on the assigned federal agency, rule evoked, and GDO species. For example, many genetically engineered animals, and presumably GDO animals, are subject to regulation by the Food and Drug Administration under the New Animal Drug (NAD) provisions of the Federal Drug and Cosmetic Act. GDOs that come under FDA as "animal drugs" would be formally reviewed for the safety and efficacy of the drug to the animal, which in the GDO case is the gene drive construct (Meghani and Kuzma 2018). Technically, broader ecosystem risks, like loss of prey for predators or impacts on ecosystem services, are not part of the legal authority of FDA's NAD review, although they can be procedurally considered under the National Environmental Policy Act. However, non-governmental actors, such as the non-profits and academics developing gene drives, would not be limited to this scope and could also consider indirect ecosystem effects, socioeconomic impacts, and cultural impacts in informal governance. Projects underway on gene drives are broadening the scope of governance questions beyond formal regulatory authority (e.g., Target Malaria 2017; Leitschuh et al. 2018).

Negative public perception is sometimes seen as a risk to be mitigated. Scientists developing GMOs in the past have expressed the need to educate the public so they do not fear genetic engineering. Sometimes the goal is to convince the public that GMOs are safe through education. These views are in line with the "deficit model" thinking of risk communication, which espouses that with more education, laypersons will be convinced of the lower risk of the technology in comparison to alternatives (e.g., Ahteensuu 2012). While it is true that public backlash and pressure could stall or even stop GDO development and deployment, most of the gene drive community recognizes the failures of deficit model thinking and unidirectional risk communication. Instead, they are turning toward public engagement and bidirectional communication to allow for the public to learn more about the risks and benefits so they can make their own decisions to support or reject GDO releases, especially in areas that they live (NASEM 2016; Target Malaria 2017; Harmon 2016; Kaebnick et al. 2016).

## Risk Governance

GDOs raise new and magnified challenges for risk governance in comparison to the deployment of other genetic engineering technologies. There is a basic mismatch between deployment of gene drives and current governance systems for first generation GEOs. Gene drives are meant to spread through a wild population, whereas regulation of GEOs has typically been based on containment or confinement in managed settings (e.g., agriculture), especially for field trial stages (Kuzma et al. 2018). Oversight systems designed for GEOs are unlikely to be sufficient for gene drives for these reasons, and greater precaution may be warranted (Kaebnick et al. 2016). The goal of spread also presents challenges to field monitoring and testing, forcing wide boundaries, and more resources for data collection. The escape of even one GDO from a limited field trial could in some cases (depending on gene drive design, Min et al. 2018) spread a gene throughout an entire population. How we mitigate the chance of escapees from the lab and how we conduct risk assessments under uncertainty come under the umbrella of risk governance.

### *Biosafety and Biosecurity Measures*

Several GDO researchers, scholars, and funders have proposed protocols and principles to ensure biosafety and biosecurity (Akbari et al. 2015; Oye et al. 2014; Emerson et al. 2017; James et al. 2018). The NASEM (2016) concluded that currently there is insufficient evidence to support deployment of GDOs into the environment but that the potential benefits justify proceeding with laboratory research and highly controlled field trials. It recommended an iterative, phased, test and release pathway that (1) gathers risk-relevant data under controlled, contained conditions, to determine whether the GDO can progress to the next phase; (2) generates release-relevant data by observing outcomes under increasingly realistic scales and conditions; and finally, (3) moves to more open and less contained field trials (NASEM 2016). However, it fell short of proposing specific decision criteria to judge the adequacy of risk information for moving to each subsequent phase. This might need to be determined on a case by case basis, taking into consideration the benefits and urgency of the problem that the GDO is designed to address. A higher risk or level of uncertainty may be tolerated under special circumstances to achieve important benefits.

More specific protocols for physical, reproductive, ecological, and molecular barriers for biosafety in laboratory studies have also been proposed (Akbari et al. 2015). Ecological barriers include performing experiments outside the habitable range of the GDO, or in areas without potential wild mates, so that in the event the GDO escapes from the laboratory, the spread would be unlikely. Reproductive strategies involve using strains in the lab that cannot reproduce with wild relatives in the surrounding area. Molecular containment methods include using strains with specific target sequences for the gene drive that do not exist in the wild population. It has been recommended that physical barriers occur at multiple levels, along with reproductive and molecular barriers. Redundant containment is important so that if one level fails, another barrier could stop an escapee from spreading (Esvelt et al. 2014; Akbari et al. 2015).

Biosecurity to prevent intentional misuse of GDOs seems almost impossible. For now, the technical complexity of successfully engineering a gene drive and getting it to work in an ecosystem provides some comfort. However, bad actors will likely have the technological capabilities in the future as the technology matures and it becomes easier to successfully deploy by a variety of bad-acting nations and miscreants. There was a time not too long ago (about 20 years) when cloning a gene or sequencing DNA was very difficult. Now one can easily order a kit over the internet, or send a gene to a DNA sequencing company, and with little training, do biotechnology. Gene drives could one day be used to deliver lethal toxins to desirable species or crop plants, and perhaps even target humans (although it would be unlikely for GDOs to affect future human generations). Some have called for the scientific research community to prevent the disclosure of exact instructions for making gene drives in scientific manuscripts or patent applications, citing the historical case in which nuclear weapons technology remained classified for 70 years after the Manhattan Project (Gurwitz 2014). Others disagree, arguing that if GDO developments were kept secret, it would prevent the progress of science not only to address important health and ecological problems in the future, but also to defend against the misuse of gene drives (Oye and Esvelt 2014).

The Defense Advanced Research Projects Agency (DARPA) has invested significant resources, upwards of $100 million through 2018, to develop tools and methodologies to "control, counter, and even reverse the effects of genome editing—including gene drives" (https://www.darpa.mil/program/safe-genes). However, DARPA's leadership in this area could be met with the suspicion that the underlying purpose is really for future weaponization (Callaway 2017). Regardless, the program explicitly prevents the release of GDOs and enforces strict biosafety conditions. In parallel, a unit of the Director of U.S. National Intelligence, the Intelligence Advanced Research Projects Agency (IARPA), is working on capabilities to detect harmful GMOs and GDOs (Callaway 2017).

Even with strict biosafety, biosecurity, and countermeasures, 100% containment or prevention of risk is never guaranteed. Furthermore, even if escape outside of lab or target area is rare, the consequences could be large given the design of many GDOs to spread in ecosystems. Risk analysis methods, such as fault-tree analysis, can be used to estimate low probability and potentially high-consequence adverse events and seem well-suited for thinking about the risks of GDOs from laboratory or confinement breakdowns (e.g., Murphy et al. 2010). However, our current ability to quantify such failures is severely limited by the significant uncertainties associated with GDOs in part stemming from a lack of relevant ecosystem and biological studies (Section "Uncertainty and irreversibility").

## *Advances in Risk Analysis and Governance*

Risk analysis can help to structure the identification of hazards from an event, the estimation of the probability of the event occurring, and characterization of the adverse consequences. For GDOs although the ultimate adverse outcomes may be the same as those associated with biocontrol or first-generation GEOs, the causal pathways that lead to those outcomes—and hence their likelihood and magnitude of these outcomes—might be unique (Hayes et al. 2018). Two approaches to hazard identification have been suggested with different strengths and weaknesses: (1) look to hazards identified for more or less similar situations like biocontrol or GEOs in a "checklist-like approach," and (2) structured hazard identification to anticipate what might go wrong, such as fault-tree analysis (Hayes et al. 2018). Table 1 captures

**Table 1** Hazardous events that could lead to adverse outcomes

| Scale | Hazardous events | Examples of potentially adverse ecological outcomes |
|---|---|---|
| Molecular | Cas9 cleaves loci with similar, but not identical, homology to the target loci | New phenotype with a different (possibly increased) capacity to spread diseases or pathogens |
| | Mutated gRNA causes Cas9 cleavage of non-target sequence | New phenotype with a different (possibly increased) capacity to spread diseases or pathogens |
| | Cas9 fails to edit or target all alleles | Changes the target organism's ability to survive, reproduce, or spread |
| | Mutations occur during repair of multiple cleavage sites | Changes the target organism's ability to survive, reproduce, or spread |
| Population | Assortative or nonrandom mating between new phenotypes | Drive is reduced and/or competitive advantage accrues to a more virulent phenotype leading to an increase in the incidence of the disease or pathogen of concern |
| | Intraspecific (admixture) and interspecific hybridization | Gene drive is acquired by, and spreads within, non-target population or non-target species leading to the suppression or modification of this population or species |
| | Unpredicted phenotypes from gene by environment interactions | Gene drive fails to produce refractory organisms in the wild but increases target organism's capacity to spread diseases or pathogens |
| Community/ Ecosystem | Population/species suppression changes competitive relationships | Release from competition allows a detrimental population or species to increase in abundance |
| | Population/species suppression causes extinction of (prey) species | Cascading effects on food web caused by decrease in abundance of predators leading to possible loss of ecosystem services |
| | Horizontal (lateral) transfer of gene drive to distant species | Gene drive is acquired by, and spreads within, non-target species, leading to suppression or modification of the non-target species |

Adapted from Hayes et al. (2018)

hazardous events at the molecular, population, and ecosystem levels from a checklist approach—many of these were discussed in sections "Types of gene drive impacts" and "Uncertainty and irreversibility".

Risk analysis is laden with assumptions and interpretations based on values. For example, the endpoints we choose to evaluate in a risk assessment are based on what we care about (e.g., certain species, certain natural resources, and certain human illnesses). Also, uncertainty in risk analysis leads to various interpretations of the data to which we bring our own experiences, cultures, and worldviews. Even in relatively straightforward cases of chemical risk assessment, the choice of a mathematical model for generating a dose–response curve from laboratory studies is an endeavor in which one can be more or less precautious about estimating risk under release conditions. Even if we do have good information on the dose–response curve, the level at which something is presumed "safe" is debatable as safety is a socially defined concept. Science gives us a guide, but what risks are acceptable are based on values, taking into consideration our experiences, culture, perceptions of the benefits, control over the situation, and trust in those managing the risks (Kuzma 2017).

GDOs present a case for risk analysis where data and information are severely limited, and therefore values will play even more of a prominent role in decision making. Quantification of risk in advance of any field releases will be nearly impossible given the uncertainties associated with GDOs in complex socio-ecological systems and the stochasticity of movements of organisms across geographic boundaries or rare genetic transfer events (Section "Uncertainty and irreversibility"). GDOs have features of "emerging risks" that are "characterized mainly by uncertainty regarding their potential consequences and/or probabilities of occurrence" which "can be due to a lack of knowledge about causal or functional relationships between new risk sources and their environment or to the insufficient application of available knowledge to the case in question" (IRGC 2015). For these situations, evaluating the "substantive validity" of risk assessments—where outcomes of the risk assessment are compared to what happens in reality—is not feasible, especially prior to any environmental release. Therefore, "procedural validity" of the risk assessment, that is *how* the risk assessment is conducted, becomes even more important than attempting to ascertain the substantive validity of particular risk evaluations prior to GDO release and field data collection.

Methods for making the process of risk assessment for GDOs more legitimate and robust have been employed under the informal governance activities of groups developing gene drives. These approaches make use of ideas from post-normal science (PNS) (Funtowicz and Ravetz 1994; Brossard et al. 2019). PNS suggests that when the decision stakes are high and the system uncertainties great, extended peer and stakeholder communities (beyond scientific researchers) should be consulted to interpret what is known and what it means for the policy decision at hand. Diverse values become an explicit part of risk assessment as the "facts" are uncertain and require interpretation for their meaning (Funtowicz and Ravetz 1994). People with "on-the-ground" knowledge, who are "interested and affected," are invited into the deliberations about risk and safety measures, along with a broader range of scholars such as ethicists and social scientists. Scientific experts and government managers still provide important technical analysis, but democratic engagement opens up the policy process for characterizing risk to communities in areas of potential GDO deployment, giving them not only a voice but also a choice in deciding what levels of risk are acceptable to them (NRC 1996).

In Australia, a deliberative engagement process was used to help assess potential risks from the deployment of *Wolbachia*-infected mosquitos to control dengue fever (Murphy et al. 2010; Murray et al. 2016). Diverse experts and stakeholders were involved in the framing and conduct of the analysis: helping to identify important parameters for fault-tree analysis, comparing the risks of the modified mosquitos to non-technological options, and estimating probabilities of a broad range of potential economic, social, cultural, ecological, and human health harms (Murphy et al. 2010). Bayesian approaches to estimating the risk, drawing on the mental models of a diverse group of experts and stakeholders, can provide important information on parameters for which little is known and thus signal areas where more research is crucial (Hayes et al. 2018).

For risk mitigation, participatory processes have also been used for GDOs. A coalition developing gene drives in mice for island conservation, the GBIRd program (Section "Purposes of gene drives and example projects") involved stakeholders in a workshop to help identify criteria to select islands that would reduce ecological risks from field trials of GDOs (Farooque et al. 2019).

A framework proposed for conducting risk analysis in support of formal regulatory decision making, the "Procedurally Robust Risk Analysis Framework" (PPRAF) draws upon principles of humility, procedural validity, inclusion, anticipation, and reflexivity (Kuzma 2019). Particular considerations for regulatory risk analysis under PPRAF are to:

(1) Assess social and behavioral foundations of vulnerability to risk
(2) Consider distributive impacts of risks amongst different groups
(3) Promote mutual learning as object of deliberation in risk analysis
(4) Engage multiple interested and affected parties in discussion of ends and means of innovation
(5) Elicit the input of interested and affected parties for scoping the risk problem and at key junctures in risk assessment
(6) Examine assumptions and framing in risk analysis
(7) Acknowledge alternative explanation to the data and analysis
(8) Reflect on quality of organizational processes used for risk analysis
(9) Reflect on meaning of any potential errors to outcomes
(10) Assess the quality of the process that led to the risk estimation
(11) Proceed with openness and transparency in conduct of risk analysis
(12) Ensure consistency in interpretation of data and information
(13) Account for changing future conditions at different timescales
(14) Consider contingencies of what is known, plausible, possible, and unknown for the future

## *Global Governance and Engagement*

Identifying possible risks is important, but ethical principles need also be integrated into processes for determining whether a field trial or release should take place. Many believe that scientists have a social responsibility for informing and engaging publics that will be affected by a gene drive (e.g., Thompson 2018). However, recommendations have been made that engagement should not be hosted by those who have a conflict of interest in seeing the technology progress, but rather should be led by local communities in areas that are candidates for deployment, while supported by global governance structures to provide the resources and expertise for deliberative engagement (Kofler et al. 2018). To date, such global governance systems for supporting engagement, conducting procedurally robust risk analysis, and comparing gene drives to other technological and non-technological alternatives are lacking.

Although GDOs will likely come under the UN CBD-BSP framework for LMOs (Section "Social, cultural, and economic impacts"), this framework is not focused on GDOs; not all countries are party to the CBD-BSP (including major actors in GDOs such as the United States); and it mainly provides for advance notice of GMO importation and risk assessment guidance.

However, the CBD is beginning to deal with risk assessment issues surrounding GDOs. In November 2018, its Subsidiary Body on Scientific, Technical and Technological Advice of the convention recommended that the Conference of the Parties (COP) as members of the parties (MOP) to the Cartagena Protocol on Biosafety consider the need for specific guidance on risk assessment of living modified organisms containing engineered gene drives at its tenth meeting of the parties (COP-MOP 10) which is likely to take place at the end of 2020. Also, the CBD's Ad Hoc Technical Expert Group (AHTEG) on Synthetic Biology has been tasked with undertaking "a review of the current state of knowledge by analysing information, including but not limited to peer-reviewed published literature, on the potential positive and negative environmental impacts, taking into account human health, cultural and socioeconomic impacts, especially with regard to the value of biodiversity to indigenous peoples and local communities, of current and near-future applications of synthetic biology, including those applications that involve organisms containing engineered gene drives" (CBD 2018). In the interim, the CBD COP calls on governments to apply a precautionary approach to introducing GDOs and to obtain the prior informed consent of indigenous and local communities where appropriate (CBD 2018).

Given that GDOs present a leap in our capabilities to engineer wild populations with great uncertainties about the potential risks, it is crucial that we give at least equal attention and resources to the development of robust and deliberative mechanisms for risk governance and decision making at the international level as we do to the development of the gene drive technologies.

# References

Ahteensuu M (2012) Assumptions of the deficit model type of thinking: ignorance, attitudes, and science communication in the debate on genetic engineering in agriculture. J Agric Environ Ethics 25(3):295–313
Akbari OS, Bellen HJ, Bier E, Bullock SL, Burt A, Church GM, Cook KR, Duchek P, Edwards OR, Esvelt KM, Gantz VM (2015) Safeguarding gene drive experiments in the laboratory. Science 349(6251):927–929
Anderson KR, Haeussler M, Watanabe C, Janakiraman V, Lund J et al (2018) CRISPR off-target analysis in genetically engineered rats and mice. Nat Methods 15(7):512–514
Baltzegar J, Barnes JC, Elsensohn JE, Gutzmann N, Jones MS, King S, Sudweeks J (2018) Anticipating complexity in the deployment of gene drive insects in agriculture. J Responsible Innov 5:S81–S97
Brossard D, Belluck P, Gould F, Wirz CD (2019) Promises and perils of gene drives: navigating the communication of complex, post-normal science. Proc Natl Acad Sci 116(16):7692–7697
Brown Z (2017) Economic, regulatory and international implications of gene drives in agriculture. Choices 32(2):1–8
Buchman A, Marshall JM, Ostrovski D, Yang T, Akbari OS (2018) Synthetically engineered Medea gene drive system in the worldwide crop pest *Drosophila suzukii*. Proc Natl Acad Sci 115(18):4725–4730
Burt A (2003) Site-specific selfish genes as tools for the control and genetic engineering of natural populations. Proc Biol Sci R Soc 270(1518):921–928
Callaway E (2017) US defense agencies grapple with gene drives. Nature News 547(7664):388
Carter S, Friedman R (2016) Policy and regulatory issues for gene drives in insects: workshop report. Available at https://www.jcvi.org/sites/default/files/assets/projects/policy-and-regulatory-issues-for-gene-drives-in-insects/report-complete.pdf
CBD (2018) Decision adopted by the conference of the parties to the convention on biological diversity 14/19. Synthetic biology. https://www.cbd.int/doc/decisions/cop-14/cop-14-dec-19-en.pdf
Champer J, Buchman A, Akbari OS (2016) Cheating evolution: engineering gene drives to manipulate the fate of wild populations. Nat Rev Genet 17:146–159
Champer J, Reeves R, Oh SY et al (2017) Novel CRISPR/Cas9 gene drive constructs reveal insights into mechanisms of resistance allele formation and drive efficiency in genetically diverse populations. PLoS Genet 13:e1006796
Champer J, Liu J, Oh SY et al (2018) Reducing resistance allele formation in CRISPR gene drive. Proc Natl Acad Sci 115:5522–5527
Cho SW, Kim S, Kim Y, Kweon J, Kim HS, Bae S, Kim J-S (2014) Analysis of off-target effects of CRISPR/Cas-derived RNA-guided endonucleases and nickases. Genome Res 24(1):132–141
Clasen FJ, Pierneef RE, Slippers B, Reva O (2018) EuGI: a novel resource for studying genomic islands to facilitate horizontal gene transfer detection in eukaryotes. BMC Genomics 19(1):323
Curtis CF (1968) Possible use of translocations to fix desirable genes in insect pest populations. Nature 218(5139):368–369
Davies K, Esvelt K (2018) Gene drives, white-footed mice, and black sheep: an interview with Kevin Esvelt. CRISPR J 1(5):319–324
Dearden PK, Gemmell NJ, Mercier OR et al (2018) The potential for the use of gene drives for pest control in New Zealand: a perspective. J R Soc N Z 48:225–244
Deredec A, Burt A, Godfray HCJ (2008) The population genetics of using homing endonuclease genes in vector and pest management. Genetics 179:2013–2026
Dhole S, Vella MR, Lloyd AL, Gould F (2018) Invasion and migration of spatially self-limiting gene drives: a comparative analysis. Evol Appl 11(5):794–808
Emerson C, James S, Littler K, Randazzo FF (2017) Principles for gene drive research. Science 358(6367):1135–1136

Esvelt KM, Smidler AL, Catteruccia F, Church GM (2014) Concerning RNA-guided gene drives for the alteration of wild populations. elife 3:e03401

Farooque MS, Barnhill-Dilling K, Shapiro J, Delborne J (2019) Exploring stakeholder perspectives on the development of a gene drive mouse for biodiversity protection on islands workshop report. Available at http://go.ncsu.edu/ges-gene-drive-workshop

Funtowicz SO, Ravetz JR (1994) Uncertainty, complexity and post-normal science. Environ Toxicol Chem 13(12):1881–1885

Gloria-Soria A, Brown JE, Kramer V, Yoshimizu MH, Powell JR (2014) Origin of the dengue fever mosquito, *Aedes aegypti*, in California. PLoS Negl Trop Dis 8:e3029

Gould F, Schliekelman P (2004) Population genetics of autocidal control and strain replacement. Annu Rev Entomol 49:193–217

Greenbaum G, Feldman M, Rosenberg N, Kim J (2019) Designing gene drives to limit spillover to non-target populations. bioRxiv. https://doi.org/10.1101/680744

Grunwald HA, Gantz VM, Poplawski G, Xu XRS, Bier E, Cooper KL (2019) Super-mendelian inheritance mediated by CRISPR–Cas9 in the female mouse germline. Nature 566(7742):105

Gurwitz D (2014) Gene drives raise dual-use concerns. Science 345(6200):1010

Harmon A (2016) Fighting Lyme disease in the genes of Nantucket's mice. The New York Times. Available at https://www.nytimes.com/2016/06/08/science/ticks-lyme-disease-mice-nantucket.html

Hayes KR, Hosack GR, Dana GV, Foster SD, Ford JH, Thresher R, Ickowicz A et al (2018) Identifying and detecting potentially adverse ecological outcomes associated with the release of gene-drive modified organisms. J Responsible Innov 5:S139–S158

IRGC (2015) Guidelines for emerging risk governance. International Risk Governance Council, Lausanne

James S, Collins FH, Welkhoff PA, Emerson C, Godfray HC, Gottlieb M, Greenwood B, Lindsay SW, Mbogo CM, Okumu FO, Quemada H (2018) Pathway to deployment of gene drive mosquitoes as a potential biocontrol tool for elimination of malaria in sub-Saharan Africa: recommendations of a scientific working group. Am J Trop Med Hyg 98(6):1–49

Jinek M, Chylinski K, Fonfara I, Hauer M, Doudna JA, Charpentier E (2012) A programmable dual-rna-guided dna endonuclease in adaptive bacterial immunity. Science 337(6096):816–821

Kaebnick GE, Heitman E, Collins JP, Delborne JA, Landis WG, Sawyer K, Taneyhill LA, Winickoff DE (2016) Precaution and governance of emerging technologies. Science 354(6313):710–711

Keeling PJ, Palmer JD (2008) Horizontal gene transfer in eukaryotic evolution. Nat Rev Gen 9(8):605–618

Kessler CC (2002) Eradication of feral goats and pigs and consequences for other biota on Sarigan Island, Commonwealth of the Northern Mariana Islands. In: Veitch CR, Clout MN (eds) Turning the tide: the eradication of invasive species. IUCN, Gland, pp 132–140

Kofler N, Collins JP, Kuzma J, Marris E, Esvelt K, Nelson MP, Newhouse A et al (2018) Editing nature: local roots of global governance. Science 362(6414):527–529

Kuzma J (2017) Risk, environmental governance, and emerging biotechnology. In: Durant R, Fiorino DJ, O'Leary R (eds) Environmental governance reconsidered: challenges, choices, and opportunities, 2nd edn. MIT Press, Cambridge, MA

Kuzma J (2019) Procedurally robust risk assessment framework for novel genetically engineered organisms and gene drives. Regulation & Governance. https://doi-org.prox.lib.ncsu.edu/10.1111/rego.12245

Kuzma J, Rawls L (2016) Engineering the wild: gene drives and intergenerational equity. Jurimetrics J Law Sci Technol 56(3):279–296

Kuzma J, Gould F, Brown Z, Collins JP, Delborne J, Frow E, Esvelt K, Guston D, Leitschuh C, Oye K, Stauffer S (2018) A roadmap for gene drives: using institutional analysis and development to frame research needs and governance in a systems context. J Responsible Innov 5(S1):13–39

Kyrou K, Hammond AM, Galizi R, Kranjc N, Burt A, Beaghton AK et al (2018) A CRISPR–Cas9 gene drive targeting doublesex causes complete population suppression in caged *Anopheles gambiae* mosquitoes. Nat Biotechnol 36(11):1062–1066

Lack JB, Greene DU, Conroy CJ, Hamilton MJ, Braun JK, Mares MA, Van Den Bussche RA (2012) Invasion facilitates hybridization with introgression in the *Rattus rattus* species complex. Mol Ecol 21(14):3545–3561

Leitschuh CM, Kanavy D, Backus GA, Valdez RA, Serr M, Pitts EA, Godwin J (2018) Developing gene drive technologies to eradicate invasive rodents from islands. J Responsible Innov 5:S121–S138

Lin AC (2013) Does geoengineering present a moral hazard. Ecol LQ 40:673–712

Lull RB, Scheufele DA (2017) Understanding and overcoming fear of the unnatural in discussion of GMOs. In: Jamieson KH, Kahan D, Dietram AS (eds) The Oxford handbook of the science of science communication. Oxford University Press, New York, pp 409–420

Maguire LA (2004) What can decision analysis do for invasive species management. Risk Anal 24:859–868

Meghani Z, Kuzma J (2011) The "revolving door" between regulatory agencies and industry: a problem that requires reconceptualizing objectivity. J Agric Environ Ethics 24(6):575–599

Meghani Z, Kuzma J (2018) Regulating animals with gene drive systems: lessons from the regulatory assessment of a genetically engineered mosquito. J Responsible Innov 5:S203–S222

Min J, Smidler AL, Najjar D, Esvelt KM (2018) Harnessing gene drive. J Responsible Innov 5:S40–S65

Mitchell PD, Brown Z, McRoberts N (2018) Economic issues to consider for gene drives. J Responsible Innov 5:S180–S202

Murphy B, Jansen C, Murray J, De Barro P (2010) Risk analysis on the Australian release of *Aedes aegypti* (L.) (Diptera: Culicidae) containing Wolbachia. Available at http://www.eliminatedengue.com/library/publication/document/csiro_report_australia_2010.pdf

Murray JV, Jansen CC, De Barro P (2016) Risk associated with the release of Wolbachia-infected *Aedes aegypti* mosquitoes into the environment in an effort to control dengue. Front Public Health 4:43

NASEM (2016) Gene drives on the horizon: advancing science, navigating uncertainty, and aligning research with public values. National Academies Press, Washington, DC

Neve P (2018) Gene drive systems: do they have a place in agricultural weed management? Pest Manag Sci 74(12):2671–2679

Nimmo DD, Beech C (2016) Genetically engineered mosquitoes in the US. Outlooks on Pest Management 26(5):207–210

Noble C, Olejarz J, Esvelt KM et al (2017) Evolutionary dynamics of CRISPR gene drives. Sci Adv 3(4):e1601964

Noble C, Adlam B, Church GM et al (2018) Current CRISPR gene drive systems are likely to be highly invasive in wild populations. elife 7:e33423

Noble C, Min J, Olejarz J, Buchthal J et al (2019) Daisy-chain gene drives for the alteration of local populations. Proc Natl Acad Sci 116(17):8275–8282

North AR, Burt A, Godfray HC (2019) Modelling the potential of genetic control of malaria mosquitoes at national scale. BMC Biol 17(1):26

NRC (1996) Understanding risk. National Academies Press, Washington, DC

Ostrom E (2009) Institutional rational choice: an assessment of the institutional analysis and development framework. In: Sabatier P (ed) Theories of the policy process, 2nd edn. Westview Press, Boulder, pp 21–64

Ostrom E (2011) Background on the institutional analysis and development framework. Policy Stud J 39(1):7–27

Oye KA, Esvelt KM (2014) Gene drives raise dual-use concerns—response. Science 345(6200):1010–1011

Oye KA, Esvelt K, Appleton E et al (2014) Regulating gene drives. Science 345:626–628

Roberts A, De Andrade PP, Okumu F, Quemada H, Savadogo M, Singh JA, James S (2017) Results from the workshop "problem formulation for the use of gene drive in mosquitoes". Am J Trop Med Hyg 96(3):530–533

Rode NO, Estoup A, Bourguet D et al (2019) Population management using gene drive: molecular design, models of spread dynamics and assessment of ecological risks. Conserv Genet 20(4):671–690

Scharenberg AM, Stoddard BL, Monnat RJ, Nolan A (2016) Retargeting: an unrecognized consideration in endonuclease-based gene drive biology. bioRxiv. https://doi.org/10.1101/089946

Scott MJ, Gould F, Lorenzen M, Grubbs N, Edwards O, O'Broutcha D (2018) Agricultural production: assessment of the potential use of Cas9-mediated gene drive systems for agricultural pest control. J Responsible Innov 5:S98–S120

Sinkins SP, Gould F (2006) Gene drive systems for insect disease vectors. Nat Rev Gen Genet 7:427–435

Target Malaria (2017) Who we are: ethics Advisory Committee and Stakeholder Engagement Team. Available at http://targetmalaria.org/who-we-are/

Thompson PB (2007) Food biotechnology in ethical perspective. Springer, Dordrecht. https://doi.org/10.1007/1-4020-5791-1

Thompson PB (2018) The roles of ethics in gene drive research and governance. J Responsible Innov 5:S159–S179

Unckless RL, Messer PW, Connallon T, Clark AG (2015) Modeling the manipulation of natural populations by the mutagenic chain reaction. Genetics 201:425–431

Unckless RL, Clark AG, Messer PW (2017) Evolution of resistance against CRISPR/Cas9 gene drive. Genetics 205:827–841

Vella MR, Gunning CE, Lloyd AL, Gould F (2017) Evaluating strategies for reversing CRISPR-Cas9 gene drives. Sci Rep 7:11038

Webber BL, Raghu S, Edwards OR (2015) Opinion: is CRISPR-based gene drive a biocontrol silver bullet or global conservation threat? Proc Natl Acad Sci 112(34):10565–10567

Zapiola ML, Mallory-Smith CA (2012) Crossing the divide: gene flow produces intergeneric hybrid in feral transgenic creeping bentgrass population. Mol Ecol 21(19):4672–4680

Zuber ST, Muller K, Laushman RH, Roles AJ (2012) Hybridization between an invasive and a native species of the crayfish genus *Orconectes* in North-Central Ohio. J Crustac Biol 32(6):962–971

# Engineered Gene Drives and their Value in the Control of Vector-Borne Diseases, Weeds, Pests, and Invasive Species

**Kathleen Hefferon and Ronald Herring**

**Abstract** Genetic engineering has created potential for moving medical and agricultural research and application frontiers forward in unprecedented ways. Despite its accepted use as a powerful tool in medical research, genetic modification and genome editing technologies remain controversial in large-scale ecological intervention and open-field agriculture. Gene drive is a technology based on genome editing that enables a trait to be pushed through a given population at a greater than expected rate. While gene drives show enormous promise as a way to address a number of challenges, such as the reduction of populations of disease-spreading pests and invasive species, they also incite great social unease because of unknown risks. The following chapter describes the mechanics of gene drives and how they could be utilized to control vector-borne diseases, weeds, and crop pests and even protect populations of endangered species. Limitations and risks associated with gene drive technologies, such as containment strategies and potential resistance, are discussed. Finally, the social impacts of gene drives with respect to international governance and public acceptance are considered.

**Keywords** Gene drive · Genome editing · Informed consent · Unwanted spread · Containment strategies · Regulation · Biosafety · Ethical considerations · Hypothetical risk · Uncertainty

## Introduction

Biotechnology is among the most scientifically promising, yet socially controversial, issues today (Lewontin 2001; Doudna and Steinberg 2017). While genetically engineered crops and livestock are steadily entering the marketplace, adoption is limited due to biosafety and risk grounds throughout much of the world. The

K. Hefferon (✉)
Department of Microbiology, Cornell University, Ithaca, NY, USA
e-mail: kathleen.hefferon@alumni.utoronto.ca

R. Herring
Department of Government, Cornell University, Ithaca, NY, USA

 
A. Chaurasia et al. (eds.), *GMOs*, Topics in Biodiversity and Conservation 19,
https://doi.org/10.1007/978-3-030-53183-6_18

introduction of genome editing has rekindled this debate by stimulating divergent world views concerning the place of humans in the natural world (Čartolovni 2017). More recently, the concept of generating synthetic gene drives to reduce or possibly even eliminate some human disease vectors and crop pests is beginning to capture the world's imagination (Wimmer 2013). Natural gene drives are selfish genetic elements that use a variety of mechanisms to ensure that they are transmitted to subsequent generations at greater than expected rates (Cutter and Jovelin 2015). Synthetic gene drives based on the CRISPR/Cas9 genome editing system could potentially alter the genetic characteristics of natural populations of organisms in ways relevant to the goals of public health, conservation, and agriculture (Zentner and Wade 2017).

Social concerns of ecology and environmental safety have been heightened because the result of such gene drives may impact wild ecosystems. (Courtier-Orgogozo et al. 2017).

It has now become possible, for example, to envision the release of a gene drive that could vastly reduce the size of mosquito populations that transmit human pathogens. The end result could be virtual elimination of disease burdens (e.g., malaria, dengue) that humans have endured for millennia. While potentially conferring such significant benefits, such alterations raise concerns that species could be pushed to extinction or that gene drive traits could be transferred to nontarget species. The social dimensions of gene drive must be explored, including the dimension of both risks and benefits to humans and, consequently, how gene drives will be governed, particularly across neighboring countries and jurisdictions.

As one recognition of this pressing issue, the US National Academies of Sciences, Engineering, and Medicine (NASEM) established the Committee on Gene Drive Research in Non-Human Organisms: Recommendations for Responsible Conduct. The committee summarized their analysis in the report, "Gene drives on the horizon: Advancing science, navigating uncertainty, and aligning research with public values" (NAS 2016). The report concluded that while attractive and not without great potential, gene drives required further study to ensure their responsible release. It was recognized that gene drive technologies could provide solutions to world problems that are difficult to address, such as vector-borne diseases, increases in pesticide and herbicide resistance of agricultural pests, and infiltration of invasive species on fragile ecosystems. However, a gene drive that is implemented and then runs out of control could eliminate certain species and change the environment as we know it permanently. This chapter concentrates on the use of gene drives to control populations of insects, weeds, and invasive or endangered species.

## *Principles of Gene Drive*

Gene drives are systems of inheritance that are biased, so that the likelihood of a sequence of DNA being passed between generations and throughout an entire population is greatly increased (Sinkins 2011). The pattern of inheritance in the presence of a gene drive becomes altered so that most or all offspring from the cross between a gene drive and a wild-type individual will inherit a particular genetic trait (Esvelt et al. 2014). The development of CRISPR/Cas9 and other synthetic genome editing tools has now greatly facilitated this process.

With the newfound ease of genome editing, the potential for gene drive technology has taken on new implications. Gene drives enabled by genome editing offer the potential to stop the spread of mosquito-borne diseases such as malaria, dengue, and Zika. Gene drives could also block the spread of weeds or even bring some species back from the endangered list through removal of invasive predators/competitors. In this case, mosquitoes containing chromosomal translocations could be mated with wild-type mosquitoes and produce heterogeneous progeny that are sterile. As a result, release of mosquitoes harboring this male-producing factor could impact the sex ratio of the mosquito population so that females reached a number below the level required for efficient disease transmission (Wieczorek 2016). This initial approach encouraged work on the use of CRISPR-Cas9 as a new tool to reduce mosquito populations (Hammond et al. 2016). For example, genes that confer a recessive female sterility phenotype can be disrupted. CRISPR-Cas9 gene drive constructs designed to target and edit each gene involved in reproduction can be inserted into the female sterility gene locus, resulting in a massive increase of sterile females. Population modelling has demonstrated that this type of gene drive could be used to effectively target female reproduction in a mosquito population. The technology could also be extended to edit mosquitoes so that they are no longer able to transmit infectious diseases (Singer and Frischknecht 2016).

## *Self-Limiting Gene Drives*

Along with gene drive technology comes significant perceived risks such as ecological damage or other unintended consequences. As a result, efforts have been made to develop gene drive systems that can spread through or be recalled from a given population (Marshall and Akbari 2018). One potential option is the use of self-limiting drives such as the Daisy drive, which would involve the development of gene drive systems that are temporally and spatially limited so that uncontrolled consequences can be easily curbed (Dhole et al. 2018). In a Daisy drive system, components of the CRISPR machinery are scattered throughout the genome in fragments so that none could "drive" on their own (Fig. 1). In spite of their spatial separation in the genome, they are functionally arranged in a Daisy chain fashion so that one element is needed to get the other one started. Local populations under drive technology can be better controlled by providing a means to limit spread based on the timing of release. Alternatively, in underdominant gene drive systems, where heterozygotes for the drive allele have a lower fitness than their homozygotes and wild-type counterparts, the rate of spread is dependent on the invasion frequency threshold, therefore requiring large release of the transgenic organism. On the other hand, this provides underdominant gene drive systems the property of improved

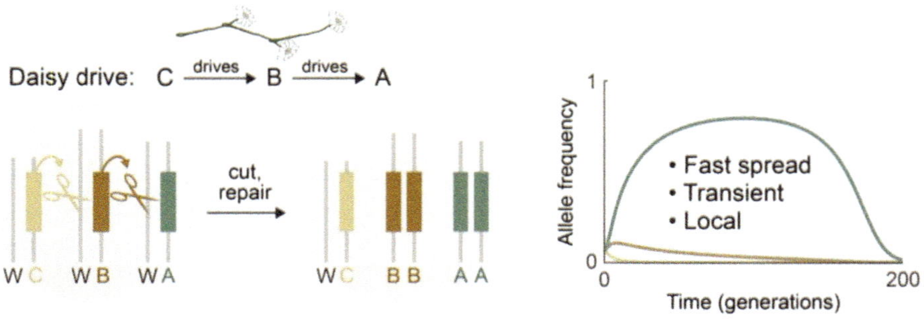


**Fig. 1** Mechanism and population-level effect of endonuclease gene drives. (Figure derived from DiCarlo et al. Nature Biotechnology volume 33, pages 1250–1255, 2015)

local confinement to a given region (Champer et al. 2018). Since the changes made do not render the organism more fit in its environment, the first gene drive unit will eventually disappear by natural selection since it does not promote its own transmission. This will create a chain reaction, and the other units will also disappear in turn. It is expected that these sorts of gene drives can be used to serve functions for local populations and either be transient or made stable by incorporating a more permanent gene drive.

One possible challenge to Daisy drives is the fact that large differences can exist with respect to the minimum release size that may be required to drive a given trait into a particular population. The release size required to make an impact in certain instances will be much greater than in others. This may result in a failure to retain Daisy drives in a geographically localized state, particularly if the cost of fitness by incorporating the gene drive trait is low (Marshall and Akbari 2018). On the other hand, drive mechanisms that require the release of individuals at higher frequencies have the potential to be highly localized and also reversible, making them attractive systems for mitigating vector-borne diseases and other pest problems (Leftwich et al. 2018).

## General Limitations of Gene Drive Technologies

There are several limitations to the use of gene drives. For example, since gene drives require multiple generations to induce change throughout a population, generation time and ability to mate with its wild-type relatives over multiple generations become integral to success of this strategy. Only organisms with a relatively short generation time and the ability to sexually reproduce, such as mosquitoes, could be managed using a gene drive strategy. Organisms which reproduce asexually, either through clonal division (banana) or through the ability to self-fertilize (dandelion), would exhibit different dynamics with a gene drive. Similarly, long-living organisms such as many tree species would not be suitable for gene drive technology. The evolution of resistance is another limitation of gene drives and will be discussed later in the chapter. A final limitation is that many of the traits spread through gene drives are in fact harmful or present reduced fitness to the organism. Eventually another gene drive will be needed to ensure that the intended driven trait is not replaced, depending on the circumstances. This suggests that gene drives may not be permanent solutions; rather they may only be transient fixtures that must be elicited periodically (Godfray et al. 2017)

In spite of these limitations, gene drives offer potential solutions to some of humanity's most pressing problems. The following section details several potential applications.

## The Potential of Gene Drives to Reduce Vector-Borne Diseases (VBD)

Vector-borne diseases, primarily spread by mosquitoes and ticks, account for 17% of all known infectious diseases (de la Fuente et al. 2017). Most vector-borne diseases are found in tropical and subtropical regions, where populations tend to be poorer and have fewer resources for surveillance and response. More recently however, infectious diseases such as Lyme disease, Zika and dengue, previously found exclusively in warmer regions, have been making their way to more temperate regions of the world (Tjaden et al. 2018). This spread stems from a complex mixture of climate change, globalization, and international trade, thus creating the emergence of new threats.

More than half of the world's populace resides in urban areas; this proportion continues to climb (United Nations 2018). This population shift may well result in the growth of concentrated populations of the poor, typically lacking in safe drinking water, waste management, and basic healthcare services. Disease burden can then be disproportionately high in poor communities. Moreover, malnourished populations typically lack essential vitamins and minerals and thus lack robust immune systems. The death rate due to vector-borne diseases is already high, and the World Health Organization (WHO) has issued a call for vector control strategies that are both unified and easy to mobilize throughout urban and rural communities alike. These strategies for vector control include a strengthened surveillance system and rapid in-field diagnostic tests so that infectious diseases such as Zika, dengue, and Chikungunya can be quickly identified and epidemics prevented. An enhanced, consistent, and unified surveillance and response infrastructure will require education, extensive communication, urban planning, availability of health services, government commitment, and strengthened policies between countries at risk of impact by vector-borne disease (Schorderet-Weber et al. 2017).

Current strategies to block disease transmission by mosquitoes and ticks include insecticides, bed nets, protective clothing, improved sanitation, and water management. While some of these approaches have proven to be effective, they are clearly not sufficient. Insecticides can be high in cost, and insects eventually may develop resistance to them. Bed nets and various forms of protective clothing are not distributed thoroughly enough to fully block mosquito and tick bites (Rakotoson et al. 2017). Urban planning to improve sanitation and still water containment strategies require social infrastructure that may not be readily available. While vaccines do exist for some infectious diseases, many do not and thus cannot detour the spread of infection. Projects such as the world Mosquito Program incorporate the use of endosymbionts such as *Wolbachia*, a bacterium that can infect mosquitoes and disrupt disease transmission (Curtis and Sinkins 1998). Ticks may also be infected with *Wolbachia* species, which can decrease their motility and thus reduce the occurrence of disease transmission (Indriani et al. 2018; Calvitti et al. 2010). Improved knowledge of microbial communities in mosquito and tick populations could potentially be controlled using endosymbionts based natural gene drives (Bull 2015).

The general concept that genetics can be used to control mosquito populations is not new; however, the use of gene drive technology has offered the possibility to generate a way to limit the spread of diseases such as Zika, dengue, and malaria through the control of insect vectors (Macias et al. 2017). In Brazil, the biotech company Oxitec is using a small laboratory, mobile production approach to generate genetically engineered mosquitoes that carry a gene that causes their offspring to die before reaching maturity. By releasing these mosquitoes into the area surrounding towns and small cities, the company hopes to reduce vector-borne diseases prevalent in the area, such as Chikungunya and dengue (Paes de Andrade et al. 2016). In this case, Oxitec generates male mosquitoes with a self-limiting gene that kills them before they are mature enough to reproduce, as well as a reporter gene whose promoter is regulated by sensitivity to tetracycline. By including the antibiotic tetracycline in the insects' water, the self-limiting gene becomes inactivated so long as they remain under caged conditions. The mosquitoes are able to reach maturity and the males are released into the environment to mate with wild females. The offspring lack access to tetracycline outside of the lab, and as a result, the self-limiting gene becomes activated, causing an early death. Using this approach, mosquito populations plummeted by over 60% in 2016 when Oxitec males were released in the Cayman Islands. It is important to note that not all of the mosquitoes released were sterile (Evans et al. 2019).

The Bill & Melinda Gates Foundation has now joined forces with a consortium known as Target Malaria to focus on removing malaria-carrying mosquitoes in West Africa. Target Malaria incorporates the CRISPR-Cas9 self-sustaining gene drive strategy into male mosquitoes of specific species that carry malaria, which, upon release, will mate with wild females. It is estimated to take multiple generations until all the modified mosquitoes will be eliminated from the population. Although the population of this specific species of mosquito may thus collapse, no impact on other mosquitoes or other insect species within that ecosystem is generally expected, although some evidence exists that this may not always be the case (McFarling 2017; Collins 2018; Fontaine et al. 2015).

Research concerning the gene editing of mosquitoes is accelerating. For example, Li et al. (2018) have successfully used the CRISPR/Cas9 system for highly efficient, site-specific mutagenesis in a diversity of malaria vectors including *Anopheles albimanus*, *A. coluzzii*, and *A. funestus* (Hammond et al. 2016). Similarly, Kyrou et al. (2018) have managed to negatively impact female *Anopheles gambiae* mosquito populations by focusing on alternatively spliced transcripts that are responsible for sex differentiation. Recently a cargo gene comprised of small RNAs which target Zika virus has been generated in *Aedes aegypti* mosquitoes that could significantly reduce Zika infection (Buchman et al. 2018).

## *Using Gene Drive to Combat Lyme Disease*

Lyme disease, transmitted by ticks, is caused by *Borrelia* bacteria and is responsible for symptoms ranging from neurological problems to arthritis. Lyme is found in both the USA and Europe, with hundreds of thousands of people diagnosed every year. Infecting humans, dogs, horses, and deer, the main reservoir of the *Borrelia* species of bacteria are mice. *Borrelia* has a highly complex genome, compared to many other bacteria, and carries multiple plasmids that provide unique pathologies, tropisms, and manifestations of disease (Casjens et al. 2017, 2018). A project that focuses on creating transgenic mice that harbor immunity to the bacteria has gained momentum, and the future possibility of employing a gene drive to carry the trait through wild populations is now under consideration (Hammond et al. 2016). Mice would either express an antibody to render them resistant to Lyme disease or else would be immunized against a protein found in tick saliva, which in turn would protect the mice against *Borrelia* and other forms of disease carried by ticks. This approach differs from a conventional vaccination as the acquired immunity would be passed on from one generation to the next. The modifications can be made using CRISPR-Cas9 genome editing technologies (Enzmann 2018). The plan is to release these edited mice into the wild, beginning with unpopulated islands and then moving on to island communities such as Nantucket and Martha's Island, where mouse populations are contained and Lyme disease is highly prevalent. The short reproductive cycle of mice enables them to be assessed for the presence of infected ticks over a relatively short time period (Bouchard 2017).

Gene drive technology in mammals such as mice, however, is far behind that of mosquitoes. Only last year has a gene drive been partially successfully implemented in mice (Grunwald et al. 2019). Researchers were able to use a CRISPR-based gene drive to change the coat color of mice from black to gray over the course of one generation (Grunwald et al. 2019). Since this initial attempt resulted in only female mice inheriting the gene drive approximately 86% of the time, more effort will be required to improve transmission efficiency before the drive could be successfully applied to a wild population. Improvements in gene drive technologies implemented on rodents such as mice could eventually be applied to reduce or eradicate rodent-borne deadly infectious diseases such as Lassa fever virus and hantavirus.

## *Gene Drive to Reduce Persistence of Weeds and Pests*

Resistance of weeds to herbicides remains a significant agricultural challenge (Godfray et al. 2017). The problem could feasibly be addressed through the use of a synthetic gene drive which could replace resistant alleles with their original, herbicide-sensitive counterparts. Similarly, a gene drive could revert insect pests who have developed resistance to commonly used pesticides such as Bt. While attractive, the technology would work only on organisms which sexually reproduce and would require that fields be kept pesticide/herbicide-free until the drive fixes. This would require cooperation from farmers who own neighboring fields that drives may spread to and thus could be difficult to implement.

Alternatively, a sensitizing gene drive could be utilized so that pest or weed populations were made vulnerable to molecules that did not affect them previously. If, for example, a gene drive provides a novel sensitivity to a chemical compound or small molecule, then an insect pest or weed species could be made vulnerable upon exposure to that small molecule. The concept is attractive as it would enable the presence of a specific species to be under strict control. It would also permit farmers to use chemicals or small molecules which are more benign to human health and the environment.

## *Gene Drives to Control Crop Pathogens*

Gene drive is under consideration as a means to address environmental problems that have not been solved by traditional conservation practices. For example, in Florida, the spread of citrus greening disease, a bacterial disease transmitted by the insect *Diaphorina citri* (psyllid species) that is destroying the citrus industry, has been particularly newsworthy. Currently, citrus growers have resorted to the spraying of antibiotics throughout their orchards to protect against the disease (McKenna 2019). While a GM-resistant citrus alternative exists, the possible use of a gene drive to control insect vectors has been explored. In this case, a self-sustaining gene drive that would spread a strain of the insect that would be incapable of transmitting the disease could replace the existing insect population (Baltzegar et al. 2018). A gene drive solution for the invasive East Asian fruit fly, *Drosophila suzukii*, which damages berries and soft-skinned fruits across the globe has also been under consideration (Li and Scott 2016). Oxitec plans to focus next on diamondback moths, a well-known crop pest and invasive species responsible for approximately $5 billion worth of damages in the USA every year. While gene drive moths are preferable to pesticides for many, others have raised concerns about the ecological impacts of their release, as well as the possibility of acquired resistance (Scharping 2017).

### *Gene Drives to Protect Invasive/Endangered Species Populations*

Invasive animal or plant species, such as cane toads, brown rats, and purple loosestrife, have caused great ecological damage (Webber et al. 2015). The top ten invasive species found in the United States are responsible for approximately $42 billion in damage every year. Alternatively, the plight of endangered species, such as amphibians sensitive to fungal diseases, is also of great concern. A synthetic gene drive could be a valuable tool to address both population types. A self-limiting drive could be implemented to target invasive organisms who sexually reproduce, such as the enzyme responsible for the production of the toxin in the saliva of the cane toad, to reduce their populations and restore natural ecosystems (Webber et al. 2015). Similarly, a gene drive, aimed at population modification, that is tailored to an endangered frog and salamander species could offer resistance to fungal pathogens and enable them to thrive once more in their natural environments. Kohl et al. (2019) conducted a survey to determine moral acceptability of gene drive technologies for the purpose of either eliminating an invasive species or protecting an endangered species. Their results suggested that the general public was more accepting of a gene drive to improve the survival of an endangered species rather than eliminate environmentally problematic wildlife populations.

## Limitations and Risks Associated with Gene Drive Organisms

### *Concerns Regarding Appropriate Containment Strategies*

Gene drives potentially promise eradication of some of humanities' worst pests. However, they also elicit concerns ranging from practical difficulties with regard to conducting field trials, unforeseen ecological changes or other complexities, and their long-term efficacy in the field (Moro et al. 2018), to the development of target site resistant populations that could prove to be unstoppable (Callaway 2017). Besides the general biology of gene drive, other challenges include governance, development, and adherence to legal structures and public acceptance (Nash et al. 2019). How the potential for ecological impacts can be addressed is discussed in the following section.

Containment measures can be administered and implemented in several ways. For example, Adelman et al. (2017) described the need for standard operating procedures (SOPs) concerning gene drive mosquitoes. While a series of SOPs will be essential for any release of gene drive organisms, implementation across multiple countries may be essential for success. Containment strategies to be set in place include an examination of land use, facilities constructed to house the target species (including labs and cages), biosafety protocols, removal of wastes, and shipping and transport precautions. International laws will be necessary to ensure that proper regulation. Staff can conduct initial trials regarding containment management by starting with non-transgenic mosquitoes, followed by transgenic varieties once competence has been demonstrated (Quinlan et al. 2018). While physical containment of gene drive organisms may be achievable in this fashion, additional safety procedures are required to ensure that gene drive strains can be identified and are trackable (Benedict et al. 2018). This will require creation of biosafety committees with methodologies to track and control gene drive organisms that are released.

Concerns about uncontrolled spread of gene drives can be addressed in other ways. For example, self-limiting drives such as the Daisy drive could generate short-lived gene drives within a given population. Alternatively, since a successful gene drive requires a particular threshold number of genome edited organisms to be released into the wild, having in reserve a large population of wild-type organisms available to be released if necessary could overcome this threshold and mitigate the impact of a gene drive within a given population.

## *Possible Resistance Developed Toward Gene Drives*

Another means by which a gene drive could overcome containment strategies and undergo uncontrolled spread is through the natural genetic variation within a given population (Zentner and Wade 2017). Sequence polymorphisms found within a given population can prevent the endonuclease from cleaving a target gene. Eventually, these naturally resistant variants will increase in abundance to the extent that the gene drive is eliminated. For example, if enough variation existed within a malaria-carrying mosquito species, a gene drive may not completely exterminate all members of a local population. Genetic variants that possessed resistance could indeed escape the effects of a gene drive and reseed a new "resistant" population of mosquitoes. This new malaria-bearing gene drive resistant mosquito population would be more difficult to control than the original population. Hammond et al. (2017) assessed the potential of emergence of resistance to gene drives in a caged mosquito population by running a female infertility-based gene drive for 25 generations. The authors observed a gradual decrease in frequency of gene drive, followed by a slow spread of mutations within the target gene that rendered it resistant to cleavage by the endonuclease. These mutations were endonuclease induced and increased at rates which were consistent with positive selection.

A strategy proposed to address this development would be the creation of a gene drive targeting multiple sites within the target gene, so that it would be statistically next to impossible to find a natural variant that harbored mutations at all sites (Godfray et al. 2017). This strategy may, however, be difficult to implement in a very large population, as it would also require a large number of guide RNAs to ensure that resistance was not selected for. Oberhofer et al. (2018) created multiple cleavage sites within a specific gene to prevent spurious mutations that conferred resistance from occurring, although homing rates were modest. Another strategy to avoid resistance would be to create several temporally successive gene drives, each targeting a select array of multiple sites. Incorporation of both temporal drives and multiple target sites might prevent resistance from emerging.

### *Other Containment Strategies*

One way to mitigate unknown ecological consequences of a gene drive could be through the simultaneous development of a reversal drive that restores the original phenotype and is ready to install at a given moment. This reversal drive could be used to overwrite any nucleotide changes that were spread by the first gene drive (Khamis et al. 2018). Alternatively, the production of sensitizing drives that are able to render the target organism vulnerable to a particular chemical could also be implemented to shut off the effect of the gene drive (Godfray et al. 2017). The presence of the chemical would have either a toxic or an inhibitory effect to the gene drive, resulting in refined control of the gene drive and greatly reduced ecological risk, although not restoring the original phenotype. Finally, the use of computational modeling to predict the outcome of a gene drive on a population of given size is a great way to identify their potential ecological impact (Edgington and Alphey 2018). This can be performed in conjunction with field trials using genetically engineered organisms that lack the gene drive function necessary to spread the trait in question and provide insight regarding properties including dispersal patterns of released insects, mating success, gene flow, and persistence. Such a plan exists for Target Malaria and is being implemented now. Currently, mathematical modeling of populations is used to determine the possibility of natural resistance taking place and taking over (Baltzegar et al. 2018). Another concern that may not be adequately captured by computational modeling is unknown behavior of the target organism in terms of mating and movement over their lifespans. For example, some mosquito species mate in swarms; others do not. Furthermore, mosquitoes raised in the lab or in cages may not serve as accurate models for their wild counterparts (Olena 2017). This can be investigated using RIDL field trials before implementing a gene drive approach (e.g., the Target Malaria phased project).

## Social Impacts: The Risk/Benefit/Uncertainty Calculus

Gene drive technology will enter an entirely new space of social acceptance and regulatory treatment: genuinely unknown terrain in which social impacts will prove decisive for progress. Might insights gained from regulation and acceptance of recombinant DNA (rDNA) pharmaceuticals and agricultural plants provide clues for coming social dynamics?

It is of great importance that a continuous assessment of how the general public perceives gene drive technologies is performed well before any gene drives are released. Jones et al. (2019) analyzed the attitudes of the US public and identified strong support for the use of gene drives to control invasive species when no other options are available. Moreover, the authors found that people who do not support GMOs within their food supply nonetheless supported the use of engineered gene drives to control invasive agricultural pests. This interesting finding suggests that concerns regarding GMO food are not necessarily transferred to other technologies that require genetic engineering. Large portions of the population remain undecided regarding their judgment of gene drives. Public attitudes are known to undergo rapid changes, as has been shown with public perception of GM crops in Europe, for example. It is interesting that the authors also found an unclear response to natural gene drives such as *Wolbachia*-based control of insect pests. While attractive to some for their "natural" component, such a gene drive would be next to impossible to control in terms of spread, in comparison to some of the synthetic gene drive implementation strategies.

A major conclusion of literature on existing rDNA organisms and products is that aggregate cost-benefit analysis under conditions of low information has driven social acceptability and regulatory response to the genomics revolution (Herring and Paarlberg 2016). Where benefits are high and demonstrable, and alternatives inferior, risks are largely accepted: rDNA pharmaceuticals illustrate this logic. These were decisive questions differentiating biotech medicine from biotech agriculture. Regulation and acceptance of rDNA pharmaceutical products was not problematic: subjected to existing regulation, such products were accepted without any exceptional stigma. There was demonstrable utility, and safety was vetted by trusted authorities – personal physicians and official science, as in the Food and Drug Administration in the USA. Risks were explicitly detailed and known to consumers; these risks were accepted for demonstrable benefits, absence of alternatives, and the extreme risks of doing nothing. Agricultural rDNA products – in marked contrast – were politically encoded as "GMOs" and restricted or blocked in much of the world. For rDNA agricultural plants there is to date no documented incremental hazard in comparison with other means of inducing new traits in plants and yet "risk" is the dominant theme in restricting spread of the technology globally (Lewontin 2001). On the benefit side of the equation, consumers of GMOs typically derived little or no benefit but are confronted with a powerful risk narrative built around the unnatural nature of "Frankenfoods."

In these existing implementations of biotechnology, the difficulty has been in coming to an appropriate *aggregate* risk/benefit analysis for society as a whole: whose risk, whose benefit? In pharmaceuticals and foods, risks are divisible and individual choices are feasible. Environmental risk management is fundamentally different: individual risk perceptions are subordinated to ecological scale dynamics, over which unanimous consent is unlikely.

For new risk/utility regulatory regimes governing gene drives to become accepted, and effective, social consensus around aggregate risk and benefit must be achieved. The history of biotechnology to date raises large cautions: such agreement assumes polities that do not exist in any meaningful sense and hard-to-conjure enforcement tools.

In a technical sense, risk is hazard multiplied by exposure or probability of hazard. Determining hazard definitively is problematic: how many "unknown unknowns" escape conventional science, how many black swans? Moreover, at the frontiers, there is no way to predict unknown *future* hazards; "risk" in this sphere can be socially constructed only in hypothetical or "anticipatory" terms, generating a distinctive politics of precaution (Gupta 2011). Even if there is no new hazard at all, proving the absence of risk is impossible for science (Giddens 1999). In these circumstances, different interpretations of risk will proliferate and parties will engage in strategic behavior to secure support for the "framing" of regulatory positions they prefer (Benford and Snow 2000). In the current environment of populist rejection of authoritative knowledge in general, and science in particular, the probability of successful risk politics increases; the plausibility of deploying gene drives widely declines.

Despite the power of risk politics evident in the global rift over GMOs, the benefit side of the equation of technology at the frontier is qualitatively greater than in previous genetic engineering episodes. Risks of uncontrollable diseases of humans, animals, and crops constitute a crisis that may produce social and political openings. Cancer patients accept the risks of powerful chemicals on the grounds that alternatives are worse. Thus gene editing technologies may resemble more the path of rDNA pharmaceutical, risk but high benefits, more than rDNA crops, perceived risk but few consumer benefits. Genome editing and gene drives promise aggregate benefits unimaginable a few decades ago, in human health, agriculture, and environmental integrity. With the dreadful risks of climate change before us, every possible tool in the toolkit for adaptation assumes even greater importance. Yet the potential for unknown hazards increases with greater potential for utility.

The second lesson from existing biotechnology in society is that rigorous monitoring and regulation are assumed for legitimacy and effectiveness, but both remain egregiously elusive. Failure of regulation in turn increases the risk side of the equation. Gene drives are especially difficult in this regard because ecological systems do not stop at artificial boundaries on a map, either locally or globally. A social compact on a scale adequate to satisfy precautionary logic is difficult to conjure; there are essentially no exemplars. More important, agreement in paper treaties has proved as often as not toothless. Scientists and officials consider precise guidelines for gene editing deployment whereas the technology itself almost uniquely permits operation outside regulation of any kind. What means of surveillance and enforcement powers can even be conjured on any meaningful scale? Climate science offers cautions. Despite wide global scientific consensus, national and local interests outweigh species interest politically and the absence of any real power to implement agreements is obvious.

We can illustrate these issues with regard to gene drives with one example from the United States, where biotechnology is widely accepted. Gene-drive mosquito technology offers great potential benefits, but democratic politics does not always accord with independent scientific assessment (Meghani and Kuzma 2018). Some groups will inevitably contest the transparency and thoroughness of the science; some organizations exist to do specifically this. Many local people in the Florida Keys remain opposed to the release of Oxitec's GM mosquitoes within their neighborhood despite extensive analysis by the USDA, FDA, and the CDC. Classified as a pesticide with potential environmental impact, the EPA is exerting regulatory oversight of the organisms. The EPA has also approved the release of *Wolbachia*-infected mosquitoes in both California and the Florida Keys.

The benefits and risks in altered mosquitoes are especially complex and hence uncertain for reasons elaborated in previous sections: likelihood for success in achieving intended outcomes, possible impact on nontarget species, risk of spreading uncontrollably, and unanticipated societal effects. Gene drives in mosquitoes then illustrate the range of social issues of greatest importance: tremendous potential benefits achievable in no other way and uncertainty that can be coded as unacceptable risk.

First of these hypothetical risks are unintended consequences for health and the environment. What is the target species and what are its habits, location, and distribution? How much certainty is enough certainty when putatively sufficient studies have been completed? What level of governance is appropriate, which is a potential chokepoint for preventing deployment: local, national, and international polities? Practical questions of social importance follow: What are best practices for delivering gene drive technology, as well as evaluations and assessment before and after the release of gene drives? How much are lab research and controlled field trial work necessary to authorize release of gene drives into the environment? How and for how long would necessary post-release surveillance studies be conducted? How would persistence of gene drive organisms be determined, to ensure unwanted spread is not occurring? For how long? How intrusive/transparent will the level of engagement be between researchers and the public? Does a community have the capacity to oversee the safe and controlled release of gene drive organisms? What containment strategies will be implemented?

Assuming answers to these questions are adequate for local and national acceptance, how will a neighboring country without adequate scientific capacity or policies be involved in decisions? Perhaps a resolution could be found within the Florida Keys, but globally scientific and regulatory capacity vary greatly. How would international spread of altered organisms be predicted adequately, halted, or mitigated? As the scope of intervention extends, how will the weight of social forces opposed to hacking evolution be addressed? Are there theological complications over space in the arrogance of intervening in "God's plan"? Will transnational advocacy networks that have powerfully mobilized against GMOs stimulate sufficient risk politics to blunt scientific consensus should it surface? What political and financial forces will mobilize on behalf of new opportunities to relieve the burden of disease in some countries, increase agricultural productivity, or sustain endangered species? What is the scope of informed consent when consequences are species-wide, not merely in one's own backyard? Who has veto power?

Unless peoples' confidence in answers to these questions can be incorporated into final decisions, regulatory uncertainty will impede investment of time, energy, and money necessary for development, blunting the effectiveness of the technology no matter how potentially beneficial. What we have learned so far is that the greatest risk of all may well be to block new technologies on the basis of ideological fundamentalism rather than rational comparison of probable risks and benefits in democratic and scientifically legitimate ways.

## Conclusions

The perception of gene drive technology presents a controversy similar to previous experiences with genetically modified organisms, yet unique in important ways. Multiple stakeholders, both pro- and anti-GMO, have created uncertainties that depress investment and complicate world trade. Specific influencers that may swing in the favor of gene drive technologies could include an increased death toll due to the rise of vector-borne diseases such as malaria in the advent of climate change, the intensity of pest pressure in a farmers' field, or the potential loss of a now endangered species to extinction in the wild. Deterrents to offset the use of gene drives could comprise of unexpected events, such as loss of control, resistance of the target species, malfeasance, or changes to an ecosystem that are detrimental to nontarget species. The costs and benefits of gene drive are distributed unevenly and at multiple levels.

Geography at all scales weighs heavily in this future. Mosquitoes do not recognize borders. Farmers who did not actively agree to the adoption of a gene drive technology to remove a crop pest may reject adamantly the presence of gene drive pest residues in their crops, even if they benefit from pest reduction generally. It is unclear how organic farmers will perceive GM insects or how they will be classified in organic certification programs. These decisions would then significantly impact international food trade between countries harboring markets that are gene drive-friendly and countries that are not. On the other hand, the removal of invasive agricultural pests will offer relief to some countries which have in the past failed phytosanitary (SPS) regulations due to the presence of an invasive insect species. Gene drive to greatly reduce or even eliminate such invasive species in produce could remove quarantines and reinstate access of these countries to the global market.

This past November, the United Nations Convention on Biological Diversity (CBD) in Sharm El-Sheikh, Egypt, rejected a proposal to temporarily ban the release of organisms carrying gene drives. While not a formal moratorium, the requirements for release of a gene drive remain vague. Part of the explanation for this result is failure to provide for community engagement, with particular emphasis on underrepresented communities that may be most affected by this technology. Community engagement is the strategy taken by the organization Target Malaria, which hopes to test gene drive mosquitoes in Africa by 2024. Although the scientific potential of gene drives as a beneficial technology continues to proceed, its progress from a social perspective will remain hindered due to regulatory uncertainty and deep-seated reservations.

## References

Adelman ZN, Pledger D, Myles KM (2017) Developing standard operating procedures for gene drive research in disease vector mosquitoes. Pathog Glob Heal 111(8):436–447

Baltzegar J, Barnes JC, Elsensohn JE, Gutzmann N, Jones MS, King S, Sudweeks J (2018) Anticipating complexity in the deployment of gene drive insects in agriculture. J Respons Innov 5(sup1):S81–S97. https://doi.org/10.1080/23299460.2017.1407910

Benedict MQ, Burt A, Capurro ML, De Barro P, Handler AM, Hayes KR, Marshall JM, Tabachnick WJ, Adelman ZN (2018) Recommendations for laboratory containment and management of gene drive systems in arthropods. Vector Borne Zoonotic Dis 18(1):2–13. https://doi.org/10.1089/vbz.2017.2121

Benford RD, Snow DA (2000) Framing processes and social movements: an overview and assessment. Annu Rev Sociol 26:611–639

Bouchard S (2017) Gene 'editing' on mice tested in war on ticks. News of Maine's Coast and Islands. http://www.islandinstitute.org/working-waterfront/%E2%80%8Bgene-%E2%80%98editing%E2%80%99-mice-tested-war-ticks

Buchman A, Marchall JM, Ostrovski D, Yang T, Akbari OS (2018) Synthetically engineered *Medea* gene drive system in the worldwide crop pest *Drosophila suzukii*. PNAS 115(18):4725–4730

Bull JJ (2015) Evolutionary decay and the prospects for long-term disease intervention using engineered insect vectors. Evol Med Public Heal 2015(1):152–166. https://doi.org/10.1093/emph/eov013

Callaway E (2017) US defence agencies grapple with gene drives. Nature 547(7664):388–389. https://doi.org/10.1038/nature.2017.22345

Calvitti M, Moretti R, Lampazzi E, Bellini R, Dobson SL (2010) Characterization of a new Aedes albopictus (Diptera: Culicidae)-Wolbachia pipientis (Rickettsiales: Rickettsiaceae) symbiotic association generated by artificial transfer of the wPip strain from Culex pipiens (Diptera: Culicidae). J Med Entomol 47(2):179–187

Čartolovni A (2017) Teilhard de Chardin's oeuvre within an ongoing discussion of a gene drive release for public health reasons. Life Sci Soc Policy 13:18. https://doi.org/10.1186/s40504-017-0064-8

Casjens SR, Gilcrease EB, Vujadinovic M, Mongodin EF, Luft BJ, Schutzer SE, Fraser CM, Qiu WG (2017) Plasmid diversity and phylogenetic consistency in the Lyme disease agent *Borrelia burgdorferi*. BMC Genomics 18(1):165. https://doi.org/10.1186/s12864-017-3553-5

Casjens SR, Di L, Akther S, Mongodin EF, Luft BJ, Schutzer SE, Fraser CM, Qiu WG (2018) Primordial origin and diversification of plasmids in Lyme disease agent bacteria. BMC Genomics 19(1):218

Champer J, Zhao J, Champer SE, Liu J, Messer PW (2018) Population dynamics of underdominance gene drive systems in continuous space. bioRxiv. https://doi.org/10.1101/449355

Collins JP (2018) Gene drives in our future: challenges of and opportunities for using a self-sustaining technology in pest and vector management. BMC Proc 19(12(Suppl 8)):9. https://doi.org/10.1186/s12919-018-0110-4

Courtier-Orgogozo V, Morizot B, Boëte C (2017) Agricultural pest control with CRISPR-based gene drive: time for public debate: should we use gene drive for pest control? EMBO Rep 18(6):878–880

Curtis CF, Sinkins SP (1998) Wolbachia as a possible means of driving genes into populations. Parasitology 116(Suppl):S111–S115

Cutter AD, Jovelin R (2015) When natural selection gives gene function the cold shoulder. BioEssays 37(11):1169–1173

de la Fuente J, Antunes S, Bonnet S, Cabezas-Cruz A, Domingos AG, Estrada-Peña A, Johnson N, Kocan KM, Mansfield KL, Nijhof AM, Papa A, Rudenko N, Villar M, Alberdi P, Torina A, Ayllón N, Vancova M, Golovchenko M, Grubhoffer L, Caracappa S, Fooks AR, Gortazar C, Rego RT (2017) Tick-pathogen interactions and vector competence: identification of molecular drivers for tick-borne diseases. Front Cell Infect Microbiol 7:114. https://doi.org/10.3389/fcimb.2017.00114

Dhole S, Vella MR, Lloyd AL, Gould F (2018) Invasion and migration of spatially self-limiting gene drives: a comparative analysis. Evol Appl 11(5):794–808

Doudna JA, Steinberg SH (2017) A crack in creation: gene editing and the unthinkable power to control evolution. FASEB J 32(1):3–4. https://doi.org/10.1096/fj.181202UFM

Edgington MP, Alphey LS (2018) Population dynamics of engineered underdominance and killer-rescue gene drives in the control of disease vectors. PLoS Comput Biol 3;14(3):e1006059

Enzmann B (2018) CRISPR and the community are teaming up to combat lyme disease. https://www.synthego.com/blog/crispr-lyme-disease

Esvelt KM, Smidler AL, Catteruccia F, Church GM (2014) Concerning RNA-guided gene drives for the alteration of wild populations. elife 3:e03401

Evans BR, Kotsakiozi P, Costa-da-Silva AL, Ioshino RS, Garziera L, Pedrosa MC, Powell JR (2019) Transgenic *Aedes aegypti* mosquitoes transfer genes into a natural population. Sci Rep 9(1):1–6

Fontaine MC, Pease JB, Steele A, Waterhouse RM, Neafsey DE, Sharakhov IV, Jiang X, Hall AB, Catteruccia F, Kakani E, Mitchell SN, Wu YC, Smith HA, Love RR, Lawniczak MK, Slotman MA, Emrich SJ, Hahn MW, Besansky NJ (2015) Extensive introgression in malaria vector species complex revealed by phylogenomics. Science 347(6217):1258524. https://doi.org/10.1126/science.1258524

Giddens A (1999) Risk and responsibility. Mod Law Rev 62(1). https://doi.org/10.1111/1468-2230.00188

Godfray HCJ, North A, Burt A (2017) How driving endonuclease genes can be used to combat pests and disease vectors. BMC Biol 15(1):81. https://doi.org/10.1186/s12915-017-0420-4

Grunwald HA, Gantz VM, Poplawski G, Xu XRS, Bier E, Cooper KL (2019) Super-Mendelian inheritance mediated by CRISPR–Cas9 in the female mouse germline. Nature 566:105–109

Gupta A (2011) An evolving science-society contract in India: the search for legitimacy in anticipatory risk governance. Food Policy 36(6):736–741

Hammond A, Galizi R, Kyrou K, Simoni A, Siniscalchi C, Katsanos D et al (2016) CRISPR-Cas9 gene drive system targeting female reproduction in the malaria mosquito vector *Anopheles Gambiae*. Nat Biotechnol 34(1):78–83

Hammond AM, Kyrou K, Bruttini M, North A, Galizi R, Karlsson X, Kranjc N, Carpi FM, D'Aurizio R, Crisanti A, Nolan T (2017) The creation and selection of mutations resistant to a gene drive over multiple generations in the malaria mosquito. PLoS Genet 13(10):e1007039

Herring R, Paarlberg R (2016) The political economy of biotechnology. Ann Rev Resour Econ 8:397–416

Indriani C, Ahmad RA, Wiratama BS, Arguni E, Supriyati ER, Sasmono RT, Kisworini FY, Ryan PA, O'Neill SL, Simmons CP, Utarini A, Anders KL (2018) Baseline characterization of dengue epidemiology in Yogyakarta City, Indonesia, before a randomized controlled trial of *Wolbachia* for Arboviral disease control. Am J Trop Med Hyg 99(5):1299–1307. https://doi.org/10.4269/ajtmh.18-0315

Jones MS, Delborne JA, Elsensohn J, Mitchell PD, Brown ZS (2019) Does the U.S. public support using gene drives in agriculture? And what do they want to know? Sci Adv 5(9):eaau8462. https://doi.org/10.1126/sciadv.aau8462

Khamis D, El Mouden C, Kura K, Bonsall MB (2018) Ecological effects on underdominance threshold drives for vector control. J Theor Biol 456:1–15. https://doi.org/10.1016/j.jtbi.2018.07.024

Kohl PA, Brossard D, Scheufele DA, Xenos MA (2019) Public views about editing genes in wildlife for conservation. Conserv Biol 33(6):1286–1295. https://doi.org/10.1111/cobi.13310

Kyrou K, Hammond AM, Galizi R, Kranjc N, Burt A, Beaghton AK, Nolan T, Crisanti A (2018) A CRISPR-Cas9 gene drive targeting doublesex causes complete population suppression in caged *Anopheles gambiae* mosquitoes. Nat Biotechnol 36(11):1062–1066. https://doi.org/10.1038/nbt.4245

Leftwich PT, Edgington MP, Harvey-Samuel T, Carabajal Paladino LZ, Norman VC, Alphey L (2018) Recent advances in threshold-dependent gene drives for mosquitoes. Biochem Soc Trans 46(5):1203–1212. https://doi.org/10.1042/BST20180076

Lewontin RC (2001) Genes in the food! The New York Review of Books 48,10. 6-21-2001. https://www.nybooks.com/articles/2001/06/21/genes-in-the-food/

Li F, Scott MJ (2016) CRISPR/Cas9-mediated mutagenesis of the *white* and *Sex lethal* loci in the invasive pest, *Drosophila suzukii*. Biochem Biophys Res Commun 469(4):911–916

Li M, Akbari OS, White BJ (2018) Highly efficient site-specific mutagenesis in malaria mosquitoes using CRISPR. G3 (Bethesda) 8(2):653–658. https://doi.org/10.1534/g3.117.1134

Macias VM, Ohm JR, Rasgon JL (2017) Gene drive for mosquito control: where did it come from and where are we headed? Int J Environ Res Public Health 14(9):pii: E1006. https://doi.org/10.3390/ijerph14091006

Marshall JM, Akbari OS (2018) Can CRISPR-based gene drive be confined in the wild? A question for molecular and population biology. ACS Chem Biol 13(2):424–430. https://doi.org/10.1021/acschembio.7b00923

McFarling UL (2017) Could this zoo of mutant mosquitoes lead the way to eradicating Zika? Stat News. https://www.statnews.com/2017/12/13/gene-drive-mosquitoes-darpa/

McKenna M (2019) Antibiotics set to flood Florida's troubled orange orchard. Nature 567:302–303. https://doi.org/10.1038/d41586-019-00878-4

Meghani Z, Kuzma J (2018) Regulating animals with gene drive systems: lessons from the regulatory assessment of a genetically engineered mosquito. J Respons Innov 5(sup1):S203–S222

Moro D, Byrne M, Kennedy M, Campbell S, Tizard M (2018) Identifying knowledge gaps for gene drive research to control invasive animal species: the next CRISPR step. Glob Ecol Conserv 13:e00369. https://doi.org/10.1016/j.gecco.2017.e00369

Nash A, Urdaneta GM, Beaghton AK, Hoermann A, Papathanos PA, Christophides GK, Windbichler N (2019) Integral gene drives for population replacement. Biol Open 8:bio037762. https://doi.org/10.1242/bio.037762

Oberhofer G, Ivy T, Hay BA (2018) Behavior of homing endonuclease gene drives targeting genes required for viability or female fertility with multiplexed guide RNAs. Proc Natl Acad Sci 115(40):E9343–E9352. https://doi.org/10.1073/pnas.1805278115

Olena, A (2017) GM mosquitoes closer to release in U.S. The Scientist. https://www.the-scientist.com/news-analysis/gm-mosquitoes-closer-to-release-in-us-30752

Paes de Andrade P, Aragão FJ, Colli W, Dellagostin OA, Finardi-Filho F, Hirata MH, Lira-Neto AC, Almeida de Melo M, Nepomuceno AL, Gorgônio da Nóbrega F, Delfino de Sousa G, Valicente FH, Zanettini MH (2016) Use of transgenic *Aedes aegypti* in Brazil: risk perception and assessment. Bull World Health Organ 94:766–771. https://doi.org/10.2471/BLT.16.173377

Quinlan MM, Mutunga JM, Diabaté A, Namountougou M, Coulibaly MB, Sylla L, Kayondo J, Balyesima V, Clark L, Benedict MQ, Raymond P (2018) Studies of transgenic mosquitoes in disease-endemic countries: preparation of containment facilities. Vector Borne Zoonotic Dis 18(1):21–30

Rakotoson JD, Fornadel CM, Belemvire A, Norris LC, George K, Caranci A, Lucas B, Dengela D (2017) Insecticide resistance status of three malaria vectors, *Anopheles gambiae (s.l.)*, *An. funestus* and *An. mascarensis*, from the South, Central and East coasts of Madagascar. Parasit Vectors 10(1):396. https://doi.org/10.1186/s13071-017-2336-9

Scharping N (2017) After mosquitoes, moths are the next target for genetic engineering. Discover. https://www.discovermagazine.com/environment/after-mosquitos-moths-are-the-next-target-for-genetic-engineering

Schorderet-Weber S, Noack S, Selzer PM, Kaminsky R (2017) Blocking transmission of vector-borne diseases. Int J Parasitol Drugs Drug Resist 7(1):90–109

Singer M, Frischknecht F (2016) Time for genome editing: next-generation attenuated malaria parasites. Trends Parasitol 33(3):202–213. https://doi.org/10.1016/j.pt.2016.09.012

Sinkins SP (2011) Invertebrate post-segregation distorters: a new embryo-killing gene. PLoS Biol 9(7):e1001114. https://doi.org/10.1371/journal.pbio.1001114

Tjaden NB, Caminade C, Beierkuhnlein C, Thomas SM (2018) Mosquito-borne diseases: advances in modelling climate-change impacts. Trends Parasitol 34(3):227–245. https://doi.org/10.1016/j.pt.2017.11.006

United Nations Department of Economic and Social Affairs (2018) https://www.un.org/development/desa/en/news/population/2018-revision-of-world-urbanization-prospects.html

US National Academies' (NAS) recent report (2016) Gene drives on the horizon: advancing science, navigating uncertainty, and aligning research with public values. The National Academies Press (US). https://doi.org/10.17226/23405

Webber BL, Raghuc S, Edwards OR (2015) Opinion: is CRISPR-based gene drive a biocontrol silver bullet or global conservation threat? PNAS 112(34):10565–10567

Wieczorek A (2016) CRISPR-Cas9 and gene drives: the risks and benefits of game-changing technology. Biotech In Focus, Cooperative Extension Service, University of Hawaii at Manoa, July, Issue 52 https://www.ctahr.hawaii.edu/biotechinfocus/images/bulletinpdf/2016_06July_I52Print.pdf

Wimmer EA (2013) Insect biotechnology: controllable replacement of disease vectors. Curr Biol 23(10):R453–R456. https://doi.org/10.1016/j.cub.2013.03.058

Zentner GE, Wade MJ (2017) The promise and peril of CRISPR gene drives: genetic variation and inbreeding may impede the propagation of gene drives based on the CRISPR genome editing technology. BioEssays 39(10). https://doi.org/10.1002/bies.201700109

# Evaluating Gene Drive Approaches for Public Benefit

**Michael R. Santos**

**Abstract** Gene drive approaches—those which bias inheritance of a genetic element in a population of sexually reproducing organisms—have the potential to provide important public benefits. The spread of selected genetic elements in wild populations of organisms may help address certain challenges, such as transmission of vector-borne human and animal diseases and biodiversity loss due to invasive animals. Adapting various naturally occurring gene drive mechanisms to these aims is a long-standing research area, and recent advances in genetics have made engineering gene drive systems significantly more technically feasible. Gene drive approaches would act through changes in natural environments, thus robust methods to evaluate potential research and use are important.

Despite the fact that gene drive approaches build on existing paradigms, such as genetic modification of organisms and conventional biological control, there are material challenges to their evaluation. One challenge is the inherent complexity of ecosystems, which makes precise prediction of changes to the environment difficult. For gene drive approaches that are expected to spread spatially and/or persist temporally, responding to this difficulty with the typical stepwise increases in the scale of studies may not be straightforward after studies begin in the natural environment. A related challenge is that study or use of a gene drive approach may have implications for communities beyond the location of introduction, depending on the spatial spread and persistence of the approach and the population biology of the target organism. This poses a particular governance challenge when spread across national borders is plausible. Finally, community engagement is an important element of responsible research and governance, but effective community engagement for gene drive approaches requires addressing complexity and uncertainty and supporting representative participation in decision making.

These challenges are not confronted in a void. Existing frameworks, processes, and institutions provide a basis for effective evaluation of gene drive approaches for public benefit. Although engineered gene drive approaches are relatively new, the necessities of making decisions despite uncertainty and governing actions with potential implications for shared environments are well established. There are methodologies to identify potential harms and assess risks when there is limited

M. R. Santos (✉)
Foundation for the National Institutes of Health, North Bethesda, MD, USA
e-mail: msantos@fnih.org



A. Chaurasia et al. (eds.), *GMOs*, Topics in Biodiversity and Conservation 19,
https://doi.org/10.1007/978-3-030-53183-6_19

experience to draw upon, and these methodologies have been applied in similar contexts. There are also laws, policies, treaties, agreements, and institutions in place across many jurisdictions that support national and international decision making regarding genetically modified organisms and the potential applications of gene drive approaches, such as public health and biodiversity conservation. Community engagement is an established component of many decision-making processes, and related experience and conceptual frameworks can inform engagement by researchers.

The existence of frameworks, processes, and institutions provides an important foundation for evaluating gene drive approaches, but it is not sufficient by itself. They must be rigorously applied, which requires resources for risk assessment, research, and community engagement and diligent implementation by governance institutions. The continued evolution of the frameworks, processes, and institutions is important to adapt to the growing understanding of gene drive approaches. With appropriate resources and diligence, it will be possible to responsibly evaluate and make decisions on gene drive approaches for public benefit.

**Keywords** Gene drive · Risk assessment · Governance · Community engagement · Biosafety · Public benefit · Decision making · Uncertainty

## Gene Drive Approaches: Potential to Provide Important Public Benefits

Gene drive is a phenomenon of biased inheritance in which the prevalence of a genetic element is increased, even in the presence of some fitness cost, leading to the preferential increase of a specific genotype that may determine a specific phenotype from one generation to the next and potentially throughout a population (National Academies of Sciences, Engineering, and Medicine (NASEM) 2016). Gene drive is a natural phenomenon observed in populations of many different organisms; there are multiple natural mechanisms that lead to gene drive, and the phenomenon has been described with various names (Burt and Trivers 2006). An important property of gene drive is that the genetic element that increases in frequency in the population can decrease individual fitness compared to a context without the driving genetic element. Normally a genetic element conveying a fitness cost would be selected against over time, because offspring that do not inherit the element will outcompete those that do. However, under some circumstances, the drive effect can outweigh a fitness cost (e.g., Burt 2003). This property of gene drive has long been recognized as potentially enabling applications that spread traits in wild populations of organisms, even if the trait does not convey a fitness advantage (e.g., Craig et al. 1960; Von Borstel and Buzzati-Traverso 1962; Curtis 1968).

The features of gene drive enable potential applications of gene drive approaches for public benefit (e.g., Esvelt et al. 2014; NASEM 2016; Australian Academy of

Science 2017). The contexts where gene drive approaches may be most applicable are those where the organism of interest reproduces sexually with a short generation time (compared to timescales of interest) and there is well-mixed mating across the relevant populations (e.g., Burt 2003; Esvelt et al. 2014). There is long-standing interest in gene drive approaches to reducing transmission of vector-borne human and animal diseases (e.g., Serebrovsky 1940; Craig et al. 1960; Curtis 1968; Ribeiro and Kidwell 1994; Burt 2003, 2014). Given the success of vector control in reducing the burden of vector-borne diseases (e.g., vector control interventions accounted for 78% of the reduction in malaria prevalence from 2000 to 2015; Bhatt et al. 2015), complementary gene drive approaches may be transformational tools to eliminate or eradicate vector-borne disease (NASEM 2016; NEPAD and the African Union 2018). There is also current interest in biodiversity preservation (e.g., through population reduction of invasive species or increased resilience for endangered species; Redford et al. 2019), agricultural pest control (e.g., Scott et al. 2018), and crop resilience (e.g., Pixley et al. 2019). Gene drive approaches can be particularly well suited to public benefit because their benefits would accrue to everyone in the area of use.

Gene drive approaches may provide a useful complement to other approaches to important public goals and should be considered in that context. The potential applications of gene drive approaches are pursued through a variety of means currently, such as insecticides/pesticides/rodenticides, conventional removal of invasive species, conventional biocontrol, breeding programs, and genetic modification (e.g., NASEM 2016). In many cases existing interventions are insufficient or undesirable (e.g., Feachem et al. 2019, for the case of malaria eradication, where cost and widespread insecticide resistance limit the impact of current tools, and Campbell et al. 2015, for the case of invasive rodents, where cost and toxicity limit the impact of toxicants). Gene drive approaches may complement existing interventions by being lower cost (because of potential spread and persistence) and species specific in their direct effect. As discussed above, gene drive approaches may also promote equity because their benefits accrue to the areas in which they are used, rather than to individuals, and thus do not depend on individual resources such as wealth and time; inequity may still exist between areas that have access to gene drive approaches and those that do not.

Given the potential for gene drive approaches to provide public benefit, there has been research into adapting or recapitulating a variety of natural gene drive mechanisms in organisms of interest, such as disease vectors. This research resulted in important advances, such as the first implementation of gene drive in a malaria vector (Windbichler et al. 2011), and has accelerated dramatically since 2015 due to the application of improved genome editing tools, such as CRISPR-Cas-based tools, to engineering gene drive (e.g., Dicarlo et al. 2015; Gantz and Bier 2015). There have been notable successes in laboratory cage experiments with malaria vectors, demonstrating proof of principle of genes that reduce the population of vectors (Hammond et al. 2016) and reduce mosquitoes' ability to transmit the parasites that cause malaria (Gantz et al. 2015). Although resistance to the gene drive mechanism may be expected over time (e.g., Burt 2003), different drive mechanisms may be more or less susceptible to the development of resistance, and techniques have been identified to delay its development (e.g., Burt 2003; Esvelt et al. 2014; Hammond et al. 2017; Kyrou et al. 2018; Champer et al. 2018); in addition, gene drive approaches will often not need to operate indefinitely to deliver their public benefit. Despite these many technical advances, gene drive is still a difficult phenomenon to engineer: highly efficient gene drive has been reported in only a handful of organisms and has been difficult to implement even in well-studied animals like mice (e.g., Grunwald et al. 2019; see also Godwin et al. 2019 and Yosef et al. 2019). To date, research into genetically engineering gene drive approaches has occurred exclusively in laboratory containment, and no genetically engineered gene drives have been introduced into the natural environment.

The advances in genome editing tools are generally expected to enable demonstrations of more varieties of gene drive. Among the principal areas of research are efforts to engineer control over the spatial spread and temporal persistence of the driving genetic element. Possible approaches include systems where the drive mechanism functions only when present above a threshold (e.g., Akbari et al. 2013; Oberhofer et al. 2019), which tends to limit the spatial spread of the genetic element, and generational limits on the persistence of the gene drive mechanism (e.g., Noble et al. 2019), after which natural selection will remove genes with fitness costs from the population. Another potential method for limiting spread is using a drive mechanism that operates on a specific genetic sequence that is prevalent only in a restricted subpopulation (e.g., Sudweeks et al. 2019). In addition to engineering a priori control into gene drive approaches, there is active research on methods to stop the spread of a driving genetic element (e.g., Esvelt et al. 2014; Vella et al. 2017; Basgall et al. 2018; Roggenkamp et al. 2018).

Given the potential for gene drive approaches to contribute to important social priorities like human and animal health and biodiversity, and the technical progress in engineering them, the potential challenges associated with gene drive approaches for public benefit have become a topic of serious consideration for stakeholders (e.g., NEPAD and the African Union 2018; Redford et al. 2019). Gene drive approaches have similar challenges to other activities that make changes to the natural environment, such as establishing a nature reserve or building a hydroelectric dam: ecosystems are complex and the consequences of interventions are difficult to predict with precision, the environment is spatially interlinked (naturally and through human-assisted transportation) so some local interventions can have effects beyond the location of intervention, and within the affected areas it is not possible for individuals to personally opt in or opt out of the effects (e.g., National Research Council 2005).

The following sections discuss the challenges of evaluating gene drive approaches for public benefit and methods for addressing them. There is an increasing diversity of potential applications and technical strategies for gene drive approaches with a wide spectrum of possible properties; thus it is not possible to evaluate them as one group (NASEM 2016). The subsequent sections highlight concepts and methods that may have applicability in case-by-case evaluation of specific gene drive approaches.

## Challenges of Evaluating Gene Drive Approaches for Public Benefit

### *Decision-Making Context*

The decision-making contexts within which gene drive approaches are evaluated are important to understanding the challenges. Because the public benefit use cases generally concern impact on communities and the shared environment, there are structures already in place for making decisions related to those goals, such as ministries of health responsible for infectious disease control and ministries of environment responsible for endangered species preservation and biodiversity. In addition, gene drive approaches, depending on their specifics, may be subject to biosafety regulations, policies, and laws, implemented by national biosafety authorities. For gene drive approaches where the responsibilities of multiple regulatory and policy interests intersect, effective integration can be difficult (e.g., NASEM 2017).

The governance of decisions on gene drive approaches—the decision-making and accountability mechanisms—includes laws and treaties as well as "soft law" tools such as guidelines, recommendations, and norms (NASEM 2016). Researchers and their institutions play a governance role through their own ethical considerations (often represented for institutions by institutional review boards or independent ethics committees) and peer- and funder-imposed norms (e.g., Akbari et al. 2015; Emerson et al. 2017). The responsibility to decide on the use of gene drive approaches is principally within governments, typically at the national level but potentially also at sub-national levels, and among governments through international treaties and agreements such as the Cartagena Protocol on Biosafety to the Convention on Biological Diversity [CBD] (Secretariat of the CBD 2000). Governance processes at all levels are generally organized for case-by-case evaluations.

There is a continuum of decisions required for potential research and use of gene drive approaches (NASEM 2016). For research, the common process is that researchers propose activities that, where necessary, are externally evaluated. Laboratory studies of gene drive approaches within containment are typically subject to biosafety review (e.g., UC San Diego Institutional Biosafety Program 2018). Governance becomes stricter as the likelihood and magnitude of potential undesirable outcomes of a decision ("harms") are judged to increase; for example, the Cartagena Protocol on Biosafety includes requirements that apply to the introduction of living modified organisms into the environment that do not apply to contained use (Secretariat of the CBD 2000; see, e.g., Maiga 2018 for the authorization of a field study of a living modified organism without a gene drive approach). Decisions regarding use, meaning introduction with direct public benefit as the primary goal, encompass potentially separate decisions about what uses are permitted (regulatory authorization decisions), what uses are actually implemented (policy and financing decisions), and what the responsibilities are for harms that are a consequence of use (liability decisions). For example, biosafety regulators and ministries of environment may be responsible for determining under which circumstances, if any, gene drive approaches to reducing malaria transmission are allowed; ministries of health and national malaria control programs may be responsible for deciding if and where gene drive approaches will be funded and delivered; and national laws and international treaties may establish liability associated with potential harm due to use of the approach (James et al. 2018).

The essential considerations for decision making on gene drive approaches for public benefit are evaluations of potential benefits, costs, and harms and their likelihoods (NASEM 2016); these considerations are elements of processes like cost-benefit analyses. Benefits, costs, and harms are subjective evaluations that depend on values, which differ among individuals and organizations (National Research Council 2005). As highlighted above, there may be multiple organizations with responsibility for authorization of a specific gene drive approach; these organizations will have different statutory responsibilities that influence the scope, scale, and weights of potential benefits, costs, and harms they consider in their evaluations (e.g., NASEM 2017). These evaluations will typically share a common objective: to establish potential outcomes associated with proposed research or use of gene drive approaches and consider the likelihoods of those outcomes.

## *Uncertainty of Potential Benefits, Costs, and Harms*

A significant challenge for the evaluation of gene drive approaches is uncertainty in the likelihoods of potential benefits, costs, and harms: ecosystems are generally complex and there is limited experience with gene drive approaches (NASEM 2016). The challenge of uncertainty is not specific to gene drive approaches or biotechnology: large uncertainties are common in the evaluation of many environmental changes, such as fisheries management (e.g., Schwaab 2014), road construction (e.g., Zhao et al. 2004), use of conventional biocontrol organisms (e.g., Benjamin and Wesseler 2016), conservation to protect endangered species (e.g., Nicholson and Possingham 2007), conventional removal of invasive species (e.g., Kessler 2011), and reintroduction of extirpated native species (e.g., Carroll et al. 2019). In contexts of high uncertainty, particularly regarding potential harms, phased testing paradigms use stepwise increases in the scale of studies to balance reduction of uncertainty, exposure to potential harms, and speed of evaluation. For example, the scale of studies for a gene drive approach might progress successively from physical containment to semi-field studies under outdoor confinement, small-scale open field studies, and, finally, larger-scale introductions (NASEM 2016). Each step provides a higher-fidelity representation of real-world use, thus further reducing uncertainty about the expected outcomes of use, but also increases the potential exposure to harms. Thus after each step the accumulated evidence informs the decision on whether to proceed to the next step; the intensity of governance also typically increases through the stepwise process.

There may be limitations to the application of the phased testing paradigm to some gene drive approaches. The paradigm is applicable for the earlier phases of research, within physical confinement. However, some gene drive approaches would be expected to spatially spread and temporally persist after the first introduction to the natural environment, even if that introduction were in the context of a field trial. Even when gene drive approaches include mechanisms to control spread and/or persistence, the efficacies of those control mechanisms won't be known until they are tested in the field. For these reasons, field trials of some gene drive approaches may be evaluated similarly to decisions about small-scale introductions, where a greater degree of environmental exposure is assumed (e.g., James et al. 2018).

The ability to detect incipient harms and respond to, eliminate, or reduce them is an important component of the evaluation of potential harms of gene drive approaches; this ability is likely to vary substantially across specific cases of gene drive approaches (NASEM 2016). In general, the complexity of ecosystems may make it difficult to determine whether any changes in the environment that happen after the introduction of a gene drive approach were caused by the gene drive approach. The potential for some gene drive approaches to spread and persist from relatively low prevalence in a population may present an obstacle to circumscribing the location where risk response is necessary, even when an effective conventional response (e.g., conventional vector control or invasive species removal) is available. Gene drive approaches designed to stop the effects of a previously introduced gene drive may have particular advantages for removing an undesired gene drive approach from the environment (e.g., Vella et al. 2017), but with the associated uncertainty of introducing another gene drive approach.

## *Potential Spread and Persistence*

The potential spatial spread and temporal persistence of some gene drive approaches create another important challenge. Depending on the scale of introduction, population biology of the target species, and specifics of the approach, communities beyond the location of introduction of a gene drive approach may be affected (e.g., Marshall 2009). Which communities will be affected, and when, will depend on properties of the gene drive approach and natural environment (e.g., North et al. 2019).

As a consequence, decision makers such as regulators could be asked to evaluate gene drive approaches expected to spread outside of their jurisdictions, including across national borders (e.g., Brown 2017). Policymakers will need to consider the legal and political implications of the potential spread of gene drive approaches; in particular, the spread of genetically modified organisms across international borders is regulated under the Cartagena Protocol on Biosafety (for countries that are signatories; Secretariat of the CBD 2000). Legal risks associated with liability and redress will be particularly consequential for decisions about potential implementation of approved gene drive approaches (e.g., Oye et al. 2014; Glover et al. 2018).

## *Community Engagement*

Consultation with stakeholders, in particular with communities where research or use of gene drive approaches is being considered, is an important element of the evaluation of gene drive approaches (e.g., NASEM 2016). For researchers, engagement with affected communities is an ethical obligation (e.g., King et al. 2014). This responsibility includes providing transparency into the research being conducted, so that concerns can be identified and addressed, and obtaining community acceptance of the research (e.g., World Health Organization 2014). In addition, community engagement by researchers creates an opportunity for co-development of innovation, where community input is expected to improve the quality of the research (e.g., NEPAD and the African Union 2018; Hartley et al. 2019). For other decision makers, such as regulators, policymakers, and implementers, community engagement is commonly an element of their decision-making process (e.g., Quinlan et al. 2016). For example, many environmental regulators include opportunity for public comment on their pending decisions, local policymakers often convene meetings of their constituents to address questions and concerns, and international governance institutions like the Convention on Biological Diversity commonly invite online comments on topics under consideration.

There are likely to be multiple elements to the challenge of successful community engagement in decision making (e.g., Kaebnick et al. 2014; Quinlan et al. 2016; NASEM 2016). Because of the potential uncertainties in spatial spread and temporal persistence of gene drive approaches, it may not be known exactly what areas and communities will be affected and when (e.g., Baltzegar et al. 2018). Communicating effectively given the scientific complexity and uncertainty of gene drive approaches may also be difficult (e.g., Brossard et al. 2019). Gene drive approaches will have area-wide effects that, like existing community interventions, do not allow for opting in or opting out at the level of the individual (e.g., Thizy et al. 2019); elements of research, such as social science research or access to private property, may still require individual consent (e.g., Kolopack and Lavery 2017). Governments routinely make decisions for communities, though the degree and mechanism for representation and participation vary; for research, community acceptance is less well defined than individual consent, and achieving representative perspectives from communities can be difficult (e.g., Kaebnick et al. 2014; Thizy et al. 2019). Given that the elements of the challenge of community engagement will vary with each individual consideration of a gene drive approach, similar to risk assessment there is unlikely to be a single prescriptive process appropriate for all gene drive approaches (e.g., Rask and Worthington 2015; NASEM 2016).

### *Conclusion*

Effective evaluation of gene drive approaches for public benefit depends on these challenges being successfully addressed. The next section discusses the groundwork that is already in place and additional efforts required to accomplish this.

## Addressing the Challenges of Evaluating Gene Drive Approaches for Public Benefit

The previous section highlighted three important challenges for evaluating gene drive approaches for public benefit. The first is that the expected environmental, social, and economic effects of the introduction of a gene drive approach will have uncertainties, a consequence of the inherent complexity of ecosystems and societies and limited experience with gene drive approaches. The second is that research or use of a gene drive approach may affect areas beyond the location of introduction, including potentially in other countries. The third is that community engagement by researchers and other decision makers is important but not straightforward. These challenges are common to other decisions about the shared environment, and thus there are existing frameworks, processes, and institutions that can help address them.

### *Managing Uncertainty in Decision Making*

Uncertainty is a common challenge in decision making, and there are frameworks to help characterize and reduce uncertainty (e.g., Aven et al. 2014). Risk assessment, a set of methods to identify and analyze potential outcomes of decisions (e.g., Rausand 2011), is one of those frameworks and is recognized as important to the evaluation of gene drive approaches (e.g., NASEM 2016; Secretariat of the CBD 2000). Risk assessment is a general and flexible framework that can identify potential harms and characterize their likelihood. Risk assessment methods can address potential health, social, cultural, economic, and environmental harms. The potential harms considered in any individual risk assessment will depend on the organization performing it. For example, when regulators perform risk assessments, the scope of potential harms considered is typically prescribed by laws and policies. Some decisions about gene drive approaches, such as policies about their use, are likely to be informed by risk assessments considering different categories of potential harms.

The first step in risk assessment is identification of potential harms within the scope that is being considered. Identifying harms can incorporate analogous prior experience (e.g., through checklists of previously experienced harms), and new potential harms can be enumerated through systematic processes: situations that could conceivably lead to harms are identified (e.g., for gene drive approaches for African malaria vectors, Roberts et al. 2017; Teem et al. 2019), and the chain of events from the decision to potential harms is articulated. The next step in risk assessment is the evaluation of the likelihoods of each potential harm (including different types and magnitudes of consequences for a given type of harm); this evaluation can be qualitative or quantitative (e.g., Rausand 2011). When there is a lack of relevant experience to inform the likelihoods (such as for potential harms that have never occurred), the likelihood of a potential harm can be inferred from the likelihoods of the events on the causal pathway to that harm: often there are data to inform the likelihoods of individual events even when the full pathway is unprecedented (e.g., Hayes et al. 2018a). For example, in fault tree analysis the events on the causal pathway to a harm are identified along with their logical relationships; when the probabilities of the individual events are estimated, they can be combined into an estimate of the probability of harm. The estimated likelihoods of harm have uncertainties, which can additionally be characterized, including quantitatively (e.g., Kaplan and Garrick 1981). Risk assessment conducted in this manner can also inform further research on gene drive approaches: by characterizing the expected likelihoods of harms and the uncertainties associated with those expectations, research studies and monitoring plans can be developed to prioritize reduction of the most consequential uncertainties.

Structured risk assessment methods have been applied across a range of complex systems, including to living modified organisms (e.g., Hayes et al. 2018a). Risk assessment is a component of many regulatory frameworks, and in particular ecological risk assessment is used by many environmental authorities (e.g., US EPA 2019; EFSA n.d.) and specifically recommended for the evaluation of gene drive approaches (NASEM 2016). The Cartagena Protocol on Biosafety requires "case-by-case," "scientifically sound" risk assessment for international transboundary movement of living modified organisms intended for use in the environment (Secretariat of the CBD 2000). No genetically engineered gene drive approach has advanced to the stage of risk assessment, but independent probabilistic risk assessments have been released for the contained use and small-scale field release of a genetically sterile malaria vector in Burkina Faso (Hayes et al. 2015, 2018b), demonstrating methods that could be applied to risk assessments of gene drive approaches.

## *Governing Gene Drive Approaches That Could Cross Borders*

Similar to gene drive approaches, many environmental decisions have implications beyond the area in which they are implemented, and there are frameworks and institutions to support those decisions. Of particular relevance are decisions in a country that can affect other countries, such as use of water from a shared source (e.g., General Assembly of the United Nations 1997). International governance institutions such as regional organizations (e.g., the European Union, the African Union) and the United Nations provide platforms for international treaties and agreements, which can inform national laws and policies. Specifically, international institutions exist to support decision making on living modified organisms (e.g., the Cartagena Protocol on Biosafety) and potential applications of gene drive approaches such as public health (e.g., the World Health Organization) and biodiversity conservation (e.g., the Convention on Biological Diversity). These institutions provide binding requirements and guidance to participating countries that apply to gene drive approaches, including provisions for liability and redress in cases where a living modified organism used within a country moves to another country and causes harm (e.g., the Nagoya-Kuala Lumpur Supplementary Protocol; Secretariat of the CBD 2011). Regional institutions can also provide guidance on the development of regional regulatory frameworks and capacity (e.g., in the African Union, Glover et al. 2018, and specifically for gene drive approaches for malaria, African Biosafety Network of Expertise 2018).

## *Effectively Engaging Communities in Decisions*

Although community engagement is expected to be challenging for research and use of gene drive approaches, there are examples to learn from and frameworks to guide future efforts. For example, foundational features of the community engagement approach of the Eliminate Dengue/World Mosquito Program have been functionally related to their impact, which may be informative for other approaches (Kolopack et al. 2015). There are relevant engagement frameworks that have been developed for genetically modified mosquitoes (e.g., Lavery et al. 2010; World Health Organization 2014; Thizy et al. 2019; Singh 2019), mice that carry tick-borne pathogens (Buchthal et al. 2019), and biodiversity (e.g., Rask and Worthington 2015). The Cartagena Protocol on Biosafety requires its signatories to promote public awareness, understanding, and participation in the decision-making process for living modified organisms (which many possible gene drive approaches would be categorized within), and there is an ongoing program of work within the Protocol to advance priority areas (UN Environment Programme 2016).

In summary, these frameworks highlight the importance of "an expansive notion of 'engagement'" (Bartumeus et al. 2019). Identifying communities, stakeholders, and publics for engagement, accounting for the uncertainties associated with spread and persistence, needs to be continuous because those identifications are likely to evolve (NASEM 2016). Engagement early and throughout the research process provides transparency and enables co-creation of approaches (e.g., Esvelt 2017; Hartley et al. 2019). Communication should use language appropriate to different audiences so that messages are understandable (e.g.,

Quinlan et al. 2016). Finally, a common refrain is the importance of financial resources and human capacity dedicated to community engagement, for research (e.g., King et al. 2014) and in government decisions (UN Environment Programme 2016).

## *Conclusion*

Collectively, the existing appropriate frameworks, processes, and institutions provide a context to address the challenges of evaluating gene drive approaches for public benefit. Established methods can characterize the expectations and uncertainties for research and use of a gene drive approach, such as the range of areas and communities that may be affected. Institutions exist to use that information to make decisions given their responsibilities and processes, and national and international governance enables decision making across communities beyond those in which a gene drive approach is introduced. Community engagement will need to be tailored for each individual case, but past experience and conceptual frameworks can guide these important activities. The next section closes with important considerations for the effective application of this context to evaluate gene drive approaches for public benefit.

## Recommendations on Decision Making for Gene Drive Approaches

Having identified the context that can help address the challenges to evaluating gene drive approaches, the effective application of that context requires resources to implement risk assessment, support research, and engage communities; diligent implementation by governance institutions; and the continued evolution of the frameworks, processes, and institutions.

Rigorous evaluation of gene drive approaches may be resource intensive, similar to other regulatory decisions, and appropriate regulatory capacity is required. Some governance institutions may already possess sufficient capacity, and in other cases capacity strengthening may be a necessary precursor to evaluation of gene drive approaches. International organizations can and should play an important role in supporting capacity development (e.g., Glover et al. 2018).

Research is an important tool to inform the potential benefits, costs, harms, and likelihoods of different outcomes from the use of a gene drive approach. Baseline environmental studies can help characterize population biology and ecosystem relationships (e.g., Moro et al. 2018) and contained use studies may reduce other uncertainties (e.g., Hayes et al. 2018a). Mathematical modeling can provide insights into potential outcomes of research or use of gene drive approaches over a range of potential contexts (e.g., Sánchez et al. 2019). Ethical, legal, and social science research will continue to inform fair and effective approaches for making decisions (e.g., National Research Council 2005; NASEM 2016). It is important that these research areas receive sufficient funding to provide informative input into the evaluation of gene drive approaches.

Decisions about gene drive approaches for public benefit, like other important public decisions, should be made with diligence, rigor, and transparency, with the understanding that a just process will produce decisions that are unlikely to satisfy every stakeholder. It is a role of governments to fairly represent the values of their constituencies in decision making, recognizing that even in processes widely recognized as good governance, such as free and fair elections, a large minority of constituents may disagree with the decision (e.g., UN General Assembly 1966). In addition, some gene drive approaches may have the potential to spread across national boundaries, requiring national authorities to act on their international obligations.

The frameworks, processes, and institutions that exist to address the challenges of evaluating gene drive approaches should continuously improve the support they provide for decision making. Given the limited current experience with gene drive approaches, it will be valuable to continue to refine the methods for identifying and characterizing potential outcomes. Governance at the national level evolves, and regular convenings of international institutions provide venues to further interpret treaties and agreements and develop guidance. Community engagement methods will continue to be informed by experience from research and governance of gene drive approaches and analogous domains. Progress in these areas, including on international liability and redress, is a necessary complement to technical progress on gene drive approaches (e.g., Oye et al. 2014). For example, the Ad Hoc Technical Expert Group on Risk Assessment of the Cartagena Protocol on Biosafety is specifically tasked with informing decisions on whether additional guidance for risk assessment of genetically engineered gene drive approaches is necessary (UN Environment Programme 2018).

Gene drive approaches have the potential to provide important public benefits by making changes in the natural environment. Evaluating them will be challenging for decision makers: ecosystems are complex, governing changes that can affect multiple communities is complicated, and there is limited experience with the use of gene drive approaches. However, these challenges are not confronted in a void. Because making decisions about the shared environment under conditions of uncertainty is a common responsibility across many domains, there are existing frameworks, processes, and institutions that can help address these challenges. If the appropriate resources and diligence are applied, it will be possible to responsibly evaluate and make decisions on gene drive approaches for public benefit.

**Acknowledgments** I gratefully acknowledge thoughtful input on this contribution from four anonymous reviewers, colleagues at the Foundation for the National Institutes of Health, Western Michigan University, EHS Consultancy, and the Commonwealth Scientific and Industrial Research Organisation.

## References

African Biosafety Network of Expertise (2018) Establishment of the West Africa integrated vector management programme. http://nepad-abne.net/establishment-of-the-west-africa-integrated-vector-management-programme-accra-19-august-2018/
Akbari OS, Matzen KD, Marshall JM et al (2013) A synthetic gene drive system for local, reversible modification and suppression of insect populations. Curr Biol 23(8):671–677
Akbari OS, Bellen HJ, Bier E et al (2015) Safeguarding gene drive experiments in the laboratory. Science 349(6251):927–929
Australian Academy of Science (2017) Synthetic gene drives in Australia: implications of emerging technologies. https://www.science.org.au/support/analysis/reports/synthetic-gene-drives-australia-implications-emerging-technologies
Aven T, Baraldi P, Flage R, Zio E (2014) Uncertainty in risk assessment: the representation and treatment of uncertainties by probabilistic and non-probabilistic methods. Wiley, Chichester, UK
Baltzegar J, Barnes JC, Elsensohn JE et al (2018) Anticipating complexity in the deployment of gene drive insects in agriculture. J Resp Innov 5(sup1):S81–S97
Bartumeus F, Costa GB, Eritja R et al (2019) Sustainable innovation in vector control requires strong partnerships with communities. PLOS Negl Trop Dis 13(4):1–5
Basgall EM, Goetting SC, Goeckel ME et al (2018) Gene drive inhibition by the anti-CRISPR proteins AcrIIA2 and AcrIIA4 in Saccharomyces cerevisiae. Microbiology 164(4):464–474
Benjamin EO, Wesseler JH (2016) A socioeconomic analysis of biocontrol in integrated pest management: a review of the effects of uncertainty, irreversibility and flexibility. NJAS – Wageningen J Life Sci 77:53–60
Bhatt S, Weiss DJ, Cameron E et al (2015) The effect of malaria control on Plasmodium falciparum in Africa between 2000 and 2015. Nature 526(7572):207–211
Brossard D, Belluck P, Gould F, Wirz CD (2019) Promises and perils of gene drives: navigating the communication of complex, post-normal science. Proc Nat Acad Sci 116(16):7692–7697
Brown Z (2017) Economic, regulatory and international implications of gene drives in agriculture. Choices 32(2)
Buchthal J, Evans SW, Lunshof J et al (2019) Mice against ticks: an experimental community-guided effort to prevent tick-borne disease by altering the shared environment. Phil Trans R Soc B 374:20180105
Burt A (2003) Site-specific selfish genes as tools for the control and genetic engineering of natural populations. Proc Biolog Sci 270(1518):921–928
Burt A (2014) Heritable strategies for controlling insect vectors of disease. Philo Trans R Soc B 369:20130432
Burt A, Trivers R (2006) Genes in conflict. Harvard University Press, Cambridge, MA
Campbell K, Beek J, Eason C et al (2015) The next generation of rodent eradications: innovative technologies and tools to improve species specificity and increase their feasibility on islands. Biolog Conserv 185:47–58
Carroll C, Lacy RC, Fredrickson RJ et al (2019) Biological and sociopolitical sources of uncertainty in population viability analysis for endangered species. Sci Rep 9:10130
Champer J, Liu J, Oh SY et al (2018) Reducing resistance allele formation in CRISPR gene drive. PNAS 115(21):5522–5527
Craig G, Hickey W, VendeHey R (1960) An inherited male-producing factor in Aedes aegypti. Science 132(3443):1887–1889
Curtis C (1968) Possible use of translocations to fix desirable genes in insect pest populations. Nature 218:368–369
Dicarlo JE, Chavez A, Dietz SL et al (2015) Safeguarding CRISPR-Cas9 gene drives in yeast. Nat Biotechnol 33:1250–1255
EFSA (n.d.). Environmental risk assessment. https://www.efsa.europa.eu/en/topics/topic/environmental-risk-assessment

Emerson C, James S, Littler K, Randazzo F (2017) Principles for gene drive research. Science 358(6367):1135–1136
Esvelt KM (2017) Precaution: open gene drive research. 355(6325):589-590
Esvelt K, Smidler A, Catteruccia F, Church G (2014) Concerning RNA-guided gene drives for the alteration of wild populations. eLife 17(3)
Feachem R, Chen I, Akbari O et al (2019) Malaria eradication within a generation: ambitious, achievable, and necessary. Lancet 394(10203):1056–1112
Gantz VM, Bier E (2015) The mutagenic chain reaction: a method for converting heterozygous to homozygous mutations. Science 348(6233):442–444
Gantz VM, Jasinskiene N, Tatarenkova O et al (2015) Highly efficient Cas9-mediated gene drive for population modification of the malaria vector mosquito Anopheles stephensi. PNAS 112(49):E6736–E6743
General Assembly of the United Nations (1997) Convention on the law of the non-navigational uses of international watercourses. United Nations, New York
Glover B, Akinbo O, Savadogo M et al (2018) Strengthening regulatory capacity for gene drives in Africa: leveraging NEPAD's experience in establishing regulatory systems for medicines and GM crops in Africa. BMC Proc 12(Suppl 8):11
Godwin J, Serr M, Barnhill-Dilling SK et al (2019) Rodent gene drives for conservation: opportunities and data needs. Proc R Soc B Biol Sci 286(1914):1606
Grunwald HA, Gantz VM, Poplawski G et al (2019) Super-mendelian inheritance mediated by CRISPR–Cas9 in the female mouse germline. Nature 566:105–109
Hammond A, Galizi R, Kyrou K et al (2016) A CRISPR-Cas9 gene drive system targeting female reproduction in the malaria mosquito vector Anopheles gambiae. Nat Biotechnol 34:78–83
Hammond AM, Kyrou K, Bruttini M et al (2017) The creation and selection of mutations resistant to a gene drive over multiple generations in the malaria mosquito. PLOS Gene 13(10):e1007039
Hartley S, Thizy D, Ledingham K et al (2019) Knowledge engagement in gene drive research for malaria control. PLOS Negl Trop Dis 13(4):e0007233
Hayes KR, Barry S, Beebe N et al (2015) Risk assessment for controlling mosquito vectors with engineered nucleases: sterile male construct. Commonwealth Scientific and Industrial Research Organisation, Hobart
Hayes KR, Hosack GR, Dana GV et al (2018a) Identifying and detecting potentially adverse ecological outcomes associated with the release of gene-drive modified organisms. J Respons Innov 5(suppl):S139–S158
Hayes K, Hosack G, Ickowicz A et al (2018b) Risk assessment for controlling mosquito vectors with engineered nucleases: controlled field release for sterile male construct. Commonwealth Scientific and Industrial Research Organisation, Hobart
James S, Collins FH, Welkhoff PA et al (2018) Pathway to deployment of gene drive mosquitoes as a potential biocontrol tool for elimination of Malaria in Sub-Saharan Africa: recommendations of a scientific working group. Am J Trop Med Hyg 98(6 Suppl):1–49
Kaebnick GE, Gusmano MK, Murray TH (2014) The ethics of synthetic biology: next steps and prior questions. Hast Cent Rep 44:S4–S26
Kaplan S, Garrick BJ (1981) On the quantitative definition of risk. Risk Anal 1(1):11–27
Kessler C (2011) Invasive species removal and ecosystem recovery in the Mariana Islands; challenges and outcomes on Sarigan and Anatahan. In: Veitch CR, Clout MN, Towns DR (eds) Island invasives: eradication and management. Proceedings of the international conference on Island invasives. IUCN and Centre for Biodiversity and Biosecurity, Gland, Switzerland/ Auckland, pp 320–324
King KF, Kolopack P, Merritt MW, Lavery JV (2014) Community engagement and the human infrastructure of global health research. BMC Med Ethic 15(1):84
Kolopack P, Lavery J (2017) Informed consent in field trials of gene-drive mosquitoes. Gates Open Res 1:14)
Kolopack PA, Parsons JA, Lavery JV (2015) What makes community engagement effective?: Lessons from the eliminate dengue program in Queensland, Australia. PLOS Neg Trop Dis 9(4):e0003713

Kyrou K, Hammond AM, Galizi R et al (2018) A CRISPR–Cas9 gene drive targeting doublesex causes complete population suppression in caged Anopheles gambiae mosquitoes. Nat Biotechnol 36:1062–1066
Lavery JV, Tinadana PO, Scott TW et al (2010) Towards a framework for community engagement in global health research. Trend Parasitol 26(6):279–283
Maiga A (2018) Arrete No 2018-453/MESRSI/SG/ANB. Ouagadougou: Ministere de l'Enseignement Superieur, de la Recherche Scientifique et de l'Innovation
Marshall JM (2009) The effect of gene drive on containment of transgenic mosquitoes. J Theor Biol 258(2):250–265
Moro D, Byrne M, Kennedy M et al (2018) Identifying knowledge gaps for gene drive research to control invasive animal species: the next CRISPR step. Glob Ecol Conserv 13:e00363
National Academies of Sciences, Engineering, and Medicine (2016) Gene drives on the Horizon: advancing science, navigating uncertainty, and aligning research with public values. The National Academies Press, Washington, DC
National Academies of Sciences, Engineering, and Medicine (2017) Preparing for future products of biotechnology. The National Academies Press, Washington, DC
National Research Council (2005) Decision making for the environment: social and behavioral science research priorities. The National Academies Press, Washington, DC
NEPAD and the African Union (2018) Gene drives for Malaria control and elimination in Africa. https://www.nepad.org/publication/gene-drives-malaria-control-and-elimination-africa.
Nicholson E, Possingham HP (2007) Making conservation decisions under uncertainty for the persistence of multiple species. Ecolog Appl 17(1):251–265
Noble C, Min J, Olejarz J et al (2019) Daisy-chain gene drives for the alteration of local populations. PNAS 116(17):8275–8282
North AR, Burt A, Godfray HC (2019) Modelling the potential of genetic control of malaria mosquitoes at national scale. BMC Biol 17:26
Oberhofer G, Ivy T, Hay BA (2019) Cleave and Rescue, a novel selfish genetic element and general strategy for gene drive. PNAS 116(13):6250–6259
Oye KA, Esvelt K, Appleton E et al (2014) Regulating gene drives. Science 345(6197):626–628
Pixley KV, Falck-Zepeda JB, Giller KE et al (2019) Genome editing, gene drives, and synthetic biology: will they contribute to disease-resistant crops, and who will benefit? Annu Rev Phytopathol 57(1):165–188
Quinlan MM, Smith J, Layton R et al (2016) Experiences in engaging the public on biotechnology advances and regulation. Front Bioeng Biotechnol 4:3
Rask M, Worthington R (2015) Governing biodiversity through democratic deliberation. Routledge, Oxon/New York
Rausand M (2011) Risk assessment: theory, methods, and applications. Wiley, Hoboken
Redford K, Brooks T, Macfarlane N (2019) In: Adams J (ed) Genetic frontiers for conservation: an assessment of synthetic biology and biodiversity conservation. IUCN, Gland
Ribeiro J, Kidwell M (1994) Transposable elements as population drive mechanisms: specification of critical parameter values. J Med Entomol 31(1):10–16
Roberts A, de Andrade PP, Okumu F et al (2017) Results from the workshop "Problem Formulation for the Use of Gene Drive in Mosquitoes". Am J Trop Med Hyg 96(3):530–533
Roggenkamp E, Giersch RM, Schrock MN et al (2018) Tuning CRISPR-Cas9 gene drives in saccharomyces cerevisiae. G3 Gene Genom Genet 8(3):999–1018
Sánchez CHM, Wu SL, Bennett JB, Marshall JM (2019) MGDrivE: a modular simulation framework for the spread of gene drives through spatially explicit mosquito populations. Method Ecol Evol 00:1–11
Schwaab E (2014) Addressing uncertainty in fisheries science and management. National Aquarium. https://www.aqua.org/Conserve/fisheries
Scott M, Gould F, Lorenzen M et al (2018) Agricultural production: assessment of the potential use of Cas9-mediated gene drive systems for agricultural pest control. J Respons Innov 5(sup1):S98–S120

Secretariat of the CBD (2000) Cartagena protocol on biosafety to the convention on biological diversity: text and annexes. Secretariat of the Convention on Biological Diversity, Montreal
Secretariat of the CBD (2011) Nagoya – Kuala Lumpur supplementary protocol on liability and redress to the cartagena protocol on biosafety. Secretariat of the Convention Biological Diversity, Montreal
Serebrovsky A (1940) On the possibility of a new method for the control of insect pests. Zool Zh 19:618
Singh JA (2019) Informed consent and community engagement in open field research: lessons for gene drive science. BMC Med Ethic 20:54
Sudweeks J, Hollingsworth B, Blondel DV et al (2019) Locally fixed alleles: a method to localize gene drive to island populations. Sci Rep 9(1):15821
Teem JL, Ambali A, Glover B et al (2019) Problem formulation for gene drive mosquitoes designed to reduce malaria transmission in Africa: results from four regional consultations 2016–2018. Malaria J 18(1):347
Thizy D, Emerson C, Gibbs J et al (2019) Guidance on stakeholder engagement practices to inform the development of area-wide vector control methods. PLoS 13(4):1–11
UC San Diego Institutional Biosafety Program (2018) Gene drives. https://blink.ucsd.edu/safety/research-lab/biosafety/gene-drives.html.
UN Environment Programme (2016) Decision adopted by the parties to the cartagena protocol on biosafety, CBD/CP/MOP/DEC/VIII/18. UNEP/CBD, Cancun, Mexico
UN Environment Programme (2018) Decision adopted by the parties to the cartagena protocol on biosafety, CBD/CP/MOP/DEC/9/13. UNEP/CBD, Sharm El-Sheikh, Egypt
UN General Assembly (1966) International covenant on civil and political rights. United Nations, New York
US EPA (2019) Risk assessment. https://www.epa.gov/risk
Vella MR, Gunning CE, Lloyd AL, Gould F (2017) Evaluating strategies for reversing CRISPR-Cas9 gene drives. Sci Rep 7:11038
Von Borstel R, Buzzati-Traverso A (1962) On the role of lethal mutants in the control of populations. Radioisotopes and radiation in entomology: proceedings of a Symposium, Bombay, 5–9 December 1960. International Atomic Energy Agency, pp 273–278
Windbichler N, Menichelli M, Papathanos P et al (2011) A synthetic homing endonuclease-based gene drive system in the human malaria mosquito. Nature 473(7346):212–215
World Health Organization (2014) Guidance framework for testing of genetically modified mosquitoes. WHO/TDR, Geneva
Yosef I, Edry-Botzer L, Globus R et al (2019) A genetic system for biasing the sex ratio in mice. EMBO Rep 20(8):e48269
Zhao T, Sundararajan SK, Tseng C-L (2004) Highway development decision-making under uncertainty: a real options approach. J Infra Syst 10(1):23–32

# Part VII
# Governance and Regulation

# Governance of Emerging Technologies/ Applications in the Bio/Life Sciences: Genome Editing and Synthetic Biology

**Krishna Ravi Srinivas**

**Abstract** This chapter discusses the issues and challenges in governing genome edited crops and synthetic biology. In case of genome editing, there are global initiatives to assess and identify principles and frameworks for governance. Regarding synthetic biology there is hardly any global initiative on governance despite concerns and Parties to Convention on Biological Diversity are discussing about regulating synthetic biology. Parallelly many ideas and proposals have been put forth on governing these technologies and some of them give emphasis to responsible research and innovation and public engagement, and changes in the regulatory regimes have been advocated. To what extent harmonization of governance at global level is not clear, but for governing these two technologies major revisions in national regulatory regimes would be needed. Given their potential to provide many promising solutions to major problems faced by human kind, harnessing that is possible only when there are effective governance systems that enjoy credibility and are based on science.

**Keywords** Genome edited crops · Genetically modified organisms · Responsible research and innovation · Biosafety · Product vs. Process · CRISPR · Cartagena protocol on biosafety · Precautionary principle · Biological weapons convention · Public engagement

## Abbreviations

| | |
|---|---|
| AHTEG | Ad-Hoc Technical Experts Group |
| BTWC | Biological and Toxic Weapons Convention |
| BWC | Biological Weapons Convention |
| CBD | Convention on Biological Diversity |
| CPB | Cartagena Protocol on Biosafety |
| CRISPR | Clustered regularly Interspaced short palindromic repeats |

K. R. Srinivas (✉)
Research and Information System for Developing Countries (RIS), New Delhi, India
e-mail: ravisrinivas@ris.org.in

 
A. Chaurasia et al. (eds.), *GMOs*, Topics in Biodiversity and Conservation 19,
https://doi.org/10.1007/978-3-030-53183-6_20

| | |
|---|---|
| DARPA | Defense Advanced Research Projects Agency (USA) |
| DIY | Do it yourself |
| EU | European Union |
| GEC | Genome edited crops |
| GMOs | Genetically modified organisms |
| iGEM | International Genetically Engineered Machine |
| IRGC | International Risk Governance Council |
| LMO | Living modified organisms |
| NIH | National Institutes of Health (USA) |
| NSF | National Science Foundation |
| OECD | Organization for Economic Cooperation and Development |
| RRI | Responsible research and innovation |
| SPS Agreement | Sanitary and Phytosanitary Agreement (WTO) |
| TAPIC | Transparency, accountability, participation, integrity, and capacity |
| WHO | World Health Organization |
| WTO | World Trade Organization |

## Introduction

As the other chapters in this volume point out we are facing challenges in governing developments in bio/life sciences which cannot be construed merely as a continuation of or variation of older ones. While there are some old issues/questions and regulatory models that are relevant for addressing the challenges in governing them, there are many novel issues too. For example, how to assess the long-term impacts of gene drives on ecosystems. Here I discuss governance issues in genome edited crops, and synthetic biology, highlighting the trends, the dilemmas, and the gaps in our understanding and in governance regimes. A usual line of argument in discussions on such technologies is that they offer unprecedented opportunities and by harnessing them we can solve problems related to hunger, agricultural productivity, health, and environment. This is equally true of these four. But now the narrative is more complex as we realize that while there are enormous opportunities, there are significant threats too.

A report from the World Bank points out that molecular biological and information technologies have enhanced rate of genetic progress by breeding and low-cost gene editing, stating that: "Biotechnology takes our ingenuity, our thirst for editing tools has created new opportunities in making targeted genetic improvements (World Bank 2019). But a recent book has an entire chapter on threats made possible by recent developments in biotechnology and synthetic biology and ends with a sombre note: "Biotechnology takes our ingenuity, our thirst for discovery—and turns it against us. It leaves us only as strong as our weakest, maddest link. It gives us promise and it gives us power, the most dangerous gifts of all" (Walsh 2019).

While it can be argued that Walsh's statement is an exaggerated assessment, as whether the gift is dangerous or not depends on how it is used. Certainly, technological determinism or dystopian visions of a technocratic society are not the right responses. Rather the challenge lies in governance of technology, particularly in shaping or directing the research and in ensuring that technology is deployed responsibly. So, I consider that it is better to pay more attention to that, than being swayed by too optimistic or dystopian views on impacts of biotechnology and synthetic biology.

From a governance perspective, as a result of features common to synthetic biology and genome editing, the following points and questions should be addressed:

1. Despite efforts to revise regulatory regimes and frameworks there are many unresolved questions, some of which are fundamental. For example, should crops developed through genome editing be treated as crops developed through traditional plant breeding or should they be treated as genetically modified organisms (GMOs)? Should gene drives be considered as living modified organisms as defined by and regulated by the Cartagena Protocol? Are the regulations meant for regulating genetic engineering adequate to regulate synthetic biology? Even if we leave out trade-related issues, the fundamental questions on governance of synthetic biology and genome edited crops are inevitable, given the advances in technology and the potential of the technologies.
2. How relevant are principles like the precautionary principle in developing regulatory regimes and are they adequate enough to regulate current applications and for future developments? Do we need new principles and ideas for this? Moreover, should applying the precautionary principle mean that only the European approach should be followed or should we revisit the idea of a "weak" vs. a "strong" precautionary principle? Can concepts like responsible research and responsible research and innovation (RRI) be integrated in governance regimes?
3. If it is argued that traditional risk assessments and environmental impact assessments are not adequate to understand the long-term environmental impacts, what are the alternatives and do we need new models and paradigms?
4. Concerns about biosafety and biosecurity are not new as they were raised in the initial years of genetic engineering and we have come a long way since then. But, in these cases, there are novel and unique features that demand new approaches and solutions. In particular there are concerns about do it yourself synthetic biology, dual use issue in synthetic biology and in case of genome editing, the potential for editing human germline raises many ethical, legal, and moral issues.
5. As these technologies have a wide range of uses in many sectors, the need for coherent governance frameworks is obvious. But will it be possible to develop frameworks that are robust enough to anticipate and be relevant in the context of technological developments? This is important because as technology advances it also shows that what was deemed to be impossible few years ago is possible now. In case of genome editing and artificial intelligence the new possibilities raise new hopes as well as concerns (Srinivas 2019).

While national regulations are being revised or reviewed to govern them, globally, there is not always a clear idea as to which treaty or convention is applicable. For example, Parties to the Convention on Biological Diversity have been discussing synthetic biology but that alone would not be sufficient. In case of genome editing the more controversial application of human germline editing is being debated by, inter alia, a WHO Commission. By now there is a consensus among scientists and science academies that governance of genome editing at the global level is necessary, although how to develop a global governance mechanism is not clear. Perhaps a combination of soft law and a convention or treaty will be necessary.

Anticipating all the impacts of a technology is not possible and should concerns about ethical aspects and/or regulations follow the innovations as and when they are adopted or can they be anticipated and planned for – this is a question for which there are no easy answers. For example, the Collingridge dilemma tells us that while it is difficult to predict the impacts till the technology is developed and used, once the technology is entrenched controlling or regulating is difficult (Collingridge 1980). It is true that technology assessment and foresight exercises and similar tools can help in understanding impacts and developing the regulatory framework. However, if the technology is radical and has wider impacts and implications in many sectors, or is a platform technology, such tools are necessary. But they may not be sufficient to fully understand and anticipate the impacts and plan accordingly. So how do we frame the technology or understand when it becomes so important? Should we take a technology/business as usual approach or try to grasp its potentials and impacts and be sensitive to them?

Highlighting the potential and wider impacts CRISPR could have, and comparing it with the Ford Model T car, Mariscal and Petropanagos (2016) argue that such an understanding of CRISPR shows there is a need for interdisciplinary, continuous, and international oversight, with participation from, inter alia, experts and stakeholders. On the other hand, Schultz-Bergin (2018) argues that to a great extent CRISPR is an ethical game-changer as it does not necessarily involve genetic materials from other species, and, thereby the question of crossing boundaries of species does not arise. While there are certainly ethical issues in applying CRISPR to animals, the significant difference from genetic engineering can result in important shifts in debates on the ethics of using CRISPR.

But from a different vantage point it has been argued, in a submission to the Biological and Toxic Weapons Convention (2018), that while genome editing technologies should be discussed, wider prospects, including aspects like "tacit knowledge," should be taken into account, as these may significantly alter the perception of risks and benefits and will pave way for a realistic understanding of impacts of these technologies for the Convention. Thus, the ramifications of these technologies go beyond issues discussed in debates on ethics, regulation, and innovation.

Highlighting the challenges in international regulation of genome editing, a pioneer in genome editing, Dounda, has cited the example of a physician in New York exporting genetically modified embryos to Mexico for implantation, particularly to evade US restrictions (Doudna and Sternberg 2017). While governance regimes are yet to emerge, different initiatives are in progress to develop regulatory frameworks and guidelines. Although there are many initiatives at global level ranging from the WHO Committee on human germline modification to discussions in CBD and CPB, from efforts by science academies to campaigns by civil society, there is currently no clarity on many issues. However, we can expect significant progress in the coming years as nations are now aware that unless these are strictly regulated, there could be many scandals like CRISPR babies in the making. After review of the development of CRISPR technology and CRISPR babies, Greely (2019) states "But, like Dolly's birth, He Jiankui's CRISPR'd babies are not the end of the world—or the beginning of the end of our species. They are a challenge both to the ability of Science to regulate itself, and to the world's trust in Science" – but trust cannot be taken granted. There are mechanisms to learn that and public engagement has to be part of it. For example, Scheben and Edwards (2018) call for transparent legislation, evaluation of potential risks, and better public engagement of the safety of genome editing. They also advocate data sharing initiatives and more collaboration among public and private sectors.

The ability to self-regulate as well as to inspire the trust and confidence of the world is a major challenge to scientists. Governance can play a key role in ensuring that science is well regulated and trustworthy, but there is no short cut in this. As academies of sciences, the WHO and other bodies deliberate on the governance of genome editing, there is hope that they will come out with guidelines and rules that explicitly permit/prohibit developing applications. Trust in science is closely linked with the ability to regulate. The WHO had issued a statement requesting regulators to disallow any human germline editing experiments.[1] Could this result in a de facto moratorium? It does not seem to have been the case.

Applying CRISPR/genome editing in agriculture has economic, ethical, and policy implications but these are beyond of the scope of this chapter on regulation. The literature on those aspects is growing (e.g., Bartkowski et al. 2018). Concerns over ethical aspects cannot be divorced from discussions on governing crop genome editing. According to Pirscher (2019), scientists working on genome editing in plants should be aware that ethics is intrinsic to their work and societal discourse on guiding values should be an integral part, right at the beginning of a research phase.

## Governance of Genome Edited Crops

### *An Overview of Regulation of Genome Edited Crops*

> But it's difficult to ignore the power of genome editing when it can be used for correcting birth defects, building resistance to disease, increasing tolerance to environmental conditions or enhancing senses or abilities. And, as history demonstrates time and again, it only takes the slightest crevice for the genie to escape the laboratory's Petri dish and become our new reality. (Carvalko 2020)

[1] https://www.who.int/news-room/detail/26-07-2019-statement-on-governance-and-oversight-of-human-genome-editing

There is now a consensus that genome editing in crops is a leap forward in plant breeding that can be effectively harnessed to develop plants with desired traits and takes relatively less time to do with genome editing. But technology is developing fast and the regulations are not in place in many countries to keep up with technological advancements. Whether GEC should be treated as GMOs for regulation is perhaps the biggest question before regulators. At the risk of oversimplification, process-based regulation and product -based regulation are the two key paradigms that guide regulatory regimes, with Europe as a prime example for the former and the USA for the latter. But beyond this preliminary level classification we have to look at the details, particularly the definitions and criteria to differentiate between a GMO and a non-GMO.

According to a survey on regulating GMOs, among the 33 countries and the EU surveyed, 15 countries and the EU used process-based regulation, and 14 countries used product-based regulation, while in four there was no well-defined regulatory framework. Of the 33 countries, 24 allowed commercial cultivation of GMOs. Argentina, the USA, Canada, the Philippines, and Bangladesh have all adopted product-based regulations, while the EU, Brazil, China, New Zealand, and Australia use process-based regulation (Ishii and Araki 2017). So, prima facie both types of regulations are in vogue among countries.

The regulatory frameworks introduced can be broadly classified as process-triggered regulation, product-triggered regulation, and new regulations for genome editing. But the picture is more complex because there are countries that allow import of GMOs and GM products as food or feed or as both, but do not permit cultivation of GM crops themselves. A recent publication comparing regulating genome editing plants and produce derived from them (Dederer and Hamburger 2019) shows that, among Argentina, Australia, Canada, the EU, Japan, and the USA, only the USA allows contained use, field trial, cultivation, and marketing as food or feed while in other countries/regions, there are restrictions. In the EU, only contained use and field trials are allowed, as is the case in Japan and Argentina. In Australia none of them is allowed and in Canada marketing as food or feed is not permitted.

Further, as Hamburger (2019) points out, while diversity and differences in regulatory regimes are based on different approaches, the differences are evident in regulatory triggers and points of entries and, in some jurisdictions, in coexistence and labelling, and differences arise also on account of legal classifications of plants and their produce. He concludes that while regulation is important, there are other factors too that matter and on account of these GEC (genome edited crops) may meet the same fate of GMOs.

Although some countries are making progress in regulating plant genome edited crops, the following can be inferred from the literature:

1. The elephant in the room is regulation within the EU on which there has been little progress.
2. Despite progress, the diversity will remain and so will legal fragmentation on account of different norms and rules being developed for different purposes.

3. Not all countries may opt for complete deregulation or treating GEC equivalent to crops developed by traditional plant breeding in all cases/circumstances.
4. There is little scope for regulatory harmonization even if countries opt for either product-based or process-based regulation.

In fact, as Hamburger (2019) points out, the differentiation in terms of product based and process based is not so helpful in understanding the complexities of regulation. The regulation of GEC is likely to be more complicated than that of GMOs, and the implications of this for development and adoption of GEC will be huge. In other words, we cannot assume that in the post-GMO world GEC will be less regulated and more accepted. Of course, there is also a view that GEC will put an end to the GMO vs. non-GMO controversy and will be better accepted.

From a technology governance perspective, regulating genome editing in crops is going to be more challenging than regulating genetic engineering because genome editing opens up new possibilities, some of which can result in plants with novel traits that were unknown before. The current frameworks certainly need revision and to bring about coherence agencies may have to work more in tandem. It is also important that we look beyond product- and process-based regulatory approaches that were developed in the context of genetic engineering.

Arguing from another vantage point, Bartkowski et al. (2018) point out that developing regulations should take into account four features, viz., non-traceability in the final product, i.e., organisms, decentralized knowledge and its use, the acceleration of breeding, and the uncertainty about off-target changes. They hypothesize that genome editing crop is likely to be confined to development of highly profitable crops.

As these authors point out, many issues ranging from ethical concerns to access to technology have to be considered to harnessing the technology in an appropriate manner and ensure that its potential is realized. In this regulation has to play a key role. However, as Zhang et al. (2020) point out, the main constraints are not the technical limitations but whether consumers would choose genome edited foods is the major question.

## *The Regulation of Genetically Modified Organisms and Crops in Europe*

To state that European Union (EU) has the most harmonized and comprehensive regulatory framework for GMOs is no exaggeration. But when it is extended to GEC it considered by many as problematic as that could delay introduction of GEC in the EU and would make the regulation of GEC complicated. But the verdict of the European Court of Justice (ECJ) given in July 2018 declared that GEC would have to be regulated as GMO (ECJ 2018).

The comprehensive regime in the EU covers authorization regarding contained use, field trials, marketing of GMO, post market monitoring, labelling, and traceability. Over the years for many reasons including regulation and consumer reluctance, the cultivation of GM crops in Europe is confined to Spain with only a single event permitted for commercial cultivation. Europe, however, imports huge quantities of GM soya and corn as animal feed. Technically the authorization is for import as food and animal feed, but due to consumer resistance, the imports are confined for feed.

Under Article 2(2) of the EU Directive on the deliberate release into the environment of genetically modified organisms of 2001 (Directive 2001/18/EC), GMO is defined as an organism in which the genetic material has been altered in a way that does not occur naturally by mating and/or natural recombination. But to make sense of this one has to read Annex 1A (parts 1 part 2), where they identify the practices that constitute genetic modification and also the practices that are deemed not to constitute genetic modification. Article 3(1) and Annex I B of the Directive exempt GMOs obtained by mutagenesis or cell fusion from the scope of the Directive. This process-based definition is at the heart of disputes over interpreting whether GEC are GMOs or not. While many scientists and academies have argued that GEC should not be considered as GMO for regulatory purposes, the European Court of Justice (ECJ) decided otherwise. It is expected that the recently appointed new commissioners will review the situation and propose a reforming of the EU regulatory framework.

Some European states have asked for a review of the GMO regulations, and in May 2019 at an Agriculture and Fisheries Council meeting, 14 member states requested the commission to undertake a review. In the judgment, the ECJ made it clear that techniques and methods of directed alteration of genetic material do constitute a genetic modification and therefore they cannot be exempted under mutagenesis exemption. This means GEC have to treated as GMOs under the EU regulation applicable to GMOs. So practically rules applicable for risk assessment and authorization applicable to GMO will be applicable to GEC.

The rationale for such a ruling has been provided in the judgment in elaborate detail, particularly in Paras 48–51 (ECJ 2018). The ruling was a surprise because the advocate general opined that the correct legal interpretation would be that the new directed mutagenesis techniques are covered by "mutagenesis techniques" and hence are exempt from the EU regulatory framework for GMO (Bobek 2018).

Many proposals have been made for review and reform of the regulations in Europe. For example, it has been suggested that constructive dialogue among key stakeholders is essential so that Europe can benefit from new developments in biotechnology without compromising on a high level of safety (Bruetschy 2019). It has been suggested that the EU framework should be revised into a flexible system that can adapt to technological advancements. In this proposed framework, there is scope for public acceptance and a role for farmers (Ricroch et al. 2016).

Medvedieva and Blume (2018) suggested that while gene edited plants with GM material should be regulated, gene edited crops that have no GM material need not be subjected regulation.

A product-based regulation in which risk regulation is triggered through an assessment of the novelty of the trait of a plant is an option. Further the novelty should be defined in terms of familiarity. Although there can be issues in defining familiarity, it is possible to define it. So, irrespective of breeding techniques, only plants with "unfamiliar" traits could then be considered novel and be subject to regulation (Dederer and Hamburger 2019). On the other hand, the irony (or tragedy) is after an extensive review of current regulations and options for reform, Voigt and Münichsdorfer (2019) are skeptical about any amendment in the near future in the regulations.

In fact they point out that the applicability of current regulatory frameworks for GMOs to GEC could prevent successful adoption of plant genome editing in Europe. A novel proposal has been put forth by Wasmer (2019), in which various options including using the flexibility available within the current regulations are exercised and the regulatory regime is revised and made up to date over a period, by undertaking step by step changes, with resultant impacts like lowering legal uncertainty and lowering the costs for innovators.

The EU can either revise the regulations or ensure that GEC are not treated as GMOs and hence they are exempted from regulation. Or it can revise it partially to treat these as GMOs for one or more purposes and not as GMOs for other purposes. Although this gives flexibility, it will disrupt the regulatory coherence based on definitions and interpretations. The other option would be to move from process-based regulation to product-based regulation. But this would mean a total revamp of the regulatory principles and oversight mechanism and is unlikely to be accepted by member states or citizens. Hence, it is likely that reform may be possible only if it is done slowly and with greater acceptance. But a consensus is not likely to be achieved given the currently polarized views and consumers' reluctance to accept GM foods. However, the consequences of not reforming the regulations can cost Europe considerably in terms of innovations and R&D. The Science Advisory Council of European Academies has made some important suggestions on regulation and has suggested that definition/exemptions of GMOs should be revised, so that Europe can capitalize on the genome editing technologies and new legal framework should be developed focusing on traits rather than on processes (EASAC 2020).

Hundleby and Harwood (2019) suggest that the EU should opt for fit for purpose regulation and this can be in harmony with regulations elsewhere so that the full potential of the new breeding technologies can be harnessed. However, it is difficult to predict how the rest of the world would regulate and whether any harmony is possible is a big question. Between total deregulation and the EU-style regulation there can be many approaches and how countries would regulate is not clear. Still, it is safe to hypothesize that although global harmonization may not be possible not many countries will prefer the EU-style regulation. Clarity may emerge in the next 5 years or so, when there is more widespread adoption of GE crops.

Hundleby and Harwood (2019) also highlight trade-related issues and point out that in Europe while GM HT soybean is not cultivated, Europe imports huge quantities of soybean every year. But the question is what will happen if the EU does not revise its regulations. This can have implications for the adoption of GE crops in countries that are dependent on the European market for export. It is likely that some countries may opt for segregating and labeling GE crops to ensure that exports are not affected. Ultimately, if consumers become convinced that GE crops are not GM crops and are safer, strict rules on labeling and segregation may become redundant. Cultivation of GM crops is very limited; for example, share of Bt corn cultivated is just 1.3% of total area of corn cultivation. In some countries which export significant quantities of feed, the EU imports significant quantities of soybeans and soybean meal, corn and corn processing by-products, and rapeseed and rapeseed meal. Regarding soybeans and soybean mean, in the major exporting countries (USA, Argentina, Brazil, Canada, and Paraguay), the share of GE soya is more than 90% (GAIN 2020).

From another vantage point after reviewing European patent law and policies on regulation of agricultural biotechnology, Jiang (2019) cautions that on account of incoherence between patent law and regulation of GMO, the European Union might lose the technological and economic advantage from adopting the technology. This point makes sense, as, while patent law facilitates innovation by incentivizing it, regulation can stifle the adoption of the same. This results in less than optimum outcomes and trade barriers.

After analyzing 11 official reports and position statements from seven countries in Europe and examining the similarities and differences in the positions taken in these seven European countries, Meyer and Heimstadr (2019) wonder whether different policy options for genome edited organisms can coexist and if so, what would be the consequences. According to them the entanglement or otherwise between technology and law will be further debated.

The earlier that entanglement and disentanglement are resolved the better as developments in Europe have implications, not only for governing GE crops in Europe and their adoption there but also in other countries in the world.

## *Thinking Beyond Product: Process Dichotomy and Issues in Regulation*

Genome editing should be considered as a tool that can be used for different applications or purposes. So it will be better to think in terms of regulating genome editing per se through a single law. In some contexts when there is no clarity on definitional categories, either new interpretations will be necessary or the definitional categories may have to be revisited. Regulation by application rather than by technology per se is a better approach when the technology can be used for many purposes on its own or in conjunction with other technologies. In the case of genome editing it is obvious that regulation for crop genome editing and human genome editing cannot be governed by the same principles and that regulation has to be sensitive to applications and outcomes. The two major regulatory frameworks in agricultural biotechnology – process-based regulation and product-based regulation – now will have to be revisited in the context of crop genome editing.

This process-based regulation, and product-based approach, could be traced to the initial years of biotechnology regulation which started with the need to regulate genetic engineering – then an emerging technology in which regulators had little experience in understanding and regulating.

Since then, we have come a long way and yet this dichotomy continues. The approaches are relevant and well entrenched, making it difficult to think beyond them, without thinking in terms of a total revision of the regulatory system. But the issue is much more than that. Ultimately what matters is the combination of the approach with other principles (e.g., the precautionary principles) and policies (e.g., import is fine for feed but not for human consumption) that determines the adoption of any technology or use of any product. The advent of GEC has necessitated a change in the regulation of agricultural biotechnology. In this context even deregulation or placing GEC beyond regulatory purview is also a way of regulating. But what matters is how consistent the regulations are and whether there is a coherent framework that is comprehensive enough to address all concerns of risk, safety, and human health.

Summarizing the presentations and talks given in an OECD conference last year, Friedrichs et al. (2019) point out that: (1) countries such as Australia, New Zealand, and India using a process-triggered regulation are reviewing the scope of current regulation and are reviewing their regulations[2]; (2) Canada and the USA are regulating on the basis of a product trigger and consider novelty of trait irrespective of the technology used to develop; and (3) in its regulations, Argentina has provided for genome editing as a subcategory. Although the revamp of the framework of biotechnology regulation in the USA is far from over, it is a positive development that many countries are moving ahead with review and revision. In case of Europe, as pointed out earlier, the probability of a revision is almost ruled out in the short term, but in the long term could be possible. But such a review and reform has to be seen light of capacities in the countries to undertake R&D in genome crop editing and commercialization. The reality is that many countries are still in the era of genetic engineering; in the Asia-Pacific region less than 10 countries have the capacity to adopt GEC (FAO 2019).

Hence, even as countries review and revise their regulations, the adoption of genome editing in agriculture may take years. Unfortunately, the countries that lack the capacity may be the ones who need it most. On the other hand, we do not know much of the regulatory costs for GEC. As regulatory regimes are being revised it is important to ensure that regulatory costs do not become a burden for developers, particularly for public sector and not for profit organizations. It is also important to ensure that the precautionary principle is used in such a way that it does not create an impression that deviation from the EU approach could compromise food safety. Applying the precautionary principle to regulating genome editing of crops need not mean that only version or interpretation is valid or accepted. Rather it is time to

---

[2] India is revising its regulatory framework and has put up a draft document, "Draft document on Genome Edited Organisms: Regulatory Framework and Guidelines for Risk Assessment" for comments (DBT 2020).

debate this in light of earlier debates on "hard" and "soft" precautionary principles for regulating agricultural biotechnology.[3]

It is not likely that all countries will switch over to product-triggered regulation. Hence it is better to explore whether the Canadian approach or that of Argentina can be tested by other countries and be adopted if found relevant. The current debates and revisions give an opportunity to think about and implement new ideas in regulation. In my view instead of stopping with regulation, it is also the time to revise the policies regarding agricultural biotechnologies in countries which are approving GEC. Given the advances in technology and the benefits, besides regulation, the enabling environment should be conducive for GEC.

Similarly, the idea of GMOs needs to be revisited. If the public were to consider that GEC was the same as GMOs, or GMOs were substantially equivalent, then their adoption might be less in countries where there is a skepticism or resistance to GMOs. Hence, if the regulatory process clears this through well-defined categories and rules and ensures that GECs are not treated like GMOs, then, the public perception may change. But to convince the public, it should be made clear that the regulation is based on science and also recognized so by scientific bodies. It is equally important that the regulatory regime is sensitive to developments in technology and science and is flexible enough to accommodate them.

Hence, current developments should be seen as opportunities to learn and revise regulations and wherever necessary revise the whole governance of agricultural biotechnology. An elephant in the room is the question of trade and regulation and how changes in regulations may impact global trade in GEC. Important questions are whether we will see another round of disputes placed before the WTO if the EU regulatory regime remains unchanged and how the status quo in European regulation will affect the countries that have adopted GEC but are also exporters of food to the EU.

Citing the case of golden rice and Hawaiian GM papaya, Hundleby and Harwood (2019) state: "Even in cases where a specific GM crop may not be envisaged as a product for a trading country, it is vital that the country's views and opinions should not negatively impact on other countries that stand to benefit from such technology." However, the picture is not so simple because countries have to balance trade concerns with other factors, and for countries that depend on exports to a specific market there are not many choices. It is equally important to understand that unless GEC have features wanted by consumers, preference for them from consumers may be less. This can result in lesser adoption.

A better solution may be the application of standards rather than regulations as a better, flexible, and more appropriate mechanism. In this the goal is to develop suitable mechanisms for managing risks and benefits, within a well-defined range of operating parameters. It may be that the adoption of renewable fuels in aviation

---

[3] For reasons of space I am not discussing this in detail. For different views on crop genome editing and precautionary principle see Dürnberger et al. (2019) and also Steel (2014) for discussion of the precautionary principle, its philosophy, and relevance in policy.

industry standards based on testing, assessment, and specifications can be an example that can be studied and adapted.

These are challenging times to regulate GECs, but they are also times to debate, revisit, learn, and revise existing regulations and to address issues in governance.

## Synthetic Biology

### *Synthetic Biology: Origins and Issues in Governance*

Synthetic biology as a discipline emerged in the early years of this millennium, and since the mid-2000s it has attracted much funding and support from the USA, Europe, and elsewhere, as reflected in securing both funding and publications. The iGEM (the International Genetically Engineered Machine) competition was held for the first time in 2004 and since then it has grown significantly. In parallel, the do it yourself biology (DIY biology) groups and initiatives also grew and diversified in terms of geography and activities. In fact, as Shapira et al. (2017) point out, funding from public research funding agencies nurtured the growth of synthetic biology and a significant share of funding and publications are from the USA and Europe.[4] While the literature highlights how synthetic biology can make positive contributions, concerns about the risks from synthetic biology are also found in the literature, for example, developing new combinations of genes without fully understanding their impacts, dual use research and applications from synthetic biology, horizontal gene transfer, impacts of synthetic biology on conservation and use of biodiversity, and recreating known pathogenic viruses. In this context it is worth noting that the World Organisation for Animal Health (OIE 2019) published "Guidelines for Responsible Conduct in Veterinary Research identifying, assessing and managing dual use" for creating awareness and to reflect upon, so that appropriate guidelines for regulating dual use, including research, can be developed by countries and institutions.

As synthetic biology diffuses the capacity to experiment with and deploy it increases. In one sense among the emerging technologies synthetic biology is the most "democratized" technology, thanks to spread of DIY culture. At the same time, this and other factors have raised concerns about biosafety and biosecurity. The governance challenges in synthetic biology stem from multiple concerns and potential risks. But the fundamental issue is synthetic biology is different from genetic engineering and hence regulatory regimes meant for genetic engineering may not be suitable. Fundamentally, through synthetic biology, it is possible to redesign the existing biological system or an organism and also possible to create totally new/novel organisms not found in nature. While the second feature has led to fears and debates about using synthetic biology to play the role of God, the first is also equally important.

---

[4] See also Raimbault et al. (2016) and https://phys.org/news/2018-09-synthetic-biology-revolution.html for a review.

The power to design DNA from scratch and writing (new) genetic code combined with tools to alter the makeup of organisms creates exciting possibilities and is an example of the allure of the technology. As synthetic biology goes beyond genetic engineering in terms of tools, scope for intervention, and potential for creating novel organisms, how adequate are the current regulatory frameworks? Although this question has been raised for many years and attempts have been made to answer this, we are yet to see the development of a comprehensive regulatory framework for Synthetic Biology. Analyzing the situation in the USA, Georgiadis and Ryznar (2019) expect that a sort of comprehensive regulatory framework might emerge, preceded by patchwork of laws.

Bruetschy (2019) points out that EU scientific opinions expressed in 2014 and 2015 acknowledged the fast developments in the field and were of the view that the first new developments could be assessed under the current regulations for GMOs. He states that the current methodologies may need to be revised and enhanced regularly to make sure that they are safe. While gene drives are also regulated under the same rules, there is a huge difference between impacts of GMOs and impacts of gene drives as "whereas for conventional (transgenic) GMOs the impact on the environment is risk assessed and risk managed with a view to limit the dissemination, the primary objective of gene drives is precisely to be disseminated in the environment to fulfil their objective." But there are many other applications of synthetic biology, and it is better to address gene drives as a separate category rather than to regulate them as GMO. Ideally they merit a separate regulatory regime different from that for synthetic biology.

## *Regulation: Old Models, New Approaches, and Proposals*

Trump (2017) points out that synthetic biology is governed by older frameworks meant for genetic engineering and they were not meant originally to govern them. Although his study examines only the USA, Europe, and Singapore, it is equally true of other countries. But irrespective of lack of relevant regulatory regimes, there could be conflicts and disconnects. Identifying them at three levels a report from OECD (2014) pointed out that, besides product vs process regulation, the applicability of the precautionary principle could be an issue. Within a country or region, on account of differences in mandates and biases, different agencies can take different views on regulation and norms. There could be regulatory conflicts among/between federal agencies and those of states or provinces. But as the IRGC (International Risk Governance Council 2009) pointed out, there could be overlaps among synthetic biology and other technologies, such as nanotechnology and biotechnology, particularly where there is convergence. Another source of contestation could be risk culture and variations in risk culture in countries. As a result, while some risk cultures may be more open to multi-stakeholder approaches and participation, some may be more biased toward centralization (Trump 2016).

Kolodziejczyk and Kagansky (2017) point out that regulations developed for GMOs are outdated with respect to synthetic biology, which needs totally new approaches for governance and risk management.

But developing a regulatory regime for synthetic biology is not easy on account of concerns over DIY biology, dual use, biosafety and biosecurity, and its impacts on environment and biodiversity. In the case of biotechnology, these were addressed, starting with self-regulation and then through frameworks. Moreover while the basic techniques in genetic engineering were developed by the mid- and late-1970s, it took time for the technology to diffuse and be adopted. Hence, when the Asilomar Conference (see above) highlighted the views of scientists on regulation, the primary concern then was biosafety and regulating genetic engineering in the initial phases.

In case of synthetic biology, the picture is not that simple. First, it has diffused at a faster rate and the number of research groups/initiatives has increased many times within a decade, as have the publications and other indicators. Given the potential of synthetic biology, it was initially supported by DARPA, NIH, and NSF in the USA but soon many other countries including China also started supporting it. Interest shown by venture capital investors and philanthropic foundations like the Gates Foundation helped synthetic biology to progress. Similarly, the growth of DIY bio has resulted in various groups and initiatives working on synthetic biology, with iGEM providing a global platform for new ideas and novel experiments. Availability of protocols and kits online has enabled a growth of the DIY culture in synthetic biology, and it is estimated that there are now about 168 DIY bio groups in the world.[5] At the global level discussions in the CBD on synthetic biology have ensured that countries are able to assess and discuss the impacts of synthetic biology for biodiversity and the environment.

All this in fact has made regulating and governing synthetic biology complex. It has been pointed out that national regulations themselves have been found inadequate and to need major revision. At the global level there is no single convention or treaty that is specifically applicable to synthetic biology. But from a biosafety perspective, the World Trade Organization's (WTO) 1995 Agreement on the Application of Sanitary and Phytosanitary Agreement (SPS Agreement) and the Convention on Biological Diversity's Cartagena Protocol on Biosafety (CPB) are most relevant. Technically, the CPB deals with living modified organisms (LMO) and the SPS Agreement sets the scope of WTO member states for restricting international trade, on grounds of food safety and animal and plant health.

The CPB was negotiated and ratified in the initial years of this millennium when synthetic biology was unheard of. Hence, its definition of LMO was based on the understanding of genetic engineering and definitions of GMO available at that time. Hence the definitions of LMO are for organisms and not for the elements that constitute them such as purified DNA. Further, the CPB is primarily concerned with the transborder movement and handling of LMOs. Hence its articles deal with physical

[5] https://www.unenvironment.org/news-and-stories/story/risks-and-potential-rewards-synthetic-biology

movement, transfer, use, and dealing with risk. This means that while the CPB may be the most relevant convention for synthetic biology, it is not adequate to handle it.

After examining the provisions of CPB and BTWC in the context of synthetic biology regulation, Rabitz (2014) puts forth the view that current institutional arrangements at the global level are not adequate to deal with potential future risks and there are gaps regarding, inter alia, transboundary movement of purified DNA and suggests that increased focus on health governance of risks associated with biological agents is desirable.

He further suggests that precautionary decision making can be used to balance risks and benefits. He takes the view that while soft law instruments like nonbinding codes of conducts may be useful, the pace of technological change may necessitate more structured and formal international regulation through amendments to existing agreements or negotiations for a new one. In 2014, the Inter-Academy Panel (IAP 2014) issued a "Statement on Realising Global Potential in Synthetic Biology: Scientific Opportunities and Good Governance" and stressed the need for global commitment in terms of "Preparing researchers for work in synthetic biology," "Engaging with the public and clarifying ethical and social concerns," "Considering alternative models for owning and sharing research outputs," and "Disseminating guidelines and calling for scientific responsibility" and highlighting the need for collaboration among stakeholders and stated: "We must collectively ensure that policy development worldwide is sufficiently flexible to encourage research and manage innovation, including those applications not yet envisaged, while suggesting sensible practices to mitigate any risks."

Currently, parties to Biological Weapons Convention (BWC) are discussing the implications of synthetic biology for the convention while parties to the CBD have deliberated on this and an Ad-Hoc Technical Experts Group (AHTEG) has been formed. The reports of AHTEG and other developments will be discussed by the convention's Subsidiary Body on Scientific, Technical and Technological Advice and by the parties to CBD in Kun Ming, in the conference of the parties to be held in 2020.

However, before discussing the global governance of synthetic biology, we need to understand in the literature on governance of emerging science and technology interventions such as risk assessment and responsible innovation, but there are not many studies that "take a more systemic view to examine how these different approaches might learn from and work alongside each other, across multiple technological domains, and at multiple levels of governance" (Chubb et al. 2018).

This is not surprising as in the literature on governance of synthetic biology, there are references to "proactive and adaptive governance," "self-governance," "anticipatory governance," and "transnational governance." But the dilemma is not that of jargon or terminology but that of governance on the basis of what principles, for what objectives, and through what mechanisms. While it is obvious to state that governance has to be dynamic and adequate to address the changes and developments in technology, that is easier said than done.

Wallach et al. (2018) point out that emerging technologies like synthetic biology pose governance challenges on account of:

1. Them tending to have multiple applications in different sectors and to be regulated by different regulatory agencies.
2. The high uncertainty about future risks and benefits is inherent in them and it is difficult to predict or anticipate them.
3. Concerns raised on account of them go beyond jurisdiction of regulatory agencies, as agencies have a limited/narrow mandate.

Hence there is a coordination problem in the governance of these emerging technologies. To address this, they suggest forming a Governance Coordination Committee (GCC) and state that a GCC approach to synthetic biology is a comprehensive model for the agile oversight of both the field as a whole and for specific applications.

On the other hand, Weik et al. (2012) suggested adopting a new approach, combining anticipatory governance and transformational sustainable science for supporting innovation in synthetic biology.

From a different perspective on innovation and governance, Stirling et al. (2018) call for a governance mechanism for synthetic biology that provides for affected parties to express their views and provide for public participation. After a comparative study of synthetic biology regulation in the USA, the EU and Singapore, Trump (2018) has proposed a TAPIC (Transparency, Accountability, Participation, Integrity and Capacity) approach to synthetic biology governance.

One way to conceptualize the regulation of synthetic biology is to focus on specific product outputs, rather than on process per se, covering broadly based and evolving methodologies. A relevant analogy will be regulation in the health care sector, where rapid innovations, including newer technologies, procedures, and devices, are the norm than an exception.[6] This idea makes sense but will require further elucidation.

After analyzing the synthetic biology policies and discourses in Europe, China, and India, Rerimassie et al. (2015) differentiate between innovation discourse and risk discourse in synthetic biology and call for a global dialogue to address the specific governance challenge in each region and to reflect regional values and concerns. They argue that an international forum is necessary for this and suggest that iGEM and UNESCO could play a role in this.

Thus, while there have been many ideas and proposals on synthetic biology governance, their impact on a revision of regulatory regimes remains unclear, nor it is becoming evident that changes envisaged in the USA and Europe will incorporate some of the values and ideas proposed. Irrespective of this, these ideas may gather traction and some may come to be used in the governance of synthetic biology.

Globally, diffusion of synthetic biology is limited and, even among countries that have some capacity in synthetic biology, there is hardly any attempt to coordinate their activities on regulation, although the OECD has been trying to bring together experts and policy makers from OECD countries and elsewhere to discuss global regulation of synthetic biology. In this context the suggestion by Kolodziejczyk and

[6] I thank an anonymous reviewer for this point and for the comparison with health care sector.

Kagansky (2017) for joint and unified research and consolidated laws and regulations among all G20 countries is worth considering.

From a different perspective, Zhang et al. (2011) identified scientific uncertainty and cross-borderness as sources of concern and mapped the issues as "Governing knowledge and non-knowing," "Cultivation of external accountability," and "Fragmentation of social authorities." According to these authors: "effective governance concerning synthetic biology may only be attained when regulators attend to these more fundamental questions. The findings of this paper indicate that scientific uncertainty and cross-borderness can be better attended to when current scientific bureaucracy is supplemented by an 'artistic' form of governance." However, as they did not elaborate this further, nor explain how to translate this into practice with case studies, this model although interesting has not moved ahead.

From the above analysis, it is clear that governance of synthetic biology is really a challenge, particularly at the global level. But, as pointed out, the pace of revisions in regulatory regimes has been slow and most countries are still using ones applicable to and developed for genetic engineering. However, any initiative on global governance or any reform of national regimes today will have to take into account:

1. The importance of synthetic biology governance for biosecurity and biodefense at the national and global level
2. The discussions in CBD and the discussions in BWC and BTWC
3. The diffusion of synthetic biology and spread of DIY culture in synthetic biology

While many of the proposals for biotechnology governance at national levels have not taken into account these three factors, global governance is unthinkable without understanding the importance of all three. Of these, the first is becoming more important and this is evident from publications like "Biodefense in the age of Synthetic Biology" (NAP 2018). The dual-use dilemma and concerns over bioterrorism are bound to impact the governance of synthetic biology. The discussions in the CBD may not result in any consensus on governance reforms nor may they result in any revision to the CBD/CPB in the short term.[7]

However, as the CBD provides a forum for parties (i.e., national governments that are signatories to the convention) and others to discuss and share ideas and also provide inputs to the various groups and for a setup by the CBD, it can facilitate a better understanding of the issues by the parties, resulting in national-level revisions or new laws. Regulating DIY synthetic biology may appear to be impossible, but DIY groups work with governments and accept the importance of self-regulation and adherence to biosafety norms. Further it is easy to cover their activities and monitor them through changes in regulatory regimes, for specific activities and for handling and acquiring designated materials.

[7] See https://bch.cbd.int/synbio/; see also Li et al. (2019).

## Conclusions

Whether it is genome edited crops or synthetic biology, governance is becoming a challenge because there are major issues with current regulatory regimes and there are new stakeholders who want to be heard and consulted. Incremental revisions/modifications are not sufficient. Governance or regulation cannot be reduced to matters which only technocrats and experts decide, strictly going by scientific risk assessment and other tools. In the last two decades there has been a realization that society cannot be taken for granted and expected to accept (with gratitude) what scientists and organized science and technology institutions and governments provide. The GM fiasco in Europe and resistance to technologies such as nuclear technology and carbon capture and storage have highlighted the need to engage with the public and to go beyond the assumption that public needs to be only educated and informed so that it appreciates the benefits and adopts them passively and need not be listened to.

In the proposals on governance we see a reflection of this experience and lessons learnt from it. But the proposals do not end with asking for more public engagement and public participation. They also suggest alternative models for assessing risks, propose incorporations of values, and try to make the innovation process a reflective one.

Responsible research and innovation (RRI), a concept and practice promoted by the European Commission, has been discussed in the literature on governance and innovation (Bruce and Bruce 2019. RRI gives emphasis to anticipation and reflexivity and takes into account inter alia ethics. RRI is one way to reconnect with society and develop innovations needed by society. The challenge lies in translating this idea into workable projects in innovation and governance in synthetic biology and genome editing. As pointed out earlier, there is an initiative that links RRI with gene editing. Although we are yet to know its impact, such initiatives will at least create awareness about responsibility in gene editing.

These initiatives complement those on ethics and governance in genome editing and on directing innovation in genome editing to meet specific objectives that are ethically sound, promote access, and facilitate responsible research. For example, the Open Plant program based in the University of Cambridge facilitates access to materials and other resources, promotes open access and open innovation, and provides a Material Transfer Agreement to promote open access and sharing in synthetic biology. Open Plant blends RRI with open source and open innovation and enables public participation and encourages DIY biology (https://www.openplant.org/).

In the case of genome editing in plants it is important to gain the confidence of people by making claims based solely on science and assuring them that these are safe as they are regulated on the basis of sound science. While the literature explains how genome editing can play a key role in the years to come in agriculture, it is important to ensure that it is not promoted through hype and unrealistic expectations from technology.

As Ricroch (2019) points out as all genes in all genomes cannot be modified by genome editing technologies, development of GMOs by transgenesis is still necessary; new traits can be produced by various tools and technologies, including through conventional breeding techniques. However, consequences in terms of risk assessment are not equivalent. He has raised an important point that has to be borne in mind in developing risk assessment as part of regulation.

So, we now need to better understand the merits and limitations of genome editing in crops and synthetic biology. Governance of these technologies has to be anticipatory, adaptive, and credible so that while society enjoys the fruits of science, it places trust in science and scientists and appreciates their contributions.

## References

Bartkowski B, Theesfeld I, Pirscher F, Timaeus J (2018) Snipping around for food: economic, ethical and policy implications of CRISPR/Cas genome editing. Geoforum 96:172–180

Bobek M (2018) Opinion of Advocate General Bobek: Case C-528/16. ECLI:EU:C:2018:20. https://curia.europa.eu/jcms/upload/docs/application/pdf/2018-01/cp180004en.pdf

Bruce A, Bruce D (2019) Genome editing and responsible innovation, can they be reconciled? J Agric Environ Ethics 32:769–788. https://doi.org/10.1007/s10806-019-09789-w

Bruetschy C (2019) The EU regulatory framework on genetically modified organisms (GMOs). Transgenic Res 28(Suppl 2):169–174. https://doi.org/10.1007/s11248-019-00149

BWTC (2018) Technical working paper on genome editing and other scientific and technological developments of relevance to the convention BWC/MSP/2018/MX.2/ WP.2 Submitted by Switzerland. https://www.undocs.org/bwc/msp/2018/mx.2/wp.2

Carvalko JM (2020) Conserving humanity at the dawn of posthuman technology Palgrave, New York. https://doi.org/10.1007/978-3-030-26407-9

Chubb J, Montana J, Stilgoe J, Stirling A, Wilsdon J (2018) A review of recent evidence on the governance of emerging science and technology. Wellcome Trust, London. https://wellcome.ac.uk/sites/default/files/evidence-review-governance-emerging-science-and-technology_0.pdf

Collingridge D (1980) The social control of technology. Frances Pinter, London. ISBN 0903804727, 9780903804721

DBT (Department of Biotechnology) (2020) Draft document on genome edited organisms: regulatory framework and guidelines for risk assessment. http://dbtindia.gov.in/sites/default/files/Draft_Regulatory_Framework_Genome_Editing-9jan2020.pdf. New Delhi

Dederer HG, Hamburger D (2019) Introduction: regulation of plants derived from genome editing—what lessons to be learned from other countries? In: Dederer HG, Hamburger D (eds) Regulation of genome editing in plant biotechnology. Cham, Springer, pp 1–17. https://doi.org/10.1007/978-3-030-17119-3_1

Doudna J, Sternberg S (2017) Preparing for the global ramifications of gene-editing technology. Global Challenges Foundation. https://globalchallenges.org/our-work/quarterly-reports/globalgovernance-in-the-age-of-disruptivetechnology/preparing-for-the-globalramification

Dürnberger C, Pfeilmeier S, Schleissing S (eds) (2019) Genome editing in agriculture: between precaution and responsibility. Nomos Verlagsgesellschaft, Baden-Baden. https://doi.org/10.5771/9783845296432-9

EASAC (European Academies Science Advisory Council) 2020 The regulation of genome-edited plants in the European Union, Brussels https://easac.eu/fileadmin/PDF_s/reports_statements/Genome_Editing/EASAC_Genome-Edited_Plants_Web.pdf

ECJ (2018) Judgement of the Court (Grand Chamber), 25 July 2018 in Case C-528/16. http://curia.europa.eu/juris/document/document.jsf;jsessionid=9ea7d0f130dcd5adc6577ba74dc9b5a

cf2530b87e485.e34KaxiLc3eQc40LaxqMbN4Pb3yRe0?text=&docid=204387&pageIndex=0&doclang=EN&mode=req&dir=&occ=first&part=1&cid=72898
FAO (2019) The status of application, capacities and the enabling environment for agricultural biotechnologies in the Asia-Pacific region. Regional background study. FAO, Bangkok, p 185. Licence: CC BY-NC-SA 3.0 IGO http://www.fao.org/documents/card/en/c/CA4438EN/
Friedrichs S, Takasu Y, Kearns P, Dagallier B, Oshima R, Schofield J, Moreddu C (2019) Meeting report of the OECD conference on "genome editing: applications in agriculture—implications for health, environment and regulation". Transgenic Res 28:419–463. https://doi.org/10.1007/s11248-019-00154-1
GAIN (Global Agricultural Information Network) of USDA (2020) Agricultural biotechnology annual: European Union report E42019-0053, March 2020, Washington, DC
Georgiadis M, Ryznar M (2019) Regulating what has yet to be created: an introduction (July 13, 2019). Texas Law Review Online 98:71–82. Available at SSRN: https://ssrn.com/abstract=3419560
Greely HT (2019) CRISPR'd babies: human germline genome editing in the 'he Jiankui affair'. J Law Biosci 6(1):111–183. https://doi.org/10.1093/jlb/lsz010
Hamburger D (2019) Comparative analysis: the regulation of plants derived from genome editing in Argentina, Australia, Canada, the European Union, Japan and the United States. In: Dederer HG, Hamburger D (eds) Regulation of genome editing in plant biotechnology. Springer, Cham. https://doi.org/10.1007/978-3-030-17119-3_8
Hundleby PAC, Harwood WA (2019) Impacts of the EU GMO regulatory framework for plant genome editing. Food Energy Secur 8:e00161. https://doi.org/10.1002/fes3.161
IAP (2014) IAP statement on realising global potential in synthetic biology: scientific opportunities and good governance. IAP, Trieste
IRGC (2009) Risk governance of synthetic biology. International Risk Governance Council, Geneva, ISBN 978–2–9700672-3-8 IRGC, Lausanne
Ishii T, Araki M (2017) A future scenario of the global regulatory landscape regarding genome-edited crops. GM Crops Food 8(1):44–56. https://doi.org/10.1080/21645698.2016.1261787
Jiang L (2019) Commercialization of the gene-edited crop and morality: challenges from the liberal patent law and the strict GMO law in the EU. New Genet Soc. https://doi.org/10.1080/14636778.2019.1686968
Kolodziejczyk B, Kagansky A (2017) Consolidated G20 synthetic biology policies and their role in the 2030 Agenda for Sustainable Development. G20 Insights, G20 Germany 2017 https://globalyoungacademy.net/wp-content/uploads/2017/08/Brief-Consolidated-G20-synthetic-biology-policies-and-their-role-in-the-2030-Agenda-for-Sustainable-Development.pdf
Li H-E, Canavan C, Cameron L et al (2019) Synthetic biology and the United Nations. Trends Biotechnol 37(11):1146–1151. https://doi.org/10.1016/j.tibtech.2019.05.011
Mariscal C, Petropanagos A (2016) CRISPR as a driving force: the model T of biotechnology. Monash Bioeth Rev 34(2):101–116. https://doi.org/10.1007/s40592-016-0062-2
Medvedieva MO, Blume YB (2018) Legal regulation of plant genome editing with the CRISPR/Cas9 technology as an example. Cytol Genet 52(3):204–212. https://doi.org/10.3103/S0095452718030106
Meyer M, Heimstadr C (2019) The divergent governance of gene editing in agriculture: a comparison of institutional reports from seven EU member states. Plant Biotechnol Rep 13(5):473–482. https://doi.org/10.1007/s11816-019-00578-5
National Academies Press (2018) Biodefense in the age of synthetic biology, Washington, DC
OIE (2019) Guidelines for responsible conduct in veterinary research. World Organisation for Animal Health. https://doi.org/10.20506/OIEguidelines.2757
Pirscher F (2019) CRISPR/Cas in crop breeding: why ethics still matter. In: Kallhoff A, Paola D, Schörgenhumer M (eds) Plant ethics, concepts and applications. Routledge, London. https://doi.org/10.4324/9781315114392
Rabitz F (2014) Regulatory gaps in the global governance of synthetic biology. In: Issue 2014/11. Institute for European Studies, Brussels. https://www.ies.be/files/PB_2014_11.pdf

Raimbault B, Cointet J-P, Joly P-B (2016) Mapping the emergence of synthetic biology. PLoS One 11(9):e0161522. https://doi.org/10.1371/journal.pone.0161522
Rerimassie V, Stemerding D, Zhang W, Srinivas KR (2015) Discourses on synthetic biology in Europe, India and China. In: Ladikas M, Chaturvedi S, Zhao Y, Stemerding D (eds) Science and technology governance and ethics. Springer, Cham. https://doi.org/10.1007/978-3-319-14693-5_11
Ricroch A (2019) Global developments of genome editing in agriculture. Transgenic Res 28:45–52. https://doi.org/10.1007/s11248-019-00133-6
Ricroch AE, Ammann K, Kuntz M (2016) Editing EU legislation to fit plant genome editing. EMBO rep 17:1365–1369. https://doi.org/10.15252/embr.201643099
Scheben A, Edwards D (2018) Bottlenecks for genome-edited crops on the road from lab to farm. Genome Biol 19:178. https://doi.org/10.1186/s13059-018-1555-5
Schultz-Bergin M (2018) Is CRISPR an ethical game changer? J Agric Environ Ethics 31(2):219–238. https://doi.org/10.1007/s10806-018-9721-z
Shapira P, Kwon S, Youtie J (2017) Tracking the emergence of synthetic biology. Scientometrics 112:1439–1469. https://doi.org/10.1007/s11192-017-2452-5
Srinivas KR (2019) Global governance of emerging biotechnologies and role of science diplomacy. Sci Dip Rev 1(3):23–30. http://fisd.in
Steel D (2014) Philosophy and the precautionary principle: science, evidence, and environmental policy. Cambridge University Press. https://doi.org/10.1017/CBO9781139939652
Stirling A, Hayes KR, Delborne J (2018) Towards inclusive social appraisal: risk, participation and democracy in governance of synthetic biology. BMC Proc 12(Suppl 8):15. https://doi.org/10.1186/s12919-018-0111-3
Trump BD (2017) Synthetic biology regulation and governance: lessons from TAPIC for the United States, European Union, and Singapore. Health Policy 121(11):1139–1146. https://doi.org/10.1016/j.healthpol.2017.07.010
Trump BD (2016) A comparative analysis of variations in synthetic biology regulation. Unpublished PhD dissertation, University of Michigan, Ann Abor
Voigt B, Münichsdorfer A (2019) Regulation of genome editing in plant biotechnology: European Union. In Dederer HG and Hamburger D (ed) regulation of genome editing in plant biotechnology. Springer, Cham, pp 137–238
Wallach W, Saner M, Marchant G (2018) Beyond cost-benefit analysis in the governance of synthetic biology. Governance of emerging technologies: aligning policy analysis with the Public's values, special report. Hastings Cent Rep 48(1):S70–S77. https://doi.org/10.1002/hast.822
Walsh B (2019) End times: a brief guide to the end of world. Hachette Books, New York
Wasmer M (2019) Roads forward for European GMO policy—uncertainties in wake of ECJ judgment have to be mitigated by regulatory reform. Front Bioeng Biotechnol 7(132). https://doi.org/10.3389/fbioe.2019.00132
Weik A, Guston D, Frow E, Calvert J (2012) Sustainability and anticipatory governance in synthetic biology. Int J Soc Ecol Sust Dev 3(2):25–38. https://doi.org/10.4018/jsesd.2012040103
World Bank (2019) Harvesting prosperity: technology and productivity growth in agriculture. World Bank, Washington, DC. https://openknowledge.worldbank.org/handle/10986/32350 License: CC BY 3.0 IGO
Zhang JY, Claire M, Nikolas R (2011) The transnational governance of synthetic biology: scientific uncertainty, cross-borderness and the 'art' of governance. For the Royal Society Science Policy Centre (UK) BIOS (Centre for the Study of Bioscience, Biomedicine, Biotechnology and Society), 36 https://kar.kent.ac.uk/36060/
Zhang Y, Pribil M, Palmgren M et al (2020) A CRISPR way for accelerating improvement of food crops. Nat Food. https://doi.org/10.1038/s43016-020-0051-8

# Part VIII
# Annex A National Legal Perspectives – Africa

# Botswana—Genetically Modified Organisms (GMOs) and Synthetic Biology: Their Potential Applications and the Legal Perspectives

**Motlalepula Pholo**

**Abstract** Genetic modification and synthetic biology are interdisciplinary areas that involve the application of engineering principles to biology. These applications offer a great potential in enhancing productivity in various industries including agriculture, medicine, energy, and the environment. However, such scientific development must be coupled with legal mechanisms that ensure a sustainable environment and protection of human health. Botswana is a landlocked country that depends heavily on food and other non-food commodities imports, even from neighboring countries with high rate of commercially released genetically modified organisms. This calls for biosafety risk assessment and management measures for both transboundary movement and natioual use of genetically modified organisms. Although Botswana acceded to the Cartagena Protocol on Biosafety in 2000, the regulatory framework and institutional capacity strengthening in the regulation of genetically modified organisms and synthetic biology technologies are in a very nascent stage. There is a need for the country to mainstream biosafety legal frameworks and institutional capacity strengthening in the national agenda.

**Keywords** Botswana · Genetic modification · Synthetic biology · Genetically modified organisms · Biotechnology · Regulatory framework · Cartagena Protocol on Biosafety

---

M. Pholo (✉)
Department of Agricultural Research, Gaborone, Botswana
e-mail: mpholo@gov.bw



A. Chaurasia et al. (eds.), *GMOs*, Topics in Biodiversity and Conservation 19,
https://doi.org/10.1007/978-3-030-53183-6_21

## Introduction

Genetic modification or engineering is the process of altering the genetic makeup of an organism, including microbes, cells, plants, and animals. This would usually involve using various methods of biotechnology, such as recombinant DNA (rDNA) technology, gene targeting, or genome editing to add, delete, or otherwise change an organism's DNA. Moreover, genetic modification can also involve moving genetic material between species (WHO 2014). The use of genetically modified organisms (GMOs) to boost agricultural productivity is often comprised of commercialized genetically modified (GM) crops with traits that address biotic stresses, such as pests, weeds, or both as well as traits that address processing and consumer issues. However, in the future, foods derived from GM microorganisms or GM animals are likely to be introduced on the market (ISAAA 2006). These could improve yield, thereby contributing positively by boosting agricultural productivity and reducing food insecurity in Africa (Cook and Downie 2010).

Although Botswana neither developed nor released GMO products to date, there have been various studies that demonstrated the presence of genetically modified sequences. Qualitative and quantitative analysis demonstrated the presence of genetically modified sequences within two brands of maize meal and soybean products in Botswana (Mpoloka and El-Kindiy 2008; Mokhawa et al. 2014). It is worth-noting that the country is landlocked, and depends heavily on food imports as a result of various constraints on the expansion of production from its own arable land. South Africa is the top partner country from which Botswana imports food products, where GM maize is highly prevalent (Cook and Downie 2010).

Furthermore, other emerging technologies like Synthetic biology extends the spirit of genetic engineering, and it focus on whole systems of genes and gene products. The goal of synthetic biology is to extend or modify the behaviour of organisms and engineer them to perform new tasks (Andrianantoandro et al. 2006). Gene editing mechanisms offer possibilities to increase performance in various sectors including health, agriculture, and environment. In the food industry and agriculture, current efforts are focused on finding novel solutions for solving limited and contaminated arable land and water resources challenges without placing further burden on the environment. The other important factor is on nutritious food in human health and the concept of food as medicine, leading to an increased demand for functional food. In that regard, synthetic biology is impacting the food and agriculture industry through engineering biosynthetic pathways and enzymes, host organisms as cell factories and traditional producers of food for various purposes, including improvement in the efficiency of existing processes of food production, strain development and improvement as well as improving disease resistance, environmental tolerance, and food quality and yield (GenScript). This could be very beneficial to Botswana in the efforts to enhance climate smart agriculture due to challenges emanating from low and poor distribution of rainfall, high temperatures, and biotic stresses such as outbreaks of pests and diseases. One of the economic challenges in the beef industry, the sector that contributes immensely to the country gross domestic products, is the frequent outbreak of foot and mouth disease (FMD). Therefore, the adoption of synthetic biology by the Botswana Vaccine Institute (BVI) in future for the development of vaccines that eliminates FMD may offer a significant breakthrough in the beef industry.

In addition to limited functional food, malaria is a major global health problem and a leading cause of morbidity and mortality. Botswana made a remarkable progress in reducing malaria cases through targeted coverage with vector control interventions, which was associated with the reintroduction of diethyl-dichloro-trichloroethane (DDT) for intensified indoor residual spraying (IRS), free mass distribution of long-lasting insecticide-treated nets (LLINs), larviciding, and intensified community mobilization campaigns to educate the public on IRS, LLINs, and early treatment (Simon et al. 2013). Despite all efforts deployed over the number of decades, complete elimination of malaria still remains a challenge. Synthetic biology, particularly the proposed gene drive mechanisms that stimulate biased inheritance of a particular gene to alter populations at the release site, changing local populations of harmful mosquitoes (Werren 1997; Johnson et al. 2006; Alphey 2014), could offer health benefits.

## Policy and Regulatory Framework

Botswana's ecosystems, species, and genetic diversity represent a huge asset to the local communities and contribute to the gross domestic product (GDP). These ecosystems and biodiversity form a basis for much of the tourism industry, which contributes to the gross domestic product (Mokaila 2007). According to the World Travel and Tourism Council (WTTC) reports, travel and tourism sector in 2018 was estimated to account for about 10.4% of the total GDP (World Travel and Tourism Council and Oxford Economics 2019). However, environmental changes and evolution as well as human-induced changes in biodiversity levels can have profound negative impacts on the functioning of ecosystems and ultimately either changes or loss of biodiversity. It is, therefore, important to have combinations of regulatory measures and activities to ensure a proper management of the country's biodiversity in order to maintain genes, species, and productive ecosystems. In that respect, it is imperative to comprehensively examine technologies such as GMOs and synthetic biology, given the potential food security, developments, health, and environmental benefits vis-a-vis the potential threat to human health and the environment (Hewett et al. 2016; Zhang et al. 2016).

## Environmental Impact Assessment

Besides application of the precautionary principle, one of the regulatory measures to address human-induced changes in biodiversity levels in Botswana is through environmental impact assessment (Environmental Impact Assessment 2012). The Environmental Impact Assessment (EIA) process and technique are intended to predict and evaluate the environmental consequences of human development activities. Furthermore, it is intended to plan appropriate measures to eliminate or reduce adverse effects and to augment positive effects. This is beneficial for the ecosystems that have a limited capacity to absorb and cope with the stress resulting from various human development activities. Moreover, it explains that elimination or reduction of cultural, social, economic, and ecological impacts is usually more costly than preventing them in the first place. However, EIA does not directly and comprehensively address issues of GMOs and synthetic biology.

## Plant, Animal, and Food Safety

Two dominant policy approaches to ensuring biosafety and/or biosecurity currently coexist at the global level. The first is the World Trade Organization's Sanitary and Phytosanitary (WTO-SPS) Agreement. This Agreement calls for national sanitary and phytosanitary measures relating to animal and plant health as well as biosafety measures to be based on scientifically sound evidence of harm, so as to prevent unnecessary restrictions on trade and avoid protectionism masquerading as risk avoidance, both key concerns of the global trade regime. The WTO-SPS Agreement also, however, allows for legitimate context-specific differences in judgments of appropriate levels of safety (WTO-SPS Agreement 1994; Christoforou 2000). Botswana is a signatory to the WTO, hence obligated to comply with the Sanitary and Phytosanitary Agreements. The main piece of legislations that governs the Sanitary and Phytosanitary (SPS) Agreements in Botswana are linked to the international standard setting organizations such as the:

1. World Organisation for Animal Health (OIE) for protection of animal health
2. CODEX: food safety
3. International Plant Protection Convention (IPPC) and Food Agricultural Organization (FAO): for the prevention and control of the introduction and spread of pests of plants and plant products

## *Plant Health*

The main piece of legislature governing the prevention of introduction, spread, and establishment of plant pests; to facilitate trade in plants; to enable Botswana to comply with its international obligations; and to provide for matters incidental thereto is the *Plant Protection Act* 2007 and the *Plant Protection Regulations* 2009. The Act contributes to ensuring environmental sustainability through application of environmentally friendly practices that include, but are not limited to, facilitating safeguarding of plant health by regulating importation and exportation of plant and plant products by issuance of phytosanitary certificates and import permits and providing technical support on the management of crop pests and diseases. The Division of Plant Protection in the Ministry of Agricultural Development and Food Security is mandated with the administration of this Act.

## *Animal Health*

Botswana has adopted the *Diseases of Animals Act* and *Regulations* 1977. The Act provides for the prevention and control of diseases of animals – to regulate the import, export, and movement of animals; to provide for the quarantine of animals in certain circumstances; and to provide for matters incidental to and connected with the foregoing. The Department of Veterinary Services has a mandate to protect the health of animals, humans, and their environment in Botswana. This is achieved through various means and one is to prevent the introduction and spread of animal and human disease through the importation of live animals and animal products.

## Food Safety

The *Food Control Act* was enacted in 1993 to ensure the provision of clean, safe, and wholesome food to consumers (Food Control Act 1993; Food Control Regulations 2003). The Act is enforced through Food Control Regulations, which have mandatory labelling requirements which include:

1. Labelling of food additives
2. Labelling of pre-packed food
3. Marketing of food for infants and young children

The Nutrition and Food Control Unit under Public Health in the Ministry of Health and Wellness is responsible to implement the provisions of the Act.

## Regulatory Policy and Framework Addressing GMOs/ Synthetic Biology

The second global dominant policy approaches to ensuring biosafety and/or biosecurity involve a mandatory disclosure by GMO producers of biosafety information and the intention to export GMOs, as a way to facilitate informed choice about import of transgenic products in diverse national contexts. This is the approach adopted by the multilaterally negotiated Cartagena Protocol on Biosafety (CPB; Secretariat of the Convention on Biological Diversity 2000) under the United Nations Convention on Biological Diversity, which was concluded in 2000 (Gupta 2013). The conclusion of the CPB has been hailed as a significant step that provides an international regulatory framework to reconcile the respective needs of trade and environmental protection with respect to a rapidly growing global biotechnology industry. The Protocol creates an enabling environment for the environmentally sound application of biotechnology, making it possible to derive maximum benefit from the potential that biotechnology has to offer while minimizing the possible risks to the environment and to human health. Therefore, genetically engineering is regulated under the Cartagena Protocol on Biosafety, whereas the medicine regulation is under the World Health Organization. Intriguingly, these regulations have an overlapping role to play for gene drive regulation (Glover et al. 2018).

### *Status of the GMOs/Synthetic Legal Legislature in Botswana*

Botswana acceded to the Cartagena Protocol on Biosafety (CPB) in 2000. In accordance with Article 19 of the Protocol, Botswana designated the Department of Environmental Affairs and Department of the Agricultural Research to fulfil the functions of the national focal point and competent national authority, respectively. The United Nations Environmental Programme/Global Environment Facility (UNEP/GEF) led to the development of the Draft National Biosafety Framework, which was completed in 2010. The draft framework comprised of an overview of background that went into developing the National Biosafety Framework (NBF), the draft Biosafety and Biotechnology Policy, and draft bill. The Consultant's Draft Bill expounds both the Policy and the Cartagena Protocol on Biosafety (CPB) by practical measures and activities that can be implemented to achieve the intended objectives of the Policy and the CPB (Department of Agricultural Reseach 2006). Consequently, parliament passed the National Biosafety Policy in June 2013 (Regonamanye 2013; Lethola and George 2016). The national policy on biotechnology and biosafety articulates appreciation of potential benefits emanating from GMOs. Other than that, it expresses the country's position regarding the different areas that can be potentially impacted by biotechnology or biosafety activities in either a positive or negative manner. These include areas of agriculture, commerce and industry, education, environment, health, and ethics.

Although the policy has been approved since 2013, the draft bill which has been reported to be ready for parliamentary approval and enactment of the national biosafety legislation is not yet in place. In support of the policy, Botswana has developed a national draft biosafety bill, which articulates all the essential elements for biosafety legislation such as the objective, the subject matter for regulation, measures to be taken for modern biotechnology application, risk assessment and management, monitoring and evaluation, and the relevant institutional framework. Though it is not yet clear on the regulation of synthetic biology, it is anticipated that the national draft biosafety bill will address both GMOs and synthetic biology, with exception to pharmaceuticals.

In that respect, Botswana has been engaged in sequential events in a biotechnology and biosafety awareness campaign project at the national level. This was conducted by the Botswana Public Awareness and Participation Innovation Platform (BOPAPIP) in partnership with the Regional Agricultural and Environmental Initiatives Network (RAEIN-Africa). The mandate of the project was to promote and facilitate public awareness and education on issues of biotechnology and biosafety in Botswana. The outcome of the process was expected to have an impact on expediting the finalization of the National Biosafety Act as well as establishing a sustainable mechanisms for public participation in biosafety decision-making, which is in line with Article 23 of the Cartagena Protocol (Ngwako et al. 2014). Despite the delay in the approval of the draft bill, the project, however, yielded a number of benefits for the country, most importantly an increased participation of stakeholders in decision-making process towards the National Biotechnology and Biosafety Policy's adoption.

## Conclusions

Genetic modification and synthetic biology are interdisciplinary areas that involve the application of engineering principles to biology. These applications offer a great potential in enhancing productivity in various industries including agriculture, medicine, energy, and the environment. However, such scientific development must be coupled with legal mechanisms that ensures a sustainable environment and protection of human health. In Botswana, modern biotechnology and synthetic biology research, regulatory framework, and institutional capacity strengthening in the regulation of these emerging technologies are in a very nascent stage. Although Botswana has enacted various legislations relating to food safety, animal health, and plant health, Biosafety Act has not yet been enacted. The scope of the approved Biosafety Policy addresses agricultural biotechnology and excludes pharmaceuticals. However, synthetic biology extends to the public health sector in areas such as gene drive technology on mosquito and malaria control. In addition, the phytosanitary certificate currently does not have a clause for GMOs/synthetic biology provisions. These are contentious cross-cutting issues on food safety, plant health, animal health, and public health that the country has to consider while enacting the National

Biosafety Act. The immediate priority is to enact the Act and mainstream biosafety issues in the national budget. In addition, there is need to build capacity on GMO research and synthetic biology as well as capacity on biosafety and/or biosecurity in order to effectively monitor and asses GMOs and synthetic biology research and products.

## References

Alphey L (2014) Genetic control of mosquitoes. Annu Rev Entomol 59:205–224

Andrianantoandro E, Basu S, Karig DK, Weiss R (2006) Synthetic biology: new engineering rules for an emerging discipline. Mol Syst Biol 2006:0028. https://doi.org/10.1038/msb4100073

Christoforou T (2000) Settlement of science-based trade disputes in the WTO: a critical review of the developing case law in the face of scientific uncertainty. NY Univ Environ Law J 8:622–648

Cook J, Downie R (2010) African perspectives on genetically podified crops – assessing the debate in Zambia, Kenya and South Africa. Report of the CSIS Global Food Security Project 24

Department of Agricultural Reseach R of B (2006) National biosafety framework. Republic of Botswana

Environmental Impact Assessment (2012) Environmental assessment CHAPTER: 65:07. In: Botswana elaws. http://www.elaws.gov.bw/law.php?id=2291

Food Control Act (1993) Government of Botswana. https://www.moh.gov.bw/Publications/acts/FOODACT1993BOTSWANA.pdf

Food Control Regulations (2003) Government of Botswana. https://www.moh.gov.bw/Publications/acts/FOODCONTROL3Regulations.pdf

GenScript. Applications of synthetic biology in food industry and agriculture. https://www.genscript.com/applications-of-synthetic-biology-in-food-industry-and-agriculture.html

Glover B, Akinbo O, Savadogo M et al (2018) Strengthening regulatory capacity for gene drives in Africa: leveraging NEPAD's experience in establishing regulatory systems for medicines and GM crops in Africa. BMC Proc 12. https://doi.org/10.1186/s12919-018-0108-y

Gupta A (2013) Biotechnology and biosafety. In: Falkner R (ed) The handbook of global climate and environment policy. Wiley, pp 89–105

Hewett JP, Wolfe AK, Bergmann RA et al (2016) Human health and environmental risks posed by synthetic biology R & D for energy applications. Appl Biosaf 21:177–184. https://doi.org/10.1177/1535676016672377

ISAAA (2006) GM approval database. 2016. http://www.isaaa.org/gmapprovaldatabase/default.asp

Johnson NA, Burt A, Trivers R (2006) Genes in conflict: the biology of selfish genetic elements. Belknap Press, Cambrdge MA

Lethola KK, George DGM (2016) Domestication of Cartagena protocol in Botswana. Botswana J Agric Appl Sci:35–39

Mokaila OK (2007) Botswana biodiversity strategy and action plan, Gaborone

Mokhawa G, Ralefala MP, Ezeogu LI, Bulawayo BT (2014) Qualitative and quantitative detection of genetically modified sequences in processed foods sold commercially in Botswana by PCR based methods Qualitative and quantitative detection of genetically modified sequences in processed foods sold commercially in. Natl Food Technol Res Cent

Mpoloka SW, El-Kindiy OK (2008) Screening for genetically modified organisms sequences in food samples in Botswana using the biosmart Allin 2.0 GMO screening system. Botswana J Technol 17:48–52

Ngwako S, Walker K, George D et al (2014) Enhancing biotechnology and biosafety public awareness and participation in Botswana: experiences of the Botswana innovation platform.

In: RAEIN-Africa (ed) Conference proceedings for the innovation Systems for Resilient Livelihoods: connecting theory to practice. RAEIN-Africa, Pretoria, pp 24–25

Regonamanye J (2013) MP's call for thorough research as biosafety policy gets green light. Sunday Standard, 21st July 2013

Secretariat of the Convention on Biological Diversity (2000) Cartagena protocol on biosafety to the convention on biological diversity. ICAO, Montreal, Canada

Simon C, Moakofhi K, Mosweunyane T et al (2013) Malaria control in Botswana, 2008-2012: the path towards elimination. Malar J 12:1–9. https://doi.org/10.1186/1475-2875-12-458

Werren JH (1997) Biology of Wolbachia. Annu Rev Entomol 42:587–609. https://doi.org/10.1146/annurev.ento.42.1.587

WHO (2014) Frequently asked questions on genetically modified foods. In: Food Saf. https://www.who.int/foodsafety/areas_work/food-technology/faq-genetically-modified-food/en/

World Travel and Tourism Council (2019) Botswana 2019 annual research: key highlights. Oxford Economics

WTO-SPS Agreement (1994) The WTO agreement on the application of Sanitary and Phytosanitary Measures (SPS Agreement). https://www.wto.org/english/tratop_e/sps_e/spsagr_e.htm

Zhang C, Wohlhueter R, Zhang H (2016) Genetically modified foods: a critical review of their promise and problems. Food Sci Hum Wellness 5:116–123. https://doi.org/10.1016/j.fshw.2016.04.002

# Democratic Republic of the Congo—GMOs/Synthetic Biology Rules/Regulations and Biodiversity: A Legal Perspective

**Freddy Bulubulu Otono, Mike Ipanga Mwaku, Aimé Diamuini Ndofunsu, and Sébastien Luyindula Ndiku**

**Abstract** Genetically modified organisms and synthetic biology products have a potential negative effect on biodiversity. The transfer of genetic material to wild populations is a major risk. The Democratic Republic of the Congo, the second largest country in Africa, has also the largest African biodiversity which is governed by a number of legal texts. Some of these legal texts prove to be ineffective; others are anachronistic in relation to the provisions of the international legal instruments to which DRC is a Party. DRC is bound by the Cartagena Protocol on Biosafety, which is an international agreement on biosafety and a supplement to the Convention the United Nations on biological diversity. But, there is still no specific law or regulation in force concerning biosafety, the lack of adequate legislation to regulate the import and monitor the introduction of GMOs and synthetic biology products. DRC proposes the revision or strengthening the legislative and regulatory on Biodiversity, in particular, updating the National Biosafety Framework and Biosafety Bill.

**Keywords** DR Congo · Biotechnology · Biodiversity · Synthetic biology · Regulation · Biosafety · Africa

## Introduction

In Article 2 of the Convention on Biological Diversity, "biotechnology" means "any technological application that uses biological systems, living organisms, or derivatives thereof, to make or modify products or processes for specific use" (CBD 1992).

F. Bulubulu Otono (✉) · A. Diamuini Ndofunsu · S. Luyindula Ndiku
Département de Biotechnologie et Biologie Moléculaire, Division des Sciences de la Vie, Commissariat Général à l'Energie Atomique, Kinshasa, République Démocratique du Congo

M. Ipanga Mwaku
Division en charge de la Biodiversité, Direction du Développement Durable, Ministère de l'Environnement, Kinshasa, République Démocratique du Congo



A. Chaurasia et al. (eds.), *GMOs*, Topics in Biodiversity and Conservation 19,
https://doi.org/10.1007/978-3-030-53183-6_22

However, the term "biotechnology" is used by many in a much narrower sense, often labeled as "modern biotechnology", that is, the application of genetic engineering whereby genes from certain species are introduced into the genetic heritage of other species. Genetic modification produces genetically modified organisms (GMOs). GMO is, however, often used interchangeably with living modified organism (LMO) (Mackenzie et al. 2003), any living organism that possesses a novel combination of genetic material obtained through the use of modern biotechnology (Cartagena Protocol on Biosafety to the Convention on Biological Diversity 2000). If they are introduced into the environment, they can potentially affect biodiversity (to local varieties, wild relatives, and non-target organisms) (Pardo 2003).

Unlike traditional genetic engineering, which typically involves the transfer of individual genes between cells, synthetic biology, an emerging field, involves the assembly of new sequences of DNA and even entire genomes (Biotechnology Innovation Organization 2016). "Synthetic biology is a further development and new dimension of modern biotechnology that combines science, technology and engineering to facilitate and accelerate the understanding, design, redesign, manufacture, and/or modification of genetic materials, living organisms and biological systems" (Wikmark et al. 2016). According to the Science for Environment Policy (2016), synthetic biology involves designing and constructing new biological parts, devices, and systems, going far beyond the modification of existing cells by inserting or deleting small numbers of genes. Cells can be equipped with new functions and entire biological systems can now be designed. Living organisms resulting from current synthetic biology techniques fall under the definition of LMOs under the Cartagena Protocol for Biosafety (CPB) and are subject to its provisions in Articles 8(g) and 19 (CBD 2015).

Although considered as one of the promising factors in achieving food security and expanding the agricultural potential of the African continent, modern Biotechnology presents, however, the risks on the environment. The introduction of GMOs and synthetic organisms may therefore have both constructive and destructive effects on the conservation and sustainable use of biodiversity. The Cartagena Protocol addresses the fact that LMOs may have biodiversity and human health impacts. The escape or release of novel organisms from synthetic biology into the environment could radically and detrimentally change ecosystems. However, the transfer of genetic material to wild populations is a major risk (Science for Environment Policy 2016).

## GMOs/Synthetic Biology Rules/Regulations and Biodiversity in the Democratic Republic of the Congo

The Democratic Republic of the Congo (DRC) is located in Central Africa. With a surface area of 2,345,000 km$^2$, DRC is the second largest country in Africa and fifth in the world by its plant and animal diversity. It has the largest African biodiversity with many species of higher plants about a third would be endemic mammals including all the major African animals, birds, fish, reptiles, and amphibians (Ministère de l'Environnement 2007; 1016).

Biodiversity is governed in DRC by a number of legal texts including Law on the Forest Code, on Basic Principles for the Protection of the Environment, Agriculture and Conservation of Nature; the Law regulating hunting and on the creation of safeguarded areas; the Decree on fishing and Decree laying down the terms of allocation of forest concessions to local communities. It should be noted that some of these legal texts prove to be ineffective because of the lack of implementing measures. Others are anachronistic in relation to the provisions of the international legal instruments to which DRC is a Party. Others are already outdated and need to be replaced.

DRC is, however, a party to a significant number of multilateral environmental agreements including the Convention on Biological Diversity, the Cartagena Protocol on Biosafety, the Nagoya Protocol on Access to Genetic Resources and the sharing of benefits from their use, the United Nations Framework Convention on Climate Change, the United Nations Convention to Combat Desertification, the Convention on International Trade in Endangered Species of Wild Fauna and Flora, the Convention on the Conservation of Migratory Species of Wild Animals and the Ramsar Convention on Wetlands. Also, DRC is Party to the Central African Forests Commission (COMIFAC). Indeed, with its various ecosystems and a rich biological and genetic diversity, DRC has been a party to the Convention on Biological Diversity (CBD) since 1994. To implement Article 6 of the CBD, DRC had developed its National Biodiversity Strategy and Action Plan in 1999, which was adopted in 2001 by the government as a National Biodiversity Policy Document.

The new National Biodiversity Strategy is based, inter alia, on the recommendations made under the CBD, including the Aichi Targets, the principle of coherence with the relevant programs in which the DRC is committed. The vision of this strategy is defined as follows:" By 2035, biodiversity is managed in a sustainable manner by its integration into all relevant national sectors, contributes to the development of the country and all Congolese are aware of its value and its contribution to their well-being" (Ministère de l'Environnement 2016).

In DRC, there is, however, the lack of adequate legislation to regulate the import and monitor the introduction of GMOs and synthetic biology products. Indeed, DRC finalized, in 2008, the process of developing the National Biosafety Framework and Draft Biosafety Bill, essentially based on GMOs regulation. Unfortunately, to date, the national biosafety framework is still not being implemented while the bill is in Parliament pending review. Otherwise, synthetic biology falls under a number of regulatory mechanisms, but most were established before the field fully developed and therefore were not intended to cope with its impacts. "Synthetic biology" as such has not been addressed specifically in the text of any multilateral treaties.

However, there are a multitude of treaties, customary rules and general principles of law, as well as other regulatory instruments and mechanisms, which could apply to all or some forms of synthetic biology.

Thus, it should be said that in DRC, there is still no specific law or regulation in force concerning biosafety, even if the laws and regulations on plant health protection, animal health (including refoulement and quarantine), and the protection of industrial property rights exist.

Let us quote some legal provisions that are related to modern biotechnology and synthetic biology:

- Law No. 11/009 of July 9, 2011, on basic principles relating to the protection of the environment, governs genetically modified organisms in section 5 of Chapter 6. Article 63 of this Law provides that a specific law must to be taken to regulate the methods of assessment and biosafety as well as the decision-making process regarding transboundary movements of GMOs.
- Law No. 14/003 of February 11, 2014 on the conservation of nature also contains provisions that can be capitalized in the context of biosecurity including the provisions relating to the environmental and social impact assessment.
- Law No. 11/022 of December 24, 2011 on Basic Agricultural Principles stipulates in Article 71 that the government shall ensure that the development, use, transfer, and release in agriculture of genetically modified organisms and pesticides are done in ways that avoid or reduce risks to the environment and health. It also ensures that certain farming practices do not have a negative impact on the environment and health.

There are therefore no provisions that take into account the requirements of the implementation of the Cartagena Protocol.

No legislative or regulatory provisions put in place an advance informed agreement procedure before triggering an export/import notification and decision-making procedure for the import of GMOs intended to be intentionally introduced into the environment of the importing party.

However, although not having developed a regulatory framework for GMOs and synthetic biology products, DRC is bound by the Cartagena Protocol on Biosafety, which is an international agreement on biosafety and a supplement to the Convention the United Nations on biological diversity.

Being given the insufficiencies met on the regulation in progress on biodiversity, the development of modern biotechnology and the impacts which GMOs as well as the products and components of synthetic biology could have on biodiversity, the DRC proposes the revision or strengthening the legislative and regulatory on biodiversity, in particular: updating the National Biosafety Framework and Biosafety Bill, effectively implementing all the provisions of the National Biosafety Framework, and developing regulatory measures on access to genetic resources and benefit sharing. This is envisaged to be achieved in a short term as the DRC has affected a part of its allocations from the Global Environment Facility (GEF 6) to a project on the effective implementation of the Nagoya Protocol. One of the components of the said project is related to the development of institutional and legal framework. With regard to the implementation of the National Biosafety Framework, the DRC is in discussion with UN Environment to propose a related project with funding, once again, from the Global Environment Facility (GEF 7). In addition, the

DRC will also have to fill other gaps in the promotion of modern biotechnology and synthetic biology; these include identifying capacity-building needs and benefiting from technology transfer.

## References

Biotechnology Innovation Organization (2016) Synthetic biology explained. Available from: https://www.bio.org/articles/synthetic-biology-explained

Cartagena Protocol on Biosafety (2000) Secretariat of the Convention on Biological Diversity, Montreal

Convention on Biological Diversity (1992) Secretariat of the Convention on Biological Diversity, Montreal

Mackenzie Rh, Burhenne-Guilmin F, LaViña A, Werksman J in cooperation with Ascencio A, Kinderlerer J, Kummer K, Tapper R (2003) An explanatory guide to the cartagena protocol on biosafety. IUCN, Gland/Cambridge.

Ministère de l'Environnement, Conservation de la Nature, Eauxet Forêts (2007) Identification et Evaluation des Besoins en Renforcement de Capacités pour la Gestion de l'Environnement en République Démocratique du Congo (Synthèse des rapports thématiquesintérimaires: Biodiversité, Désertification & Déforestation et Changement Climatique), p 92

Ministère de l'Environnement, Conservationde la Nature et Développement Durable (2016) Stratégie et Plan d'Action Nationaux de la Biodiversité (2016–2020), p 91

Pardo CA (2003) Genetically modified organisms and biodiversity: assessing the threats. Biotecnología Aplicada 20:1–8

Science for Environment Policy (2016) Synthetic biology and bidiversity, Future brief 15. Produced for the European Commission DG Environment by the Science Communication Unit, UWE, Bristol. Available at: http://ec.europa.eu/science-environment-policy

Secretariat of the Convention on Biological Diversity (2015) Synthetic biology, Montreal, Technical Series No. 82, 118 pages

Wikmark OG, Brautaset T, Agapito-Tenfen SZ, Okoli AS, Myhr AI, Binimelis R, Ching LL (2016) Synthetic biology – biosafety and contribution to addressing societal challenges, Biosafety report 2016/02. GenOk – Centre for Biosafety, Tromso

# Kenya—A Review of Regulation of Genetically Modified Organisms (GMOs)—Case Study of Kenya

**Josphat N. Muchiri, Theophilus M. Mutui, and Dorington O. Ogoyi**

**Abstract** Globally, national regulation of Genetically Modified Organisms (GMOs) is largely consistent with provisions of the Cartagena Protocol on Biosafety (CPB) to the Convention on Biological Diversity (CBD), which covers transboundary movement of Living Modified Organisms (LMOs). Kenya signed the Cartagena Protocol in 2000, ratified it in 2003, and developed the National Biotechnology Development Policy in 2006. The policy led to the enactment of the Biosafety Act No. 2 of 2009 that provides for legal, institutional, and regulatory framework for harnessing the benefits of modern biotechnology. This Act established the National Biosafety Authority (NBA). The Authority facilitates responsible research in modern biotechnology while minimizing potential risks that may be posed by GMOs to human and animal health as well as adequate protection of the environment. The Authority ensures adequate level of protection for safe transfer, handling, and use of GMOs in Kenya by establishing a transparent, science-based and predictable process for review of applications. The authority has also published four biosafety regulations, namely, Contained Use (2011), Environmental release/Placing on the Market (2011), Import, Export, and Transit (2011), and Labeling (2012) and other enabling tools to enable it exercise its mandate. A number of applications for Import, Export, and Transit of humanitarian food products (maize/soybean blend) and contained use as well as confined field trials have been approved by the Authority. The crops under research include cotton, maize, sorghum, cassava, bananas, sweet potato, yam, cowpea, beans, and gyphsophila. For environmental release applications, the four key areas considered during the decision-making process include risk assessment (Food Safety Assessment as well as Environmental Risk Assessment), socioeconomic considerations, public participation, and consultations among relevant regulatory agencies. Whereas the process has been progressive, it has not been devoid of challenges that include low public awareness on biosafety matters, absence of regulations and inadequate expertise in New Breeding Techniques (NBTs) and socioeconomic issues which are being addressed through continuous capacity building to ensure seamless implementation of biosafety regulatory framework in Kenya. This chapter provides an overview of the legal, institutional, and

J. N. Muchiri · T. M. Mutui · D. O. Ogoyi (✉)
National Biosafety Authority, Nairobi, Kenya
e-mail: dogoyi@biosafetykenya.go.ke



A. Chaurasia et al. (eds.), *GMOs*, Topics in Biodiversity and Conservation 19,
https://doi.org/10.1007/978-3-030-53183-6_23

administrative mechanism for regulation of GMOs, steps in decision processes, experiences, synergies, and challenges faced in the biosafety regulatory processes as well as opportunities for improvements.

**Keywords** Biosafety · Cartagena Protocol · GMO · Modern Biotechnology · Regulatory framework · Risk Assessment · Kenya

## Introduction

### *General*

The world's population is set to reach 9 billion by 2050, and Africa is projected to contribute the biggest proportion of this increase. As a result, increased food production is a priority in efforts to feed and empower the population. At the national level, the Kenyan Government in its short- term plan referred to as the "Big Four Agenda" identifies food security and nutrition as one of the pillars to eradicate poverty by 2022 (Government of Kenya 2018). The ability to produce more food through expanding the current area under cultivation, increasing the application of agrochemicals and use of irrigation, are limited in small-scale farming situations which form the bulk of farming systems in Kenya. As such, the use of modern biotechnologies has been identified as one of the possible additional tools that can increase agricultural production, and reduce production costs as well as manage post-harvest losses, which in some instances account up to 30–40%. Biotechnology is believed to hold great promise for increasing food production (Karembu et al. 2009; Juma 2011; Chambers 2013), thus considerable effort has been expended in many African countries to set up regulatory frameworks to support the responsible utilization of this technology (Karembu et al. 2009). Tapping into the potential of modern biotechnology whilst ensuring that the health of humans, other animals, and the environment is safeguarded requires a dynamic and functional regulatory regime (Kinyua et al. 2014).

### *The Biosafety Legal and Regulatory Regime in Kenya*

Genetically engineered (GE) also referred as genetically modified (GM) food is becoming an increasing part of the global food supply (James 2012). As countries do cross border and international trade, movement of GE foods between countries becomes inevitable. This calls for the enactment of legislation to support trade while ensuring sanitary and phytosanitary standards are maintained.

The need to have biosafety frameworks and laws to govern the safe use of biotechnology has its genesis in the provisions of the Convention on Biological

Diversity (CBD), specifically articles 8(g) and 19(3). One of the supplementary agreements to the CBD is the Cartagena Protocol on Biosafety (CPB). Indeed, Kenya was the first country to sign the Protocol in 2000 and ratified it in 2002 (Wafula 2009) just before it entered into force on September 11, 2003. Recognizing the potential benefits of modern biotechnology and cognizant of possible potential risks, the Kenyan Government established the National Biosafety Authority (NBA) in 2010 pursuant to the recommendations of the National Biotechnology Development Policy of 2006 and subsequent enactment of Biosafety Act No. 2 of 2009. NBA is mandated to exercise general supervision and control over the development, transfer, handling, and use of genetically modified organisms (GMOs) to ensure safety of human and animal health as well as provision of adequate protection of the environment. NBA is therefore the competent national authority on matters of GMOs in Kenya.

The Biosafety Act defines GMO as: "an organism that possesses a novel combination of genetic material obtained through the use of modern biotechnology techniques; that includes the application of; (a) in-vitro nucleic acid techniques including the use of recombinant deoxyribonucleic acid (DNA) and direct injection of nucleic acid into the cells or organelles; or (b) fusion of cells beyond the taxonomic family, that overcome natural, physiological, reproductive and recombinant barriers and which are not techniques used in traditional breeding and selection." This definition fits very well with the description of living modified organism (LMO) in the CPB. Pharmaceuticals for human use are exempt in the context of the Biosafety Act which are regulated in other public health laws.

To further support implementation of the Biosafety Act, Kenya has published four sets of regulations, namely The Biosafety (Contained Use) Regulations, 2011, Biosafety (Environmental Release) Regulations, 2011, Biosafety (Import, Export and Transit) Regulations, 2011, and the Biosafety (Labelling) Regulations, 2012. These regulations have been instrumental in the conduct of genetic engineering research at laboratory and field trials (Table 1), import, export and transit of GM products into Kenya and neighboring countries, as well as environmental release of GMOs in the country.

## Decision-Making Process for Environmental Release Applications

The Biosafety Act anticipates three types of activities, namely contained use research; import, export and transit; and environmental release of GMOs. The process below focuses on environmental release review procedures.

The decision-making process for environmental release applications entails the following steps: (1) the applicant/developer fills a prescribed form and submits to the National Biosafety Authority accompanied by applicable fees; (2) the application is screened for administrative completeness and acknowledged within 30 days;

**Table 1** Status of GM crop research in Kenya as on December 2018

| Crop | Modified trait | Stage | Remarks |
|---|---|---|---|
| Maize | Drought tolerance (MON 87460) | Confined field trial (CFT) | CFT completed |
| | Insect resistance (MON 810) | Environmental release (limited) | Pending National Performance Trials(NPTs) |
| | Stack (insect resistance and drought tolerance) | Confined field trial | CFT ongoing |
| | Maize lethal necrosis disease resistance | Greenhouse trials | Ongoing |
| Cotton | Insect resistance (MON 15985) | Environmental release (limited) | Currently undergoing second season National Performance Trials and distinctness, uniformity and stability (DUS) tests |
| Gypsophila cut flowers | Pink coloration of petals | Confined field trial | CFT trial completed. Environmental release request declined |
| Cassava | Virus resistance | Confined field trial | CFT ongoing |
| | Bio-fortification | Confined field trial | CFT completed |
| | Stress tolerance | Greenhouse trial | Ongoing |
| Sweet potato | Virus resistance | Confined field trial | Ongoing |
| | Weevil resistance | Greenhouse trial | Ongoing |
| Irish potato | Late blight disease resistance | Greenhouse trial | Ongoing |
| Banana, plantains and enset | Bacterial disease resistance | Confined field trial | Completed |
| | Viral disease resistance | Greenhouse trial | Ongoing |
| | Nematode resistance | Greenhouse trial | Ongoing |
| | Double haploidy | Greenhouse trial | Ongoing |
| Sorghum | Bio-fortification | Confined field trial | CFT ongoing |
| Yam | Nematode resistance | Greenhouse trial | Ongoing |
| Cowpea | Drought tolerance | Greenhouse trial | Ongoing |
| Pigeon pea | Insect resistance | Greenhouse trial | Ongoing |
| Beans | Virus resistance | Greenhouse trial | Ongoing |

(3) engagement of independent biosafety experts to review food/feed safety, environmental and ecological safety as well as socio-economic data on the application; (4) review of the application by other relevant government bodies; (5) public notice of non-confidential information of the application; (6) consolidation of public and expert's review comments by the NBA Secretariat; and (7) review of application by NBA Board Technical Committee and finally a decision by NBA Full Board.

This entire review process from the time of receiving an administratively complete application to the time the decision is communicated to applicant takes 90–150 calendar days. NBA decisions on environmental release applications are informed by the: risk assessment report, technical experts review comments, Regulatory Agencies review comments, socio-economic impact report, and actionable comments from the general public. It is important to emphasize that NBA assesses the safety of a transgenic event in a GM crop relative to its conventional counterpart. Other developmental and regulatory processes such as variety release, seed certification schemes, and routine market surveillances follow similar steps as those for non-GM crops. To date, Kenya has reviewed and made decisions on three applications for environmental release as illustrated in Table 2. In regard to product identification, all products whose GM content is above 1%, once approved for commercial release, are clearly labeled as "Approved GM Product" for consumers' information and ease of traceability. Additionally, environmentally released GM products are monitored for the first 10 years and another 10 years upon renewal (cumulative 20 years of post-release monitoring) after which the product is no longer regulated under the Biosafety Act (Fig. 1). However, existing surveillance programs continue as is the practice with other non-modified crops or products. If the Board's decision is to reject the application, then clear reasons for such a decision are communicated to the applicant who may wish to appeal.

The four key areas in the decision-making process for environmental release applications include risk assessment, socio-economic considerations, public participation, and consultations, among regulatory agencies.

## Risk Assessment and Risk Management

Article 15 of the CPB requires that risk assessments be undertaken in a scientifically sound manner, in accordance with Annex III and taking into account recognized risk assessment techniques. The objective of this risk assessment is to identify and evaluate the potential adverse effects of living modified organisms on the conservation and sustainable use of biological diversity in the likely potential receiving environment, taking also into account risks to human health. Assessment of risk is based on the logical definition of risk being a function of hazard and exposure.

The assessment process focuses on both food/feed safety as well as environmental safety. The global practice for safety assessment of GM crops is that the developer or applicant bears the primary responsibility for demonstrating product safety through conducting laboratory and field trials using established and approved

**Table 2** Summary of public comments received for the environmental release applications of Bt maize, Bt cotton, and gypsophila cut flower

| No. | 1 | 2 | 3 |
|---|---|---|---|
| Project name/crop | Bt maize (MON 810) | Bt cotton (MON 15985) | Gypsophila cut flower |
| Modified trait | Insect resistance | Insect resistance | Modified flower color |
| Total number of submissions received from the public | 15,096 | 11,719 | 69 |

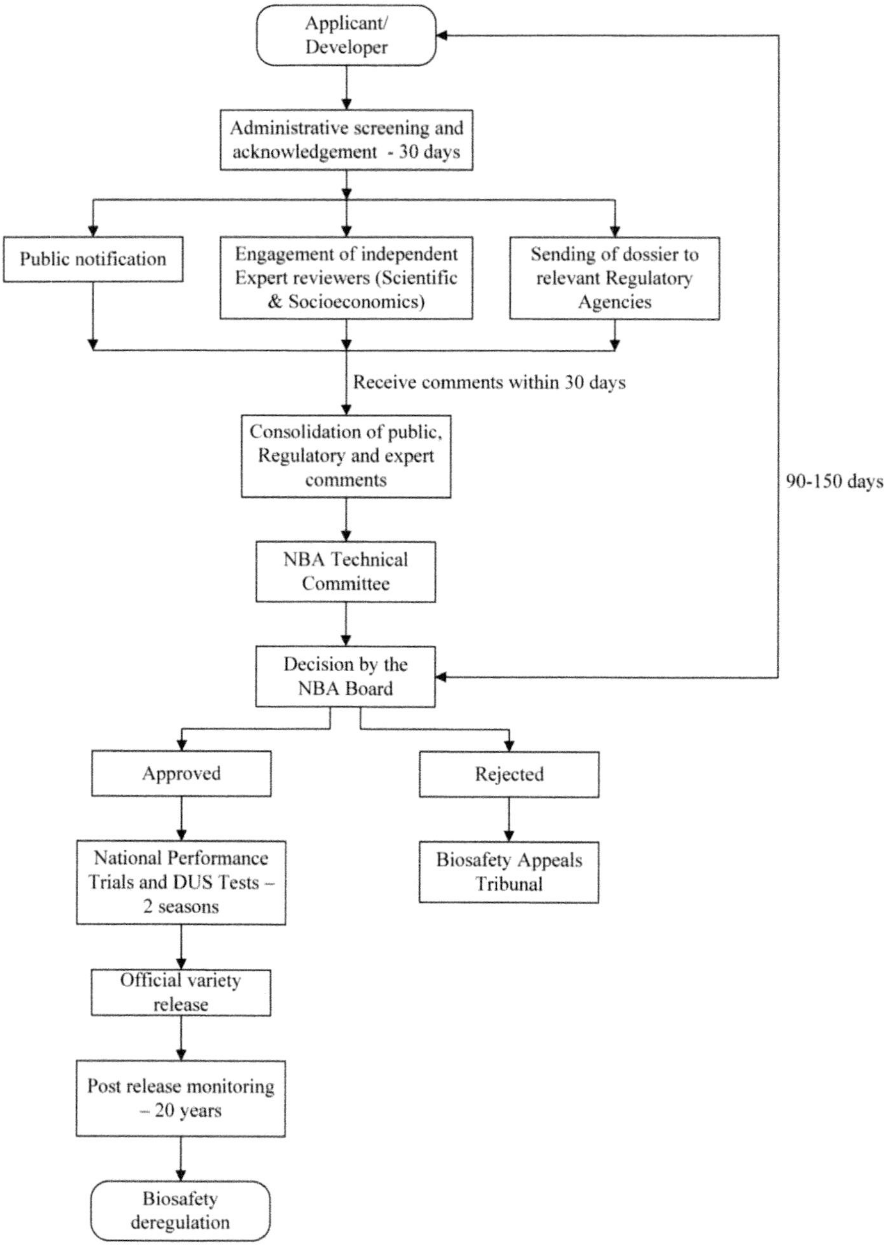


**Fig. 1** Flowchart for the environmental release process in Kenya

testing protocols. When this biosafety data is submitted to competent national authorities, it is peer reviewed by these authorities who may, in most cases, engage independent experts in appropriate fields (e.g., molecular biologists, entomologists, statisticians, immunologists, ecologists, socio-scientists). Based on the review of submitted data and experts' opinions, competent authorities such as NBA can seek additional information from applicants or require additional tests to be carried out before a decision is made. The review of submitted data by independent experts who do not work for the regulatory agency gives technical expertise to the NBA while also providing assurance and confidence to the public. Overall, the safety assessment of GM crops is geared towards establishing whether the modified crop is as safe as the conventional crop, normally referred to as the "substantial equivalence principle." Among the food safety parameters that NBA considers in an application are: the nature of the unmodified organism, the introduced trait, molecular characterization, compositional analysis, toxicity, allergenicity, and any unintended effects. Upon conducting a food/feed safety assessment, it will be determined whether the new product is substantially equivalent to the conventional counterpart except for the introduced trait, and whether the toxicity, allergenicity, and nutritional data submitted raise any safety concerns. As Kenya is a member to several international bodies such as Codex Alimentarius Commission (CAC), World Health Organization (WHO), CPB, and Food and Agriculture Organization (FAO), the risk assessment process follows internationally agreed standards and guidelines. These are now domesticated into national biosafety laws, Standards, and operational manuals. Biosafety data transportability is permissible on case by case basis.

When the NBA makes a decision, it considers the potential risks posed by the GMO and risk management measures proposed by applicants. If the Authority determines that the proposed risk management is inadequate, it institutes additional mitigation measures communicated through approval conditions.

One of the challenges Kenya has experienced in the recent past is the integration of Environmental Impact Assessment (EIA) required under the Environmental Management and Coordination Act into the Environmental Risk Assessment (ERA) required by the Biosafety Act, these laws being implemented by different agencies. This causes delays in subsequent steps required by applicants before a final decision on product release. There is a need for a critical analysis of underlying issues with other countries such as South Africa which may have similar legal provisions, and how they synchronize the two processes.

## *Socio-economic Considerations*

Article 26 of the CPB provides for Parties to take into account socio-economic considerations in reaching a decision on the import of LMOs, but only to the extent consistent with that country's other international obligations.

By definition, socio-economic assessments are *ex-ante* (before the fact) for products in the regulatory approval process. There may be cases where a biosafety regulatory system may require post-release monitoring and evaluation of socio-economic impacts, but falls under *ex-post* assessment, where there is a long and well-established literature and experience for assessments after environmental release (Kiplagat 2009). The CPB does not define socio-economic considerations, so the interpretation of this Article is left to individual Parties.

In the Kenyan context, over and above the risk assessment, all environmental release applications require consideration for socio-economic impact assessment before arriving at the final decision. As the Biosafety Act is not explicit on what entails socio-economic considerations in environmental release applications, a guideline was developed through national stakeholder consultations that identified eight key socio-economic issues key to the country. These include: food security and sustainability; access to the technology; income to farmers; cost of seeds and other inputs; co-existence between GM and non-GM counterparts (conventional and organic); trade implications at the national, regional and international level; benefits of the technology and freedom of choice; as well as biosafety and stewardship plans put in place by technology developers. This framework is still being tested as the country has so far had three environmental release applications to consider with varying degree of success. One challenge is that Kenya has not commercialized any GM product, as such socio-economic data are therefore hypothesized or taken from other countries with a history of commercialization. The thresholds for "accept or reject" are also not clearly defined.

## *Public Participation*

Article 23 of the CPB encourages Parties to promote and facilitate public awareness, education, and participation in the safe transfer, handling and use of LMOs in relation to the conservation and sustainable use of biodiversity, taking also into account risks to human health (SCBD 2000). The Protocol further provides for consultation with the public to be part of the decision-making but necessarily unique to each country's legal system and regulations. The Protocol does not give guidance on the public participation procedures to be used. When engaging the public, each country should consider the level of education, language of communication, and the medium to be used (Mugwagwa and Kiplagat 2014).

Public participation in decision-making is anchored in the Kenyan Constitution. Additionally, the Biosafety Act Para 54 and the implementing Biosafety (Environmental release) Regulations 2011 obliges the NBA to engage the public before arriving at a decision. Through stakeholder consultations, a guideline for public participation on GMO projects was developed and adopted in 2015. The guideline provides for: publishing a notice in at least two newspapers with nationwide circulation, publication of a non-confidential dossier on the NBA website, an official Government Gazette Notice, and holding at least one public forum in English and/or Kiswahili (the national languages in Kenya). Members of the public are allowed to submit any comments to the NBA within 30 days. All received public comments are reviewed by NBA and the applicant is requested to provide any additional information needed if it was not provided in the submitted application before a final decision is made.

In the three applications so far reviewed for environmental release (Bt maize, Bt cotton, and gypsophila cut flowers), the Authority received comments either supporting or opposing the commercialization of GM technology, while others were uncertain. In the case of Bt maize, the Authority received 15,096 comments from the public (Table 2); 14,956 (99.1%) were in support, 37 (0.2%) against, and 103 (0.7%) uncertain. Most of the submissions amounting to 14,986 were just in support or against the application without any reasons. However, in 110 of the submissions, the public had indicated the issues that informed their decisions, categorized as food/feed safety, environmental, and socio-economic concerns (Table 3). Whereas people are concerned about food safety as well as the environment, socio-economics also played a key role in the acceptance of GM technology and should not be ignored. For the Bt-cotton and transgenic gypsophila applications, 11,719 and 69 comments were received, respectively (Table 2). The low number of public submissions in regard to transgenic gypsophila application could be attributed to the

**Table 3** Issues raised by the public in regard to the Bt maize environmental release application

| S. no. | Issue of concern by the public |
|---|---|
| | **Food/feed safety concerns** |
| 1. | Toxicity issues |
| 2. | Possibility of the Bt maize causing allergenic properties on the populations |
| 3. | Nutritional, dietary and compositional changes |
| 4. | Antibiotic resistance |
| | **Environmental concerns** |
| 1 | Impacts on non-target organisms |
| 2. | Mechanisms for monitoring released Bt maize |
| 3. | Gene flow leading to increased fitness |
| 4. | Effect on target organism leading to development of resistance |
| 5. | Possible loss of biodiversity including soil and water micro fauna |
| 6. | Environmental impact assessment (EIA) yet to be conducted |
| 7. | Issues of climate change |
| | **Socio-economic concerns** |
| 1. | Co-existence framework between modified and non-modified maize |
| 2. | Food security and sustainability |
| 3. | Access to the technology including intellectual property rights |
| 4. | Possibility that cost of Bt maize seeds may be too high thus unaffordable to most farmers |
| 5. | Social-ethical issues |
| 6. | Trade implications at regional and international level |
| 7. | Freedom of choice—labelling of the Bt maize for consumer information |
| 8. | Stewardship program for variety purity throughout the production cycle |
| 9. | Possible monopolization of the Kenyan seed industry by foreign companies |
| 10. | Low level of public awareness of the technology |
| 11. | Religious beliefs |

low public interest on the application as it is an ornamental plant not a food related crop that elicits a lot of public interest.

Whereas the current mechanism of public engagement is working, it could be further optimized by advertising through the radio. According to a survey conducted in Kenya, radio scored 72% as the best channel for communication on new GMOs to the public, while newspapers scored only 50% (Sang et al. 2014). The engagement could perhaps be further enhanced by holding public fora at regional or county levels, and using local languages as literacy levels on GMOs remain low in the country.

## *Consultation Among Regulatory Agencies in the Country*

Recognizing that GMOs cut across a broad spectrum of disciplines, the Biosafety Act in the First Schedule lists eight regulatory agencies that NBA consults (on a need to basis) before making a decision. These include the: Kenya Plant Health Inspectorate Service (KEPHIS); Kenya Bureau of Standards (KEBS); National Environment Management Authority (NEMA); Directorate of Veterinary Services (DVS); Department of Public Health, Pests Control Products Board (PCPB); Kenya Wildlife Service (KWS); and Kenya Industrial Property Institute (KIPI). Five of these agencies, KEPHIS, DVS, KEBS, NEMA, and the Department of Public Health, also sit in the NBA Board, and so participate fully in the decision making process. During the review process, the applicant submits the application to NBA to identify relevant Regulatory Agencies from among the eight; the Authority then sends the dossier to the relevant agencies for review, comments, and/or sets conditions for approval or rejection. Any concerns raised by the Agencies are addressed before the NBA Board makes a final decision. There is a challenge of integrating the Environmental Impact Assessment (EIA) required by the National Environment Management Authority (NEMA) into the Environmental Risk Assessment (ERA) provided for by CPB and required by NBA, thus delaying the process. When a decision is made, the monitoring of research projects and commercial activities is jointly by the NBA and relevant Regulatory Agencies. To streamline the working between NBA and these Agencies, a coordination framework has been developed and the agencies meet at least annually to keep abreast of recent developments.

The central management of GMO applications at NBA has greatly reduced duplicity where applicants would be required to submit parallel applications to different Agencies. The process has, however, not been without challenges, including: limited human capacity for review of GMO applications, and the extended time for submitting comments to NBA that negatively affect when NBA can communicate to applicants. There are instances where NBA makes a decision which is then vetoed by another regulatory agency citing their statues, resulting in delays in arriving at a decision. The existence of a number of regulatory agencies serves as a "third-eye" and checks that NBA is not abdicating from its legal mandate of exercising due diligence.

## Discussion and Conclusions

Kenya has made great strides in establishing an efficient and effective biosafety system for the regulation of GMOs and their derived products. Indeed, it has the necessary policy, legal, institutional, administrative, and public participation mechanisms anchored into enforceable laws. Using the existing framework, Kenya has made several regulatory decisions on contained use, confined field trials, environmental release, imports, exports, and transit.

However, during the implementation of biosafety laws, some gaps have emerged that include limited expertise in some areas of risk and socio-economic impact assessments, lack of clarity in socio-economic parameters to be considered and thresholds for approvals, limited capacity among respective government agencies involved in evaluation of applications, limited technical knowledge of food safety assessment principles by policy makers and the general public, and lengthy consultations for environmental release applications that makes it difficult to make decisions within set legal timelines.

In regard to existing expertise at the national level, the NBA Secretariat has limited permanent technical staff of about 13 and a national pool of 42 independently sourced experts. Experts are engaged on a "need" basis depending on the nature of application. Whereas these experts offer invaluable expertise in the review process, there are limitations in the number of available experts with in-depth knowledge in particular fields such as toxicology, allergenicity, emerging technologies, and socio-economic experts. Additionally, despite existence of guidelines on socio-economic considerations, the thresholds for approval or rejection still remain ambiguous in the decision-making process involving environmental release of GMOs especially when a country has no historical data on their commercialization.

The review process in Kenya entails applicants submitting an application centrally to the NBA, although other agencies will be involved in review of the submitted information and ultimately in decision-making. As such, when the NBA receives an application, it identifies all relevant regulatory agencies and forwards copies to them to give opinions based on their mandate. The administrative process of sharing application dossiers has been largely smooth. However, most of the agencies lack a dedicated biosafety desk with standby biosafety officers who would be responsible for reviewing regulatory dossiers submitted to them. Consequently, staff who perform risk assessment may not have previously performed this task. While there are efforts by the NBA to capacity-build the key agencies and encourage them to designate biosafety officers, this hardly materializes in practice due to high staff turnover and other priorities within their institutions. Another challenge is that whereas the NBA expects review comments from agencies within 30 days, on average it takes 45 days, negatively impacting on the time taken to reach a decision by the NBA, and subsequent communication to the applicants.

Limited awareness on GMOs by policy makers and the general public also remains a challenge. Negative public perception of the technology remains a major impediment to full exploitation of the technology despite the existence of a regulatory framework. This is compounded by myths and misconceptions on GM technology compared to conventional methods and organic agriculture. One core mandate of NBA is creation of public awareness on biosafety matters. As such, various fora involving policy makers, media engagements, farmers, and consumers have been established, although not on large-scale and robustness needed due to budgetary constraints. Such efforts would greatly enhance informed public participation in decision-making process. Related to this is inadequacy in the number of effective biosafety risk communicators with ability to use simplified language as opposed to scientific jargon in their communication.

At the regional level in Africa, there is currently limited collaboration among the regional biosafety agencies due to the different stages of development of their regulatory frameworks and lack of a harmonized risk assessment process. Efforts to harmonize risk assessments in the region have now been initiated at the East African Community (EAC) and Common Markets for Eastern and Southern Africa (COMESA) levels, but little progress has so far been made in the approval and implementation of proposed plans. There is a need to fast track this harmonization process as well as establishing both formal and informal links between the competent biosafety authorities in the region. This would be beneficial through a regular exchange of information and experiences.

Lastly, whereas the Kenyan biosafety laws stipulate that decisions should be communicated to applicants within 90–150 days, this has proved impractical for the three environmental release applications received due to the lengthy, consultative review process, and public engagements that led to decisions being made beyond the legally mandated time. A review of the Biosafety Act is now needed to provide a more practical review period, but one still within the 270 days provided for in the CPB.

Whereas the review process explained in this chapter focuses on GM crops, the review process for GM animals and microorganisms is similar.

## References

Chambers JA (2013) Biosafety of GM crops in Kenya, Uganda and Tanzania. An evolving landscape of regulatory Progress and retreat. CSIS Global Food Security Project Report Ebook

Government of Kenya (GoK) (2018) Unpacking the big four. Kenya School of Government (KSG) Weekly Bull 7(20)

James C (2012) Global status of commercialized biotech/GM crops, ISAAA brief no.44. ISAAA, Ithaca/New York

Juma C (2011) The New Harvest; agricultural innovation in Africa. Oxford University Press, New York

Karembu M, Nguthi F, Ismail H (2009) Biotechnology crops in Africa, The final frontier. ISAAA Africenter, Nairobi

Kenya Biosafety Act (2009) http://www.biosafetykenya.go.ke/index.php?option=com_content&view=article&id=16&Itemid=121

Kenya Biosafety Regulations (Contained use) (2011) http://www.biosafetykenya.go.ke/index.php?option=com_content&view=article&id=17&Itemid=122

Kenya Biosafety Regulations (Environmental release) (2011) http://www.biosafetykenya.go.ke/index.php?option=com_content&view=article&id=17&Itemid=122
Kenya Biosafety Regulations (Import, Export and Transit) (2011) http://www.biosafetykenya.go.ke/index.php?option=com_content&view=article&id=17&Itemid=122
Kenya Biosafety Regulations (Labelling) (2012) http://www.biosafetykenya.go.ke/index.php?option=com_content&view=article&id=17&Itemid=122
Kinyua MG, Otieno DF, Tonui W, Ogoyi DO (2014) Case Study – Kenya: biosafety in Africa; Experiences and best practices. (Keetch DP, Makinde D, Weebadde CK, Maredia KM, eds), Michigan State University. pp 89–93. http://www.agriknowledge.org/file_downloads/x633f103b
Kiplagat B (2009) Implementation of the Cartagena protocol on biosafety at National Level- Legal Issues . ABNE Policy Brief no.1
Mugwagwa J, Kiplagat B (2014) Legislative and policy issues associated with GE crops in Africa. Biosafety in Africa; experiences and best practices, pp 48–53
Sang H, Ogoyi DO, Mavua A, Tonui W (2014) An evaluation of communication strategies for introducing controversial biotechnology seeds in Kenya. African Commun Res 7(3):365–385
Secretariat of the Convention on Biological Diversity (SCBD) (2000) Cartagena protocol on biosafety to the convention on biological diversity: text and annexes, Montreal
Wafula D (2009) Kenya Biosafety Law- Congratulations! Bivision News 11:1–4

# South Africa—Synthetic Biology Regulatory Considerations and Biodiversity – A Legal Perspective for South Africa

James Ian Rhodes and Wadzi Mandivenyi

**Abstract** The South African government recognizes science and technology as essential in achieving South Africa's development goals. Accordingly a number of key policies and initiatives have been put in place to support innovation. The South African national Bio-economy Strategy provides a policy framework that proposes resources and support for the development of competencies and infrastructure to support the use of technologies, including synthetic biology. Although synthetic biology has not been formally defined nationally, the products of synthetic biology will include genetically modified organisms. South Africa already has a well-established GMO regulatory system which provides a robust framework to regulate activities with synthetic organisms and their products. This includes measures for the responsible development, production, and use of synthetic organisms and their products. This framework is described, and specific concerns for synthetic biology are discussed.

**Keywords** Innovation · Development goals · Biotechnology · Policy · GMO · Genetically modified · Regulation · Sustainable · Risk analysis · Risk assessment

## Background

The South African government recognizes science and technology, including biotechnology, as essential in achieving South Africa's development goals (DST 2013). Support for science and technology and how innovation must be harnessed to address development is outlined in a number of key government policies (DED

J. I. Rhodes (✉)
Biosafety South Africa, Cape Town, South Africa
e-mail: james@biosafety.org.za

W. Mandivenyi
Department of Environment, Forestry and Fisheries, Pretoria, South Africa
e-mail: wmandivenyi@environment.gov.za



A. Chaurasia et al. (eds.), *GMOs*, Topics in Biodiversity and Conservation 19,
https://doi.org/10.1007/978-3-030-53183-6_24

2011; DTI 2014; NPC 2011; RSA 2013). The South African national Bio-economy Strategy approved by cabinet in 2013 provides a policy framework that proposes resources and support for the development of competencies and infrastructure to support the use of synthetic biology in order to underpin the competitiveness of South Africa's bio-economy. Although synthetic biology has not been formally defined nationally, it is acknowledged that synthetic biology is a broad term that incorporates a number of different disciplines and technologies across varying applications from relatively minor modifications of existing genetic material, to the design/creation of new life forms (DEA 2018). Irrespective of the exact definition, the products of synthetic biology will include genetically modified organisms (GMOs). The discussions on synthetic biology are therefore considered in the context of biotechnology and in the legislative framework of biotechnology and GMOs.

## Legal Instruments

The Genetically Modified Organisms Act 15 of 1997 (GMO Act), as amended by Act 23 of 2006 and read together with its accompanying regulations, is the principal legal instrument for regulating all activities involving GMOs in South Africa. In addition, South Africa is a Party to the Cartagena Protocol on Biosafety (CPB) which provides an international set of rules that determine the import, export, transit, handling, and activities related to the use of GMOs in order to protect the environment, biodiversity, and human health.

The overall objective of the GMO Act is to provide for measures to promote the responsible development, production, use, and application of GMOs. This encompasses the entire pipeline of GMO development including research and development (contained use and confined field trial activities), production (contained use and general release activities), import and export, transport, use, and application of GMOs. Accordingly, the act aims to ensure that any GMO-related activity in South Africa is conducted so as to limit potential risks to the environment, to human, and other animal(including human) health, and takes socio-economic considerations into account. The GMO Act and the relevant regulations govern all activities with GMOs according to permits issued in terms of this Act. Different types of permits can be applied for relating to a particular GMO activity. These include permits for import, commodity clearance, general release, field trials, and contained use.

The definition of a GMO under the act is "an organism, the genes or genetic material of which, have been modified in a way that does not occur naturally through mating or natural recombination or both." This includes GMO viruses and bacteriophages and the use of gene therapy, but excludes the regulation of human gene therapy. South Africa supports an approach focused on the GMOs and not the process through which they have been created. In line with this approach, the current South African regulatory framework is designed to regulate GMOs, irrespective of the process through which it was developed, and regulation is based on two guiding principles: (1) a product-based trigger because that is the source of potential harm; and (2) a threshold of genetic variation beyond that which may also occur naturally (DEA 2018). Accordingly the Executive Council of the GMO Act, which is the regulatory authority for South Africa, considers synthetic biology as fundamentally similar to genetic modification contemplated in the GMO Act, and activities involving synthetic biology will fall under the definition of a GMO (DEA 2018). In this regard, synthetic biology is considered as an extension of the current biotechnology techniques and would therefore primarily be regulated under the provisions of the GMO Act. This regulatory framework is well established with extensive experience and regulatory expertise at all stages of the development pipeline in the regulation of activities with GM plants, animals, and microorganisms.

The framework uses a case-by-case risk analysis process to regulation that is iterative in nature and takes into consideration the trait, the organism, and the receiving environment (DAFF 2004). The risk analysis framework that is followed to assist in decision-making composes four main steps. These include: (1) setting the context and scope; (2) the scientific risk assessment; (3) risk decision-making, including risk management; and (4) risk communication. This approach provides a rigorous framework to assist with decision-making with activities that involve synthetic biology in South Africa. This includes all stages of the decision-making process including assessing, managing, and communicating risks on biodiversity and to human health. This is not to say that there may not be new approaches that may be necessary in managing and accessing specific traits or characteristics of organisms produced through synthetic biology, for example, to address issues of appropriate comparators for synthetic organisms, and South Africa supports horizon scanning efforts on the impact of synthetic biology (CBD 2018). This will require a responsive regulatory system in order to respond to this evolving technology; however, this will occur within the broader risk analysis framework.

Two other acts have specific provisions for GMOs. The National Environmental Management Biodiversity Act (NEMBA 2004) confers to the South African National Biodiversity Institute the responsibility to monitor and report on the environmental impacts of GMOs released into the environment in South Africa and also establishes a mechanism whereby the Minister of Environmental Affairs may request an environmental impact assessment of a GMO under the National Environmental Management Act (NEMA 1998). The definition of a GMO under both NEMA and NEMBA is as defined under the GMO Act. In addition, the GMO regulatory system is complemented by other national legislation such as Promotion of Access to Information Act (2000), Agricultural Pests Act (1983), Animal Diseases Act (1984), Fertilizers, Farm Feed, Agricultural Remedies and Stock Remedies Act (1947), Medicines and Related Substances Amendment Control Act (1997), Promotion of Administrative Justice Act (2000), and the Foodstuffs, Cosmetics and Disinfectants Act (1972).

## Additional Considerations

The South African GMO regulatory framework also includes socio-economic considerations as part of the decision-making system. This framework therefore provides sufficient scope for addressing the ethical and social aspects associated with synthetic biology applications. This current regulatory framework enables decision-making that addresses potential risks while taking into consideration the potential benefits of synthetic biology as contemplated in the bio-economy strategy. The potential applications of synthetic biology to develop efficient and effective ways to respond to challenges associated with bioenergy, agriculture, health, and chemical production, among others, are recognized in the bio-economy strategy (DST 2013). The strategy suggests striking a balance between recognizing the potential benefits of biotechnology and the ethical considerations of the technology while at the same time being responsive to the significant social and economic development goals of South Africa (DST 2013). This is in alignment with the CPB as in its previous decisions, the convention invited Parties to take into account appropriate socio-economic, cultural, and ethical considerations when identifying the potential benefits and adverse effects related to synthetic biology. This will require a constantly evolving ethical and regulatory framework. Specifically additional guidelines will likely be required in establishing protection goals to assess the benefits and risks of synthetic biology during the setting of the context and scope step of the risk analysis framework as well as additional guidance on how environmental and social benefits are to be weighed up during the decision-making step. However, by ensuring constant engagement between the scientists and regulatory authorities, it is possible to advance the safe use of synthetic biology as part of a developmental agenda.

## Discussion

Some concerns for the regulation of synthetic biology remain. In terms of South Africa's participation in the CPB, there are concerns that the CPB's focus on process, in this instance synthetic biology, is creating unnecessary complications, duplications, and confusion (DEA 2018). Irrespective of the exact definition that may be used for it, "synthetic biology" falls under the definition of "modern biotechnology." The products of synthetic biology will comprise LMOs which are already subject to the CPB. Deliberating on different techniques separately gives the impression that: (1) modern biotechnology and synthetic biology has its own set of risks; and (2) each technique cannot be regulated using the same broad risk analysis process. This can also lead to developing countries determining that they are not sufficiently competent at assessing the risks of new techniques such as synthetic biology when, in actuality, the different techniques may have the same outcomes/ products and regulators are equipped with the skills and training to assess applications for activities with GMOs.

South Africa supports an integrated approach to LMO risk governance under the CPB that focuses on the LMO and not the process through which it was produced. This aligns with the South African approach to the regulation of synthetic biology. There are also concerns that current regulations for GMOs will not adequately address DIY citizen scientist's use of synthetic biology; however, the GMO Act provides for penalties for those that contravene it Act. Communication and education are likely to play an important role to ensure compliance of citizen scientists.

The GMO regulatory system for South Africa therefore provides a guiding framework that would regulate activities with synthetic organisms and its products. This includes contained and confined activities; general (environmental) release; use for food and feed; and import and export.

## References

CBD (Convention on Biological Diversity) (2018) Decision adopted by the conference of the parties to the Convention on Biological Diversity. Fourteenth meeting, Sharm El-Sheikh, Egypt, 17–29 November 2018, Agenda item 27. https://www.cbd.int/doc/decisions/cop-14/cop-14-dec-19-en.pdf

DAFF (Department of Agriculture, Forestry and Fisheries) (2004) Guideline document for work with genetically modified organisms. Directorate: Genetic Resources, Department of Agriculture, Forestry and Fisheries, Pretoria. https://www.nda.agric.za/doaDev/sideMenu/biosafety/doc/GUIDELINE4WORKWORKwit.pdf

DEA (Department of Environmental Affairs) (2018) South Africa's submission to the convention on biological diversity on information pertaining to experiences on synthetic biology. http://bch.cbd.int/database/attachment/?id=17739

DED (Department of Economic Development) (2011) The new growth path: framework. http://www.economic.gov.za/communications/publications/new-growth-path-series/download

DST (Department of Science and Technology) (2013) The bio-economy strategy. https://www.gov.za/documents/bio-economy-strategy

DTI (Department of Trade and Industry) (2014) Industrial policy action plan: economic sectors and employment cluster IPAP 2014/15–2016/17 http://www.gov.za/sites/www.gov.za/files/IPAP2014.pdf

NPC (National Planning Commission) (2011) National Development Plan: vision for 2030. http://policyresearch.limpopo.gov.za/bitstream/handle/123456789/941/NDP%20Vision%202030.pdf?sequence=1

RSA (Republic of South Africa) (2013) Medium-term strategic framework (MTSF) 2014–2019. http://www.gov.za/sites/www.gov.za/files/MTSF_2014-2019.pdf

# Tunisia—The Use of Modern Biotechnology in Tunisia – Regulatory Framework

**Maher Chaouachi, Raja Chalghoumi, Mohamed Ilyes Kchok, Sana Jaballah, Hazar Belli Abelkafi, Hatem Ben Belgacem, Saloua Ben Fredj, Raoudha Hellal, and Maher Medini**

**Abstract** Regulation of genetically modified organisms (GMOs) started in the 1990s in the United States, and European Union (EU) adopted two different approaches for GMO regulations: one based on the "substantial equivalence" and the other on the "precautionary approach" and the "right to know" of the consumer. Other countries developed their regulations in between these two concepts. However, despite the underlying opposite approaches, both countries recognized some common aspects in GMO regulation that cover different aspects of the cultivation and commercialization of GM crops, such as approval, risk assessment, labeling, traceability, and coexistence; but also aspects related to the development of new GM

M. Chaouachi (✉)
Laboratoire de Recherche "Bioressources: Biologie Intégrative & Valorisation", Institut Supérieur de Biotechnologie de Monastir, Avenue Tahar Hadded, BP 74, Université de Monastir, Monastir, Tunisia

R. Chalghoumi
Laboratory of Improvement & Integrated Development of Animal Productivity & Food Resources, Higher School of Agriculture of Mateur, University of Carthage, Bizerte, Tunisia

M. I. Kchok
Centre of Biotechnology at Borj Cedria Technopark, Hammam-Lif, Tunisia

S. Jaballah
Agri-food Technical Center (CTAA), Ariana, Tunisie

H. B. Abelkafi · H. B. Belgacem
Ministry of local affaire and Environment, Direction Générale de l'Environnement et de la Qualité de la Vie (DGEQV), Cité administrative, rue de développement, Tunis, Tunisie

S. B. Fredj
Laboratory of seed Analysis and Plants, Ministry of Agriculture, Water Resources and Fisheries, Tunis, Tunisia

R. Hellal
Central Laboratoiry of Analyses and Assays (LCAE), Tunis, Tunisie

M. Medini
Laboratory for GMO detection, National Genes Bank of Tunisia. Boulevard du Leader-Yasser-Arafat, Tunis, Tunisia



A. Chaurasia et al. (eds.), *GMOs*, Topics in Biodiversity and Conservation 19,
https://doi.org/10.1007/978-3-030-53183-6_25

crops, such as rules for laboratory and field trials and intellectual property rights (IPR) protection (Vigani and Olper, AgBioforum 18:44–54, 2015). Regulation of biotechnology and GMO has a direct effect on trade and market (Gruère, An analysis of trade related international regulations of genetically modified food and their effects on developing countries, EPT discussion paper 147. International Food Policy Research Institute, Environment and Production Technology Division (EPT), Washington, DC, 2006; Gruère et al., Rev Int Econ 17:393–408, 2009; Vigani and Olper, Food Policy 43:32–43, 2013, AgBioforum 18:44–54, 2015). Despite the efforts of the Codex Alimentarius and of the Biosafety Protocol in searching for international agreement on labeling and rules for the trans-border movements of GMOs, to date there is no consistent and harmonized set of rules to regulate GMOs. This is partially due to the different food security strategy in developing and developed countries (Vigani and Olper, AgBioforum 18:44–54, 2015). Hence the "wait and see" attitudes of most developing countries, including Middle East and North African (MENA) countries, which fear that the implementation of any particular regulations may have a direct effect on their current and future agricultural exports to countries with stringent regulations (Zarrilli, International trade in GMOs and GM products: National and multilateral legal frameworks, Policy issues in international trade and commodities study series, 29. United Nations Conference on Trade and Development, New York/Geneva, 2005).

**Keywords** Tunisia · Biotechnology · Biosafety · GMO · Synthetic biology · Regulation · Cartagena protocol

## Background

Today, modern biotechnology has made huge progress towards new techniques and uses targeting genomic modifications including synthetic biology that aims to exercise control in the design, characterization, and construction of biological parts, devices, and systems to create more predictable biological systems. The areas of research that are considered "synthetic biology" include DNA-based circuits, synthetic metabolic pathway engineering, synthetic genomics, protocell construction, and xenobiology. Synthetic biology presents potential challenges to biosecurity, as well as potential tools to aid in security efforts. Biosecurity concerns related to biodiversity include the use of synthetic biology to create destructive pathogens targeting agriculture or other natural resource bases. Existing livestock and crop diseases could be made more lethal, and novel pathogens designed to impact agricultural biodiversity predict that biological weapons customized to attack specific groups are highly likely in the long term (10 or more years). Hence the international concerns on synthetic biology.

Tunisia ratified the Cartagena Protocol on Biosafety (CPB) that entered into force on September 11, 2003. The implementation of the protocol started in Tunisia with the establishment of a National Biosafety Framework (NBF) in the frame of the United Nation Environment Programme/Global Environment Facility[1] (UNEP/GEF) project (2007–2015) coordinated by the Ministry of Local Affairs and Environment (MLAE). However, the legal basis to NBF implementation is still totally absent. Hence the confusion established among stakeholders to go further. The MLAE updated its national strategy in order to overcome these misunderstandings based on a collective awareness of the importance of local genetic resources, the precautionary principle, and adherence to a code of conduct. Thus, different actors, i.e., scientists, policy makers, producers, or consumers, need to coordinate their efforts to bring this community around the same concepts. This contribution describes the state of art regarding advancements in terms of regulation in the case of modern biotechnology in Tunisia.

## Modern Biotechnology: State of Art in Tunisia

In Tunisia, the evaluation of the status of the biotechnologies points out that GMO-related biotechnology laboratories, and research units are making steady progress in research in agriculture, health, environment, and agribusiness. In fact, Tunisia is one of the countries that have a particular potential for research development in the MENA region and Africa, and have many research laboratories, biotechnology centers, and institutes involved in the establishment of Biotech crop development and also their detection and quantification (Chaouachi et al. 2013; Nabi et al. 2016). Tunisia has succeeded in developing many biotech products with traits such as biotic and abiotic stress tolerance (Gargouri-Bouzid et al. 2006; Gouiaa et al. 2012; Feki et al. 2013). All these transgenic crops are, however, still confined in laboratories and not yet authorized for cultivation in field trials due to a lack of regulation regarding the risk assessment of such activities and the level needed to sufficiently protect against harm to the environment and consumers.

In terms of trade, Tunisia is a major importer of corn products (Americas: 36% market share), soy products (Americas: 84%), and sugar products (Brazil: 67%) while a lesser importer of those of alfalfa (EU: 99%), rapeseed (EU: 100%), and cotton (EU: 73%). Imported feed ingredients are a necessity for Tunisia's livestock and poultry production (Ahmed and Chahed 2012).

In addition, Tunisia today has built the capacity of detection and quantification of GMOs in different matrixes due to the implementation of the activity in four official centers and laboratories: (1) National Gene Bank (http://www.bng.nat.tn); (2) Technical Center of Agrifood (http://www.ctaa.com.tn/accueil/); (3) Central Laboratory of Analysis and Assays (https://lcae.nat.tn/); and (4) the Laboratory of

[1] Actually "United Nations Environment.

Analysis of Seeds and Plants. Furthermore, a Network convention signed in 2016 between these four institutions has been implemented to better coordinate and exchange scientific and technical information regarding GMO detection. These core institutions are coordinated by the Ministry in charge of the Environment and supported by different Biotechnology institutes, such as the High Institute of Biotechnology of Monastir (http://www.isbm.rnu.tn/) and research centers (CBS: http://www.cbs.rnrt.tn/fra/home and CBBC: http://www.cbbc.rnrt.tn/) and is part of the MENA regional Network (MENANGL) and the Global GM Network coordinated by the Joint Research Center (JRC) under the European Commission. In addition, these testing laboratories are already or undergoing accreditation to ISO17025. However, while these laboratories are operational technically, they are not yet legally and administratively entitled to perform such analyses pending adoption of the national Biosafety Law.

## Establishment of the National Biosafety Framework (NBF)

First, it is important to differentiate between Biosafety and Biosecurity. According to definitions established by the WHO (2006), "laboratory Biosafety" concerns the containment principles, technologies and practices that are implemented to prevent the unintentional exposure to pathogens and toxins, or their accidental release, while "laboratory Biosecurity" concerns the protection, control, and accountability for valuable biological materials (VBM) within laboratories, in order to prevent their unauthorized access, loss, theft, misuse, diversion, or intentional release. FAO, however, uses the term biosecurity in relation to sanitary, phytosanitary, and zoosanitary measures applied in food and agricultural regulatory systems. As such, biosecurity is seen as a strategic and integrated approach that encompasses policy and regulatory frameworks (including instruments and activities) that analyze and manage risks in the sectors of food safety, animal life and health, and plant life and health, including associated environmental risk. Thus, biosecurity covers the introduction of plant pests, animal pests and diseases, and zoonoses, the introduction and release of genetically modified organisms (GMOs) and their products, and the introduction and management of invasive alien species and genotypes. Biosecurity is a holistic concept of direct relevance to the sustainability of agriculture, food safety, and the protection of the environment, including biodiversity (FAO 2003).

Two main international agreements are pertinent, the Codex Alimentarius and the Cartagena Protocol on Biosafety (CPB); these are the most diffused and developed agreements on GMOs. The purpose of the Codex Alimentarius is to define international standards to protect consumer health and promote fair relationship in trade practices. It has successfully reached an agreement on safety assessment procedures for GMOs, but no formal labeling standard has been yet achieved. The CPB is part of the United Nations Convention on Biological Diversity (CBD), and introduced a procedure for risk assessment, risk management, and trans-boundary movements of living modified organisms (LMOs). The CPB requires a comprehensive risk assessment and risk management framework provided by the exporter before the introduction of any LMO into the importer territory. The CPB was proposed as a primary policy for those countries without domestic regulations on GMOs and to protect countries holding most of the global biodiversity, typically located in the south of the world. To comply with CPB requirements is costly, and developing countries could benefit from collective funds provided by the agreement (Vigani and Olper 2013).

Here we describe the Tunisian legal framework for Biosafety as described, including the outputs of UNEP/GEF projects on the development of the NBF and the state of the art of the Biosafety strategy in Tunisia (Fig. 1).

The Conference of the Parties to the CBD requested the Global Environment Facility (GEF) to provide support to eligible countries for building capacity to implement the CPB, including the MENA countries. Although the Protocol entered into force in 2003, the GEF actually began supporting capacity building activities for biosafety in 1997 with pilot projects in 18 countries. The evaluation of this support, submitted to the GEF Board in November 2005, found that: "the GEF has responded very expeditiously and systematically to the request from the CBD for support to the Cartagena Protocol. UNEP, UNDP, and the World Bank have remained neutral in this dynamic debate among the various interest groups, and have succeeded in doing so." The report also found that: "the GEF has contributed to considerable progress toward implementation of the Protocol by enhancing capacity on scientific, administrative, legal and information management matters, as well as promoting cross-sectorial collaboration and collaboration between the public and the private sectors as well as the civil society." The GEF support enabled some 120 countries to prepare their own National Biosafety Framework (NBF). One of the projects supported by the UNEP/GEF was conducted between 2007 and wrapped up in the 18th Biosafety National Project Coordinator (NPC) meeting for Africa in Tunisia (2015). In the frame of these projects, three national commissions have been created to manage three main pillars of the NBF: (1) Legal Framework Commission; (2) Technical Commission; and (3) Communication, Awareness and Public Participation Commission. The project outputs and actions put in place are described in Fig. 2. In addition, these commissions developed a draft law on biosafety that includes living modified organisms (LMOs), invasive species, and pathogens. The country profile can be found on the Biosafety Clearing House (BCH) of the CPB (Biosafety Clearing House 2019). Through these UNEP/GEF projects, Tunisia has ensured the training of several stakeholders on the use of the BCH to be in line with the new LMOs actualities and authorizations all over the world. Tunisia also established the national subcommittee on communication, education, and public awareness in 2014 to, among others, facilitate the exchange of information on LMOs in the BCH; and to operationalize the national BCH including the development and validation of information and outreach materials. In 2016, in line with its communication plan, an NGO was established, the Tunisian Association for Biosafety and Environmental Education (ATB2E) (www.atb2e.tn) whose objectives are, among others, to raise awareness among various social groups regarding issues related to Biosafety. Tunisia also shares various outreach materials and other information on a

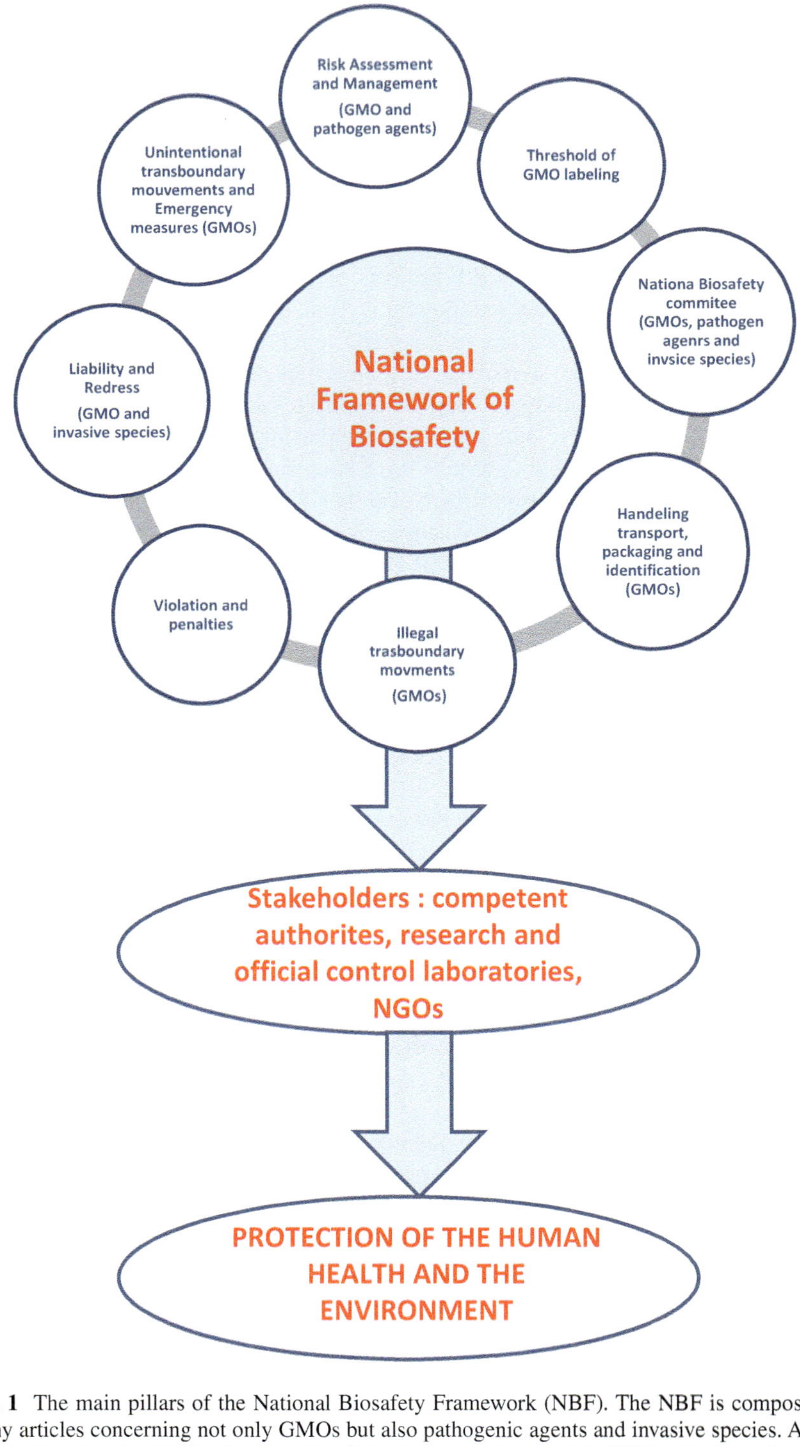


**Fig. 1** The main pillars of the National Biosafety Framework (NBF). The NBF is composed of many articles concerning not only GMOs but also pathogenic agents and invasive species. All the stakeholders involved in the framework will implement and control the NBF with the main aim of protecting human health and the environment

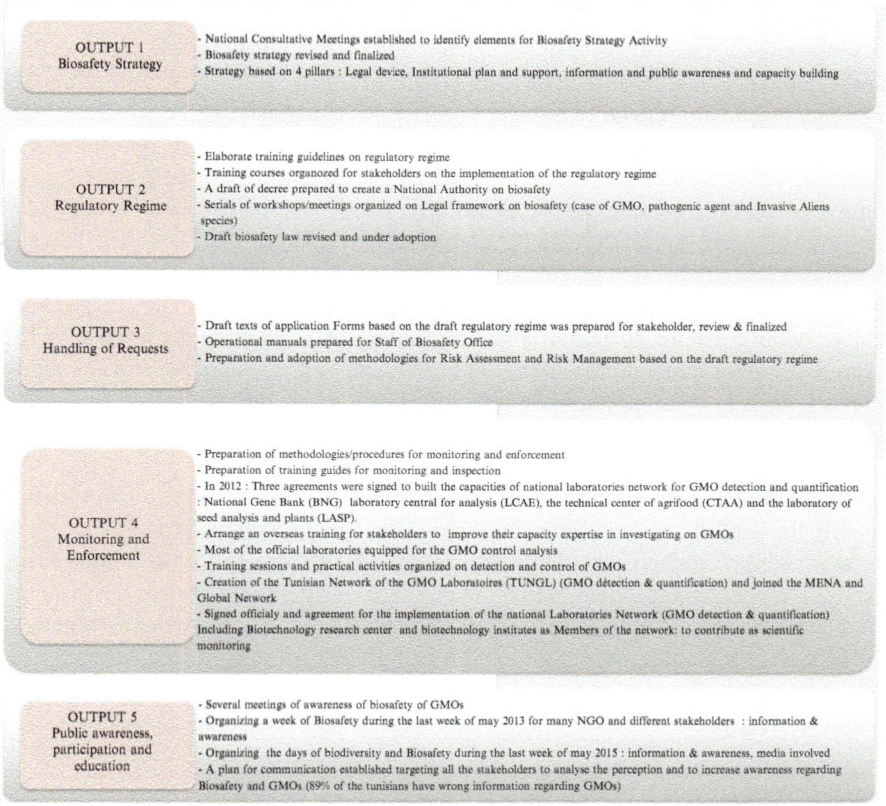


**Fig. 2** The state of art of the main pillars and outputs of the Biosafety National framework to be implemented in Tunisia as a result of the UNEP/GEF project

regular basis that are publicly available on USB keys, CD-ROMs in French, Arabic, and other national languages (Ben Belgacem et al. 2017–2018).

## GMO Regulation and Legal Perspectives?

A common problem in many countries was inadequate coordination of roles and responsibilities among regulatory bodies. The lead responsibility for implementation of the CPB often rests with the Ministry in charge of the environment or a similar such body, presumably because of the link between the Protocol and the CBD. Yet the actual management of the approval process of GMOs tends to rest with ministries of Agriculture, Higher Education and Scientific Research, Science and Technology, or Foreign Trade. All of these tend to be more powerful ministries than that responsible for enforcing the regulations, and this imbalance may lead to some challenges in implementing the Protocol in some countries. Interestingly, several countries have treated biosafety in conjunction with wider issues of biosecurity, agrobiodiversity, invasive alien species, and illegal transboundary movement of endangered species. This indicates that GMOs are seen, by at least some countries, as being intimately related to broader biodiversity concerns.

In developing the national strategy for biosafety in Tunisia, many questions were raised and complicated more and more the establishment of a specific law targeting only GMOs. This complexity stems from modern biotechnology being applied in different domains, agriculture, industry, health, and environment. Thus, the Tunisian lawmaker is challenged to reach a consensus on GMO regulation to address the serious conflict between two groups: (1) agri-biotech investors and their affiliated scientists who consider agricultural biotechnology as a solution to food shortage, scarcity of environmental resources, and weed and pest infestations; and (2) opposing scientists, environmentalists, farmers, and consumers who insist that genetically modified food introduces new risks to food security, the environment, and human health.

After the examination of most of the international regulations in existence, mainly those of the EU and based on the process for obtaining GM products arising from modern biotechnology (Christiansen et al. 2019), and that of the United States focused on a final product policy using the substantial equivalence principal (Eckerstorfer et al. 2019), Tunisia preferred to create a new regulatory framework on Biosafety including not only the GMOs, but also pathogenic agents/toxins and invasive species. Tunisia opted to take up this challenge, and the emerging legal and regulatory project turns around five strategic, legal, and institutional orientations:

(1) A cautionary orientation based on a priori control
(2) Transparency
(3) Ongoing vigilance, strategic, and, specific control
(4) Integration of the national control and evaluation system of risks into the international system
(5) A strategic authority separation

This framework summarizes the steps to be taken starting from laboratory risk assessments, containment levels and traceability, environment dissemination, the authorization process of and notifications for import and export of GMOs, up to the determination of the threshold of labeling on products derived from GMOs. Moreover, this law is planning to include articles focusing on violation of the present law and the penalties to be undertaken. However, the framework has not involved as yet the definition of the new emerging technologies regarding "gene editing techniques" and synthetic biology.

## International Regulations and Synthetic Biology: State of the Art and Challenges

Synthetic biology is a collection of emerging and enabling technologies which utilize advances in genetic and systems engineering. As with more established research in genetic engineering, such research is aimed at selectively altering genotypic (genetic) information to trigger a desirable shift in an organism's phenotype or physical characteristics (Trump 2017). The term synthetic biology is used increasingly, but without a clear definition. Most of the recent research carried out in this field is genetic engineering, as defined by current GMO-legislation in the European Union (Christiansen et al. 2019). Synthetic biology has developed its own language. In vitro synthesis of DNA also carries the label synthetic biology. It is important to analyze whether present and future activities of synthetic biology are within the scope of existing national and international legislation and whether it will eventually be introduced as a specific domain or presented as amendments of previous biosafety regulations already adopted·and will follow the same procedures and directives as GMOs.

If we analyze the actual perspective of regulation at the international level, we note that first the CBD had published its opinion regarding synthetic biology, and considering its decision XI/1124, the Conference of the Parties (COP) notes: "based on the precautionary approach, the need to consider the potential positive and negative impacts of components, organisms, and products resulting from SynBio techniques on the conservation and sustainable use of biodiversity" and also "recognises the development of technologies associated with synthetic life, cells or genomes, and the scientific uncertainties of their potential impact on the conservation and sustainable use of biological diversity and urges Parties and invites other Governments to take a precautionary approach" (Convention on Biological Diversity 2019). The preparatory work encompasses two notes which compile relevant information on components, organisms, and products resulting from SynBio techniques that may have an impact on the conservation and sustainable use of biological diversity and associated social, economic, and cultural considerations.

On the other hand, the European Union consulted many Committees of the European Commission (EC) and also working groups and in 2014 drafted Scientific Opinions on synthetic biology that provide an operational definition and address risk assessment methodology, safety aspects, environmental risks, knowledge gaps, and research priorities (European Commission 2014). The main conclusion was uncertainty whether Directives 2009/41/EC and 2001/18/EC in the European GMO regulatory framework were the appropriate legislation to cover synthetic genomics and Synbio. However, the United States considered that the existing policy and regulatory framework for biotechnology might apply, with minor adaptations, to synthetic organisms. In fact, for synthetic nucleic acids, the US National Institutes of Health (NIH) Recombinant DNA Advisory Committee concluded that, in most cases, biosafety risks were comparable to recombinant DNA research and that the current risk assessment framework could be used to evaluate synthetically produced nucleic acids with attention to the unique aspects of this technology (NIH guidelines for research involving recombinant or synthetic nucleic acid molecules (NIH 2019).

## Conclusion

In the framework of commitments to the CPB, Tunisia is now waiting for the adoption of the legal framework for biosafety and the implementation of the updated strategy. The implementation of the legal texts was carried out within a large consultation among the different actors involved, especially the concerned ministries and the public, and in harmony with the protocol provisions. This will allow all the key actors at the national level to rely on this framework in their future works and projects involving the three domains (GMOs, pathogenic agents, and invasive species) and make operational the application of the CPB and CBD requirements. Nonetheless, biotechnology regulations need the total support of agricultural and agri-food groups to balance the need for biodiversity protection with trade and market constraints. Finally, one of the major challenges that the MENA region and Tunisia particularly will face in the near future is how the emerging technologies will be regulated and whether the awaited framework will be amended after adoption to include the new terminologies and policies regarding "gene editing" and "synthetic biology."

## References

Ahmed HF, Chahed Y (2012) USDA, gain REPORT number TS1205. Agricultural Biotechnology Annual. http://www.agriexchange.apeda.gov.in/MarketReport/Reports/Agricultural%20Biotechnology%20Annual_Tunis_Tunisia_7-2-2012.pdf

Ben Belgacem H, Chalghoumi R, Jaballah (2018) Experiences and lessons learned from Tunisia. *Biosafety*. Protocol Newsletter 13:18–19

Biosafety Clearing House (2019) http://bch.cbd.int/about/countryprofile.shtml?country=tn)-

Chaouachi M, Nabi N, Ben Hafsa A et al (2013) Monitoring of genetically modified food and feed in the Tunisian market using qualitative and quantitative real-time PCR. Food Sci Biotechnol 22:1–10

Christiansen AT, Andersen MM, Kappel K (2019) Are current EU policies on GMOs justified? Transgenic Res 28:267–286

Convention on Biological Diversity (2019) http://www.cbd.int/decision/cop/default.shtml?id=13172

Eckerstorfer MF, Engelhard M, Heissenberger A et al (2019) Plants developed by new genetic modification techniques-comparison of existing regulatory frameworks in the EU and Non-EU countries. Front Bioeng Biotechnol 19:7–26

European Commission (2014) https://ec.europa.eu/health/scientific_committees/consultations/public_consultations/scenihr_consultation_21_en

FAO (2003) Biosecurity1 in Food and Agriculture. Committee on agriculture. 17th session. Rome, 31 March–4 April 2003

Feki K, Quintero FJ, Khoudi H et al (2013) A constitutively active form of a durum wheat $Na^+$/$H^+$ antiporter SOS1 confers high salt tolerance to transgenic Arabidopsis. Plant Cell Rep 33:277–288
Gargouri-Bouzid R, Jaoua L, Rouis S et al (2006) PVY-resistant transgenic potato plants expressing an anti-NIa protein scFv antibody. Mol Biotechnol 33:133–140
Gouiaa S, Khoudi H, Leidi EO et al (2012) Expression of wheat Na(+)/H(+) antiporter TNHXS1 and H(+)- pyrophosphatase TVP1 genes in tobacco from a bicistronic transcriptional unit improves salt tolerance. Plant Mol Biol 79:137–155
Gruère GP (2006) An analysis of trade related international regulations of genetically modified food and their effects on developing countries, EPT discussion paper 147. International Food Policy Research Institute, Environment and Production Technology Division (EPT), Washington, DC
Gruère GP, Carter CA, Farzin YH (2009) Explaining international differences in genetically modified food labeling regulations. Rev Int Econ 17:393–408
International Service for the Acquisition of Agri-biotech Applications (ISAAA): http://www.isaaa.org/
Nabi N, Chaouachi M, Zellama MS et al (2016) A new QRT-PCR assay designed for the differentiation between elements provided from *Agrobacterium* sp. in GMOs plant events and natural *Agrobacterium* sp. bacteria. Food Chem 196:58–65
NIH (2019). http://oba.od.nih.gov/oba/rac/Guidelines/NIH_Guidelines.htm
Trump BD (2017) Synthetic biology regulation and governance: lessons from TAPIC for the United States, European Union, and Singapore Health Policy, vol 121, pp 1139–1146
Vigani M, Olper A (2013) GMO standards, endogenous policy and the market for information. Food Policy 43:32–43
Vigani M, Olper A (2015) Patterns and determinants of GMO regulations: an overview of recent evidence. AgBioforum 18:44–54
WHO (2006) WHO biorisk management: laboratory biosecurity guidance, 2006, WHO/CDS/EPR/20066
Zarrilli S (2005) International trade in GMOs and GM products: National and multilateral legal frameworks, Policy issues in international trade and commodities study series, 29. United Nations Conference on Trade and Development, New York/Geneva

# Zimbabwe—The Status of Biosafety in Zimbabwe – A Legal Perspective

**Deckster T. Savadye, Annah R. Takombwa, and Abisai Mafa**

**Abstract** Zimbabwe, like most countries in Africa, has gone through difficult times in trying to balance the judicious use of biotechnology for socioeconomic development and addressing the possible threats of some aspects of modern biotechnology on the national economy, human health, socio-cultural interests, and the environment. Additionally, there are strong stakeholder differences in perception over the utility and safety of modern biotechnology techniques and products. Consequently, a barrage of measures was put in place by the government and the legislator as far as 1998 to ensure safety in biotechnology practice in the country. These include revision of existing laws, enactment of new ones, and developing regulations and guidelines. The existing frameworks, although comprehensive, need to be updated to cater for advances in the field of biotechnology. Additionally, there is need to harmonize institutional arrangements to close gaps and reduce duplication. This chapter, therefore, provides a snapshot of the current status of Zimbabwe's legal and institutional arrangements for biotechnology research, development, and application.

**Keywords** Biosafety · Policy · Law · Synthetic biology · GMOs · Biodiversity · Zimbabwe · Biotechnology · Regulations

## Background

During the early 1990s, Zimbabwean scientists approached the then Scientific Liaison Officer in the Office of the President and Cabinet (OPC), requesting the Government to put in place a legal framework for regulating modern biotechnology. The scientists were aware of the potential benefits of harnessing modern

---

D. T. Savadye (✉) · A. R. Takombwa · A. Mafa
National Biotechnology Authority, Harare, Zimbabwe
e-mail: dsavadye@nba.ac.zw



A. Chaurasia et al. (eds.), *GMOs*, Topics in Biodiversity and Conservation 19,
https://doi.org/10.1007/978-3-030-53183-6_26

biotechnology but also realized that there was need to ensure the judicious application of the technology. In a bid to fast-track the regulatory process, the Research Act [Ch.10:22] of 1986 was amended in 1998 to provide for the management of potentially harmful technologies and undertakings through Safety Boards.

In order to implement the new law, the Research (Biosafety) Regulations were enacted in 2000 and the Biosafety Board was formed (Sithole 2006). The Biosafety Board operated under the Research Council of Zimbabwe (RCZ), an institution that reports directly to the OPC. The Biosafety Board also developed Biosafety Guidelines for use by researchers in all work involving genetically modified organisms (GMOs) in a bid to ensure safety in biotechnology practice. Under these interim arrangements, the country conducted field trials of *Bacillus thuringiensis (Bt)* maize and cotton from 2001 to 2005. In addition, the framework was also used to regulate the import of genetically modified (GM) food and feed (Keeley and Scoones 2003). For instance, during the drastic food shortages that hit Southern Africa in 2002, Zimbabwe together with Malawi and Mozambique requested that the maize grain donated by the USA be milled before distribution to prevent farmers from planting it (Zerbe 2004).

In 2005, Zimbabwe became a party to the Cartagena Protocol on Biosafety (CPB) to the Convention on Biological Diversity (CBD). In the same year again, the country received a grant from the United Nations Environment Programme (UNEP)/ Global Environment Facility (GEF) project for developing national biosafety frameworks. Therefore, Zimbabwe proceeded to implement the CPB through the development of a National Policy on Biotechnology in 2005 and the enactment of the National Biotechnology Authority Act [Ch.14.31] of 2006. The new policy provided for a set of measures to promote the deployment of biotechnology in all fields of human development, including a commitment to set aside a proportion of national Gross Domestic Product (GDP) for supporting biotechnology research, development, and commercialization. The new law established the National Biotechnology Authority (NBA) which replaced the Biosafety Board. The law also repealed the Research (Biosafety) Regulations of 2000.

The Research (Biosafety) Regulations had provisions for the regulation of research or undertakings involving recombinant DNA technology or genetic modification. The NBA Act of 2006 provides a holistic approach, encompassing regulation, promotion, and training in both conventional and modern biotechnology. The NBA Act in its present form places the responsibility of regulating and promotion on the National Biotechnology Authority to allow directed establishment and growth of the biotechnology industry. It is envisaged that once there is a vibrant biotechnology industry in Zimbabwe, the Act will be amended to remove the promotion role.

## Biosafety Legal Instruments

The following are the functions of the pieces of legislation that govern biotechnology activities in Zimbabwe:

1. National Biotechnology Authority Act [Ch.14.31] of 2006 – to support and manage biotechnology research, development, and application
2. Food and Food Standards (Food Labelling) Regulations, Statutory Instrument 265 of 2002 under the Ministry of Health and Child Care (MoHCC) – states that all genetically modified (GM) foods should be clearly labelled
3. National Biotechnology Authority (Food, Feed, Food and Feed Additives and Seed) (Import, Export and Transit) Regulations, Statutory Instrument 157 of 2018 – biosafety regulation of the transboundary movement of food, feed, food and feed additives, and seed
4. National Biotechnology Authority (Genetically Modified Food and Feed) (Labelling), Regulations, Statutory Instrument 159 of 2018 – regulations for the compulsory labelling of food and feed which contains at least 1% of products of genetic modification. Guidelines for operators who wish to label food or feed which contains less than 1% of products of genetic modification are also contained in the regulations
5. National Biotechnology Authority (Agricultural Biotechnology Products), Regulations, 2018, Statutory Instrument 160 of 2018 – to regulate the import, export, transit, handling, use, and application of biofertilizers, biopesticides (bioinsecticides, biofungicides, and bioherbicides), and biostimulants

## Synthetic Biology

The 13th Conference of Parties (CoP) to the CBD decided that the operational definition of synthetic biology is: “a further development and new dimension of modern biotechnology that combines science, technology and engineering to facilitate and accelerate the understanding, design, redesign, manufacture and/or modification of genetic materials, living organisms and biological systems” (Secretariat of the Convention on Biological Diversity XE "Biological diversity" 2017). Although synthetic biology is rather a relatively new technique, the broadness of the NBA Act of 2006 makes it possible to regulate this technology. Subsection 3 (2) of the NBA Act of 2006 states that the Act shall apply to: “(c) any activity involving biological and molecular engineering technologies such as metabolic engineering, proteomics, metabolomics, nanotechnology, genetic modification, cloning, DNA-chip technology and bioinformatics; and such other technologies as may be declared by the Authority to constitute potentially harmful research or undertaking.” However, given that scientists across the globe are yet to agree on a definition of synthetic biology (Secretariat of the Convention on Biological Diversity 2015) and that the technology keeps on advancing, there is need for the NBA to gazette a Statutory

Instrument for comprehensive risk analysis (The Parliamentary Office of Science and Technology 2015).

## Biodiversity

The Ministry of Environment, Tourism and Hospitality Industry is responsible for biodiversity issues. It delivers its duties through its State agencies like the Environmental Management Agency (EMA), Forestry Commission of Zimbabwe (FCZ), Zimbabwe Parks and Wildlife Management Authority, Non-Governmental Organizations (NGOs), and a committee of key biodiversity stakeholders, known as the National Biodiversity Forum.

Though the current National Biodiversity Strategic Action Plan of 2014 incorporates biosafety issues, there is need for more work to be done as far as consideration of biosafety issues under biodiversity programs, action plans, or policies is concerned. This is because if, for instance, the country decides to adopt GM crops, data on the potential impact of the particular GM crop on biodiversity should be available to enable informed decision-making.

## Current Status on GMOs and Way Forward

Currently there is no research on GM crops taking place in the agriculture sector in Zimbabwe. The Government of Zimbabwe had been taking a precautionary approach on GM crops. Highly polarized views on the matter used to appear in the media, raising health fears among the general public (Makotamo et al. 2015). The situation has, however, changed as the new government is willing to listen to how biotechnology can solve the problems currently bedevilling the country's agricultural, health, environmental, and industrial sectors. In the health sector, though, there are some clinical trials of recombinant human immunodeficiency virus (HIV) vaccines under way.

Some seed company players are keen on introducing GM crops and already have products ready in other markets. Awareness activities on GMOs such as workshops and field visits to other countries have been done by the NBA in partnership with some seed companies and regional partners.

Fluctuations in the international cotton lint prices have contributed to low production in Zimbabwe (Buka 2016), though the Government is giving free inputs to small-scale farmers (USDA 2019). Adoption of *Bt* cotton may bring about sustainability in this sector. It should, however, be noted that increased public awareness, education, and training will be a key ingredient for the successful adoption of *Bt* cotton. In addition, GM cotton for lint production seems to be more acceptable than GM food products, which are rumored to be associated with health effects among the general populace. The experience of other countries who have produced *Bt* cotton should, however, makes it less difficult for the government of Zimbabwe to make a decision.

## Conclusion

A solid regulatory framework for the regulation of GMOs exists in Zimbabwe though there is need to improve certain elements. Currently, the country is not engaged in GM crop trials or commercialization, a situation which may leave the country as a net importer of agricultural produce from its neighbors which are either doing trials or growing GM crops commercially. To this end, there is need for the country to revisit its GM crop stance and come up with decisions which boost agricultural productivity.

## References

Buka G (2016) Cotton and its By-Products Sector in Zimbabwe. UNCTAD, [Online]. UNCTAD/WEB/SUC/2017/3, 1–32. https://unctad.org/en/PublicationsLibrary/websuc2017d3_en.pdf

Food and Food Standards (Food Labelling) Regulations, Statutory Instrument 265 of 2002

Keeley J, Scoones I (2003) Contexts for regulation: GMOs in Zimbabwe. Institute of Development Studies Brighton, Sussex. ISBN 1 85864 494 1

Makotamo L, Takombwa AR, Muvhunzi F, Mufandaedza J (2015) Public perception towards transgenic crops and genetically modified foods in Zimbabwe. Paper presented at the Conference on Research and Innovation in Technical and Vocational Education: The 21st Century Scenario for Zimbabwe, Hwange Safari Lodge, Hwange, 12–13 March 2015, 83–97

National Biotechnology Authority (Food, Feed, Food and Feed Additives and Seed) (Import, Export and Transit) Regulations, Statutory Instrument 157 of 2018

National Biotechnology Authority (Agricultural Biotechnology Products), Regulations, (2018) Statutory Instrument 160 of 2018

National Biotechnology Authority (Genetically Modified Food and Feed) (Labelling), Regulations, Statutory Instrument 159 of 2018

National Biotechnology Authority Act [Ch.14.31] of 2006

Secretariat of the Convention on Biological Diversity (2015) Synthetic biology, Montreal, Technical series no. 82, 118 pages

Secretariat of the Convention on Biological Diversity (2017) Mainstreaming biodiversity for well-being. Decisions of the thirteenth meeting of the conference of the parties to the convention on biological diversity, Cancun, Mexico, 2–17 December 2016. Montreal: Secretariat of the Convention on Biological Diversity, 379 pages

Sithole SZ (2006) Biosafety regulations and guidelines in Zimbabwe. In: Status and risk assessment of the use of transgenic arthropods in plant protection. International Atomic Energy Agency (IAEA-TECDOC--1483). https://inis.iaea.org/search/search.aspx?orig_q=RN:37048132

The Parliamentary Office of Science and Technology (2015) Regulation of synthetic biology, London: PostNote. https://researchbriefings.parliament.uk/ResearchBriefing/Summary/POST-PN-0497

United States Department of Agriculture (USDA) (2019) Zimbabwe: cotton production and consumption in Zimbabwe. https://www.fas.usda.gov/data/zimbabwe-cotton-production-and-consumption-zimbabwe

Zerbe N (2004) Feeding the famine? American food aid and the GMO debate in Southern Africa. Food Policy 29:593–608

# Part IX
# Americas

# Argentina—Regulatory Framework for Modern Biotechnology

**Martin A. Lema**

**Abstract** Argentina has a full-fledged and seasoned regulatory system for modern biotechnology. After almost three decades of existence, its experience includes thousands of field trials, dozens of commercial authorizations, as well as practical experience with GM plants, animals, and microorganisms. Besides, it is able to cope with the latest innovations, such as genome-edited organisms.

The present chapter summarizes the history of this regulatory system, as well as its institutional and legal frameworks. A focus is made on issues of current interest such as liability and redress and synthetic biology. Finally, a forecast of future developments is included.

**Keywords** Biotechnology · Regulation · CONABIA · Argentina · GMO · Genome-editing · Synthetic-biology · Recombinant-DNA

## History and Current Regulations

The Argentine regulatory framework began functioning in 1991, after the creation of the National Commission on Agricultural Biotechnology (CONABIA). To date, Argentina has issued 52 commercial authorizations for GM crops (SGAI 2019). The present chapter is a summary of a more extended description, which is available for further reading (Whelan and Lema 2019).

Activities involving GM crops are regulated under Law 20.247 on Seed Trade (INFOLEG 2019a), Law 27.233 on Plant and Animal Health (INFOLEG 2019b),

M. A. Lema (✉)
Biotechnology Directorate, Secretariat of AgroIndustry, Buenos Aires, Argentina

Av. Paseo Colón 982, Buenos Aires, Argentina

National University of Quilmes, Bernal, Argentina

Roque Sáenz Peña 352, Buenos Aires, Argentina
e-mail: mlema@unq.edu.ar

 
A. Chaurasia et al. (eds.), *GMOs*, Topics in Biodiversity and Conservation 19,
https://doi.org/10.1007/978-3-030-53183-6_27

and Law 22.520 on Ministries of the Executive as implemented in Decree 802/2018 (INFOLEG 2019c).

Ministerial Resolution no. 763/11 establishes the implementation aspects of general laws applied to the regulation of GMOs (INFOLEG 2019d). In turn, multiple subordinate regulations from the Ministry of Agroindustry, the National Seeds Institute (INASE), and the National Agrifood Health and Quality Service (SENASA) rule different activities involving GM plants, animals, and microorganisms (INFOLEG 2019e).

As regards International Law, the Argentine Republic is a signatory of the treaty on Sanitary and Phytosanitary (SPS) measures of the World Trade Organization (WTO). As a consequence, its GMO regulatory framework is kept in line with standards of SPS reference organizations: *Codex Alimentarius* Commission, International Plant Protection Convention (IPPC), and the World Organisation for Animal Health (OIE). Argentina has also applied the WTO Dispute Settlement Understanding to challenge the earlier functioning of the European Communities' GMO regulatory system (WTO 2010).

Argentina was a founder signatory country of the Cartagena Protocol on Biosafety. Currently, the Argentine regulatory system is compatible with the treaty, but the country has not ratified the Protocol yet.

## Regulatory Agencies and Authorities

The Biotechnology Directorate under the Ministry of Production and Labor is the leading regulatory bureau. It chairs CONABIA and coordinates activities with other relevant agencies. The Directorate is under the Secretariat of Foodstuff and Bioeconomy, which is the competent authority regarding the authorization of environmental and market release of GMOs of agro-industrial use.

CONABIA is a commission of specialists in different fields of expertise, which act as representatives of different institutions. Its main role is to perform biosafety assessments and evaluation of confinement measures. It also advises broadly in scientific and technical issues related to biotechnology.

Noteworthy, CONABIA is recognized as the Center of Reference in GMO Biosafety by the United Nations Organization for Food and Agriculture (FAO 2014). It was chosen for such purpose on the basis of its level of expertise and ability to perform capacity building in biosafety.

Inspectors of INASE and SENASA are in charge of controlling biosafety measures. INASE is focused on seeds or other viable plant propagation material, while SENASA controls grains and plant-derived foodstuff, as well as microorganisms and animals.

## Confined Use

Laboratories manipulating GMOs must have a Biosafety Level 2, in accordance with the Manual on Laboratory Biosafety of the World Health Organization (WHO 2004). If not developed locally, the importation of GM plant propagative material requires a special clearance by SENASA under Res. 498/2013 (INFOLEG 2019f).

Those performing activities outside of laboratories (including contained use in a greenhouse or confined field release) are required to enroll in the Registry of Operators of Genetically Modified Plant Organisms INFOLEG (2019g), before requesting separate permit for each activity.

Permit categories for confined activities include (a) activities under biosafety greenhouses, which are regulated by Resolution 241/2012 (INFOLEG 2019h); (b) open field trials, regulated under Res. 763/11 abovementioned; and (c) production of seeds for export or plant biomass for special uses, which is regulated by Res. 17/2013 (INFOLEG 2019i).

## Commercial Release

The approval for marketing of GM crops and derived products is granted by the Secretariat of Foodstuff and Bioeconomy. In this case, the decision is based, among other studies, on a case-by-case biosafety assessment performed by CONABIA according to the abovementioned Res. 701/11.

## Liability and Redress

The duty to preserve the environment, as well as the responsibility to redress its damage, is mandated by Article 41 of the Constitution (INFOLEG 2019j). In addition, its Article 43 establishes the possibility of appealing justice for expedited measures to avoid or mitigate damage to the environment. Then, the liability regime for environmental damage in Argentina is established in law no. 25.675 (INFOLEG 2019k). Finally, if a GMO would hypothetically cause damage to the environment, human health, or trade, broad provisions on civil liability could be appealed to obtain restoration or monetary compensation.

## Synthetic Biology

Argentina issued an ad hoc explicit regulation regarding "new breeding techniques" in 2015 (Whelan and Lema 2015). It requires that products obtained with the aid of any recombinant DNA (therefore including synthetic biology, gene drives, gene editing, etc.) to be assessed by CONABIA; the outcome of that assessment is to determine if the organism is under the GMO regulatory system. This is done case by case, on the basis of the LMO definition of the Cartagena Protocol.

There is no technically sound and universally accepted definition of the term "synthetic biology." However, considering examples of its use in the technical literature hand in hand with the current operational definition being used in discussions under the Convention on Biological Diversity (CBD 2019), it's clear that products of synthetic biology involving novel combinations of genetic material would be regulated as GMOs.

## Expected Changes in the Near Future

The Argentine regulatory framework is updated frequently to cope with technological advances and the refinement of criteria for the discipline of GMO risk assessment. It was thoroughly revised in 2011, and several amendments were conceived from then, at an average rate of two new norms per year. For instance, recent regulatory amendments include special treatment for the repeated use of the same genetic construct, for field trials of microbial inputs for agriculture, opportunity for the public to comment on CONABIA's decision documents, etc.

In 2019, another complete update was in progress. Expected changes include modifications to allow for a more transparent and expedited assessment, simplification of the regulatory texts (i.e., redundancies, obscure or obsolete sections, number of dispersed pieces of regulation), and measures for reducing procedural bureaucracy.

It also includes modifications to make regulations even more explicitly in line with the Cartagena Protocol on Biosafety text and COP-MOP decisions. This is made in line with keeping the Argentine regulatory framework as harmonized as possible with international standards as well as keeping the door open for Argentina to ratify the Protocol in the near future.

Finally, Argentina will complete the framework by incorporating the possibility of applying for the commercial release of GM microorganisms (beyond recombinant vaccines, which is currently the only option available) and GM animals.

**Disclaimer** The information and views are attributable to the author and do not necessarily represent those of the organizations where he works.

## References

CBD (2019) Portal on synthetic biology. https://bch.cbd.int/synbio/
FAO (2014) La FAO y Argentina refuerzan la seguridad en biotecnología. http://www.fao.org/director-general/newsroom/news/detail/es/c/264211/
INFOLEG (2019a) Ley de Semillas y creaciones Fitogenéticas. http://servicios.infoleg.gob.ar/infolegInternet/anexos/30000-34999/34822/texact.htm
INFOLEG (2019b) Sanidad de los Animales y Vegetales Ley 27233. http://servicios.infoleg.gob.ar/infolegInternet/anexos/255000-259999/257451/norma.htm
INFOLEG (2019c) Ley de Ministerios 22.520. http://servicios.infoleg.gob.ar/infolegInternet/anexos/45000-49999/48853/texact.htm
INFOLEG (2019d) Resolución 763/2011. http://servicios.infoleg.gob.ar/infolegInternet/anexos/185000-189999/185806/norma.htm
INFOLEG (2019e) Subsidiary regulations to Resolución 763/2011. http://servicios.infoleg.gob.ar/infolegInternet/verVinculos.do?modo=2&id=185806
INFOLEG (2019f) Resolución N° 498/2013. http://servicios.infoleg.gob.ar/infolegInternet/anexos/220000-224999/224698/norma.htm
INFOLEG (2019g) Resolución 241/2012. http://servicios.infoleg.gob.ar/infolegInternet/anexos/195000-199999/198153/norma.htm
INFOLEG (2019h) Organismos Vegetales Geneticamente Modificados. Resolución 46/2004. http://servicios.infoleg.gob.ar/infolegInternet/anexos/90000-94999/92241/norma.htm
INFOLEG (2019i) Resolución 17/2013. http://servicios.infoleg.gob.ar/infolegInternet/anexos/220000-224999/224874/norma.htm
INFOLEG (2019j) Constitución de la Nación Argentina. Ley N° 24.430. Publicación del texto oficial de la Constitución Nacional (sancionada en 1853 con las reformas de los años 1860, 1866, 1898, 1957 y 1994). http://servicios.infoleg.gob.ar/infolegInternet/anexos/0-4999/804/norma.htm
INFOLEG (2019k) Política Ambiental Nacional Ley 25.675. http://www.infoleg.gob.ar/infolegInternet/anexos/75000-79999/79980/norma.htm
SGAI (2019) List of GM events approved in Argentina. https://www.agroindustria.gob.ar/sitio/areas/biotecnologia/ogm/
Whelan AI, Lema MA (2015) Regulatory framework for gene editing and other new breeding techniques (NBTs) in Argentina. GM Crops & Food 6:253–265
Whelan AI, Lema MA (2019) Regulation of genome editing in plant biotechnology: Argentina. In: Regulation of genome editing in plant biotechnology. Springer, Cham, pp 19–62
WHO (2004) Laboratory biosafety manual. http://www.who.int/csr/resources/publications/biosafety/WHO_CDS_CSR_LYO_2004_11/en/
WTO (2010) Measures affecting the approval and marketing of biotech products. https://www.wto.org/english/tratop_e/dispu_e/cases_e/ds293_e.htm

# Colombia—GMOs/Innovative Biotechnology Regulations

**Elizabeth Hodson de Jaramillo**

**Abstract** The Government of Colombia has shown its commitment to biosafety issues by implementing several measures since 1998 and was a leading country in formulating the Cartagena Protocol on Biosafety. The biotechnological advances are moving much faster than the mechanisms that govern the diverse products such as the regulatory status of organisms or products derived from new molecular technologies such as synthetic biology or gene editing. Recent regulations include a procedure for the analysis of cultivars obtained by innovative biotechnological techniques such as gene editing, in order to determine if the final product contains or does not contain foreign genetic material, corresponding or not to a Living Modified Organism (LMO), and, consequently, determine whether the regulation of LMOs should apply.

**Keywords** Biosafety · New breeding techniques NBTs · Gene editing · Agrobiotechnology

## Introduction

The Government of Colombia has shown its commitment to biosafety issues by implementing several measures since 1998. As a major hot spot of biodiversity, it is also an important center of origin or diversification of several agricultural crops (Sandoval-Sierra and Chaves-Servia 2014). As a megadiverse country and a signatory member country of the Convention on Biological Diversity (CBD), Colombia has a global responsibility to design and implement a strong, unambiguous, and effective legal framework. Because of this, it is not surprising that Colombia was a leading country in formulating the Cartagena Protocol on Biosafety (CPB) to the Convention of Biological Diversity, was the host of the sixth meeting of the Open-Ended Ad Hoc Working Group on Biosafety in 1999, and was one of the first

E. Hodson de Jaramillo (✉)
Facultad de Ciencias, Pontificia Universidad Javeriana, Bogotá, Colombia

Academia Colombiana de Ciencias Exactas, Físicas y Naturales, Bogotá, Colombia

 
A. Chaurasia et al. (eds.), *GMOs*, Topics in Biodiversity and Conservation 19,
https://doi.org/10.1007/978-3-030-53183-6_28

signatory countries. With the ratification of the CP through the Colombian "Law 740" in May 2002, it became imperative for the country to adjust and adopt a national legislative framework in order to regulate and harmonize activities related to living modified organisms (LMOs) across different sectors and national institutions (Hodson and Carrizosa 2006).

Regarding other international agreements, Colombia plays an active role in the discussions of the Nagoya-Kuala Lumpur Protocol on redress and liability and the CPB Conference of the Parties, as a signatory. In addition, Colombia is also a signatory to the International Treaty on Plant Genetic Resources for Food and Agriculture and the International Plant Protection Convention (IPPC) and attends Codex meetings to discuss issues on biotechnology. In 2017, Colombia joined the Global Low Level Presence Initiative to develop international approaches to manage LLP (USDA 2018).

## GMOs

As part of the CPB implementation process, the "GEF-WB project Colombia: Capacity-Building for the Implementation of the Cartagena Protocol on Biosafety" was executed, looking forward to develop recommendations for a national biosafety legislative and policy framework and to formulate scientific-based guidelines for the protection of human, animal, and plant health, the environment, and the economic well-being and to consolidate the national scientific and technical capacity in biosafety of living modified organisms (LMOs), including risk assessment, management, monitoring, and communication. The outcome of the project proved to be an important scenario for the different institutions involved to interact between them, seeking to articulate and complement the national normative related to the implementation of the CPB.

The current legal framework for biosafety is largely derived from preexisting, sector-specific legislations (especially environment, agriculture, and health), which have been adjusted to encompass and update applications of new scientific developments, mainly taking into consideration the social responsibility of the competent authorities not only to guarantee the safety of the technologies protecting human, animal, and environmental health but also to guarantee opportune access of the users to technological developments without raising the cost of the required processes.

The guiding principles of the Colombian regulations related to biosafety are responsibility, safety, and protection, based in a scientifically solid, efficient, and reliable system (Hodson 2018). In 2005, Decree 4525 was issued in order to implement the CPB, in which three competent authorities were clearly defined, and designed responsible for the evaluation and assessment of risks related with the applications of recombinant DNA biotechnology (rDNA including GMOs), prior to their approval for commercial use and release. These are the Ministry of Agriculture and Rural Development, the Ministry of Environment and Sustainable Development, and the Ministry of Health. Indirectly involved are the Ministry of Trade, Industry, and Tourism and the Ministry of Foreign Affairs (National Chancellery) acting as the National Focal Point for the CPB. Also, three interagency committees were established to carry out the risk assessments and to deliver their technical concept.

The biosafety committees are in charge of conducting risk assessments, on a case-by-case basis, proposing the studies required for the evaluations and defining actions to be taken for monitoring and minimizing any possible risk related to modern biotechnology application, in agreement with the protocol. The competent authorities along with their regulatory instances are constantly revising, adjusting, and complementing the normative currently in force, providing opportunities to engage the governmental regulatory agencies with technical outreach that facilitates the adoption of science-based regulatory policies.

## Synthetic Biology and Gene Editing

Currently, the vertiginous biotechnological advances are moving much faster than the regulatory mechanisms that govern the diverse products. In several countries around the world, governments are concerned and stressed in relation to the regulatory status of organisms or products derived from new molecular technologies such as synthetic biology or gene editing. For the responsible competent authorities, it is clear that many of the current innovations need new standards and provisions. For example, gene editing is a very powerful tool in the continuum of plant breeding innovations, which offer enormous opportunities, but at the same time represents challenges to adjust the regulations, which need to be updated and adapted to the astounding biotechnological advances. One of the main concerns is if designated authorities and evaluators are sufficiently updated and really understand the rapid advances of current techniques of modern biotechnologies, i.e. molecular biology and gene editing, in order to define what are the real scientific-based risks of new biotechnology products. Additional concerns are the social and economic costs involved, as well as the benefits of future developments and the possible impact of the delay in their adoption or opportunity cost. The issue is how to make very rigorous, transparent, efficient, and accurate risk analysis evaluations for the safe use of new biotechnology products, facilitating the timely access and potential benefits of the application to the society. As a matter of fact, for the applications of these new technologies in biomedical research and human health including gene therapy, there were no updated regulations or new standards available in the country at the time of writing. In relation to plants, ICA Resolution 3168 of 2015 regulates commercial activities related to seeds obtained from plant breeding techniques (conventional or nonconventional) and the registration of plant breeding units that, once authorized, can follow specific regulations to import genetically modified material in order to carry out experimental trials, with a more simplified procedure of importation.

For agricultural applications, the question that arises is if new technology products (i.e. gene edited organisms) truly present a hazard or a risk that demands regulatory oversight, for which the criterion is to compare the new development, on a case-by-case basis, with the corresponding products obtained by conventional breeding and with GMOs. One of the considerations, especially in plant manipulation, is that many of the products of new technologies show small known, directed genetic alterations such as single-nucleotide substitutions, deletions, and frameshift mutations, alterations that could happen spontaneously in nature and are found in many cells. These modifications contribute to the genetic variation within species that is regularly used in conventional as well as in modern plant breeding. There is often little or no distinguishable difference between the newly developed products and what have previously been considered natural products. Therefore, the doubt is whether or not genetic alterations, which can also occur in natural conditions, should be under special legislations and scrutiny.

## Benefits to Agricultural Production

In the specific case of plants and agricultural production, these various new technological developments offer innovative opportunities that facilitate responses to imperative social and environmental challenges such as facing climate change, intensification of sustainable agricultural practices, improving resilience, food security (food and feed supply), increasing agrodiversity, and product variety, all of them related to the transition toward a circular bioeconomy, promoting social, environmental, and economic sustainability. When applied to plants and crops, these emerging technologies allow faster, more exact and more directed results than the ones obtained by conventional crop breeding methods.

Colombia has been favorable to the adoption of new technological developments and, as mentioned, had regulations related to biotechnological crops in place since 1998. The first approval for GM crops in the country was for the blue carnation for export only in greenhouse containment in 2000, and in 2002, GM cotton was the first biotechnological plant cultivated on a nonrestricted commercial basis in Colombia (Hodson and Carrizosa 2006). Since then, several GM crops and their products have been approved for cultivation (cotton, maize, soybean, blue roses, blue gypsophila, and blue chrysanthemum), some are under development in research institutions, and there are several approvals of GM products for food or feed. Currently, labelling, low-level presence (LLP), and approval synchronicity issues, as well as internal discussions around new agrobiotechnological developments and their respective regulations, are causing some regulatory uncertainty and potentially delaying the adoption of new technologies (USDA 2018). In relation to more innovative agrobiotechnologies, there are currently two research groups working on genome editing. The International Center for Tropical Agriculture (CIAT) Research Center is working on several agronomic or nutritional interesting traits: herbicide-tolerant cassava, increased yields in rice, viruses and bacteria resistance in rice, high zinc and iron in rice, nutritional quality in bean, and, most recently, cadmium absorption in cacao. EAFIT University is conducting research on castor bean oleic acid content (USDA 2018).

## Recent Developments and Prospects

In relation with recent regulations, the Instituto Colombiano Agropecuario (ICA), the designed competent authority for agriculture, forestry, and fisheries, issued Resolution 29299 (ICA 2018), which establishes a procedure for the evaluation of an improved cultivar obtained by innovative biotechnological techniques, and the final product does not contain foreign genetic material, in order to determine if the given cultivar corresponds to a living modified organism (LMO) or not, and, consequently, determine whether the regulation of LMOs should apply. The interested party submits an application to ICA for review, and within a period of 60 business days, if no further information is required, ICA will determine whether the new cultivar is considered LMO or not, and therefore, it is within or beyond the scope of regulation for LMOs. If it is considered to be LMO, the cultivar will have to go through the existing regulatory LMO framework. Otherwise, it will be treated under existing conventional crop legislation and regulation (USDA 2018).

Over the next 10 years, the world can expect a proliferation of abundant and diverse biotechnological products obtained through various innovative technologies, for multiple uses and applications in all fields and beyond the known traditional and industrial uses. This profusion of new developments can overwhelm the regulatory systems and regulators. Evaluators will have to face complicated challenges and respond with rigorous, clear, reliable, reproducible, and at the same time highly efficient and safe regulations for health, the environment, and development.

## References

Hodson de Jaramillo E (2018) GM crops situation in Latin America. Frontiers in science and engineering. International Journal Edited by the Hassan II Academy of Science and Technology, Morocco, vol 8 – Supplement (1):73–81

Hodson de Jaramillo E, Carrizosa MS (2006) Agrobiotechnology and biosafety in Colombia. CAB Rev. 1:N° 016

ICA – Instituto Colombiano Agropecuario (2018) Resolución 0029299 Procedimiento para el trámite ante el ICA de solicitudes de un cultivar mejorado con técnicas de innovación en fitomejoramiento a través de Biotecnología moderna. Available at: https://www.ica.gov.co/getattachment/2d02cc52-d1c5-4123-8a5a-aea9ad2ce926/2018R29299.aspx

Sandoval-Sierra CL, Chavez-Servia JL (2014) Uso alimenticio de especies vegetales por las comunidades indígenas de Colombia: una revisión de literatura. Agroecología: Ciencia y Tecnología 2(1):1–6

USDA Foreign Agricultural Service (2018) Colombia. Agricultural biotechnology annual. Colombia continues to work through regulatory challenges. GAIN report number: CO1821

# Ecuador—Modern Biotechnology in Ecuador—Development and Legal Framework

María Torres and Efrén Santos-Ordóñez

**Abstract** In this chapter, we acknowledge the slow development of modern biotechnology in Ecuador. Some research projects have used molecular tools mainly to study the genetic diversity of several plant and animal species of importance for conservation or agriculture. To our knowledge, there could be a few cases, or none at all, in which the use of modern biotechnology is applied for industrial purposes. In this context, we describe an example of a research project related to the genetic transformation of bananas, an important agricultural crop for the country. The current regulations related to this subject are analyzed, and the lack of a National Biosafety Framework that ensures the development and proper use of these technologies in Ecuador is highlighted. The lack of political decision and the correct understanding of modern biotechnology and its implications for various sectors of society represent the greatest challenges that Ecuador has to face in order to be able to handle this issue adequately, promote the development of this type of biotechnology, and preserve the country's biodiversity.

**Keywords** Ecuador · Megadiverse country · Conservation · GMOs · Synthetic biology · Legal framework · Political support

M. Torres (✉)
Laboratorio de Biotecnología Vegetal, Universidad San Francisco de Quito. Colegio de Ciencias Biológicas y Ambientales, Quito, Ecuador
e-mail: ltorres@usfq.edu.ec

E. Santos-Ordóñez
ESPOL, Escuela Superior Politécnica del Litoral, ESPOL, Centro de Investigaciones Biotecnológicas del Ecuador (CIBE), Guayaquil, Ecuador

Facultad de Ciencias de la Vida (FCV), ESPOL, Escuela Superior Politécnica del Litoral, ESPOL, Guayaquil, Ecuador

 
A. Chaurasia et al. (eds.), *GMOs*, Topics in Biodiversity and Conservation 19,
https://doi.org/10.1007/978-3-030-53183-6_29

## Introduction

It is quite a challenge to speak on the situation regarding GMOs and synthetic biology in Ecuador. Although there have been several attempts to establish a logical framework to discuss these issues, political interests have always prevailed over technical and scientific ones. Here, we first briefly consider the development of these technologies in the country and then try to understand the regulatory framework that applies to them.

## Development of the New Biotechnologies

Within the past 20 years, several institutions of higher education in Ecuador have offered programs, mostly undergraduate, in biotechnology. There is no doubt that during those years, several research groups, with greater or lesser funding, have developed projects related to molecular biology and some genetic transformation. The term "synthetic biology," however, is just starting to be mentioned in Ecuador, and cases of research being carried out that imply the development of this field are few.

Ecuador is recognized as a megadiverse country, with numbers of plant, amphibian, and bird species, among living macroscopic organisms, that reflect a unique richness within different Ecuadorian ecosystems (INEC 2015; MAE 2015). Recently, microbiome richness has also begun to be analyzed, as well as microbial consortia that might lead to bioremediation, medical, or industrial applications. The results obtained up to now reveal a great diversity in microorganisms as well—which have been insufficiently studied until now.

In general terms, when we hear of molecular tools being used in local research projects, this mainly implies the use of technologies to determine genetic diversity and population structure of species, as well as molecular characterization of microorganisms. Studies in phylogeny, evolution, and systematics are the ones leading in use of molecular techniques.

A different scenario appears when attempting to analyze applied investigations being carried out in Ecuador and how these use molecular or genetic transformation technologies. For example, there are very few national breeding programs for plants or animals, which comprehensively use molecular tools for making progress in agriculture or livestock fields. The National Institute of Agricultural Research carries out breeding projects in several crops such as cacao (*Theobroma cacao*), chocho (*Lupinus mutabilis*), and beans (*Phaseolus vulgaris*), where molecular characterization of the germplasm has been performed. It is important to mention that the present analysis does not take into account the area of human health—where other interests and regulations exist but where modern technologies have been barely implemented.

The reality is that in Ecuador, modern biotechnology, especially related to GMOs, has not yet evolved. Box 1 reports a case study on the only project that has been developed here in relation to genetic transformation which has come to our

**Box 1: Case Study: Banana and Plantain (*Musa* spp.) Genetic Engineering in Ecuador**

**Background**

In Ecuador, banana and plantain (*Musa* spp.) are important crops for the export market and also for local consumption. Genetic improvement in *Musa* could be performed through conventional breeding; however, a banana breeding program has not yet been established in Ecuador, mainly due to the high levels of sterility in most cultivars, polyploidy, and long production cycles. On the other hand, a biotechnology improvement program for banana and plantains has been established at ESPOL University covering different topics.

**Establishment of Suitable Explants for Genetic Modification**

The ideal explant for genetic modifications is embryogenic cell suspension (ECS), which avoids the risk of chimera formation (Sowmya et al. 2016). Two main methodologies were performed on the development of ECS, including male inflorescence and scalps (Korneva et al. 2010). Furthermore, banana tissue culture has been established for mass propagation by using temporary immersion systems (Korneva et al. 2013). Some major drawbacks are the time to develop ECS and the regeneration of in vitro plants after genetic transformation. Protocols need to be adjusted according to the genotype; therefore, other explants have been used including meristematic regions from in vitro plants.

***Agrobacterium*-Mediated Transformation**

Two major methodologies have been used in banana for genetic modifications worldwide: biolistics and *Agrobacterium tumefaciens* (Khanna and Deo 2016). However, *Agrobacterium*-mediated transformation has been used preferentially, due to the low copy number of integrated transgenes and higher efficiency of transgenic lines obtained. In Ecuador, *Agrobacterium*-mediated transformation has been performed in the banana "Williams" (genotype AAA) and the plantain "Dominico" (genotype AAB) using ECS (Santos et al. 2016b).

Currently, genetically modified plantain plants of the cultivar "Dominico" transformed with the GTPCHI gene from *Musa* and the ADCS from *Arabidopsis thaliana* driven with the banana expansin promoter are in the greenhouse under confined conditions. These genetically modified plantains were produced within the framework of a research project in collaboration with the Ghent University in Belgium. The modified plantains should be tested for increased levels of folate in the fruit. Folates (water-soluble vitamin B9) are molecules consisting of a pterine ring, a para-aminobenzoate moiety and a $\gamma$-linked tail with one or more L-glutamates (Strobbe and Van der Straeten 2017). The developed protocols for the generation of genetically

(continued)

**Box 1:** (continued)

modified banana and plantain plants resulted in the establishment of a GMO detection platform. Furthermore, food analysis confirms the presence of products containing transgenic ingredients (Pacheco Coello et al. 2017; Santos et al. 2016c).

**Promoter Characterization and Gene Identification**

In a banana and plantain improvement program using genetic engineering, gene function and regulation characterization should be performed. Promoter sequences are needed to drive constitutive or specific expression of the transgenes. Furthermore, native promoters should imply a better acceptance from the consumers than heterologous sequences. Isolation of promoters could be performed through a T-DNA tagging approach by fusing a promoterless reporter gene next to a border of the T-DNA (Santos et al. 2007, 2009). Promoter characterization could be performed in silico, by analyzing the expression of the corresponding gene by RT-qPCR and/or by fusing the promoter to a reporter gene (Santos et al. 2016a; Villao et al. 2019). For gene identification, different strategies could be performed. Candidate resistant genes for black sigatoka disease were identified by using a suppression subtractive hybridization technique of the natural resistant banana "Calcutta 4" (Sánchez Timm et al. 2016). The candidate genes could be used in the generation of a cisgenic banana resistant to black sigatoka disease (BSD).

**Economic Prospects in the Ecuadorian Banana Industry if Modern Biotechnology Could Be Adopted**

Although at the moment, because of the legal framework situation, it is difficult to envisage the impacts of introducing a GMO product in the Ecuadorian economy, we need to reflect on how such a development could affect its economy. The impact of a transgenic crop will be largely dependent on the actual trait improved. For instance, while developing a biofortified plantain may have an impact in human nutrition, crop management improvement is not expected to occur, and therefore, wide cultivation in Ecuador will be limited to plantain farms unless an international demand of biofortified fruit is established at a higher product price than conventional non-modified plantains. This scenario could present itself similarly to the present situation with organic banana cultivation, where a high price of fruit is obtained in international markets (Pascal 2008). On the other hand, an improvement in crop management will attract banana and plantain farmers, which could lead to an early adoption of transgenic banana. For instance, a BSD-resistant plant could lead to a decrease in production costs, as the investment to control BSD is estimated to be 10–20% of the current total cost, averaging US$ 1300 $ha^{-1}$ each year (FAO 2016). In 2017, 158,000 and 99,000 ha have been used for banana and plantain cultivation in Ecuador, respectively (FAO 2019); therefore, at least in banana production, US$ 2 million could be saved each year. To further implement a modern biotechnology development, these economic data are important to keep in mind together with a well-established biosafety framework.

attention; this concerns banana, a species of considerable economic importance in Ecuador. However, the possibility of other projects being run in this field, with unsuccessful results or inadequate communication, cannot be ruled out.

While a profound analysis of the reasons that have brought about this state of affairs might be conducted, in order to simplify things, it is easy to say that at every level, this technology's development has faced economic limitations, scant investment, and probably meager interest in related areas, such as agriculture and livestock. But there is no doubt that the high levels of politicization and vilification that this topic brings out in our country are crucial factors.

## The Regulatory Framework

In our view, extremist positions have been taken by a small group within Ecuadorian society, but one that has had influence and access to the spheres of political decision-making. An important example of this situation was the approval of a new Ecuadorian Constitution, which includes Article 401, with a text based on misinformation and with no objective or open discussions on this topic. Article 401, effective since 2008, states the following:

> Se declara al Ecuador libre de cultivos y semillas transgénicas. Excepcionalmente, y sólo en caso de interés nacional debidamente fundamentado por el Presidente de la República y aprobado por la Asamblea Nacional, se podrán introducir semillas y cultivos genéticamente modificados. El Estado regulará bajo estrictas normas de bioseguridad, el uso y el desarrollo de la biotecnología moderna y sus productos, así como su experimentación, uso y comercialización. Se prohíbe la aplicación de biotecnologías riesgosas o experimentales.

English Translation:

> Ecuador is declared free from transgenic cultures*[1] and seeds. Exceptionally, and only in cases of national interest properly substantiated by the President of Ecuador and approved by the National Assembly, can genetically modified seeds and crops be introduced. The State shall regulate, under strict biosafety norms, the use and development of modern biotechnology and its products, as well as its experimentation, use, and marketing. The application of experimental or hazardous biotechnologies is forbidden. (translated by us)

Our country's conduct in the field of genetically modified organisms has been determined by Article 401 and its interpretations.

Some points in this constitutional provision merit comment:

1. *Ecuador is declared free from transgenic cultures and seeds:* this sentence was interpreted by the last government as referring to the prohibition of agricultural crops and transgenic seeds in Ecuador, but not of the use of GMOs or their devel-

[1] The Spanish term "cultivos" loosely translates in English as "crops." However, "cultivos" is a nonspecific term which, in this legal context, should be correctly described with a specific adjective: "cultivos agrícolas," meaning "agricultural crops." We have used the English term "cultures" to emphasize this omission.

opment. The question that begs to be asked is, why develop genetically modified plants that later cannot be cultivated? Thus, even though research is not forbidden, what is the sense of investigating when there is a legal lock written into the highest law, the country's constitution, which is very difficult to amend.

2. In the same way, this first sentence has shed light on poor interpretations of what *free from transgenic cultures* refers to. Why include only agricultural crops and not animal or microorganism cultures? Certain groups maintain that, alluding to the term *culture* and its lack of qualifiers, all kinds of cultivation are forbidden in the country. This represents a very obvious problem of interpretation and legal clarity that could be the cause of many inconveniences.
3. *Exceptionally, and only in cases of national interest properly substantiated by the President of Ecuador and approved by the National Assembly, can genetically modified seeds and crops be introduced.* How? When? Who is to determine what constitutes national interest? This is extremely subjective and open; if anyone wanted to establish this exception, months of debate would follow just in order to see if national interest could be defined. We feel this exception does not follow technical criteria and falls prey to political and economic interests, unrelated to this technology's development.
4. *The State shall regulate, under strict biosafety norms, the use and development of modern biotechnology and its products, as well as its experimentation, use, and marketing.* In itself, this sentence is adequate and implies that the government of Ecuador should act in a responsible manner and—in order to comply with what is stated—should establish a national framework to regulate, control, and monitor activities pertaining to modern biotechnology. However, the fact is that Ecuador is far from having a national framework of this kind.
5. *The application of experimental or hazardous biotechnologies is forbidden.* A sentence completely out of context. What experimental or hazardous biotechnologies are being referred to? As a discipline, biotechnology implies experimentation; is this forbidden in Ecuador? What is the scope of this provision? No one can provide a clear answer to these doubts, and once again, this reflects the inconsistency with which this topic has dealt within our country.

We could list other laws also pertaining to GMOs in the country, such as the organic health law, the organic consumer law, the food sovereignty organic law, and the new environmental code, but an analysis of these regulations' provisions dealing with GMOs—and their lack of articulation—would only evidence the same confusion and lack of clarity on how to proceed in those areas.

In this context, to consider that a rational conversation might be had on how to regulate or manage new technological advances such as the use and application of synthetic biology only anticipates doubts and uncertainty.

The corrections necessary in this legal labyrinth in order to oversee modern biology, GMOs, and forthcoming developments such as synthetic biology require a decision for dialogue between the interested stakeholders, in which more objective and non-ideologized viewpoints could be presented. The pros and cons on how these technological developments might be addressed in a megadiverse country such as Ecuador should be put on the table for discussion, including the uses that these technologies might have in the conservation of our ecosystems and species diversity and toward new opportunities for our sustainable development.

Although several attempts to establish a national biosafety framework have been made, specific regulations on how to oversee GMOs and their risk analyses have been proposed, and policy guidelines for the adequate management of biotechnology have been put forward; in the country's history, such initiatives have regrettably always failed due to lack of adequate political support. Ecuador lives in the confusion generated by a profound legal void in this area, in which developing new technologies must face not only economic limitations but also legal barriers than cannot be interpreted.

Hopefully, myths, and extremisms might someday leave some space to rationally establish real regulatory systems and guide the adequate implementation of these topics in Ecuador. Education surely plays a predominant role in this scenario. Currently, there are some efforts, mainly at the university level, to introduce biosafety concepts into courses related to the development of modern biotechnology. We need some time to see if this approach achieves a more scientific and technical management of new biotechnology developments in a country such as Ecuador, where we need to balance conservation efforts and sustainable development.

## References


Código Orgánico del Ambiente, Registro Oficial Suplemento 983, 12 April 2017

Constitución de la República del Ecuador, Registro Oficial 449, 20 October 2008

FAO (2016) Ecuador's banana sector under climate change: an economic and biophysical assessment to promote a sustainable and climate-compatible strategy, by Elbehri A, Calberto G, Staver C, Hospido A, Roibas L, Skully D, Siles P, Arguello J, Sotomayor I, Bustamante A. Rome, Italy. http://www.fao.org/3/a-i5697e.pdf. Accessed 18 May 2019

FAO (2019) FAOSTAT database. FAO, Rome, Italy. http://www.fao.org/faostat/en/#data/QC Accessed 18 May 2019

INEC (Instituto Nacional de Estadística y Censos) (2015) Cifras de biodiversidad del Ecuador. http://www.ecuadorencifras.gob.ec/documentos/webinec/Sitios/pagina%20ambientales/

Khanna HK, Deo PC (2016) Novel gene transfer technologies. In: Mohandas S, Ravishankar K (eds) Banana: genomics and transgenic approaches for genetic improvement. Springer, Singapore

Korneva S, Ortega N, Santos E, Peralta E (2010) Obtención de suspensiones celulares embriogénicas de *Musa* spp. utilizando diferentes explantes. Memorias XIX Acorbat, Medellín, Colombia, pp 351–356

Korneva S, Flores J, Santos E, Piña F, Mendoza J (2013) Plant regeneration of plantain 'Barraganete' from somatic embryos using a temporary immersion system. Biotecnol Apl 30(4):267–270

Ley Orgánica de Defensa del Consumidor, Registro Oficial Suplemento 116, 10 July 2000

Ley Orgánica de Salud, Registro Oficial Suplemento 423, 22 December 2006

Ley Orgánica del Régimen de la Soberanía Alimentaria, Registro Oficial Suplemento 583, 5 May 2009

Ministerio del Ambiente (2016) Estrategia Nacional de Biodiversidad 2015–2030. Quito, Ecuador

Ministerio del Ambiente del Ecuador (2015) Quinto Informe Nacional para el Convenio sobre la Diversidad Biológica. Quito, Ecuador

Pacheco Coello R, Pestana Justo J, Factos Mendoza A, Santos Ordoñez E (2017) Comparison of three DNA extraction methods for the detection and quantification of GMO in ecuadorian manufactured food. BMC Res Notes 10:758

Pascal L (2008) Certification in the value chain for fresh fruits. Food and Agriculture Organization of the United Nations (FAO). http://www.fao.org/3/a-i0529e.pdf

Sánchez Timm E, Hidalgo Pardo L, Pacheco Coello R, Chávez Navarrete T, Navarrete Villegas O, Santos Ordóñez E (2016) Identification of differentially- expressed genes in response to *Mycosphaerella fijiensis* in the resistant Musa accession 'Calcutta-4' using suppression subtractive hybridization. PLoS One 11(8):e0160083

Santos E, Remy S, Thiry E, Windelinckx S, Swennen R, Sági L (2007) Tagging low-temperature responsive promoters in banana using the luciferase reporter gene. Acta Hortic 763:99–105

Santos E, Remy R, Thiry E, Windelinckx S, Swennen R, Sági L (2009) Characterization and isolation of a T-DNA tagged banana promoter active during in vitro culture and low temperature stress. BMC Plant Biol 9:77

Santos E, Pacheco R, Villao L, Galarza L, Ochoa D, Jordán C, Flores J (2016a) Promoter analysis in banana. In: S. Mohandas and K. V. Ravishankar Banana: genomics and transgenic developments for crop improvement. Springer. 157-179

Santos E, Sánchez E, Hidalgo L, Chávez T, Villao L, Pacheco R, Flores J, Korneva S, Navarrete O (2016b) Advances in banana transformation through *Agrobacterium tumefaciens* in Ecuador: progress, challenges and perspectives. Acta Hortic 1114:197–202

Santos E, Sánchez E, Hidalgo L, Chávez T, Villao L, Pacheco R, Navarrete O (2016c) Status and challenges of genetically modified crops and food in Ecuador. Acta Hortic 1110:229–235

Sowmya HD, Usharani TR, Mohandas S (2016) Somatic embryogenesis as a tool in genetic transformation. In: Mohandas S, Ravishankar K (eds) Banana: genomics and transgenic approaches for genetic improvement. Springer, Singapore

Strobbe S, Van Der Straeten D (2017) Folate biofortification in food crops. Curr Opin Biotechnol 44:202–211

Villao L, Sánchez E, Romero C, Galarza L, Flores J, Santos E (2019) Activity characterization of the plantain promoter from the heavy metal-associated isoprenylated plant gene (MabHIPP) using the luciferase reporter gene. Plant Gene. https://doi.org/10.1016/j.plgene.2019.100187

# Honduras—GMOs/Synthetic Biology Rules/Regulations and Biodiversity – A Legal Perspective from Honduras

**Carolina Alduvín**

**Abstract** Honduras was the first country in the Mesoamerican region to regulate the introduction of GMO crops. This chapter describes how the Earth Summit of 1992, the conventions and agreements derived from it, paved the way to create the legal frame and government structures in Honduras to make possible the assessment of new biotechnological crops. How government, academia, and productive sectors propitiate the adoption of new biotechnologies in order to improve crops and yields for the benefit of small farmers and big producers alike are also described. Honduras has an up-to-date legal frame even for the newest precision biotechnologies.

**Keywords** DiBio · Zamorano · Mesoamerica · CertiSem · Olancho · Comayagua · UNAH · SAG · SERNA · CNBBA

## Background

United Nations Conference on Environment and Development, also known as Earth Summit, was held at Rio de Janeiro in 1992 and led to several international agreements including the *Declaration on Environment and Sustainable Development*, *Agenda 21*, *Forest Principles*, *Framework Convention on Climate Change*, *United Nations Convention to Combat Desertification*, and *Convention on Biological Diversity* (CBD).

The CBD was ratified by Honduras in 1995, and in order to accomplish its recommendations, the Biodiversity Directorate (DiBio) was created in 1997 within the Ministry of Natural Resources and the Environment (SERNA). Among the CBD recommendations were the creation of a National Commission on Biodiversity,

C. Alduvín (✉)
National Committee on Agricultural Biotechnology and Biosafety (CNBBA), Tegucigalpa, M.D.C., Honduras

Biotechnology Group, Honduran National Academy of Sciences (ANCH), Tegucigalpa, M.D.C., Honduras

Universidad José Cecilio del Valle, Tegucigalpa, M.D.C., Honduras



A. Chaurasia et al. (eds.), *GMOs*, Topics in Biodiversity and Conservation 19,
https://doi.org/10.1007/978-3-030-53183-6_30

integrating five Committees: wildlife, genetic resources, bioethics, biotechnology, and biosafety. Since the country has very few specialists both in biotechnology and biosafety, in its inaugural meeting, the assembly decided to join these two disciplines in one committee.

At the end of October 1997, the National Committee on Biotechnology and Biosafety (CNBB) was created with representatives from the government, academia, and private entrepreneurs; in the first session, the members wrote and signed its internal rules and a confidentiality agreement. All founder members agreed that the membership should be exclusively by invitation and based on previous assessment of the scientific capabilities of individuals as opposed to organizations.

The CNBB was inaugurated by representatives from SERNA through DiBio, the Ministry of Agriculture and Livestock (SAG), and the National Service for Animal and Plant Health (SENASA), specifically the Seed Certification Office; the Ministry of Health through the Codex Alimentarius Office; the Ministry of Planning; the National Autonomous University of Honduras (UNAH); Zamorano University; Standard Fruit Company; the National Association of Seed Producers; and the National Association of Sugarcane Producers.

## GMO Regulation

At the first session of the CNBB held at Zamorano in November 1997, the representative from SAG announced the first requests to introduce to the country living modified organisms. He explained that SAG, under the Phytozoosanitary Law (Art. 9.1, Decree 157–94 of 4 November 1994), states that the sanitary and phytosanitary regulations for organisms and products of modern biotechnology fell within the legal competence of SENASA. Therefore, every person or society who imports, researches, exports, experiments on, moves, liberates, multiplies, or deals commercially with living genetically modified organisms or their products, processes, or agents for biological control and other types of organisms for agricultural or livestock purposes produced inside or outside the country must have prior approval from SENASA.

The first LMOs to be considered were *Bacillus thuringiensis* corn and glyphosate-tolerant corn. The next year, 1998, the CNBB began to set procedures, study the dossiers, and distribute responsibilities among members according to expertise. An unanimous decision was that all field essays were to be conducted by experts at Zamorano University, near our capital city, Tegucigalpa. Simultaneously, an application of genetic improvement in banana was reviewed.

By June 1998, the CNBB had received official recognition from the government by means of a Presidential Decree as a consultative group of experts to advise the Honduran government on biotechnology and biosafety matters. Then, we made our formal presentation before public and private and national and international institutions in our society. There were some protest groups, and a few media tried to discredit our work. We therefore initiated an educational program on scientific literacy targeted to journalists and community organizations; we counted on national and international experts to spread the new concepts, rules, and procedures.

Also, in September 1998, the official government gazette published Agreement No. 1570–98, Biosecurity Regulation with Emphasis on Transgenic Plants, the legal instrument on which all our regulatory work is now based. This document establishes general principles to be taken into account to ensure regulated use of GMOs. Under Article 8, it orders the creation of national committees with advisory functions to enable national authorities to assess possible risks to human health, animal and plant production, and protection to the environment due to the use of GMOs.

By 2001, the CNBB had concluded all necessary studies to confirm that the first GMO plant presented no harm to animal or human health or to the environment and recommended pertinent authorities to permit semicommercial usage, and commercial use was granted in 2003. This was the first event of this kind in the Mesoamerican region, and Honduras therefore was recognized as giving regional leadership in the regulation of GMOs. Several other requests for introduction and liberation came in the following years, with new versions of modifications in queues.

## LMO Regulation

In 2008, Honduras ratified the Cartagena Protocol on Biotechnology of the Convention on Biological Diversity (CPB) which had been adopted by the United Nations in 2000; this international instrument regulates transboundary movements of LMOs promoting biosafety on their manipulation and use. Its focal point is also SAG–SENASA. Since then, all information regarding requests, approvals, and liberation of LMOs is reported to the Biosafety Clearing House (BCH) mechanism on information exchange regarding biosafety of this biotechnology. The BCH contains detailed information on the LMO, its modifications, products, risk assessment procedures, risk management measures, and results as a reference for all parties and users of the CPB.

The secretariat of the CBD has subsequently created special working groups on controversial issues such as risk assessment, risk management, synthetic biology, and digital sequences information on genetic resources, among others, through Ad Hoc Technical Expert Groups (AHTEGs). After several rounds of discussions, mostly online, they have presented guideline-like documents. These guidelines serve as a tool for those parties and organizations that at the moment have not implemented their own regulations and procedures. The delegates from Honduras, however, are strongly opposed to the mandatory adoption of such guidance.

By 2017, Honduras had systematized its procedures in a document and shared it in the BCH. This is in the form of the handbook *Proposal of Norms and Procedures for Use of Living Modified Organisms in Honduras* and is organized in the following sections:

1. Internal Rules
2. Authorization Procedures
3. Judgment Format
4. Use Authorization
5. Decision Format for BCH
6. Risk Assessment Publication Format
7. Model for Appointment Letter
8. Model for CNBBA Meeting Agenda
9. Model for CNBBA Acts
10. Model for No Divulgation Agreement
11. Model for Communication to SENASA

Also in 2017, the CNBB turned to CNBBA (the National Agricultural Committee on Biotechnology and Biosafety), and its members formally took an oath in a special ceremony presided over by the SENASA director. This change was legally backed by Agreement No. 177–2017. This Agreement was based on Article 245, No. 11 of the Constitution of the Republic; Articles 7, 29, and 36, Nos. 2,5 and 6 of the General Law of Public Administration; Articles 1, 17, and 22 as reformed by Decree No. 344–2005 of the Phytozoosanitary Law; and also Chapter II, Article 4 of the Phytozoosanitary Law, according to Decree No. 157–94.

The 2017 appointment of the CNBBA as an advisory body to SENASA in matters of biotechnology and biosafety led to a forum for discussion, harmonization, and consensus for policies related to the production, productivity, and competitivity of the agricultural sector, establishing priorities such as surveillance to ensure compliance with international conventions, national laws, and rules on both subjects and the proposal of criteria and procedures to be followed.

Since the end of the twentieth century, a biodiversity law has been on draft and subject to several rounds of public consultation among groups of experts, users, civil society organizations, government officers, academics, entrepreneurs, indigenous peoples, and local communities. It includes a chapter on biotechnological issues, on which members of the CNBBA have been consulted and given their opinions and advice. As far as we know, the responsible lawyers have prepared a final version which is currently somewhere in the long procedure toward approval by our Congress. So, currently, this remains a project, and so is no interest for our legislators.

## Synthetic and Precision Biology

On September 12, 2019, the official government gazette published the Agreement SENASA 008-2019 which regulates procedures to authorize agricultural and livestock products obtained by novel techniques of genetic improvement or precission biotechnology. The new techniques of synthetic biology and precision biology are not considered in any law project in Honduras at this time. Biotechnology is still an underdeveloped subject in our academic institutions. Nevertheless, UNAH and Zamorano University have been equipping laboratories to international standards for genetic research with various tools of molecular biology. In industry, these approaches are seldom used in production processes, only in selected agricultural enterprises and a few diagnostic procedures in hospitals or other sanitary facilities. At this time, two academic projects have been developed in international fora: in 2014, a team of students from Zamorano University and the National Autonomous University of Honduras (UNAH) presented a proposed project at the International Genetically Engineered Machine (iGEM) competition in Boston, MA, USA.

The team emphasized legal aspects of their synthetic biology project and received a bronze medal award. In 2015, a team from UNAH, sponsored by the Honduran Institute of Science, Technology, and Innovation (IHCieTI), presented a project on production of a vaccine to prevent dengue fever, fulfilling all necessary legal requirements and procedures.

## References

Government of Honduras (1993) General law on environment. La Gaceta Magazine, June 8th, Tegucigalpa, Honduras

Government of Honduras (1994) Phytozoosanitary law, decree 157–94. La Gaceta Magazine, November 4th, Tegucigalpa, Honduras

Government of Honduras (1998) Biosafety regulations with emphasis on transgenic plants, agreement 1570–98. La Gaceta, September 1998, Tegucigalpa, Honduras

Government of Honduras (2005) Modified Phytozoosanitary law, decree 344–2005. La Gaceta Magazine, December 29th, Tegucigalpa, Honduras

Government of Honduras, Ministry of Environment, Biodiversity Directorate (n.d.) Anteproject of Biodiversity Law [draft]

Government of Honduras (2019) SAG-SENASA. Board of Directors Agreement 008-2019

Interamerican Institute for Agricultural Cooperation (2017) Proposal of norms and procedures for living modified organisms use, Tegucigalpa, Honduras

United Nations Environment Programme (1992) Convention on biological diversity. United Nations Environment Programme, Nairobi

United Nations Organization (1993) Agenda 21. Rio de Janeiro, Brazil

United Nations Organization (2000) Cartagena protocol on biosafety of genetically modified organisms. United Nations, Montreal

# Venezuela—GMOs/Regulations and Biodiversity – A Legal Perspective in Venezuela

**María Eugenia Cavazza, Luis Plata, María Correnti, Gysell Plata, Juan Fernando Marrero, and Carliz Díaz**

**Abstract** The Bolivarian Republic of Venezuela has a long and important tradition in the conceptual, legislative, and regulatory development on environmental affairs. The creation of the Ministry of the Environment and Renewable Natural Resources in the 1970s, as a predecessor for the current Ministry of the Popular Power for Ecosocialism was a milestone. The Biological Diversity Management Law was promulgated in 2008. It's aims were to establish notions, such as the provisions for the biological diversity management. The National Action Plan details the main causes of biodiversity loss and then identifies seven strategic lines. These concern information management; conservation of endangered species, strategic areas for conservation, sustainable use of biological diversity, prevention, control, and eradication of exotic species, control and supervision of GMOs, species trafficking or illicit trade's prevention and management.

M. E. Cavazza (✉)
Biomedicine Institute. Ministry for Popular Power of Health. Molecular Microbiology Laboratory, J.M.Vargas School, Medicine Faculty, Central University of Venezuela, Caracas, Venezuela

L. Plata
Economic and Social Science, Central University of Venezuela, Caracas, Venezuela

M. Correnti
Oncology anda Hematology Institute, Ministry for Popular Power of Health. Molecular Genetic Laboratory, Caracas, Venezuela

G. Plata
Biomedicine Institute. Ministry for Popular Power of Health. Molecular Microbiology Laboratory, Caracas, Venezuela

J. F. Marrero
Agronomy Faculty, Central University of Venezuela, Caracas, Venezuela

C. Díaz
Ministry for Ecosocialism.Biodiversity Direction, Caracas, Venezuela

 
A. Chaurasia et al. (eds.), *GMOs*, Topics in Biodiversity and Conservation 19,
https://doi.org/10.1007/978-3-030-53183-6_31

**Keywords** National Strategy · Sustainable · Exotic species · Law of seeds · Biodiversity · Venezuela · Community · Social

## Background

The Bolivarian Republic of Venezuela has a long and important tradition in the conceptual, legislative, and regulatory development on environmental affairs. The creation of the Ministry of the Environment and Renewable Natural Resources in the 1970s, as a predecessor for the current Ministry of the Popular Power for Ecosocialism, was a prominent institutional contribution, together with the promulgation of the Organic Law for the Environment (promulgated in 1976 and last amended in 2006) and the Environmental Penal Code (promulgated in 1992 and revised in 2012), among other laws and numerous regulations, standards, and resolutions, all with strong emphasis on environmental protection, including natural resources and biodiversity.

In addition, this law considers different geographical areas under the Special Administration Regime (Áreas Bajo Régimen de Administración Especial (ABRAE)) among which 43 National Parks and 36 Natural Monuments are recognized and which are part of a continuing trend by the environmental authority and the nation's Executive.

The promulgation of the Constitution of the Bolivarian Republic of Venezuela in 1999 was a historic landmark in relation to the express enunciation of innovative principles, rights, and obligations regarding the environment. Thus, the *Carta Magna* noted that: "environmental protection is a State duty, in co-responsibility with the citizens. Established notions, such as sustainable development, the non-patentability of living beings' genomes, the obligatoriness of environmental and sociocultural impact studies for all projects that are likely to generate environmental damage, including consultation and community participation".

The Bilogical Diversity Management Law (BDML) was promulgated in 2008. This law aims to "Established notions,such the provisions for the biological diversity management in its various components, comprising: natural or manipulated genomes, genetic material and their derivatives, species populations, communities and ecosystems" present in spaces including continental, island, lake, and river, territorial sea, inland sea areas, soil, subsoil and airspace, in warranty of security and sovereignty of the Nation and to achieve greater collective well-being within the framework of sustainable development."

Beyond the scope of regulating and protecting biodiversity, there is an aim to mitigate human-activity-associated risks; the standard is very accurate in relation to the hazards associated with products of modern biotechnology; Article 2.11 points out "that the management of biological diversity comprises the adoption of actions and measures in the field of biosafety relating to genetically modified organisms to prevent adverse effects on biological diversity."

The Biological Diversity Management Law gives the State a leading role in the implementation of these principles, which regulates biotechnology, biosecurity, and this type of technology. In regard to this:

- The State must establish measures to prevent and avoid threats to biodiversity resulting from the use of biotechnology, especially those risks associated with the use, marketing, transportation, and release of products or organisms resulting from modern biotechnology processes. This includes LMOs and genetically modified organisms (GMOs).
- The State should promote biotechnological development within the country as an instrument for sustainable development, emphasizing on environmental conservation, biodiversity protection, nutritional security, and human health. This means that there is no denial or prohibition for the advancement of biotechnology in the country, in the broad sense, including progress in modern biotechnology, as long as the protection of the environment is always taken into account and guarantied.
- Persons performing activities in the country involving organisms resulting from modern biotechnology (e.g., LMOs) are subjected to State supervision by the Environmental National Authority, and they must fully comply with all rules, mechanisms, and instruments for control as established in this law, as well as other applicable regulations. Regarding biosafety, it should be mentioned that most aspects of this law refer to chapter III: Biosafety (Articles 50–55):

  - The environmental National authority will regulate research activities, import, export, release, confined management, production, distribution, trade, mobilization, and storage of the genetically modified organisms (GMOs), their derivatives, and any products containing them with the aim of preventing risks on biodiversity.
  - As long as safety is not proven in regards to conservation and sustainable use of biodiversity, GMO's transfer, handling, and use is prohibited. It should be noted that, in accordance with the precautionary principle, the lack of scientific information on the safety of a GMO should not prevent the possibility of taking measures in favor of safeguarding and protecting against impending risks arising from GMO use, handling, or transfer.
  - The management of GMOs will be confined and handled in conditions according to what is established in this law. Biological Diversity Management, therefore, provides "closed management", under conditions of isolation for GMOs.
  - Natural or legal persons, public or private agents, and national or foreign entities, who intend to carry out GMO-related activities within the country, should follow the Biosafety Protocol by the National Environmental Authority, according to national regulations.
  - The National Environmental *Authority* will carry out risk assessments, in coordination with the community and other agencies, according to the prior informed procedures and agreements, in accordance with Cartagena's Biosafety Protocol on Biotechnology and the Biological Diversity Convention.

## Homeland and National Strategy for the Conservation of Biological Diversity and the National Action Plan 2010–2020

The second Socialist Plan for Social and Economic Development of the Nation, 2013–2019, better known as the Homeland Plan, in its historic objective No. 5 sets the goal of contributing to the preservation of planetary life and the safeguarding of the human species. National target 5.1 aims to: "build and promote an Ecosocialist productive economical model, based on a harmonious relationship between humans and nature, which guarantees the sustainable, rational and optimal use/exploitation of natural resources, respecting nature's processes and cycles."

Likewise, in its programs and policies, there are environmental guidelines: policy "No. 5 promoting a different relationship between human beings and mother earth; promoting an alternative development model based on the ecological, cultural, social and political sustainability."

Guidelines for carrying out the National Action Plan don't merely mention dialogue promotion and sustainable use of biological diversity but give the main direct boost to generate actions in this field, including the following:

> It is this sense, complying with this mandate, the Bolivarian Republic of Venezuela in May 2012 collectively elaborated the document on the National Strategy for the Conservation of Biological Diversity and its National Action Plan, for the period 2010 2020; it was drawn up by the Ministry of the Popular Power for Ecosocialism, built with active participation of organized communities, students, groups, teachers, officials, and militants. It took place in two phases, the first began with problem identification associated to biological diversity loss and the study of its causes and consequences; the second phase was aimed at the development of the National Action Plan.

The National Action Plan details the main causes of biodiversity loss and then identifies seven strategic lines. These concern information management; conservation of endangered species; strategic areas for conservation; sustainable use of biological diversity; prevention, control, and eradication of exotic species; control and supervision of GMOs; and species trafficking or illicit trade's prevention and management. The strategic line on GMOs is the one pertinent to the theme of this book.

This line aims to strengthen regulation mechanisms, procedures, and actions related to GMOs, in order to avoid adverse effects on biological diversity and promote food sovereignty and the enduring supreme socialhappiness of the Venezuelan folk. The strategy recognizes the need for regulation, transit control and monitoring of GMOs, and protection of autonomous genetic diversity and sovereignty and therefore aims to promote debate and knowledge exchange about the potential ecological, economic, social, and cultural effects of GMOs.

GMO control and supervision is needed in order to develop and strengthen an integrated policy on biotechnology safety through the creation of a National Biosafety System, with accredited laboratories capable of detection, quantification, and monitoring of GMO samples and derived products. The strategy also promotes access to the official and alternative information on GMOs, ending information hijacking by implementing a report system for biotechnology and biosafety activities, which include GMO labeling (of products and derivatives), and allowing community participation in the decision-making process.

The National Action Plan also includes a series of transverse axes cutting across the seven strategic lines: education for conservation, environmental legislation, management for conservation, and international management and policy.

## National Strategy for Biological Diversity'sConservation and the Cartagena Protocol on Biosafety

An important aspect of the Convention on Biological Diversity (CBD) corresponds to explicit and implicit principles that preexist the international agreements and conventions, specifically that of caution (Principle 15 of the Rio Declaration). This is also acknowledged in the Preamble to the Cartagena Protocol on Biosafety (CPB) in Articles 1, 10.6, and 11. (j), as well as in Annex III, which are intended to ensure an adequate level of protection in the safe transfer, handling, and use of LMOs, as well as the prevention of serious or irreversible damage. The possible lack of scientific certainty, however, is ambiguously handled in the CPB.

The National Strategy for Biological DiversityConservation, developed by the Bolivarian Government, is the resulting instrument from a collective elaboration of national actors concerned with public spaces, with direct involvement from the Government, for conservation and research of the country's biological diversity. This provides a public mandate for actions that will rule the State Agencies' performance in the field of biodiversity, to carry out the protection and conservation of biological diversity and its sustainable use, in terms of equity. At the same time, the strategy responds to the mandate created by the CBD Article 6 on General Measures for Conservation and Sustainable Use that indicates "Each Contracting Party shall, in accordance with its particular conditions and capabilities … develop national strategies, plans or programmes for the conservation and sustainable use of biological diversity or adapt for this purpose existing strategies, plans or programs which shall reflect, *inter alia*, the measures set out in this Convention ….".

The CPB is an executive arm that establishes how to carry out the mandates contained in the aforementioned Convention, specifically in the area of transboundary movements, being its goal to "… ensure an adequate level of protection in the field of safe transfer, handling and use of living modified organisms resulting from modern biotechnology, that may have adverse effects on the conservation and sustainable use of biodiversity, taking also into account human health risks and focusing in particularly, on cross-border movements …." As outlined in the Protocol, the precautionary approach established at the meetings which gave way to the CBD (Rio 1992) is present throughout the entire document (Tables 1 and 2).

**Table 1** Comparison of the principles and objectives of the National Strategy with those of the CBD and CPB

| National strategy (ENCDB) | CBD | CPB |
|---|---|---|
| *Principles:*<br>1) Eco socialist ethics<br>2) Sovereignty<br>3) Inclusion and social justice | *Objectives:*<br>1) Conservation<br>2)Sustainable use of biological diversity<br>3) Participation in the benefits derived from natural resources | *Objective:*<br>"... Contribute to ensuring an adequate level of protection in the field of safe transfer, handling and use of living modified organisms (LMOs) resulting from modern biotechnology that may have adverse effects on the conservation and sustainable use of biological diversity, taking into account potential risks on human health and specifically focusing on transboundary movements ..." |

**Table 2** Provisions on sovereignty

| National strategy (ENCDB) | CBD | CPB |
|---|---|---|
| *Principle 2: Sovereignty:*<br>"To promote conservation and sustainable use of biological diversity in Venezuela in a regional, continental and world-oriented scale. Sovereignty as a way toward the integration of solidarity and enduring social supreme happiness as well as a dignified management by communities (...).<br>2.1 To recognize biological diversity and its management as an object of sovereignty (...)<br>2.2 To recognize biological diversity and its management as a source of sovereignty and as a medium for popular development and integration (...)." | *Article 3:*<br>"In accordance with the Charter of the United Nations and the principles of international law, every State has, ..., the sovereign right to exploit their own resources pursuant to their own environmental policy, and the responsibility to ensure that, the activities carried out within their jurisdiction or control, do not cause damage to the environment of other States or of areas beyond the limits of national jurisdiction ..." | *Article 2:*<br>"... Nothing in this Protocol shall affect in any way the sovereignty of States over their territorial sea established in accordance with international law, and the sovereign rights and the jurisdiction which the States have in their exclusive economic zones and their continental shelves in accordance with international law, and the exercise by ships and aircraft of all States of navigational rights and freedoms as provided for international law and as reflected in the relevant international instruments ..." |

## Governing Body and National Authorities for Modern Biotechnology'sBiosafety: Biosafety National Commission (NCB)

In 2003, a National Commission was created by the National Executive Decree (Decree No. 37.733, 16 July 2003, published in the Official Gazette of the Bolivarian Republic of Venezuela). This Decree takes into consideration that Venezuela is a Party to the CBD and CPB. It has the purpose of forming a technical-scientific consultative body to advise the National Executive Power on LMO-related activities and its derivatives and products, allowing it to establish guidelines and dictate regulations that will guarantee safe use and handling of modern biotechnology's resulting organisms. According to this Decree, the governing body on biosecurity is the Ministry of the Popular Power for Ecosocialism, as expressed in Article 2. The NCB is composed of 11 members each representing a different body or sectors:

- Ministry for the Environment and Natural Resources (now Ministry of the Popular Power for Ecosocialism), which will coordinate this commission
- Ministry for Agriculture and Lands (today Ministry of the Popular Power for Productive Agriculture and Lands)
- Ministry for Light industry and Trade (today Ministry of the Popular Power for the Basic, light industries and Commerce)
- Ministry for Health (today Ministry of the Popular Power for Health)
- Ministry for Science and Technology (now Ministry of the Popular Power for University Education, Science, Technology and Innovation)
- Ministry for Food (today Ministry of the Popular Power for Feeding)
- Universities and public institutes of higher education
- Industrial and commercial sector for agriculture, food, medicinal, and pharmaceutical products
- Agricultural sector for small- and medium-sized producers (subsector plant and animal)
- Agricultural sector for large producers (subsector plant and animal)
- The organized community

Moreover, the NCB is the Competent National Authorities (CNA) for the implementation of risk control and biosafety measures derived from LMO use and handling according to its area of competence, as expressed in Article 13 of the Decree:

(a) Ministry of the Popular Power for Ecosocialism
(b) Ministry of the Popular Power for Productive Agriculture and Lands
(c) Ministry of the Popular Power for Feeding
(d) Ministry of the Popular Power for Health
(e) Ministry of the Popular Power for University Education, Science, Technology and Innovation

In accordance with the purpose of this Decree, each CNA must ensure each area is within its particular competence, in order to establish mechanisms and procedures conducive to reducing GMO use- and handling-associated risks. The CNA should dictate measures to safeguard the environment and diversity due to the associated risks of GMO use. The CNA on agricultural production will do the same to protect the sustainability of primary production considering potential dangers derived from use of genetic materials (e.g., seeds) of transgenic origin.

The CNA in nutritional matters has the mission to ensure food safety and is obliged to enact measures to guarantee the safety of GM foods for consumers and also toward, through correct labeling, all potential risks and in any necessary case, to prohibit consumption. The CNA on health must take measures that ensure people's overall health related to GMO use, consumption, and handling. Finally, the

CNA, in the field of education, science, and technology, should take steps to educate the public on GMO's potential risks and promote associated research and development.

However, the main function of the NBC is to advise the National Executive Power and particularly the CNA referred above in the following aspects:

- Recommend GMO-related policies, plans, projects, and actions that include their derivatives and any products containing them.
- Propose a National Strategy to coordinate all public institutions that participate in the management of GMOs and biosafety.
- Participate in the elaboration of the biosafety regulation.

It should also propose effective implementation mechanisms for biosafety laws according to CPB and CBD to:

- Issue opinions on the chosen authorizations for activities related to GMOs and their derivatives.
- Issue opinions on evaluation studies and risk analysis for the purpose of authorizing GMO-related activities.
- Advise on risk management measures and its follow-up.
- Promote the strengthening of specialized human resources and institutional capacity on the matter of biosecurity.
- Collaborate in the development of a Biosafety regulatory framework and public management guidelines.
- Provide the support required, as scientific and technical advisor, to the national executive body.

Other national existing regulations to take into consideration for State-level actions related to GMOs include: the Organic Law of National Security, Law of Science, Technology and Innovation (LOCTI), Organic Law on Safety and Agrifood, Integral Agricultural Health Law, Law of Seeds, and Environmental Organic Law.

In general, these instruments all have the principles of precaution, equity, justice and self-determination of the people, non-maleficence, and finally responsibility and stewardship of the State, as a common denominator. The State is jointly responsible, together with the natural and legal actors, for the risks associated with GMOs, in the field of environmental protection, biological diversity, and human health, defining the institutional competence in this area, especially for the National Biosecurity Agency.

## References

Asamblea Nacional Constituyente (1999) Constitución de la República Bolivariana de Venezuela. Gaceta Oficial No. 36860 de fecha 30 de diciembre. Venezuela. In March 2018: http://www.inpsasel.gob.ve/moo_doc/ConstitucionRBV1999-ES.pdf

Asamblea Nacional de la República Bolivariana de Venezuela (2000) Ley de Diversidad Biológica. Gaceta Oficial 5.468 de fecha 24 de mayo del 2000. In April http://www.acnur.org/t3/fileadmin/Documentos/Pueblos_indigenas/ley_diversidad_biologica_ven.pdf?view=1

Asamblea Nacional de la República Bolivariana de Venezuela (2006) Constitución de la Comisión Nacional de Bioseguridad. Gaceta Oficial 38.392 de fecha 7 de marzo del 2006. In July 2018: http://avisa.org.ve/wp-content/descargas/gaceta38392.pdf and http://virtual.urbe.edu/gacetas/38392.pdf

Asamblea Nacional de la República Bolivariana de Venezuela (2013) Plan de la Patria: Segundo Plan Socialista de Desarrollo Económico y Social de la Nación 2013–2019. Publicado en Gaceta Oficial de la República Bolivariana de Venezuela No 6.118 Extraordinario, 4 de diciembre de 2013. In April 2018: http://www.mppp.gob.ve/wp-content/uploads/2013/09/ley_plan_patria.pdf

CBD – Convention of Biological Diversity (1992) Convenio sobre Diversidad Biológica. Organización de las Naciones Unidas. In March 2018: https://www.cbd.int/doc/legal/cbd-en.pdf

De Cózar Escalante J (2005) Principio de Precaución y Medio Ambiente. Revista Scielo Rev Esp Salud Publica 79(2) Madrid mar./abr. 2005. In April 2018: http://scielo.isciii.es/scielo.php?script=sci_arttext&pid=S1135-57272005000200003

Greiber T y Col (2013). Guía Explicativa del Protocolo de Nagoya sobre Acceso y Participación en los Beneficios. UICN Serie de Política y Derecho Ambiental No. 83. In April 2018: https://www.propiedadintelectual.gob.ec/wp-content/uploads/downloads/2017/micrositio/articulos-tecnicos/guia-explicativa-del-protocolo-de-nagoya.pdf

Ministerio del Poder Popular para el Ambiente (2014) Estrategia Nacional para la Conservación de la Diversidad Biológica y su Plan de Acción Nacional (2010–2020) ISBN: 978-980-04-1483-5. In March 2018: http://diversidadbiologica.minamb.gob.ve/media/bibliotecas/biblioteca_636.pdf

United Nations Conference on Environment and Development (UNCED or the Earth Summit)Rio de Janeiro, Brazil (1992).In April 2018: https://www.un.org/esa/dsd/resources/res_docukeyconf_eartsumm.shtml

# Part X
# Asia

# India—GMOs/Synthetic Biology Rules/Regulations and Biodiversity – A Legal Perspective from India

**Brajesh Barse and Syed Shams Yazdani**

**Abstract** In this chapter, we highlight the current scenario of Indian biological research in the area of genetically modified organisms (GMOs) and synthetic biology and describe national policies and the biodiversity act that are regulating the GMOs and synthetic biology-related research activities in India. We also review here a recent draft document released by the Department of Biotechnology, Government of India on "Genome Edited Organisms: Regulatory Framework and Guidelines for Risk Assessment" considering the recent developments in the field of genome editing (GEd) technologies. It has been evident from the geographical location and ancient history of India that it would have a wide variety of biological diversity along with the associated traditional knowledge available for exploitation. India is a member of Convention on Biological Diversity (CBD) and signatory to the Cartagena Protocol on Biosafety (CPB) and Nagoya Protocol on Access to Genetic Resources and the Fair and Equitable Sharing of Benefits on access (ABS). Government of India has set up a National Biodiversity Authority (NBA) to regulate the use of biological resources and associated knowledge occurring in India for commercial or research purposes or for the purposes of bio-survey and bio-utilization. The biological resources available in the country and their associated knowledge could be of great help in developing new technologies for human use with the help of modern genetic engineering and synthetic biology tools. Along with the legal framework to work on GMOs and GEd, this chapter also discloses various initiatives of Govt of India to promote synthetic biology research in the country.

**Keywords** GMOs · Genetic engineering · Genome editing · Biodiversity · Synthetic biology · Regulations · Risk assessment · Biological diversity · National Biodiversity Authority · Policies and guidelines

---

B. Barse · S. S. Yazdani (✉)
Microbial Engineering Group, International Centre for Genetic Engineering and Biotechnology (ICGEB), New Delhi, India

DBT-ICGEB Centre for Advanced Bioenergy Research, International Centre for Genetic Engineering and Biotechnology (ICGEB), New Delhi, India
e-mail: shams@icgeb.res.in



A. Chaurasia et al. (eds.), *GMOs*, Topics in Biodiversity and Conservation 19,
https://doi.org/10.1007/978-3-030-53183-6_32

## National Policies on GMOs

India has a systematicregulatory framework for biosafety evaluation of genetically modified organisms (GMOs) and products thereof. India was one of the early movers in development of a biosafety regulatory system for GMOs in the year 1989. The Ministry of Environment and Forests (MoEF) enacted an environment (protection) act in 1986 to provide for the protection and improvement of the environment and related matters (The Environment (Protection) Act 1986). Under this act, the rules and procedures for the manufacture, import, use, research, and release of GMOs, as well as products made by the use of such organisms, were notified by MoEF through notification in Official Gazette of Govt. of India on 5 December 1989 (Rules for the manufacture, use/import/export and storage of hazardous micro-organisms/genetically engineered organisms or cells 1989). These rules and regulations cover the areas of research as well as large-scale applications of GMOs and products throughout India. The rules also cover the application of hazardous microorganisms which may not be genetically modified. Hazardous microorganisms include ones pathogenic to humans and other animals as well as plants. The rules cover activities involving manufacture, use, import, export, storage, and research. Besides hazardous naturally occurring microorganisms, the target substances covered are all genetically engineered organisms including microorganisms, plants, and animals (Rules for the manufacture, use/import/export and storage of hazardous micro-organisms/genetically engineered organisms or cells 1989). The Rules of 1989 are supported by guidelines on research activities in contained environments, confined field trials, food safety assessment, environmental risk assessment, etc.

### *Definitions*

The definition of genetic engineering in the "Rules, 1989," which was further updated in 2017 (Regulations and Guidelines for Recombinant DNA Research and Biocontainment 2017), is rather broad and includes the "modification, deletion or removal of parts of heritable material." Therefore, the new genome engineering technologies, including gene editing and gene drives, may also be covered under the Rules. To be more precise, the Rules specifically defines the terms biotechnology, cell hybridization, gene technology, genetic engineering, and microorganisms.

### *Applications*

These rules are applicable to the manufacture, import, and storage of microorganisms and gene-technological products and applicable to specific cases such as sale, exportations, and importation of genetically engineered cells and manufacturing-related activities of genetically engineered products and pharmaceutical/drug, food stuff distilleries and tanneries etc., which make use of microorganisms/GMOs

**Table 1** Competent authorities dealing with various aspects of GMOs in India

| | |
|---|---|
| Recombinant DNA Advisory Committee (RDAC) | This committee shall review developments in biotechnology at national and international levels and shall recommend suitable and appropriate safety regulations for India in recombinant research, use, and applications from time to time |
| Institutional Biosafety Committee (IBSC) | Committee constituted at institutions handling risk-inherent microorganisms or GE organisms. On-site emergency plan and update from time to time according to the manuals/guidelines of the RCGM and make available as required copies to the District Level Committee/State Biotechnology Coordination Committee and the Genetic Engineering Appraisal Committee |
| Review Committee on Genetic Manipulation (RCGM) | DBT committee that monitors the safety-related aspects in respect of ongoing research projects and activities involving genetically engineered organisms/hazardous microorganisms |
| Genetic Engineering Appraisal Committee (GEAC) | Under MoEF, it appraises activities involving large-scale use of hazardous microorganisms, GE organisms, or cells in research and industrial production from the environmental angle and appraises proposals relating to release of GE organisms and products into the environment including experimental field trials. It has powers to take punitive action under the "Environment (Protection) Act" of 1986 |
| State Biotechnology Coordination Committee (SBCC) | Inspect, investigate, and take punitive action in case of violations of statutory provisions through the Nodal Department and the State Pollution Control Board/Directorate of Health/Medical Services |
| District Level Committee (DLC) | Monitor the safety regulations in installations engaged in the use of genetically modified organisms/hazardous microorganisms and their applications in the environment |

(Rules applicable to the manufacture, import and storage of micro-organisms and Gene-Technological products 1989).

## *Competent Authorities*

The implementation of the Rules of 1989 is the responsibility of the Department of Biotechnology (DBT), Ministry of Science and Technology (MoS&T), and the MoEF. The rules define the competent authorities and composition of six such authorities for handling of various aspects of the rules (Table 1).

While the RDAC has an advisory role, IBSC, RCGM, and GEAC are involved in regulations and SBCCs and DLCs in monitoring.

## *Approvals and Prohibitions*

The approvals and prohibitions under Rules, 1989, related to GMOs (Rules applicable to the manufacture, import and storage of micro-organisms and Gene-Technological products 1989) are summarized as follows. Under this rule, no person shall import, export, transport, process, and use or sell any hazardous or pathogenic microorganisms without the approval of GEAC. The use of pathogenic microorganisms can only be allowed in laboratory with proper measures mentioned in the MoEF Environment (Protection) Act of 1986. For any scale-up or pilot operations, the license must be obtained from GEAC. Certain experiments for education purpose within the field of gene technology may be carried out outside the laboratories or in laboratory areas with the approval of IBSC committee.

## Guidelines and Regulatory Framework for Genome Editing Technologies

A new draft guideline has been made available by DBT on genome editing technologies (GEdT) (Draft Document on Genome Edited Organisms: Regulatory Framework and Guidelines for Risk Assessment2020). Genome editing is a precise molecular method of mutation leading to deletion or addition or substitution of target base pair(s) in the native genes/nucleic acid sequences. Genetically engineered (GE) organisms (GMOs or LMOs) typically contain sequences from other organisms to modify an existing trait or introduce a new trait. Genome editing similarly facilitates the introduction of a foreign gene(s) to introduce a new trait(s), but the site of integration is more precise in genome edited (GEd) organisms. Within GEd organisms, there are differences depending on the type or nature of site-directed nuclease (SDN) or oligo-directed mutagenesis (ODMs) used in genome editing process.

GEdT represent the latest innovation and its potential applications in a wide range of sectors covering human and animal health, food, agriculture, microbial biotechnology, bio-economy, etc. As with all new technologies, GEdT have dual-use potential leading to safety and security issues.

### *Governing Genome Editing Technologies and Procedures*

GEdT have implications to international treaties/agreements such as the Cartagena Protocol on Biosafety (CPB), Biological Weapons Convention, Wassenaar Arrangement on Export Controls for Conventional Arms and Dual-Use Goods and Technologies, and Australia Group (AG). India as a party to these treaties/agreements remains committed to the fulfilment of obligations under them and will take necessary steps to regulate GEd whenever required.

Export of hazardous microorganisms or toxins listed in Special Chemicals, Organisms, Materials, Equipment and Technologies (SCOMET) and developed using GEdT requires prior approval from the Directorate General of Foreign Trade (DGFT) as specified under the Foreign Trade Policy of India. Food Safety and Standards Authority of India (FSSAI) under the Food Safety and Standards Act of

2006 is responsible to assess the safety of food and its ingredients where food contains or consists of genome edited products.

## *Risk Assessment of Genome Edited (GEd) Products/Organisms*

The risk evaluation matrix, in line with globally followed risk assessment for any new technology, has been developed to determine the overall risk levels. According to the matrix, the risk level will be determined based on the extent of complexity/modification introduced and the risk category into which an organism falls. The categories are essentially based on the complexity of modification and prior knowledge/familiarity with the modification in natural/existing population. Risk levels can range from low, to moderate, to high, and as the level increases, the data requirement and biosafety assessment levels increase. Three risk category groups are recognized:

1. *GEd Group I*– Single or few base-pair edits/deletions/insertions leading to least complexity (phenotype/genotype). Changes leading to knockdown/knockout of protein/RNA that result in a new trait which may be familiar with prior knowledge. Chances of off-target effects.
2. *GEd Group II*– Several base-pair edits leading to certain degree of complexity in phenotype/genotype (leading to improvement of an existing attribute or creation of a new attribute). Changes leading to gain of function with a new protein or RNA. May or may not be familiar with prior knowledge. Chances of off-target effects.
3. *GEd Group III*– Insertion of foreign gene/DNA sequence leading to high degree of complexity in phenotype/genotype (leading to creation of a new attribute, new metabolic pathways, etc.). Changes leading to gain of function with new protein or RNA. May not have prior knowledge. Chances of off-target effects.

## *Regulatory Consideration for Genome Edited (GEd) Organisms/ Products Derived Thereof*

The regulatory process and granting of approvals by IBSC/RCGM/GEAC for GEd products/organisms/processes will depend on the purpose for which approvals are sought and the extent of modification(s) introduced and risk levels of the resulting products/organisms/processes. Table 2 describes the regulatory approval and regulatory pathways for approval of GEd organisms and/or products derived thereof in India.

The regulatory consideration pathways for GEd plants, animals, and human GEd stem cells and products derived thereof are given in Figs. 1, 2, and 3 respectively.

**Table 2** Regulatory framework for genome edited organisms/products derived thereof in India

| Statutory committee for authorization | GEd researchand product development | **Toward regulatory approvals** for release of GEd organism/cells/products | | |
|---|---|---|---|---|
| | | **GEd plants** | **GEd animals:** laboratory animals and livestock | **Human stem cells:** Gene therapy (Somatic stem cells) |
| **IBSC** | All research and product development experiments related to GEd Group I (Plants, animals/human stem cells) | GEd plants and products derived from Group I experiments (plants) | IBSC to recommend to RCGM after evaluation of molecular characterization data of Group I, Group II, and Group III (Animals/human stem cells) | |
| **RCGM** | All research and product development experiments related to GEd Groups II and III (Plants, animals/ human stem cells) | RCGM to recommend to GEAC based on molecular characterization data and contained/confined trial data of GEd plants or product(s) of Group II and III experiments and GEd animals falling under Group I, II, and III experiments | | RCGM to recommend to CDSCO based on PCT studies |
| **GEAC** | | GEd organisms and products derived from Group II and III experiments on plants and Group I, II, and III experiments on animals/human stem cells for environmental release | | |
| **Statutory market authorization agency** | | MoA&FW, GoI, FSSAI | CPCSEA, FSSAI, DAHR, MoA&FW, GoI | CDSCO, MoA&FW, GoI |

## National Biodiversity Act

The National Biodiversity Act 2002 (National Biodiversity Act of India 2002) regulates the use of biological resources of India, including genes used for improving crops and livestock through genetic interventions. The Act covers conservation, use of biological resources, and associated knowledge in India for commercial or research purposes or for the purposes of bio-survey and bio-utilization. It provides a framework for access to biological resources and sharing the benefits arising out of such access and use. The Act also includes in its ambit the transfer of research results and application for intellectual property rights (IPRs) relating to Indian biological resources.

The Act covers foreigners, nonresident Indians (NRI), body corporate, associations, or organizations that are either not incorporated in India or incorporated in India with non-Indian participation in their share capital or management. These

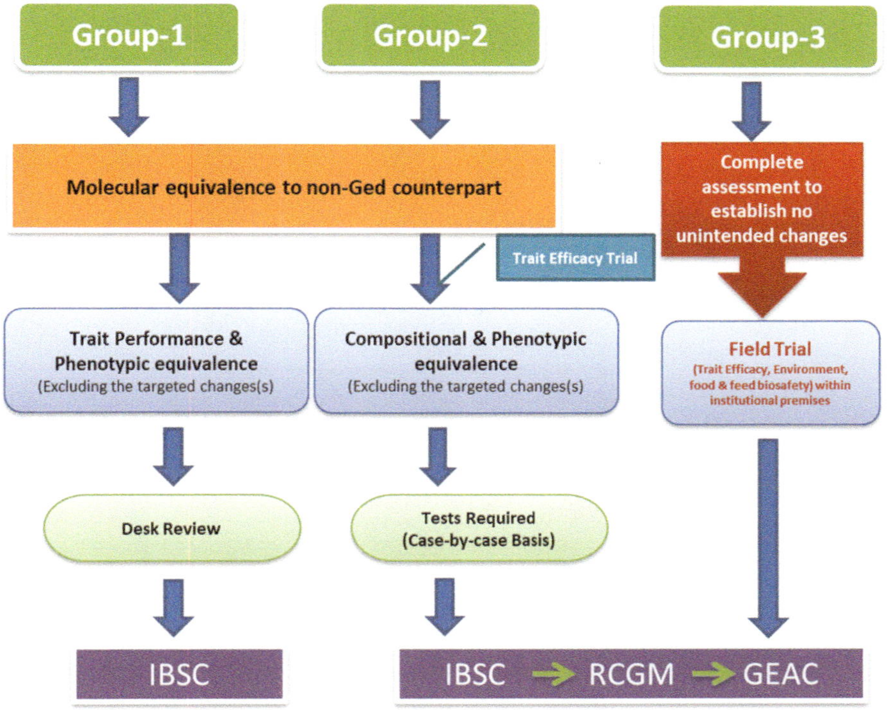


**Fig. 1** Regulatory consideration for genome edited plants and products derived thereof in India

individuals or entities require the approval of the National Biodiversity Authority (NBA) when they use biological resources and associated knowledge occurring in India for commercial or research purposes or for the purposes of bio-survey or bio-utilization.

Indians and Indian institutions, however, do not require the approval of the NBA when they engage in the above activities. Nevertheless, they would need to inform the State Biodiversity Boards prior to undertaking such activities. Any commercial application related to use of biological resources must still be approved by the authority (Approval requirements of different entity in India 2002).

## *NBA Application Process*

Several application processes are required to fulfil the NBA requirements for approval (Application forms for obtaining approval of NBA2002), which are summarized in Table 3.

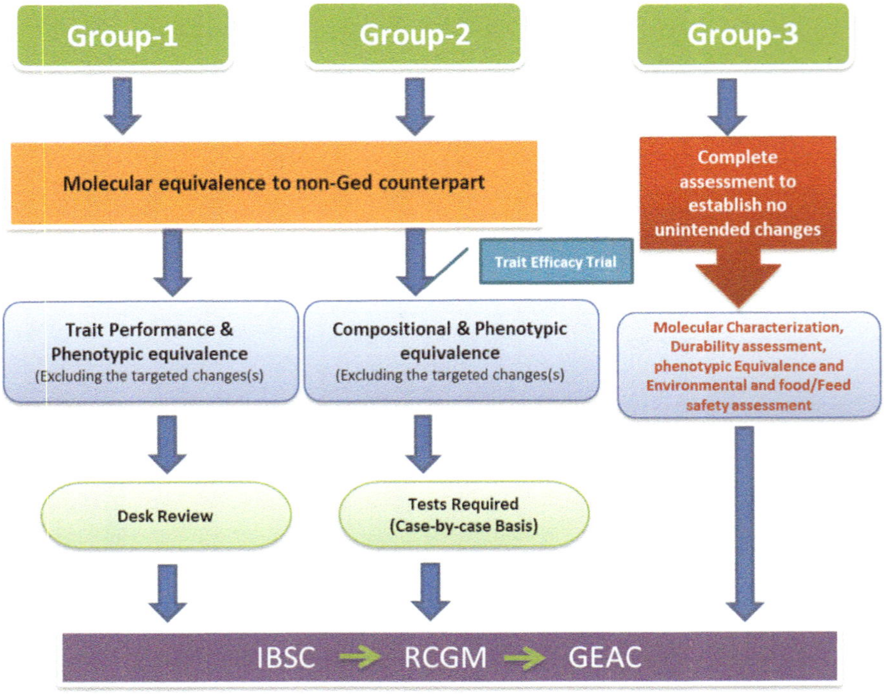


**Fig. 2** Regulatory consideration for genome edited animals and products derived thereof in India

## *Sections of the Indian Biodiversity Act*

For the purpose of access to Indian biological resources and their commercialization, the National Biodiversity Act has regulations defined in various sections (National Biodiversity act 2002 and Biological Diversity rules 2004) Some of the important sections are the following:

- *Section 3*, which defines who should be taking prior approval from the National Biodiversity Authority for accessing biological resources occurring in India for commercial utilization, research, or bio-survey and bio-utilization
- *Section 4*, which indicates that prior approval is also needed if any person wants to transfer the results of the research relating to Indian biological resources to the person falling under Section 3
- *Section 5*, which defines rules for exemption related to transfer or exchange of biological resources or associated information for the research institutions in India having collaborative research project with similar institution(s) in/of other countries
- *Section 6*, which defines rules for obtaining prior approval from the National Biodiversity Authority for applying intellectual property right (IPR) or any

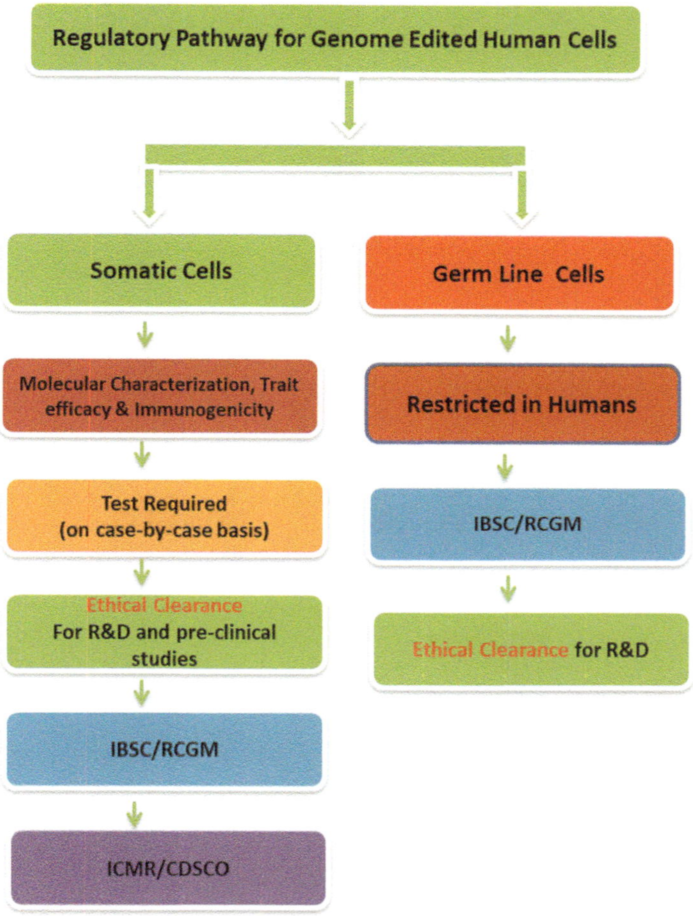


**Fig. 3** Regulatory consideration for genome edited cells (gene therapy product)

invention based on any research or information on a biological resource obtained from India

- *Section 7*, which indicates that an Indian entity, except few who are under exemption category, needs to give prior intimation to the relevant State Biodiversity Board for commercial utilization and/or bio-survey and bio-utilization of Indian biological resources
- *Section 40*, which states that pursuant to central government notifying any item including biological resources normally traded as commodities under this Section, the same becomes exempted from the purview of the Act

**Table 3** Application process for getting approval from NBA

| Application forms | Purpose of application | Who should apply |
|---|---|---|
| Form I | Access of biological resources occurring in or obtained from India and/or associated traditional knowledge for research, commercial utilization, bio-survey, or bio-utilization | Non-Indian, NRI, Foreign entity, or Indian entity having non-Indian participation in share capital or management |
| Form II | Transfer the results of research | Any Indian/non-Indian or entity to any non-Indian, NRI, foreign entity, or Indian entity having non-Indian participation in share capital or management |
| Form III | Applying for intellectual property rights for inventions based on any research or information on a biological resource obtained from India | Any Indian/non-Indian or entity |
| Form IV | Transfer of biological resources/knowledge already accessed to a third party | Any person who obtained approval of NBA in Form I, to Indians/ non-Indians entities |
| Form A | If the applicant is a trader/manufacturer/ company, he/she shall submit along with Form I, as per regulation 2 of ABS Guidelines (2014) | Trader/manufacturer/company |
| Form B | Conducting of noncommercial research or research for emergency purpose outside India by Indian researchers/government institutions, as per regulation 13 of ABS Guidelines (2014) | Indian researchers/government institutions |
| Form C | Deposition of microorganism in non-Indian repository for claim of novel species | Indian scientist/researchers |

## Convention on Biological Diversity and Synthetic Biology Regulation Under the Nagoya Protocol

The Convention on Biological Diversity (CBD), known as the Biodiversity Convention, is a multilateral treaty, which entered into force on 29 December 1993. The Convention has three main objectives (Draft Document on Genome Edited Organisms: Regulatory Framework and Guidelines for Risk Assessment 2020):

1. Conservation of biological diversity or biodiversity
2. Sustainable use of its components
3. Fair and equitable sharing of benefits arising from genetic resources

These objectives are meant to develop national strategies for the conservation and sustainable use of biological diversity. CBD has two supplementary agreements – Cartagena Protocol and Nagoya Protocol. Cartagena Protocol on Biosafety (CPB) and the Nagoya Protocol dealing with Access to Genetic Resources and the Fair and Equitable Sharing of Benefits Arising from their Utilization (ABS).

The Nagoya Protocol on ABS was adopted on 29 October 2010 (Draft Document on Genome Edited Organisms: Regulatory Framework and Guidelines for Risk Assessment2020) and covers traditional knowledge (TK) associated with genetic resources that are covered by the CBD and the benefits arising from its utilization, providing a transparent legal framework for the effective implementation of one of the key objectives of the CBD. The Nagoya Protocol creates incentives to conserve and sustainably use genetic resources and therefore enhances the contribution of biodiversity to development and human well-being.

As synthetic biology research drew attention at the global level, representatives of 196 nations joined and discussed its regulation at a CBD Conference of the Parties 13 (COP 13) meeting in 2015; the Parties focused on a critically important question: was the use of digital sequence information from genetic resources in foreign countries subject to the access and benefit-sharing requirements of the Nagoya Protocol and the fair and equitable sharing of benefits from their utilization? Based on the precautionary approach adopted by the meeting, the representatives considered the potential positive and negative impacts of components, organisms, and products resulting from synthetic biology techniques on the conservation and sustainable use of biodiversity (Manheim 2016).

## Synthetic Biology Research and Development in India

Synthetic biology research and development is at an initial stage in India. Currently, there is no product developed through such research in India that has reached the market. However, the Government of India, through the Department of Biotechnology (DBT), initiated several schemes to encourage research and development in this field, so there is a great interest in India to promote the field using cutting-edge research for the development of advanced technologies. Major initiatives will fund research in clean energy, health, agriculture, and bioremediation (Fig. 4); some of

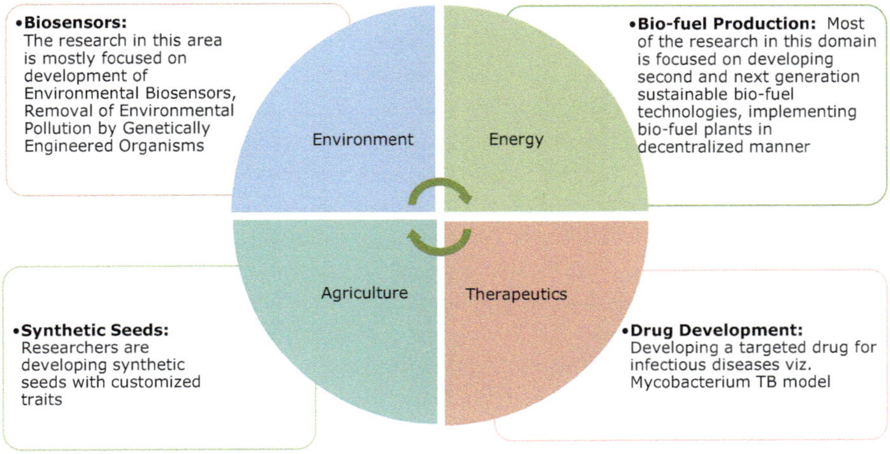


**Fig. 4** Major research initiatives in India in the field of synthetic biology

the academia (Supplementary Table S1) and industries (Supplementary Table S2) are involved in the synthetic-biology-related areas in India, and few special centers of excellence have been developed to perform dedicated research in the area of clean energy using synthetic biology (centers of excellence developed by DBT to perform dedicated research in the area of clean energy), namely, the DBT-ICGEB Centre for Advanced Bioenergy Research, DBT-ICT Centre for Energy Biosciences, and PAN-IIT Centre for Bioenergy. Several workshops and training program related to synthetic biology activities have also been funded (B4 Young Scientist Program, Workshop on Synthetic Biology2018; DBT Sponsored Training Program in Synthetic Biology2018). DBT has also been supporting seed funding to undergraduate/postgraduate students to participate at the popular iGEM's synthetic biology competition (DBT's support for iGEM's Giant Jamboree Synthetic Biology Event; DBT's initiative for Indian Biological Engineering Competition (iBEC)). Several international exchange programs have been supported to give exposure to students and researchers to advance the field (DBT's Indo-US Genome Engineering/Editing Technology Initiative (GETin) program; DBT and Indo-US Science and Technology Forum's joint initiative Bioenergy-Awards for Cutting Edge Research (B-ACER)). The Biotechnology Industry Research Assistance Council (BIRAC), a Public Sector Undertaking of DBT, which supports translational and industrial projects to public or private enterprises, has also encouraged technology development in the area (Biotechnology Industry Research Assistance Council (BIRAC) initiative for technology development in the area of synthetic biology).

## Synthetic Biology Rules and Regulations in India Related to Biological Diversity

The intense funding efforts from the government are likely to bear fruit in the near future with synthetic biology products starting to appear in the market. India currently believes the existing rules and regulations for research, development, deployment, and commercialization of GMOs are exhaustive enough to be applied to organisms, components, and products of synthetic biology at various stages of their development. These rules and regulations are governed by a three-tier mechanism (see Section on "Competent Authorities") in order to approve research and development on recombinant DNA products, environmental release of genetically engineered (GE) crops, and monitoring and evaluation of research activities involving recombinant DNA technology (regulations with respect to research activities involving recombinant DNA technology). Nevertheless, various meetings and workshops are organized and supported to take account of current developments and their impact on biological diversity (Special Seminar Series "Synthetic Biology – Policy and Implementation Issues" 2017). A draft regulatory framework has also been made for genome edited organisms (Draft Document on Genome Edited Organisms: Regulatory Framework and Guidelines for Risk Assessment 2020). India is

**Supplementary Table S1** List of academic players in India working on synthetic biology research

| S. no. | Institute/university | Technology | Application | Resource link |
|---|---|---|---|---|
| 1 | Centre of Energy Biosciences, ICT Mumbai | Synthetic metabolic pathways construction | Synthesis of drop in biofuel | http://www.ictmumbai.edu.in/DepartmentHome.aspx |
| 2 | Center of Innovative and Applied Bioprocessing, Mohali, Punjab | Metabolic engineering | Ethanol and fatty acid production | https://www.researchgate.net/profile/Ashok_Pandey5/info |
| 3 | Centre for Biosystems Science and Engineering (BSSE) – IISc, Bangalore | Biological-synthetic interface | Drug development | http://www.be.iisc.ernet.in/~siddharth/ |
| 4 | DBT-ICGEB Centre for Advanced Bioenergy Research, New Delhi | Synthetic biology/metabolic engineering | Biofuel production | https://www.icgeb.org/pavan-jutur.html |
| 5 | Dept. of Zoology, Delhi University | Genetic-synthetic strategy | Drug development | http://www.jbc.org/content/289/30/21142.long |
| 6 | Delhi University | Genetic engineering/synthetic gene | Strain development | http://dbtepromis.nic.in/bindcurrentyear.aspx |
| 7 | IISER, Pune | Synthetic biology | Building networks of genes-proteins and simulating their behavior | http://blogs.plos.org/collections/igem-report-005/ |
| 8 | IIT Delhi | Protein engineering/nano-engineering | Custom-designed protein for genome editing/joint replacements | http://web.iitd.ac.in/~sundar/ |
| 9 | IIT Kanpur | Synthetic chaperones | Drug discovery | http://www.iitk.ac.in/bsbe/arun-k-shukla |
| 10 | IIT Madras | Oscillators and network design | Metabolism | https://home.iitm.ac.in/kraman/lab/research/ |
| 11 | IIT Bombay | Genetic engineering | Biofuel | http://dbtepromis.nic.in/bindcurrentyear.aspx |
| 12 | Invertis University, Bareilly, Uttar Pradesh | Gene circuit | Biosynthetic pathway | https://www.omicsonline.org/editor-profile/Vijai_Singh/ |
| 13 | JNU | Synthetic biology/metabolic engineering/system biology | Biomolecule production/health/energy/environment | http://dbtepromis.nic.in/bindcurrentyear.aspx |

(continued)

**Supplementary Table S1** (continued)

| S. no. | Institute/university | Technology | Application | Resource link |
|---|---|---|---|---|
| 14 | Madurai Kamaraj University | Metabolic engineering | Biofuel | http://dbtepromis.nic.in/bindcurrentyear.aspx |
| 15 | National Chemical Laboratory (NCL), Pune | Synthetic biology | Biosynthesis of isoprenoids | http://academic.ncl.res.in/hv.thulasiram |
| 16 | National Institute for Interdisciplinary Science and Technology (NIIST), Trivandrum | Metabolic engineering | Ethanol and fatty acid production, biofuel | http://journal.frontiersin.org/article/10.3389/fenrg.2017.00008/full |
| 17 | National Institute for Interdisciplinary Science and Technology (NIIST) | Metabolic engineering | Ethanol and fatty acid production, biofuel | http://journal.frontiersin.org/article/10.3389/fenrg.2017.00008/full |
| 18 | NCBS, Bangalore | Genetic networks | – | http://rsta.royalsocietypublishing.org/content/roypta/371/1984/20110548.full.pdf |
| 19 | NCCS, Pune | Synthetic signaling circuits | Pathology | http://www.venturecenter.co.in/nccs/capabilities/436#more-436 |
| 20 | Pune University | Bio-bricks | Construction of new biological system | http://www.unipune.ac.in/snc/institute_of_bioinformatics_and_biotechnology/ibb_webfiles/pdf/iGEM1.pdf |
| 21 | Rajiv Gandhi Centre for Biotechnology, Trivandrum | Metabolic engineering | Ethanol and fatty acid production, biofuel | http://journal.frontiersin.org/article/10.3389/fenrg.2017.00008/full |
| 22 | Saha Institute of Nuclear Physics, Kolkata | "Bio-bots" | | https://sites.google.com/site/sbaghsyntheticbiology/ |
| 23 | University of Kerala, Thiruvananthapuram | Programming languages | Creating synthetic organisms | https://link.springer.com/article/10.1007/s11693-011-9070-y |
| 24 | International Crops Research Inst. For the Semi-Arid Tropics | Metabolic engineering | Food fortification | – |
| 25 | University of Hyderabad | Metabolic engineering | Biomolecule production | http://dbtepromis.nic.in/bindcurrentyear.aspx |
| 26 | Tamil Nadu Agricultural University | Genetic engineering | Pest resistance | – |
| 27 | University of Calcutta | Synthetic biology | Nanotechnology | http://dbtepromis.nic.in/bindcurrentyear.aspx |

**Supplementary Table S2** List of industrial players in India working on synthetic biology research

| S. no. | Company | Technology | Application | Resource link |
|---|---|---|---|---|
| 1 | FIB-SOL Life Technologies, Chennai | Nanofiber substrates | Organic farming | http://www.genomecompiler.com/bio-incubator-synthetic-biology/ |
| 2 | Indian Oil Corporation R&D Centre, Faridabad | Metabolic engineering | Biobutanol production | https://www.ris.org.in/sites/default/files/pdf/DP%20194%20Ravi%20Srinivas.pdf |
| 3 | Innovation Labs, TCS, Hyderabad | Programming languages | Design and construction of organisms | https://www.ncbi.nlm.nih.gov/pmc/articles/PMC3065592/ |
| 4 | Pondicherry Biotech Private Limited, Puducherry | Biodesign | DNA synthesis techniques | http://science.sciencemag.org/content/344/6179/55 |
| 5 | Praj Industries Limited, Pune | Synthetic biology | Cellulosic ethanol production | http://greeneconomypost.com/synthetic-biology-biofuel-biochemical-company-17244.htm |
| 6 | Purius Nanosystems, Chennai | Synthetic biology | Drug discovery, simulation, molecular modeling | http://www.genomecompiler.com/bio-incubator-synthetic-biology/ |
| 7 | Travancore Analytics, Technopark, Trivandrum, India | Programming languages | Design and construction of organisms | https://site2corp.com/in/travancore-analytics-private-limited/news |
| 8 | Vital Bioscientific Solutions, Chennai | Synthetic biology | Drug discovery, simulation, molecular modeling | http://www.genomecompiler.com/bio-incubator-synthetic-biology/ |
| 9 | Yaathum Biotech, Chennai | Computational and synthetic biology | Diagnosis and drug development | http://www.genomecompiler.com/bio-incubator-synthetic-biology/ |
| 10 | Sea6 Energy | Synthetic biology | Biofuel | https://patentscope.wipo.int/search/en/detail.jsf?docId=WO2013050860&recNum=44&docAn=IB2012001967&queryString=ALL:(algae%20production)&maxRec=35225 |
| 11 | Evolva | Synthetic biology | Health, wellness, and nutrition | http://www.evolva.com/mission-strategy/ |

represented in the Ad Hoc Technical Expert Group (AHTEG) on Synthetic Biology of the CBD to discuss the impact of synthetic biology on biological diversity (Convention of Biological Diversity Portal on Synthetic Biology). The country was represented in meetings in Montreal in 2015 and 2017, contributed to the synthesis of meeting documents, and was proactive in providing feedback (India's submission on synthetic biology in response to CBD Notification No. 86375, 2017). The Indian Council of Medical Research (ICMR), the apex body in India for the formulation, coordination, and promotion of biomedical research under the Ministry of Health and Family Welfare, has given special considerations in guidelines for synthetic-biology-related research and development in terms of (a) precautionary principle to prevent harm to humans, the environment, and ecosystem; (b) biosecurity to scrutinize the product of dual use, one beneficial use for a particular purpose and the other for harmful use, for example, as a biological weapon; (c) observing GLP, GMP, and GCP when conducting clinical trials; (d) contained and stepwise release into the environment after taking clearance from appropriate authority for its safety; etc. (ICMR's Ethical Guidelines 2017). The larger consensus within the government bodies is to evaluate research activities pertaining to synthetic biology more intensively and assess the gravity of potential risks on case-by-case basis.

## Conclusions

Research and development in genetic engineering and biotechnology in India took a giant leap when in 1986 a decision was made to create Department of Biotechnology (DBT) within the Ministry of Science and Technology (MoS&T). India now stands on a solid platform to innovate in various social areas, such as health care, food and agriculture, energy, and environmental security. With the efforts of DBT and the Ministry of Environment and Forest (MoEF), a regulatory framework has been established for R&D work on GMOs and their exploitation and release.

In new field of genome editing technologies, draft guidelines on genome edited organisms have been prepared by DBT. This comprises applicable laws, acts, and procedures governing genome editing and general considerations and takes a tiered approach to risk assessment of genome edited organisms and products, a regulatory approval road map, data requirements for risk assessment, and institutional mechanisms for governance and oversight. Further, India became a Party to the CBD in 1994, and in 2003, a National Biodiversity Authority was established under MoEF to facilitate, regulate, and advise the government on conservation, the sustainable use of biological resources, and fair and equitable sharing of benefits arising out of the use of those resources. Although synthetic biology research in India is at a preliminary level, it is possible that with the intense funding support and with the several regulatory frameworks now in place, the field will advance quickly in India.

However, it is important that discussions in various fora take place time to time to evaluate the benefits and potential risks of research outcomes so that development in the field is not hindered.

## References

Application forms for obtaining approval of NBA (2002). http://nbaindia.org/content/26/59/1/forms.html

Approval requirements of different entity in India (2002). http://nbaindia.org/text/12//TheBiologicalDiversityAct2002.html

B4 Young Scientist Program, Workshop on Synthetic Biology (2018). http://dbtindia.gov.in/sites/default/files/B4-Synthetic-Bio-workshop-Flyer_Final-1.pdf

Biotechnology Industry Research Assistance Council (BIRAC) initiative for technology development in the area of Synthetic Biology. http://www.birac.nic.in/cfp_view.php?id=15&scheme_type=16

Centres of excellence developed by DBT to perform dedicated research in the area of clean energy. http://dbtindia.gov.in/schemes-programmes/research-development/energy-environment-and-bio-resource-based-applications-1

Convention of Biological Diversity Portal on Synthetic Biology. https://bch.cbd.int/synbio/

DBT and Indo-US Science and Technology Forum's joint initiative Bioenergy-Awards for Cutting Edge Research (B- ACER). https://www.iusstf.org/program/bioenergy-awards-for-cutting-edge-research

DBT's Indo-U.S. Genome Engineering/ Editing Technology Initiative (GETin) program. http://dbtindia.gov.in/whats-new/call-for-proposals/indo-us-genome-engineeringediting-technology-initiative-getin31st

DBT's initiative for Indian Biological Engineering Competition (iBEC) competition. http://dbtindia.gov.in/schemes-programmes/building-capacities/awards/indian-biological-engineering-competition-ibec-pre

DBT support for iGEM's Giant Jamboree Synthetic Biology Event. http://www.dbtindia.nic.in/student/

DBT Sponsored Training Program in Synthetic Biology (2018).http://www.ncaah.org/component/content/article/54-news-updates/231-trainingprograminsynteticbiologymarch2018

Draft document on genome edited organisms: regulatory framework and guidelines for risk assessment (2020). http://dbtindia.gov.in/sites/default/files/Draft_Regulatory_Framework_Genome_Editing-9jan2020.pdf

ICMR's Ethical Guidelines (2017). https://www.icmr.nic.in/sites/default/files/guidelines/ICMR_Ethical_Guidelines_2017.pdf

India's submission on synthetic biology in response to CBD Notification No. 86375 (2017). https://bch.cbd.int/database/attachment/?id=17595

Manheim B (2016) Regulation of synthetic biology under the Nagoya Protocol. Nat Biotechnol 34:1104–1105. About the Nagoya Protocol. https://www.cbd.int/abs/about/

National Biodiversity act 2002 and Biological Diversity rules, 2004 (2004). http://nbaindia.org/uploaded/act/BDACT_ENG.pdf

National Biodiversity Act of India (2002). http://nbaindia.org/content/25/19/1/act.html

Regulations and Guidelines for Recombinant DNA Research and Biocontainment (2017). http://dbtindia.gov.in/sites/default/files/uploadfiles/Regulations_%26_Guidelines_for_Reocminant_DNA_Research_and_Biocontainment%2C2017.pdf

Regulations with respect to research activities involving recombinant DNA technology. http://dbtindia.gov.in/guidelines-biosafety#

Rules applicable to the manufacture, import and storage of micro-organisms and gene- technological products (1989). http://geacindia.gov.in/resource-documents/biosafety-regulations/acts-and-rules/Rules-for-the-manufacture-use-import-export-and-storage-1989.pdf

Rules for the manufacture, use/import/export and storage of hazardous micro-organisms/genetically engineered organisms or cells (1989). http://geacindia.gov.in/resource-documents/biosafety-regulations/acts-and-rules/Rules-for-the-manufacture-use-import-export-and-storage-1989.pdf

Special Seminar Series "Synthetic Biology – Policy and Implementation Issues" (2017). http://tdu.edu.in/wp-content/uploads/2017/07/seminar1_21july_report.pdf

The Environment (Protection) Act (1986). https://indiacode.nic.in/bitstream/123456789/4316/1/ep_act_1986.pdf

# Malaysia—Genetic Modifications and Synthetic Biology Regulations and Biodiversity – A Legal Perspective for Malaysia

**Mohamad Faiz Foong Abdullah and Mohana Anita Anthonysamy**

**Abstract** Biodiversity is a major source of wealth for Malaysia, and various laws and regulations are in place to protect this wealth. The use of genetically modified organisms (GMOs) in agriculture and other areas however may pose certain threats to biodiversity. The legal framework to regulate the use of GMOs and to manage the risks arising from their introduction into the environment is discussed from the Malaysian perspective. Future challenges are also discussed.

**Keywords** Biodiversity protection · Regulation of biotechnology · Living modified organisms · Biosafety law · Risk assessment · Contained use · Open release · Synthetic biology · Socioeconomic impacts

## Introduction

Malaysia is recognized as one of the 17 megadiverse countries in the world by the Convention of Biological Diversity (CBD). It has an extremely rich and highly diverse biological resource, with about 25,000 plant species, 15,000 of which are endemic, and 2795 vertebrate species, of which 1103 are endemic. This biodiversity plays a large role in the country's socioeconomic development, providing food, materials, natural products for sustenance, and a large gene pool for future food plant research. It also acts as a buffer to climate change. Protecting and conserving natural biodiversity remain a top priority for Malaysia (National Policy on Biological Diversity 2016–2025).

Advances in plant breeding and agrotechnology, however, may have a disruptive effect on this biodiversity. The monoculture approach of modern agriculture, if not

M. F. F. Abdullah (✉)
Faculty of Applied Sciences, Universiti Teknologi MARA, Shah Alam, Selangor, Malaysia
e-mail: mohdf184@uitm.edu.my

M. A. Anthonysamy (✉)
Department of Biosafety Malaysia, Ministry of Environment and Water, Putrajaya, Malaysia

 
A. Chaurasia et al. (eds.), *GMOs*, Topics in Biodiversity and Conservation 19,
https://doi.org/10.1007/978-3-030-53183-6_33

managed properly, can replace existing biodiversity with an environment much reduced in variety species. Malaysia has in place various measures to ensure that these effects are minimized, including strict regulations on plantation planning and management. The use of genetically modified (GM) crops raised further concerns on their possible invasiveness and impact on existing ecosystems (Carpenter 2011). GM crops engineered with advantageous traits, such as disease resistance, may displace local species, reducing the available gene pool.

There is yet little information on the long-term impacts of GM crops on biodiversity and the ecosystem. The history of GM crop use has some notable success stories, e.g., GM cotton in India, and equally controversial issues, e.g., effect on non-target organisms and possible cause of chronic disease (Raman 2017). Globally, there is collective agreement among almost all countries to protect the earth's biodiversity as described in the Cartagena Protocol on Biosafety (CPB). Countries, however, draw their own standards with regard to how much protection is necessary, depending on the level of "biotechnology friendliness" of their populace. Implementation of the provisions in the CPB has also been problematic both technically and politically (Lim 2007). Together with the advent of a slew of "new breeding techniques", the future regulation of GM crops and organisms remains highly challenging.

Here, we discuss the implementation of the CPB in Malaysia, the challenges they posed, the success achieved, and possible measures for improvement.

## The Malaysian Biosafety Act 2007

Biodiversity conservation in Malaysia is guided by the National Policy on Biological Diversity for sustainable use. Target 12 of this policy aims to have a comprehensive biosafety system (inclusive of a liability and redress regime) which is operational to manage potential adverse impacts of modern biotechnology on biodiversity and human health. The responsible agency for biodiversity is the Division of Biodiversity, under the Ministry of Energy and Natural Resources. Modern biotechnology is regulated by the Biosafety Act 2007 with the objectives of protecting human, plant, and animal health, the environment, and biological diversity. Under this law, any release (e.g., commercial use, planting, field trial, and disposal), contained use, and import or export activities of living modified organisms (LMOs) or their products must obtain approval of the National Biosafety Board (NBB). The NBB comprises representatives from six relevant ministries and is headed by the secretary-general of the Ministry of Environment and Water. In addition, there are four expert NBB members who have the knowledge and experience of biosafety to strengthen the decision-making capabilities of the NBB.

## Regulation of LMOs and Modern Biotechnology in Malaysia

Under the Biosafety Act 2007, LMOs are defined as "any living organism that possesses a novel combination of genetic material obtained through the use of modern biotechnology." Modern biotechnology is defined as the application of (a) in vitro nucleic acid techniques, including recombinant DNA and direct injection of the nucleic acid into cells or organelles, or (b) fusion of cells beyond the taxonomic family of the organism that overcome natural physiological reproductive or recombination barriers and that are not techniques used in traditional breeding and selection. This replicates the definition used under the CPB. The Act covers all activities that involve LMOs. Products derived from LMOs are also regulated by it, with the exception of products classified as pharmaceuticals. The Act, however, empowers the minister-in-charge to exempt certain categories of products or activities which are considered to be of very low risk. In general, the Biosafety Act 2007 is very much in line with the provisions of the CPB and its guidelines.

Release and contained use of LMOs and products of modern biotechnology are subjected to a process of risk assessment, and appropriate risk mitigation measures are required to be in place to prevent, reduce, or control the risks and any possible adverse effects that LMOs and its products will have (or likely to have) on human, plant, and other animal health, the environment, and biological diversity. Activities involving LMOs intended to be in "contained use" situations (e.g., laboratory research, greenhouse trials) are prevented from coming into contact with the external environment. The risk assessment must show that the risk of unintentional release and to human health is minimal before the NBB grants approval. LMOs intended for open release (e.g., open-field trials, plantations) go through a more stringent risk assessment to consider all possible impacts to the environment in a holistic manner. Risk assessments are conducted by a Genetic Modification Advisory Committee (GMAC), a panel of experts comprising academicians, scientists, and representatives from industry and nongovernmental organizations. The GMAC provides scientific and technical advice to the NBB on the risks of an application and whether sufficient mitigation measures are being taken. A decision for open release takes into account the views of all stakeholders and also includes a public consultation process. The NBB makes the final decision for approval, after taking into consideration input from GMAC and also other aspects, i.e., socioeconomic and existing policy factors. The NBB can also impose terms and conditions where necessary to further mitigate any residual risks.

Until the first quarter of 2019, Malaysia had approved the importation of 36 LMOs for the purpose of food, feed, and processing (FFP). These include 18 cultivars ("varieties") of GM maize, 11 of GM soybean, 4 of GM cotton, 2 of GM canola, and 1 of GM potato (Table 1). A number of these contained stacked events, i.e., two or more genetic modifications combined in a single cultivar. Risk assessment for stacked events includes an additional step of evaluation for possible interactions among the events. Besides these, two LMO products were approved for open release as a biopesticide, and eight cultivars of carnation to be sold for

**Table 1** GM crop events in Malaysia approved for the purpose of food, feed and processing (to Feb. 2019)

| Crop | Identifier |
|---|---|
| Corn | DP4114, MON 87427, MZIR098, MZHG0JG, DAS-59122-7, 3272, GA21, MIR162, MIR604, 5307, MON88017, MON89034, TC1507, T25, SYN-Bt11-1, MON 863, MON 810, MON 603 |
| Soybean | DAS-68416-4, DAS-44406-6, DAS-81419-2, 305423, 305423, SYHT0H2, FG72, A5547-127, CV127, MON 89788, ACS-GM5-3, MON 4032 |
| Canola | DP73496 |
| Oilseed rape | MS8RF3 |
| Cotton | GHB119, LLCotton25, T304-40, GHB614 |
| Potato | Y9, E12 |

ornamental purposes in the market, two open-field trials of GM plants, and one field trial of GM mosquitoes have been approved. Besides approvals for open release, researchers using GMO in their work are also required to file a notification for contained use. The numbers of such notifications have increased substantially since the Biosafety Act was enforced, indicating an encouraging sign that the local research community has accepted the regulations and oversight as part of their scientific responsibility (Fig. 1). This has resulted in a challenge for the regulators to meet the proscribed timeframe for reviewing and approving an increasing number of applications.

## New Biotechnology Techniques

A number of new technologies for genetic manipulation and plant breeding have emerged in the past decade, and some of these hover in the gray area of existing regulations (Lassoued et al. 2018). Such technologies, including genome editing, RNA inhibition, transient expression, intragenesis, reverse and accelerated breeding, and synthetic biology, make use of conventional genetic engineering techniques, but the end products may or may not contain foreign genes or recombinant DNA fragments. Consequently, these technologies and their products may not fall within the scope of the Biosafety Act. The time required for producing, for example, a gene-edited crop is also considerably shorter than conventional techniques, posing further challenges to timely regulatory processes.

Currently, applications to conduct activities with these technologies are considered on a case-by-case basis, whereby the GMAC will advise applicants as to whether their activities are regulated under the Biosafety Act. In general, genome editing and other innovative techniques that result in the insertion of foreign DNA fragments, or require the use of exogenous DNA templates, are likely to fall into regulatory overview. Techniques that do not insert foreign DNA or require DNA

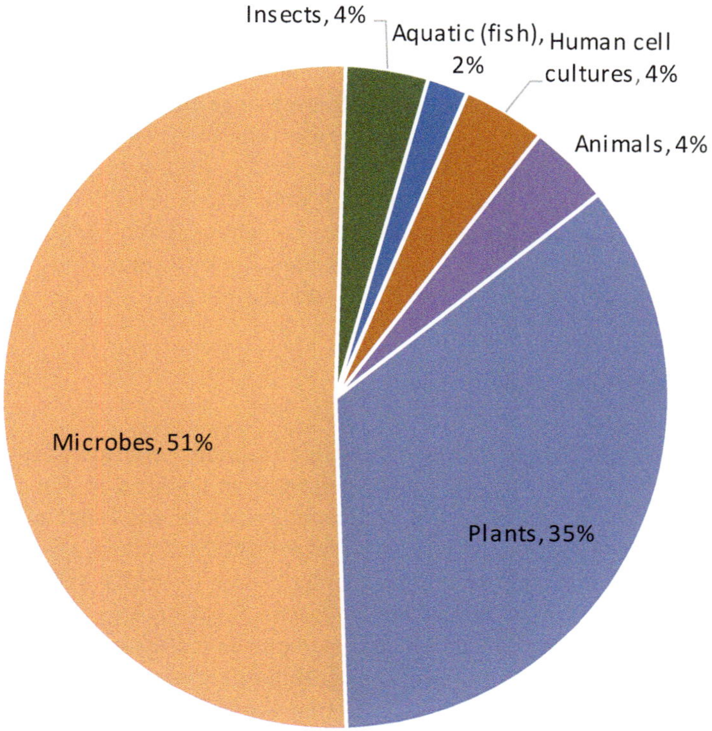


**Fig. 1** Contained use activities notified to the National Biosafety Board (to Feb. 2019)

templates, or the product can be shown to degrade and retain none of the exogenous nucleic acids, go through a preliminary assessment to determine if they are regulated under the current law. Discussions have been initiated to develop a more comprehensive workflow, and this is expected to be in place by the end of 2020 to guide applicants on the next steps.

## Synthetic Biology

Synthetic biology is an emerging technology that utilizes biotechnology and system engineering concepts to build new biological systems with novel or specialized functions. The technology has the potential to bring benefits to many fields, but the uncertain nature of the genetic techniques involved may necessitate a regulatory framework to reduce potential risks (Trump 2017). Current synthetic biology techniques are considered an extensions of modern biotechnology. If it is used to produce an LMO, however, it is clearly regulated under the Biosafety Act 2007. Nevertheless, recent advances have propelled synthetic biology to a state where small genomes can be synthesized and transformed into protoplasts that are able to self-replicate or even evolve. This poses a challenge for legislation, which is bound by the definition and scope described under the Biosafety Act 2007. Synthetic biology products pose a special challenge, as the notion of substantial equivalence may be blurred in the absence of a suitable comparator. The environmental impact of an organism heavily modified by synthetic biology is also elevated, especially in the form of gene drives that can rapidly and irreversibly push a species or variant to extinction. However, as the scope of the legislation also covers in vitro nucleic acid techniques, synthetic biology is deemed to be bound by the Biosafety Act 2007 and will be managed under the current laws. Similarly, an LMO produced via synthetic biology can be managed with the measures already in place.

## Socioeconomic Impacts

The Biosafety Act 2007 empowers the NBB to take into consideration local and regional socioeconomic impacts when evaluating an application. For instance, a technology that is highly disruptive to indigenous social and economic activities may be subjected to further scrutiny and management. Malaysia is also very careful to ensure there is minimal risk of cross-border escape for LMOs released into the environment. Both positive and negative impacts of a technology will be carefully considered for a sustainable scientific and entrepreneur ecosystem. One case which drew international attention was the use of GM mosquitoes as a control measure for vectors of dengue fever. A restricted field trial was carried out for data collection before a full release could be considered (Subramaniam et al. 2012).

For trade transparency, all approved events are listed in the official website of the Department of Biosafety (DoB 2019a) and two international databases – the Biosafety Clearing House (BCH 2019) under the Convention of Biological Diversity (CBD) and the FAO GM Foods Platform (FAO 2019). Food products containing GM ingredients are required to be labelled under the Food Act 1983 under the Ministry of Health, and guidance is provided through the "Guidelines on Labelling of Foods and Food Ingredients Obtained through Modern Biotechnology" (DoB 2019b). This ensures that consumers are given choices and prevents trade disputes.

National security is also a concern. Regulatory oversight for biosecurity and biodefence concerns is covered under a bioweapon and toxin bill. However, the anticipated scope of this law would be to cover microorganisms, including viruses, that have the potential to be used as bioweapons. This may not, however, cover all types of organisms that can be used in synthetic biology.

## Conclusion and Future Directions

The current risk assessment framework needs to be reviewed and/or adapted and/or a new framework be provided if it proves needed to assess organisms produced via new biotechnology techniques and synthetic biology. The assessment should be robust enough to assess not only work that involves the incorporation of heterologous genes into existing organisms but also the construction of novel life-forms where there are no parent organisms as comparators.

The major issue will be to determine if new biotechnologies and their products fall within the remit of the Biosafety Act and decide if there is a need to review or amend the Act and associated regulations. It is also imperative to consult the public and other stakeholders if there is a need to revise or expand the current protection goals.

A stepwise approach is needed whereby safety data and characteristics of the organisms produced through synthetic biology can be more properly understood, for example, in potential interactions with other organisms, impact of any horizontal gene transfer, and unforeseen evolution so that these data can be used for risk assessment and as a basis for making decisions.

Regardless of these concerns, we consider that the spirit of the Biosafety Act provides enough avenues to achieve a balance to safeguard biodiversity, human, plant, and animal health, and the environment without creating undue barriers to scientific innovation and entrepreneurship. Any new review of the laws and regulation will surely need to take these factors into consideration.

## References

BCH (2019) Biosafety clearing house Malaysia Country's decisions. https://bch.cbd.int/database/results?searchid=776061

Carpenter JE (2011) Impact of GM crops on biodiversity. GM Crops 2:7–23

DoB (2019a) Department of Biosafety Malaysia Official Website. http://www.biosafety.gov.my/en-my/Pages/default.aspx

DoB (2019b) Department of Biosafety Malaysia Official Website – publications/guidelines by other agencies. http://www.biosafety.gov.my/en-my/media-library/Pages/Publication.aspx

FAO (2019) FAO GM foods platform. http://www.fao.org/food/food-safety-quality/gm-foods-platform/browse-information-by/country/country-page/en/?cty=MYS

Lassoued R, Smyth SJ, Phillips PWB, Hesseln H (2018) Regulatory uncertainty around new breeding techniques. Front Plant Sci 9:1291–1301

Lim LL (2007) Cartagena protocol on biosafety. In: Travik C, Lim LC (eds) Biosafety first. Tapir Scad Press, Penang, pp 1–18

National Policy on Biological Diversity 2016–2025. Ministry of Energy and Natural Resource Malaysia

Raman R (2017) The impact of genetically modified (GM) crops in modern agriculture: a review. GM Crops Food 8(4):195–208

Subramaniam TSS, Lee HL, Ahmad NW, Murad S (2012) Genetically modified mosquito: the Malaysian public engagement experience. Biotechnol J 7:1321–1327

Trump BD (2017) Synthetic biology regulation and governance: lessons from TAPIC for the United States, European Union, and Singapore. Health Policy 121:1139–1146

# Pakistan—Synthetic Biology – Challenges and Opportunities from a Biodiversity Perspective in Pakistan

**Faisal F. Khan**

**Abstract** Synthetic biology is an emerging field that is set to revolutionise the global life sciences and biotechnology landscape. There are several, often broad, ways of defining this new field, but one of the more popular definitions is provided by the UK Royal Academy of Engineering which defines it as an emerging field that 'aims to design and engineer biologically based parts, novel devices and systems as well as redesigning existing, natural biological systems' (The Royal Academy of Engineering: Synthetic biology: scope, applications and implications. https://www.raeng.org.uk/publications/reports/synthetic-biology-report, 2009).

**Keywords** Biodiversity · GMOs · Conservation · Synthetic Biology · Pakistan · Genetic devices · Ecosystems · Engineered Organisms

## Background

Synthetic biology is an emerging field that is set to revolutionise the global life sciences and biotechnology landscape. There are several, often broad, ways of defining this new field, but one of the more popular definitions is provided by the UK Royal Academy of Engineering which defines it as an emerging field that 'aims to design and engineer biologically based parts, novel devices and systems as well as redesigning existing, natural biological systems' (RAE 2009).

From a biodiversity and conservation perspective, there is an emerging discourse on the interplay of conservation and synthetic biology and whether both disciplines undermine each other's efforts or whether there can be a mutually beneficial alliance. Several arguments are being made in support and against this idea (Redford et al. 2014; Piaggio et al. 2017). Firstly, the question of unintended outcomes, especially horizontal gene transfer (HGT), is posited as an environmental argument,

F. F. Khan (✉)
Institute of Integrative Biosciences, CECOS University, Peshawar, Pakistan
e-mail: faisal@cecos.edu.pk



A. Chaurasia et al. (eds.), *GMOs*, Topics in Biodiversity and Conservation 19,
https://doi.org/10.1007/978-3-030-53183-6_34

which means that synthetic genes (and therefore traits) may transfer, through the normal course of reproduction, from engineered organisms to wild-type ones in nature. Proponents argue that since these engineered organisms are exposed to the forces of natural selection, they will not be able to survive out in the wild, but the counter argument challenges by using the fact that there is a chance they may become invasive disturbing the ecosystem and potentially even resulting in biodiversity loss. Secondly, the threat posed by private ownership of engineered organisms and their products to the principle of access and benefit sharing is another argument made against the use of engineered organisms. If any of the engineered components, processes or approaches are legally protected as intellectual property (such as patents), it denies anyone else the right to use, produce or share it. Moreover, organisms can also be engineered to have 'genetic switches' that inhibit the use of, for example, their seeds beyond one generation which means the farmer needs to go back to the company and buy such seeds every year. This is sometimes very alarming, especially in indigenous, low-income communities, where something which belonged to nature thus far, such as seeds, and was open for all and available to everyone is now private property and cannot be used. With more than 80% of farmlands being smallholder in Pakistan, for example, this becomes a question of subsistence and a struggle for survival in low-income communities.

The custodianship of traditional knowledge on genetic material with regard to biological diversity and the ownership of local seeds, especially cereal crops, is a special concern from a perspective of a developing country. Another concern is related to land-use policies and the potential increase in demand for land for scale-up industrial biotechnology applications and increasingly synthetic biology applications. For example, biofuel production demands high volumes of sugar from sugarcane fields or uses microalgae and other cultivated plants. This not only poses a dire risk to natural habitats and biodiversity but also brings in the risks of land extension. This exacerbates the pressures of food security and to increase crop yields.

However, it can also be argued that HGT is a concern for engineered organisms meant for environmental release that can only be guarded against by adopting a precautionary principle and an exhaustive case-by-case risk assessment and mitigation process. There is also an issue in an increasingly open-access community that the new discipline of synthetic biology is evolving in a *democratic* manner and is being disseminated and propagated far and wide. This open culture might be challenging for companies which intend to capture and protect the intellectual property, as there is the potential to create a whole generation of *open-access* versions of seeds. Synthetic biology methods are, however, more specific when compared to traditional biotechnology, so decreasing the chance of unwanted and unintended effects.

Both sides of the argument for and against the use of engineered organisms and/or their products need an unbiased, evidence-based discourse particularly in a local context, looking at local realities, culture and ethical values. Decisions taken after such a discourse would then lead to more informed and contextualised decisions in the best interest of not just our flora and fauna but also human life. This will also ensure that we do not miss out on the potential economic revolution emerging technologies like synthetic biology bring with them which is also crucial for the socio-economic development of the country.

## State of Synthetic Biology in Pakistan

The current biotechnology landscape in Pakistan has been in place for a few decades now with the key institutes being the National Institute for Biotechnology and Genetic Engineering (NIBGE est. 1994), the Centre for Excellence in Molecular Biology (CEMB) in Lahore (est. 1981) and the International Center for Chemical and Biological Sciences (ICCBS), which is a complex of multiple institutes in Karachi (est. 1967). Although very nominal, NIBGE has the most documented commercialised products, mainly non-GM and GM crop cultivars ('varieties'). Amongst the more recently established departments and institutions, the Department of Biology in the Lahore University of Management Sciences, the Atta-ur-Rehman School of Biological Sciences at the National University of Science and Technology and the Department of Biotechnology at the Quaid-i-Azam University are the more prominent. In 2014, there are not more than 500 active biotechnology research groups in the country but with none of them working in synthetic biology (Malik 2014).

Despite the infrastructure being in place, synthetic biology has only emerged very recently in Pakistan and also where it was least expected. In spite of not being one of the major cities known for biotechnology, Peshawar, the capital city of the north-western province of Khyber Pakhtunkhwa, has pioneered synthetic biology in the country beginning with two successful participations in the International Genetically Engineered Machine (IGEM) competition in 2016 and 2017. IGEM is a flagship international student competition that began as a module at the Massachusetts Institute of Technology (MIT) in 2004 and quickly evolved into a global student competition for young synthetic biologists, with an annual Giant Jamboree in Boston every year. The IGEM Peshawar team in 2016 (IGEM Peshawar 2016), which included undergraduate students from all four provinces of the country, worked on a biosensor – a bacterial cell that could detect carbon monoxide and oxides of nitrogen and produce a coloured pigment; it won a bronze medal at the competition. In 2017, a second IGEM team from Peshawar developed a biosensor for arsenic in freshwater and won a silver medal (IGEM Peshawar 2017). The two teams and several outreach and training programmes were funded by the Directorate of Science and Technology in the provincial government of Khyber Pakhtunkhwa and CECOS University under the banner of the SynBioKP project (SynBioKP 2015). The project has won numerous awards and has managed to train several hundred university students and over 16,000 school and high school students in synthetic biology. Synthetic biology also appears as a track, recently for the third time in a row, at the National STEM School organised by the Pakistan Innovation Foundation (The National STEM School 2018) and hosted by the Lahore University of Management Sciences (LUMS) and other institutions. The winter school attracts around 30 of the brightest children from around the country for a 10-day residential science school. In 2017, some students from the synthetic biology track were inspired to go on and create the first IGEM high school team from Pakistan. The LACAS_Biobots team (IGEM LACAS Biobots 2018) from the Lahore College of Arts and Sciences (LACAS) worked on developing a synthetic version of 'mother of pearl' and attended the Giant Jamboree in Boston in 2018.

IGEM Peshawar alumni are also responsible for catalysing a movement of community biospaces, labs with mainly do-it-yourself (DIY) infrastructure set up outside an academic setting and open to public. Examples include Codon Corps (Rawalpindi), AbroBios (Hyderabad), House of Interdisciplinary Interaction (Karachi) and the Undivided BioArt Community (Peshawar).

With respect to foundational technologies such as next-generation DNA sequencing technologies, there are several institutes in the country which house a MiSeq machine (Illumina) including the Aga Khan University in Karachi, Lahore University of Management Science in Lahore and Rehman Medical Institute in Peshawar. However, no DNA synthesis facility yet exists in the country.

## Existing Regulations

The Pakistan Biosafety Rules of 2005 was notified by the government of Pakistan under the Pakistan Environmental Protection Act in 1997 (Ministry of Climate Change, Pakistan Biosafety Rules 2005b). National biosafety guidelines were also issued in the same year (Ministry of Climate Change, National Biosafety Guidelines 2005a). The biosafety rules and guidelines recommended setting up (a) an Institutional Biosafety Committee (IBC), (b) a Technical Advisory Committee (TAC) and (c) a National Biosafety Committee (NBC). All institutions, both public and private involved in research and development in biotechnology, are required to establish an IBC, which is responsible for giving clearance for initiating research according to the biosafety guidelines. Up till now, nearly 40 IBCs belonging to both the private and public sectors have been registered with the National Biosafety Center (NBC) established by the federal government.

Pakistan has been a signatory to the UN Convention on Biological Diversity (CBD) since 1992 (ratified by the Cabinet in 1994) and has been making reasonable progress at different policy level planning and mapping different interventions. To fulfil Pakistan's obligations to the CBD, the Government of Pakistan approved a Biodiversity Action Plan which not only provides a roadmap but also helps the country to monitor the progress in achieving the Aichi Biodiversity Targets.

The sixth National Biodiversity Strategy and Action Plan (NBSAP) was approved by the government of Pakistan in 2018 which also covers the issue of emerging new technologies including synthetic biology and stressed the need to develop a national position based on a an unbiased, inclusive and evidence-based discourse around the topic to weigh both its advantages and disadvantages and the need for more transparency, improved implementation and use of technology for surveillance and documentation of the local biodiversity. The government has recently established a Directorate of Biodiversity dedicated to work on implementation of the provisions of the NBSAP. It is expected that the Directorate of Biodiversity will be able to meet this by 2020.

## The Way Forward

In view of the current pace of developments in the area of synthetic biology, the following are proposed interventions that need to be made both at the strategy and execution level in a time-sensitive manner to address the challenges at the interface of synthetic biology, biodiversity and conservation in Pakistan.

### *Strong Political Will and a National Discourse*

The incumbent government, under Prime Minister Imran Khan, has shown increasing interest in tackling the challenge of climate change, as evidenced, for example, by the Billion Tree Tsunami project (IUCN News 2017). This trend in political will towards issues of environment is refreshing but not enough to deal with the challenges of synthetic biology. Serious concerted efforts need to be put in place, and a national discourse has to be developed around the issues we face. Although the Government of Khyber Pakhtunkhwa launched the SynBioKP project with a strong forward-looking approach, it still has to reach scale and do that fast in order to develop and build capacity in order to be able to understand and engage the challenges that accompany powerful, dual-use technologies.

### *Address Serious Capacity Issues*

The Ministry of Climate Change and the Biodiversity Unit both need serious investments at the federal level and also need to establish provincial wings that can achieve objectives in a devolved fashion. Concerned government departments and agencies are in dire need of highly qualified young professionals as the older ones either have retired or are close to retiring. Strong research and policy teams need to be present and have the expertise to (1) engage all stakeholders, (2) convene regular meetings and awareness events, (3) engage the media for public awareness and a national discourse, (4) publish position papers and primary research regularly and (5) represent Pakistan and share its position at international meetings such as the annual Conference of the Parties (COP) of the CBD.

## *Upgrade Curricula*

The curriculum in universities, schools and colleges would benefit from an upgrade. With tectonic developments in biology and emerging challenges such as climate change and biodiversity loss, a fresh perspective needs to be brought in on how we teach biology at all levels. Currently, for example, there is very limited textbook content in government and most private schools on the rich biodiversity of Pakistan, the challenges we face and emerging technologies such as synthetic biology.

## *Well-Planned Research Interventions*

Research needs to be conducted to identify, catalogue, biobank and protect the biota of the country using the latest of technologies available. For example, countries like Pakistan that are very rich in biodiversity can begin with fairly easier yet economically significant species such as spices, herbs, medicinal plants and local races of crops, fruits and vegetables.

**Disclosures** I was the founding director of the Institute of Integrative Biosciences, CECOS University, remain a faculty member there and have been the principal investigator of the SynBioKP and IGEM projects in Peshawar. This contribution is condensed from a paper submitted to the Department of Climate Change, Government of Pakistan, as an input to the sixth National Biodiversity Strategy and Action Plan in 2017.

# References

IGEM LACAS_Biobots (2018) Official wiki-page. http://2018.igem.org/Team:LACAS_BioBots

IGEM Peshawar (2016) Official wiki-page. http://2016.igem.org/Team:Peshawar

IGEM Peshawar (2017) Official wiki-page. http://2017.igem.org/Team:Peshawar

IUCN News (2017) Pakistan's Billion Tree Tsunami restores 350,000 hectares of forests and degraded land to surpass Bonn challenge commitment, August 2017. https://www.iucn.org/news/forests/201708/pakistan%E2%80%99s-billion-tree-tsunami-restores-350000-hectares-forests-and-degraded-land-surpass-bonn-challenge-commitment

Malik KA (2014) Biotechnology in Pakistan: status and prospects. Report by the Pakistan Academy of Sciences. http://paspk.org/wp-content/uploads/2015/12/Biotechnology-Report-2014.pdf

Ministry of Climate Change (2005a) National Biosafety Guidelines. Government of Pakistan. http://www.environment.gov.pk/desc.php?page=guidelines

Ministry of Climate Change (2005b) Pakistan biosafety rules 2005. Government of Pakistan. http://www.environment.gov.pk/desc.php?page=rules_regulations

Piaggio AJ, Segelbacher G, Seddon PJ et al (2017) Is it time for synthetic biodiversity conservation? Trends Ecol Evol 32(2):97–107

Redford KH, Adams W, Carlson R et al (2014) Synthetic biology and the conservation of biodiversity. Oryx 48(3):330–336. https://doi.org/10.1017/S0030605314000040

SynBioKP (2015) Official website. http://www.synbiokp.com

The National STEM School (2018) Pakistan Innovation Foundation website. http://pif.org.pk/stem-school/

The Royal Academy of Engineering (2009) Synthetic biology: scope, applications and implications. https://www.raeng.org.uk/publications/reports/synthetic-biology-report

# Thailand—GMOs and Synthetic Biology Regulations – Thailand Perspective

**Chalinee Kongsawat and Weerawat Runguphan**

**Abstract** Thailand has become a party to the Cartagena Protocol on Biosafety (CPB) since 8 February 2006. This protocol aims to ensure the safe handling, transport, and use of living modified organisms (LMOs) or genetically modified organisms (GMOs) derived from modern biotechnology that may have adverse effects on biological diversity. The protocol takes into account risks to human health and particularly those that arise from transboundary movements (Secretariat of the Convention on Biological Diversity 2000). The ratification ensures that Thailand can participate fully in this new multilateral approach to managing the potential risks of LMOs. Even though Thailand is in the process of enacting a specific law that details provisions for working with LMOs, there are existing legislations and guidelines to regulate the activities of LMOs as follows.

**Keywords** Thailand · GMOs · GM plant · GM microorganism · GM animal · Synthetic biology · Biosafety regulation

## GMOs Regulations in Thailand

Thailand acceded to the Cartagena Protocol on Biosafety (CPB) on 10 November 2005, and the agreement entered into force on 8 February 2006. The Office of Natural Resources and Environmental Policy and Planning (ONEP) in the Ministry of Natural Resources and Environment (MONRE) has been nominated as the National Focal Point and Biosafety Clearing House for the Protocol (Technical Biosafety Committee 2010). This protocol aims to ensure the safe handling, transport, and use of living modified organisms (LMOs) or genetically modified organisms (GMOs) derived from modern biotechnology that may have adverse effects on biological diversity. The Protocol takes into account risks to human health and particularly those that arise from transboundary movements (Secretariat of the

C. Kongsawat (✉) · W. Runguphan
National Center for Genetic Engineering and Biotechnology (BIOTEC), National Science and Technology Development Agency (NSTDA), Ministry of Higher Education, Science, Research and Innovation (MHESI), Khlong Luang, Pathumthani, Thailand
e-mail: chalinee@biotec.or.th

 
A. Chaurasia et al. (eds.), *GMOs*, Topics in Biodiversity and Conservation 19,
https://doi.org/10.1007/978-3-030-53183-6_35

Convention on Biological Diversity 2000). The ratification ensures that Thailand can participate fully in this new multilateral approach to managing the potential risks of LMOs. Even though Thailand is in the process of enacting a specific law that details provisions for working with LMOs, there are existing legislations and guidelines to regulate the activities of LMOs as follows.

## *Genetically Modified (GM) Plants*

In Thailand, genetically modified (GM) plants are prohibited by the Plant Quarantine Act B.E. 2507 (1964). The Act was amended by the Plant Quarantine Act (No. 2) B.E. 2542 (1999) and, subsequently, the Plant Quarantine Act (No. 3) B.E. 2551 (2008). The Act contains 27 sections. In order to prevent the invasion of plant pests and diseases from outside the country, a number of regulation notifications and orders are applied to imported plants and plant products under the Plant Quarantine Act. The Act is administered by the Department of Agriculture (DOA), Ministry of Agriculture and Cooperatives (MOAC). Under the Plant Quarantine Act, in the Notification of Ministry of Agriculture and Cooperative on specification of plant from certain sources as prohibited articles, of exceptions and conditions under the Plant Quarantine Act B.E. 2507 (1964) (No. 10) B.E. 2553 (2010) of the importation of 33 species, 51 genera, and one family of GM plants as prohibited materials, exceptions are permitted only for cases approved in advance by the DOA for the experimentation or research to conduct risk assessments as granted by the Director General of the DOA in compliance with the Notification on Criteria, Procedures, and Conditions for the importation or bringing in transit of prohibited, restricted and, unprohibited materials (B.E. 2551 (2008)).

The Notification of DOA on Guidelines for importation or transit of prohibited articles under the Plant Quarantine Act B.E. 2507 (1964) (No. 3) B.E. 2544 (2001) establishes a step-by-step approach for the importation of GM plants that complies with the Advance Informed Agreement (AIA) procedure of the CPB. First, the applicant has to submit relevant information about the GM plant to the DOA Biosafety Committee where two subcommittees evaluate and examine all technical information and make a recommendation. The Risk Assessment Sub-Committee evaluates the risks and safety of the GM plant, while the Field Inspection Sub-Committee inspects and monitors the laboratory/greenhouse, closed containment, or field trial where the research is being conducted. Based on the advice and recommendation of the Biosafety Committee, the Director General (DG) of DOA then prepares and submits an opinion to the MOAC concerning any possible adverse effects related to the GM plant and then submits to the Cabinet to consider the approval for field trial.

## Genetically Modified (GM) Microorganisms

Microorganisms including bacteria, molds, viruses, and parasites are regulated by the Pathogens and Animal Toxins Act B.E. 2558 (2015), administered by the Department of Medical Sciences (DMSC) in the Ministry of Public Health (MOPH). The scope of the Act includes only microorganisms that are capable of causing diseases in humans, livestock, beasts of burden, or other animals prescribed in the additional Notification. The Act regulates production, import, export, sale, transit, or possession of both GM and non-GM pathogens. The Act classified pathogens into four risk groups, based on their relative risks for causing diseases or hazards. The risk groups are consistent with the characteristics of the place of production or possession of pathogens, tools, equipment, accompanying documents, labels, containers, or packages for each group of pathogens. All pathogens must remain in contained use and exterminated by an appropriate method before disposing of them into the environment. When the technologies used in the production of a pathogen may increase disease severity with any genetic change, the Act requires that MOPH be notified about the procedures and conditions in relation to the safety assessment of technologies used in the production of pathogens and animal toxins.

The use of GM microorganisms at large scale is regulated by the voluntary Biosafety Guidelines for Contained Use of Genetically Modified Microorganisms (GMMs) at Pilot and Industrial Scales, which is published by National Science and Technology Development Agency (NSTDA), Ministry of Higher Education, Science, Research, and Innovation (MHESI). The objective of these guidelines is to provide guidance for the contained use of GMMs at the pilot and industrial scales to ensure safety to operators, the community, and the environment. Risk assessment of work using GM microorganisms is classified according to the degree of safety taking into account both the nature of GM microorganisms and the relevant working procedures in order to achieve appropriate levels of containment. All contaminated liquid or solid waste from containment must be inactivated by validated means before disposal. The treated waste shall not contain any transferrable gene to ensure that it will not be disseminated into the environment.

## Genetically Modified (GM) Animals

Aquatic animals are regulated by the Emergency Decree on Fisheries B.E. 2558 (2015), which is administrated by the Department of Fisheries (DOF), MOAC. The objective of the Decree is to supervise the import of aquatic animals and aquatic plants in order to protect rare aquatic animal species and prevent danger caused by epidemic disease. Although the Decree does not specifically control GM aquatic animals, Section 65 of the Decree states that the MOAC has the power to prohibit the importation, exportation, bringing in transit, culturing, or possession of any kind of aquatic animals unless a license from the Director General of the DOF has been obtained. In addition, DOF also regulates the importation of aquatic animals, marine algae, and amphibious and aquatic plants by the Royal Decree on Prohibiting Importation of some Aquatic Animals B.E. 2547 (2004). A DOF license is required for the importation of aquatic animals specified in the Royal Decree. In the case of GM aquatic animals, the advice of the Institutional Biosafety Committee (IBC) must also be included in the license application.

Non-GM livestock is regulated by the Animal Epidemic Act B.E. 2558 (2015) which is administrated by the Department of Livestock Development (DLD), MOAC. For prevention and control of epidemics, any person who imports, exports, or transits an animal or carcass through Thailand is required to obtain a license from the Director General of DLD. Although the Act does not specifically control GM livestock, Section 31 of the Act states that the application for a license and the procedures on import, export, or transit through Thailand of livestock, including GM livestock, shall follow the criteria, procedures, and conditions prescribed in the Notifications.

## Draft National Biodiversity Law

Thailand currently has no specific law regulating GMOs. There are existing laws and regulations concerning specific family of organisms such as plants, animals, and microorganisms, as well as bacteria, molds, viruses, and parasites. Given the lack of the specific biosafety regulation, this leads to gap especially in decision-making. Therefore, MONRE has prepared a draft Biosafety Act to ensure appropriate handling of all organisms including GMOs. Moreover, MONRE has included in the draft Biological Diversity Act the core of biological diversity for sample access and benefit sharing of genetic resources as well as invasive alien species (IAS) to provide the effective framework that was relevant the three objectives of the Convention on Biological Diversity (CBD), namely, the conservation of biodiversity, the sustainable use of the component of biodiversity, and the fair and equitable sharing of the benefit arising out of the utilization of genetic resources. The Biodiversity Act will bridge the gap associated with regulating GMOs in Thailand.

## Synthetic Biology Regulations in Thailand

A key question with regard to synthetic biology is whether or not LMOs developed through synthetic biology (e.g., via genome editing techniques) that do not harbor any foreign DNA should be considered LMOs. In decision XIII/17, the Conference of the Parties to CBD took note of the conclusion of the Ad Hoc Technical Expert Group on Synthetic Biology (AHTEG) that living organisms developed through synthetic biology are similar to LMOs as defined in the CPB. Moreover, general principles and methodologies for risk assessment under that Protocol and existing biosafety frameworks provide a good basis for risk assessment of living organisms developed through synthetic biology, but such methodologies might need to be updated and adapted (CBD 2017). However, some organisms developed through gene editing or total genome synthesis may contain only a single or few base-pair changes that could have been obtained through traditional breeding techniques. In such cases, it is still unclear whether these organisms would be considered as LMOs (CBD 2017).

As a party to both the CBD and CPB, Thailand has set up the national legislation to comply with these international agreements. At present, most existing laws that regulate LMOs do not have clearly defined definitions for "LMOs," "modern biotechnology," and "synthetic biology." Even though the draft Biological Diversity Law for Thailand defines LMOs and modern biotechnology as in Article 3 of the CPB, a consensus among the international community on what these definitions are is essential for proper enforcement of the law in the future.

## References

Animal Epidemics Act BE 2558 (2015) The Royal Gazette 132 (14A): 22–41

CBD/SYNBIO/AHTEG/2017/1/2 (2017) Overview of Work Done in Response to Decision XIII/17 and Background Information to Facilitate Deliberations by the Ad Hoc Technical Expert Group on Synthetic Biology

CBD/SYNBIO/AHTEG/2017/1/3 (2017) Report of the Ad Hoc Technical Expert Group on Synthetic Biology

Emergency Decree on Fisheries BE 2558 (2015) The Royal Gazette 132 (108 A): 1–45

Notification of Ministry of Agriculture and Cooperatives Re: Specification of plant from certain sources as prohibited articles, of exceptions and conditions under the Plant Quarantine Act BE 2507 (1964) (No. 10) BE 2553 (2010) The Royal Gazette 127 (Sp. Part 60 D): 62–63

Pathogen and Animal Toxin Act BE 2558 (2015) The Royal Gazette 132 (80 A): 9–31

Plant Quarantine Act BE 2507 (1964) Amended by Plant Quarantine Act (No. 2) B.E. 2542 (1999) Plant Quarantine Act. (NO. 3) B.E. 2551(2008) The Royal Gazette 116 (39 A): 1–9

Royal Decree on Prohibiting Importation of some Aquatic Animals BE 2547 (2004) The Royal Gazette 121 (Sp. Part 35 A): 4–6

Secretariat of the Convention on Biological Diversity (2000) Cartagena protocol on biosafety to the convention on biological diversity: text and annexes. Secretariat of the Convention on Biological Diversity, Montreal

Technical Biosafety Committee (2010) White paper: updated status and perspective of Thailand on Research and Development of Modern Biotechnology and Biosafety Regulation. National Center for Genetic Engineering and Biotechnology, Bangkok

# Part XI
# Australasia

# Australia—Biodiversity Considerations as Part of the Regulation of GMOs, Including Synthetic Organisms

**Andrea Robold and Heidi Mitchell**

**Abstract** Australia regulates GMOs under nationally consistent legislation which came into effect in 2001. The scope of the legislation is broad and covers all GMOs – microorganisms, plants and animals – both in contained facilities and when released into the environment. This broad coverage also encompasses organisms developed using synthetic biology.

The legislation requires consideration of the risks to people and the environment from work involving GMOs. The protection of environmental biodiversity is achieved using robust risk analysis methodology and the use of consultation with experts on the release of GMOs into the environment. This process and how it is used to protect environmental biodiversity from possible impacts of synthetic organisms is discussed further.

**Keywords** Biodiversity · Australia · Gene technology · Risk analysis · Environment · Synthetic biology · GT Act

## Specific Legislation to Regulate Activities with GMOs

The Commonwealth of Australia is a federation of states and territories. An agreement between the state, territory and Commonwealth governments formed the basis for Australia's national regulatory system for gene technology, implemented through the *Gene Technology Act 2000* (GT Act; Commonwealth of Australia 2000) and the Gene Technology Regulations 2001 (GT Regulations; Commonwealth of Australia 2001), together with corresponding state and territory legislation.[1] The GT Act and

[1] In this document, reference to the Commonwealth Act or Regulations or gene technology legislation also includes corresponding law enacted in other Australian jurisdictions.

A. Robold · H. Mitchell (✉)
Office of the Gene Technology Regulator, Canberra, ACT, Australia
e-mail: heidi.mitchell@health.gov.au



A. Chaurasia et al. (eds.), *GMOs*, Topics in Biodiversity and Conservation 19,
https://doi.org/10.1007/978-3-030-53183-6_36

the GT Regulations are regularly reviewed and amended to keep up to date with scientific advances. The legislation regulates certain activities with all GMOs, including microorganisms, plants and animals both in contained facilities and when released into the environment. The scheme is set up so that all work with GMOs is prohibited unless authorised.

The Australian gene technology legislation is administered by the Gene Technology Regulator (the Regulator) who is an independent decision maker. The Regulator is supported by the Office of the Gene Technology Regulator.

The definition of the term gene technology in the GT Act was deliberately worded broadly to prevent it becoming outdated through the development of new gene technologies. In fact, the definition would have captured technologies which were, even at the beginning of the regulatory scheme, considered to have a history of safe use, were it not for specific exclusions in the regulations, e.g. radiation and chemical mutagenesis.

While the GT Act contains no specific definition for the term synthetic biology, it falls within the broad definition of gene technology. Individuals, organisations and governments both in Australia and overseas use a variety of definitions for these terms and so provide no clarity on the subject. However, the 2018 review of the GT Act found that there is wide agreement among stakeholders that synthetic biology includes techniques for producing novel nucleic acids, protein sequences or a combination thereof. The review concluded that these techniques are covered by the GT Act and that the current risk analysis approach remains appropriate (Commonwealth Department of Health 2018).

## Regulated Activities with GMOs

The GT Act acknowledges that certain activities with a GMO may provide a pathway to harm to people or the environment. The regulated activities (*dealings*) with a GMO are to experiment with it; make, breed or grow it; and import, transport or dispose of it (see GT Act, Section 10, for more information). Specifying these activities provides for their regulation both in contained laboratory research and upon environmental release.

## Regulatory Framework for GMOs and GM Products

While the Regulator is responsible for decisions on activities with live GMOs, Australian product regulators administer other laws that may be applicable to GMOs or their products. For example, while growing a GM plant in the field is subject to regulation under the GT Act, use of the GM plant in commercially available food requires a pre-market safety assessment and approval by Food Standards Australia New Zealand. Other product regulators include those responsible for human medicines, animal medicines, agricultural chemicals, industrial chemicals and biosecurity of imports. In addition, the environment minister administers the *Environment Protection and Biodiversity Conservation Act* 1999 which provides for the protection of the environment, with an emphasis on matters of national environmental significance (Commonwealth of Australia 1999). The Regulator consults with the environment minister and the product regulators when assessing a GMO for environmental release. As discussed below, as part of a risk assessment, the Regulator will assess whether a GMO may have adverse effects on biodiversity.

## The Object of the GT Act

The object of the GT Act (GT Act, Section 3):

> […] is to protect the health and safety of people, and to protect the environment, by identifying risks posed by or as a result of gene technology, and by managing those risks through regulating certain dealings with GMOs.

Thus, the GT Act requires the Regulator to consider potential adverse effects on people or the environment. However, considerations of possible benefits; product efficacy; and social, cultural, economic or marketing implications are out of scope.

## Protecting the Environment, Including Biodiversity

The environment is defined in the GT Act as including ecosystems and their constituent parts, natural and physical resources and the qualities and characteristics of locations, places and areas. This broad definition is considered to encompass biodiversity.

The GT Act provides for protection of the environment through, for example:

1. Prohibiting all activities with GMOs unless authorised under the GT Act
2. Conducting risk assessments
3. Identifying and applying effective risk management or refusing to issue a licence if risks cannot be managed
4. Consulting widely with external stakeholders
5. Maintaining awareness of overseas regulation of GMOs

The Risk Analysis Framework (OGTR 2013) explains the Regulator's approach to risk analysis. It guides risk evaluators to consider effects on the biotic and abiotic components that may lead to harm to the environment. Harm to the environment may result from impaired health of organisms due to toxicity or disease; displacement of organisms; predation/altered predator-prey cycles; reduced quality of abiotic components, such as soil, water or air; or disruption of ecosystem processes through, e.g. altered nutrient levels or fire regimes. Any one or a combination of these may result in harm to biodiversity.

The phrase ‘risks posed by or as a result of gene technology’ in the object of the GT Act indicates a comparative assessment, in which the impacts of a GMO and particularly those characteristics that were altered by gene technology are compared to those of a similar non-GM organism, often referred to as the ‘parent organism’. For some organisms created through synthetic biology, there may not be a parent organism to use as a comparator. However, the above risk analysis approach can still be applied because both characteristics of organisms that cause harm to the environment and the nature of harmful effects on the biotic and abiotic environmental components are known.

The perception of what constitutes ‘harm’ to the environment is value-based and can vary between people. It can also change over time and differ according to other factors such as variations in the vulnerability of individuals or type of land use. International standards such as those of the International Plant Protection Convention (IPPC) and World Organisation for Animal Health (OIE) and national health and environmental legislation can provide guidance on the values to be protected from harm. In addition, the Regulator adopts values such as the risk categorisation of pathogens (Standards Australia/New Zealand 2010) or those associated with good agricultural management practices for managing weeds, pests or diseases. These values are taken into account in the risk assessment of GMOs. For example, insect-resistant *Bacillus thuringiensis* (Bt) cotton is known to be toxic to certain lepidopteran pests. In the risk assessment conducted for this GMO, toxicity to these pests was not considered harm as these pests are deliberately killed in conventional cotton management (OGTR 2006).

A stepwise development process is typically followed for intentional release of GMOs (OECD 1986): data from initial contained research, overseas release/s or release of a similar GMO inform authorisations for small, short-term, confined trials where the GMO is removed from the environment once the trial is finished. This provides the Regulator with the information necessary to enable the assessment of a larger or unconfined release and to address any uncertainty before the Regulator authorises any large-scale release, thus facilitating protection of the environment.

For regulated activities that require a licence, which includes any intentional release of a GMO into the environment, the Regulator must not issue a licence unless satisfied that risks are able to be managed (Section 56 of the GT Act). Also, the Regulator can impose specific requirements on GMO releases of any size to ensure the object of the GT Act is met.

For a GMO that is used in contained laboratory research and not intended for environmental release, protection of the environment largely occurs through applying the appropriate level of physical containment.

## Concluding Remarks

Australia has dedicated gene technology legislation which provides science-based regulation of activities with GMOs. The broad definition of gene technology in the GT Act, and regular legislation reviews, enables new technologies, including synthetic biology, to be captured. The regulatory system facilitates the development and use of gene technologies while ensuring protection of people and the environment, including biodiversity.

## References

Commonwealth Department of Health (2018) The third review of the National Gene Technology Scheme. October 2018 final report. http://www.health.gov.au/internet/main/publishing.nsf/Content/gene-technology-review

Commonwealth of Australia (1999) Environment Protection and Biodiversity Act. Available from the Federal Register of Legislation. http://www7.austlii.edu.au/cgi-bin/viewdb/au/legis/cth/consol_act/epabca1999588/

Commonwealth of Australia (2000) Gene Technology Act 2000. Available from the Federal Register of Legislation. https://www.legislation.gov.au/Details/C2016C00792

Commonwealth of Australia (2001) Gene Technology Regulations 2001. Available from the Federal Register of Legislation. https://www.legislation.gov.au/Details/F2016C00615

OECD (1986) Recombinant DNA safety considerations. Available from the OECD. https://www.oecd.org/env/ehs/biotrack/Recombinant-DNA-Safety-Considerations.pdf

OGTR (2006) Risk Assessment and Risk Management Plan for DIR 066/2006: commercial release of herbicide tolerant and/or insect resistant cotton lines north of latitude 22° South. Available at http://www.ogtr.gov.au/internet/ogtr/publishing.nsf/Content/DIR066-2006

OGTR (2013) Risk analysis framework 2013. Available at http://www.ogtr.gov.au/internet/ogtr/publishing.nsf/Content/risk-analysis-framework

Standards Australia/New Zealand (2010) Safety in laboratories – microbiological safety and containment AS/NZS 2243.3:2010, Standards Australia/New Zealand

# New Zealand—GMO Rules and Regulations in New Zealand

**Jack A. Heinemann, Dorien S. Coray, and Brigitta Kurenbach**

**Abstract** The New Zealand legal and regulatory landscape for genetically modified organisms is split between food and environmental safety. Different regulators take primary responsibility. New Zealand and Australia jointly regulate GMOs used as food primarily through the Food Standards Australia New Zealand Act. Each country has its own environmental regulator. The apex law for research and release is the Hazardous Substances and New Organisms (HSNO) Act. The New Zealand courts have confirmed that products of synthetic biology and gene/genome editing techniques are GMOs for purposes of regulation under the HSNO Act. In some parts of the country, regulation also involves local government where districts or regions have opted to regulate under the Resource Management Act. There have been calls from some sectors to revise the definitions of GMOs and approach to regulation. However, the current government has indicated that this is not a priority and that doing so would harm the agricultural sector. The law as currently framed has contributed to there being no evidence of harm to human health or the environment, and no economic losses either. However, a recent determination by the Environmental Protection Agency to allow open air use of nucleic acids on eukaryotic organisms is too fresh to speculate on future effects.

**Keywords** Hazardous Substances and New Organisms Act · Biosecurity Act · Resource Management Act · Environmental Protection Authority · dsRNA · Genome editing · New organisms

J. A. Heinemann (✉) · D. S. Coray · B. Kurenbach
School of Biological Sciences, Centre for Integrated Research in Biosafety, and Centre for Integrative Ecology, University of Canterbury, Christchurch, New Zealand
e-mail: jack.heinemann@canterbury.ac.nz; brigitta.kurenbach@canterbury.ac.nz



A. Chaurasia et al. (eds.), *GMOs*, Topics in Biodiversity and Conservation 19,
https://doi.org/10.1007/978-3-030-53183-6_37

## Introduction

The legislative framework governing genetically modified organisms (GMOs) follows New Zealand's international obligations to the agreements into which it has entered. For the purposes of this article, those agreements are relevant to GMOs used as food or in the environment. New Zealand does not formally regulate GMOs that have food safety approval, including those imported for animal feed, provided that any GMO ingredient is in a nonviable state. This article discusses the regulation of GMOs first as food and then provides more detail on the legislative framework for GMOs in the responsible development of safe biotechnologies, according to the country's obligations as a Party to the Convention on Biological Diversity (CBD) and its family of treaties.

In New Zealand's legal framework, GMOs are considered to be "new organisms." Their status and regulation is consistent with the country's strict biosecurity laws intended to reduce threats to its agriculture and unique biodiversity from invading organisms. The special feature of GMOs is that they are considered new even if they are created in New Zealand.

The use of genetic engineering technologies is widespread in biological research in the country, including at universities, polytechnical institutions, Crown Research Institutes, and hospitals. This use is contained. In addition, between 1988 and 1997, there were 53 approved outdoor field experiments involving GMOs, including those developed as potential vaccines, although not all proceeded (EPA). A further 20 field tests have been approved by the Environmental Protection Authority since it became the responsible agency. Tested GMOs have been food plants, flowers and trees, and bacteria and animals (EPA). With the exception of provision for emergency use of a vaccine for horses, there are no GMOs approved for outdoor use.

## Food

GMOs for food fall under the 1991 Food Standards Australia New Zealand Act (Food Standards Australia New Zealand Act 1991). This legislation creates a binational food regulator, Food Standards Australia New Zealand (FSANZ). Standard 1.5.2 of the FSANZ Act defines food produced using gene technology as "food which has been derived or developed from an organism which has been modified by gene technology" and gene technology as "recombinant DNA techniques that alter the heritable genetic material of living cells or organisms" (Food Standards Australia New Zealand Act 1991).

A treaty between the two countries, the Trans-Tasman Mutual Recognition Agreement (MPI), implemented through the Trans-Tasman Mutual Recognition Act of 1997, binds New Zealand to the decisions made by FSANZ.

The treaty Agreement also provides a basis under exceptional circumstances for New Zealand to opt out of decisions made by FSANZ. These are described as "exceptional health, safety, third country trade, environmental, or cultural grounds." To our knowledge, New Zealand has only once exercised the opt out provisions of the Agreement (Terry 2007). In that case, it was on the basis of third-country trade implications.

Oversight of gene technology for use in food ends with a premarket assessment by FSANZ. After that, both countries rely on a non-codified "duty of care" from manufactures, suppliers, or others connected to the products of gene technology to adequately monitor any harm if it should later eventuate and to report and/or mitigate, as appropriate (Brent et al. 2003).

FSANZ recognizes guidance provided by Codex Alimentarius, a joint WHO and FAO body, and other high level international authorities (Brent et al. 2003). Because both Australia and New Zealand are members of the same organizations and subscribe to the same treaties on this matter, a joint regulator is harmonious for meeting both countries' international obligations.

Domestic implications are not as clear cut. The FSANZ Act is entirely a product of Australian law, and the regulator is answerable only to the Australian minister in charge (Scott 2003). It is perceived to limit the normal powers of New Zealanders to hold public entities to account (Scott 2003). For example, Australians, but not New Zealanders, can request information under the Australian Freedom of Information Act (FOIA); FSANZ is not subject to the New Zealand equivalent 1982 Official Information Act (OIA). To our knowledge, FSANZ has never denied a request from a New Zealander for information. Nevertheless, it is unclear whether the information provided would have been gathered to the same standards used when a public agency was officially responding to a request made through the OIA.

## National-Level Regulation of GMOs in the Environment (Nonfood Uses)

Australia and New Zealand diverge significantly in their international obligations for the use, exchange, and benefit sharing of GMOs. The remainder of this article will therefore be specific to New Zealand.

New Zealand is the home of both unique and rare species and is a country that is dependent on agriculture for much of its export income. Several laws recognize and protect both these attributes. The main legislative instruments of the national government for GMOs are the 1996 Hazardous Substances and New Organisms Act (HSNO Act, pronounced locally as "has no act") and the 1993 Biosecurity Act that empowers authorities to act on illegal or potentially unsafe GMOs (Biosecurity Act 1993). In addition, the 1991 Resource Management Act (Resource Management Act 1991) provides for regulation of GMOs by local governments.

## *New Organisms*

The HSNO Act encompasses all organisms whether or not they are a priori deemed to be biosecurity threats. Their status as new organisms places the burden of proof that the GMOs may be safely used in or released into New Zealand upon those who wish to use or release the organisms.

GMOs are a specific category of new organism in the HSNO Act. It states that "genetically modified organism means, unless expressly provided otherwise by regulations, any organism in which any of the genes or other genetic material—(a) have been modified by *in vitro* techniques; or (b) are inherited or otherwise derived, through any number of replications, from any genes or other genetic material which has been modified by *in vitro* techniques" (Hazardous Substances and New Organisms Act 1996).

New Zealand's definition is similar, but not identical, to that used by the Cartagena Protocol on Biosafety (CPB) to define living modified organisms (LMOs), which includes living genetically modified organisms. The Protocol definition of a living modified organism is "any living organism that possesses a novel combination of genetic material obtained through the use of modern biotechnology." Both the Protocol and the HSNO Act refer to in vitro techniques, but the Protocol emphasizes the use of nucleic acids, whereas the HSNO Act emphasizes modification of genes and other genetic material.

In many respects, the HSNO Act resembles European Union Directive 2001/18/EC. The similarity of language may be no accident because the frameworks have a common origin in those developed by the United Kingdom from the late 1970s onward (Heinemann 2015). Thus, it is also perhaps unsurprising that both New Zealand and the EU courts have arrived at similar interpretations of the scope of processes that create GMOs.

The EU and New Zealand adopted similar approaches in their legislative frameworks. Both regulate using what is called "process-based" legislation (Steinbrecher and Paul 2017), capture a broad category of processes used in making GMOs, and then use specific criteria in their regulations to determine what is and is not to be regulated (Heinemann 2015). For example, plants altered through the use of some chemical and radiation mutagenesis techniques are technically defined as GMOs but specifically excluded from GM assessment provisions (Atanassova and Keiper 2018).

The regulator created by the HSNO Act is the Environmental Protection Authority (EPA; previously the Environmental Risk Management Authority). New organisms not already covered by the Biosecurity Act are governed by this Authority. The Ministry for Primary Industries (MPI) is the enforcement agency for both.

## *When Is a New Organism a GMO?*

In what is regarded as a world-first decision, New Zealand's High Court ruled that organisms made using the techniques of gene/genome editing, including CRISPR/Cas9, ZFN, and Talens, are omitted from the list of exclusions to the HSNO Act (Kershen 2015). Thus, most if not all living applications of synthetic biology (Secretariat 2015) probably would be subject to regulation. This decision was followed later by a similar one of the European Court of Justice (Callaway 2018).

The High Court ruling was prompted by an EPA determination under Section 26 of the HSNO Act that some applications of these techniques were sufficiently similar to excluded techniques that they were not to be subject to its GMO provisions. This determination conflicted with internal staff advice and interpretation of the legislation (Kershen 2015).

The Sustainability Council of New Zealand initiated a High Court appeal of the EPA determination. The Court quashed the EPA's determination because "the Authority erred in its interpretation of the regulation because it considered that the regulations did not set out an exhaustive list and that techniques that are comparable and sufficiently similar to those listed in the Regulations should also be excluded" (Mallon 2014).

More recently, the EPA issued another Section 26 determination that eukaryotic organisms "treated with externally applied double-stranded RNA molecules to induce a small interfering RNA (siRNA) response do not fulfil the definition of genetically modified organisms detailed in the Act and therefore are not new organisms for the purposes of the HSNO Act" (EPA 2018a). This determination applies only to organisms exposed to RNA molecules, not organisms created by alteration of DNA to produce new RNA molecules (Heinemann 2019).

In this case, the determination by the New Zealand EPA's Section 26 Committee was similar to the advice received from EPA staff, who also anticipated that these techniques would soon be relevant to open air applications of double-stranded RNA-based pesticides (EPA 2018b). This determination is being challenged because it relied upon knowledge about only a few kinds of eukaryotes, contradicted existing knowledge of them and other eukaryotes, and failed in other respects such as properly considering the risk of harmful viruses being released (Heinemann 2019).[1]

Here again, New Zealand was at the forefront of setting regulations on new biotechnological applications. Few if any other parties to the CPB have regulations on this open air (externally applied) use of nucleic acids (Heinemann 2019; Heinemann and Walker 2019). Internationally, at least one other CPB member country, Mauritania, has requested the Ad Hoc Technical Expert Group on Risk Assessment and Risk Management for advice on conducting risk assessments and management of this kind of technology, referred to as "environmental application of in vitro nucleic acid techniques" (AbdelKawy 2016; CBD 2016). Presently, the New

[1] For example, as a vector for SARS-CoV-2 should batches become contaminated during manufacturing or post-sale.

Zealand EPA has called in the decision to deregulate the use of externally applied double-stranded RNAs but has yet to make a final determination.

## Local-Level Regulation

New Zealanders have been exercising their rights to participate in the regulation and safe use of GMOs for many decades. Among other manifestations of this engagement are numerous districts and regions that have adopted their own regulations through local planning processes. This mirrors similar trends in Europe (USDA 2017). Presently, the Hastings, Auckland, Whangarei, and Far North Districts have implemented provisions for local decision-making on the release of GMOs. The Northland Regional Council is considering the same after being prompted by the Whangarei and Far North District Councils to include provisions for the region's coastal and marine areas.[2] These districts comprise a significant proportion of the area of the North Island of New Zealand and cover the areas in which a majority of New Zealanders reside.

A large number of public and private research organizations and some universities, as well as private citizens and experts, have participated in one or more local planning processes. For example, submissions for the Auckland District planning process included among others the University of Otago, the government Centre of Research Excellence the Maurice Wilkins Centre, the Crown Research Institute Scion, experts from the consortium Pastoral Genomics, the industry groups NZBIO and Federated Farmers, and indirectly the Royal Society of New Zealand through a report commissioned from them by Federated Farmers.

A fundamental disagreement over the proper jurisdiction for the regulation of GMOs arose early in the process. Local governments relied upon the Resource Management Act (RMA) 1991 to justify adopting local regulations. Objections to this were heard by the Environment Court followed by an appeal to the High Court. Both courts ruled unambiguously that the RMA gave local bodies the power to regulate the use and release of GMOs in their jurisdictions.

A further argument against the use of the RMA was that it was duplicating work already performed by the national regulator, the EPA. However, the approach taken at the local level was found to not only be legally valid under the RMA; any actual duplication was ruled irrelevant to the RMA (Mathias 2018).

During each of the quasi-judicial proceedings used by local governments in the adoption of their 10-year plans as required by the RMA, the proposed forms of regulation were opposed by some sectors and private experts. In general, the nature of the opposition was that there existed a scientific consensus on the safety of GMOs and that there was a significant economic risk should additional regulations inhibit their use.

[2] One of us (Heinemann) was a formal expert witness in all these processes.

The hearing judges found in all four cases so far (with one still pending) that the basis for asserting a scientific consensus on safety was unconvincing. In particular, the case for safety was largely based on use as food, which was irrelevant to environmental regulation. Moreover, the evidence provided was specific to crop plants and the regulations applied to all kinds of organisms including viruses, bacteria, fungi, other plants, and animals. Finally, there was no counter evidence to an expert economist who found that adoption of the additional local regulations would have no adverse economic effects.

## Summary and Future Directions

GMOs fall under the scope of several laws in New Zealand. Reflecting New Zealand's international obligations, it has laws for both food safety standards and the environmental and contained research use of GMOs. The category of regulated organisms includes the products of new techniques such as gene editing and other forms of mutagenesis, with some older techniques specifically excluded by the regulations. These laws determine the activities of a variety of regulators at or above the national level, including FSANZ, EPA, and the MPI, and local government bodies.

New Zealand law appears aligned with at least the EU for the foreseeable future. New Zealand also appears most closely aligned to the EU in its approach to regulation, especially for the management of environmentally released GMOs. It has a "process trigger" for its legislation and requires specific categories of processes to be excluded from provisions.

Beginning with the High Court ruling confirming that new techniques of mutagenesis are within the coverage of the HSNO Act, there have been calls from some to revise the law. Advocacy for revision reappeared following the similar European Court of Justice ruling. Interestingly, the language used both in and outside of New Zealand is very similar, possibly homologous, with frequent calls in various countries to amend legislation to that which is "fit for purpose" (Devuyst 2018; Jones 2018; Manhire 2018). In the recent past, the RMA has been amended, but those changes did not affect local government from asserting rights to regulate GMOs not intended for use as medicine (Davison 2017).

Taking into consideration both environmental and economic safety, the current government is signaling that changes to GMO regulation are not coming in the foreseeable future (Dreaver 2018). The topic of GMOs and other kinds of biotechnology is as polarized and fraught in New Zealand as elsewhere. However, to date, the combination of approaches taken by the country to regulate GMOs has prevented any known irreversible harm to the environment or human health.

## References

AbdelKawy O (2016) Submission by Mauritania for suggestions on Risk assessment guidance on additional topics. Le Ministère de l'Environnement et du Développement Durable, La Mauritanie. https://bch.cbd.int/database/attachment/?id=16832

Atanassova A, Keiper F (2018) Plant breeding innovation: a global regulatory perspective. Cereal Chem 95:8–16

Biosecurity Act (1993). http://www.legislation.govt.nz/act/public/1993/0095/latest/DLM314623.html

Brent P, Bittisnich D, Brooke-Taylor S, Galway N, Graf L, Healy M, Kelly L (2003) Regulation of genetically modified foods in Australia and New Zealand. Food Control 14:409–416

Callaway E (2018) CRISPR plants now subject to tough GM laws in European Union. Nature 560:16

CBD (2016) UNEP/CBD/BS/cop-mop/8/8. Convention on Biological Diversity. https://www.cbd.int/doc/meetings/bs/mop-08/official/bs-mop-08-08-en.pdf

Davison I (2017) RMA reforms pass into law with support from Maori Party. New Zealand Herald. https://www.nzherald.co.nz/nz/news/article.cfm?c_id=1&objectid=11833888

Devuyst P (2018) GMO directive is no longer fit for purpose, say EU scientific advisors. Informa. https://iegpolicy.agribusinessintelligence.informa.com/PL218317/GMO-Directive-is-no-longer-fit-for-purpose-say-EU-scientific-advisors

Dreaver C (2018) 'The core technology of the future': time to rethink NZ's GMO-free status?. The Spinoff. https://thespinoff.co.nz/science/20-11-2018/the-core-technology-of-the-future-time-to-rethink-nzs-gmo-free-status/

EPA (2018a) Decision. Environmental Protection Authority. https://www.epa.govt.nz/assets/FileAPI/hsno-ar/APP203395/APP203395-Decision-FINAL-.pdf

EPA (2018b) EPA staff report. Determining whether eukaryotic cell lines treated with double-stranded RNA are genetically modified organisms: Environmental Protection Authority. https://www.epa.govt.nz/assets/FileAPI/hsno-ar/APP203395/APP203395-EPA-Advice-Document-FINAL.pdf

EPA. GM field tests. https://www.epa.govt.nz/industry-areas/new-organisms/rules-for-new-organisms/gm-field-tests/

Food Standards Australia New Zealand Act. 118 of 1991 (1991). https://www.legislation.gov.au/Details/C2016C01118

Hazardous Substances and New Organisms Act (1996). http://www.legislation.govt.nz/act/public/1996/0030/93.0/DLM381222.html

Heinemann JA (2015) Expert scientific opinion on the status of certain new techniques of genetic modification under Directive 2001/18/EC. University of Canterbury, Christchurch. https://www.canterbury.ac.nz/media/documents/science-research/inbi/new-techniques-of-genetic-modification.pdf

Heinemann JA (2019) Should dsRNA treatments applied in outdoor environments be regulated? Environ Int 132:104856

Heinemann JA, Walker S (2019) Environmentally applied nucleic acids and proteins for purposes of engineering changes to genes and other genetic material. Biosaf Health 1:113–123

Jones M (2018) Making the Gene Technology Act fit-for-purpose. Murdoch University. https://www.murdoch.edu.au/news/articles/opinion-making-the-gene-technology-act-2000-fit-for-purpose

Kershen DL (2015) Sustainability Council of New Zealand Trust v The Environmental Protection Authority: gene editing technologies and the law. GM Crops Food 6:216–222

Mallon J (2014) Sustainability Council of New Zealand Trust v The Environmental Protection Authority. High Court of New Zealand

Manhire T (2018) 'I literally covered my wall in post-it notes': meet NZ's new chief scientist. The Spinoff. https://thespinoff.co.nz/science/13-11-2018/i-literally-covered-my-wall-in-post-it-notes-meet-nzs-new-chief-scientist/

Mathias GJ (2018) WDC and FNDC – GE Supplementary Opening Submissions. https://northlandregionalcouncil-my.sharepoint.com/:x:/r/personal/evaniaa_nrc_govt_nz/_layouts/15/Doc.aspx?sourcedoc=%7B3fb622a1-b3a9-45dc-8124-e66876c5d994%7D&action=default
MPI. Australia-New Zealand co-operation. https://www.mpi.govt.nz/law-and-policy/legal-overviews/food-safety/australia-new-zealand-co-operation/
Resource Management Act (1991). http://www.legislation.govt.nz/act/public/1991/0069/211.0/DLM230265.html
Scott C (2003) The creation of trans-Tasman authorities – Food Standards Australia New Zealand: public sector reform implications. N Z Pub Sec 26:14–19
Secretariat (2015) Meeting of the Ad Hoc Technical Expert Group on Synthetic Biology. https://www.cbd.int/meetings/SYNBIOAHTEG-2015-01
Steinbrecher RA, Paul H (2017) New genetic engineering techniques: precaution, risk, and the need to develop prior societal technology assessment. Environ Sci Policy Sus Dev 59:38–47
Terry S (2007) Food safety credibility. Sustainability Council, Wellington. http://www.sustainabilitynz.org/docs/FoodSafetyCredibility_GMLysineCorn.pdf
USDA (2017) EU-28. GAIN report: USDA Foreign Agricultural Service

# Part XII
# Europe

# Czech Republic—GMO Regulations and Biodiversity: A Legal Perspective

**Slavomír Rakouský and Zuzana Doubková**

**Abstract** The Czech Republic adopted first regulation of the use of genetically modified organisms (GMOs) in 2001. Nowadays, the national regulatory framework is set up by the Act No. 78/2004 Coll., on the Use of Genetically Modified Organisms and Genetic Products, as amended, and by the implementing Decree No. 209/2004 Coll. The Czech GMO Act transposes two EU Directives: 2001/18/EC and 2009/41/EC, covering thus all three types of GMO use: (1) contained use, (2) deliberate release into the environment for any other purpose than placing on the market and (3) placing on the market of GMOs as products or contained in products. EU Regulations 1829/2003 and 1830/2003 concerning the authorisation of genetically modified food and feed, traceability and labelling of genetically modified organisms and genetically modified food and feed, and Regulation 1946/2003 implementing the Cartagena Protocol on Biosafety, have been directly applicable in the Czech Republic since its accession to the EU in May 2004.

The Competent Authority for handling the notifications and for regulation of the use of GMOs in the Czech Republic (except for GM food and feed) is the Ministry of the Environment, while the Ministry of Agriculture is the Competent Authority under Regulation 1829/2003, on genetically modified food and feed, and it is responsible for the rules of coexistence of GM and non-GM crops.

**Keywords** Regulation · Biodiversity · Czech Republic · Legislation · Biosafety · Policy · Resilient ecosystem · Environment · Synthetic biology

---

S. Rakouský (✉)
Czech Commission for the Use of Genetically Modified Organisms and Genetic Products, Prague, Czech Republic

Z. Doubková
Czech Commission for the Use of Genetically Modified Organisms and Genetic Products, Prague, Czech Republic

Ministry of the Environment, Prague, Czech Republic
e-mail: zuzana.doubkova@mzp.cz



A. Chaurasia et al. (eds.), *GMOs*, Topics in Biodiversity and Conservation 19,
https://doi.org/10.1007/978-3-030-53183-6_38

## GMO Regulations

The first regulation of the use of genetically modified organisms (GMOs) in the Czech Republic came into force in 2001 (Act No. 153/2000 Coll., on the Use of Genetically Modified Organisms and Genetic Products). Although the Czech Republic was not a member of the European Union (EU) at that time, the content of the Act already followed the EU GMO legislation. Nowadays, the national regulatory framework is set up by the Act No. 78/2004 Coll., on the Use of Genetically Modified Organisms and Genetic Products, as amended, and by the implementing Decree No. 209/2004 Coll., on Detailed Conditions for the Use of Genetically Modified Organisms and Genetic Products, as amended (EU Official Bulletin 2019). Both documents are harmonized with the EU GMO legislation.

The Czech GMO Act transposes two EU Directives, 2001/18/EC and 2009/41/EC, covering thus all three types of GMO use: (1) contained use, (2) deliberate release into the environment for any other purpose than placing on the market and (3) placing on the market of GMOs as products or contained in products. The Act on the Use of Genetically Modified Organisms and Genetic Products has been amended several times. The last amendment to the Act came into force on 1 January 2017 and deals with:

1. Simplification of the administration regarding contained use of GMOs.
2. Transposition of EU Directive 2015/412 providing the possibility for EU member states to restrict or prohibit the cultivation of GMOs. However, the Czech Republic has not imposed any ban on GM crops yet (Biosafety Clearing-House, Czech Republic 2019).

EU Regulations 1829/2003 and 1830/2003 concerning the authorization of genetically modified food and feed and the traceability and labelling of genetically modified organisms and genetically modified food and feed and Regulation 1946/2003 implementing the Cartagena Protocol on Biosafety (CPB) have been directly applicable in the Czech Republic since its accession to the EU in May 2004.

## Competent Authorities and Advisory Bodies

Pursuant to EU Directive 2001/18/EC and the Czech Act No. 78/2004, the Competent Authority for handling the notifications and for regulation of the use of GMOs in the Czech Republic (except for GM food and feed) is the Ministry of the Environment. It closely cooperates with the Ministry of Agriculture in agricultural aspects, such as seed issues, animal health, food and feed, and with the Ministry of Health as regards human health aspects. The Ministry of the Environment is also the National Focal Point for the CPB and for Regulation (EC) 1946/2003 (Biosafety Clearing-House, Czech Republic 2019). The Ministry of Agriculture is the Competent Authority under Regulation (EC) 1829/2003, on genetically modified food and feed and it is responsible for the rules of coexistence of GM and non-GM crops (Trnková et al. 2015).

Based on Act No. 78/2004, the Ministry of the Environment established its' expert advisory body, the Czech Commission for the Use of GMOs and Genetic Products, consisting of scientists and representatives of administrative authorities and NGOs. The activities of the Commission cover especially environmental risk assessment, and it is authorized to:

1. Assess the information contained in notifications of the use of GMOs and issue opinions on these notifications.
2. Check and assess reports on the use of GMOs and other documents submitted by the users.
3. Carry out environmental risk assessments and comment on the notifications for placing GMOs on the market under EU Directive 2001/18/EC and Regulation 1829/2003.
4. Issue its expert positions and statements on specific topics or documents, including the international exchange of information.
5. Inform the public on scientific developments and its own activities.
6. Prepare documents, identify emerging biosafety issues and provide ad hoc consultations for the authorities (Doubková 2011a).

The Ministry of Agriculture has its own group of experts, serving as its advisory body, the Scientific Committee for Genetically Modified Food and Feed. Activities of the Committee are focused especially on the risk assessment of GM food and feed and actual problems in this area.

## Supervision and Enforcement

The Czech Environmental Inspectorate is the main competent authority on state supervision of the use of GMOs as regards contained use and deliberate release into the environment. It cooperates with other state supervision bodies responsible for various products where GMOs are used or could be present as unauthorised admixtures, the:

1. Czech Agriculture and Food Inspection Authority, in charge of food inspections and control
2. Central Institute for Supervising and Testing in Agriculture, in charge of seeds, feed and plant protection products
3. State Veterinary Administration concerning animal-related supervision
4. State Institute for Drug Control concerning medicinal products
5. Custom Authorities in charge of export and import

Four authorized detection laboratories are available to these authorities in the Czech Republic, and a National Reference Laboratory for GMOs has been established at the Crop Research Institute in Prague.

## Cultivation of GM Crops and Coexistence

Genetically modified products consumed or commercially grown in the Czech Republic do not differ from other products with regard to their potential risks to human and animal consumption or to the environment. However, special rules apply to their sale (obligatory labelling) and field production (rules of coexistence) (Křístková 2010).

The commercial cultivation of GM crops falls within the authority of the Ministry of Agriculture. Simultaneously, the following authorities' expert opinions are requested: the Czech Agriculture and Food Inspection Authority (CAFIA); the Czech Environmental Inspectorate (CEI); the Central Institute for Supervising and Testing in Agriculture (CISTA), together with its departments of the former State Phytosanitary Administration; the Crop Research Institute-National Reference Laboratory for GMO identification and DNA fingerprinting (CRI); and the State Agricultural Intervention Fund (SAIF). Altogether these institutions form a network, which processes all information on GM crops. The GMO issue is evaluated thoroughly and in a complex way. Both points of view are taken into account, i.e. the legislative one and the possible risks of release into the environment (field trials) and placing on the market (Trnková et al. 2015).

Czech growers of GM crops can only cultivate such GM varieties that contain genetic modification(s) approved at the EU level and that have been registered in the National Plant Variety Register of the Czech Republic or in the Common Catalogue of Varieties of Agricultural Plant Species of the EU (Trnková et al. 2015).

The concept of coexistence aims at the parallel existence of different agricultural production systems (conventional, organic and that based on GM crops) and the separation of these systems and their products to avoid unwanted admixtures of genetic modifications in conventional and organic products. The coexistence concept in the Czech Republic has been obligatory for every farmer growing GM crops since the first year of GM crop cultivation in 2005. Coexistence measures as general rules are incorporated in Act No. 252/1997 on Agriculture, as amended. More detailed conditions for the cultivation of GM varieties are given by Decree No. 58/2010 (originally Decree No. 89/2006). In line with the above-mentioned legislation, the Ministry of Agriculture has laid down basic principles for GM crop growers (so far for maize, potatoes and soybean) (Trnková et al. 2015).

Based on the experience obtained up to now in the country with GM maize and GM potato cultivation, it has been shown that simple and well understandable rules on coexistence enable the easy, reliable and fully transparent performance of GM field production, reducing thus potential risks to human and animal health, the environment and other agricultural production systems (Křístková 2010).

At present only one GM crop is authorised for cultivation in the EU – Bt maize line MON810 (resistant to the European corn borer). In the Czech Republic, Bt maize had been commercially cultivated on only a limited acreage from 2005 to 2017, when its area dropped (from a maximum of 8.380 ha in 2008) to zero, mainly due to higher administrative demands on farmers and problems with the marketing of GM production.

## Biosafety Policy

The Czech Republic was among the first countries which signed the Cartagena Protocol on Biosafety (CPB) in May 2000, on the occasion of the Fifth Meeting of the Conference of the Parties to the Convention on Biological Diversity in Nairobi, and it ratified the Protocol on 8 October 2001.

Contrary to the legislative framework for the use of GMOs, biosafety policy as a stand-alone document has not been developed in the Czech Republic. Instead, the Ministry of the Environment, as the main responsible body in this area, decided to incorporate biosafety principles into relevant strategic documents. The reasons for this decision were, among others, the negative experience of some other countries with similar political and economic conditions and their difficulties with approval of such a governmental document (Doubková 2011b). Therefore, biosafety principles are reflected in the following documents:

1. Strategy for Sustainable Development
2. State Environmental Policy
3. State Programme of Nature Conservation and Landscape Protection
4. National Biodiversity Strategy
5. Food Safety Strategy
6. Action Plan on Health and the Environment
7. State Programme of Environmental Education and Public Awareness
8. Reports on the Environment in the Czech Republic

## Sustainable Development

The first Sustainable Development Strategy of the Czech Republic was approved by the Government of the Czech Republic in 2004 as a long-term framework for political decision-making. Currently, the Strategic Framework of the Czech Republic up to 2030 represents a new key document for sustainable development of the country and society in the decades to come. Czech Republic 2030 was being formed at the time when the global community formulated its vision of the future world at the United Nations into 17 Objectives of Sustainable Development, and when the international community also adopted a new Paris Agreement under the United Nations Framework Convention on Climate Change, with its ambitious goals (Kárníková 2017). The priorities and objectives of sustainable development are classified in this document into six priority axes. Of these, Axis 3 “Resilient ecosystems” is pertinent and further structured into five subchapters concerning: landscape and ecosystem services, biodiversity, water in the landscape, soil care and strategic objectives.

Although GMOs are not directly mentioned throughout the Strategic Framework document, all five subchapters of Axis 3 are relevant for biosafety, especially the subchapter on biodiversity.

## State Environmental Policy

The State Environmental Policy of the Czech Republic 2012–2020 sets a framework for effective protection of the environment in the country until 2020. The document was updated in 2016 after its midterm evaluation.

The main objective of this policy is to ensure a healthy and high-quality environment for citizens living in the Czech Republic, to significantly contribute to a more effective use of resources and to minimize the negative impacts of human activities on the environment, including cross-border impacts, and thus to contribute to the improvement of the quality of life both in Europe and globally. The Policy focuses on the following areas:

1. Protection and sustainable use of resources
2. Climate protection and improvement of ambient air quality
3. Protection of nature and landscape
4. Safe environment

"Safe Environment", in the chapter on "Risk prevention", deals with chemicals, dangerous waste and GMOs, including relevant objectives and indicators (Ministry of the Environment 2016).

## National Biodiversity Strategy

The National Biodiversity Strategy of the Czech Republic, based on the Convention on Biological Diversity (CBD), was formulated shortly after the accession of the Czech Republic to the EU. The first document outlined biodiversity conservation and management for the period 2005–2015, and in March 2016 the Government endorsed the National Biodiversity Strategy for the years 2016–2025.

Biotechnology is mentioned in the Strategy in connection with the conservation of genetic resources. GMOs are specifically dealt with in the chapter on the impact of agriculture on biodiversity.

## Emerging Biosafety Issues: New Gene Techniques and Synthetic Biology

In the Czech Republic, new gene techniques (genome editing) have so far been applied only in contained use situations and mostly for basic research. Ongoing projects use the CRISPR-Cas or TALEN techniques which have been regulated in the same way as GMOs. The European Court of Justice (ECJ) in its ruling from July 2018 endorsed this approach (ECJ 2018).

## References

Biosafety Clearing-House, Czech Republic (2019). https://www.mzp.cz/Biosafety/index.htm

Doubková Z (2011a) III. Administrative system. In: Roudná M (ed) Genetic modifications and biosafety measures – Czech Republic. Ministry of the Environment, Prague, pp 10–23

Doubková Z (2011b) I. Biosafety policy. In: Roudná M (ed) Genetic modifications and biosafety measures – Czech Republic. Ministry of the Environment, Prague, pp 3–7

ECJ (2018) Organisms obtained by mutagenesis are GMOs and are, in principle, subject to the obligations laid down by the GMO directive. Court of the European Union, Judgement in Case C-528/16, Luxembourg, 25 July 2018. Available via CURIA https://curia.europa.eu/jcms/upload/docs/application/pdf/2018-07/cp180111en.pdf

EU Official Bulletin (2019) Available via EUR-Lex. https://eur-lex.europa.eu/homepage.html?locale=en

Kárníková A (ed) (2017) Strategic framework Czech Republic 2030. Office of the Government of the Czech Republic, Department of Sustainable Development, Prague

Křístková M (2010) Experience with Bt maize cultivation in the Czech Republic 2005–2009. Ministry of Agriculture, Prague

Ministry of the Environment (2016) State Environmental Policy of the Czech Republic 2012–2020. Ministry of the Environment, Czech Republic. https://www.mzp.cz/en/state_environmental_policy

Trnková J et al (2015) Organisation and inspection of GM crops cultivation in the Czech Republic, 2nd edn. Ministry of Agriculture, Plant Commodities Department, Prague. Available via Ministry of Agriculture, Czech Republic. http://eagri.cz/public/web/mze/zemedelstvi/gmo-geneticky-modifikovane-organismy/

# Hungary—Hungary's GMO-Free Policy and Its Legal Background

**Rita Andorkó and Levente Kőrösi**

**Abstract** Keeping agriculture free from genetically modified organisms is a priority of the Hungarian government. Hungary's GMO-free agricultural strategy is based on the precautionary principle and takes into consideration the interest of future generations, contributes to the protection of biodiversity, and creates economic benefits for farmers and food producers. Due to the country's strong commitment and its consistent GMO-free policy, no GM plants have ever been cultivated in Hungary.

**Keywords** Hungary · GMO-free policy · Fundamental Law · Pannonian biogeographical region · Safeguard clause · Amflora case · GMO-free labelling · European Soy Declaration

## Background

Keeping agriculture free from genetically modified organisms (GMOs) is among the top priorities of the Hungarian Government. Hungary is one of the strongest opponents of agricultural gene technology in the European Union (EU), excluding the possibility of cultivation of any genetically modified crops in the country.

The first piece of legislation that made provisions on genetically modified organisms was the Nature Conservation Act.[1] Paragraph 9 of the Act declares that the creation of such organisms, the experiments with them, their cultivation and their import to or export from the country may only occur in line with certain conditions specified by a separate law. As a follow-up, Hungary was the first in Central-Eastern

[1] Act No. LIII of 1996 on Nature Conservation

R. Andorkó (✉) · L. Kőrösi
Department for Biodiversity and Gene Conservation, Ministry of Agriculture, Budapest, Hungary
e-mail: rita.andorko@am.gov.hu; levente.korosi@am.gov.hu



A. Chaurasia et al. (eds.), *GMOs*, Topics in Biodiversity and Conservation 19,
https://doi.org/10.1007/978-3-030-53183-6_39

Europe to adopt a specific legal framework[2] in 1998, more than 5 years before EU accession. According to this Act, all activities related to gene technology such as trade, production, distribution, use, transportation, cultivation and also related to research activities, such as contained uses, are subject to authorisation procedure.

The responsibilities for the authorisation of gene technology activities in Hungary are divided between two competent ministries, depending on the field of the respective gene technological activities related to the agricultural and food sector (including process additives used in food production), contained use and other industrial gene technological activities, or activities related to human health, to the production of human pharmaceutical products and to cosmetics in direct contact with the human body. The competent authorities are supported by the Gene Technology Advisory Board, an independent biotechnology committee, which consists of nominees of the Hungarian Academy of Sciences, of the competent ministries as well as of non-governmental organisations. Authorisation decisions are made on a case-by-case basis and the potential risks of the release or of the contained use of the GMOs and the management of risks associated with those are all considered.

The importance of maintaining the GMO-free status of the Hungarian agriculture was recognised already in 2006, when all five parliamentary parties at that time mutually agreed on the country's GMO-free agricultural strategy and the process of implementation aimed at its realisation.[3] Since then, this strategy has not been changed (Darvas and Székács 2011); to the contrary, 6 years later, the Hungarian Parliament adopted a new Fundamental Law, which entered into force on 1 January 2012, which includes the pursuit of a GMO-free agriculture[4] (see below). Certain objectives related to this issue are integrated into the Hungarian Government's Program and also into several national strategies such as the National Strategy for the Conservation of Biodiversity (2015–2020),[5] the fourth National Environmental Protection Programme (2015–2020)[6] and the National Rural Development Strategy (NRDS 2012–2020).

Due to this strong commitment and the consistent GMO-free policy of the Hungarian Government, with the exception of a few experimental releases into the environment in the past, no GM plant could have ever become cultivated in Hungary.

---

[2] Act No. XXVII of 1998 on gene technology activities

[3] Parliamentary Resolution 53/2006. (XI.29) on various issues relating to gene technology activities, their use in agriculture and food production and the Hungarian strategy concerning them

[4] Article XX: (1) Everyone shall have the right to physical and mental health. (2) Hungary shall promote the effective application of the right referred to in paragraph (1) through agriculture free of genetically modified organisms, by ensuring access to healthy food and drinking water, by organising safety at work and healthcare provision and by supporting sports and regular physical exercise as well as by ensuring the protection of the environment.

[5] Parliamentary Resolution 28/2015. (VI. 17) on the National Strategy for the Conservation of Biodiversity (2015–2020)

[6] Parliamentary Resolution 27/2015. (VI. 17) on the National Environmental Protection Programme (2015–2020)

## Consequences of the EU Membership

Since 2004 Hungary has been a Member State (MS) of the European Union; therefore, its legislation is in line with the legal framework of the EU (National Reports on the implementation of the Cartagena Protocol on Biosafety 2005, 2007, 2011, 2015, 2019). Due to its different environmental features, the EU was enriched with the accession of Hungary by a new ecological region, the Pannonian biogeographical region. This biogeographical region significantly differs from the Western and other Central European areas with intensive land use, different climate and vegetation. Agricultural areas and agroecosystems in Hungary are much more diverse than most other biogeographical regions referred to. There are numerous protected species in the country that have an important role in the grassland ecosystem, vegetation and soil maintenance and exist primarily or exclusively in the Carpathian Basin. After the accession to the EU, these differences raised strong doubts about the applicability to Hungary of the risk analysis applied in other MSs with different ecosystems.

As a MS of the EU, the safeguard clause was the only legal instrument available for several years to follow and maintain the respective strategies of the individual member countries. In line with this legal possibility, in 2005 the Minister of Agriculture and Rural Development introduced a cultivation ban of Monsanto's MON810 maize in Hungary.[7] This safeguarding measure was based on national research and new scientific evidence which resulted in the exclusion of MON810 from cultivation in Hungary (Darvas 2006; Darvas and Székács 2011).

A ban on cultivation of BASF's Amflora GM potato was introduced in 2010. The same year, Hungary, supported by several other MSs, challenged the European Commission's decision on the authorisation of the same GM potato, claiming that it was a threat to human and animal health. The European Court of Justice[8] annulled the decisions approving and authorising the general production of Amflora GM potato in its judgement, which practically meant that no further cultivation was possible in the EU.

Besides the protection of our environment and the safety of future generations, the market advantage provided, and related economic interests are also part of the reasons of having a GMO-free strategy. Therefore, in 2015 Hungary immediately applied the new EU Directive[9] providing freedom for MSs to decide whether they want to cultivate GM crops in their territory or not and transposed it into its national legislation. According to this Directive and the relevant domestic legislation Hungary exempted

[7] Decree 53/2013. (VI. 17.) of the Ministry of Rural Development on the safeguard clause on the seeds of stems and hybrids of maize MON810

[8] Judgment of the General Court of 13 December 2013 — Hungary v Commission (Case T-240/10) (2014/C 39/26)

[9] Directive (EU) 2015/412 of the European Parliament and of the Council of 11 March 2015 amending Directive 2001/18/EC as regards the possibility for the Member States to restrict or prohibit the cultivation of genetically modified organisms (GMOs) in their territory

its entire territory from cultivation of several GM maize events (MON810, 1507, 59122, 1507x59122, Bt11, GA21, MIR604, and Bt11xMIR604xGA21).

## Penalties

Illegal activities with GM plants and propagating material are subject to criminal liability in Hungary. According to the Criminal Code of Hungary[10] any person who unlawfully imports, stores, transports or places on the market in the territory of Hungary the propagating materials of genetically modified plant cultivars ("varieties") which have not been authorised in the EU, or releases such into the environment; or unlawfully releases into the environment the propagating materials of genetically modified plant varieties which have not been authorised in the EU for cultivation purposes; or violates the prohibitive measures imposed for the duration of the safeguard procedure is guilty of a misdemeanour punishable by a maximum of 2 years of imprisonment.

## GMO-Free Labelling

In 2016 a new element of the Hungarian GMO-free strategy (European Parliament, Greens/EFA conference 2013) entered into force establishing the legal framework of GMO-free labelling.[11] The Decree provides the possibility of using specific labelling for food derived from GMO-free raw materials (including plants, meat, fish, egg, milk from animals fed with GMO-free feed and GMO-free honey and other apiary products). The legal framework was complemented by a GMO-free trademark system in 2018 with the aim of providing proper and sufficient information about the respective product, ensuring that it is from GMO-free production and gives freedom of choice to the consumers. The use of GMO-free labelling is, however, voluntary.

## Hungary's GMO-Free Policy

Hungary's GMO-free policy is largely based on the precautionary approach addressing the existing gaps and uncertainties in risk assessment and on scientific results that have proven adverse effects of GMOs. In addition, studies and surveys have clearly indicated that most Hungarians including farmers reject the use of GM plans

[10] Act No. C of 2012 on the Criminal Code, under Section 362 Violation of Legal Liabilities Relating to Genetically Modified Plant Varieties

[11] Decree No. 61/2016 (15 September) of the Minister of Agriculture on indicating the absence of genetically modified organisms

in agriculture and the food industry (i.e. Research Survey Report 2016). The majority consider GM food "unnatural", raising possibly health and economic concerns, and consider GM plants harmful to the environment and biodiversity (Research Survey Report 2016). GMO-free products have great advantages on the sowing seed and food markets. Consequently, the GMO-free strategy is highly important and advantageous for Hungary not only in terms of biodiversity protection but also as an economic incentive, because that can boost the competitiveness of our products on the global market and help us accessing new markets.

Since Hungary is amongst the five largest sowing-seed exporters in the world (Seed Exports 2017), proving and protecting the genetic purity and GMO-free status of seeds is an issue of primary importance for both producers and foreign trading partners. Therefore, it is crucial to guarantee that crops as well as seeds and propagating materials produced in Hungary are free from any GMOs, since no cultivation of any kind of genetically modified plant cultivar/hybrid is authorised in Hungary according to the legislation in force.

In addition, the Hungarian authorities strictly control the possible GMO content of sowing seeds. Official controls are applied to both Hungarian crops and imported seed lots coming from the EU and non-EU countries. All contaminated shipments are destroyed. Hungary has a zero-tolerance policy to the presence of GMOs in propagating materials and sowing-seeds.

In 2015, as part of this policy, Hungary launched the "Alliance for the GMO-free Europe" in order to preserve the GMO-free status of agriculture and food production. This initiative intends to reach and/or maintain the GM-free status of countries with concrete steps at European, regional, national and local levels in order to achieve GM-free agriculture and food production and contribute to the protection of biodiversity. Hungary is also a signatory to the "Danube-Soy Declaration" which indicated dedication to providing excellent quality GMO-free soy for the consumers. Building on this cooperation, Hungary, together with Germany, initiated the European Soy Declaration in 2017. Its goal is to recognise that European agriculture and food production need a comprehensive protein policy to counterbalance the amount of GM soy mainly imported from outside the EU and used in feed and food production (Tikász and Varga 2017). As part of the Hungarian Protein Strategy, the issue of primary importance is to have enough GMO-free alternatives to the imported GM soya. Several other steps have already been taken in order to decrease the import dependency on soybean and soya meal (i.e. designation of a new agricultural support scheme on the use of alternative protein sources and certain plant genetic resources).

## Importance of Genetic Resources

In recent decades Hungary has worked extensively to collect and preserve its plant genetic resources. The country's central gene bank, the National Biodiversity and Gene Conservation Centre, preserves both wild and agricultural genetic resources for future generations and is considered to be amongst the 20 largest agricultural gene banks from the more than 1750 in the world (FAO 2010). This immeasurable treasure includes several species and cultivars rich in proteins potentially suitable for use as soy bean alternatives.

## Summary

The Hungarian GMO-free agricultural strategy encompasses a wide range of measures and activities. It is based on the precautionary principle and takes into consideration the interest of future generations, contributes to the protection of biodiversity, creates economic benefits for farmers and food producers and possibly also brings economic benefits in the fields of health care and tourism. For Hungary, being a major sowing-seed producer, it is crucial to keep its agriculture free from genetically modified organisms. GMO-free seeds, propagating materials and other GMO-free products have a specific added value in international markets and may help to improve the country's export position.

## References

Darvas B (ed) (2006) Mezőgazdasági géntechnológia – elsőgenerációs GM-növények. Agricultural Committee of the Hungarian Parliament, Budapest

Darvas B, Székács A (eds) (2011) Hungarian background on views of 1st generation genetically modified plants. Agricultural Committee of the Hungarian Parliament, Budapest

European Parliament (2013) Greens/EFA conference. GMO-free labelling of food products. https://www.greens-efa.eu/en/article/gmo-free-labelling-of-food-products/

FAO (2010) The second report on the state of the World's plant genetic resources for food and agriculture. FAO, Rome

NRDS (2012–2020) National Rural Development Strategy http://videkstrategia.kormany.hu/download/6/ae/40000/DIT2_angol_t%C3%B6rdelt_120910.pdf

Research Survey Report (2016) Századvég Foundation. Ordered by the Ministry of Agriculture

Seed exports (2017) International Seed Federation https://www.worldseed.org/resources/seed-statistics/

National Reports on the implementation of the Cartagena Protocol on Biosafety (2005, 2007, 2011, 2015, 2019). https://bch.cbd.int/

Tikász IE, Varga E (2017) Perspectives of non-GM soybean based value chains in Hungary's livestock industries. In: Szendrő K, Horváthné KB, Barna R (eds) Proceedings of the 6th international conference of economic sciences. Kaposvár University, Kaposvár, pp 391–404

# Italy—GMOs and Synthetic Biology Rules/Regulations and Biodiversity: The Legal Perspective of Italy

**Valentina Rastelli, Giovanni Bacaro, Simona Baima, Marzia De Giacomo, Valeria Giovannelli, Vincenza Ilardi, Matteo Lener, Stefano Mocali, Annalisa Paternò, Biancamaria Pietrangeli, Giovanni Staiano, and Elena Sturchio**

**Abstract** The Italian regulatory framework on genetically modified organisms (GMOs) follows the cardinal principles of the European Union legal order. The incomplete implementation of the Italian legislation has led to a de facto moratorium of the deliberate releases of GMOs into the environment, also for experimental purposes, thus slowing down the Italian research in this field. The most recent techno-

---

V. Rastelli (✉) · V. Giovannelli · M. Lener · G. Staiano
Dipartimento per il monitoraggio e la tutela dell'ambiente e per la conservazione della biodiversità, Istituto Superiore per la Protezione e la Ricerca Ambientale (ISPRA), Rome, Italy
e-mail: valentina.rastelli@isprambiente.it

G. Bacaro
Dipartimento di Scienze della Vita, Università di Trieste, Trieste, Italy

S. Baima
Consiglio per la ricerca in agricoltura e l'analisi dell'economia agraria, Centro di ricerca Genomica e Bioinformatica – Sede di Roma, Rome, Italy

M. De Giacomo
Dipartimento di sicurezza alimentare, nutrizione e sanità pubblica veterinaria, Istituto Superiore di Sanità (ISS), Rome, Italy

V. Ilardi
Consiglio per la ricerca in agricoltura e l'analisi dell'economia agraria, Centro di ricerca Difesa e Certificazione, Rome, Italy

S. Mocali
Consiglio per la ricerca in agricoltura e l'analisi dell'economia agraria, Centro di ricerca Agricoltura e Ambiente – Sede di Firenze, Florence, Italy

A. Paternò
National Reference Laboratory for GM Food and Feed (NRL-GMFF), Istituto Zooprofilattico Sperimentale del Lazio e della Toscana (IZSLT) M. Aleandri, Rome, Italy

B. Pietrangeli · E. Sturchio
Dipartimento innovazioni tecnologiche e sicurezza degli impianti, prodotti e insediamenti antropici, Istituto nazionale Assicurazione Infortuni sul Lavoro (INAIL), Rome, Italy



A. Chaurasia et al. (eds.), *GMOs*, Topics in Biodiversity and Conservation 19,
https://doi.org/10.1007/978-3-030-53183-6_40

logical developments opened new perspectives for research and applications, posing new challenges for the regulatory system. Synthetic biology is one of these new challenges: even if in Italy there is still a growing debate on whether the application of the existing legislation on GMOs to some of the organisms resulting from the applications of synthetic biology is possible, training and research activities are already under way. We would like to emphasize that, although the present GMOs regulatory framework is effective to preserve biodiversity, further improvements could be needed and should be focused on simplifying the authorization procedure for certain products. It is also necessary to promote and guarantee research and experimentation, in order to provide policy makers with science-based decision support system, and not to keep Italy out of the opportunities offered by technological advances.

**Keywords** Italy · Genetically modified organisms · Biodiversity · Regulatory framework · Precautionary approach · Risk assessment and risk management · Biosafety · Synthetic biology · Modern biotechnology · Research

## GMOs: Existing Regulations and How They Are Addressing Biodiversity Issues

In Italy the regulatory framework for the authorization of genetically modified organisms (GMOs) follows the European Union (EU) legal order, which is composed of two main instruments: (1) Directive 2001/18/EC on the deliberate release into the environment of GMOs, for experimental purposes and placing on the market, including cultivation (transposed into national law with Legislative Decree n° 224 dated 8 July 2003), and (2) Regulation (EC) 1829/2003 on genetically modified food and feed, including cultivation of plants for food and feed uses.[1] A further legislative instrument made compulsory the traceability and labelling of GMOs and the traceability of food and feed products produced from GMOs (Regulation (EC) 1830/2003).

The National Competent Authority (NCA), responsible for complying with the requirements of Directive 2001/18/EC on the deliberate release into the environment of GMOs, is the Italian Ministry of the Environment, Land and Sea; the NCA responsible for implementing Regulation (EC) 1829/2003 at national level is the Italian Ministry of Health, which also carries out management and coordination activities of the official controls on the presence of GMOs in food and feed, envisaged by the Italian local authorities. The NCAs can be supported by Advisory Committees: for the deliberate release, the Italian Institute for Environmental Protection and Research (ISPRA) has taken on this role from September 2018.

[1] Within the European Union, regulations are binding legislative acts that must be applied in their entirety across the EU, while directives are legislative acts that set out goals that all EU Member States must achieve, but it is up to the individual States to devise their own laws on how to reach these goals. For this reason, directives have to be transposed into national law.

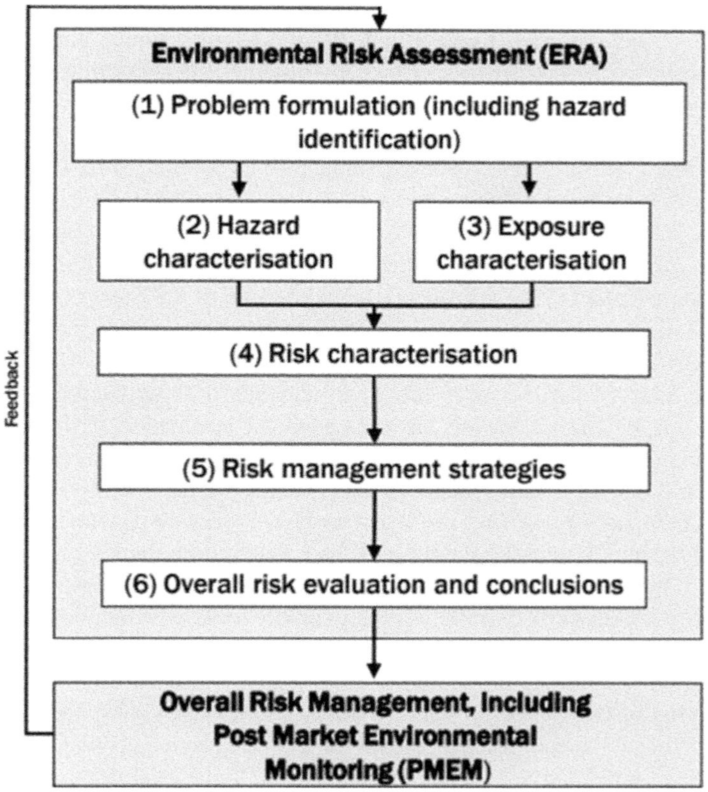


**Fig. 1** Six steps within the environmental risk assessment (ERA) and relationship to risk management, including monitoring, according to Directive 2001/18/EC and Regulation (EC) No. 1829/2003. (Reproduced from EFSA 2010)

The main objectives of this regulatory framework on GMOs are to ensure a high level of protection of the environment and of human and animal health and to protect the consumers' interests while ensuring the internal market functioning. The key points are as follows: (1) a precautionary approach; (2) a case-by-case approach; (3) an environmental risk assessment (ERA) of the deliberate release of a GMO (the assessment should not only address the possible direct and immediate effects of releasing the GMO but also any indirect and delayed effects on human health and the environment, as well as cumulative long-term effects) and related monitoring activities; (4) a step-by-step approach, used for the ERA (assessment based on six steps) (Fig. 1); and (5) a stepwise approach, used for the introduction of GMOs into the environment, which means that the containment of a GMO is gradually reduced and the scale of the release increased, but only if the evaluation of the earlier steps in terms of the human health and environment protection indicates that the next step can be undertaken.

Directive 2001/18/EC was modified over the years and consequently also the Italian national law. In 2015, in order to meet the demands of several Member States and to grant the freedom of choice of consumers, farmers and operators, Directive 2001/18/EC was amended by Directive (EU) 2015/412, which entitles Member States to have the possibility to adopt legally binding acts restricting or prohibiting the cultivation of GMOs in their territory, without affecting the risk assessment provided in EU authorization procedures for GMOs. In line with this new Directive, Italy enacted Legislative Decree n° 227 dated 14 November 2016 and to date has requested the restriction of the cultivation for six events of GM maize (MON810, 1507, 1507x59122, 59122, Bt11, GA21). Directive (EU) 2018/350 further amended Directive 2001/18/EC, updating the technical Annexes on the environmental risk assessment in order to take into account the guidance of the European Food Safety Authority on the environmental risk assessment of genetically modified plants (EFSA 2010). This Directive has been transposed into national law with the Decree of the Ministry of the Environment, Land and Sea n° 108 dated 18 June 2019. Regulation (CE) 1829/2003 was also amended, by Regulation (EU) 503/2013, in order, among other things, to take into account the abovementioned EFSA guidance.

Further Italian legal tools are as follows: (1) Decree of the Ministry of Agriculture, Food and Forestry Policies and Tourism dated 19 January 2005, which provides that the environmental risk assessment has to be integrated to also assess the potential impact on agro-biodiversity, that specific crop management measures are put in place and that field trials must be limited to public local sites identified by authorities; (2) Law n° 5 dated 28 January 2005 that establishes the necessary measures to ensure coexistence among transgenic, conventional and organic agriculture in case of cultivation of GMOs for commercial purposes, also providing the possibility to identify dedicated areas for each cultivation; and (3) Decree of the Ministry of the Environment, Land and Sea dated 8 November 2017 that establishes a general plan for the environmental surveillance, including field controls, of the deliberate release into the environment of GMOs. Finally, a system of controls to verify the presence of unauthorized GMOs, both for cultivation and for marketing, has been set up, including the national network of veterinary institutes (IZS), the National Environmental Protection System (SNPA), customs agencies, local plant health services and centres for seed certification.

Italy has implemented international agreements strictly related to biodiversity and biosafety: the Convention on Biological Diversity (CBD) was ratified by Law n. 124 dated 14 February 1994 and the Cartagena Protocol on Biosafety (CPB) to the Convention on Biological Diversity with Law n. 27 dated 15 January 2004. Regarding the Nagoya-Kuala Lumpur Supplementary Protocol on Liability and Redress to the Cartagena Protocol on Biosafety, Italy finalized the ratification process in 2019 (Law n. 7 dated 16 January 2019).

In conclusion, Italian legislation on GMOs is particularly stringent. Unfortunately, this legislation was not properly implemented, thus leading to a de facto moratorium of experimental field trials since 2005. Despite this, several research activities were still carried out, under field (without using GMOs but plants with variety-specific molecular/genetic markers) and contained greenhouse conditions, focused on the evaluation of GMOs impacts on biodiversity, taking into account the territorial, landscape and agro-biodiversity peculiarities of Italy. Some examples of these activities, before and after the moratorium, are as follows: Camastra et al. 2014 (who propose a software tool, TÉRA, based on a fuzzy inference engine, for the environmental risk assessment of genetically modified plants); Canfora et al. 2014 (within the Life+ Project MAN-GMP-ITA, this subgroup analysed soil in order to obtain soil health and fertility indicators to be used as baselines in the environmental risk assessment model developed by the Project); Castaldini et al. 2005 (a polyphasic approach, with microcosm and greenhouse experiments, has been developed to gain knowledge of suitable key indicators for the evaluation of environmental impact of genetically modified Bt 11 and Bt 176 corn lines on soil ecosystems); Ilardi and Barba 2002 and Tomassoli et al. 2004 (GM tomato lines resistant to a virus were produced and assessed for transgene flow); Lener et al. 2013 (within the Life+ Project MAN-GMP-ITA, this group validated and improved an existing methodology, developed always by Italian researchers, for the environmental risk assessment of GM plants, in order to achieve a decision support system); Manachini et al. 2018 (evaluation of the potential exposure of butterflies as results of possible cultivation or naturalization of spilled seed of oilseed rape in Sicily); Mocali et al. 2009 (a multidisciplinary approach used to assess the effects of GM eggplants on soil quality and microbial diversity after two different treatments: cutting up and extirpation); Turrini et al. 2004 (an experimental model system developed to monitor the impact of genetically modified plants, events Bt11 and Bt176 and aubergine plants expressing Dm-AMP1 defensin on arbuscular mycorrhizal fungi). All the aforementioned surveys showed that the experimental model systems and multimodal approach used in greenhouse and field trials were suitable assays of the impact of GMOs on biodiversity and pointed out the importance of evaluation on a case-by-case basis.

We also highlight that the possible application of the present legislation on GMOs, including also risk assessment and risk management procedures, to organisms produced by the new techniques of genetic engineering is under debate and discussions, at the European and Italian levels.

## Synthetic Biology: Existing Regulations and How They Are Addressing Biodiversity Issues

Synthetic biology, a new interdisciplinary branch of biology, can include diverse fields of research and a broad range of applications. For these reasons, there is an ongoing wide scientific and socio-economic debate on its definition, at both national and international levels.

Italy supported the operational definition of synthetic biology agreed during the Thirteenth Conference of the Parties to the CBD in 2016: "Synthetic biology is a further development and new dimension of modern biotechnology that combines science, technology and engineering to facilitate and accelerate the understanding, design, redesign, manufacture and/or modification of genetic materials, living organisms and biological systems". This operational definition was approved as a starting point with the clear aim of facilitating further discussions and resolutions on this matter, until a clear and unambiguous definition can be globally accepted. This definition is more comprehensive than that proposed by three EU Scientific Committees (SCENIHR et al. 2014): "Synthetic Biology is the application of science, technology and engineering to facilitate and accelerate the design, manufacture and/or modification of genetic materials in living organisms", as it includes techniques involving cell-free systems not resulting in the development of living organisms.

While technological developments are advancing at an accelerated rate, to date in Italy and in the EU there is no relevant or specific legislation regarding synthetic biology. Nevertheless, in examining whether and how organisms, components and products resulting from the applications of synthetic biology could be regulated, a first analysis confirms that most living organisms already developed using these techniques and approaches can be evaluated within the GMOs legislative framework. Conversely, components and products resulting from the applications of synthetic biology and entirely new organisms that will be developed in the future with these applications may require a further assessment to clarify if there is a need to develop specific regulations according to the intended use or if other existing legislative frameworks, such as those applied to chemicals or plant protection products, can be adapted. Concerning the evaluation of the potential risks posed by organisms, components and products resulting from synthetic biology, Italy, in agreement with the EU position, highlights the importance of applying a precautionary approach, as well as carrying out risk assessments on a case-by-case basis and following the step-by-step approach.

The need of a precautionary approach to synthetic biology was agreed during the last Conference of the Parties to the CBD in 2018 (decision CBD/COP/DEC/14/19). Negotiation on the text of this decision was very intense: one of the most discussed points was related to the introduction of organisms containing engineered gene drives into the environment. The final agreed text calls upon the Parties to apply a precautionary approach and to only consider introducing organisms containing engineered gene drives into the environment, including for experimental releases and research and development purposes, when scientifically sound case-by-case risk assessments have been carried out and risk management measures are in place. Such a decision foresees the need to perform a regular horizon scanning, monitoring and assessment of the most recent technological developments for reviewing new information on the potential positive and negative impacts of synthetic biology. The EU was one of the main actors at the negotiating table on this decision, strongly promoting the insertion of a text as cautious as possible, and Italy supported this position.

Italy is not only participating and following these discussions at the international level but is also promoting a debate at the national level, involving stakeholders, industries, research institutes, universities, and decision makers. At present, most initiatives carried out in Italy on this topic focus on communication and/or training, but research activities are also taking place. Two examples are the International Synthetic and Systems Biology Summer School, with its sixth course in 2019 (organized by the Scuola Normale Superiore, Pisa), and Synthetic and Systems Biology for Biomedicine (a research line carried out by the Italian Institute of Technology).

## Final Considerations

The Italian legislation on GMOs, which follows the European Union one, can be considered an exhaustive but rather cumbersome legislation. Indeed, while it guarantees the preservation of biodiversity from potential negative impacts related to the deliberate releases of GMOs into the environment, at the same time its incomplete application led to an undesired slowdown of Italian research in this field.

In the past 15 years, the technological developments in genetic engineering advanced at an accelerated rate and the related research activities have been focused also on safety aspects, to meet some of the concerns raised on GMOs. Indeed, these new developments may allow us to limit and/or reduce unintended or side effects and thus to facilitate the risk management. For this reason, simplified procedures, based on evidence and experience accumulated in these years, could be the correct way to facilitate authorization procedures, and the restart of scientific research and experimentation in the field in Italy would be both welcome and advisable. Indeed, one of the excellences of Italian research is the genomics of cultivated plants and the genomes of several typical Italian crops were sequenced. This knowledge and the restart of research activities are among our most important resources to enable Italian traditional and local varieties, characterized by an extraordinary diversity and high quality, to cope with emerging biotic and abiotic stresses, within a sustainable agriculture.

Another consideration is that the rapid evolution of biotechnology is often associated with expectations and fears regarding possible impacts on human health and on the environment. Italian governmental and research institutions are working on strengthening communication among researchers, risk assessors, decision makers and the general public. In addition, Italian research scientists are deeply involved in the dissemination of the acquired knowledge, trying to make their research known to the general public and to point out possible practical applications of their research and related tools.

Research projects in synthetic biology and interest in the economic potential of bioproducts have exponentially increased in recent years. Although these applications and products may bring benefits to society, there remain many scientific uncertainties over the development of synthetic life, cells and genomes, especially in terms of their impact on biodiversity (Science for Environment Policy 2016).

In the future, synthetic organisms could be developed in such a way that they will fundamentally differ from the naturally occurring ones. Thus, for a timely environmental risk assessment, existing methodologies and guidance may need to be adapted and improved. For example, it could be difficult to identify an appropriate comparator, to gather relevant information in order to perform characterization of potential hazards, to identify routes of exposure and adverse effects arising from the integration of protocells into living organisms, to predict the behaviour and impacts of new xenobiological organisms (SCENIHR et al. 2015). Furthermore, as synthetic biology is a rapidly evolving technology, risk assessment methodology should be revised at regular intervals.

Lastly, we emphasize that further improvements, also necessary for GMOs risk assessments, should generally be focused on developing instruments, such as predictive and/or simulation models, databases and networks, with the aim of simplifying, standardizing and harmonizing data production and collection.

The key messages from our experience are highlighted in Box 1.

**Box 1: Key Messages from Italian Experience**

- Technological developments within the biotechnology domain are advancing at an accelerated rate and have been focused on safety aspects, also to meet some of the concerns raised over the past years regarding GMOs. These developments allow us to limit and/or reduce unintended or side effects. For this reason, we highlight the need to work towards simplified procedures within the regulatory . This simplification should not lead to deregulation of future products and organisms, but should rather allow the strengthening of case-by-case assessment (i.e. regulation commensurate with the scientifically assessed level of risk), strictly related to the product/organism characteristics.
- Taking into account that we are facing rapidly evolving technologies, a fast revision of risk methodologies and risk management procedures could be advisable, together with guidance for the development of monitoring plans.
- Further improvements should also focus on developing instruments, such as predictive and/or simulation models, and networks, with the aim of simplifying, standardizing and harmonizing data production and collection.
- Cost-benefit analysis should be improved to guarantee protection of and full consideration of alternative options.
- Strengthening the communication among researchers, risk assessors, decision makers and the general public and increasing the dissemination of the acquired knowledge are crucial points to gain the trust of the general public and institutions.
- Research and experimentation in the field must soon be resumed in , in order to provide policy makers with science-based decision support systems and so as not to keep out of opportunities offered by technological advances.
- A first analysis confirms that most of the living organisms already developed within can be evaluated using the GMOs legislative framework, while components, products and new future organisms could require a further assessment to clarify if there is a need to develop specific according to the intended use or if other existing legislative frameworks can be adapted.
- For evaluation of the potential risks posed by organisms, components and products resulting from the applications of , highlights the importance of applying a , as well as carrying out risk on a case-by-case basis and following the step-by-step approach.

**Disclaimer** This text presents the authors' views that are not necessarily those of the Institutions they belong to nor of the Italian government.

## References

Camastra F, Ciaramella A, Giovannelli V et al (2014) TÉRA: a tool for the environmental risk assessment of genetically modified plants. Eco Inform 24:186–193

Canfora L, Sbrana C, Avio L et al (2014) Risk management tools and the case study *Brassica napus*: evaluating possible effects of genetically modified plants on soil microbial diversity. Sci Total Environ 493:983–994

Castaldini M, Turrini A, Sbrana C et al (2005) Impact of Bt corn on rhizospheric and soil eubacterial communities and on beneficial mycorrhizal symbiosis in experimental microcosms. Appl Environ Microbiol 71(11):6719–6729

European Food Safety Authority (EFSA) - Panel on Genetically Modified Organisms (2010) Guidance on the environmental risk assessment of genetically modified plants. EFSA J 8(11):1879

Ilardi V, Barba M (2002) Assessment of functional transgene flow in tomato fields. Mol Breed 8(4):311–315

Lener M, Giovannelli V, Arpaia S et al (2013) Applying an operating model for the environmental risk assessment in Italian sites of community importance (SCI) of the European Commission habitats directive (92/43/EEC). Bull Insectol 66(2):257–267

Manachini B, Bazan G, Schicchi R (2018) Potential impact of genetically modified Lepidoptera-resistant *Brassica napus* in biodiversity hotspots: Sicily as a theoretical model. Insect Sci 25(4):562–580

Mocali S, Dentice A, Marcucci A et al (2009) The impact of post-harvest treatments of transgenic eggplant residues on soil quality and microbial diversity. Agrochimica 53(5):296–307

SCENIHR (Scientific Committee on Emerging and Newly Identified Health Risks), SCCS (Scientific Committee on Consumer Safety), SCHER (Scientific Committee on Health and Environmental Risks) (2014) Synthetic biology I definition. Opinion, 25 September, 2014. https://ec.europa.eu/health/scientific_committees/emerging/docs/scenihr_o_044.pdf.

SCENIHR (Scientific Committee on Emerging and Newly Identified Health Risks), SCHER (Scientific Committee on Health and Environmental Risks), SCCS (Scientific Committee on Consumer Safety) (2015) Synthetic biology II – risk assessment methodologies and safety aspects. Opinion, May 2015. https://ec.europa.eu/health/scientific_committees/emerging/docs/scenihr_o_048.pdf.

Science for Environment Policy (2016) Synthetic biology and biodiversity. Future brief 15. Produced for the European Commission DG Environment by the science communication unit, UWE, Bristol. http://ec.europa.eu/science.

Tomassoli L, Lumia V, Zaccaria A et al (2004) Genetically modified tomato: chance of gene transfer by cross pollination. Petria 14(1):29–39

Turrini A, Sbrana C, Nuti MP et al (2004) Development of a model system to assess the impact of genetically modified corn and aubergine plants on arbuscular mycorrhizal fungi. Plant Soil 266:69–75

# Norway—The Norwegian Gene Technology Act: Presenting Case Studies to Illustate the Act's Advances in Protecting Biodiversity

**Anne Ingeborg Myhr, Idun Merete Grønsberg, and Arinze Stanley Okoli**

**Abstract** Here we present the scope and administrative practice of the Norwegian Gene Technology Act. The main focus is on the objective of the environmental risk assessment which acknowledges both direct and indirect effects as well as immediate, delayed and cumulative effects on the environment. We describe the prohibition of two vaccines and seven applications of GM rapeseed. These applications are approved in Europe, but forbidden in Norway after taking environmental and biodiversity risks into consideration. The only approved GMOs in Norway are the import of five carnations, where the risk to the environment was not relevant. With the emergence of gene-edited organisms Norway does, as many other countries, discuss how to regulate, monitor and trace these organisms which represent a challenge to present regulative framework.

**Keywords** Norwegian Gene Technology Act · GMO · GM crops · Biodiversity · Environmental risk assessment · Gene-edited plants · GM vaccine · GM oilseed rape

## The Norwegian Gene Technology Act

The Norwegian Gene Technology Act came into force in 1993 (Norwegian Gene Technology Act 1993). According to Chapter 1 General provision Para 1: "The purpose of this Act is to ensure that the production and use of genetically modified organisms and the production of cloned animals take place in an ethically justifiable and socially acceptable manner, in accordance with the principle of sustainable development and without adverse effects on health and the environment".

The Act state that genetically modified organisms (GMOs) may only be approved when there is no risk of adverse effects on human or other animal health or the environment, and that "considerable weight shall be given to whether the deliberate

---

A. I. Myhr (✉) · I. M. Grønsberg · A. S. Okoli
GenØk-Centre for Biosafety, SIVA Innovation Centre, Tromsø, Norway
e-mail: anne.i.myhr@uit.no

 
A. Chaurasia et al. (eds.), *GMOs*, Topics in Biodiversity and Conservation 19,
https://doi.org/10.1007/978-3-030-53183-6_41

release will be of benefit to society and is likely to promote sustainable development" (Para 10,2). The societal utility criterion in the Act are only relevant for impacts of the product within Norway, whereas sustainability may take into consideration factors that evolves over longer term, with an impact on global developments.

The Act is separated in one section for contained use and one for deliberate release. The contained section governs GMOs produced, grown, stored, destroyed or used in physically, chemically or biologically contained environments. Activities considered as deliberate release include field experiments, bioremediation and disposal of waste containing GMOs. Deliberate release also concerns commercial environmental use of GMOs and marketing of products consisting of or containing GMOs. The Act was extended in 1996 to cover import and transport of GMOs.

Norway is not a part of the European Union (EU) but due to Norway's obligations through the European Economic Area (EEA) agreement, the EU Directive 2001/18/EC is applicable to Norway. Consequently, this means that an approval of a GMO in the EU, automatically leads to an approval in Norway, unless it becomes prohibited. Both the EU and Norway requires that before a GMO can be approved a health and environmental risk assessment must be carried out. In addition, it is required in Norway that effects on sustainability, societal benefit and ethical aspects are evaluated for GMOs that are regulated by the Act. The application of the Directive 2001/18/EC was based on a compromise, that Norway can reject authorizations on the basis of concerns other than health and the environment (Rogers 2015). This compromise is similar to the later implemented EU Directive 2015/412/EU which allows EU member states to restrict or prohibit GMOs on social grounds. As of June 2020, the Norwegian authorities have approved five GM plant applications. These GM plants are GM carnations with changed colours. Fourteen applications approved in the EU have been forbidden in Norway. This includes 11 GM plants, two vaccines and one test kit that contains GM microorganisms (Lovdata 2020). In this chapter, we will describe how environmental and biodiversity issues were acknowledged in some of the prohibited cases: the two vaccines (against rabies and pseudorabies) and the four applications for deliberate use and release of GM oilseed rape.

## *The Administrative Process Under the Act*

When applications are considered under the Act, certain information must be made publicly available. Such information concerns description of the GMOs, the identity of the user, the purpose of the release and target area of release. In addition, monitoring strategies and the applicant's assessment of foreseeable consequences must be included. According to the Act (Section 4(b)), GMOs are defined as: organisms altered through the use of gene technology or cell technology.

The administrative practice is based on a case-by-case review and a step-by-step approach. During its form of procedure, the Ministry of Climate and Environment handles the recommendation from the Norwegian Environment Agency. The Agency coordinates the decision-making process which is based on advice from the Norwegian Biotechnology Advisory Board and the Norwegian Scientific Committee for Food and Environment (VKM) as well as a public hearing. The Board has the responsibility for assessment of the GMOs contribution to sustainability, the social utility and if there are any ethical aspects. VKM has the responsibility to carry out environmental risk assessments (ERA) for the Agency and health risk assessment for the Norwegian Food Safety Authority.

An important section of the Act concerns liability. When activities that are regulated by the Act cause damage, inconvenience or loss, the person responsible for the activity has liability for the damage. This would, for example, include liability for changes in an ecosystem due to reduced biodiversity after introduction of a GMO. This clause has not yet been implemented in Norway since no damage by GMO has been reported.

## Impact Assessment Under the Norwegian Gene Technology Act

In Para 11 of the Act it is stated: "Applications for the approval of deliberate release pursuant to section 10 shall include an impact assessment setting out the risk of adverse effects on health and the environment and other consequences of the release".

The objective of the impact assessment is according to Appendix 2: "On a case by case basis, to identify and evaluate potential adverse effects of the genetically modified organism, either direct or indirect, immediate or delayed, on human health and the environment which the deliberate release or the placing on the market of genetically modified organisms may have." The same objective can be found in Article 4(2) of Directive 2001/18/EC. The Norwegian Act contains stronger requirements than the Directive 2001/18/for how to handle uncertainties, as the preparatory work of the Act emphasizes that the precautionary principle should apply in such situations. The Norwegian authorities has also acknowledged the principle in restrictive decisions. The Norwegian regulative system does also request information about how the GMO contributes to sustainability, the societal benefits and whether the production and use can take place in an ethically and socially justifiable way. However, until now it has been difficult to achieve such information from the applicants. The implementation of Directive 2015/412/EU do not involve a request of such non-safety information in the EU regulative system.

## Differences in Decisions on GMOs Based on Biodiversity Considerations

To highlight the importance of environmental issues in decisions on the deliberate release and use of GMOs, we have choosen to focus on cases where decisions based on ERA differs from decisions done in other regions or countries. For Norway the most appropriate comparator is the EU, especially since Norway considers the same applications. As per June 2020 there are 14 cases with different decisions to those taken in EU (see Table 1). Several of these cases concerns GM plants that contains antibiotic resistance genes that represent a risk to health, one GM maize has been forbidden due to ethical reasons, while risks to the environment was important in two vaccines and in several of the GM rapeseed applications.

### *Environmental Risk Assessment (ERA) of GM Virus-Vector Applications*

GM virus vector applications includes GM virus-based vaccines (GM vaccines) and gene therapy products. With GM vaccines the aim is to provide a prophylaxis to protect healthy individuals from the disease that would normally be caused be a more virulent wildtype strain, therefore, a large group of individuals (humans and animals) are often targeted during a vaccination program. In GM therapy, the modified virus is used to cure an existing disease disorder, e.g. cancer or congenital diseases, and hence the recipients of the GM products are often isolated individuals that are confined in special facility. Therefore, there is a higher risk of introduction and subsequent spread of GM vaccine viruses in the environment compared to GM therapy products. The ERA of GM vaccines is focused on the GM virus-vector, i.e. the changes on the characteristics of the virus as a result of the gene modification(s). Viruses most commonly used as GM vectors are *Herpesvirus*, *Adenovirus* and *Poxvirus* (Okoli et al. 2016; Lundstrom 2018). The environment in the ERA is considered as the entire ecosystem surrounding the recipient of the GM vaccine/therapy product. i.e. both humans, animals and microorganisms other than the human/animal being vaccinated. The same stringency in ERA is required for both clinical trials and products seeking market approval. However, higher stringency is placed on the ERA of replication competent GM virus vector (RCVV) than replication incompetent vectors, the former having higher capacity to impact biodiversity if introduced into the environment. Other biodiversity-related considerations in ERA of GM vaccines include genetic and genome stability, phenotypic stability of expressed transgene, host range of modified virus, ability of the GM virus to be shade by host, survivability of the GM virus in the environment, ability of the GM virus to recombine with the wild-type virus, and potential spread between target and non-target

**Table 1** GMOs prohibited in Norway (based on Lovdata and EU Register of Authorized GMOs)

| GMO case (Event/ Name) | Approved in EU | Prohibited in Norway | Reason for prohibition in Norway |
|---|---|---|---|
| GM rabies vaccine (RABORAL) | 19.10.1993 | 01.10.1997 | Risk to health and the environment. Risk for recombination with naturally circulating viruses and potentially non-target effects and long-term effects. No social utility since rabies is not found in Norway. |
| GM tobacco (C/F/93/08–02) | 08.06.1994 | 01.10.1997 | Risk to health for humans and animals. Contains antibiotic resistance genes that can spread to pathogenic bacteria. No social utility and access to alternative production systems. |
| GM pseudo-rabies vaccine (Nobi-Porvac Aujeszky) | 18.07.1994 | 01.10.1997 | Risk to health and the environment. Risk for recombination with naturally circulating viruses and potentially non-target effects and long-term effects. No social utility since pseudorabies is not found in Norway. |
| GM oilseed rape (MS1 x RF1 (PGS1)) | 06. 02.1996 (authorization is expired) | 01.10.1997 | Risk to health for humans and animals. Contains antibiotic resistance genes that can spread to pathogenic bacteria. No social utility and access to alternative production systems. |
| GM chicory (RM3–3, 3–4, 3–6) | 20.05.1996 | 01.10.1997 | Risk to health for humans and animals. Contains antibiotic resistance genes that can spread to pathogenic bacteria. No social utility and access to alternative production systems. |
| GM maize (Bt176) | 23. 01.1997 (authorization is expired) | 01.10.1997 | Risk to health for humans and animals. Contains antibiotic resistance genes that can spread to pathogenic bacteria. No social utility and access to alternative production systems. |
| GM oilseed rape (MS1 x RF2 (PGS2)) | 06. 06.1997 (authorization is expired) | 01.10.1997 | Risk to health for humans and animals. Contains antibiotic resistance genes that can spread to pathogenic bacteria. No social utility and access to alternative production systems. |
| GM test kit from Valio Oy | 14.07.1997 | 15.12.2000 | Risk to health for humans and animals. Contains antibiotic resistance genes that can spread to pathogenic bacteria. |
| GM oilseed rape (Topas 19/2) | 22.04.1998 (authorization is expired) | 14.12.2012 | Risk to health for humans and animals. Contains antibiotic resistance genes that can spread to pathogenic bacteria. No social utility and access to alternative production systems. |

(continued)

**Table 1** (continued)

| GMO case (Event/ Name) | Approved in EU | Prohibited in Norway | Reason for prohibition in Norway |
|---|---|---|---|
| GM maize (1507) | 08.08.2005 (authorization is expired) | 02.06.2017 | Ethical reason. The GM plant is tolerant to the herbicide glufosinate ammonium, which is prohibited in Norway. No social utility and access to alternative production systems. |
| GM oilseed rape (GT73) | 31.08.2005 | 14.12.2012 | Risk to environment by spread of the GM plant or by the transgene. No social utility and access to alternative production systems. |
| GM oilseed rape (Ms8) | 26.03.2007 | 02.06.2017 | Risk to environment by spread of the GM plant or by the transgene. No social utility and access to alternative production systems. |
| GM oilseed rape (Rf3) | 26.03.2007 | 02.06.2017 | Risk to environment by spread of the GM plant or by the transgene. No social utility and access to alternative production systems. |
| GM oilseed rape(Ms8xRf3) | 26.03.2007 | 02.06.2017 | Risk to environment by spread of the GM plant or by the transgene. No social utility and access to alternative production systems. |

species (Okeke et al. 2017). Non-target organism is the organism that is not the target of vaccination, but which may inadvertently be infected or affected by the vaccination program.

Although the rabies and pseudorabies GM vaccines are approved for use in the EU, they are prohibited for use in Norway since they were considered to represent risks to the environment (see Table 1). In particular, the GM vaccines are assessed as having the ability to recombine with naturally circulating viruses. Further, the Norwegian authorities pointed out that the modified virus vectors in GM rabies and pseudorabies vaccines can potentially spread to non-target organisms and can also potentially have delayed or long-term effects. Since no baseline study has been conducted on the naturally circulating vaccine-relevant viruses in Norway, it is difficult to predict if recombination between the vaccine strains and wildtype strains can happen in the field or if the use of GM vaccines can have other impacts on the biodiversity. Rabies and pseudorabies do not constitute a problem in Norway, there is no incidents of rabies in Norwegian wildlife or of pseudorabies (Aujeszky's disease) in domesticated animals, and thus, the utility of the vaccination program was considered low. The Ministry of Climate and Environment therefore concluded that there was no need to take any risk since there where low utility and that the approval of the vaccines would be in conflict with the precautionary principle. In some of the EU countries rabies in foxes and other animals is a serious problem necessitating the need for approval of the oral vaccine. There is, however, no available data showing that baseline studies were conducted prior to vaccination in the EU by the GM rabies vaccine, but surveillance and monitoring are being conducted post vaccination (Müller et al. 2009, SGE RAB 2019).

Domesticated and pet animals, especially cats and dogs, are regularly taken on visits to Europe and other parts of the world where they may be vaccinated and/or be in contact with GM virus vaccines which are not approved in Norway. Hence, there may be a need of a Norwegian register of pets and domesticated animals that have been vaccinated with vaccines not approved in Norway.

## *Environmental Risk Assessment (ERA) of GM Rapeseed Applications*

In a Norwegian perspective, many of the GM plant applications as GM soy, maize and cotton are not of high importance as cultivation plants due to climatic reasons. However, oilseed rapes (*Brassica napus*) are cultivated at a small scale in Norway, making GM rapeseed most relevant, together with GM potato. According to the Norwegian Agriculture Agency, production of oilseed as a whole has varied between 6.0 (2013-2014) to 18.7 tons (2001-2002). In the last 5 years it has increased from 6.0 to 10.2 tons/year. The agency does not distinguish between seeds like soy, cotton, sunflower, linen, sesame and rapeseed etc. Approximately 2% (40,000 acres) of the grain harvested is oilseed rape (Yara 2014). We will focus on GM oilseed rape further in this section.

All applications under the 1829/2003 directive (food and feed purposes) of GM oilseed rape has been forbidden in Norway (Lovdata 2020). Three of these GM plants was forbidden since they contained antibiotic resistance genes, while four has been forbidden to avoid spread of the transgene (herbicide tolerance) to wild relatives. GM oilseed rape belongs to a family of plants that has relatives in Norway, growing as far north as Finnmark. Since wild relatives, for example the *Brassica rapa* (including its subspecies), have a larger growth and climatic span than the cultivated oilseed rape, the potential for hybridization and spread of unwanted traits, are of interest.

The relevant transgenes that are of special concern are those that introduce tolerance against herbicides and antibiotics, but potentially also other important traits as resistance to fungi. It has therefore been important to look at the potential consequences for incorporated traits potentially being transferred to populations of relatives. A scenario of spread of herbicidetolerance genes creates concerns for a potential build-up of tolerance in non-GM oilseedrape and in feral populations (or wild relatives). The spread of transgenes or escaping GM oilseed rape plants have been investigated in Canada which has similar climatic conditions as Norway, where it was found that escaped GM oilseed rape have established themselves outside fields of cultivation, and some of these has also acquired multiple herbicide traits (Beckie et al. 2003, 2006). This distribution and the persistence have been found to be dependent of agricultural transport and cropping patterns (Knispel and

McLachlan 2010). Spread to conventional fields may affect co-existence, and buildup of resistance may make it difficult to use herbicides to kill weeds. The Norwegian authorities did in their decision consider it important to avoid spread of such genes to wild relatives (Lovdata 2020).

Environmental monitoring of GM plants, and especially the case of GM oilseed rape, represents a scenario that we have described(GenØk 2013, 2015) as a "worst-case scenario" due to the distinctive traits of the plant and seeds: high survival rate, small and readily dispersible seeds that have long survival, long distance travel potential and natural relatives, among others.

The features of rapeseed plants, their seed (amount, size and viability) and pollen (spread), makes it extremely important to have a stringent plan focusing on how to avoid spread and consequently loss of biodiversity. At present, Norway lacks a total overview of naturally occurring relatives, which is information of high importance as this will provide baseline data for an estimate of potential for spread and hybridization in the environment in and outside agricultural practices. Also, a plan for handling of equipment when harvesting, for transport routes and routines as well as handling throughout the whole plant's life and storage of it needs to be established, The issue of a monitoring plan (co-existence issues, buffer zones, separation areas etc.) is therefore necessary.

## Future Challenges

New breeding technologies (NBTs), as Genome-editing technologies, raise new challenges to the regulative framework of different products derived from such technologies, including GMOs (Agapito-Tenfen et al. 2018). A common argument is that these technologies are comparable to mutagenesis and other traditional forms of breeding in terms of risk to health and environment. On the other hand, one could say that there is a difference regarding the history of safe use for these different technologies. In 2018, the Norwegian Biotechnology Advisory Board, recognizing the challenges for regulating NBTs as GMOs, presented a report where the majority of the board proposed a tiered approval system for the deliberate release of GMOs based on the degree of genetic change (NBAB 2018). This system is based on four levels based that are differentiated based on the degree of genetic change. At the lowest level, a notification to the authorities may be sufficient. At higher levels of genetic change, organisms would require approval before release is authorized, but may be subject to differentiated risk assessment and approval requirements under the NGTA. This suggestion has been submitted to the Ministry of Climate and Environment for further consideration.

## References


Agapito-Tenfen S, Okoli AS, Bernstein M et al (2018) Revisiting risk governance of GM plants: the need to consider new and emerging gene-editing techniques. Front Plant Sci 9:1874

Beckie HJ, Warwick SI, Nair H et al (2003) Gene flow in commercial fields of herbicide-resistant canola (*Brassica napus*). Ecol Appl 13:1276–1294

Beckie HJ, Harker KN, Hall LM et al (2006) A decade of herbicide-resistant crops in Canada. Can J Plant Sci 86:1243–1264

EU (2019) Register of authorized GMOs. http://ec.europa.eu/food/dyna/gm_register/index_en.cfm

GenØk (2013) Monitoring of GMOs released into the Norwegian environment: a case study with herbicide-tolerant GM rapeseed. GenØk report 2013/02, Tromsø, Norway. http://genok.no/wp-content/uploads/2014/07/GenOk_Biosafety_Report_2013_02.pdf

GenØk (2015) Environmental risk of fungus resistant oilseed rape. GenØk biosafety report 2015/1, Tromsø, Norway. http://genok.no/wp-content/uploads/2015/03/raps.pdf

Knispel AL, McLachlan SM (2010) Landscape-scale distribution and persistence of genetically modified oilseed rape (Brassica napus) in Manitoba, Canada. Environ Sci Pollut Res Int 17:13-25

Lovdata (2020) Forskrift om forbud mot omsetning i Norge av bestemte genmodifiserte produkter. https://lovdata.no/dokument/SF/forskrift/2000-12-15-1268?q=Forskrift%20om%20forbud%20mot%20omsetning (in Norwegian)

Morgera E, Roger A (2015) In the Public Interest? A comparative analysis of Norway and EU GMO regulations. Rev European Comp Int Environ Law 24(3):264-277

Lundstrom K (2018) Viral vectors in gene therapy. Diseases 6:42

Müller T, Batza HJ, Beckert A et al (2009) Analysis of vaccine-virus-associated rabies cases in red foxes (*Vulpes vulpes*) after oral rabies vaccination campaigns in Germany and Austria. Arch Virol 154:1081–1091

NBAB (Norwegian Biotechnology Advisory Board) (2018) The Gene Technology act – invitation to public debate. http://www.bioteknologiradet.no/filarkiv/2010/07/genteknologiloven-engelsk-hele-for-web-v-2.pdf

Norwegian Gene Technology Act, Act of 2 April 1993 No. 38 Relating to the Production and Use of Genetically Modified Organisms, etc. (Gene Technology Act). https://www.regjeringen.no/en/dokumenter/gene-technology-act/id173031/

Norwegian Agriculture Agency (Landbruksdirektoratet). Kornleveranser (only in Norwegian), https://www.landbruksdirektoratet.no/no/leveransedata/kornleveranser?leveranser=korn&leveransetype=Oljefr%C3%B8&varekategori=Oljefr%C3%B8&leveranseaar=17%2F18&fylke=

Okeke MI, Okoli AS, Diaz D et al (2017) Hazard characterization of modified vaccinia virus Ankara vector: what are the knowledge gaps? Viruses 9:318

Okoli AS, Asare N, Gjøen T et al (2016) Knowledge base for the assessment of environmental risks by the use of genetically modified virus-vectored vaccines for domesticated animals. Scientific Opinion of the Panel on Microbial Ecology of the Norwegian Scientific Committee for Food Safety (VKM). Oslo, Norway. http://hdl.handle.net/11250/2472697

SGE RAB (Standing Group of Experts on Rabies in Europe) (2019) First meeting (SGE RAB1) Brussels, Belgium, 13 February 2019, https://web.oie.int/RR-Europe/eng/Regprog/en_GF_TADS%20-%20Standing%20Group%20Rabies.htm

Yara (2014) Mer oljevekster vil styrke norsk kornproduksjon (only in Norwegian), https://www.yara.no/gjoedsel/fagmateriell/gjodselaktuelt/gjodselaktuelt-host-2014/mer-oljevekster-vil-styrke-norsk-kornproduksjon/

# Slovenia—GMOs and Synthetic Biology Regulations and Biodiversity: A Slovenian Legal Perspective

**Martin Batič, Ruth Rupreht, Jelka Šuštar-Vozlič, and Marko Dolinar**

**Abstract** Modern biotechnology and synthetic biology are globally recognized as the most rapidly evolving fields in life sciences. In Slovenia, most of the activities in these fields are linked to contained use of GMOs, and to a minor extent to experimental release into the environment. In compliance with the legislative requirements, 81 systems for contained use of GMOs (laboratories, production sites or other premises) are currently recorded in the Slovenian GMOs Register. This corresponds to 3.4 contained systems per 100,000 inhabitants, which is comparable to other highly developed EU countries. The majority (78.8%) of contained systems is located in universities and research institutes; the remaining 19.7% are located in companies (18.2%) and high schools (1.5%), respectively. The vast majority of activities with GMOs (82%) pose no or negligible risk to humans or the environment. The remaining 18% are conducted in Containment Level II facilities as their risk to the environment and biodiversity was concluded to be low. Slovenian research institutes and universities are actively involved in the progress of modern biotechnology and synthetic biology. Genome editing techniques such as oligonucleotide-directed mutagenesis (ODM), zinc finger nucleases (ZFNs), transcription activator-like effector nucleases (TALENs) and CRISPR-Cas9 (clustered regularly interspaced short palindromic repeats associated (Cas) nucleases) are used. All genetic modifications are strictly adhered to national legislation which fully implements the European Union (EU) directives and guidelines.

**Keywords** Slovenia · Legislation · Biosafety · Genome editing techniques · Synthetic biology · Risk assessment · Genetically modified organisms (GMOs) · Contained use · Deliberate release

---

M. Batič (✉) · R. Rupreht
Republic of Slovenia, Ministry of the Environment and Spatial Planning, Ljubljana, Slovenia
e-mail: martin.batic@gov.si

J. Šuštar-Vozlič
Agricultural Institute of Slovenia, Ljubljana, Slovenia

M. Dolinar
Faculty of Chemistry and Chemical Technology, University of Ljubljana, Ljubljana, Slovenia

 
A. Chaurasia et al. (eds.), *GMOs*, Topics in Biodiversity and Conservation 19,
https://doi.org/10.1007/978-3-030-53183-6_42

## Introduction

Modern biotechnology and synthetic biology are globally recognized as the most rapidly evolving fields in life sciences. With the rapid development in these fields, an increasing number of biotechnology products are introduced to the global market. In the near future, the same can be expected for products of synthetic biology. Accordingly, international collaboration is one of the key issues in preventing dual use of modern biotechnology and its products. It also ensures mutual understanding of the importance of risk assessment and risk management processes. Slovenian research institutes and universities are actively involved in the progress of modern biotechnology and synthetic biology. All genetic modifications are strictly adhered to national legislation which fully implements the European Union (EU) directives and guidelines.

## Biosafety[1] in Slovenia

In order to ensure the safe use of modern biotechnology and synthetic biology, the EU and Slovenia, as a Member State, established a legislative framework that is based on the risk assessment of human, other animal and plant health and the environment.

The framework refers to principles of containment as well as technologies and practices used to prevent accidental exposure and unintentional release of GMOs into the environment. The framework thus sets out measures to prevent and minimize potential adverse effects to the environment, in particular with regard to biodiversityconservation, and to health of those who may be exposed to GMOs in contained use, deliberate release into the environment or placing on the market.

The legislative framework used in Slovenia (Table 1) is tasked with protecting the health of humans, animals and the environment before GMOs are released and placed on the market. The legislation also imposes a post-market monitoring of the environment for each authorized GMO. In addition, traceability of origin and labelling are required for any authorized GMO in order to provide consumers with information and freedom of choice.

Procedures are based on case-by-case risk assessment and authorization procedures for GMOs, which are effective, transparent and time-limited.

Authorization procedures for all GMOs are subject to the precautionary approach. The risk assessment for a particular GMO is based on harmonized criteria that are believed to be among the most demanding ones in the world. In the EU, the European Food Safety Authority (EFSA) and professional bodies in the Member States are

[1] Biosafety is a term used to describe efforts to reduce and eliminate the potential risks to the environment, biodiversity and human health resulting from modern biotechnology and synthetic biology and its products.

**Table 1** Legislative framework relating to GMOs

| |
|---|
| **Slovenia** |
| Management of Genetically Modified Organisms Act (Official Gazette of RS, No. 23/05 – current official consolidated text: No. 90/12) (MESP 2005) |
| Act on Coexistence of Genetically Modified Plants with Other Agricultural Plants (Official Gazette of RS, No. 41/09) (MAFF 2009) |
| Restriction or Prohibition of the Cultivation of Genetically Modified Plants Act (Official Gazette of RS, No. 69/15) (MAFF 2015) |
| Act Ratifying the Convention on Biological Diversity (Official Gazette of RS - International Treaties, No. 7/96) (MESP 1996) |
| Act Ratifying the Cartagena Protocol on Biosafety to the Convention on Biological Diversity (Official Gazette of RS - International Treaties, No. 23/02) (MESP 2002) |
| Act Ratifying the Nagoya–Kuala Lumpur Supplementary Protocol on Liability and Redress to the Cartagena Protocol on Biosafety (Official Gazette of RS - International Treaties, No. 4/14) (MESP 2014) |
| **European Union** |
| Directive 2009/41/EC of the European Parliament and of the Council on the contained use of genetically modified micro-organisms (EC 2009) |
| Directive 2001/18/EC of the European Parliament and of the Council on the deliberate release into the environment of genetically modified organisms and repealing Council Directive 90/220/EEC (EC 2001) |
| Directive (EU) 2015/412 of the European Parliament and of the Council amending Directive 2001/18/EC as regards the possibility for the Member States to restrict or prohibit the cultivation of genetically modified organisms (GMOs) in their territory (EC 2015) |
| Commission Directive (EU) 2018/350 amending Directive 2001/18/EC of the European Parliament and of the Council as regards the environmental risk assessment of genetically modified organisms (EC 2018) |
| Regulation (EC) No 1829/2003 of the European Parliament and of the Council on genetically modified food and feed (EC 2003) |
| Regulation (EC) No 1830/2003 of the European Parliament and of the Council concerning the traceability and labelling of genetically modified organisms and the traceability of food and feed products produced from genetically modified organisms and amending Directive 2001/18/EC (EC 2003a) |
| Regulation (EC) No 1946/2003 of the European Parliament and of the Council on transboundary movements of genetically modified organisms (EC 2003b) |
| Regulation (EC) No 882/2004 of the European Parliament and of the Council on official controls performed to ensure the verification of compliance with feed and food law, animal health and animal welfare rules (EC 2004) |
| Regulation (EC) No 178/2002 of the European Parliament and of the Council laying down the general principles and requirements of food law, establishing the European Food Safety Authority and laying down procedures in matters of food safety (EC 2002) |
| Commission Regulation (EC) No 65/2004 establishing a system for the development and assignment of unique identifiers for genetically modified organisms (EC 2004a) |
| Commission Recommendation of 13 July 2010 on guidelines for the development of national coexistence measures to avoid the unintended presence of GMOs in conventional and organic crops (EC 2010) |
| **International** |
| Convention on Biological Diversity (SCBD 1992) |
| Cartagena Protocol on Biosafety to the Convention on Biological Diversity (SCBD 2000) |
| Nagoya-Kuala Lumpur Supplementary Protocol on Liability and Redress to the Cartagena Protocol on Biosafety (SCBD 2010) |

responsible for the risk assessment. In all cases, the risk assessment must demonstrate that, under the intended conditions of use of the GMO-containing product, this is safe for human health, as well as for animals, plants and the environment; otherwise the consent will not be granted.

## Risk Assessment of GMOs

Risk assessment is an obligatory part of each application for authorization. In Slovenia, three types of authorization exist:

1. Contained use of GMOs including research and development of GMOs or application of genetic engineering techniques in the laboratory, as well as industrial production in a contained setting (e.g. bioreactor, greenhouse)
2. Deliberate release of GMOs for field trials
3. Deliberate release of GMOs for cultivation or placing on the market

In the first two cases consent is granted at the national level, while in the third approval is given at the EU level. Risk assessments are prepared by notifiers. In Slovenia, these assessments are evaluated by one or both scientific committees, which are established on the requirements and criteria set up in the Management of GMOs Act (MESP 2005), namely, the Scientific Committee for work with GMOs in contained use and the Scientific Committee for the deliberate release of GMOs into the environment and placing products on the market. The assessment of risks is the starting point for decision making in the process of authorization of GMOs. After authorization of deliberate release, compulsory monitoring of GMOs allows for the establishment of correct, appropriate and proportionate safeguards if unexpected, unforeseen and/or delayed effects are detected during the release.

The aforementioned precautionary and stepwise licensing of GMOs intended for release into the environment stipulates, first of all, testing in containments, followed by testing under limited environmental release conditions (e.g. field experiments, gene therapy experiment). After safety to human health and the environment is proved under limited conditions, approval of such a GMO product for placing on the market can be obtained. However, in the process of authorization at the EU level, the EU Member States until now have never approved a GMO for market introduction with the absolute majority because several concerns raised by the EU Member States were not discussed.

At the international level, modern biotechnology and GMOs are addressed by the Cartagena Protocol on Biosafety (CPB) to the Convention on Biological Diversity (SCBD 2000). This international agreement aims at ensuring the safe handling, transport and use of living modified organisms (LMOs). Slovenia and the EU are amongst the 172 countries who are signatories of the CPB (SCBD 2000). The requirements of the Protocol regarding transboundary movement of GMOs have been adopted by the EU Implementing Regulation 1946/2003 (EC 2003b) and by the Slovenian Management of GMOs Act (MESP 2005).

All applications for the placing of GMOs on the market must also include a post-release monitoring plan. GMO monitoring depends on the intended use of the GMOs. They may be intended for food/feed, processing and cultivation or for any other purpose. Monitoring plans for the GMOs intended for release into the environment must meet the requirements of the Annex VII of the Directive 2001/18/EC (EC 2001) and the Slovenian Management of GMOs Act (MESP 2005).

In Slovenia, GMOs are mostly used for research purposes in contained use where the containment measures are tight. Experiments that include limited release of GMOs into the environment for research purposes are only related to gene therapy in humans and animals. Any kind of cultivation of GMOs in Slovenia is currently prohibited according to the Restriction or Prohibition of the Cultivation of Genetically Modified Plants Act (MAFF 2015).

## Potential and Likelihood of Risk

Products of modern biotechnology and synthetic biology do not have a long history of safe use. Therefore, legislative framework for the management of such GMOs should determine measures to prevent and minimize possible negative and harmful effects to the environment, in particular with regard to preserving biodiversity and human health. Currently, experimental release into the environment in Slovenia is limited to gene therapy. There was no application for other types of GMO release so far in Slovenia. Also, we recorded no applications for placing of GMOs on the market up to now.

### *Contained Use of GMOs*

In the EU, contained use of GMOs falls entirely within the competence of responsible authorities of individual Member States. In Slovenia, the competent authority is the Ministry of the Environment and Spatial Planning.

In contrast to the EU Directive 2009/41/EC (EC 2009), genetically modified microorganisms in Slovenian legislation are not designated differently from other GMOs; i.e. the same term is used for microorganisms and higher eukaryotes. According to Slovenian legislation, the term "contained use" includes all activities with genetically modified organisms (including viruses, viroids, microorganisms and animal and plant cells). In the containment, such GMOs are cultured, stored, transported, destroyed, disposed of or used in any other way. Specific containment measures are used to limit their contact with and to provide a high level of safety for the general population and the environment.

In compliance with the legislative requirements, 81 systems for contained use of GMOs (laboratories or production sites or other premises) are currently recorded in the Slovenian GMOs Register. This corresponds to 3.4 contained systems per

100,000 inhabitants, which is comparable to some other highly developed EU countries. The largest percentage (78.8%) of contained systems is in universities and research institutes, the remaining 19.7% is located in companies (18.2%) and high schools (1.5%) (Fig. 1).

Based on the adopted opinions of the Scientific Committee for work with GMOs in contained use, the vast majority of activities with GMOs (82%) pose no or negligible risk to humans or the environment. The remaining 18% are conducted in containment level II facilities as their risk to the environment and biodiversity was concluded to exist, but is low (Fig. 2). No GMOs are retained in containment levels III and IV in Slovenia.

Considering only containment level II laboratories, GMO activities take place at different institutions. Thirty-five percent of applications were notified from universities, 39% from research institutes and 25% from companies. These R&D activities with GMOs allow innovations in the broad field of modern biotechnology in Slovenia.

At universities and research institutes, genome editing techniques such as oligonucleotide-directed mutagenesis (ODM), zinc finger nucleases (ZFNs), transcription activator-like effector nucleases (TALENs) and CRISPR-Cas9 (clustered regularly interspaced short palindromic repeats associated (Cas) nucleases) are used. These new techniques, whose products are classified as GMOs in Slovenia and in the EU, are currently used for research in six laboratories. The use of these new techniques began in Slovenia soon after they were invented and accessible.

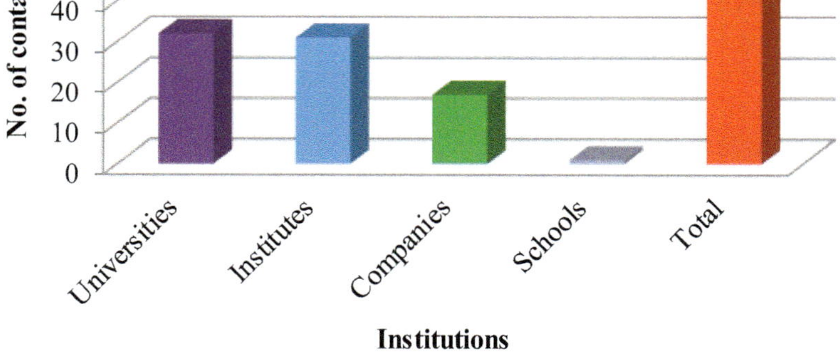


**Fig. 1** Number of systems in which contained use of GMOs is reported in Slovenia (MESP 2017)

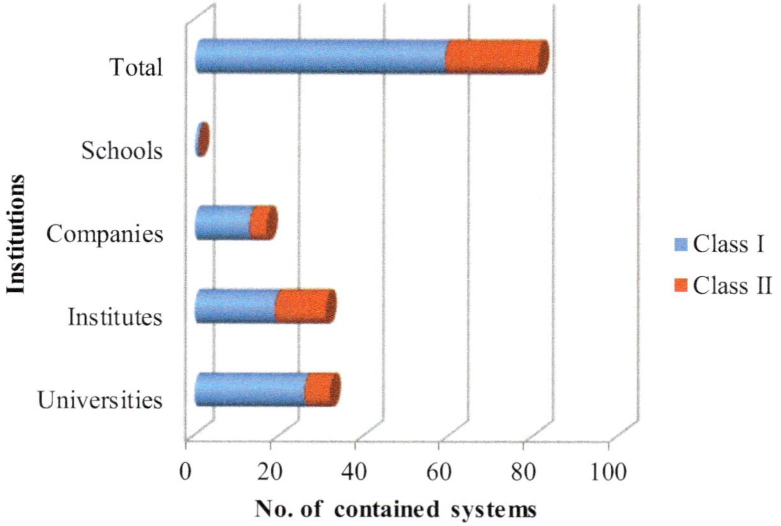


**Fig. 2** Number of GMO contained systems and containment level by institution type in Slovenia (MESP 2017)

## *Modern Biotechnology and Synthetic Biology*

Since modern biotechnology and synthetic biology rely on the generation of GMOs and their modification, they are legally covered by the management of GMOs Act (MESP 2005). Most of the activities are for the contained use of GMOs and to a minor extent for experimental release into the environment. In the forefront are activities in medicine and human health. Specifically, the first activities in the synthetic biology in Slovenia were recorded in 2006 with a project on a genetic device that could prevent sepsis development in patients; this was research conducted mainly at the National Institute of Chemistry (Ciglič et al. 2007). In the field of plant biotechnology, in the forefront are experiments performed at the Biotechnical Faculty (University of Ljubljana), National Institute of Biology, and others.

In parallel to experimental achievements, we are observing a growing awareness of the importance of the social, ethical and legal aspects of the use of modern biotechnological approaches. A recent evaluation of the field (2016–2018) performed by a group of experts examined several aspects of modern biotechnology and synthetic biology (Šuštar-Vozlič et al. 2019), an overview of techniques, their possible impact on the environment, coverage by current legislation and recommendations are presented. The applicability of existing approaches for environmental risk assessment was specifically addressed and appropriate amendments were proposed where needed. Moreover, socio-economic aspects of novel technologies were assessed and a set of socio-economic factors associated with the wide use of new techniques has been defined. Results will be important for deciding on the necessity of adjustments to the existing biosafety system in Slovenia.

## Future Regulation

In Slovenia, we are aware that due to the rapid development and innovation in the field of biotechnology and synthetic biology, the existing biosafety system and its legislative and administrative framework must be constantly adjusted and complemented. This can effectively ensure safe use of new techniques and products and prevent and/or reduce possible short- and long-term harmful impacts on the environment, biodiversity and human health. In this respect, solutions are sought which ensure safe use of new techniques and activities in the field of modern biotechnology and synthetic biology in Slovenia.

## References

Ciglič M, Fekonja O, Kovač J et al (2007) Engineered human cells: say no to sepsis. IET Synth Biol 1(1–2):13–16

EC (2001) Directive 2001/18/EC of the European Parliament and of the Council of 12 March 2001 on the deliberate release into the environment of genetically modified organisms and repealing Council Directive 90/220/EEC. https://eur-lex.europa.eu/legal-content/EN/TXT/?uri=celex%3A32001L0018

EC (2002) Regulation (EC) No 178/2002 of the European Parliament and of the Council of 28 January 2002 laying down the general principles and requirements of food law, establishing the European Food Safety Authority and laying down procedures in matters of food safety. https://eur-lex.europa.eu/legal-content/EN/ALL/?uri=celex%3A32002R0178

EC (2003) Regulation (EC) No 1829/2003 of the European Parliament and of the Council of 22 September 2003 on genetically modified food and feed. https://eur-lex.europa.eu/legal-content/en/ALL/?uri=CELEX%3A32003R1829

EC (2003a) Regulation (EC) No 1830/2003 of the European Parliament and of the Council of 22 September 2003 concerning the traceability and labelling of genetically modified organisms and the traceability of food and feed products produced from genetically modified organisms and amending Directive 2001/18/EC. https://eur-lex.europa.eu/legal-content/en/ALL/?uri=CELEX%3A32003R1830

EC (2003b) Regulation (EC) No 1946/2003 of the European Parliament and of the Council of 15 July 2003 on transboundary movements of genetically modified organisms. https://eur-lex.europa.eu/legal-content/EN/TXT/?uri=CELEX%3A32003R1946

EC (2004) Regulation (EC) No 882/2004 of the European Parliament and of the Council of 29 April 2004 on official controls performed to ensure the verification of compliance with feed and food law, animal health and animal welfare rules. https://eur-lex.europa.eu/legal-content/EN/ALL/?uri=celex%3A32004R0882

EC (2004a) Commission Regulation (EC) No 65/2004 of 14 January 2004 establishing a system for the development and assignment of unique identifiers for genetically modified organisms. https://eur-lex.europa.eu/legal-content/EN/ALL/?uri=CELEX%3A32004R0065

EC (2009) Directive 2009/41/EC of the European Parliament and of the Council of 6 May 2009 on the contained use of genetically modified micro-organisms (Recast). https://eur-lex.europa.eu/legal-content/EN/TXT/?uri=celex%3A32009L0041

EC (2010) Commission Recommendation of 13 July 2010 on guidelines for the development of national co-existence measures to avoid the unintended presence of GMOs in conventional and organic crops. https://eur-lex.europa.eu/legal-content/EN/ALL/?uri=CELEX%3A32010H0722

EC (2015) Directive (EU) 2015/412 of the European Parliament and of the Council of 11 March 2015 amending Directive 2001/18/EC as regards the possibility for the Member States to restrict or prohibit the cultivation of genetically modified organisms (GMOs) in their territory. https://eur-lex.europa.eu/legal-content/EN/TXT/?uri=CELEX%3A32015L0412

EC (2018) Commission Directive (EU) 2018/350 amending Directive 2001/18/EC of the European Parliament and of the Council as regards the environmental risk assessment of genetically modified organisms. https://eur-lex.europa.eu/legal-content/EN/TXT/PDF/?uri=CELEX:32018L0350&from=EN

MAFF (2009) Act on coexistence of genetically modified plants with other agricultural plants. https://www.uradni-list.si/glasilo-uradni-list-rs/vsebina?urlurid=20091980

MAFF (2015) Restriction or prohibition of the cultivation of genetically modified plants act. http://www.pisrs.si/Pis.web/pregledPredpisa?id=ZAKO7213

MESP (1996) Act ratifying the convention on biological diversity, http://www.pisrs.si/Pis.web/pregledPredpisa?id=ZAKO785

MESP (2002) Act ratifying the Cartagena protocol on biosafety to the convention on biological diversity. http://www.pisrs.si/Pis.web/pregledPredpisa?id=ZAKO3465

MESP (2005) Act on management of GMOs. Slovene biosafety portal. http://www.biotechnology-gmo.gov.si/eng/slovenia/legislaton

MESP (2014) Act ratifying the Nagoya - Kuala Lumpur supplementary protocol on liability and redress to the Cartagena protocol on biosafety. http://pisrs.si/Pis.web/pregledPredpisa?id=ZAKO6726

MESP (2017) Porocilo o okolju v Republiki Sloveniji 2017. http://www.mop.gov.si/fileadmin/mop.gov.si/pageuploads/pomembni_dokumenti/porocilo_o_okolju_2017.pdf

SCBD (1992) Convention on biological diversity. https://www.cbd.int/convention/text

SCBD (2000) The Cartagena protocol on biosafety to the convention on biological diversity. https://bch.cbd.int/protocol

SCBD (2010) The Nagoya – Kuala Lumpur supplementary protocol on liability and redress to the Cartagena protocol on biosafety. http://bch.cbd.int/protocol/supplementary/

Šuštar-Vozlič J, Dolinar M, Milavec M, Juvančič L (2019) Procedures for ensuring safety and social acceptability of new techniques and applications of synthetic biology and modern biotechnology. ARRS research project Code V4-1650, 2016–2018. https://www.sicris.si/public/jqm/prj.aspx?lang=eng&opdescr=search&opt=2&subopt=400&code1=cmn&code2=auto&psize=1&hits=1&page=1&count=&search_term=%C5%A1u%C5%A1tar%20vozli%C4%8D&id=10371&slng=&order_by=

# Index





A. Chaurasia et al. (eds.), *GMOs*, Topics in Biodiversity and Conservation 19,
https://doi.org/10.1007/978-3-030-53183-6